AA002464

2012 24th International Conference on Microelectronics

(ICM 2012)

Algiers, Algeria
16 – 20 December 2012

IEEE Catalog Number: CFP12473-PRT
ISBN: 978-1-4673-5289-5

Copyright © 2012 by the Institute of Electrical and Electronic Engineers, Inc
All Rights Reserved

Copyright and Reprint Permissions: Abstracting is permitted with credit to the source. Libraries are permitted to photocopy beyond the limit of U.S. copyright law for private use of patrons those articles in this volume that carry a code at the bottom of the first page, provided the per-copy fee indicated in the code is paid through Copyright Clearance Center, 222 Rosewood Drive, Danvers, MA 01923.

For other copying, reprint or republication permission, write to IEEE Copyrights Manager, IEEE Service Center, 445 Hoes Lane, Piscataway, NJ 08854. All rights reserved.

This publication is a representation of what appears in the IEEE Digital Libraries. Some format issues inherent in the e-media version may also appear in this print version.

IEEE Catalog Number: CFP12473-PRT
ISBN 13:   978-1-4673-5289-5

Additional Copies of This Publication Are Available From:

Curran Associates, Inc
57 Morehouse Lane
Red Hook, NY 12571 USA
Phone: (845) 758-0400
Fax: (845) 758-2633
E-mail: curran@proceedings.com
Web: www.proceedings.com

Table of contents

Digital signal and data processing

A secure and robust audio watermarking system for copyright protection 1
Himeur Yassine and Boudraa Bachir

HOG Based Fast Human Detection 5
Mouloud Kachouane, Safia Sahki, Mustapha Lakrouf and Noureddine Ouadah

Using the DAPGF filter for speech analysis, Application for cochlear implants 9
Boucherit Ismail and Guerti Mehania

Wavelet based segmentation and time-frequency caracterisation of some abnormal heart sound signals 13
Nayad Kouras, Daoud Boutana and Messaoud Benidir

A Comparative Evaluation of VoWLAN Call Capacity With and Without Codec's Silence Suppression 17
Yacine Harkat and Abderrahmane Amrouche

Image denoising using selective neighboring Quaternionic wavelet coefficients 21
Mohammed Kadiri, Mohamed Djebbouri and Philippe Carré

Color image segmentation by a Genetic Algorithm based Clustering and Connected Component Labeling 25
Fatima Zohra Bellala and Feryel Souami

An Application of ICA-based Detection to OFDM-IDMA System 29
Meftah El Hadi, Anou Abderrahman and Bensebti Messaoud

Automatic Detection Method of R-wave Positions in Electrocardiographic signals 33
Fatiha Bouaziz and Daoud Boutana

Use of Nakagami statistical model in ultrasonic tissue mimicking phantoms characterization 37
Nardjess Bahbah, Hakim Djelouah and Ayache Bouakaz

Neutralization Technique Model Using Artificial Neural Networks   41

Rafik Addaci, Nazih Hamdiken, Tarek Fortaki and Robert Staraj

Selection of a Suitable Mother Wavelet For Microemboli Classification Using SVM And RF Signals   45

Karim Ferroudji, Nabil Benoudjit, Mohamed Bahaz and Ayache Bouakaz

INFLUENCE OF G722.2 SPEECH CODING ON TEXT-INDEPENDENT SPEAKER VERIFICATION   49

Meriem Fedila and Abderrahmane Amrouche

EVALUATION OF THE PERFORMANCE OF BLOCK MATCHING ALGORITHMS IN WAVELET-BASED T+2D VIDEO CODING   53

Mehrez Marzougui

Analog circuits

Automated Design Technique for Constant-gm Rail-to-Rail for Input Stage   57

Ahmed Reda and Fathi Farag

Inductor Implementation using CMOS Current Conveyor Integrator for Low Voltage Low Power Applications   61

Fathi A. Farag

Power Recovery From Data Line in Avionic Applications   65

Jinying Zhang, Saeid Hashemi, Masood Karimian, Zied Koubaa and Mohamad Sawan

Control of Chaotic Behaviour in Buck-Boost DC-DC Converters   69

Ammar Natsheh and J. Gordon Kettleborough

Modeling of Op- Amp Nonlinearities in Pipelined ADC   73

Samir Barra, Souhil Kouda, Abdelghani Dendouga and Nour-Eddine Bouguechal

High Bandwidth 0.18μm CMOS Transimpedance Amplifier for Photoreceiver Circuit   77

Escid Hammoudi

Power Electronics Circuit for Speed Control of Experimental Wind Turbine   81

Adel Merabet, John Kerr, Vigneshwaran Rajasekaran and Derek Wight

Energy Monitoring System for Security and Power Management Applications 85

Sepideh Shariati, Radu Muresan and Anthony Vannelli

Studies on the influence of corona on overvoltage surges by simulation using the ATP/EMTP 89

Zahira Anane and Abdelhafid Bayadi

A High-resolution DCO with MOS Capacitors 93

Zixuan Wang, Cheng Huang and Jianhui Wu

Device characterization and modeling

A microcontroller-based pulse generator for isothermal I-V measurements 97

Maurizio Costagliola, Vincenzo D'Alessandro, Grazia Sasso and Niccolò Rinaldi

A New RADFET dosimeter design for environment monitoring applications 101

Mohamed Meguellati, Djeffal Faycal and Abelmalek Kadri

The thermal effect on the output conductance in AlGaN/GaN HEMT's 105

Bellakhdar Aissa, A. Telia , L. Semra and A. Soltani

Compact modeling of long channel DoubleGate MOSFET transistor 109

Billel Smali, Saida Latreche, Samir Labiod

The Electro-Thermal Sub Circuit Model for Power MOSFETs 113

Messaadi Lotfi, Smail Toufik

New method for determination of the diode parameters in the presence of the leakage currents 117

Mahi Khaled, Bahous Messani and Saad Machraoui

TSV Impact on Circuit Performance and Recommended Design Methodologies 121
Khaled Salah, Alaa El Rouby, Hani Ragai, Yehea Ismail

RF circuits and integrated antenna

New Tapered Slot Vivaldi antenna for UWB Applications 125

Djalal Ziani Kerarti and Fatima Zahra Marouf

MMIC Doherty Power Amplifier for a 5W Pico-Cell Base Station in GaN HFET Technology 129

Sashieka G. Seneviratne, Mustapha C. E. Yagoub and Rony E. Amaya

A Novel Ultra-Wide Stopband Microstrip Low-Pass Filter for Rejecting High Order Harmonics and Spurious Response 133

Mouloud Challal, Arab Azrar and Danielle Vanhoenacker-Janvier

Wireless Microelectronic Nose Network for Gas Classification 136

Muhammad Hassan and Amine Bermak

A 0.9 V High Gain and High Linear Bleeding CMOS Mixer for Wireless Applications 140

Sid Ahmed Tedjini-Bailiche, Abdelhalim Slimane, Mohand Tahar Belaroussi, Mohamed Trabelsi and Sami Beghami

Improved antenna diversity system by low correlation between elements exploiting Single-Negative Metamaterials 144

Aouadi Belgacem

EC-CPW Fed Elevated Patch Antenna 148

Adel Emhemmed

Multi-Band Metamaterial Structures Based On Hexagonal Shaped Magnetic Resonators 152

Hayet Benosman and Noureddine Boukli Hacene

Design and Implementation of passive UHF RFID system 156

Rafik Khelladi, Mustapha Djeddou and Mustapha Bensalah

Pattern Nulling of Linear Array Antennas with Mutual Coupling Effects using Taguchi Method 160

Abdelmadjid Recioui, Hamid Bentarzi

Fast and Accurate Analysis Method of a Circular Patch Antennas using Neurospectral Method 164

Nazih Hamdiken, Rafik Addaci and Tarek Fortaki

Efficient ECC Implementation Architecture Suitable for RFID Technology 167

Mustapha Benssalah, Mustapha Djeddou and Karim Drouiche

R F MEMS CIRCUIT FOR SPACE COMMUNICATIONS SYSTEMS 171

Oukil Souad, Boudjamai Abdelmadjid and Boughenmi Nabil

Digital circuits and systems

A Highly Efficient Substitution Matrix Loader for Pairwise Sequence Alignment 175

Mohd Nazrin Md Isa, Khaled Benkrid and Thomas Clayton

Implementation of Spectral Subtraction Method on FPGA using High-Level Programming Tool 179

Mohammed Bahoura and Hassan Ezzaidi

FPGA Design of Real-Time Obstacle Detection System Using Stereovision 183

Hamza Bendaoudi and Abdelhakim Khouas

Design of a C-Element Based Clock Domain Crossing Interface 187

Zaid Al-Bayati, Otmane Ait Mohamed, Syed Rafay Hasan and Yvon Savaria

Design and Implementaiton of a Fall Detection System using Compressive Sensing and Shimmer Technology 191

Hassan Rabah, Abbes Amira and Afandi Ahmad

Virtual Shared Memory Architecture for Inter-Task Communication in Partial Reconfigurable Systems 195

Chuan Hong, Khaled Benkrid, Ali Ebrahim and Xabier Iturbe

Memristors-based NMOS Logic Circuits 199

Zine-Eddine Abid, Dirar Homouz, Baker Mohammad and Wei Wang

FPGA Implementation of a Feature Extraction Technique based on Fourier Transform 203

Mohammed Bahoura and Hassan Ezzaidi

Modified Circular Hough Transform using FPGA 207

Messaoudi Khadidja

Montgomery Modular exponentiation on FPGA 211

Nadjia Anane, Mohamed Anane and Mohamed Issad

GA-based Modeling of a Non-ideal Second Order low-pass Modulator   215

Abdelghani Dendouga, Samir Barra, Souhil Kouda and Nour-Eddine Bouguechal

Embedded Implementation of an IP-PBX /VoIP Gateway   219

Faroudja Abid

A Fast Hardware/Software Co-Verification Method using a real hardware acceleration   223

Ben Ayed Mossaad, Bouchhima Faouzi and Abid Mohamed

Design and Implementation of an FPGA-based Fuzzy MPPT Controller for PV Standalone Systems   227

Messai Adnane, Mellit Adel and Guessoum Abderrezak

A Comparative Performance Evaluation Study of The Basic Binary Tree and Aloha Based Anti-collision Protocols For Passive RFID System   231

Ahmed Rennane, Hadjer Saadi, Rachida Touhami and Mustapha C.E. Yagoub

Wireless and communication systems

Power Conservation For Wireless Mesh Network   235

Sarra Mamechaoui, Fedoua Didi and Guy Pujolle

A Novel Cluster-based Fault-tolerant Scheme for Wireless Sensor Networks   239

Mohamed Lehsaini and Chifaa Tabet Hellel

Successive Interference Cancellation Receiver (SIC) in DS-OCDMA System   243

Kada Biteur and Malika Kandouci

Coverage and Connectivity Protocol for Wireless Sensor Networks   247

Rachid Baghdad and Abdelkader Khalil

Differential Dual Amplitude-Width PPM for Wireless Optical Communication Systems   251

Mehdi Rouissat, Riad Borsali and Mohammad Chikh-Bled

The Influence of Noisy Channel on Acoustic Echo Cancellation in Mobile Communication   255

Mahfoud Hamidia and Abderrahmane Amrouche

CAD and simulation VLSI

Fault injection for verifying testability of fault tolerant structures at the Verilog level   259
Ghania Ait Abdelmalek, Rezki Ziani and Mourad Laghrouche

TSV Impact on Circuit Performance and Recommended Design Methodologies   N/A
Khaled Mohamed, Alaa El-Rouby, Yehea Ismail and Hani Ragai

Design and Simulation Methodology for Switch-Cap Circuits Used in Data Converter Applications   263
Sanad Kawar, Mahmood Mohammed and Khaldoon Abugharbieh

Particle Swarm Optimization for the Identification of Worst Case Test Vectors of Total-Dose Induced Leakage Current Failures in ASICs   267
Mostafa Abdel-Aziz, Hamzah Abdel-Aziz, Ahmed Abou-Auf and Amr Wassal

Study and modeling of defects in integrated circuits for their reliability analysis   271
Ghania Ait Abdelmalek, Rezki Ziani and Mourad Laghrouche

A New Computer Based Auto-recloser Framework   275
Hamid Bentarzi, Mahfoud Chafai and Boubakour Harhati

A Non-Convex Classifier Support for Abstraction-Refinement Framework   279
Samir Ouchani and Otmane Ait Mohamed

Numerical modeling of MOS transistor using implicit finite different-time domain method   283
Samir. Labiod, Saida. Latreche, Billel. Smali, M. R. Beghoul, C. Gontrand

Study the self- heating effects in SOI MOSFET transistor and Numerical Simulation Using Silvaco Software   287
Fatima Zohra Rahou, Mohamed Rahou and Ahlem Guen Bouazza

Multi-Objective Genetic Algorithm Optimization of CMOS Operational Amplifiers   291
Samir Barra, Abdelghani Dendouga, Souhil Kouda and Nour-Eddine Bouguechal

Simulation of Ion Implantation for CMOS 1µm Using SILVACO Tools   295
Boubaaya Mohamed, Hadj Larbi Fayçal and Oussalah Slimane

Modeling CMOS PIN Photodiode using COMSOL 298

Mohamad Hamady, Ehsan Kamrani and Mohamad Sawan

Device physics and novel structures

A Novel Method based On Capacitance-Voltage for Negative Bias Temperatures Instability Studies: Concept and Results 302

Abdelmadjid Benabdelmoumene, Boualem Djezzar, Leonard Trombetta, Hakim Tahi, Amel Chenouf and Mohamed Kechouane

Impact of the Inhomogeneous Structure of the Active Layer on the Transfer Characteristic of Polysilicon TFT's 306

Hadjira Tayoub

ANFIS-based approach to study the subthreshold swing behavior for nanoscale DG MOSFETs including the interfacial traps effect 310

Toufik Bentrcia, Djeffal Faycal and Elasaad Chebaki

Influence of vertical scaling and temperature on impact-ionization effects in SiGe HBTs 314

Grazia Sasso, Vincenzo D'Alessandro, Maurizio Costagliola and Niccolò Rinaldi

A New Eye on NBTI-Induced Traps up to Device Lifetime Using on the Fly Oxide Trap Method 318

Boualem Djezzar, Hakim Tahi, Abdelmadjid Benabdelmoumene and Amel Chenouf

An Accurate Extraction Methodology for NBTI Induced Degradation Using Charge Pumping Based Methods 322

Hakim Tahi, Boualem Djezzar, Abdelmadjid Benabdelmoumene and Amel Chenouf

A study of volatile organic compounds diffusion in thin plasma polymerized TEOS thin film 326

Ali Bougharouat, Azzedine Bellel and Salah Sahli

Oxide Trap Annealing by H2 cracking at E' Center under NBTI Stress 330

Cherifa Tahanout, Hakim Tahi, Boualem Djezzar, Abdelmadjid Benabdelmoumene, Amel Chenouf and Becharia Nadji

Deep Experimental Investigation of NBTI Impact on CMOS Inverter Reliability 334

Amel Chenouf, Boualem Djezzar, Abdelmadjid Benadelmoumene and Hakim Tahi

Electromechanical Response Simulation of Film Bulk Acoustic Wave Resonator   338

Serhane Rafik, Boutkedjirt Tarek and Hassein Bey Abdelkader

Process and technology

Safe Operating Area of a 0.15µm GaAs PHEMT in Overdrive Operating Conditions   342

Naoufel Ismail and Adel Kalboussi

Standard CMOS Implementation of Schottky Barrier Diodes for Biomedical RFID   346

Sebastião Cabral, Leonardo Zoccal, Paulo Crepaldi and Tales Pimenta

Comparative Study of Current Mode and Voltage Mode Sense Amplifier for sub-28nm SRAM   350

Baker Mohammad, Percy Dadabhoy and Paul Basseett

A Compact Analytic Expression of the Oscillation Amplitude in MOS LC-Oscillators   356

Bassem Fahs, Adnan Harb, Hassan Bazzi and Mohammad Mansour

Morphology and structural properties of nano-PZT thin films deposited on unheated substrate by RF sputtring system   360

Mahdi Mohammed and Kadri Mohammed

Characterization by SEM and FTIR of B-LPCVD polysilicon films after thermal oxidation   363

Moufida Bouzerdoum and Boubekeur Birouk

A low power thermal protection topology   367

Alex Pivoto, Paulo Crepaldi and Tales Pimenta

Boron Redistribution in Strongly Doped Silicon Thin Bi-layers Gates   371

Salah Abadli and Farida Mansour

Special session: solar energy

FPGA-Based implementation of IncCond algorithm for photovoltaic applications   374

Adel Mellit, Nadjwa Chettibi and Mahmoud Drif

Comparative study on Water Max A 64 DC pump performances based Photovoltaic Pumping System design to select the optimum heads in arid area   378

Azzedine Boutelhig and Yahia Bakelli

Optimal Model Selection For PV Module Modeling 382

Idriss Hadj Mahammed, Y Bakelli, S Hamid Oudjana, A Hadj Arab and S Berrah

Photovoltaics in Italy: toward grid parity in the residential electricity market 386

Alessandro Massi Pavan and Vanni Lughi

Modeling and Simulation of a PV Pumping System Under Real Climatic Conditions 390

Douadi Bendib, Achour Mahrane, Mohmaed Ayad and Madjid Chikh

Additional Papers:

New Approach for Modeling Effect of Phosphorous Diffusion on Minority Carrier Lifetime in Multicrystalline Silicon Wafers 394

Bentalhoda Soleimany, Anahita Shojai Hashemi, Ebrahim Asl-Soleimani

Compact Multi-band Rectangular Slotted Antenna for Global Navigation Satellite Systems (GNSS) 398

Mustapha Djebari, Amine Abdelhadi

Secure and robust audio watermarking system for copyright protection

HIMEUR Yassine
Department of Electronics
University of Jijel
Jijel, Algeria
him.yassine@gmail.com

BOUDRAA Bachir
LCPTS
USTHB
Algiers, Algeria
b.boudraa@yahoo.fr

Abstract—
Transform-domain digital audio watermarking has a performance advantage over time-domain watermarking by virtue of the fact that frequency transforms offer better exploitation of the human auditory system (HAS). In this paper, an innovative watermarking scheme for audio signal based on double insertion of the watermark image in DWT-DST domain of the host signal is proposed. Subjective and objective tests reveal that the proposed watermarking scheme maintains high audio quality and is simultaneously highly robust to different attacks, including MP3 compression, low-pass filtering, amplitude scaling, additive Gaussian noise, reacquisition, cropping, sampling, high pass filtering. Comparison of the proposed algorithm with similar techniques such as Cox et al and Dhar et al, shows the superiority of the proposed scheme in term of robustness and imperceptibility.
Key words: audio watermarking, DWT, PN-sequence, DST,FRIT, copyright protection, robust watermarking.

I. INTRODUCTION

Information hiding techniques have recently become important in a number of application areas. Digital audio, video, and pictures are increasingly furnished with distinguishing but imperceptible marks, which may contain a hidden copyright notice or serial number or even help to prevent unauthorised copying directly [1].
The traditional information security technology based on cryptography theory mostly has its limitations. In order to resolve the shortcomings of traditional information security technology, more and more researchers has been starting to study the digital watermarking technology because it can effectively compensate for the deficiencies of the security and protection application of traditional information security technology. The watermark information can be copyright information, authentication information or controlling information so as to determine the copyright owner of the digital works, certify the authenticity and integrity of multimedia works, control copying according to the embedded control information, achieve the purpose of copyright protection. Digital watermarking technology has many applications in protection, certification, distribution, anti-counterfeit of the digital media and label of the user information. It has become a very important study area in information hiding [2].
Existing audio watermarking systems exploit the irrelevant properties of the human auditory system (HAS) [3]. In

particular, HAS is insensitive to small amplitude changes in the time and frequency domains, allowing the addition of weak noise signals (watermarks) to the host audio signal such that the changes are inaudible [4-9]. Frequency-domain techniques, in particular, have been more effective than time-domain techniques since watermarks are added to selected regions in the transformed domain of the host audio signal, such that inaudibility an robustness are maintained [10-12].

II. TRANSFORMATION TECHNIQUES

II. 1 Difference between the DCT and the DST

By applying the DCT to 8 identical values *(x(i) = 100, i=1,2,...,8)* we can see that the DCT compacts all the energy of the data into the single DC coefficient whose value is identical to the values of the data items. Applying the DST to the same eight values, on the other hand, results in seven AC coefficients whose sum is a wave function that passes through the eight data points but oscillates between the points [13].

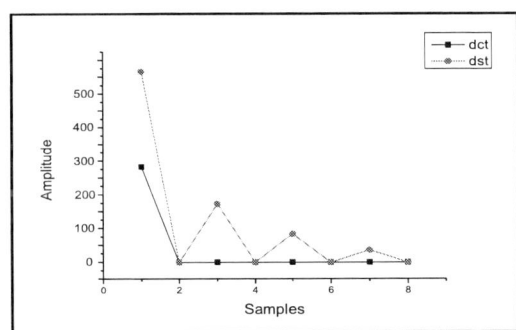

Figure 1. DCT and DST of eight identical values

So we can say that the DST has compacted the energy input values by using more AC coefficients than the DCT. This example, with the fact that the DCT produces highly decorrelated coefficients, argues strongly in favor of using the DCT in data compression in contrast to DST.
In watermarking, the insertion of the watermark in the DC coefficients gives a better robustness, but degrades the quality of the watermarked signal. In addition, the DCT does not have many choices for insertion of the watermark, while the DST, as a result of compacting the signal energy over several coefficients gives more choices and therefore more security to watermarking system. In addition, it ensures a good compromise between robustness and imperceptibility.

II.2 The Finite Ridgelet Transform

978-1-4673-5289-5/12 $31.00 © 2012 IEEE

Finite Ridgelet Transform, also known as FRIT, is a kind of invertible, non-redundant transform [14]. To overcome the periodization effect of a finite transform, by using a novel ordering of the FRAT coefficients. Taking the one-dimensional wavelet transform on the projections of the FRAT in a special way results in the finite ridgelet transform (FRIT), which is invertible, non-redundant, and computed via fast algorithms. In 2-D, points and lines are related via the Radon trans- form, thus the wavelet and ridgelet transforms are linked via the Radon transform. More precisely, denote the Radon transform as :

$$R_f(\theta, t) = \int_{R^2} f(x)\delta(x_1 \cos\theta + x_2 \sin\theta - t)dx \qquad (6)$$

Then the ridgelet transform is the application of a 1-D wavelet transform to the slices (also referred to as projections) of the Radon transform,

$$CRT_f(a, b, \theta) = \int_R \psi_{a,b}(t)R_f(\theta, t)dt \qquad (7)$$

So, the advantage of the Finit Ridgelet Transform, is invertible and to compute its inverse transform you must have two information on the original image. Namely; the normalized mean and the structure of wavelet decomposition that are needed for reconstruction. These two information are generated during the transformation process and without this information we cannot do the reconstruction of the original image. In this paper, the FRIT is used as an encrypting system to secure the watermarking algorithm. So, the transformed coefficients in are inserted in the host audio signal.

III. WATERMARK EMBEDDING AND EXTRACTION PROCESS

The proposed watermark embedding process is implemented in the following seven steps:

1) The original audio signal is first decomposed with the wavelet transform.

2) The approximation (A_1) and detail (D_1) coefficients are then segmented into non-overlapping frames and transformed with the DST. The watermark is first transformed with the FRIT and the result coefficients are double inserted into the approximation (low frequencies) and detail coefficients (high frequencies).

3) Each sample of the watermark is then inserted into a coefficient of each frame using equation (8) [11]. Generally samples of the watermark are inserted in the first five coefficients of each frame. This ensures that the mark is added to the components of the most important perception.

$$S_w = S + \alpha \times W \qquad (8)$$

The value of α denotes the strength of the watermarking and it is called embedding strength.

5) Take the inverse DST of the modified DST coefficients to calculate the watermarked high and low frequencies A^w_1 and D^w_1 respectively.

6) At the end, we reconstruct the watermarked audio signal S_w using the inverse discrete wavelet transform.

The proposed watermark extraction is implemented in the following three steps:

1) Calculate the DWT of the attacked watermark audio signal.

2) The approximation and detail coefficients A^w_1 and D^w_1 are then segmented into non-overlapping frames and transformed with the DST.

3) The N coefficients of the watermark inserted are extracted from the first frames coefficients of Sw have been selected during the insertion phase using the following relation.

$$W' = \frac{S_w - S}{\alpha} \qquad (9)$$

The obtained coefficients are than transformed with the inverse FRIT. Thus, the extraction and transformation operations retrieve the two watermarks included in the coefficients of low and high frequencies. We only keep the watermark that has suffered the least damage. It will be used as proof of ownership of copyright.

IV. SIMULATION RESULTS AND DISCUSSION

In this section, we evaluate the performance of our watermarking system using eight different types of 16 bit mono audio signals sampled at 44.1 kHz for α=7 by considering the frame size of 512 samples. A comparison study has been done between the proposed algorithm, cox's algorithm [11] and Dhar's algorithm [12]. In this paper, we propose to use a binary image as a watermark instead of the PN-sequence used in [11] and [12], to allow a subjective validation of the extracted watermark.

IV.1 Imperceptibility Test

In order to evaluate the quality of watermarked signal, the following signal-to-noise ratio (SNR) equation is used:

$$SNR = 10.\log\frac{\sum_{i=0}^{N-1} s(i)^2}{\sum_{i=0}^{N-1}[s(i) - s^w(i)]^2} \qquad (10)$$

where $s(n)$ and $s^w(n)$ are original audio signal and watermarked audio signal respectively. After embedding watermark, the SNR of all selected audio signals using the proposed method are above 25 dB (TABLE 1) which ensures the imperceptibility of our proposed system. Which satisfies the IFPI requirement (20 dB) described in [15]. These results have clearly demonstrated the superiority of the proposed scheme over Cox and Dhar methods.

TABLE 1. SNR comparison between proposed system , Cox's and Dhar's systems

Audio signals	Title	proposed	DWT-DCT	Dhar	Cox
S1	listening	**28.91**	26.90	21.81	17.23
S2	Male Voice	**25.15**	23.29	18.49	16.03
S3	Fifth symphony	**25.89**	23.63	21.07	16.23
S4	arabic music	**26.96**	25.50	20.25	18.33
S5	Let it be	**31.07**	29.44	19.07	16.63
S6	Piano	**27.12**	22.49	20.93	17.33
S7	Hey Jude	**27.13**	23.99	20.71	16.74
S8	Classical music	**26.65**	22.29	21.46	18.25
Mean	/	**25.51**	22.06	19.92	14.77

IV.2 Robustness Test

In order to evaluate the performance of the proposed watermarking system, the following Normalized Correlation (NC) is used to evaluate the correlation between the extracted watermark and the original watermark:

$$NC(w, w') = \frac{\sum_{i=1}^{M_1} \sum_{j=1}^{M_2} w(i,j) w'(i,j)}{\sqrt{\sum_{i=1}^{M_1} \sum_{j=1}^{M_2} w(i,j)^2} \sqrt{\sum_{i=1}^{M_1} \sum_{j=1}^{M_2} w'(i,j)^2}} \quad (11)$$

Where i is the index in the watermark image w. Table 3 shows the performance comparison in terms of NC between the proposed system and Cox's and Dhar's methods when no attack is applied to four different types of watermarked audio signals. Various common signal-processing attacks are used to assess the robustness of our scheme are given below (TABLE 2). The presented algorithm is simulated by programming in Matlab 7.3. The size of the original signal S is 512 × 1024 and the size of image watermark w is 32 × 32.

TABLE 2. Robustness comparison of proposed algorithm, Cox's and Dhar's algorithms

	Audio signals	Proposed	ALG2	ALG1
MP3 compression (64 kbps)	S1	**0.999**	0.971	0.960
	S3	**0.999**	0.966	0.975
	S5	**0.999**	0.977	0.991
	S7	**0.999**	0.955	0.992
Resampling (44.1→22.05→44.1)	S1	**1**	0.507	0.125
	S3	**1**	0.507	0.152
	S5	**1**	0.977	1
	S7	**1**	0.977	0.967
Requantification 16→8→16	S1	**1**	0.961	0.991
	S3	**1**	0.977	0.999
	S5	**1**	0.977	1.000
	S7	**1**	0.973	0.999
Additive White Gaussian Noise (20 dB)	S1	**0.999**	0.977	0.819
	S3	**0.999**	0.977	0.978
	S5	**0.998**	0.977	0.997
	S7	**0.999**	0.976	0.996
Low-pass filtering (fc=4 kHz)	S1	**1**	0.856	0.437
	S3	**1**	0.977	0.987
	S5	**1**	0.977	0.983
	S7	**1**	0.960	0.982
Cropping (10%)	S1	**0.995**	0.977	0.999
	S3	**0.994**	0.977	0.999
	S5	**0.995**	0.977	0.999
	S7	**0.995**	0.977	0.999
High-pass filtering (fc=400 Hz)	S1	**1**	0.503	0.777
	S3	**1**	0.376	0.810
	S5	**1**	0.528	0.810
	S7	**1**	0.346	0.808
Amplification (150 %)	S1	**0.992**	0.846	0.491
	S3	**0.992**	0.846	0.491
	S5	**0.992**	0.846	0.491
	S7	**0.992**	0.846	0.491

A comparaison between watermarks which were extracted after applying various attacks using the proposed algorithm, cox's algorithm and Dhar's algorithm are given in Figure 2. In addition to the extracted watermark quality and also the SNR of the watermarked audio signal which is described in the previous section, another privilege of our method can be depicted in robustness performance especially in case of MP3 compression with different bit rates and Additif White Gaussian Noise with different SNR, as shown in Figures 3 and 4.

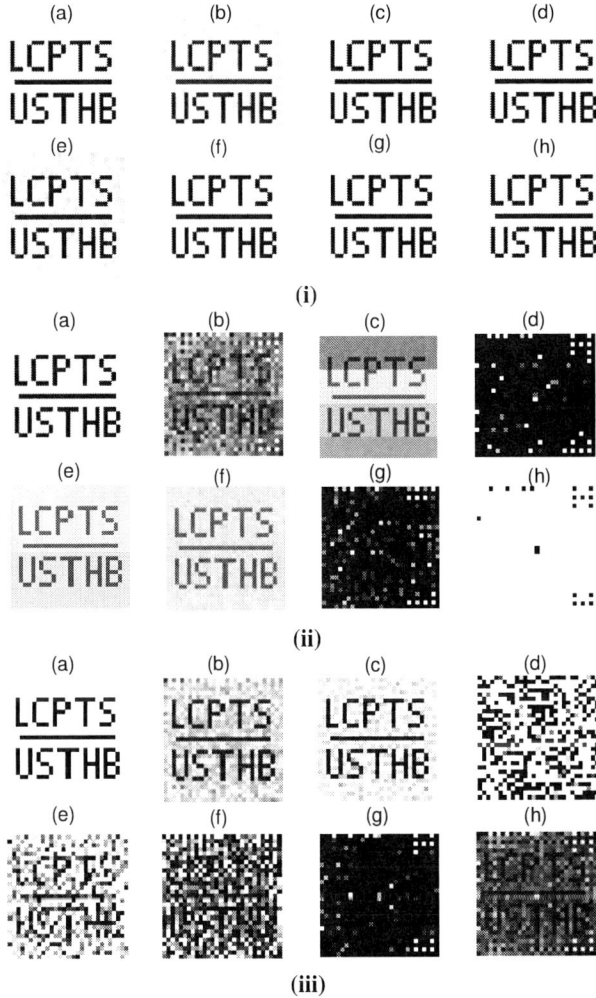

Figure 2. Extracted watermarks using: i) proposed algorithm, ii) cox's algorithm, iii) Dhar's algorithm after different attacks a) original watermark, b) MP3 compression, c) requantization, d) resampling, d) AWGN, e) low pass filtering, f) High-pass filtering, g) amplification

It is clear from Figures 6 and 7, the efficiency of the proposed algorithm to preserve the watermarks inserted into the audio signals to prove the copyright protection. It shows a high superiority against other algorithms of comparison.

V. CONCLUSION

A novel audio watermarking method is proposed based on double embedding in the hybrid domain, DWT-DST. The method uses the DWT variety of time-frequency decomposition for audio signals, and modifies its DST coefficients with a watermark image before re-constituting the signal. The effectiveness of the DWT-DST based algorithm was demonstrated by watermarking eight different audio signals. Moreover, it is also a secure scheme, the watermark cannot be extracted without knowing the keys generated during the insertion process.

Future work will concentrate on making the method more practical by modifying the technique such that the original audio signal is not required to extract the watermark and this approach may also be extended to image and video watermarking.

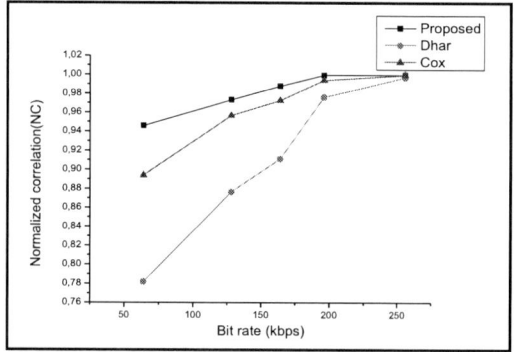

Figure 3. Comparison of robustness against MP3 compression with different bit rates

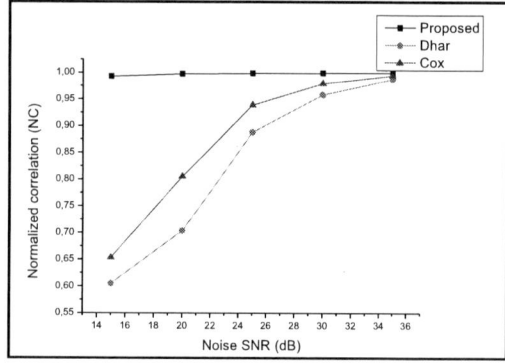

Figure 4. Comparison of robustness against AWGN with different SNR

References

[1] H.X. Wang, Z. Z. Diao, "A Novel Real-Time Information Hiding Technique in VLC Domain," 2009 Fourth International Conference on Innovative Computing, Information and Control 2009.

[2] J. X. Liu, Z. M. Lu ; J. S. Pan "A Robust Audio Watermarking Algorithm Based on DCT and Vector Quantization," Eighth International Conference on Intelligent Systems Design and Applications, 2008.

[3] M. Arnold, "Audio watermarking: features, applications and algorithms," in Proc. of IEEE International Conference on Multimedia and Expo, vol. 2, pp. 1013-1016, 2000.

[4] P. Bassia, I. Pitas, N. Nikolaidis, "Robust Audio Watermarking in the Time Domain". IEEE Transactions on Multimedia, vol 3, No. 2, 2001.

[5] W. Bender, D.Gruhl, and N. Morimoto, "Techniques for data hiding, " IBM Systems journal, 1996, 35(3/4): 131-336.

[6] J. Huang, Y. Wang, and Y.Q. Shi, "A blind audio watermarking algorithm with self-synchronization," Proc. Of IEEE, Int. Sym. On Circuits and Systems, vol. 3 pp. 627-630, 2002.

[7] D. Gruhl, A. Lu and W. Bender, "Echo Hiding," Proc. Of 1st Information Hiding Workshop, LNCS vol. 1174, Berlin, Germany: Springer – Verlag, pp. 295 -315, 1996.

[8] Kim, H.J., and Choi Y.H., "A novel echo-hiding scheme with backward and forward kernels," IEEE Transactions on circuits and Systems for video and technology, vol. 13, 2003.

[9] A.N. Lemma, J. Aprea, W. Oomen, L.v. Kerkhof, "A temporal domain audio watermarking technique, " IEEE Trans. Signal Process., 51 (4) (2003), pp. 1088–1097.

[10] Boney, L., Tewfik, A. H., and Hamdy, K. N., "Digital watermarks for audio signal," International Conference on Multimedia Computing and Systems, pp. 473-480, 1996.

[11] I. J. Cox, J. Kilian, F. T. Leighton, and T. Shamoon. "Secure spread spectrum watermarking for multimedia. IEEE Trans. Vol. 6, No. 12, Dec 1997.

[12] P. K. Dhar, M. I. Khan, C. H. Kim, and J. M. Kim, "An Efficient Audio Watermarking Algorithm in Frequency Domain for Copyright Protection, " CCIS, 2010, Vol 122, pp. 104-113, Springer

[13] D. Salomon. G. Motta and D. Bryant, "Data Compression: The Complete Reference," Fourth Edition. Springer-Verlag 2007.

[14] Minh N. Do, Martin Vetterli, "The Finite Ridgelet Transform for Image Representation, " IEEE Trans on Image Processing, Vol. 12, No. 1, pp. 16 – 28, 2003.

[15] S. Wu, J. Huang, D. Huang, Y. Q. Shi, "Efficiently Self-Synchronized Audio Watermarking for Assured Audio Data Transmission, " IEEE Trans on Broadcasting, Vol. 51, No. 1, pp. 69 – 76, 2005.

HOG Based fast Human Detection

M. Kachouane, S. Sahki
Electronics dept.
Université Saad Dahlab de Blida
Blida – Algeria
Kachouane.mouloud@hotmail.fr

M. Lakrouf, N. Ouadah
NCRM team, Robotics & Automation dept.
Centre de Développement des Technologies Avancées
Baba Hassen, Algiers, Algeria
mlakrouf@cdta.dz

Abstract—**Objects recognition in image is one of the most difficult problems in computer vision. It is also an important step for the implementation of several existing applications that require high-level image interpretation. Therefore, there is a growing interest in this research area during the last years. In this paper, we present an algorithm for human detection and recognition in real-time, from images taken by a CCD camera mounted on a car-like mobile robot. The proposed technique is based on Histograms of Oriented Gradient (HOG) and SVM classifier. The implementation of our detector has provided good results, and can be used in robotics tasks.**

Human detection; HOG descriptor; SVM classifier; Real- time detection.

I. INTRODUCTION

The problem of object recognition is to decide whether a specific object or object within a class of objects is contained in an image or not. This problem can be seen as a match between the target model and a set of descriptors, which are extracted from an image test. This generalization, as simple as it seems to be, is able to explain the existence of multiple approaches that depends on the choice of object descriptors, type and complexity of its model, and the methods used for the learning and matching object model. Beyond being a general problem in computer vision, object recognition is an important tool for many applications. It is used in video surveillance, digital image databases, and largely for complex robotic tasks as in our case.

Person detection is particularly difficult, mainly because of the high variability of appearances and possible situations. The problem is to find a representation of a human that is both sufficiently generic to cover all types of situations, and sufficiently discriminative for humans. For this, we generally use an intermediate representation, based on the computation of one or more features, taken from the information contained in the only values of the image pixels.

In this paper, we address the problem of person detection from images taken by a CCD camera, embedded on an outdoor mobile robot. Several approaches have been implemented in real-time, to perform an efficient and fast detection, in order to be used in autonomous navigation tasks. This work has been done in the context of autonomous transportation system project of NCRM team, in CDTA.

The paper is organized as follow: in section II we present the HOG descriptor combined with the SVM classifier, used in this work. Section III is dedicated to show experimental tests and results discussion.

II. HUMAN DETECTION APPROCH

Early works on people detection date from the late 1990s [1]. In one of the first proposed methods, stereovision is used to detect objects using a Hough transform. The method can detect pedestrians, but is not exclusive to this type of object. In 1998, Wöhler and Heisele use leg movement to achieve the pedestrian detection and classification, with constraints on the location of pedestrians to the ground [1]. However, these methods are not generic.

From the 2000s, research advances focused more on face detection and in particular the method of Viola & Jones, which has been extended in 2005 to target detection using motion [2]. The method allows a more generic detection, requiring no prior information about the structure of the scene, with a processing time close to the real-time. In 2005, INRIA researchers propose a new technique based on the histograms of oriented gradient (HOG) [3]. The good performances obtained by this technique make it few years after a standard method [4]. In 2008, researchers from Rutgers University have introduced a new descriptor built as a covariance matrix, which provides better performance in known environment [5]. Also in 2008, Mu & al. have successfully used local binary patterns (LBP), a type of features that had proven a good effectiveness especially when applied to face detection [6]. In this work, we focus our interest on the HOG based technique, which is detailed bellow:

A. HOG descriptor

HOG descriptors have been introduced by Dalal and Triggs in [7] and [8]. The main idea behind the histogram of oriented gradient is that the local appearance and shape of object in an image can be described by the intensity distribution of gradients or direction of the contours. The implementation of these descriptors can be obtained by dividing the image into small connected regions, called cells. Then, for each cell we compute a histogram of gradient directions or edge orientations for all pixels of the cell. The combination of these histograms is the descriptor.

The HOG descriptor has some key advantages. Since it operates on localized cells, the method maintains the invariance to geometric and photometric transformations. We implemented the HOG descriptor on C/C++ and using OpenCV [9] libraries following four steps as shown below:

978-1-4673-5289-5/12 $31.00 © 2012 IEEE

1) Training database conception: the database used was the "INRIA pedestrian database" which contains images covering a wide variety of pedestrians. We have then updated it by adding images of pedestrians in different states (running, standstill, in front, side and back) and some images of pedestrians in clothes from Islamic society: women wearing veil of different colors and shapes.

2) Luminance normalisation: The input images of our system are initially converted to grayscale. Then their luminance is normalized. We used then the Gamma correction.

3) Image gradient : The gradient is a key step for the descriptors formation. The accuracy of computed orientations and histograms and the results are closely related to the method used to calculate the gradient of the image. Fast computation of the gradient can be done using, for example, 1-D simple derivation masks, 2-D operators like Sobel or recursive operators like Deriche.

In our case, we have used the algorithm of the first derivative which is one of the most simple and fast operators. This operator uses convolution matrix to calculate an approximation of the horizontal and vertical derivative. Let I be the source image. Images which contain at each point the derivative approximations respectively horizontal and vertical are calculated as follows:

$$G_x = [-1\ 0\ 1] \times I \qquad (1)$$

$$G_y = [-1\ 0\ 1]' \times I \qquad (2)$$

At each point, the approximations of horizontal and vertical gradients are combined as follows to obtain an approximation of the gradient norm:

$$G = \sqrt{Gx^2 + Gy^2} \qquad (3)$$

We also compute the gradient direction as follows:

$$\theta = arctan\left(\frac{Gy}{Gx}\right) \qquad (4)$$

4) Gradient orientation histograms: A histogram is an array of numbers where each element corresponds to the frequency of occurrence of a range of values for a set of data. In a part of an image, for example, each box of the histogram may represent the pixels with the same color. A histogram is a transformation of the data space to the positive real numbers. From a statistical point of view, a histogram provides distribution of a certain type of data set. The image is divided into cells of 8x8 pixels size, and for each cell we compute the gradient orientation histogram. Each pixel of the cells participates in the vote. For each pixel of coordinates (x, y), the associated value in the histogram H is given by:

$$H(a) = H(a) + NG(x, y) \qquad (5)$$

We have created 9 sub-images called binaries images, as we have chosen the size of the histogram box equal to 20° (180°/20° = 9). All pixels in these images are set to zero except the pixels in the original image for which the values of the gradient orientation correspond to the particular case. These 9 images constitute the full histogram. For each cell of size 8 x 8 pixels, we compute the values of the 9 boxes of gradient orientation histogram. For a block of 2 x 2 cells, we calculate the HOG descriptor for each cell, then the obtained arrays are assembled into a single array of 36 components, and it is normalized in accordance with standard L1 defined by the normalization factor f given by:

$$f = \frac{v}{(\|v\|_1 + \epsilon)} \qquad (6)$$

were v represents the non-normalized vector containing all the histograms of a single block, $\|v\|k$ its k-norm and ϵ is a constant.

This normalized vector corresponds to the HOG descriptor for a block. An image 64x128 pixels, contains 7 x 15 blocks with multiple overlap. We assemble the normalized vectors obtained for all blocks in a single 1-D vector of 3780 components, which represents the final HOG descriptor for our case.

B. SVM Classifier

The set of descriptors (105 x 36 = 3780 values) is used to feed the SVM classifier, which generates a model (a set of support arrays). During the decision phase, the descriptors are calculated in an identical manner as in the learning phase. Decision making, regarding the class membership is made directly by the decision function of SVM. In our experiments, we use the linear kernel function to classify the descriptors given by the learning database.

III. EXPERIMENTAL RESULTS

We have implemented the HOG based detection technique in real-time on the mobile robot "Robucar" available in CDTA, under two different environments: a wide corridor (indoor environment) and a parking (outdoor environment). For that, we have used our database (called CDTA base), composed by 112 images of 320x240 pixels, and containing 216 human targets. In this case, we have considered as being a good detection if the frame covers more than the half of the human body. Images acquisitions and processing management are performed within a modular LAAS-CNRS architecture called "GenoM" [10], using C/C++ and the library OpenCV on a laptop with 2 GHz Core-i5 processor and 4 GB Ram.

Detection parameters of the proposed technique have been tested: Gamma correction of the image, gradient filter type, blocks and cells sizes and the global threshold. The goal is to identify the best parameters to increase the detection rate, and decrease the processing time.

A. Image Gamma correction

TABLE I.   DETECTION RATE & GAMMA CORRECTION

	Detection	False positive	Processing time (ms)
With Correction	125 (58%)	17 (8%)	200
Without Correction	105 (49%)	11 (5%)	180

Figure 1. Human detection without (left) and with (right) Gamma correction

We notice that introducing the Gamma correction of the image enhances the detection rate.

B. Gradient filter

We have tested two gradient filters, and here are the results:

TABLE II.   DETECTION RATE & GRADIENT'S FILTERS

Gradient filter	Detection	False positive	Processing time (ms)
Sobel	127 (59%)	16 (7.5%)	240
1-D Derivator	125 (58%)	17 (8%)	180

Note that the Sobel filter gives slightly better results, but with less processing time. We chose then to use the 1-D differentiator, because it is faster.

C. Cells& Blocks sizes :

TABLE III.   DETECTION RATE & CELLS SIZE

Cells size (pixel)	Detection	False positive	Processing time (ms)
6 x 6	162 (75%)	20 (9%)	700
7 x 7	163 (75%)	17 (8%)	600
8 x 8	158 (73%)	17 (8%)	150
9 x 9	158 (73%)	16 (7%)	1000
10 x 10	138 (64%)	13 (6%)	547

The cells size selected affects the quality of detection, and even more on the detection time. We chose the 8x8 size for a reasonable amount of time. The sizes below are slightly better, but their computation time does not allow a real-time application.

TABLE IV.   DETECTION RATE & BLOCKS SIZE

Block size	Detection	False positive	Processing time (ms)
24 x 24	160 (74%)	16 (7%)	135
28 x 28	168 (78%)	17 (8%)	143
30 x 30	172 (80%)	16 (7%)	145
32 x 32	173 (80%)	17 (8%)	150
36 x 36	177 (82%)	18 (8%)	160

We have then tried several combinations of blocks size and cells size. The one which gives the best detection is: blocks of 36x36 pixels, 9x9 pixels cell, but it constrains execution time to be near to 3 seconds. For real-time use (as for robotics tasks), we have chosen the combination: 32x32, 8x8, as it gives a quite good detection, with low processing time (135ms ≈ 7 frames/second).

D. Global threshold

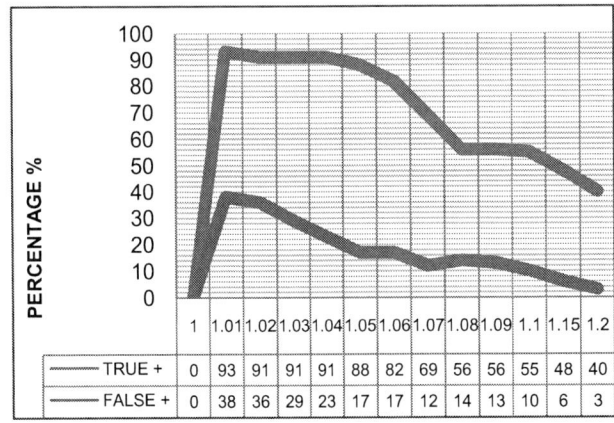

Figure 2. Detection rate according to global threshold

From this figure, we notice that the best detection rates (in bleu) are between thresholds values of 1.01 and 1.06, and the rate of false detection is quite low (in yellow) from 1.04.

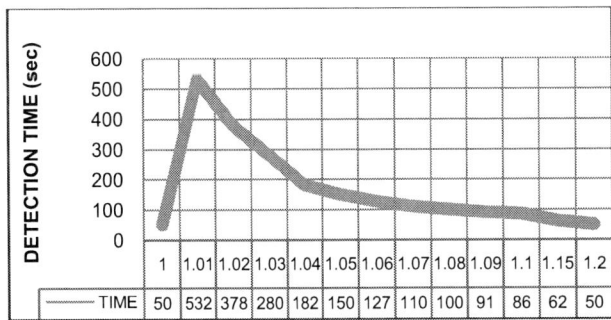

Figure 3. Detection time according to global threshold

In the second figure, we can see that processing time takes lowest values when the threshold is high. Combining both results, we have chosen the value of 1.05 which gives a fairly high detection rate. The false positive rate is low where detection time is close to be on real-time.

After parameters evaluations and tunings, we have tested our programs on several databases. Among them, one downloaded from the Internet (INRIA database). Here are presented some detection results applied on images taken from this database.

Figure 4. HOG based detector on INRIA database

Figure 5. HOG based detector on CDTA database

Figure 5 presents detection results applied on images taken from our base. Are summarized in this table statistics on detection rates and false detections taken from tests on the two cited databases.

Databases	Image number	Target number	Detected target	False detection	Rate detection
INRIA	900	1178	1012	170 (17%)	86%
CDTA	1689	4031	3491	712 (17%)	87%
All	2589	5209	4503	822 (17%)	86%

TABLE V.   TABLE OF RESULTS

IV.   CONCLUSION

In this paper, we present fast humans detection system implemented, with our own database, using histograms of oriented gradient (HOG) combined with the SVM classifier. Several tests have been performed to identify optimal parameters. Global detection rate is quite interesting, form results of detector application on two different databases. Processing time is suitable for embedded applications, but can be enhanced by detector combination. Future works will focus on integrating this technique in complex robotic task.

ACKNOWLEDGMENT

The authors are very grateful to Anthony Mallet from LAAS-CNRS, for its involvements in this work.

REFERENCES

[1] B. Heisele and C. Wöhler, *"Motion-based recognition of pedestrians"*, Proceedings of the 14th International Conference on Pattern Recognition, vol. 2, p. 1325-1330, Brisbane, Australia, 16-20 août 1998.

[2] P. Viola, M. Jones and D. Snow, *"Detecting Pedestrians using Patterns of Motion and Appearance"*, In IJCV, vol. 63, no 2, p. 153-161, 2005.

[3] N. Dalal and B. Triggs, *"Histograms of Oriented Gradients for Human Detection"*, Conference on Computer Vision and Pattern Recognition, p. 886 - 893, San Diego, USA, June 2005.

[4] M. Enzweiler and D. Gavrila, *"Monocular Pedestrian Detection: Survey and Experiments"*, TPAMI, Vol 31, n° 12, p. 2179-2195, December 2009.

[5] O. Tuzel, F. Porikli and P. Meer, *"Pedestrian detection via classification on Riemannian manifolds"*, IEEE Transactions on Pattern Analysis and Machine Intelligence, vol. 30, no 10, p. 1713-1727, October 2008.

[6] Y. Mu, S. Yan, Y. Liu, T. Huang, and B. Zhou, "Discriminative local binary patterns for human detection in personal album", CVPR, p. 1–8, 2008.

[7] N. Dalal and B. Triggs. "Histograms of oriented gradients for human detection ". In CVPR, p. 886-893, 2005.

[8] N. Dalal. "Finding People in Images and Videos". Phd Thesis, L'institut National Polytechnique de Grenoble, 2006.

[9] Sara Fleury, Matthieu Herrb. "Genom: Manuel d'utilisation", LAAS CNRS, France, 2003.

[10] R. Alami, R. Chatila, S. Fleury, M. Ghallab, and F. Ingrand. "An architecture for autonomy". Int. Journal of Robotic Research, vol. 17(4), p. 315–337, 1998.

Using the DAPGF Filter for Speech Analysis : Application for cochlear implants

Boucherit ismail

Electronics Department
Yahia Fares University - Médéa Algeria
And Signal and Communications Laboratory
Electronics Department
Ecole Nationale Polythechnique
Algiers Algeria
boucheritw@yahoo.fr

Guerti Mhania

Signal and Communications Laboratory
Electronics Department
Ecole Nationale Polythechnique
Algiers Algeria
mhania.guerti@enp.edu.dz

Abstract— **A cochlear implant is an electronic device that replaces the function of the damaged inner ear. The recent developments for cochlear implant are especially at the stage of the analysis of the sound and the strategies of stimulation. In this paper, we explore the use of a particular auditory filter-bank based on the Differentiated All-Pole Gammatone Filter (DAPGF) for frequency decomposition. The results show that the shape of the DAPGF filter has a better contribution at low frequencies compared to the conventional filter in noisy conditions.**

Keywords- Cochlear Implant ; Auditory Filter ; Gammatone.

I. INTRODUCTION

Cochlear implants (CIs) are surgically implanted prosthetic devices, which are used to provide a sensation of hearing in profoundly deaf people [1]. Contemporary cochlear implants consist of a microphone, a sound processor, a transmitter, a receiver, and an electrode array that is positioned inside the cochlea. The sound processor is responsible for decomposing the input signal into a small number of bands (16 - 22) via the Fast Fourier Transform or a bank of bandpass filters, and the envelopes are extracted from each band. The envelopes are used to modulate biphasic pulses which are in turn sent to the electrodes for stimulation [2].

Auditory filtering is closely related to some specific parts of the auditory system, especially the Basilar Membrane (BM). In time-domain auditory models, the spectral analysis performed by the BM is often simulated by a bank of Gammatone Filters (GTF) [3]. Auditory filter banks are non-uniform band-pass filter banks designed to imitate the frequency resolution of human hearing. The cochlea transfer function at a particular place in the BM is neither purely low-pass nor purely band-pass [4].

Speech intelligibility in noise environments might be limited for cochlear implant users since the sound fidelity perceived by CI users is not as high as normal hearing listeners. A number of solutions for the problem of Speech Recognition in noise have been proposed [5], [6]. They can be classified in two categories : speech enhancement and model compensation. In this paper we are interested in the second category in order to exploit an auditory filter-bank that is not much affected by

the noise. The auditory filter-bank shapes are a function of level, so we fix the level to study only the particularity of the shape.

II. AUDITORY FILTER

A. Gammatone Filter

The Gammatone Filter, developed by Patterson and al, is one such filter. Its name is due to the nature of its impulse response, which is a gamma envelope modulated by a tone carrier centered at f_c Hz [7].

$$g(t) = at^{n-1} \exp(-2\pi b B(f_c)t) \exp(j2\pi f_c t) \qquad (1)$$

$B(f_c)$ is the Equivalent Rectangular Bandwidth (ERB) of the center frequency f_c

$$B(f) = 01039 . f + 24.7 \qquad (2)$$

The GTF is inherently nearly symmetric in the pass band, while physiological measurements show a significant asymmetry in the biological cochlea transfer function. In addition, it is not easy to use the parameterization of the GTF to model level-dependent changes in the auditory filter [8].

The Gammachirp filter was derived by Irino and Patterson [3] as a theoretically optimal auditory filter which includes a chirp parameter in the impulse response of each filter. It can be expressed as the cascade of a Gammatone filter, with an asymmetric compensation filter. The Gammachirp auditory filter was developed to extend the domain of the gammatone auditory filter and simulate the changes in filter shape that occur with changes in stimulus level. Another cascade filter system namely the Differentiated All-Pole Gammatone Filter [4], provide a robust foundation for modeling cochlea transfer functions.

B. Differentiated All-Pole Gammatone Filter (DAPGF)

The DAPGF response is attractive because it exhibits certain characteristics suitable for modeling a variety of auditory data : level-dependent gain, linear tail for frequencies well below the centre frequency, asymmetry, etc. The DAPGF can be considered as a cascade of (N–1) identical

978-1-4673-5289-5/12 $31.00 © 2012 IEEE

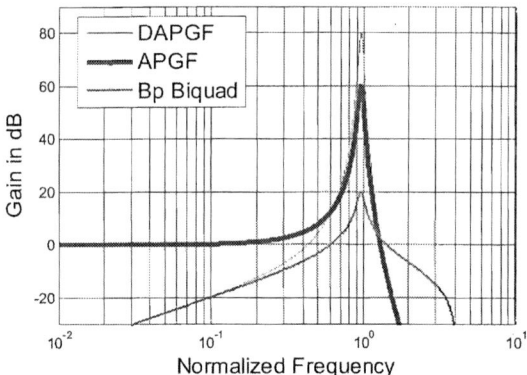

Figure 1. Transfer function of the DAPGF of N = 4 and Q = 10 and its decomposition to a 3rd-order APGF and a scaled BP biquad with a gain of 20 dB

Low Pass biquads (i.e. a (N–1) th-order APGF) and a rightly scaled Band Pass biquad [4]. The DAPGF transfer function is :

$$H_{DAPGF}(s) = \frac{w_0^{2N-2}}{[s^2 + \frac{w_0}{Q}s + w_0^2]^{N-1}} \times \frac{w_0 s}{[s^2 + \frac{w_0}{Q}s + w_0^2]} = \frac{w_0^{2N-1}s}{[s^2 + \frac{w_0}{Q}s + w_0^2]^N} \quad (3)$$

$$K_1 = \omega_0^{2(N-1)} \text{ and } K_2 = \omega_0$$

III. SPECTRAL ANALYSIS

A. FilterBank Analysis

The simulation consists on a filter-bank Analysis. We use at first the DAPGF Filter-bank with 21 channels for frequency decomposition. The Center Frequencies (CF) of the filter-bank were chosen according to critical-band spacing [9], we note here that the peak of the transfer function shifts to the left as the parameter Q decreases, so to get 0 dB DAPGF gain we take Q=0.912 and N=4. Then we use another conventional filter-bank (6th order Butterworth) with the same center frequencies and bandwidth at -3 dB (see Fig. 2).

Input speech was sentence "they hear" of 75 dB Sound Pressure Level (SPL), and Speech-Shaped Noise (SSN) was added so that the SNR was 5 dB.

Table I shows center frequencies of the conventional filter-bank with 21 channels.

TABLE I.   CENTER FREQUENCIES OF THE CONVENTIONAL FILTER-BANK

Chanals	Ch1	Ch2	Ch3	Ch4	Ch5	Ch6	Ch7
CF (Hz)	85	155	230	310	397	483	595
Chanals	Ch8	Ch9	Ch10	Ch11	Ch12	Ch13	Ch14
CF (Hz)	705	830	980	1140	1320	1533	1775
Chanals	Ch15	Ch16	Ch17	Ch18	Ch19	Ch20	Ch21
CF (Hz)	2060	2400	2810	3320	3950	4725	5650

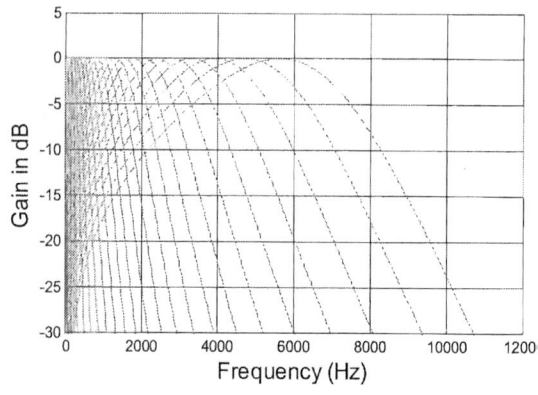

(a)

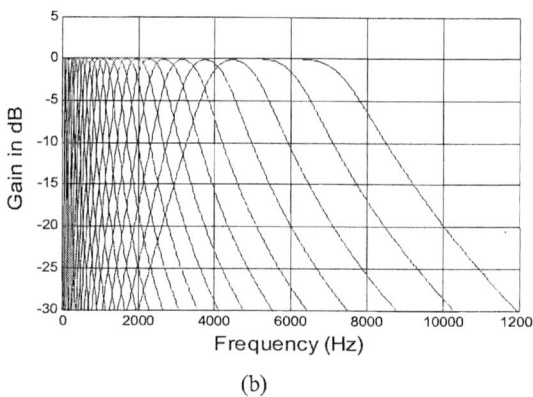

(b)

Figure 2.   Frequency response of the : (a) DAPGF filter-bank with Q=0.912 (b) conventional filter-bank

Fig. 3 and Fig. 4 show quite a few results with the present model. Two situations are indicated : filtered speech in noise (reconstructed after filtering). These results show that the formant trajectories are well represented when the two filter-banks were used.

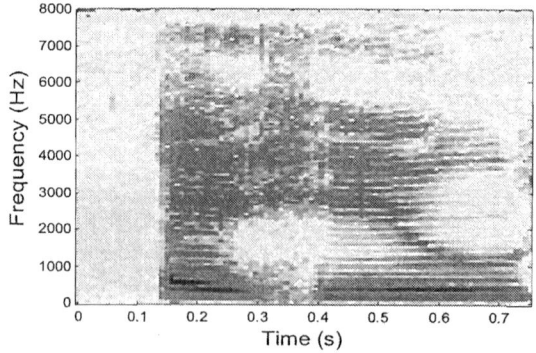

Figure 3.   Spectrograms of sentence "they hear". Speech processed by the conventional bandpass filter-Bank. SSN was added so that the SNR was 5 dB

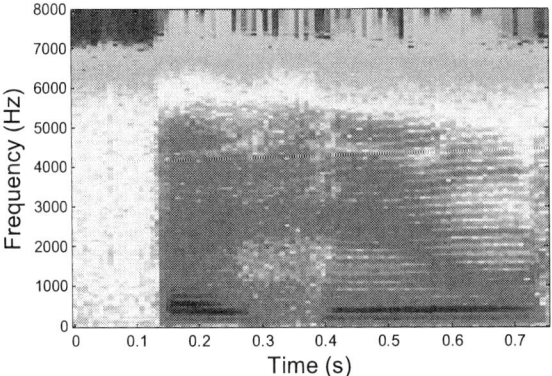

Figure 4. Spectrograms of sentence "they hear". Speech processed by the DAPGF filter-Bank. SSN was added so that the SNR was 5 dB.

B. Acoustic measurements for vowels

As the spectrograms in the above figure cannot show visibly the difference between the two types of filter-banks, we employ acoustic measurements as used in [10].

1) Spectral contrast measurements

Estimates of the vowel contrast were made on vocalic segment of the vowels (containing the 20% - 80% of the vowel duration) by computing the Root-Mean-Square energy (RMSe). Measurements on spectral contrast were made every 10 ms, using the 21 filter-bank values (Fig. 5). Spectral contrast (in dB) was defined to be the difference between the spectral peak and the spectral valley [11], and was computed as follows:

$$SC_{dB} = 20 \log_{10} \frac{F_{max}}{F_{min}} \qquad (4)$$

Where :

- $F_{max} = max \ F_i$, and $1 \leq i \leq 16$ is the spectral peak magnitude.
- $F_{min} = min \ F_i$ is the spectral valley amplitude, and F_i is the i-th RMSe filter-bank value.

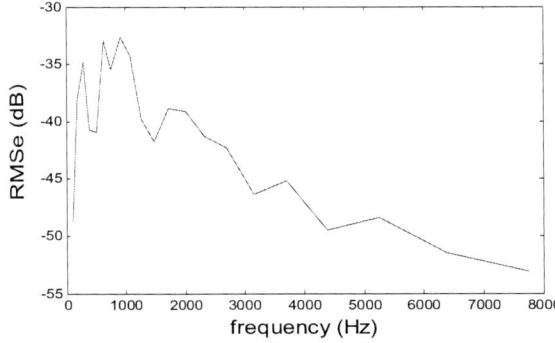

Figure 5. RMS energy vs. frequency for a 21 channels filter-bank . Plot of one sample frame 10 ms of vowel [a].

The spectral peak and valleys were sought only within 0-3 kHz region, corresponding to critical bands 1-16.

2) Spectral distance measurements

Spectral distance is used to quantify the effect of noise in individual frequency bands [10]. We use Euclidean distance metric of the filter-Bank energies without normalization. The vocalic segment of the vowels was first divided into frames of 10 ms. Each frame of 10 ms was first filtered through a 21 channel filter-bank. Three spectral distance measurements were made, one of each band, every 10 ms : 0-1 kHz (this is the band where the formant F1 resides for most vowels), 1 - 2.7 kHz (this is the band where the formant F2 resides for most vowels) and 2.7 - 6 kHz.

$$SB_1 = \sqrt{\frac{1}{9} \sum_{i=1}^{9} \left(F_i^c - F_i^n \right)^2} \qquad (5)$$

$$SB_2 = \sqrt{\frac{1}{6} \sum_{i=10}^{15} \left(F_i^c - F_i^n \right)^2} \qquad (6)$$

$$SB_3 = \sqrt{\frac{1}{6} \sum_{i=16}^{21} \left(F_i^c - F_i^n \right)^2} \qquad (7)$$

where :

- F_i^c denotes the i-th filter-bank energy of the clean vowel.
- F_n^c denotes the i-th filter-bank energy of the noisy vowel.

IV. RESULTS

The spectral contrast was measured for noisy vowels after adding noise at -5, 0, 5 and 10 dB SNR. Fig. 6 shows the spectral contrast results of individual vowel [a] for additive SSN at different SNRs. We note that the spectral contrast increase as the SNR increase.

The DAPGF filter-Bank has a narrow band at the first region (Fig. 2). Transfer function of the auditory filter is neither purely low-pass nor purely band-pass.

The output of the first filter contains a part of the power of following harmonics. Similarly, the output of the second filter contains a portion of the power of the third. It goes on and on, until the last filter that contains most of the power of other filter. That's why the DAPGF filter-Bank has more spectral contrast compared to the conventional filter-Bank.

The spectral distance gives an idea about spectral difference between the noisy and clean vowel spectra. Fig. 8 shows the case of SNR= 0 dB.

In the first region (0-1 kHz bandwidth), it is clear that the DAPGF filter-Bank have the less spectral values. In the middle region both filter-Banks have approximately the same values for vowel [a] and less spectral values for vowel [i] and [o]. The spectral difference between the two filter-Bank occurs in the high-frequency band (f > 2.7 kHz). This difference can be explained by the differences of the shape of the transfer function between the Butterworth and the DAPGF filter,

978-1-4673-5289-5/12 $31.00 © 2012 IEEE

especially below -3 dB. As the transfer function has a linear tail for frequencies well below the centre frequency. This shape make it more selective at low-frequency band, but at high-frequency band the DAPGF filter becomes more larger spanning all the spectrum bandwidth, the DAPGF filter as explain above passes a portion of the power which allowed the noise to exceed. We obtained the same results for vowels [i] and [o] at -5, 5 and 10 dB SNR.

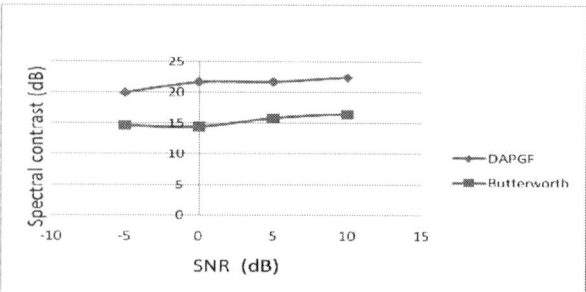

Figure 6. Spectral contrast results of vowel [a]

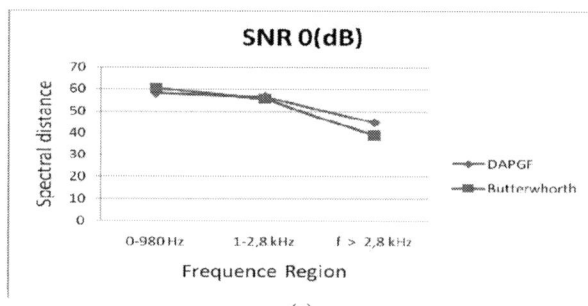

(a)

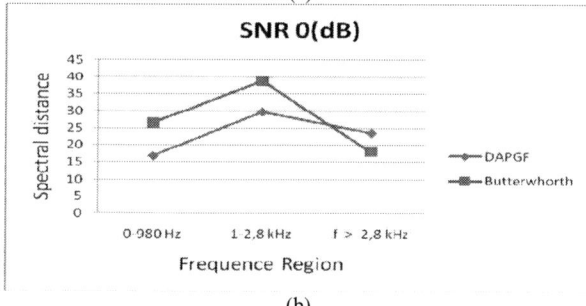

(b)

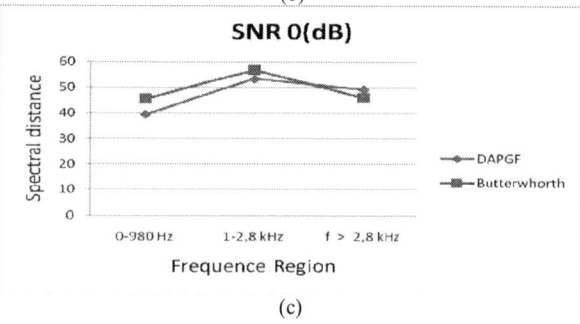

(c)

Figure 7. Spectral distance results for vowels corrupted by speech-shaped noise at SNR = 0 dB. (a) for vowel [a], (b) for vowel [i], (c) for vowel [o].

V. CONCLUSION

In this paper we describe an efficient speech analysis filter-bank based on the properties of the peripheral auditory system.

The characteristics of the Cochlea as a bank of filters performs frequency analysis and sound discrimination are closely linked to the notion of auditory filters that determine the powers of perception and extraction of information in our high-performance hearing system. It therefore seems worthwhile to consider an analysis and estimate of the noisy signal in agreement with the inner workings of the auditory system. We verify the advantage of the strategy based on the filter DAPGF for robust representation of the formant from a spectral analysis and measures.

The present study evaluated the performance of the DAPGF auditory filter as well as good filter-bank in terms of enhancing speech intelligibility in noisy conditions, particularly at low-frequency band (< 3 kHz).

In general, the results from the present study suggest that the DAPGF filter can replace the contemporary Butterworth filter at low frequencies. The DAPGF filter bank is intended for use as an input to CI speech processor.

REFERENCES

[1] V. Gopalakrishna, N. Kehtarnavaz, and P C. Loizou, "A Recursive Wavelet-Based Strategy for Real-Time Cochlear Implant Speech Processing on PDA Platforms," IEEE. Trans. Biomed. Eng, vol. 57, no. 8, pp. 2053–2063, August 2010.

[2] Y. Hu and P C. Loizou, "A new sound coding strategy for suppressing noise in cochlear implants," J. Acoust. Soc. Am., vol. 124, no. 1, pp. 498–509, July 2008.

[3] T. Irino and R. D. Patterson, "A time-domain, level-dependent auditory filter: the gammachirp," J. Acoust. Soc. Am., vol. 101, no. 1, pp. 412-419, January 1997.

[4] A.G. Katsiamis, E. M. Drakakis, and R. F. Lyon, "Practical Gammatone-like Filters for Auditory Processing," EURASIP Journal on Audio, Speech, and Music Processing, vol. 2007, 15 Pages, 2007.

[5] Y. Hu and P C. Loizou, "A comparative intelligibility study of single-microphone noise reduction algorithms," J. Acoust. Soc. Am., vol. 122, no. 3, pp. 1777–1786, September 2007.

[6] F. Toledo, P C. Loizou and A. Lobo, "Subspace and Envelope Subtraction Algorithms for noise Reduction in cochlear Implants," 25 Annual International Conference of the IEEE EMBS, Cancun, Mexico September 17-21, 2003.

[7] R. D. Patterson, K. Robinson, J. W. Holdsworth, D. McKeown, C. Zhang, and M. Allerhand, "Complex sounds and auditory images," in Auditory Physiology and Perception, Y. Cazals, L. Demany, and K. Horner, Eds., pp. 429-446. Pergamon, Oxford, 1992.

[8] A. G. Katsiamis , E. M. Drakakis, and F. Richard Lyon, "Introducing the Differentiated All-Pole and One-Zero Gammatone Filter Responses and their Analog VLSI Log-domain Implementation," In Proceedings of the 49th IEEE International Midwest Symposium on Circuits and Systems, vol 1, pp. 561–565, Augus 6-9, San Juan, Puerto Rico, 2006 .

[9] E. Zwicker and H. Fastl, "Psychoacoustics: Facts and Models," Springer Verlag: Berling, 1999.

[10] G. K. Parikh, and P C. Loizou, "The effect of noise on the spectrum of speech," J. Acoust. Soc. Am., vol. 118, no. 6, pp. 3874–3888, December 2005.

[11] P C. Loizou and O. Poroy, "Minimum spectral contrast needed for owel identification by normal hearing and cochlear implant listeners," J. Acoust. Soc. Am., vol. 110, no. 3, pp. 1619–1627, September 2001.

Wavelet based segmentation and time-frequency caracterisation of some abnormal heart sound signals

Nayad Kouras, Daoud BOUTANA

dept.of electronic, faculty of sciences and technology
university of Jijel ,
jijel, Algeria
nayadkouras@yahoo.fr, daoud.boutana@mail.com

Messaoud BENIDIR

Laboratoire des signaux et systèmes
Supelec, université Paris –Sud
91192 Gif sur-Yvette,France
benidir@LSS.supelec.fr

Abstract— **Biomedical signal recordings are so complex and nonstationnary, they are also affected by different kinds of noise that make their interpretation difficult. The major goal of the paper consists of two ideas. In the first one, we present the results of segmentation method followed by the time frequency caracterisation of some phonocardiogram (PCG) signals. The paper using the Discrete Wavelet Transform (DWT) in conjunction with Shannon entropy provided a useful tool for phonocardiogram (PCG) segmentation. In the segmentation technique, we calculate the entropy of the details coefficients at each level and threshold it in order to detect the murmur of heart sound signals. Several real-life signals were used: Early Aortic Stenosis (EAS), Late Aortic stenosis (LAS), Mitral Regurgitation (MR), Aortic Regurgitation (AR) .The results of the method illustrate clearly the detection of the main components S1, S2, S3, pathological murmurs and the identification of the valves disease.**

Keywords- **wavelet analysis, time-frequency representation, heart sound signal, pathological murmurs.**

I. INTRODUCTION

Heart sound segmentation is considered as a helpful operation for detection and partitioning in main component and especially pathological murmur in each cardiac cycle. The two main audible heart sounds in a normal cardiac cycle are the first (S1) and second (S2) heart sound. The first heart sound S1 consists of two major components M1 and T1, corresponding, respectively to the mitral and tricuspid valves. However the second heart sound S2 consists of two main components noted A2 and P2 corresponding respectively to the aortic and pulmonary valves. Information given by auscultation only, is still insufficient to characterize and diagnose some heart diseases. In [1,2] the authors used Discrete Wavelets Transform (DWT),implemented by Mallat algorithm [3] and thresholding to denoise PCGs signals. In [4] the authors present a comparative study of some wavelet functions (the Daubechies of order 4, 6, 8, 10 and 15, Coiflets of order 3, 4, 5 and finally Symlets 10 in the denoising of (PCG) signals. The authors presents in [5] the performances of the DWT and the packet wavelet transform in the PCG signal analysis using some wavelet function. So, several studies indicate that Daubechies order 10 (db10) outperform other wavelet functions. Consequently, db10 wavelet may be considered such as potential candidates in the PCG signal segmentation. So, the work presented in this paper will focus

mainly on the analysis of abnormal PCG signals using wavelet analysis that allows viewing information in different filter banks. Wavelet decomposition using discrete time filter banks make the identification and segmentation of HS signal easier. It permits to locate efficiently in time and frequency domain the different components of the HS signal. The segmentation method developed in [6] is based on normalized average Shannon energy to compute the PCG envelope. Another paper present a successfully method for heart sound segmentation [7] based on Shannon energy. However, the method fails to segment heart pathological murmurs. Another segmentation method based on windowing the HS signal using the DWT and selecting the wavelet detail coefficients at several bands by an adaptive peak detector, is proposed [8]. Samjin Choi et al. used [9] the normalized average Shannon energy, the envelope information of Hilbert transform, and the cardiac sound characteristic waveform in order to extract the envelope for the cardiac sound signal segmentation based. Recently, Boutana et al in [10] used the time-frequency distribution in conjunction with Rényi entropy measure for identification and segmentation of some pathological heart sound signals. In this paper, we achieve the segmentation of some abnormal real-life phonocardiogram (PCG) signals using wavelet analysis. Then, we use the time-frequency distribution [11] such as the spectrogram for characterizing the separated murmur signals.

The paper is organized as follows. In Section 2, we give a brief overview of wavelets analysis. We present in section 3 the theoretical aspect of the time-frequency representation. In section 4, we present the segmentation method based on wavelet analysis using entropy measure. Section 5 summarizes and concludes the paper.

II. WAVELET ANALYSIS

The Continuous Wavelet Transform (CWT) of a signal x (t) is given by:

$$T(a,b) = \frac{1}{\sqrt{a}} \int_{-\infty}^{+\infty} x(t)\psi^* \left(\frac{t-b}{a} \right) dt \qquad (1)$$

Where $\psi^*(t)$ is the complex conjugate of the analyzing wavelet function or the mother wavelet $\psi(t)$, a is the scale parameter, b is the translation parameter of the wavelet and t is the time variable. Equation (1) may be re-written as:

$$T(a,b) = \int_{-\infty}^{+\infty} x(t)\psi_{(a,b)}^*(t) dt \qquad (2)$$

978-1-4673-5289-5/12 $31.00 © 2012 IEEE

$$\psi_{(a,b)}(t) = \frac{1}{\sqrt{a}} \psi\left(\frac{t-b}{a}\right), \quad a > 0, \text{and } b \in R \qquad (3)$$

represents the scaled and translated versions of the wavelet basis function [3, 4]. For the implementation of the wavelet transform, the parameter a and b need to be discretized. When the discretization of the parameters a and b is dyadic (i.e., $a = 2^j$ and $b = k2^j$), an efficient implementation may be obtained by using quadrature mirror filter (QMF) decomposition at each scale level j. The family of wavelets generated by dyadic scaling and translation forms an orthonormal basis [3, 7]. So, Equation (3) can be re-written as:

$$\psi_{j,k}(t) = 2^{-j/2} \psi\left(2^{-j}t - k\right) \qquad (4)$$

Similarly, the scaling function (or father wavelet) is given by :

$$\varphi_{j,k}(t) = 2^{-j/2} \varphi\left(2^{-j}t - k\right) \qquad (5)$$

The above discrete version of the CWT is called the Discrete Wavelet Transform (DWT) and can be implemented using Mallat's algorithm, which uses filter bank pairs of low pass and high pass filters [3].

The approximation (or coarse) wavelet coefficients are obtained by projecting the signal on scaled and translated versions of the scaling function as:

$$A_j = \sum_k a_{j,k} \varphi_{j,k} \quad \text{with} \quad a_{j,k} = \langle x(t), \varphi_{j,k}(t) \rangle \qquad (6)$$

On the other hand, the details wavelet coefficients are obtained by projecting the signal on scaled and translated versions of the basic wavelet function as:

$$D_j = \sum_k d_{j,k} \psi_{j,k} \quad \text{with} \, d_{j,k} = \langle x(t), \psi_{j,k}(t) \rangle \qquad (7)$$

In the above expressions, $\varphi_{j,k}$ and $\psi_{j,k}$ represent the scaling and wavelet functions, respectively, at the j-th level of decomposition, for j=0,...., J. At each level of decomposition, the high pass filter produces detail information in the signal, Dj, while the low pass filter produces coarse approximations, Aj. The original signal may be written as:

$$x(t) = A_j(t) + \sum_{j \leq J} D_j(t) \qquad (8)$$

III. TIME-FREQUENCY REPRESENTATION

Another useful tool in the analysis of non-stationary signals is called the Time-frequency distributions (TFDs). These representations considered as energy distribution in the time-frequency plane can be expressed as [12]:

$$C_x(t,f) = \iiint e^{j2\pi(\xi t - \xi f - f\tau)} \Phi_x(\xi,\tau) x\left(u+\frac{\tau}{2}\right) x^*\left(u-\frac{\tau}{2}\right) du \, d\tau d\xi \qquad (9)$$

Where x(t) represents the analytical form of signal under consideration (x*(t) is the complex conjugate of x(t)) and $\Phi_x(\xi,\tau)$ is called the kernel of the distribution. All the integrals are from $-\infty$ to $+\infty$. A choice of a particular kernel function yields a particular quadratic TFD with its own specificities [1]. In particular, the kernel of the Wigner-Ville distribution (WVD) is unity $\Phi_x(\xi,\tau)=1$. The kernel of the spectrogram is the Wigner distribution of the analysis window itself so $\Phi_x(\xi,\tau) = W_h(\xi,\tau)$ and the spectrogram is defined as the squared magnitude of the short-time Fourier transform (STFT), that is:

$$|S(t,f)|^2 = \left| \int_{-\infty}^{+\infty} x(\tau) h(\tau-t) e^{-j2\pi f\tau} \, d\tau \right|^2 \qquad (11)$$

The spectrogram has been used in this paper in order to resolve the time-frequency localization of the pathological murmurs in consideration.

IV. METHOD OF SEGMENTATION

For this experiment we tested the efficiency of the proposed method implemented in MATLAB on four groups of subjects: EAS, LAS, MR and AR. The valvular heart disorders are detected by using the DWT at the selected level. The entropy of detail's coefficients may be used; we employed a thresholding scheme to distinguish between murmur and the main component. The proposed segmentation method includes the following steps:

Step 1(DW decomposition): the first step is to decompose the heart signal into six levels using Daubechies (db10) as a wavelet filter. The choice of the best level of decomposition is based on the Shannon entropy values at each level of decomposition.

Step 2 (the entropy of details coefficients): Calculation of the normalized Shannon entropy of details coefficients as follow:

$$E_j = -d_j \log(d_j) \qquad (12)$$

Where d_i is the detail coefficient at a level of decomposition.

Step 3(thresholding entropy values): Distinguish between the main component (S1-S2) of the PCG signal and the murmur by thresholding the entropy values. We propose the mean value of the Shannon entropy and investigate its use as an appropriate threshold to determine the boundary between the main components and the pathological murmur. Then, samples of the main component were detected for points of the entropy values below the threshold and those of the murmur were detected for everything above the threshold.

Step 4 (reconstruction of signals): In the synthesis step we obtained separated signals using the inverse DWT.

V. EXPERIMENTAL RESULTS AND DISCUSSION

The segmentation study was conducted on normal and abnormal real-life PCG signals [12]. Each PCG signal was sampled at a frequency : EAS and LAS signals with 11 KHz , MR and AR with 8 KHz.The frequency rangs of the details coefficients vary for each order level decomposition and depend to the sampling frequency following the values given in TABLE 1.

A. Example1: Case of the early aortic stenosis

The aortic stenosis is defined as a narrowing of the aortic valve, it's impedes the delivery of blood in the body through the aorta and makes the work of heart difficult. In this case, the murmur is located in early systole. The applied method on the EAS signal has given the results showed by the Fig. 1.It can be seen that the murmur signal is well separated. The method using the entropy values such as a threshold has given the 5th level of the decomposition allowing a good reconstruction of the murmur signal and the main components (S1 and S2) of the HS signal. The 5th level of wavelet decomposition give the detail d4 corresponding to the frequency ranges situated from 172 Hz to 344 Hz. Also, the time-frequency representation obtained by the spectrogram in Fig. 2 show the same frequency band with a mean frequency value located at 250 Hz.

B. Example 2: Case of the late aortic stenosis

The murmur of the LAS is located in late systole just before the start of the second heart sound signal. The results obtained by the proposed method on the LAS signal are showed by the Fig. 3 It can be seen that the pathological murmur is well separated and arriving just before the second S2 signal. Using the detail coefficients d5 in the reconstruction of the murmur signal and the main components of the HS signal, we obtain the results showed in Fig.3. According to the value of frequency sampling, the frequency band of the murmur signal is span from 172Hz to 344Hz. The results are confirmed by the spectrogram such as contour plot in Fig.4. From the figure, the murmur signal occupied the same frequency band where the mean frequency is situated at 250 Hz.

TABLE 1 FREQUENCY RANGES OF THE DETAILS COEFFICIENTS

Sampling Frequency	Level decomposition					
	d1	d2	d3	d4	d5	d6
8000 Hz	2000-4000	1000-2000	500-1000	**250-500**	125-250	62,5-125
11 000 Hz	2750-5500	1375-2750	687,5-1375	344-687,5	**172-344**	86-172

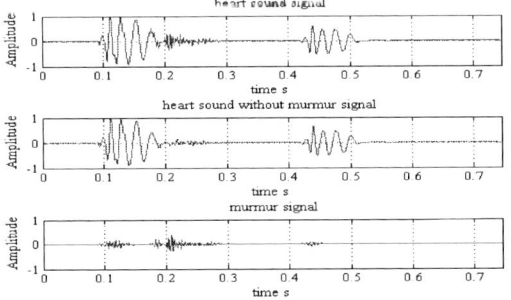

Figure 1. Segmentation of the early aortic stenosis signal : from top to bottom: original signal ,separated signal without murmur and pathological murmur.

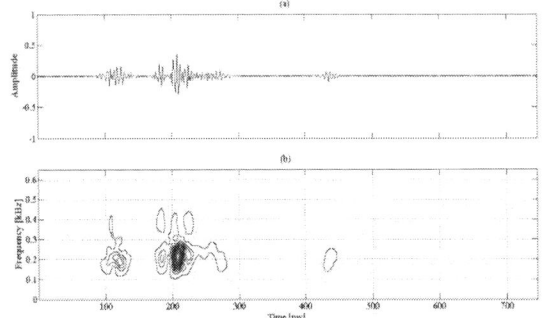

Figure 2. (a) separated murmur and (b) Time-frequency representation of the separated murmur signal for early aortic stenosis signal.

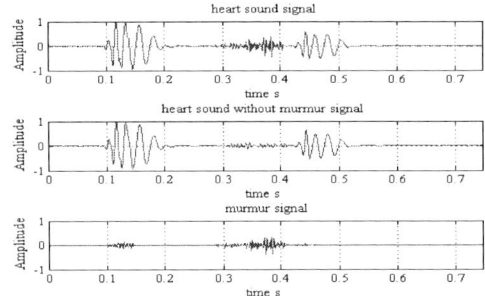

Figure 3. Segmentation of the late aortic stenosis signal: from top to bottom: original signal, separated signal without murmur and pathological murmur.

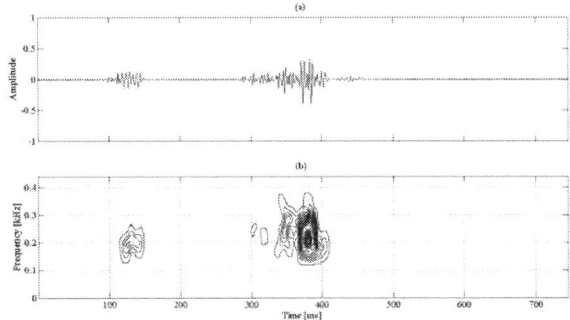

Figure 4. (a) Separated murmur and (b) Time-frequency representation of the separated murmur signal for late aortic stenosis signal.

C. Example 3: Case of the mitral regurgitation

Another pathological HS signal in order to validate the method is used called the Mitral Regurgitation (MR). In the case of the MR, the valve does not close properly and causes blood to leak back (regurgitate) into the left atrium when the left ventricle contracts. MR is mainly characterised by systolic murmurs starting just after the first HS S1 and the presence of the third HS S3 as can be seen in the bottom of the Fig. 5. The best level of wavelet decomposition obtained is the 4th order So, the applied method has permit to separate the systolic murmur and the main components S1, S2 and also S3 illustrated in Fig 5.The observation of the corresponding frequency ranges to the 4th order of decomposition and the time frequency representation in Fig. 6 show that the murmur of MR occupies a frequency band from 250 Hz to 500Hz.

D. Example 4: Case of the aortic regurgitation

We study also the case of the Aortic Regurgitation (AR) produced by the backflow of blood across the aortic valve and caused a diastolic murmur. The signal of AR is represented by the top of the Fig. 7 where the murmur is located at the end of the second HS signal. The applied method has given the 3rd order as the best level of decomposition for segmentation. The separated murmur characterising this disease of HS signal is represented by the bottom part of the Fig. 7. The time frequency representation of the separated murmur signal is illustrated in Fig. 8. From the figure, we can observe that the frequency band of the murmur caracterising the AR signal span from 200Hz to 600 Hz with a mean frequency located at 400 Hz having multicomponents configuration.

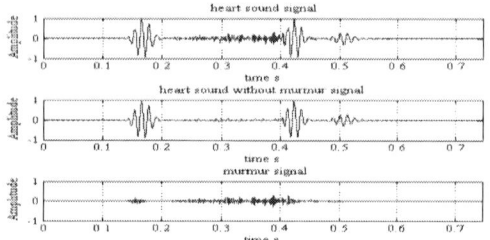

Figure 5.Segmentation of the mitral regurgitation signal : from top to bottom: original signal ,separated signal without murmur and pathological murmur.

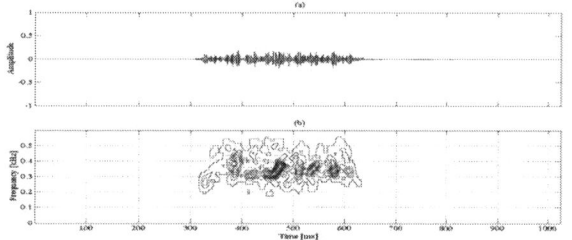

Figure 6. (a) Separated murmur and (b)Time-frequency representation of the separated murmur signal for mitral regurgitation signal.

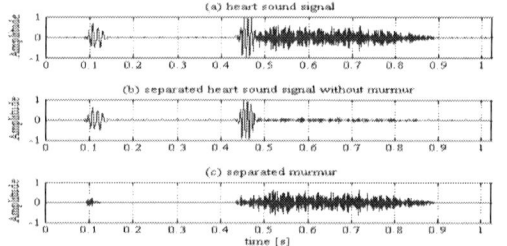

Figure 7. Segmentation of the aortic regurgitation signal : from top to bottom: original signal,separated signal without murmur and pathological murmur.

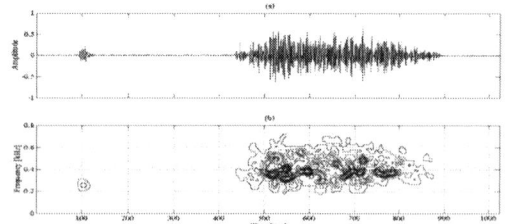

Figure 8. (a) Separated murmur and (b)Time-frequency representation of the separated murmur signal for aortic regurgitation signal.

VI. CONCLUSION

In this paper, we presented a segmentation method based on wavelet decomposition and reconstruction. The wavelet analysis is conducted by Daubechies db10 considered as the best wavelet function in earlier works. The best selection of the level decomposition and the details coefficients is achieved on thresholding the entropy values of the details coefficients.The performances of the method have been assessed on real-life abnormal heart sound signals such as EAS, LAS, MR and AR. For these complicated diseases, the method has shown good results. Time frequency characterisation has been achieves on these separated murmur signals. These features characterising the signals formed by the first and the second heart sound signals and abnormal murmurs can be used for diagnosis of characteristic valvular or dysfunction diseases of the cardiac heart.

In our future work, we will investigate very large number of pathological heart sound signal and introduce another criterion for features extraction.

REFERENCES

[1] David L Donoho and I.M. Johnstone, "Ideal spatial adaptation by wavelet shrinkage," Biometrika, Vol 81, pp 425-455, 1994.

[2] David L Donoho, "De-noising by soft thresholding. IEEE Trans. on Information Theory, 41(3), pp 613–627, 1995.

[3] S. Mallat, "A theory for multiresolution signal decomposition: the wavelet representation," IEEE Trans. on Pattern Anal. Mach. Intell., vol.2, N°7,pp.674-693, July 1989.

[4] Nayad Kouras, D. Boutana, M. Benidir and B. Barkat, "A comparative study of some wavelet functions in the denoising of phonocardiogram signals," in Proc. of the 2nd International Conference on Advanced Computer Theory and Engineering (ICACTE 2009), September 25 - 27, 2009, Cairo, Egypt.

[5] L. Hamza Cherif, S.M. Debbal and F. Bereksi-Reguig, "Choice of the wavelet analyzing in the phonocardiogram signal analysis using the discrete and the packet wavelet transform," Expert Systems with Applications 37 (2010) 913–918

[6] Liang H, Lukkarinen S, Hartimo I., "A heart sound segmentation algorithm using wavelet decomposition and reconstruction Proceedings of the IEEE EMBS, 19th annual international conference 1997;pp 1630-1633.

[7] D. Kumar, P. Carvalho, M. Antunes, J. Henriques, L. Eug´enio, R. Schmidt, J. Habetha, "Detection of S1 and S2 Heart Sounds by High Frequency Signatures", Engineering in Medicine and Biology Society, 2006. EMBS '06. 28th Annual International Conference of the IEEE Aug. 30 2006-Sept. 3 2006 1410 – 1416

[8] Zümray Dokur and Tamer Ölmez, " Feature determination for heart sounds based on divergence analysis", Digital Signal Processing 19 (2009) 521–531

[9] Samjin Choi , Youngkyun Shin , Hun-Kuk Park, "Selection of wavelet packet measures for insufficiency murmur identification", Expert Systems with Applications: An International Journal, v.38 n.4, p.4264-4271, April, 2011

[10] D. Boutana, M. Benidir and B. Barkat, "Segmentation and identification of some pathological phonocardiogram signals using time-frequency analysis",Signal Processing, IET Volume: 5 , Issue: 6 , 2011 , Page(s): 527 -537

[11] B. Boashash, "Time Frequency signal analysis and Processing: A Comprehensive Reference," in Elsevier Publications, First edition 2003, ISBN 0- 08- 0443354.

[12] http://www.egeneralmedical.com/

A Comparative Evaluation of VoWLAN Call Capacity With and Without Codec's Silence Suppression

Yacine Harkat

Theory and Computer Systems Engineering, DTISI
CERIST Research Center
Algiers, Algeria
yacine.harkat@gmail.com

Abderrahmane Amrouche

Speech Communication and Signal Processing Laboratory,
Faculty of Electronics and Computer Sciences, USTHB,
P.O. Box 32, El Alia, Bab Ezzouar, 16111, Algiers, Algeria
namrouche@usthb.dz

Abstract— **This paper presents the findings of experiments that were performed on a VoWLAN (Voice over WLAN) cell using the IEEE 802.11b standard. The experiments were simulated using OPNET Modeler 14.0 software. The experiments were aimed at finding the maximum number of simultaneous Voice over Internet Protocol (VoIP) users the VoWLAN cell would support under the G.711 and G.729 codec standards with and without silence suppression. The packetization interval was 10 milliseconds (ms). To determine the capacity of the model under various experiments, we checked three metrics: jitter, delay and packets loss. The findings of the first experiment (without silence suppression) indicated that, the maximum number of simultaneous VoIP users the VoWLAN cell was able to support was 6, using the IEEE 802.11b mode under G.729A codec, which is consistent with recent research findings. The finding of the second experiment (with silence suppression) however, demonstrated that with codec's silence suppression enabled, the maximum number of VoIP users the VoWLAN cell was able to support was 13, using the IEEE 802.11b mode under G.729A codec.**

WLAN, IEEE 802.11b, Codec, VoIP, OPNET, Call capacity, Silence Suppression, and VoWLAN.

I. INTRODUCTION

Voice over IP (VoIP) or IP telephony is a very popular way of communication not only for single users but also for big enterprises. Due to fast-growing wireless technology and ease of use of wireless networks, VoIP is now being deployed over Wireless LANs (VoWLANs). Capacity planning is an essential factor to consider at the time of developing VoIP network. Many works have been done to study the performance of transmitting voice over WLAN. In [1] Hole and Tobagi give out the analysis and simulation results of the capacity of VoIP over WLAN under different codecs, varying delay constraints, channel conditions and voice call quality requirements. Many other related researches are also done in the literature [2, 3, 4, and 5]. However, none of them consider the codec's silence suppression. In the first experiment of this paper we evaluated the VoWLAN call capacity. Generally a VoIP application consists of the codecs (Coder-decoder), the Real-Time Protocol

(RTP), the Real-Time Control Protocol (RTCP) and the User Datagram Protocol (UDP). The codec is usually a Digital Signal Processor based system employing one of the compression algorithms. Recent encoders can be used in order to reduce bandwidth utilization. Further reduction in the data rate can be achieved by means of Silence Suppression. In the second experiment of this paper we evaluated the VoWLAN call capacity under codecs with silence suppression performed.

The rest of this paper is organized as follows: Section II gives an overview about the VoIP system, section III highlights the silence suppression technique, Section IV covers the WLAN technology, section V describes the scenarios we tested, section VI and VII analyses and discusses the results, and the conclusion is drawn in section VIII.

II. VOIP SYSTEM OVERVIEW

The commonly used VoIP codecs are G.711 [6], G.729 [7] and G.723.1 [8], which are standardized by the ITU-T in its G-series recommendations. The reduced bandwidth utilization is at the expense of additional complexity and encoding delay as well as slightly lower quality. Further reduction in the data rate can be achieved by means of Voice Activity Detection (VAD), in which case no signal is encoded during silence periods. When silence suppression is employed, the codecs then operate in two states: a silent state at zero bit-rate and an active state at the compressed bit-rate. Regardless of the state, the frame period and frame size are still fixed. More discussions on silence suppression can be found in Section III.

As the voice packets are sent over an IP network, they are subject to variable delays and network drops. Even if a lot of voice codecs can tolerate some small packet loss without severe degradation, voice traffic has unacceptable performance if long delays are incurred. It is recognized that the end-to-end delay has a great impact on the perceived quality of interactive conversations with a threshold effect around 150 ms [9]. Voice traffic is also highly sensitive to the other network Quality of Service (QoS) factors such as jitter and packet loss. In this paper, we measured delay, jitter and packet loss for voice traffic. We based our observations on the values in Table I.

978-1-4673-5289-5/12 $31.00 © 2012 IEEE

TABLE I. FACTORS ACCEPTED VALUES

Factor	Range	
	Accepted range	Unaccepted range
Delay (ms)	0 to 150	Above 150
Jitter (ms)	0 to 50	Above 50
Packets Loss %	0 to 2%	Above 2%

III. SILENCE SUPPRESSION

Typical voice conversations contain up to 65% of silence [10]. Reduction in the codec bit rate can be achieved by means of silence suppression, in which case no signal is encoded during silence periods. This feature saves more than the half of required bandwidth. Silence suppression includes three main components: voice activity detection (VAD), discontinuous transmission (DTX), and comfort noise generation (CNG). VAD is responsible for determining when the talker is silent. DTX stops transmitting frames when the VAD has detected a silent period. CNG is used to recreate low-level background noise to the receiver. CNG is recommended for user confidence in the call connection: if the call appears too quiet, users may anticipate that the call has been disconnected. When silence suppression is employed, the codec then operate in two states: a silent state at zero bit-rate and an active state at the compressed bit-rate. Unlike the PSTN, which generally does such silence suppression across trans-continental links, IP telephony performs silence suppression at the end-points. Hence, no network support is required to take advantage of end system silence suppression. This leads to a reduction in cost to perform the silence suppression (as it is distributed to end systems where it can be done cheaply).

IV. WIRELESS LAN OVERVIEW

The architecture of this type of network is the same as Local Area Network (LAN)'s except that the transmission happens via radio frequency (RF) or Infrared (IR) and not through physical wires/cables, and at the MAC sub-layer, as uses different standard protocol. The main characteristics of the WLAN technologies are mobility, simplicity, scalability, edibility and cost effectiveness. The IEEE 802.11-1999 standard [11], known as a wireless Ethernet, specifies two channel access mechanisms: A mandatory contention-based distributed coordination function (DCF) and an optional polling-based point coordination function (PCF). DCF uses Carrier Sensing Multiple Access/Collision Avoidance CSMA/CA as the access method [12-15]. CSMA/CA deploys Interframe Space (IFS), Contention Window (CW) and Acknowledgment. IFS is waiting period of transmission over IP-based network. In DCF access method, STA senses traffic before send or share packet over IP-based network. If DCF found medium is sensed busy, sender would be waiting until the medium ready for the transmission. This process is called DCF interfram space (DIFS). Then the sender STA will send Request-to-Send (RTS) to get permission from receiver by sending Clear-to-Send (CTS). CW is a number of slots which ranging from 0 to 1. CW stops its timer while station finds the channel busy due to overflow or bursty traffic over networks. And restart timer when the channel is sensed as idle. STA can calculate the random counter interval backoff and choose from CW. The senders then enter the backoff phase, in which every sender chooses a random backoff counter from (0, CWmin).

Then CW as: Backoff_Phase= rand (0, CW) * slot_time, where CWmin<CW<CWmax. The backoff counter is reduced by 1 for each idle time slot and stop until busy period. When the backoff counter reaches 0, then sender starts the transmission [16]. Then the receiver sends an ACK to sender [17].

V. SIMULATED SCENARIOS

Using OPNET Modeler 14.0, we tested the VoWLAN call capacity under different scenarios. Our first experiment aimed at finding the maximum number of simultaneous Voice over Internet Protocol (VoIP) users a VoWLAN cell would support using the physical modes IEEE 802.11b under the G.711 and G.729A codec standards without enabling the silence suppression when the packetization interval was 10ms. The second experiment examined the VoWLAN voice call capacity under the G.711 and G.729 codec with the silence suppression enabled using the physical mode 802.11b and the same packetization interval (i.e. 10 ms) as the first experiment.

We consider a network in infrastructure mode. As the IEEE 802.11 includes a mechanism for adjusting the flow can switch between the physical modes 1, 2, 5.5, 11Mbps, the capacity of the cell depends on the spatial distribution of users. In this paper, all users (N stations) are assumed to be at equal distance (d = 2.5m) of AP, and that to ensure the system operates with maximum speed of the standard used. See Figure 1. In OPNET, we simulated each experiment's scenario for duration of 150 seconds. Every 2 seconds a VoIP call was added. Each VoIP user was to call continuously till the end of the simulation. The inter-repetition time was constant and set to 0 seconds (i.e. no offset between calls). The operation mode of all calls was set to concurrent. All VoIP calls were performed in both directions (STAs are divided into N / 2 pairs, each two STAs in the same group send and receive VoIP traffic to and from each other). A summary of the parameters and their values is listed in Table II and Table III. Notice that all other simulation values that are not shown indicate that the default OPNET values were left the same.

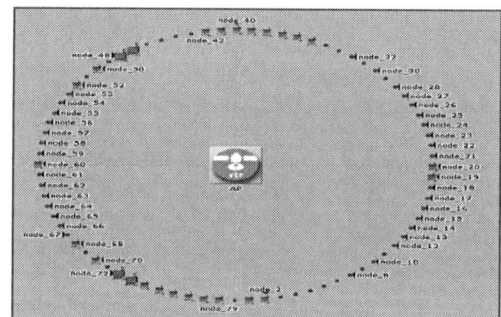

Figure 1. Measurement setup

TABLE II. EXPERIMENT'S PARAMETERS (CODECS)

Codec	Parameter	Value
G.711	Frame Size (seconds)	10 ms
	Lookahead Size (seconds)	0
	Coding Rate (bits/seconds)	64 Kbps
	Speech Activity Detection	Disabled/Enabled
G.729A	Frame Size (seconds)	10 ms
	Lookahead Size (seconds)	5 ms
	Coding Rate (bps)	8 Kbps
	Speech Activity Detection	Disabled/Enabled

978-1-4673-5289-5/12 $31.00 © 2012 IEEE

TABLE III.   EXPERIMENT'S PARAMETERS (PHYSICAL MODES)

Parameter	Value		
	802. 11b	802. 11g	802. 11a
SLOT	20 μs	20 μs	9 μs
SIFS	10 μs	10 μs	16 μs
DIFS (SIFS + 2xSLOT)	50 μs	50 μs	34 μs
PHY header length	192 μs	20 μs	20 μs
Minimum data rate (Mbps)	1	1	6
Maximum data rate (Mbps)	11	54	54
RTS (Octets)	20	20	20
CTS (Octets)	14	14	14
ACK (Octets)	14	14	14
CWmin (SLOT)	31	15	15
CWmax (SLOT)	1023	1023	1023

VI.   RESULTS ANALYSIS

We start by looking at the end-to-end delay. Figure 4 shows the evolution of the end-to-end delay in function of the time of simulation. The delay increased as the time of simulation increased. Without silence suppression and under G.729A codec, for example, the delay stayed acceptable (less than or equal to 150 ms) as long as the time of simulation was 12 seconds or less. At t=13 seconds, however, the delay jumped to 210 ms. It can be concluded that the last successful call was added at t=10 seconds). This helps in saying that in term of end-to-end delay, the VoWLAN cell can accommodate:

$$\left(\frac{\text{Time of last successful call}}{\text{inter}-\text{repetition time}} + 1 \right) = \left(\frac{10}{2} + 1 \right) = 6 \text{ VoIP Calls or}$$

less, which is consistent with findings in related literature.

Second, we study the packets loss. Figure 2 and 3 shows the percentage of data loss calculated as the difference between the packets sent and the packets received. It can be concluded that, that without silence suppression and under G.729A codec, for example, when the time of simulation is 14 seconds or less, the packets loss is acceptable (less than or equal to 2%). However, when time of simulation increases to 15 seconds or more, the data loss becomes unacceptable. It can be concluded that the last successful call was added at t=12 seconds. In this case, VoWLAN call capacity in term of parquets loss is equal to $\left(\frac{12}{2} + 1 \right) = 7$ calls. Third, we examine the jitter. Figure 5 shows that without silence suppression and under G.729A codec, for example, the jitter is within the acceptable range (less than 50 ms) until the time of the simulation is less than or

equal to 24 seconds, which gives a VoWLAN capacity equal to 12 calls $\left(\frac{22}{2} + 1 \right)$. Therefore, it is clear that the jitter is not the primary deciding factor in determining the VoWLAN call capacity of the cell. Table IV summarizes the results achieved under G.711 and G.729A codec with and without silence suppression enabled. From first experiment, we have concluded that when the number of simultaneous calls is low in the network, the voice performance is not affected very much. However when the number of simultaneous calls is getting higher, the voice performance will degrade rapidly. These results can be surprising with respect to the available physical data rate available in the cell, especially at 2.5 m, where the data rate of the physical mode is at the maximum (11Mbps for 802.11b). In fact, IEEE 802.11b suffers from a huge overhead, due to the RTS/CTS handshake, the acknowledgment, the MAC header, the backoff window, and the basic rate of 1Mbps used to transmit the control packets and the physical header. Moreover, for each voice frame, a RTP/UDP/IP header has to be added. The proportion of this overhead is particularly high for small data packets.

VII.   DISCUSSIONS

The results of simulations that we conducted allowed us to show that from point of view VoWLAN call capacity, the use of codecs with low bit rate (small frames) has advantages over the use of codecs with high bit rate (large frames). In the second experiment we compared the VoWLAN call capacity using IEEE 802.11b standard under G.711 and G.729A codec with and without silence suppression, to observe how the silence suppression impact the VoWLAN call capacity. In OPNET, when silence suppression is enabled, there will be no data send during the silence periods. The findings of this experiment indicated the VoWLAN call capacity with silence suppression is two times higher than the VoWLAN call capacity without silence suppression. This result was expected because, as discussed in section III, when silence suppression enabled no signal is encoded during silence periods. This feature saves more than the half of required bandwidth. The bandwidth saved compensates the overhead for a higher number of concurrent calls. We believe that the results conducted in this paper are important because they show that the WLANs scheduling cannot be treated as wired, and a simple calculation of basic data rate required and available,

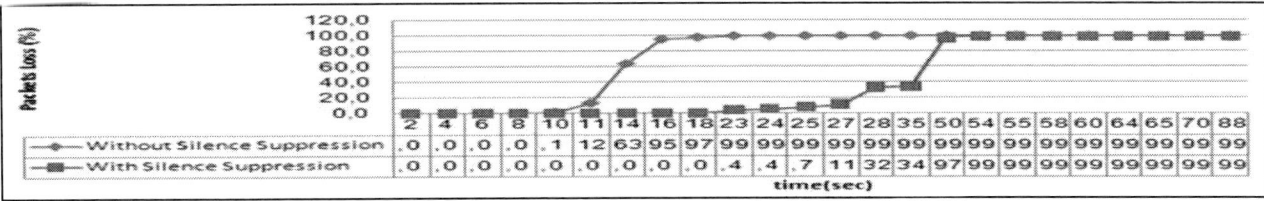

Figure 2.   Packets Loss With and without silence suppression 802.11b_G.711/G.729A

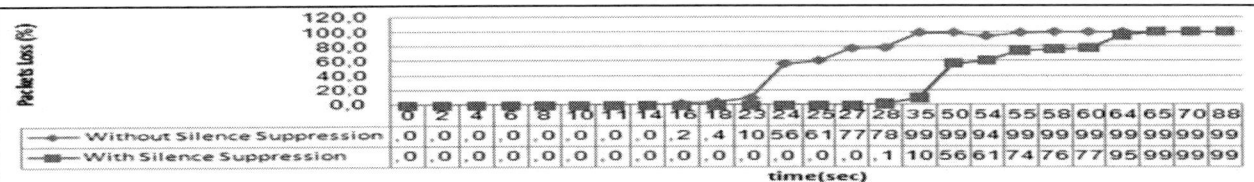

Figure 3.   Packets Loss With and without silence suppression 802.11b_G.711/G.729A

would be totally inoperable without taking into account the overhead induced by the different layers and access mechanisms.

VIII. CONCLUSION

In this paper, we evaluated the VoWLAN call capacity with and without codec's silence suppression. We showed that enabling the silence suppression increase considerably the voice call capacity of a WLAN cell. We summarize our findings in Table IV. In the first experiment, using different scenarios, we were able to show the impact of encoder schemes on the VoWLAN call capacity. We based our observations on the values in Table I. The findings of the first experiment indicated that the minimum number of simultaneous VoIP users the VoWLAN cell was able to support was 5 using IEEE 802.11b mode under G.711 codec. This result is consistent with recent research findings. The maximum number of simultaneous VoIP users using IEEE 802.11b mode under G.729A, without silence suppression, codec was 6.

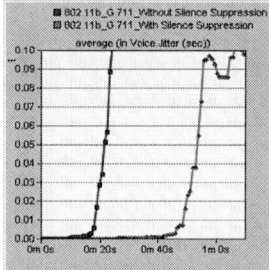 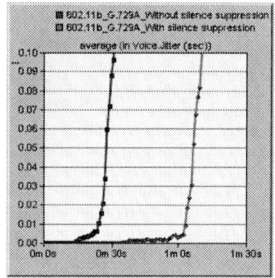

Figure 4.   End-to-End Delay With and without silence suppression 802.11b_G.711/G.729A

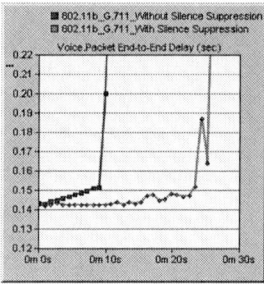 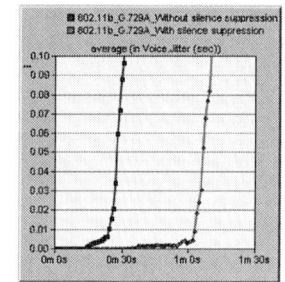

Figure 5.   Jitter With and without silence suppression 802.11b_G.711/G.729A

TABLE IV.   RESULTS SUMMARY

VoWLAN call capacity	IEEE 802.11b			
	Under G.711		Under G.729A	
	Without silence supp.	With silence supp.	Without silence supp.	With silence supp.
In term of End-to-End Delay ($\leq$ 150ms)	5	11	6	13
In term of Packets Loss ($\leq$ 2%)	5	11	7	14
In term of Jitter ($\leq$ 50ms)	10	26	12	32
VoWLAN call capacity	5	11	6	13
Decision Factors	Delay& P.Loss	Delay & P.Loss	Delay	Delay

In the second experiment we presented statistical and graphical analysis to enable us find out comparison evaluation. This is very important and useful way as the comparison evaluation can tell us what is happening in VoIP conversation when we modify parameters (codec's silence suppression in this case). The findings of this experiment demonstrated that the maximum number of VoIP users the VoWLAN 802.11b cell under G.711 codec with silence suppression enabled was able to support was 13. Similar experiment without silence suppression indicated a maximum of only 6 users.

REFERENCES

[1] David P. Hole and Fouad A. Tobagi, "Capacity of an IEEE 802.11b Wireless LAN supporting VoIP", to appear in Proc. IEEE Conference on Communications (ICC) 2004.

[2] Garg, S., and Kappes, M., "Can I add a VoIP call?" Proc.of IEEE ICC'03, vol. 2, pp. 779--783, May 2003.

[3] Garg S, Kappes M. "An experimental study of throughput for UDP and VoIP traffic in IEEE 802.11b networks", Proceedings of the IEEE WCNC'03, vol. 3, New Orleans, LA, U.S.A., March 2003; 1748–1753

[4] Coupechoux, M., Kumar, V., and Brignol, L., "Voice over IEEE 802.11b Capacity," Proc. of the 16th ITC Specialist Seminar on Performance Evaluation of Wireless and Mobile Networks, Sept. 2004.

[5] L. Cai and X. Shen, "Voice Capacity Analysis of WLAN With Unbalanced Traffic", IEEE Transactions on Vehicular Technology, Vol. 55, No. M3, May 2006.

[6] ITU-T Recommendation G.711. Pulse Code Modulation (PCM) of voice frequencies. ITU-T, November 1988.

[7] ITU-T Recommendation G.729 Annex A. C source code and test vectors for implementation verification of the G.729 reduced complexity 8 kbit/s CS-ACELP speech codec. ITU-T, November 1996.

[8] ITU-T Recommendation G.723.1. Speech coders : Dual rate speech coder for multimedia communications transmitting at 5.3 and 6.3 kbit/s. ITU-T, March 1996.

[9] ITU-T Recommendation G.114. One way transmission time. ITU-T, May 2003.

[10] O. Hersent, D. Gurle, and J-P. Petit, IP Telephony: Packet-based multimedia communications systems, Addison-Wesly, Harlow, England, 2000.

[11] IEEE Std 802.11-1999, Part 11: Wireless LAN MAC and physical layer specifications, Reference number ISO/IEC 8802-11:1999(E).

[12] Q. Cao, T. Li, Tianji and D. Leith. Achieving fairness in Lossy 802.11e wireless multi-hop Mesh networks. In: Third IEEE International Workshop on Enabling Technologies and Standards for Wireless Mesh Networking MESH. Macau SAR, P.R. China pp. 1-7, 10 November, 2009.

[13] P. Dini, O. Font-Bach and J. Mangues-Bafalluy. Experimental analysis of VoIP call quality support in IEEE 802.11 DCF. Communication Systems, Networks and Digital Signal Processing. 6th International Symposium, pp. 443 – 44729 August 2008.

[14] T. Li, Q. Ni, T. Turletti, Xiao. Performance analysis of the IEEE 802.11e block ACK scheme in a noisy channel. Broadband Networks, 2005. 2nd International Conference. Hamilton Inst. Ireland, Vol 1, pp. 511-517, 13 February 2006.

[15] Q. Ni, T. Li, T. Turletti, and Y. Xiao. Saturation Throughput Analysis of Error-Prone 802.11 Wireless Networks. Wiley Journal of Wireless Communications and Mobile Computing (JWCMC). INRIA, France, Vol. 5, Issue 8, pp. 945-956. December 2005.

[16] W. Wang, S. Liew and V. Li. Solutions to performance problems in VoIP over a 802.11 wireless LAN. Journal of IEEE transactions on vehicular technology, New York, USA, Vol. 54, pp. 366-384, 2005.

[17] S. Ahmed, X. Jiang, P. Ho and S. Horiguchi. Wireless Access Point Voice Capacity Analysis and Enhancement Based on Clients' Spatial Distribution. Multi- media Communication. San Diego, CA, Vol 58, pp. 2597 – 2603, 11 June 2009.

Image denoising using selective neighboring Quaternionic wavelet coefficients

Mohammed kadiri
Telecommunication and Signal
Processing Laboratory
Djillali Liabes University
Sidi Bel Abbes, Algeria

Mohamed Djebbouri
Telecommunication and Signal
Processing Laboratory
Djillali Liabes University
Sidi Bel Abbes, Algeria

Pilippe Carré
XLIM-SIC Laboratory
CNRS UMR 6172
University of Poitiers, France

Abstract— Removing noise for image is a classical problem in image processing but it is still relevant issue. In this paper, we adopt an innovative and original approach, which is combined between a recent quaternionic wavelet transform and selective neighbouring coefficients to image denoising. The choice of quaternionic transformation is justified by the fact that it gives a very good separation of the coefficients in terms of amplitude and 3-angles phase. This representation generalizes better the concept of analytic signal to the image. Quaternionic transformation retains property of shift invariant but it allows the introduction of a true 2D analysis compared with the traditional complex wavelet transform. After decomposition, selective neighbouring coefficients are used for denoising. Interesting results are obtained by comparing in term of PSNR with the complex wavelet transform and the discrete wavelet transform.

Index Terms—Image denoising, Complex Wavelet Transform, Quaternionic Wavelet Transform, Neighboring coefficients, Thresholding.

I. INTRODUCTION

Over the past decade, wavelet transforms have shown great promises in such diverse fields as pattern recognition, image denoising, image compression, and computer graphics, to name only a few. For this work, we address the problem of removing additive Gaussian noise from a corrupted image. The goal of any denoising method is to eliminate the noise parts while retaining as much as possible of the important signal characteristics, such as, for example, the edges and sharp features. More precisely, we transform image in wavelet domain, threshold wavelet coefficients, and perform the inverse wavelet transform. In the decomposition-reconstruction part, the Complex Wavelet Transform (CWT) and the new Quaternionic Wavelet Transform (QWT) are very powerful tools compared to discrete wavelet transform (DWT). In fact, the later has a serious problem which is shift dependant. A small shift in the input of the process generates large variations in the output.

Several authors have studied the CWT and its application to image denoising. Kingsbury introduced a very elegant computational structure, the dual - tree complex wavelet transform (DT-CWT) [1][3], and he examines its property of the shift invariant and filtring of signals[4]. In [2], the same author incorporates the DT-CWT into the image restoration and enhancement. Chen and kégl [5] used the complex ridgelets transform for image denoising. The DT-CWT remedy two drawbacks of DWT: shift invariant and the lack of a notion of phase to encode signal location information.

Recently, research interest is oriented to the quaternionic wavelet transform [7][8][9][10][11]. The only phase of the CWT can lead to ambiguity when translating the image so that the phase of the QWT is represented by three angles. The two first angles encode horizontal and vertical orientations while the third encode texture information's and edge. R. Soulard and P. Carré study the QWT. They use it in texture classification [10]. In [11] Shan Gai et al apply the dual tree QWT, then he uses the bivariate shrinkage function to model the dependencies between QWT coefficients and its corresponding parents coefficients. This method is based on a probabilistic estimator that seeks the relationship between the coefficients of two successive scales for the noise variance and the threshold value where its complexity.

In our work, the quaternionic wavelet is performed on the image. Then, we threshold the wavelet coefficients by exploiting the relationship between the coefficients in a neighboring window localized in the same scale [12] [13]. This approach was used on rolling bearing fault diagnosis [14]. We find its extension to image denoising in [15][16]. The contribution that we seek is the use of a transformation adapted to 2D or image. This transformation is an extension of modern approaches to thresholding. We also compare the proposed method with other traditional algorithms.

The rest of the paper is organized as follows: Section 2 gives a short description of quaternionic wavelet transform and section 3 explains how to incorporate neighboring wavelet coefficients into image denoising and threshold. Section 4 discusses the implementation results and Section 5 concludes the paper.

II. QUATERNIONIC WAVELET TRANSFORM

Because the real and imaginary parts of the coefficients from the complex wavelet transform are in quadrature, its magnitudes are almost shift invariant with moderate redundancy. But, the CWT does not give a perfect localization signal because its phase is encoded on a single angle. We have a better location with Quaternionic Wavelet Transform whose phase is encoded in three angles.

The QWT is based on an alternative theory for the 2-D Hilbert transform and can be computed using a dual-tree filter bank with linear computational complexity.

- Quaternion Hilbert transform

There are several alternatives to the 2-D analytic signal; we focus here on one due to Bulow [6]. It combines the Hilbert transforms to form an analytic signal comprising a real

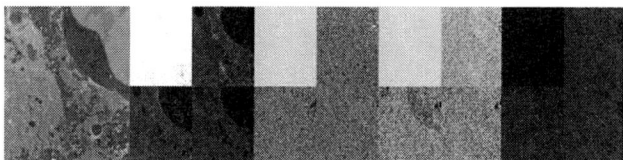

Figure1. One level decomposition by Quaternionic Wavelet Transform.From left to right: Original image, magnitude |q|, and phase angles: φ , θ, ψ

part and three imaginary components that are manipulated using quaternion algebra [6]. Quaternion is a generalization of complex number with three imaginary parts defined as:

$$q = a + bi + cj + dk \qquad (1)$$

and $i^2 = j^2 = k^2 = -1$.

In polar form, it's represented by its magnitude and tree phase angles:

$$q = |q|e^{i\varphi}e^{j\theta}e^{k\psi} \qquad (2)$$

Where (φ, θ, ψ) are the quaternion phase angles defined within the range $[-\pi, \pi] \times \left[-\frac{\pi}{2}, \frac{\pi}{2}\right] \times \left[-\frac{\pi}{4}, \frac{\pi}{4}\right]$. These angles are computed using the following formulas:

$$\varphi = \tfrac{1}{2} arctan\left(\frac{2(bd+ac)}{a^2+b^2-c^2-d^2}\right) \qquad (3)$$

$$\theta = \tfrac{1}{2} arctan\left(\frac{2(cd+ab)}{a^2-b^2+c^2-d^2}\right) \qquad (4)$$

$$\psi = \tfrac{1}{2} arcsin\big(2(ad - bc)\big) \qquad (5)$$

Bulow introduces an alternative definition of 2-D analytic signal based on the quaternion Fourier transform (QFT) [6]. The QFT of a 2-D signal f is given by:

$$F^q(u, v) = \int_{R^2} e^{-i2\pi ux} f(x,y)\, e^{-j2\pi vy}\, dxdy \qquad (6)$$

In space domain, the 2-D quaternion analytic signal is defined as:

$$f_A^q(x,y) = f(x,y) + if_{Hi1}(x,y) + jf_{Hi2}(x,y)$$
$$+kf_{Hi}(x,y) \qquad (7)$$

Where, f_{Hi1}, f_{Hi2}, f_{Hi} are, respectively, the total and the partial Hilbert transform of a real 2D signal f. For more details on the construction of these components, see [6].

The 2-D dual-tree QWT rests on the quaternion definition of 2-D analytic signal. We obtain a 2-D analytic wavelet and its associated quaternion wavelet transforms (QWT) by organizing the four quadrature components of a 2-D wavelet (the real wavelet and its 2-D Hilbert Transforms) as a quaternion. The scaling function is ϕ and wavelets are ψ^D, ψ^V, ψ^H in the diagonal, vertical and horizontal directions[10].
To compute the QWT coefficients, we can use a separable 2-D implementation of the dual-tree filter bank in Figure 2 (this figure is proposed by R. Soulard[10]. The link with complex transform is shown). During each stage of filtering, we independently apply the two sets of h and g wavelet filters to each dimension (x and y) of a 2-D image; for instance, applying the set of filters h to both dimensions yields the scaling coefficients, and the diagonal, vertical, and horizontal wavelet coefficients. Therefore, the resulting 2-D dual-tree implementation comprises four independent filter banks

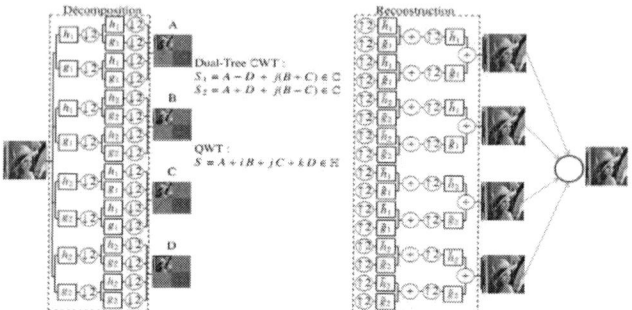

Figure 2. Filterbank for compute the QWT and CWT

applied to each dimension (hh, hg, gh, and gg)) operating on the same 2-D image. We combine the wavelet coefficients of the same sub-band from the output of each filter bank using quaternion algebra to obtain the QWT coefficients. So we have a multi-scale representation of analytic signal with module and phase information (Figure1 shown magnitude and tree phase angles obtained after QWT decomposition of spot image).

III. INCORPORATING NEIGHBORING COEFFICIENTS IN IMAGE DENOISING

We have done a 2-D quaternionic wavelet transform for image denoising. At every decomposition level, we get magnitude of four frequency sub-bands, namely, LL, LH, HL, and HH. The next level should be applied to the low frequency sub-band LL only. This process is continued until a pre-specified level is reached. Since the Gaussian noise will be nearly averaged out in the low frequency wavelet coefficients and we want to keep small coefficients in these frequencies, only wavelet coefficients in the high frequency levels need to be threshold i.e. we need to threshold all LH, HL, and HH within these high frequency sub-bands.

Due to the linearity of the wavelet transform the additive noise model in the image domain remains additive in the wavelet transform domain as well [16]:

$$W_{k,d}(x,y) = Y_{k,d}(x,y) + N_{k,d}(x,y) \qquad (8)$$

$W_{k,d}(x,y)$, $Y_{k,d}(x,y)$, and $N_{k,d}(x,y)$ are noisy, noise free wavelet coefficient and noise component. Scale and orientation of wavelet coefficients are denoted by k and d.

The difficulty of QWT is the information thresholding strategy. Several methods based on the relationship between amplitude and phase interscale such as gaussian scale mixture and derotated complex wavelet coefficients [17] were used in image denoising in the complex domain, but in quaternionic domain very few methods are developed. We here try a first strategy by working on the energy information.

The basic motivation of neighbor thresholding is that if the current coefficient contains information, it is likely that the neighbor coefficients also do (Wavelet coefficients are correlated in a small neighborhood). For this reason, at each location we threshold every coefficient by using it as much as the coefficients in its neighborhood. For every coefficient of our interest we choose local windows around it. The size of this window is predefined as a function of the size of the image to denoising. For the wavelet coefficient to be thresholded, we shrinkage it according to the following formula [15]:

$$W_{k,d}(x,y) = W_{k,d}(x,y) \times T(x,y) \qquad (9)$$

T(x,y) is the threshold factor defined as :

$$T(x,y) = 1 - \frac{\lambda^2}{S^2(x,y)} \qquad (10)$$

$T(x,y) = T(x,y)$ if $\lambda^2 < S^2$ and $T(x,y) = 0$ if $\lambda^2 > S^2$

$\lambda^2 = 2\sigma^2 log n^2$

$$S^2(x,y) = \sum_{(x,y)\in n^2} W^2(x,y) \qquad (11)$$

λ: universal threshold.

σ : standard deviation of coefficients.

n^2 : size of local neighborhood window

$S^2(x,y)$: Summation of squared coefficients of local window.

It can be seen that this procedure chose constant size of neighboring window at every level of wavelet decomposition. However, this method is not accurate enough. The size of window is fixed while de correlation between coefficients is variant at different levels of decomposition. In [16], To solve this problem and after studying the regularity of wavelet coefficients dependency, the author changes equation (11) by:

$$S_j^2(x,y) = \sum_{(x,y)\in n^2} W_j^2(x,y) \quad , \; n=n_0\text{-}j \quad (12)$$

Where: j is the level of decomposition and n_0 is a constant defined according the size of noisy image and the support of wavelet filter.

It should be noted that during this stage only the magnitude of the wavelet coefficients were thresholded. The probabilistic model adapted to the magnitude is the Rayleigh distribution. The information from the phase has not been taken into account.

The image denoising algorithm (called QWT-NeighShrink algorithm) by considering the correlation between wavelet coefficients can be described as follows:

1. Perform quaternionic wavelet decomposition for noisy image.
2. Threshold the selective wavelet coefficients using neighborhood window and universal threshold.
3. Reconstruct the image by inversing decomposition scheme.

IV. EXPRIMENTAL RESULTS

The experimentation has been conducted on the gray scale image, Lena of size 512×512. Our programs are written in free Scilab and Matlab calling the free software package WaveLab developed by Donoho et al. The performance has been evaluated by denoising image corrupted by white gaussian noise at different levels. To decompose the image, tree levels of the DT-QWT are applied. For comparison, in first, we implement the universal hard- thresholding and soft-thresholding denoising techniques, and our method QWT-NeighShrink. All details scales are thresholded. The results of this test are showed in figure 3. The objective results in PSNR are given in table I. The PSNR is defined as:

$$PSNR = -10 log_{10} \frac{\sum_{x=1}^{N}\sum_{y=1}^{M}(I(x,y)-I1(x,y))^2}{255^2 NM} \qquad (13)$$

The original image is I ; its size is N×M and $I1$ is the denoised image.

It is shown that the neighbor thresholding method is better than the term-by-term (hard or soft) thresholding method. In fact, it has averaged a gain of 5dB for a low level of noise and 1dB for high level, compared to hard thresholding and soft thresholding. We can also see that, hard and soft shrink does not have any denoising power when the noise level is low.

TABLE I.
PSNR for the tree threshold methods with DT-QWT

Threshold method	Input PSNR-Noisy Image					
	34.12	28.10	24.58	22.08	20.14	18.56
Hard shrink	32.69	30.25	28.65	27.82	26.89	26.21
Soft shrink	30.31	28.23	27.17	26.53	25.94	25.68
Neigth shrink	**37.41**	**33.37**	**30.92**	**29.42**	**28.23**	**27.18**

TABLE II.
PSNR for DT- QWT , COMRIDGELET, DT-CWT, and DT-DWT

Wavelet Type	Input PSNR-Noisy Image					
	34.12	28.10	24.58	22.08	20.14	18.56
DT-DWT	33.03	29.46	26.58	25.50	23.79	23.02
DT-CWT	37.08	33.06	30.67	29.13	27.82	26.96
ComRidgelet	37.19	33.13	30.79	29.21	28.00	27.03
DT-QWT	**37.41**	**33.37**	**30.92**	**29.42**	**28.23**	**27.18**

Then, we compared the DTQWT-NeighShrink denoising method with denoised image by complex Ridgelet Shrink (ComRidgelet) proposed in [5] , dual tree complex wavelet transform (DTCWT-NeighShrink) and duel tree discrete wavelet transform (DTDWT-NeighShrink) (see figure 4). Like the comparison in table I, table II reflects efficiency of the adopted method in this last test.

In this experiment we have noticed that quaternionic and complex transformation can capture more local features than discrete wavelet transformation. A gain difference of 4 dB is obtained in comparison to denoising by the DT-DWT. But, between denoising by DT-QWT and complex methods (DT-CWT and ComRidgelet), we see that we average gain 0.25 dB. Some of these results are produced by the redundancy introduced by the transformations. Besides, only the amplitude is taken into account while phase which represents the major difference between the two transformations is not considered in this paper. This work is a first approach and a study of feasibility.

V. CONCLUSION AND FUTUR WORK

We have presented a new approach to image denoising by introducing a 2-D multiscale wavelet representation, the quaternionic wavelet transform, that is particularly efficient for coherent processing of relative location information in images. This transform generalizes complex wavelets to higher dimensions and inspires new processing and analysis methods for wavelet phase. QWT used here is based on an alternative definition of the 2-D Hilbert Transform and 2-D analytic signal and on quaternion algebra. We have demonstrated the efficacy of the consideration of neighborhood in the thresholding in comparison to term by

978-1-4673-5289-5/12 $31.00 © 2012 IEEE

term thresholding hard or soft. The choice of local windows size is very important in this step. In fact, a very large size introduces pixels that are not directly correlated with the pixel of our interest and we return to the term by term if this size is very small.

We have also demonstrated that the QWT provides gain in quality of denoising compared to the CWT when the thresholding is applied only to the amplitude. To increase this gain we can apply the denoising jointly, on QWT amplitude and phase. The information contained in the QWT phase, which is encoded in three angles, and particularly the third angle that represents the texture in the image, is very important in signal localization. This phase information can also be studied by considering the relationship between scales, so called the inter-scale relationship. All these points will be the subject of our future work.

Noisy – Free   Noisy Image PSNR= 22

Soft Shrink PSNR= 26.50   Hard Shrink PSNR= 27.78

Neigth Shrink PSNR= 29.42

Figure 3. Noisy image with PSNR = 22 dB denoising by DT-QWT using different threshold methods

REFERENCES

[1] N. G. Kingsbury, "The dual-tree complex wavelet transform: a new technique for shift invariance and directional filters", in the Proceedings of the IEEE Digital Signal Processing Workshop, 1998.

[2] N.G. Kingsbury, "The dual-tree complex wavelet transform: a new efficient tool for image restoration and Enhancement", in: Proceedings of EUSIPCO'98, Rhodes, pp.319–322, September 1998.

Noisy Image PSNR= 22   DWT PSNR = 25.50

DT-CWT PSNR=29.13   DT-OWT PSNR=29.42

Figure. 4. Comparison between DT-DWT, DT-CWT and DT-QWT Image denoising.

[3] N.G. Kingsbury, "A dual-tree complex wavelet transform with improved orthogonality and symmetry Properties", in: Proceedings of IEEE ICIP, Vancouver, September 11–13, 2000.

[4] N.G.Kingsbury," Complex wavelets for shift invariant analysis and filtering of signals". Journal of Applied and Computational Harmonic Analysis,vol. 10, no3, pp.234-253, 2001.

[5] G.Y. Chen and B. Kégel," Image denoising with complex ridgelets". The journal of Pattern Recognit, 40, pp: 578-585, 2007

[6] T.Bulow," Hypercomplex Spectral Signal representation for the Processing and analysis of Images", PhD Thesis, Christian Albrechts.

[7] E.B. Corrochano, "Multi-resolution image analysis using the quaternion wavelet transform", The journal of Nnumerical Algorithms, 39(1), pp: 35-55, 2005.

[8] E.B. Corrochano, "The theory and use of quaternion wavelet transform", The journal of Mathematical   Imaging and Vision, 24(1), pp: 19-35, 2006.

[9] R. Soulard, « Quaternions et algèbres géométriques pour le traitement d'images », Master's thesis, University of Poitiers, 2009.

[10] R. Soulard and P. Carré, "Qaternionic wavelets for texture classification", Pattern Recognition Letters 32, pp: 1669-1678, 2011.

[11] S. Gai, P.Liu, J.Liu,X.Tang, "A new image denoising algorithm via bivariate shrinkage based on quaternion wavelet transform", Journal of Computational Information System 6:11, pp:3751-3760, 2010.

[12] T.T.Cai, B.W.Silverman, "Incorporating information on neighboring coefficients into wavelet estimation", Sankhya Series , B63(2), pp:127-148, , 2001.

[13] G. Y. Chen and T. D. Bui, "Multiwavelets Denoising Using Neighboring Coefficients", IEEE Signal Processing Letters, Vol. 10, No. 7, July 2003

[14] S.Hailiang1, ZI Yanyang1, HE Zhengjia, W. Xiaodong and Y. Jing," Translation-invariant multiwavelet denoising using improved neighbouring coefficients and its application on rolling bearing fault diagnosis", 9th International Conference on Damage Assessment of Structures   (DAMAS2011), Journal of Physics: Conference   Series 305 (2011) 012012.

[15] B. Chinna Rao, M. Madhavi Latha," Selective neighbouring wavelet coefficients approach for image denoising", International Journal of Computer Science and Communication Vol. 2, No. 1,pp. 73-77, January 2011.

[16] G.Y. Chen, T.D. Bui, A. Krzy´zak, "Image denoising with neighbour dependency and customized wavelet and threshold", Pattern Recognition 38, pp: 115 –124, (2005).

[17] M. Miller and N. Kingsbury," Image Denoising Using Derotated Complex Wavelet Coefficients", IEEE Tran. On Image Processing, vol 17, N 9, September 2008.

Color Image Segmentation by a Genetic Algorithm based Clustering and Connected Component Labeling

Fatima Zohra BELLALA BELAHBIB, Feryel SOUAMI

LRIA Département d'Informatique FEI Université des Sciences et de la Technologie Houari Boumediene
Alger, Algérie
fzbellala@usthb.dz, fsouami@usthb.dz

Abstract—**In this paper, we present a two steps segmentation method. The first consists in color image quantization by genetic algorithm based clustering method where variable string length and fitness function with a smallest number of variables are used. Once pixels are classified in 3D color space, a Connected Component Labeling adapted to color image is proposed and applied to the 2D image space to get separate regions. Results are discussed using Berkeley color images base in RGB and Lab color spaces.**

Keywords- Segmentation, Genetic algorithm, Connected Component Labeling, FCM Clustering, Quantization

I. INTRODUCTION

Image segmentation is one of the important issues in image processing and computer vision. It is defined as process of partitioning an image into constituent parts of objects [10]. It's the first stage and its results affect the results of other stages in image analysis.

Several methods of segmentation are proposed in literature. They can be broadly classified into seven groups [8] applied initially, for grey level images and then extended to color ones: Histogram thresholding; Clustering (Fuzzy and Hard); Region growing, region splitting and merging; Edge based; Physical model base; Fuzzy approaches and Neural network and Genetic Algorithm (GA) approaches.

Genetic algorithms are optimization techniques and randomized search [16] inspired by natural species evolution. They converge towards global solution even in presence of complex data and are parallelized. In genetic algorithms, population of solutions or individuals is encoded in the form of chromosomes (strings). Individuals are randomly initialized from a large search space and coded often in binary. This population is then evolved from generation to other using genetic operators: crossover, mutation and selection. The selection operation depends on evolutionary function called fitness (objective function) which measures the goodness of the solution. The process of selection, crossover and mutation are continually applied for individuals till termination condition is satisfied or a fixed number of generations is reached.

Genetic algorithms are applied in image analysis, especially in segmentation where satisfactory results have been obtained [3, 18, 5]. GA can be used in image segmentation by two different approaches. Parameters optimization [3] and pixel level segmentation [18, 5]. In the former, genetic algorithms are applied to adjust parameters of a known segmentation method to improve its output and in the latest, GA are applied to perform regions labeling so that each pixel is classed in a particular region depending on its characteristics.

In [3], Bhanu, Lee and Ming consider image segmentation as an optimization problem and they present a closed loop image segmentation system in which genetic learning system algorithm is used to chose adequate parameters for well known segmentation algorithms. In [18], Ramos and Muge consider image segmentation as a clustering problem and use GA to partition image pixels in their proper cluster among a set of a given clusters. The k-Means clustering model is used as a GA evaluation function. The main drawback of this method is that the number of clusters must be known a priori. And in [5], Chun and Yang use firstly a clustering algorithm to obtain large number of small regions. A GA is then used to ameliorate the obtained result. They define a fitness function as product of the compact and separation and the magnitude of the gradient. Even though the algorithm is simple to implement, it requires much computer storage and computation time.

In this paper, we present a segmentation algorithm consisting of two steps as in [6]. In the first one, the color image is quantized by a genetic algorithm based on fuzzy clustering model and using a variable string length [17] without prior knowledge of exact regions number. Once pixels are classified into set of classes according to their color characteristic, a Connected Component Labeling algorithm is applied in the second step to get separate regions (spatial segmentation). The benefit of this separation is double. First methods of classification and spatial segmentation can be used. In the other part, separation permits to avoid the difficulty of analyzing the similarity of colors and their distribution at the same time.

The rest of this paper is organized as follows: In the second section, we describe the genetic algorithm used for color image quantization. In the section three, we present the extended CCL algorithm for color images used to separate clusters obtained by the genetic algorithm to get image regions. Results are discussed in section four and we finally, conclude our work.

II. GENETIC ALGORITHM CLUSTERING FOR COLOR QUANTIZATION

Clustering is unsupervised classification of data where elements in a same class (cluster) are as similar as possible and elements in different classes are as dissimilar as possible.

978-1-4673-5289-5/12 $31.00 © 2012 IEEE

The Fuzzy C-Means (FCM), introduced by Ruspini in 1969 [20] and improved by Bezdek [2], is the most popular fuzzy clustering algorithm [13]. FCM is very sensitive to initialization parameters and requires prior knowledge of cluster centers number and can lead to local minima instead of global one. However, genetic algorithms effectively minimize the problem of converging to the local minima [3] if the initial population is distributed throughout the search space. We use genetic algorithm to perform clustering process instead of FCM.

In the other part, the aim of quantization is to reduce the amount of data of color image (256^3 for 8 bits RGB color space) and it consists of two steps [21]: palette design in which a reduced number of palette colors is specified and pixel mapping in which each color pixel is assigned to one of the colors in the new palette. Quantization can be considered as clustering technique applied for color image in the 3D color feature space and genetic algorithm clustering method is used to perform this quantization.

We use the GA proposed by Bellala and Souami in [1] which is concisely described in the following subsections.

A. Chromosomes Encoded and Initialization

Solutions (individuals of population) are feasible palettes and one gene chromosome is a three dimensional vector where each coordinate is coded in real codification which is currently more suitable [9] than the binary representation. The string length is variable from individual to other so that the palette size can be found without any prior knowledge. Just a maximal possible colors number must be known and it depends on the precision desired by the quantifier. This number is then adjusted by the algorithm to give the colors final number adequate for the quantization. String items are initialized to colors of randomly chosen pixels of a given image (which permits convergence in fewer generations) and some items are made to an invalid value to get different lengths of strings. The population size P of individuals is taken to be 100 (P=100) in our work.

B. Fitness Function

We use the fitness function F proposed in [1] and defined by:

$$F = 1/(\sum_{j=1}^{N}(1/\sum_{i=1}^{C}(1/d(x_j, \beta_i))) * (1 - 5 * \exp(-t/10))^2) \quad (1)$$

Where: - t is the generation number of population solution;

- $d(x_j, \beta_i)$ is the distance from the color point x_j of a given image to the color point β_i of a given individual. In our work, we use Euclidian distance.

- N is the dimension of the image to be quantized;

- C is the palette color numbers (maximal individual length);

The fitness function F depends just on data to be classified (pixels colors in our case) and the cluster centers (palette colors) which means that the fitness function is more promptly evolved comparing to the objective function in the FCM case [2].

C. Genetic Operators

The selection operation determines the survival individuals for the next generation according to their fitness value. In this paper, we used the roulette wheel selection method [9] enriched by some bad individuals as in [1] to choose the individuals to be crossed and mutate.

The crossover operation consists of interchange parts of two randomly chosen parent chromosomes to generate child chromosomes. In our algorithm, a one-point crossover is applied with a probability p_c=0.82.

Mutation is the operation which alters a gene at a random position in chromosome chosen with a small probability p_m. p_m=0.01 in our work.

The process of selection, crossover and mutation is iterated for a finite number. The best chromosome is then taken from the last generation and is used as solution for the clustering problem. Therefore the length of this chromosome gives the palette length and genes the palette colors values. Each pixel is assigned to its new color by the argmin function.

III. CONNECTED COMPONENT LABELING

Clusters generated by the genetic algorithm stage can't present exactly image regions. Indeed, the GA classification is carried in the 3D color space so pixels which are not necessarily connected in the image plan can have the same color value and than belonging to the same cluster.

To achieve spatial image segmentation using clustering, two different methods can be applied: -Redefine the clustering objective function to have at the same time the color as feature information and the spatial information; -Analyze the obtained image (the quantized image) to separate set of pixels from the same class following their connectivity.

In our case, we use the second method to get separate regions from the quantized image. Thus we focus on the connected problem in image pixels. Two pixels P and Q are connected (4-connected or 8-connected) if there exists a path of pixels $(p_0, p_1, ..., p_n)$ such that p_0=P, p_n=Q and $\forall i$ $1 \leq i \leq n$, p_{i-1} and p_i are neighbors [7]. To extend this definition for color images, we add the condition that pixels p_{i-1} and p_i must have the same color to get a connected color pixels set and then regions.

We use Connected Component Labeling (CCL) process to separate pixels according to their connectivity and their color. CCL is an approach that labels regions within a binary image [14] introduced in the year of 1966 by Rosenfeld and Pfaltz. [19]. Many algorithms have been proposed for addressing this issue using ordinary computer architecture [7, 14, 12] or a specific one [11].

Using ordinary computer architecture, the two raster scan algorithm is the classical and the most commonly applied for binary images. It consists in assigning a provisional label to each no background pixel, based on the value of its neighbors already visited and then replacing the provisional labels by their equivalent labels using an equivalent table stored in the first scan. Points with the same label form one component of connected pixels candidate of region.

Instead of converting color image to binary one and then applied a given CCL algorithm, we propose an extension to the classical algorithm for color images. Note that the objective

978-1-4673-5289-5/12 $31.00 © 2012 IEEE

desired by this work isn't to present a new CCL algorithm but our aim is to show the ability to apply the CCL algorithm for color images to get separate regions.

A pixel is assigned a label checking its color value and the ones of visited neighbor pixels and their labels. The C code of the proposed algorithm is given by the figure 1 using 4-connectivity. Where x is the checking point, p and q are the neighbors already visited, lx, lp and lq are their respectively assigned labels and I is the labeled image. If the lp and lq labels of the visited neighbors are different for the same color of the points p and q, we maintain the smallest one and we store the equivalence between the two labels.

As in [7], we propose to process the equivalence during the first scan. The equivalence is stored in the T table with two fields. One denotes the equivalent label value R and other denotes the occurrence (size) S of the label value in the image which can be very useful for the processing step which can succeed segmentation.

```
// FIRST SCAN:
NewLabel=1;
for(i=1; i<NROWS-1; i++)
for(j=1; j<NCOLS-1; j++)
{lp = I[i-1,j];  lq = I[i,j-1];
if((Color(p)==Color(q))  && (Color(p)== Color(x))
if(lp==lq) {lx=lp; T[lx].S++; }
else
  {
  e=min(lp,lq);
  a=max(lp,lq);
  lx=T[e].R;
  k=a;
  while(T[k].R!=k)
    {(T[T[k].R].S)+=T[k].S;
    T[k].S=0;
    k=T[k].R;}
  T[a].R=k;
  }
else if (Color(p)==Color(x)) lx=lp;
else if (Color(q)==Color(x)) lx=lq;
else {lx=NewLabel; NewLabel++;}
I[i,j]=lx; T[lx].S++;
}
//SECOND SCAN
```

Figure 1. CCL adapted to color images

IV. APPLICATION AND DISCUSSION

The proposed color image segmentation algorithm was applied for Berkeley images base [15]. The images are 24bits/pixel real color images. The RGB and Lab color spaces are used and results are compared for the two spaces. Better results are obtained by more genes and generations number. But these two parameters affect the run time and convergence is observed at around 17 generations and 31 colors as maximal string length.

The number of regions obtained by the algorithm can exceed 4000 regions. To overcome the problem of small regions, a supplementary (post processing) phase of merging small regions with the nearest ones according to their color values is required. Examples of the obtained results with the previous fixed parameters using RGB and Lab color spaces are shown in figure 2 where regions with a surface less than 0.1% of the image surface are merged with neighbors regions in column (c) and

(d). Note that the merged percentage depends on the precision desired by the segmentation process. The higher this percentage is the lower number of regions is.

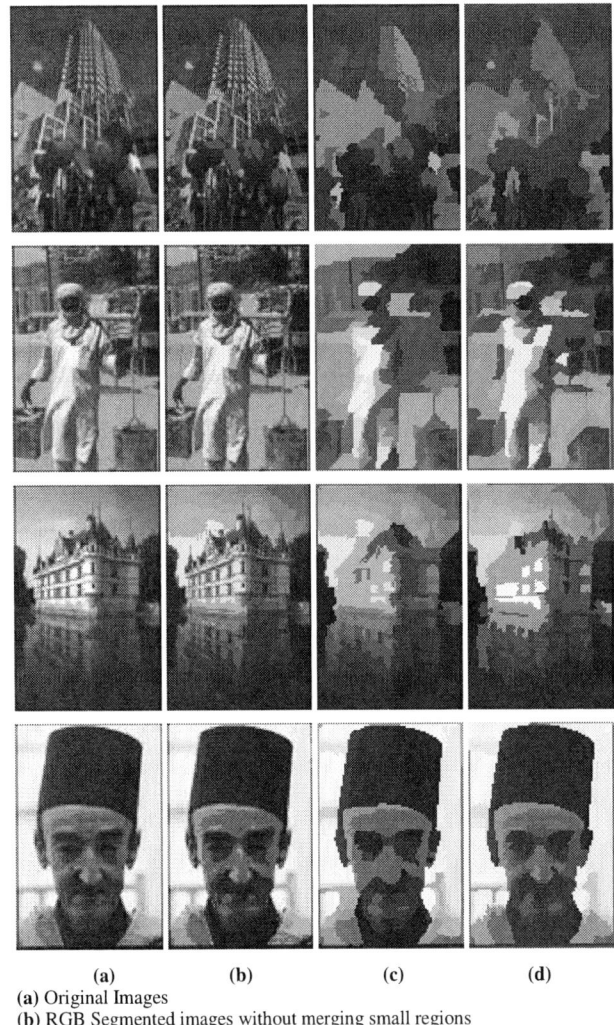

(a) Original Images
(b) RGB Segmented images without merging small regions
(c) RGB Segmented images after merging small regions
(d) Lab Segmented images after merging small regions

Figure 2. Original and segmented images in RGB and Lab color spaces

To evaluate the obtained results, two unsupervised evaluation measures are considered: Zeboudj and Borsotti quantitative measures [4]. The corresponding Zeboudj and Borsotti values for the presented images are given in table1 for the RGB color space and in table 2 for the Lab color space. Moreover, we present in these tables the labeled initial regions number obtained by the two steps proposed segmentation algorithm and the labeled final regions number obtained by the supplementary process of merging small regions.

For the Lab color space, even if the Zeboudj values, more adapted to the homogenous images, haven't relatively been changed comparing to the RGB color space, due to the images nature, but remarkable amelioration can be observed in the Lab segmentation with the Borsotti values which are less important than the ones obtained in RGB segmentation.

TABLE I. RGB SEGMENTATION RESULTS EVALUATION FOR THE PRESENTED IMAGES

	1st image	2nd image	3rd image	4th image
Initial regions number	2872	4071	2123	2457
Final regions number	111	110	149	188
Zeboudj	0,0818	0,0855	0,0763	0,1115
Borsotti	76,6143	82,5423	32,7402	9,9854

TABLE II. Lab SEGMENTATION RESULTS EVALUATION

	1st image	2nd image	3rd image	4th image
Initial regions number	2010	3149	1329	1709
Final regions number	161	148	169	193
Zeboudj	0,0958	0,1088	0,0843	0,0951
Borsotti	4,1046	3,8199	1,7013	1,6856

The length of the palette obtained by the genetic algorithm is adjusted according to the initial maximal string length and to colors of the image to be segmented. Table 3 presents, for some initial string length, the final palette length obtained for the first image after 17 generations in Lab color space.

TABLE III. INITIAL AND FINAL STRING LENGTH

Initial string length	10	20	40	60
Final string length	10	19	37	53

For the presented images, segmentation is carried in about 7mn22s in Intel Celeron Processor unit with 733MHz CPU. The fitness evaluation function is certainly the more time consuming (even if it contains fewer variables). This one requires scanning the whole image as many as population size is to get distance from each individual. One solution is to define an appropriate structure to save for each chromosome gene, its distance from each color point. Even if this proposition is largely more storage space consuming it may reduce significantly the run time.

V. CONCLUSION

Segmentation is the process of dividing image into separate regions and commonly precedes all image analysis. Genetic algorithms are a heuristic technique optimization and CCL is an approach that labels regions within a binary image. In this paper, a two steps segmentation method is presented based on clustering genetic algorithm and on the CCL approach generally used in pattern recognition.

In the proposed algorithm, real-coded genes is adopted and fewer variables for the fitness function are required and variable string length is used to adjust at the same time the palette size and the colors values. Furthermore, the idea of extending the CCL algorithm to color image avoid color image conversion and then a considerable gain in run time.

Application of the algorithm to Berkeley images base proves its efficiency and the use of more powerful processor may improve the run time. Currently works are in progress to

optimize this run time so that the algorithm can be applied as part of CBIR (Content Based Image Retrieval) to index color images.

REFERENCES

[1] F. Z. Bellala Belahbib and F. Souami, "A genetic algorithm based clustering for color image quantization", 3rd European Workshop on Visual Information Processing euvip'2011 Paris, 2011.

[2] J. C. Bezdek, J. K. Raghu Krisnapuram and Nikhil R. Pal, "Fuzzy models and algorithms for pattern recognition and image processing", Kluwer Academic Publishers, 1999.

[3] B. Bhanu, S. Lee and J. Ming, "Adaptive image segmentation using a genetic algorithm", IEEE Transactions on Systems, vol. 25, no. 12, 1995.

[4] S. Chabrier, B. Emile, C.Resenberger and H. Laurent, "Unsupervised performance evaluation of image segmentation", Special Issue on Performance Evaluation in Image Processing, EURASIP Journal on Applied Signal Processing, pp. 1-12, 2006.

[5] D. N. Chun, H. S. Yang, "Robust image segmentation using genetic algorithm with a fuzzy measure", Pattern Recognition, vol. 29, no. 7, pp. 1195-1211, 1996.

[6] Y. Deng and B. S. Manjunath, "Unsupervised segmentation of color-texture regions in image and video", IEEE Transaction on Patern Analysis and Machine Intelligence, vol. 23, pp. 800-810, 2001.

[7] L. Di Stefano, A. Bulgarelli, "A simple and efficient connected labeling algorithm", In 10th International Conference on Image Analysis and Processing, pp. 322-327, 1999.

[8] S. Dutta, B. B. Chaudhuri, Fellow, "Homogenous region based color image segmentation", Proceeding of the WordCongress on Engineering and Computer Science, vol. II WCECS 2009, San Francisco, USA, 2009.

[9] D. B. Fogel, Evolutionary computation. Toward a new philosophy of machine intelligence, 3rd ed., IEEE Press, 2006.

[10] T. Gevers, Color image invariant segmentation and retrieval. PhD thesis, University of Amsterdam, The Netherlands, May 1996.

[11] C. Grana, D. Borghesani and R. Cucchiara, "Connected component labeling techniques on modern architechures", LNCS 5716, ICIAP 2009, pp. 816-824, 2009.

[12] L. He, Y. Chao and K. Suzuki, "A linear-time two-scan labeling algorithm", Proceedings of the IEEE International Conference on Image Processing, ICIP 2007, pp. 241-244, 2007.

[13] A.K. Jain, "Data clustering: 50 years beyond K-means", Pattern Recognition Letters, vol. 31, vo. 8, pp. 651-666, June 2010.

[14] U. L. Jau and C. S. The, "Real-time object-based video segmentation using colour segmentation and connected component labeling", LNCS 5857, IVIC 2009, pp. 110-121, 2009.

[15] D. Martin, C. Fowlkes, D. Tal and J. Malik, "A database of human segmented natural images and its application to evaluating segmentation algorithms and measuring ecological statistics", ICCV, vol. 2, pp. 416-423, 2001.

[16] U. Maulik and S. Bandyopadhyay, "Genetic algorithm based clustering technique", Pattern Recognition, vol. 33, no. 9, pp. 1455-1465, 2000.

[17] U. Maulik and S. Bandyopadhyay, "Fuzzy partitioning using real coded variable length genetic algorithm for pixel classification", IEEE Transactions on Geosciences and Remote Sensing, vol. 41, no. 5, pp. 1075-1081, 2003.

[18] V. Ramos and F. Muge, "Image colour segmentation by genetic algorithm", Proceedings of the 11th Portuguese Conference on Pattern Recognition", pp 125-129, May 2000.

[19] A. Rosenfeld and J. L. Pfaltz, "Sequential operations in digital picture processing", Journal of the Association for Computing Machinery vol. 13, no. 4, pp. 471-494, 1966.

[20] E. H. Ruspini, "A new approach to clustering". Information and Control 15, pp. 22–32, 1969.

[21] P. Scheunders, "A genetic C-Means clustering algorithm applied to color image quantization", Pattern Recognition, vol. 30 no. 6, pp 859-866, 1997.

An Application of ICA-based Detection to OFDM-IDMA System

E.H.Meftah , A.Anou and M.Bensebti

Electronics Department, university SAAD DAHLAB of BLIDA, ALGERIA

BP 270 BLIDA, ALGERIA

Email: meftah2elhadi@yahoo.fr

Abstract—In this paper, an ICA-based blind multiuser detection approach for OFDM-IDMA system is presented. ICA is a signal processing technique that aims to detect a set of unknown mutually independent source signals from their observed mixtures without prior knowledge of the mixing coefficients. The proposed algorithm is applied as post processors attached to IDMA receiver. In this study, the performance of the proposed detector is compared to the conventional turbotype iterative OFDM-IDMA multiuser detector in which chip interleavers are the only means of user separation. Validity and performances of the described approach are demonstrated by numerical simulations based on examining the bit error rate (BER) criterion in the quasi-static Rayleigh fading channel.

Keywords—OFDM-IDMA;ICA;BSS;MUD;wireless communication.

I. INTRODUCTION

Interleave-based systems are broadly used in various wireless applications. In order to use the capacity of multi-access systems, multiuser detection techniques are necessary. A large number of schemes and algorithms have been devised to enhance the performance, and also to reduce the involvement of a receiver in a multiuser environment [1].

To suit the increase in demands on wireless services, future wireless communication systems have to structure new technologies that support high capacity. In [2], [3], it has been shown that interleave division multiple access (IDMA) systems support higher information capacity over orthogonal frequency division multiplexing (OFDM). Furthermore, in a multi-user wireless communication system over multipath channels where the multiuser access interference (MAI) and the inter-symbol interference (ISI) are major sources of disability to performance, it can be proves that OFDM-IDMA systems accommodate resistance to frequency selective fading and interferers by virtue of correlation inter user.

Independent component analysis (ICA) has received immense interest from the research society in signal processing, mainly due to its ability to solve the problem of blind source separation (BSS) which consists of converting the mixed random signals into source signals or components using the presupposition that the source signals are mutually independent [4].ICA aims to separate independent blind sources from their observed linear mixtures without any prior knowledge of the channel state information (CSI). This technique has been widely used in the past decade to extract

useful features from observed data in many fields (e.g. telecommunications).

In the case of CDMA system application, for instance, it is not meaningful to apply ICA on its own because of the ambiguities about which extracted source is the desired signal. Note that in CDMA communication, the source signals are well defined by the spreading codes [5][6]. However, if the interference suppression property of ICA is used with available code sequence information of the wanted user, it would then be meaningful of associating ICA with CDMA [7].

In this paper, we propose to detect and separate the transmitted OFDM-IDMA symbols, without estimating the channel, by using the ICA algorithm as post processor attached to IDMA receiver. The motivation behind this proposed method is to avoid transmitting the bits of the channel parameters, and consequently, it permits to optimize the spectral effectiveness as well as the computation time at reception.

This paper is organized as follows. In the next section, the system model and performance measures of OFDM-IDMA systems are presented. Then, in Section III, the ICA-based detector structure is described. Next, in Section V, the simulation results are presented. Finally, in Section VI, a conclusion

II. SYSTEM MODEL OF OFDM-IDMA

IDMA is considered as a special case of code division multiple access (CDMA) in which IDMA applies random inter-leavers to distinguish different users in wireless communication system [8]. The main advantage of IDMA is that it allows the use of a low-complexity iterative multi-user detection (MUD) strategy to systems with a large number of users and the computational complexity of the iterative MUD in IDMA systems is a linear function of the number of users, and is much lower than that in CDMA systems.

The IDMA receiver complexity over multi-path channels is related to the channel length. In [2], OFDM-IDMA was proposed as multi-user system combining OFDM and IDMA in the multi-path environment, in which OFDM are adopted to resolve the ISI.

A K-user OFDM-IDMA system is shown in Fig. 1, the block diagram of the transmitter and (iterative) receiver structures follows the principles propose in[1].At the transmitter part, the information bits d_k for k^{th} user,k = 1,2,...,K , are spread with a speading sequence c_k , after the speading bits are interleaved by a user-specific interleaver π_k, we assume that interleaver are generated randomly and independently , In addition , Quadrature Phase Shif Keying (QPSK) mapping is considered ,the complex result sequence $\{X_k(n) = X_k^{Re}(n) + iX_k^{Im}(n)\}$ are operated by Inverse Fast Fourier Transform (IFFT) operation. Furthermore, a guard interval (GI), whose duration is longer than the channel maximum delay, is inserted between adjacent OFDM symbols in order to prevent ISI and the inter-carrier interference (ICI). We assume the complex channels coefficient of each user $h_k(l), = [h_k(0), h_k(1), \dots, h_k(L-1)]$, are stationary in a frame period and mutually independent, where L is the number of paths.

At the receiver part, the received signals after OFDM demodulation (includes a guard interval elimination followed by FFT) can be expressed as:

$$R(n) = \sum_{k=1}^{K} H_k(n)X_k(n) + W(n) , n = 0,1, \dots, N_c - 1 \quad (1)$$

Where N_c is the number of OFDM subcarrier and $H_k(n) = \sum_{l=0}^{L-1} h_k(l)e^{-j2\pi l/N_c}$ denotes the channel frequency response on the nth subcarrier for the kth user, $\{W(n)\}$ are samples of complex additive white Gaussian with zero mean and σ^2 variance.

Let $H_k(n) = |H_k(n)|e^{-i\theta}$, the equation (1) became as:

$$Re\left(e^{-i\theta}R(n)\right) = |H_k(n)|X_k^{Re}(n) + \xi_k^{Re}(n) \quad (2)$$

Where:

$\xi_k^{Re}(n) = Re\left(e^{-i\theta} \sum_{k'\neq k} H_{k'}(n)X_{k'}(n) + e^{-i\theta}W(n)\right)$ represents MAI and ISI, Using the central limit theorem, $\xi_k^{Re}(n)$ can be approximated by Gaussian random variable.

The turbo-type iterative OFDM-IDMA multiuser detector consists of an iteration process composed by an elementary signal estimator (ESE) and K single-user a posteriori (APP) decoders (DECs). The next sub-section gives brief details of the iteration process, for more details refer to [2], [9]. According to (1), the coded block often occupies several OFDM symbols; we can rewrite the received signal corresponding to the mth symbol of coded block as:

$$R_m(n) = \sum_{k=1}^{K} H_{k,m}(n) X_{k,m}(n) + V_m(n)) \quad (3)$$

If we want to express the equation (3) in matrix notation, we can replace the user signature sequences and channel coefficient with a matrix G, by some abuse of notation and we can express the mth symbol OFDM as:

$$R_m = \sum_{k=1}^{K} H_{k,m} d_{k,m} c_k + V_m \quad (4)$$

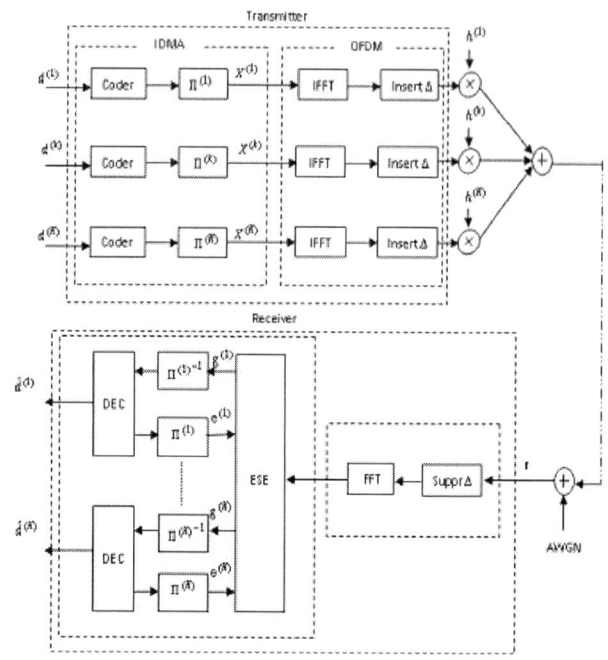

Fig. 1. Block diagram of the transmitter and (iterative) receiver structures of OFDM-IDMA.

$$\bar{R}_m = H_{1,m} d_{1,m} \begin{bmatrix} c_{1,1} \\ c_{2,1} \\ \vdots \\ c_{S,1} \end{bmatrix}_{Sx1} + \cdots + H_{K,m} b_{K,m} \begin{bmatrix} c_{1,K} \\ c_{2,K} \\ \vdots \\ c_{S,K} \end{bmatrix}_{Sx1} + \bar{V}_m$$

$$= [\bar{c}_1, \bar{c}_2 \cdots \bar{c}_k]_{SxK} \begin{bmatrix} H_{1,m} & 0 & .. & 0 \\ 0 & H_{2,m} & .. & 0 \\ \vdots & \vdots & \ddots & \vdots \\ 0 & 0 & .. & H_{K,m} \end{bmatrix}_{KxK} \begin{bmatrix} d_{1,m} \\ d_{2,m} \\ \vdots \\ d_{K,m} \end{bmatrix}_{Kx1}$$

$$+ \begin{bmatrix} V_1 \\ V_2 \\ \vdots \\ V_S \end{bmatrix}_{Sx1}$$

$$\bar{R}_m = [\bar{c}_1 H_{1,m}, \bar{c}_2 H_{2,m} \cdots \bar{c}_k H_{K,m}]_{SxK} \ \bar{d}_{m(Kx1)} + \bar{V}_m$$

$$\bar{R}_m = G \ \bar{d}_m + \bar{V}_m \quad (5)$$

Where all the vectors are assumed column vectors and they are indicated by ' $\bar{\ }$ ', and m is the index to the position of the coded symbol in the coded block, which is very similar to that of IDMA [10], except that $H_{k,m}$ is the frequency domain factor of the corresponding subcarrier, rather the time domain fading factor, c_k is the Sx1 vector representation of kth user's signature sequence and $\bar{V}_m$ is the Sx1 AWGN vector, G is an SxK matrix with elements $G = H_{k,m}^{(n)} \cdot c_k^{(n)}$, $\forall$ k and n

A. Elementary Signal Estimator (ESE)

The ESE computes the extrinsic log-likelihood ratio (LLR) about $X_k^{Re}(n)$, given by:

$$g(X_k^{Re}(n)) = \frac{2|H_k(n)|}{\text{Var}\left(\xi_k^{Re}(n)\right)} \left(\text{Re}\left(e^{-i\theta}R(n)\right) - E(\xi_k^{Re}(n))\right) \quad (6)$$

The method for calculating $E(\xi_k^{Re}(n))$ and $\text{Var}\left(\xi_k^{Re}(n)\right)$ can be found in [9], and the $X_k^{Im}(n)$ can be generated in a similar way.

B. The decoder (DEC) operation

The DECs carry out APP decoding using the output of the ESE as the input. With QPSK signaling, outputs of the DEC-k are the extrinsic LLRs for $X_k^{Re}(n)$ and $X_k^{Im}(n)$. We use the extrinsic information to update the mean and variance of each chip in the next iteration.

.

III. ICA MODEL

ICA is a statistical technique based on higher order statistics, which has applications in many areas where the model of ICA is satisfied, such as DS-CDMA communications. It is a blind technique, because besides source signals, the mixing coefficients are also unknown. Thus, we suppose that the sources (symbols in our case) are statistically independent. This is a fundamental assumption for using ICA that is generally verified in communication systems. Also, the extraction of the sources can be done by ICA by exploiting the essential features of the sources and system [11].

A. ICA estimation

The simplest case of ICA model deals with linear instantaneous mixing .In this model, the observed data vector s is modeled by the linear transformation as given below [12]:

$$x = As \quad (7)$$

Where x is a vector variable, of dimension P, in which each variable is an observed signal mixture and s is a vector variable, of dimension M, in which each variable is a source signal. We assume that $P \geq M$.The mixing matrix A defines a linear transformation on S, which can usually be reversed in order to recover an estimate vector U of s from x, i.e.:

$$S \approx U = Ws \quad (8)$$

Where the separating matrix $W = A^{-1}$ is the inverse of A. However, A is an unknown matrix and cannot therefore be used to find W.

B. Proposed ICA-based detector

Let us observe k scalar random variables $r_1, r_2, \ldots, r_k$, which are assumed to be linear combinations of u unknown independent components (ICs) $d_1, d_2, \ldots, d_u$ that are statistically independent and zero-mean. In the vectorial notation, let us consider the two following vectors: $\bar{r} = \left(r_1, r_2, \ldots, r_k\right)^T$ and $\bar{d} = \left(d_1, d_2, \ldots, d_u\right)^T$; then the linear relationship between vector $\bar{r}$ and vector d is given by:

$$\bar{r} = G\bar{d} \quad (9)$$

In the presence of the noise, the model in (9) becomes [14]:

$$\bar{r}_m = G\bar{d}_m + \bar{n}_m \quad (10)$$

where $\bar{r}_m =$ is m^{th} observed data vector, G is an unknown full rank mixing matrix, $\bar{d}_m$ is a realization of an unknown non-Gaussian source and $\bar{n}_m$ is a realization of additive noise process. The subscript m in the equations refers to the observed variable at the mth sampling interval. The model in (10) is then similar with the OFDM-IDMA model given in equation (5).The purpose is to estimate $\bar{d}_m$ given only $\bar{r}_m$, and the assumption of the independence of the sources, which can be expressed as follows:

$$y_m = \bar{w}^T \bar{r}_m \quad (11)$$

So, y_m can be used to estimate the data bit of the wanted user at the mth interval. As the binary antipodal modulation is used, the symbol of the desired k^{th} user can be obtained simply by taking the $signum$ function of y_m:

$$\hat{d}_m^k = sgn(y_m) = sgn(\bar{w}^T \bar{r}_m) \quad (12)$$

IV. SIMULATION RESULTS

In this section, numererical results concerning the BER performance of the proposed ICA-based detector are presented and compared with those obtained by classical conventional detectors in the case of the IDMA and OFDM-IDMA systems. For all simulation, only the spreading code of the wanted mobile user is known, while the codes of the interfering users are unknown (i.e. blind detection), the interleavers used in simulations are generated randomly and independently, the modulation BPSK constellation and the quasi-static Rayleigh fading channel are assumed ,the iteration number is fixed at $it = 10$ to guarantee convergence [3].

First, we tested the ICA-based detector in IDMA system, the result is compared with ICA-assisted minimum mean square error labelled MMSE-ICA detector for CDMA system, a simulated system with $K = 20$ users, spread with short Gold Codes of length $S = 31$ are considered, The results so obtained are given by Fig. 2. It can clearly be seen that IDMA-ICA detector outperforms CDMA MMSE-ICA detector despite its simplicity and that the performance advantage increases at high signal-to-noise ratio (SNR) values and at high MAI communication environment, attributed to the use of user specific interleavers, which greatly facilitates the convergence of iterative MUD [15].

Secondly, we simulated the performance of the proposed ICA-based detector in an uplink OFDM-IDMA context. Let $N_{info} = 256$ $bits$ be the number of information bits in a frame per user, number of user $K = 8$, random sign-alternating spreading codes are used with spread length $S = 8$ (loaded scenarios), the BPSK symbols of each user are modulated on to subcarriers by IFFT, the number of subcarriers is set to be $N_c = 64$, cyclic prefix whose duration is longer than the maximum channel delay is inserted between adjacent OFDM symbols. According to Fig. 3, It can be seen that the ICA-based detection schemes can also effectively detect the transmitted signals. However, there is a large gap between the ICA-based detection and perfect CSI based detection using chip-by-chip (CBC) detection in the OFDM-IDMA scheme.

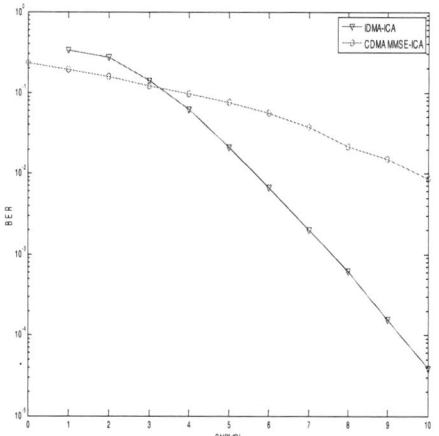

Fig. 2. BER vs. SNR performance of the ICA based blind detector for IDMA system.

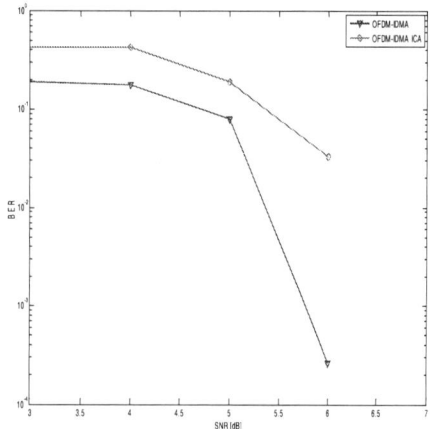

Fig. 3. BER vs. SNR performance of the ICA based blind detector for OFDM-IDMA system

V. CONCLUSION

In this paper, we have introduced to an ICA-based detection approach for uplink OFDM-IDMA system. We started only with the synchronization information and the spreading code of the wanted user, simulation results indicated that the performance of the proposed detector was much better than the conventional ICA-MMSE in CDMA context and it was comparable to OFDM-IDMA with CBC iterative MUD technique based on prior knowledge of spreading codes, delays and amplitudes of all the interfering users. In a real OFDM-IDMA wireless communication, the interfering users keep on changing, this makes the wireless environment remarkably dynamic, in which case, the MUD requires accurate information about the interfering users that reduces the spectral efficiency. The proposed ICA-based detector was attempted to estimate the original source signals without knowing the parameters of mixing process [6], and thus can handle changing wireless conditions. As perspective, we extend work that was reported in [16], in which we will envisage to detect and separate the transmitted symbols in noisy environment [14].

REFERENCES

[1] K. Kusume, G. Bauch, and W. Utschick, . "IDMA vs. CDMA: analysis and comparison of two multiple access schemes,". *Wireless Communications, IEEE Transactions on*, vol. 11, no. 1, pp. 78-87, 2012.

[2] I. M. Mahafeno, C. Langlais, and C. Jego, . "OFDM-IDMA versus IDMA with ISI cancellation for quasistatic rayleigh fading multipath channels ,". *in Turbo Codes&Related Topics; 6th International ITG-Conference on Source and Channel Coding (TURBOCODING), 2006 4th International Symposium on*, 2006, pp. 1-6.

[3] L. Ping, Q. Guo, and J. Tong, . "The OFDM-IDMA approach to wireless communication systems," Wireless Communications, IEEE, vol. 14, no. 3, pp. 18-24, 2007.

[4] R. X. Tian, X. Hui, H. Z. Tao, W. Fenghua, and L. F. Bo, . "Fast-ICA based blind estimation of the spreading sequences for down-link multirate DS/CDMA signals," *in Intelligent Computation Technology and Automation (ICICTA), 2012 Fifth International Conference on*, 2012, pp. 501-504.

[5] T. Ristaniemi and J. Joutsensalo, . "On the performance of blind source separation in CDMA downlink," *in Proc. Int. Workshop on Independent Component Analysis and Signal Separation (ICAâAZ'99)*, 1999, pp. 437-441.

[6] W. Y. Leong and J. Homer, . "Implementing ICA in blind multiuser detection,". *in Communications and Information Technology, 2004. ISCIT 2004. IEEE International Symposium on*, vol. 2, 2004, pp. 947-952.

[7] T. Ristaniemi and J. Joutsensalo, . "Advanced ICA-based receivers for DS-CDMA systems, ". *in Personal, Indoor and Mobile Radio Communications, 2000. PIMRC 2000. The 11th IEEE International Symposium on*, vol. 1, 2000, pp. 276-.281.

[8] L. Ping, L. Liu, K. Wu, and W. K. Leung, . "Interleave division multipleaccess, *Wireless Communications, IEEE Transactions on*, vol. 5, no. 4,pp. 938-947, 2006.

[9] L. Ping, K. Y. Wu, L. Liu, and W. K. Leung, . "A simple, unified approach to nearly optimal multiuser detection and space-time coding, ". *in Information Theory Workshop, 2002. Proceedings of the 2002 IEEE*, 2002, pp. 53-56.

[10] H. Bie and Z. Bie, . "A hybrid multiple access scheme: OFDMA-IDMA, ". *in Communications and Networking in China, 2006. ChinaCom'06. First International Conference on*, 2006, pp. 1-3.

[11] A. Hyvärinen, J. Karhunen, and E. Oja, Independent component analysis.2001. Wiley, New York, 2001.

[12] A. Hyvärinen and E. Oja, . "Independent component analysis: algorithms and applications, ". Neural networks, vol. 13, no. 4, pp. 411-430, 2000.

[13] A. Hamza, E. H. Meftah, S. Chitroub, and R. Touhami, . "ICA-based blind symbol detection for compound system MIMO-OFDM in CDMA context, ". *in Telecommunications (ICT), 2010 IEEE 17th International Conference on*, 2010, pp. 345-349.

[14] O. Ekici and A. Yongacoglu, . "Application of noisy-independent component analysis for CDMA signal separation,. in Vehicular Technology Conference, " 2004. *VTC2004-Fall. 2004 IEEE 60th*, vol. 5, 2004, pp.3812-3816.

[15] A. Hamza, G. Salut, and S. Chitroub, . "Independent component analysis in IDMA systems, ". *Proceedings of IEEE NEWCAS-TAISA*, pp. 64-68, 2009.

[16] E. H. Meftah, A. Anou, and M. Bensebti, . "Noisy ICA-based detection method for compound system MIMO-OFDM in CDMA context, ". *in I/V Communications and Mobile Network (ISVC), 2010 5th International Symposium on*, 2010, pp. 1-4.

Automatic Detection Method of R-wave Positions in Electrocardiographic signals

Fatiha Bouaziz, Daoud Boutana
Department of electronic,
Faculty of science and technology,
University of Jijel, Algeria
bouhali.fatiha_@yahoo.fr, daoud.boutana@mail.com

Messaoud Benidir
Laboratoire des signaux et systèmes Supelec,
Université Paris-Sud, 91192 ;Gif-Sur-Yvette-France.
benidir@Lss.supelec.fr

Abstract—In this study, we have presented a robust method of R-wave detection in electrocardiographic signals (ECG) using multiresolution wavelet transform. Daubechies wavelets (db4 and db6) were used for this detection. The presented method requires a pre-processing by using a Median filter which is an effective tool to remove the baseline of the signal and correct it. Then, the ECG signals under test have been decomposed to the required level using the selected wavelets. Finally, the algorithm has been tested and evaluated on MIT-BIH arrhythmia database. Signals from lead II have been only analyzed. In this validation, the QRS detection algorithm achieved the very good detection performance using db6 wavelet, because db6 wavelet gives the higher values of the overall sensitivity and predictivity than to db4 wavelet.

Index Terms—ECG, R-wave, multi-resolution wavelet transform, MIT arrhythmia database, QRS complex, Daubechies wavelet.

I. INTRODUCTION

The Electrocardiogram is a graphical representation of the potential difference between two points on the body surface, versus time. It has become one of the most valuable diagnostic tools in modern medicine. The diagnosis and treatment of many cardiovascular diseases relies mainly upon the visual inspection of the recorded ECG signal. The recognition and analysis of ECG signal are difficult, since their size and form may eventually change and not only that, the presence of noise from different sources can also affect the true form of the signal. Therefore the processing of the ECG signal is a very important in facilitating the evaluation of ECG by physicians.

The QRS complex is the most prominent waveform within the electrocardiographic (ECG) signal, with normal duration from 0.06 s to 0.1 s [1]. It reflects the electrical activity within the heart for total ventricular muscle depolarization. Its shape, duration and time of occurrence provide valuable information about the current state of the heart. Because of its specific shape, the QRS complex serves as an entry point for almost all automated ECG analysis algorithms and detection of the QRS complex is the most important task in automatic ECG signal analysis [2].

The QRS detection is not a simple task, due to the varying morphologies of normal and abnormal complexes and because the ECG signal experiences different types of disturbances with

complex origin. The various researchers are conducted to explore effective signal analysis techniques for the detection of QRS complex [3-4]. Wavelet transform is a powerful tool for analyzing the bio medical signals [5,6], because it gives good estimation of time-frequency localization.

The method presented in this paper is robust and simple to implement. The R-wave positions have been identified using the selective coefficients method described by S. Pal and al [7].

II. MATERIAL

A. Discrete Wavelet Transform

In wavelet transform, a linear operation transforms the signal to decompose it at different scales. The Discrete Wavelet Transform (DWT) is a special case of the Wavelet Transform (WT) that provides a compact representation of a signal in time and frequency that can be computed efficiently. Basically, wavelet transform is the convolution operation of the subject signal x(t) and the family of wavelet functions, $\Psi_{j,k}(t)$ [8] :

$$T_{j,k} = \int_{-\infty}^{+\infty} x(t)\psi_{j,k}(t)dt \qquad (1)$$

The approximation coefficient of the signal $x(t)$ is represented as:

$$A_{i,k} = \int x(t)\phi_{j,k}(t)dt \qquad (2)$$

where $\phi(t)$ is the scaling function, j and k are scale and location factors respectively.

For a range of scale m, the original signal $x(t)$ under discrete wavelet transform can be represented as:

$$x(t) = x_m(t) + \sum_{j=1}^{m} d_j(t) \qquad (3)$$

Where

$x_m(t)$: is the mean signal approximation at scale *m* and is given by:

$$x_m(t) = A_{m,k}\phi_{m,k}(t) \qquad (4)$$

and $d_j(t)$: is the detail signal corresponding to scale *j*.

In DWT, filters of different cut-off frequencies are used for analyzing the signal at different scales. For this purpose, the signal is passed through a series of highpass and lowpass filters in order to analyze low as well as high frequencies in the signal.

B. Wavelet Selection

The selection of the analyzing function in Wavelet Transform, which is called the mother wavelet, has a significant effect on the result of analysis and should be selected carefully based on the nature of the signal [9]. But, there is no absolute way to choose a certain wavelet. There are several wavelet families like: Harr, Daubechies, Biorthogonal, Coiflets, Symlets, Morlet,...Daubechies (db4 and db6) wavelets have been selected for extracting ECG features in our application, because they have been found to give details more accurately than others. Moreover, the above two mentioned wavelets show structural similarity with QRS complexes and energy spectrum is concentrated around low frequencies [10].

C. Database

For the analysis of detector performance, it is necessary that a standard database must be chosen so that the obtained results can be interpreted with respect to that manually annotated database. In order to test the performance of the proposed method, real electrocardiogram signals have been used, collected from the MIT-BIH database [11]. Annotated ECG signals are noted by file (.dat).

III. METHOD OF DETECTION

The ECG signals (.dat files) downloaded from Physionet are first converted in to MatLab readable format (.mat files). The signals from lead-II sampled at 360 Hz are only taken for our analysis. Detection of the R waves and the elimination of the abnormalities in the ECG signal is the important step in localizing the aim of this paper. Detection process is summarized by the following steps:

A. Pre-processing

Before preceding the features detection method of the ECG signal, it is first necessary to remove baseline wandering from an ECG signal. For this purpose, we have used the correction process proposed by Z.K.Chalabi and al[12].

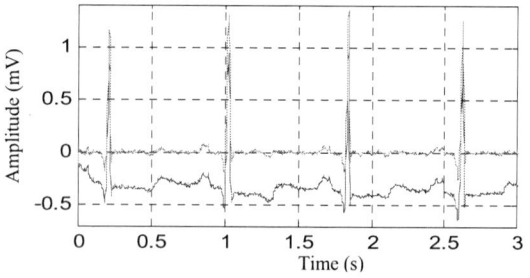

Fig. 1. The effect of the Median filter on the ECG signal (100.dat)

B. Decomposition of the Signal

The corrected ECG signal is decomposed for eight wavelet scales which are selected for better illustration to ensure the presence of some low frequency components of original signal (Fig. 2). It is clear that small scales represent the high frequency components and large scales represent the low frequency components of the signal.

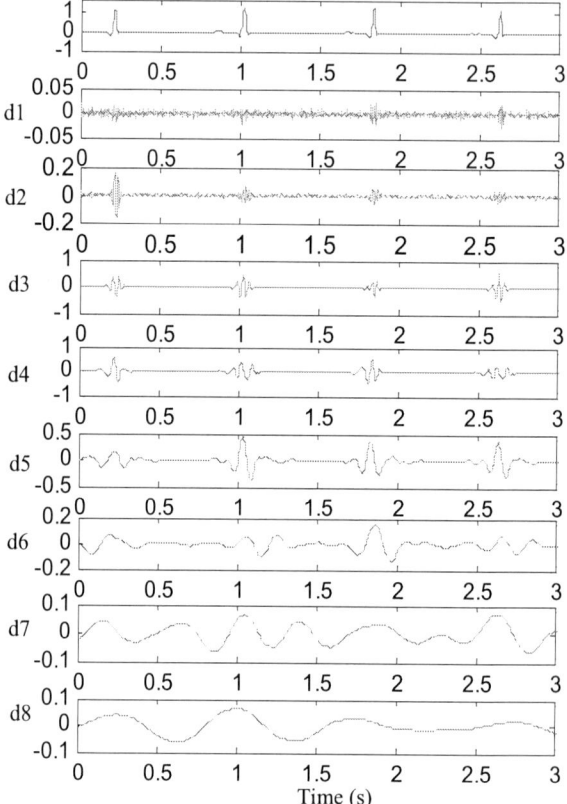

Fig. 2. ECG signal decomposition using db6 wavelet

C. Detail Selection for R-Wave Detection

The selection of specific detail coefficients which are used for R-wave detection is based on the energy and cross correlation analysis [13]. So, we have first calculated the energy contained within the detail coefficients at each scale using equation (5) then the cross-correlation coefficients between all the decomposed signals individually with the original ECG signal were calculated. The results show that the most energy of the QRS complex is concentrated at decomposition levels 3, 4 and 5 and also the highest values of correlation coefficients were observed at those levels. Thus, detail coefficients d_3, d_4 and d_5 are chosen for the detection of QRS complex.

$$E_m = \sum_{i=1}^{m} [d_i]^2 \qquad (5)$$

D. R-Wave Positions

Figure 3 shows the plot of the original signal along with the reconstructed signal using the selected detail coefficients (d_3, d_4, d_5).

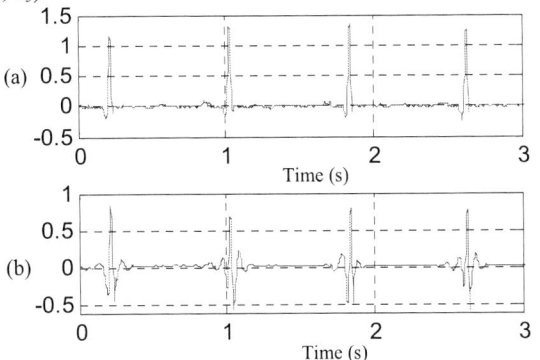

Fig. 3. Original signal (a) and reconstructed signal (b)

It is clear that the QRS region is properly captured but it is difficult to identify R peaks due to its oscillatory nature. For this reason, we have used the detection method developed by S. Pal and al[7] in order to correctly extract the R-Waves positions in ECG signals. They are found by the determination of maximum amplitude points of the modulus of two functions e_1 and e_2, which have been defined as [7]:

$$e_1 = d_3 + d_4 + d_5 \qquad (6)$$

$$e_2 = \frac{d_4(d_3 + d_5)}{2^n} \qquad (7)$$

where n is the level of decomposition.

Then, the modulus of $e_1 \times e_2$ is taken for the detection of R-wave positions. Figure 4 shows the plot of this modulus along with the original ECG signal. It is evident that the QRS complex may have less amplitude, but it is much closely spaced in time. Hence the R peaks are identified as the maximum amplitude points.

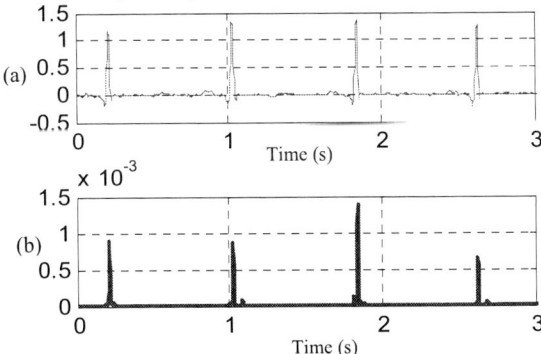

Fig. 4. Original signal (a) and plot of modulus of $e_1 \times e_2$ (b)

IV. RESULTS AND VALIDATION

In the present work, physionet MIT-BIH database is used to test and evaluate the algorithm. As discussed in the previous section. The database contains 48 records and each recording is of 30 minutes duration. Here, 10 records are used to evaluate the R-wave detection algorithm. In our evaluation of the proposed technique, we have calculated the following parameters [14]:

- the Sensitivity (*Se*) defined as:

$$Se = \frac{TP}{TP + FN} \qquad (8)$$

- the Positive predictivity (p^+), which has been written as:

$$P^+ = \frac{TP}{TP + FP} \qquad (9)$$

Where:
TP: is the total number of beats under test;
FN: is the number of the false negative beats;
FP: the number of the false positive beats.

A False Negative beat (*FN*) occurs when the algorithm fails to detect an actual QRS quoted in the corresponding annotation file of the MIT-BIH record and a false positive beat (*FP*) represents a false beat detection.

Table I shows the results obtained from the test of 10 records, using two wavelet functions of Daubechies family (db4 and db6).

Figure 5 shows two examples of the detection of R-wave positions using the proposed technique. We have used the symbol '*' to illustrate the R-wave positions on the original signal.

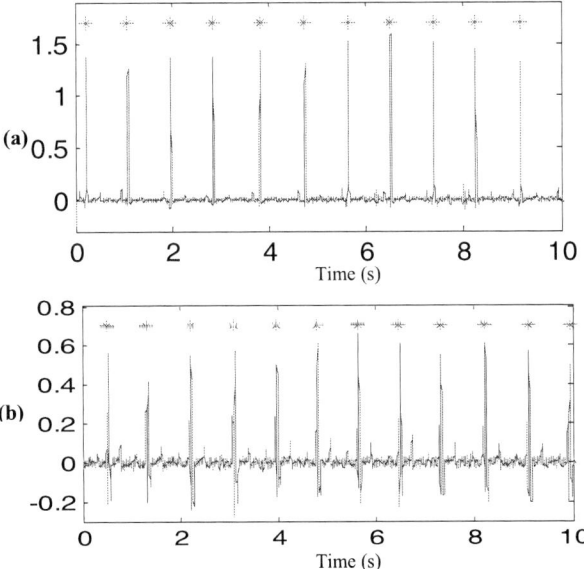

Fig. 5. Original signal and R-peaks position : (a) 101, (b) 111

From the obtained results, it is clear that the proposed method gives the good detection performances (Sensitivity and Positive predictivity). We have also obtained that db6 wavelet shows the better detection parameters than to db4.

TABLE I. VALIDATION RESULTS OF THE PROPOSED DETECTION ALGORITHM USING TWO TYPES OF DAUBECHIES WAVELETS

Record	Total N° beats	db4 wavelet				db6 wavelet			
		FP beats	FN beats	Se (%)	P^+ (%)	FP beats	FN beats	Se (%)	P^+ (%)
100	2273	0	3	99.86	100	0	0	100	100
101	1865	1	4	99.78	99.94	0	2	99.89	100
103	2084	1	2	99.90	99.95	0	0	100	100
105	2572	1	5	99.80	99.96	0	0	100	100
107	2137	3	2	99.90	99.85	2	1	99.95	99.90
109	2132	2	2	99.90	99.90	1	1	99.95	99.95
111	2124	10	4	99.81	99.53	0	1	99.95	100
112	2532	2	8	99.68	99.92	1	2	99.92	99.96
115	2545	4	2	99.92	99.81	0	1	99.96	100
200	2021	6	3	99.85	99.70	0	1	99.90	100
Total	**22285**	**30**	**35**	**99.84**	**99.86**	**4**	**9**	**99.96**	**99.98**

V. CONCLUSION

In this paper, an algorithm based on multiresolution wavelet transform has been presented for the R-wave detection in ECG signals. Wavelet decomposition of ECG signal up to the eight level using Daubechies wavelets (db4 and db6) generates eight scales of detail coefficients. The smaller scales correspond to high frequency components and higher scales correspond to low frequency components of the signal. According to the energy of the signal, it has been clear that the most energy of the QRS complex is concentrated at decomposition levels 3, 4 and 5. This result has been further justified by calculating cross correlation coefficients between the original ECG signal and each detail coefficient. Thus, detail coefficients d_3, d_4 and d_5 were identified for the detection of QRS complex. But, it has been proved that the R-wave extraction using reconstructed signal comprising detail coefficients (d_3-d_5) is difficult due to its oscillatory nature. Finally, the presented R-wave detection algorithm has been validated using MIT-BIH standard database. We have seen that this method achieved very good detection performance (Se, P^+). The result of the comparison between the two used wavelets shows that db6 wavelet gives the better values of the overall sensitivity and predictivity than to db4.

REFERENCES

[1] M. Goldman, "Principle of Clinical Electrocardiography", eleventhed., Lange Medical Publication, Drawer L., Los Altos, California 94022, 1982.

[2] C. Li, C. Zheng, C. Tai, Detection of ECG characteristic points by wavelet transforms, IEEE Transactions on Biomedical Engineering 42 (1) (1995) 21–28.

[3] A. Ruha, S. Sallinen and S. Nissila, "A real-time microprocessor QRS detector system with a 1ms timing accuracy for measurement of ambulatory HRV", IEEE Transactions on Biomedical Engineering, vol.44, no. 3, pp. 159-167, March,1997.

[4] R. Legarreta, P.S. Addison and N. Grubb, "R-wave detection using continuous wavelet modulus maxima", IEEE proceedings on Computers in Cardiology, Vol. 30, pp. 565-568,Feb 2003.

[5] M. Bahoura, M. Hassani and M. Hubin, "DSP implementation of wavelet transform for real time ECG wave forms detection and heart rate analysis", Computer methods and programs in biomedicine (1997), pp. 35–44

[6] P.S. Addison, "Wavelet transforms and the ECG: a review", Physiological measurement, vol. 26, no. 5, pp.155, 2005.

[7] S.Pal, M. Mtra, « Detection of ECG characteristic points using Multiresolution Wavelet Analysis based Selective Coefficient Method »,Measurement 43 (2010) 255–261.

[8] R. Polikar,"The Wavelet Tutorial Part IV, Multiresolution Analysis: The Discrete Wavelet Transform", www.cs.ucf.edu/courses/cap5015/WTpart4.pdf, (2008).

[9] S. Sumathi, M. Y. Sanavullah, "Comparative study of QRS complex detection in ECG based on discrete wavelet transform", International Journal of Recent Trends in Engineering, Vol 2,No. 5, November 2009.

[10] S. Z. Mahmoodabadi, A .Ahmadian , M. D.Abolhasani, "ECG Feature Extraction using Daubechies Wavelets" Proceedings of the fifth IASTED, International Conference VIisualization, Imaging, and Image Processing, Benidorm, Spain, September 7-9, 2005.

[11] http://www.physionet.org/mitdb

[12] Z.K. Chalabi, A. Boudjemaoui, L. Saadia, and N. Berrached, "Détection et classification Automatiques d'Arythmies Cardiaques", the fifth International Conference. Sience of Elecyronic, Technologies of Information and Telecomunications, March 22-26, 2009-Tunisia.

[13] A. Pachauri, and M. Bhuyan, "Robust Detection of R-Wave Using Wavelet Technique", World Academy of Science, Engineering and Technology 56 2009.

[14] M. Talbi, A. Aouinet, L. Salhi, and A. Cherif, "New Method of R-wave Detection by continuous Wavelet Transform", Signal processing, An international journal (SPIJ),volume (5), issue (4), 2011.

Use of Nakagami statistical model in ultrasonic tissue mimicking phantoms characterization

Nardjess BAHBAH, Hakim DJELOUAH

Faculty of Physics, Material Physics Laboratory
U.S.T.H.B
Algiers, Algeria
nardjess_bahbah@yahoo.fr,
djelouah_hakim@yahoo.fr

Ayache BOUAKAZ

Université F. Rabelais
INSERM U930
Tours, France
bouakaz@med.univ-tours.fr

Abstract— In order to improve tissue characterization, the probability density function of ultrasonic backscattered echoes which may be treated as random signals, is modeled using Nakagami statistical distribution. Recently, it has been shown that Nakagami statistical model constitutes a quite good model for tissue characterization due to its simplicity and general character. Experiments were performed using a 5MHz linear array connected to a 128 individual channels open research platform (M2M, Les Ulis, France) equipped with individual analog transmitters. Ultrasound images and corresponding RF signals were acquired at different transmit frequencies ranging from 2 to 6 MHz. Transmitted signals had frequency bandwidths of 80 to 90 %. A commercially available phantom was used to mimic tissue backscattering. The analysis was performed on different sizes and locations of the sampling window. The received echoes have been filtered around the transmitted center frequency and around twice the transmitted frequency (2nd harmonic). Acquired RF signals have been analyzed in order to evaluate Nakagami parameter (m), the scaling parameter (Ω) and the probability density function (PDF). These latter results have been compared to those obtained by using Field II software.

Keywords- Nakagami distribution, Tissue characterization, Ultrasonic backscattering, Ultrasonic imaging.

I. INTRODUCTION

Many researchers have used stochastic models to describe the probability density function (PDF) of the ultrasonic signals envelope backscattered by tissues which may be treated as random signals. The parameters of these distributions depend on some characteristics such as the density (number of scatterers within the transducer resolution cell) and scattering amplitude related to the size of the scatterers. Among the commonly used distributions, we can quote Rayleigh distribution (square root of an exponential distribution), K-distribution [1] (square root of the product of a Gamma distribution with an exponential distribution), and Nakagami distribution (square root of a Gamma distribution). Rayleigh model which is commonly used [2], needs some conditions such as the presence of a large number of randomly located scatterers. Wagner [3] classifies the other models according to their Signal to Noise Ratio (SNR) as compared with the SNR of Rayleigh distribution. The first class called pre-Rayleigh

(SNR < 1.91) describes heterogeneous textures. The second, called Rayleigh (SNR=1.91), defines the homogenous texture class. The third corresponding to the periodic texture is the post-Rayleigh class (SNR > 1.91).

It has been shown that K-distribution constitutes a quite good model for pre-Rayleigh and Rayleigh textures [4, 5]. The two K-distribution parameters provide information on the number of scatterers, the variation in the scattering amplitude and the average scattering amplitude. But they are not enough general to describe the statistics of the backscattered echo from range cells containing a periodic alignment of scatterers giving rise to post-Rayleigh. Recently Nakagami statistical model, initially proposed to describe the statistics of radar echoes, has been considered to be able to quantitatively characterize biological tissues thanks to its two parameters, Nakagami parameter (m) and scaling parameter Ω [6]. In addition to the scattering amplitude and density, this model can take into account the regularity of the scatterers spacing [7]. Nakagami statistical model has comparatively less computational complexity than the other models and is enough general to describe a wide range of scattering conditions in medical ultrasound, including pre-Rayleigh, Rayleigh and post-Rayleigh distributions. Although the Nakagami distribution can fit well with the PDF of the ultrasonic envelope, a multiple statistical distribution may be more appropriate to model the envelope statistics, since the ultrasonic signals returned from the tissues may contain contributions from more than one mechanism [8].

This paper is organized as follows. First, we introduce Nakagami model and its parameters. Secondly, computer simulations and experiments on phantoms are processed with the estimation of the PDF and the two Nakagami parameters. The comparison, the discussion, and some concluding remarks close up the paper.

II. STATISTICAL MODEL: NAKAGAMI DISTRIBUTION

The probability density function, of the envelope f(R) of the backscattered signal can be described in terms of the Nakagami distribution and, in this case, it is defined by:

$$f(R)=[\,2m^{m}R^{2m-1}/\,\Gamma\,(m)\,\Omega(m)]\exp(-mR^{2}/\,\Omega)U(R) \quad (1)$$

where $\Gamma(\ .\ )$ and $U(\ .\ )$ are the Gamma function and the unit step function, respectively. Nakagami parameter (m) and scaling parameter (Ω) can be calculated as follows:

$$m=[E(R^2)]^2/E[R^2- E(R^2)]^2 \qquad (2)$$

and

$$\Omega=E(R^2) \qquad (3)$$

where $E(\ .\ )$ is the statistical mean. The scaling parameter refers to the average power of the backscattered envelope. Moreover, Nakagami parameter is particularly useful for characterizing the probability distributions of ultrasonic backscattered envelopes, including the statistical conditions for pre-Rayleigh, Rayleigh, and post-Rayleigh distributions. When the resolution cell of the ultrasonic transducer contains a large number of randomly distributed scatterers, the envelop statistics of the ultrasonic backscattered signals obeys the Rayleigh distribution. If the resolution cell contains the scatterers with randomly varying scattering cross sections having comparatively high degree of variance, the envelop statistics are pre-Rayleigh distributions. If the resolution cell contains periodically located scatterers in addition to randomly distributed scatterers, the envelop statistics are post-Rayleigh distributions. Because the values of m ranging between 0 and 1 reflect statistics ranging from pre-Rayleigh to Rayleigh distributions and larger values correspond to the PDFs of post-Rayleigh or Rician distributions, thus the Nakagami parameter can be used to classify the properties of tissues. This has been validated in computer simulations on phantoms [6, 8] and clinical measurements [9].

It is possible to see that the Nakagami distribution can be identified as belonging to the Gamma distribution class. If we define a new random variable A of Gamma distribution with parameters (α; β) and $R=\sqrt{A}$, the probability density function f(R) can be written as:

$$f(R)= [\ 2\ \beta^{\alpha}\ R^{2\ \alpha -1}/\ \Gamma\ (\alpha)\ ]\exp(-\beta\ R^2)U(R) \qquad (4)$$

which is a Nakagami distribution with parameter (m =α; Ω = α/β). For convenience, we use the density given by (4). as the density function of the Nakagami distribution with parameters (α; β). The second order moment of this distribution is then given by:

$$E(R^2)= \alpha/\beta \qquad (5)$$

III. EXPERIMENTAL AND THEORETICAL RESULTS ON PHANTOMS

Both experimental measurements and computer simulations were carried out to explore the effect of some physical parameters on the estimation of the Nakagami statistical parameters.

The experimental arrangement includes an ultrasonic phantom a 5MHz linear array connected to an open research platform which is connected to a computer in order to visualize the machine interface where RF signals and B-scan ultrasound images are displayed, as well as the different windows allowing the selection of various physical parameters of the system such as the transmission frequency, the bandwidth, the number of time samples of the RF signals, etc.... For different positions of the sampling window, RF signals have been acquired at different frequencies f0 (2, 3, 4, 5, 6 MHz) and bandwidths of the excitation waveforms of 80 to 90 %. The received RF waveforms were then filtered around the center frequency and around twice the center frequency 2f0 (4, 6, 8, 10, 12 MHz) in order to detect an eventual non linear effect. Tissues are known to be nonlinear propagation media which can generate new frequency components. To better characterize tissue properties, it is then necessary to take into account the nonlinear behavior in the statistical model.

For the simulations, we used Field II software, developed by Jorgen Jensen [10] at the Technical University of Denmark, which is dedicated to the calculation of the pressure field at any point and whose major interest is to simulate probes of complex shape. By extension it allows the simulation of RF signals. This software has the advantage of simulating the contributions of the probe characteristics (shape, shooting strategy, excitation ...) on a 2D or 3D digital phantom and obtains the corresponding RF signals [11]. Ultrasound images are obtained by examining tissues defined by their reflectivity, with a probe itself characterized by its impulse response. Ultrasonic RF signals thus result from summing the responses of the diffusers during the ultrasonic wave propagation.

The results obtained from measurements and simulations were compared in order to study the effect of some physical parameters of the phantom on the sensitivity of Nakagami parameters, in particular Nakagami parameter (m).

The theoretical and experimental results were obtained for two square sampling windows of size 0.3 cm2 located at different positions inside the phantom, and for two different transmitting bandwidths (80% and 90%). The results show the same evolution of the Nakagami parameters according to the physical parameters. Changes in location of the sampling windows and bandwidth, have not a great influence on this evolution for the two transmitting ultrasonic frequency (Fig. 1 and Fig. 2). For the central frequency of 5 MHz, in most cases, the values of the Nakagami parameter are very close.

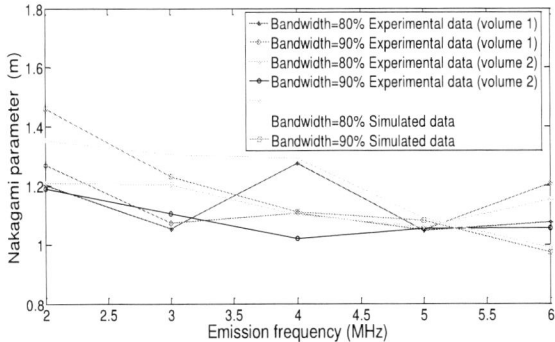

Figure 1. Comparison of theoretical and experimental values of Nagakami parameters versus the transmitting frequency f_0 (2, 3, 4, 5, 6 MHz) for different transmitted bandwidths and without filtering the detected waveforms.

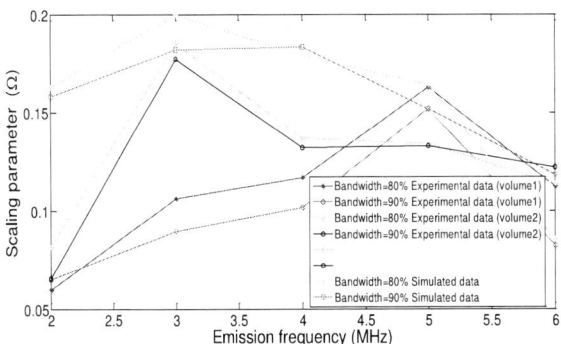

Figure 2. Comparison of theoretical and experimental values of the scale parameters versus the transmitting frequency f0 (2, 3, 4, 5, 6 MHz) for different transmitted bandwidths and without filtering the detected waveforms.

Variations of Nakagami parameter according to the transmission frequency and the position of the sampling window and of the bandwidth, increases with the filtering and it is more important in the case of a filtering around 2f0 (Fig. 3 and Fig. 4).

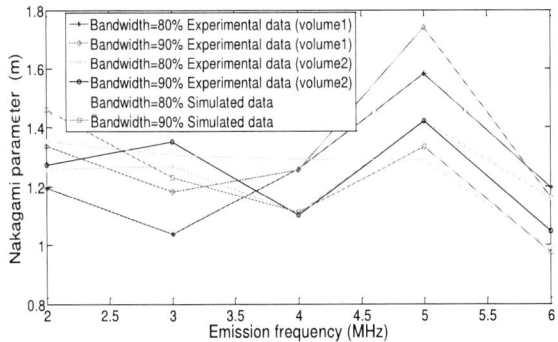

Figure 3. Comparison of theoretical and- experimental values of Nagakami parameters versus the transmitting frequency for different transmitted bandwidths and with a filtering of the detected waveforms around 2f0 (4, 6, 8, 10, 12 MHz).

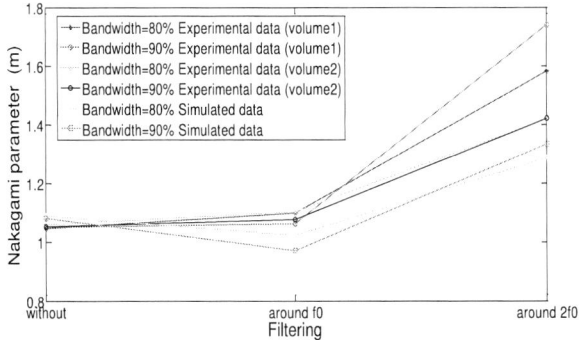

Figure 4. Comparison of theoretical and experimental values of Nakagami parameters versus the filtering in the case for a transmitted frequency f0=5 MHz.

Fig. 5 and Fig. 6 represent the probability density function (PDF) of the backscattered envelope for theoretical and experimental envelopes, respectively.

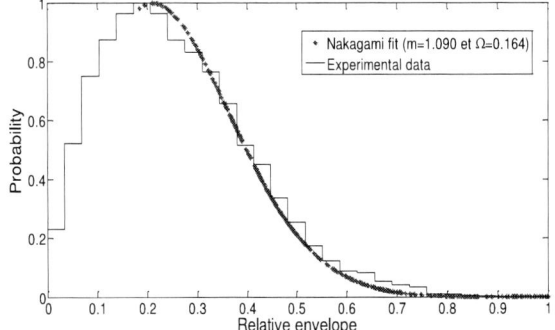

Figure 5. The probability density function (PDF) of the theoretical backscattered envelope.

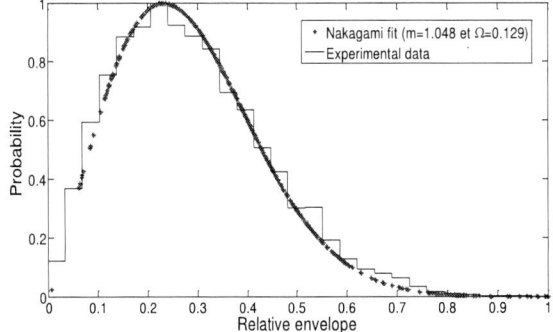

Figure 6. The probability density function (PDF) of the experimental backscattered envelope.

IV. CONCLUSION

The statistical analysis of the RF harmonic signals (twice the transmitted frequency) showed that Nagakami parameters (m, Ω) and the probability density function (PDF) depend ddc on the applied scanning conditions (frequency bandwidth, transmitted frequency and the location of the sampling window in the phantom). In conclusion, these results demonstrate that tissue characterization using Nagakami statistical model is more sensitive to the variation of the physical parameters of the system when performed in the harmonic mode than in the conventional fundamental mode.

REFERENCES

[1] E. Jakeman and P. N. Pusey, "A model for non-Rayleigh sea echo", IEEE Trans On antennas and Prop, vol. 24(6) , 1976, pp. 806–814.

[2] J.W. Goodman, "Statistical properties of laser speckle patterns", J.C. Dainty (Ed.), Laser Speckle and Related Phenomena, 2nd Edition, Springer, Berlin, vol. 9 of Topics in Applied Physics, 1984.

[3] R. F. Wagner, S. W. Smith, J. M. Sandrik, and H. Lopez, "Statistics of speckle in ultrasound B-Scans", IEEE Trans. Sonics Ultrason, vol. 30, 1986, pp. 156–163.

[4] J.M. Reid, F. Forsberg, E.J. Halpern, C.W. Piccoli, R. C Molthen, "Comparisons of the Rayleigh and k-distribution models using in vivo breast and liver tissue", Ultrasound Med. Biol, vol. 24(1), 1998, pp. 93–100.

[5] P. M. Shankar, V. A. Dumane, J. M. Reid, V. Genis, F. Forsberg, C.W. Piccoli, and B.B Goldberg, "Use of the k-distribution for classification of breast masses", Ultra. Med. Biol, vol. 26(9), 2000, pp. 1503–1510.

[6] Shankar PM, "A general statistical model for ultrasonic backscattering from tissues", IEEE Trans. Ultrason Ferroelec Freq Contr, vol. 47, 2000, pp. 727–736.

[7] Robert M. Cramblitt and Kevin J. Parker, "Generation of non-Rayleigh speckle distribution using marked regularity models", IEEE Transactions on Ultrasonics, Ferroelectrics and Frequency Control, vol. 46(4), 1999, pp. 867–873.

[8] Tsui PH, Wang SH., "The effect of the transducer characteristics on the estimation of Nakagami parameter as a function of scatterer concentration", Ultrasound Med Biol, vol. 30, 2004, pp. 1345– 1353.

[9] Shankar PM, Dumane VA, Reid JM, Genis V, Forsberg F, Piccoli CW, Goldberg BB, "Classification of ultrasonic B-mode images of breast masses using Nakagami distribution", IEEE Trans. Ultrason Ferroelec Freq Contr, vol. 48, 2001, pp. 569–580.

[10] J. Jensen and P. Munk, "Computer phantoms for simulating ultrasound B mode and cfm images",In Boston, Massachusetts, USA, 3rd Acoustical Imaging Symposium, vol. 23, 1997, pp. 75–80.

[11] J. Jensen and N. Svensen, "Calculation of pressure fields from arbitrarily shaped, apodized, and excited ultrasound transducer", IEEE Transactions on Ultrasonics, Ferroelectrics and Frequency Control, vol. 39, 1992, pp. 262–267.

Neutralization Technique Model Using Artificial Neural Networks

R. Addaci, R. Staraj
LEAT, University of Nice-Sophia Antipolis-CNRS,
Bât. Forum, Campus Sophi@Tech,
930 Route des Colles, 06903 Sophia Antipolis, France
rafik.addaci@unice.fr

N. Hamdiken, T. Fortaki
Advanced Electronics Laboratory
University of Batna
05000, Batna, ALGERIA
hamdikenazih@yahoo.fr

Abstract—In this paper, a neuronal model based on artificial neural network (ANN) is developed to model the so called neutralization line isolation technique. This approach allows a faster analysis and design of multi-antenna systems based on two PIFAs mounted on a mobile phone type printed circuit board (PCB) with very significant time gain and accurate results compared to other electromagnetic simulation tools whether in analysis or design sense.

Keywords-component; Antenna modeling, artificial neural networks, multi-antenna systems, neutralization technique.

I. INTRODUCTION

During the last two decades, the use of wireless communications has grown significantly. However, all the new wireless technologies require a high data transfer rate to offer high quality of service to users. Furthermore, such transmissions have to deal with fading channels and/or multipath environments. Therefore, new concepts of antennas have emerged to cope with large number of users, high transmission rates demand, but also with possible difficult propagation environments. Among these concepts, the multi-antenna systems have attracted particular interest for the scientific community, because they can efficiently take advantage of multipath phenomenon to fight the fading of the signal caused by the environment and to achieve significant spectral efficiencies. Several kinds of systems have been developed but they all have in common to provide multiple de-correlated copies of the transmitted signal [1-2].

Diversity and MIMO systems have been used in WLAN standards and will inevitably be implemented in the future wireless communications such as WiMAX and 4G. This is due to their ability to significantly increase the transmission rate without increasing power and spectrum of the transmitted signal and to solve the fading problems appearing in multipath environments [3-5]. The advantages offered by these systems are however conditioned by an adequate design allowing to obtain the maximum isolation between the feeding ports of the antennas which is a critical parameter for this type of systems [6-7].

Recent advances in the miniaturization of electronics devices have generated a growing need for small-sized antennas. Indeed, the use of miniature antennas tends to spread in the various systems of modern telecommunications. Therefore, the design of multi-antenna systems is a challenge even more difficult to reach because the required performance and features such as matching, isolation, small size and compact structure have to be maintained over a wide frequency band and/or multiple frequency bands. The fact that the antennas are very close may degrade the isolation, and then the total efficiency, the diversity performance and the transmitting capacity of the system.

It has been shown in several studies that the distance between antennas must be at least a half wavelength to obtain a maximum of isolation. This condition is very difficult to satisfy in increasingly small communication devices. Many researchers have studied this problem and several solutions have been developed [8-10] based on the insertion of electromagnetic band-gap (EBG) cells [11-12], or slots in the PCB/ground plane of the device [13-14], or using a defected ground plane structure [15], lumped circuit networks [16] or neutralization technique [17-19]. Several researchers [20-21] have demonstrated that this last one can be successfully applied to address specific port-to-port isolation problems.

II. NEUTRALIZATION TECHNIQUE

The so-called neutralization technique previously proposed by A. Diallo and all is an efficient isolation technique of two closely spaced PIFAs in small handsets. The originality of this technique consists in linking the two radiating elements by a suspended microstrip line to enhance the isolation between their feeds. The basic idea of the neutralization technique is to consider that the electromagnetic coupling between two PIFA (antennas) positioned on the top corners of a PCB is mainly capacitive because of the two metal plates placed facing each other, and thus, that a judicious insertion of an inductance between the two metal plates will allow to obtain a notch filter behavior for the $|S_{21}|$ parameter characterizing the coupling between the two feeding ports of the multi-antenna system.

This work was supported under the BMCI project, funded by the French Government (DGE-DGCIS).

The position of the zero deep of this notch filter can be tuned to the desired frequency by changing the width and/or the length of the inserted neutralization line. However, if the desired frequency band changes, by modifying for example the lengths and/or the widths of the PIFA, or if the distance between the two PIFAs or their relative positions compared to the edge of the printed circuit board (PCB) have to be modified to answer a new consideration design, a new parametric study should be carried out again to optimize the new system which generates a significant time losses.

To overcome these limits, a model developed using artificial neural networks (ANN) dedicated to the analysis and the design of multi-antenna systems based on two planar inverted F antennas (PIFA) connected by a neutralization line and placed on a mobile phone printed circuit board (PCB) type is presented in this paper. Our objective is to allow a faster and more accurate implementation of the neutralization technique in the case of different applications in different frequency bands, without needing to realize the parametric studies to optimize the system for each application and each frequency band and so to extremely reduce the designing time. Thus, we have tried to model this technique using a model based on artificial neural networks (ANN) approach.

To test the validity of this modeling method on our multi-antenna system, we have chosen the configuration presented in Fig. 1 as an initial design of our study.

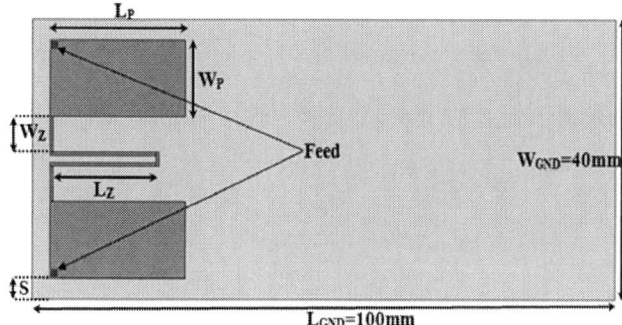

Fig. 1. Initial multi-antenna system.

In this configuration, the neutralization line of width 0.5mm is connected between the two short circuits (0.5mm) of the two PIFA placed on a ground plane of given dimensions $L_{GND} \times W_{GND} = 100 \times 40 mm^2$, which represent the standard dimensions of a PCB of a mobile phone PCB. The height of the PIFAs is 8mm. The other variables such as the length and width of the PIFAs (L_P, W_P), the length of the neutralization line (L_Z), the lengths (W_Z) and the distance of the antenna from the edge of the PCB (S) shown in Fig.1, will be considered as the network inputs parameters.

III. ARTIFICIAL NEURAL NETWORKS (ANN)

ANNs are nonlinear structures inspired by observed process in natural networks of biological neurons in the brain [22-23]. They are parallel computational systems consisting of a set of simple processing units connected together in a specific way

and communicating by sending signals to each other over a large number of weighted connections in order to perform a particular task. Their particular properties such as the massive parallelism that makes them very efficient, robust and noise tolerant, and their capability of learning from training data and generalizing to new situations, makes them as a powerful tool for modeling, especially when the underlying data relationship is unknown. This is why the ANNs are now being increasingly recognized in the area of classification and prediction.

Within ANN systems, it is useful to distinguish three types of layers: the input layer, the output layer and the hidden layers. These networks can have one or more hidden layers. Each layer for this structure receives inputs from the external sources and uses them to compute an output signal which is then propagated to other layers.

By using a set of data samples called training (learning) database, for which each sample consists of an unique input signal and the corresponding desired response, the network modifies and adjusts the weights and repeats the training for many samples in the database in order to reduce as much as possible the error, which means the difference between the output value of ANN and the target (attribute). The training is stopped when there are no further significant changes in the synaptic weights or in accordance with an appropriate criterion. During the training, the ANNs identify and learn correlated patterns between input data sets and corresponding target values by constructing an input-output mapping for the considered problem. After training, ANNs can be generalized and used to predict outputs for inputs not encountered during training. The multilayer perceptrons (MLP) network model, which is among the simplest and the most used architecture, was adapted for the calculation of the desired outputs of the proposed multi-antenna system. In this work, the most widely known learning algorithm, the back-propagation algorithm, based on gradient descent to tune network parameters to best fit a training set of input-output pairs, is used to train the MLP network in a supervised manner.

IV. OPTIMIZED MODEL AND PERFORMANCE

A. Optimized Analysis Model and its Performance

The first idea in this work is to propose an analysis model that gives the performance of the system shown in Fig. 1, in a faster and accurate way compared to the existing electromagnetic software. This means that the inputs parameters of the network are the length "L_P" and the width "W_P" of the PIFAs, the length "L_Z" and the width "W_Z" of the neutralization line and the distance "S" of the PIFA from the edge of the PCB. The principal desired performance as the resonant frequency, the level of the adaptation parameter ($|S_{11}|$) at this frequency, the -6dB bandwidth and the level of the isolation parameter ($|S_{21}|$) at the resonant frequency and at the edges of the obtained bandwidth, constitute the outputs of the network, Fig. 2.

We have tried to create, using the electromagnetic simulation software ANSYS HFSS, a training database that represents a large variation of the inputs parameters of the network, with the respect however of the physical realization

978-1-4673-5289-5/12 $31.00 © 2012 IEEE

limits. The minimal and maximal values of these parameters considered in this work are grouped in Table I. These values limit and affect the resonant frequency of the neutralized multi-antenna system.

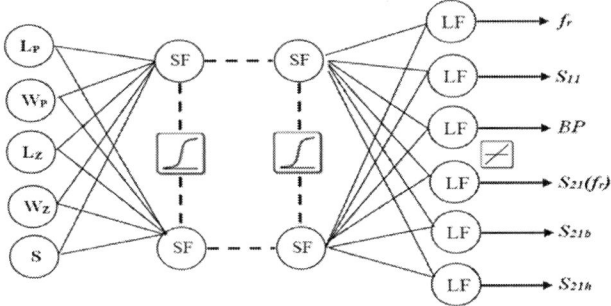

Fig. 2. Architecture of the proposed ANN model: SF (sigmoid function), LF (linear function).

TABLE I
VARIATION RANGE OF THE INPUTS PARAMETERS FOR THE ANALYSIS MODEL

Parameters (mm)	L_P	W_P	L_Z	W_Z	S
Min	15	10	1	1	1
Max	30	15	26	7	6

In order to show the performance of the optimized analysis neuronal model, a comparison between the results obtained by this model and those obtained by HFSS, which is used to create the learning database, for some randomly selected configurations was therefore necessary. The comparison is made on the outputs parameters of the network, for the two configurations. The results obtained by HFSS and the neuronal model for the studied system, are in very good agreement (Fig. 3).

Fig. 3. Comparison of HFSS and ANN model results for two different

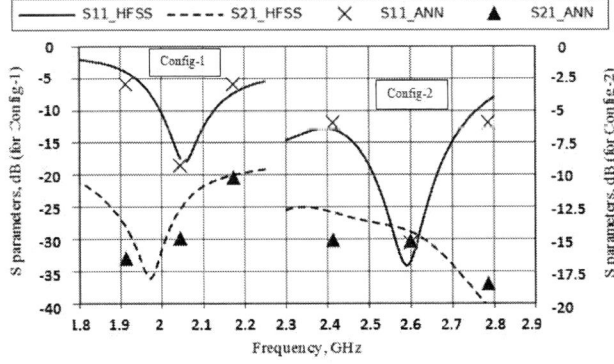

configurations.

Besides the time needed to create the database which is used for the training of the neural network and the time required to optimize the final ANN model, the advantage of the analysis neuronal model presented in this paper, is the very substantial reduction in CPU time compared to the existing electromagnetic softwares, such as HFSS considered as

reference in this study. Effectively, on a personal computer having a 2.13GHz processor and 8GB installed RAM, HFSS takes an average time of 4minutes for one given configuration, while on the same computer the optimized neuronal model gives the results instantly.

B. Design Code and its Performance

In the previous section, an efficient analysis model for the neutralized multi-antenna system shown in Fig. 1 is presented. This model offers a very important time gain, by giving simultaneously the performance in terms of resonant frequency, level of matching ($|S_{11}|$) at this frequency, -6dB bandwidth and level of the isolation parameter ($|S_{21}|$) at the resonant frequency and at the edges of the obtained bandwidth for any configuration for which its designed parameters (the inputs of the network showed in Fig. 2) are comprised in Table I. To give more facilities to the designer, it would be interesting, in the second step of this paper, to develop a synthesis model allowing to give in this case the designed parameters of the system (the inputs of the network showed in Fig. 2) for desired performance (the outputs of the network showed in Fig. 2) fixed, in this case, by the user.

This is possible by carried out the inverse learning; the outputs of the network shown in Fig. 2 become the inputs and vice versa. By developing this new model, we found that this model does not take into account the physical limits of the feasibility, and can produce some configurations which are unrealizable on the considered ground plane. Effectively, the model offers sometimes a combination of these parameters that causes a physically unrealizable system, like for example overlapping between the elements constituting the system or the exceeding of the dimension of the considered ground plane. However, in this case these parameters are the outputs of the network, and we have not the ability to set their variation ranges, as in the analysis model.

To remedy to this problem, and having always in mind the objective of facilitating the design of the considered system, we have developed, in addition to the analysis model, a code based on Matlab software, where the users have the possibility to set both the variation range and the variation step of the inputs parameters of the analysis network model (Fig. 2) but also, allowing the designer to give the desired performance in terms of resonant frequency, bandwidth, level of $|S_{11}|$ and $|S_{21}|$ at the resonant frequency and the level of $|S_{21}|$ for the frequency limits of the bandwidth. The code performs a parametric study and displays the values of the inputs parameters and their performance only for the configurations giving the desired output performance.

This solution allows thus to eliminate the problem encountered previously in the designing model, and offers especially an important time gain over the parametric studies, therefore, when designing the system. It also gives only the configurations giving superior performance to those set by the designer in the code, and gives directly their corresponding performance, unlike the results of the other software that require more CPU time for the selection of the configurations giving the desired performance. Without taking into account this CPU time, Table II shows the comparison of the necessary time to carried a different number of parametric studies on the considered system between HFSS and the proposed model. We can observe that the difference in the time calculation is impressive.

TABLE II
COMPARISON OF THE SIMULATION TIME BETWEEN HFSS AND ANN MODEL

		Numbers of parametric study		
		156	33852	203112
Necessary Time	HFSS	624 min	94 days	564 days
	ANN_code	6 sec	9 min	32 min

V. CONCLUSION

In this paper, an analysis neuronal model was developed for the isolation technique called neutralization line. This model gives instantly the performance in terms of resonant frequency, $|S_{11}|$ at the resonance, bandwidth, $|S_{21}|$ at the resonance and at the limits of the bandwidth, of any configuration. By using this analysis model in parallel with a code developed on Matlab software, the time needed to design a multi-antenna system with desired performance has become very small compared to the time required by others electromagnetic softwares. Due to the considerations taken into account during the creation of the training database, the validity of the model, in the designing sense, remains true for any system for which the resonant frequency is comprised in the [1.75-3.65GHz] band. Considering the speed of the neuronal model, the size of the ground plane can be now considered in our future works to allow the design of multi-antenna systems giving the best possible performance for different applications other than the mobile phone case considered in this study.

REFERENCES

[1] M. A. Jensen and J. W. Wallace, "A review of antennas and propagation for MIMO wireless communications", *IEEE Trans. On Antennas and Propagation*, vol. 52, no. 11, pp. 2810-2824, Nov. 2004.

[2] Z. Tang, A.S. Mohan, "Characterize the indoor multipath propagation for MIMO communications," *Microwave Conference Proceedings, 2005. APCM 2005. Asia-Pacific Conference Proceedings*, 4-7 Dec 2005, vol. 4, pp. 4.

[3] Y. Zhang, Z. Ye, C. Liu "Estimation of Fading Coefficients in the Presence of Multipath Propagation," *IEEE Trans Antennas Propagat.*, vol. 57, no.7, pp. 2220-2224, July 2009.

[4] G. Foschini and M. Gans, "On limits of wireless communications in fading environment when using multiple antennas," *Wireless Personal Communications*, vol. 6, no.3, pp. 331–335, March 1998.

[5] J. Winters, "On the capacity of radio communication systems with diversity in a Rayleigh fading environment," *IEEE Journal on Selected Areas in Communications*, vol. 5, no. 5, pp. 871-878, June 1987.

[6] A.P. Feresidis, T. Kokkinos, Qian Li, "Isolation enhancement of monopole antennas and PIFAs on a compact ground plane", *Antennas & Propagation Conference, Loughborough*, 16-17 Nov. 2009, pp.465-468.

[7] S. Zhang, J. Xiong, S. He, "MIMO antenna system of two closely positioned PIFAs with high isolation", *Electronics Letters*, vol. 45, no.15, pp. 771-773, 16th July 2009.

[8] D. Pozar, "Input Impedance and Mutual Coupling of Rectangular Microstrip Antennas," *IEEE Trans. Antennas Propag.*, vol.30, no.6, pp.1191-1196, Nov. 1982.

[9] F. Yang, Y. Rahmat-Samii, "Mutual Coupling Reduction of Microstrip Antennas Using Electromagnetic Band-Gap Structure," *IEEE Antenna and Propagation Society International Symposium*, Boston, 8-13 Jul. 2001, vol.2, pp.478-481.

[10] M.K. Ozdemir, H. Arslan, E. Arvas, "Mutual Coupling Effect in Multiantenna Wireless Communication System," *IEEE Global Telecommunications Conference GLOBECOM*, San Francisco, USA, 1-5 Dec. 2003, vol.2, pp.829-833.

[11] R. Makinen, V. Pynttari, J. Heikkinen, M. Kivikoski, "Improvement of Antenna Isolation in Hand-Held Devices Using Miniaturized Electromagnetic Band-Gap Structures", *Mic. & Opt. Tech. Lett.*, vol. 49, no. 10, pp. 2508-2513, Oct. 2007.

[12] C-C. Hsu, K-H. Lin, H-L. Su, H-H. Lin, C-Y. Wu, "Design of MIMO Antennas with Strong Isolation for Portable Applications", *IEEE Antennas & Prop. Int. Symp.*, Charleston, 1-5 June 2009, pp. 1-4.

[13] K.J. Kim, W.G. Lim, J.W. Yu, "High Isolation Internal Dual-Band Planar Inverted-F Antenna Diversity System with Band-Notched Slots for MIMO Terminals", *36th European Microwave Conference*, Manchester, 10-15 Sep. 2006, pp.1414-1417.

[14] K.J. Kim, K.H. Park, "The high isolation dual-band inverted F antenna diversity system with the small N-section resonators on the ground plane", *Asia-Pacific microwave conference*, Yokohama, 12-15 Dec. 2006, pp. 195-198.

[15] F. Zhu, J.D. Xu, Q. Xu, "Reduction of Mutual Coupling Between Closely-Packed Antenna Element Using Defected Ground Structure," *3rd IEEE International Symposium on Microwave, Antenna, Propagation and EMC Technologies for Wireless Communications*, Beijing, 27-29 Oct. 2009, pp.1-4.

[16] S-C. Chen, Y-S. Wang, S-J. Chung, "A decoupling technique for increasing the port isolation between two strongly coupled antennas," *IEEE Trans. ant. & Prop.*, vol. 56, no. 12, pp. 3650-3658, Dec. 2008.

[17] A. Diallo, C. Luxey, P. Le Thuc, R. Staraj, G. Kossiavas, "Study and Reduction of the Mutual Coupling Between Two Mobile Phone PIFAs Operating in the DCS1800 and UMTS Bands", *IEEE Ant. & Prop*, vol.54, no. 11, pp.3063-3074,Nov. 2006.

[18] A. Diallo, C. Luxey, P. Le Thuc, R. Staraj, G. Kossiavas, "An efficient two-port antenna-system for GSM/DCS/UMTS multi-mode mobile phones", *Elect. Lett.*, vol. 43, no. 7, pp. 369-370, 29th March 2007.

[19] R. Addaci, A. Chebihi, A. Diallo, P. Le Thuc, C. Luxey and R. Staraj, "Multi-antenna systems for clamshell mobile phones," *Proceedings of the Fourth European Conference on Antenna and Propagation (EuCAP)*, Barcelona, 12-16 April 2010, pp. 1-4.

[20] J. Byun, J-H. Jo, B. Lee, "Compact dual-band diversity antenna for mobile handset applications", *Microwave and Optical Technology Letters*, vol. 50, no. 10, pp. 2600-2604, October 2008.

[21] K. Chung, J.H. Yoon, "Integrated MIMO antenna with high isolation characteristic," *Electronics Letters*, vol. 43, no. 4, pp. 199-201, 15th February 2007.

[22] R.E. Uhrig, "Introduction to artificial neural networks," *Proceedings of the 21st international conference on industrial electronics, control, and instrumentation*, Orlando, FL, 6-10 Nov. 1995, vol. 1, pp. 33-37.

[23] G. Antonini and A. Orlandi, Gradient evaluation for neural-networks-based electromagnetic optimization procedures, IEEE Transactions on microwave theory and techniques, vol. 48, no. 5, pp. 874-876, May 2000.

Selection of a Suitable Mother Wavelet For Microemboli Classification Using SVM And RF Signals

K. Ferroudji, N. Benoudjit, M. Bahaz.

Electronics Department, University of Batna, Algeria
Email: karim_hab@hotmail.com, nbenoudjit@gmail.com, med_bahaz@yahoo.fr.

A. Bouakaz.

UMR Inserm U930 CNRS ERL 3106,
Tours, France
Email: bouakaz@med.univ-tours.fr

Abstract—**Fourier transform (FT) based techniques have been traditionally used for microemboli classification. We suggest in this study exploiting the Radio-Frequency (RF) signal backscattered by the emboli since they contain additional information on the embolus than the Doppler signal. A recent alternative approach of Wavelet Transform is applied for this purpose. The performances of such classification system are highly dependent upon the selection of an appropriate wavelet mother. Many mother wavelets have been used for this analysis such as Biorthogonal, Coiflet, Daubechies, and Symlet. This experimental study describes a strategy to choose a suitable mother wavelet for microemboli classification using SVM and RF signals.**

Index Terms—**Mother wavelet; Daubechies; Coiflet; Symlet; Biorthogonal; Classification; Support Vector Machine; Microemboli; Gaseous embolus; Solid embolus; RF signals.**

I. INTRODUCTION

Embolism is intravascular migration of an insoluble body such as gas bubble, a fat globule, a blood clot, an atheromatous plaque, or a piece of thrombus. Embolus formation inside the body could be attributed due to different physiological, physical, and intervention mechanisms [1], it can travel to any part of the body, accounting for many serious (and sometimes life-threatening) disorders thus the importance of an automatic classification system.

Recently, a new approach based on the analysis of radio frequency (RF) signal and using the nonlinear characteristics of gaseous bubbles and Fourier transform to classify emboli was investigated [2, 3].

In this study, we employ a number of wavelet functions within a microemboli classification system based on discrete wavelet transform (DWT) and support vector machine (SVM), and try to determine how the classification result is influenced by the choice of the wavelet function.

The discrete wavelet transform (DWT) presents a multiresolution analysis in the form of coefficient matrices which can be used in a manner similar to Fourier series coefficients. This DWT representation can be thought of as a form of "feature extraction" on the original signal. The signal,

for feature extraction [4–7], can be studied considering its approximations and its details. Various frequencies can be extracted separately to be studied, making signal analysis easier. The use of a wavelet transform requires the selection of the mother wavelet [9-11].

Support Vector Machine (SVM) is a new promising pattern classification technique proposed recently by Vapnik and co-workers [12]. Unlike traditional methods which minimize the empirical training error, SVM aims at minimizing an upper bound of the generalization error through maximizing the margin between the separating hyperplane and the data.

Experimental results are obtained for two different concentrations (5μl and 10μl) of microbubbles at low and high Mechanical Index (MI).

II. WAVELET TRANSFORM

Wavelets are functions that satisfy certain mathematical requirements and are used in representing data or other functions. The basic idea of the wavelet transform is to represent any arbitrary signal $x(t)$ as a superposition of a set of such wavelets or basis functions. These basis functions are obtained from a single photo type wavelet called the mother wavelet by dilation (scaling) and translation (shifts). The discrete wavelet transform for one-dimensional signal can be defined as follows [9].

$$c(a,b) = \int_R x(t) \frac{1}{\sqrt{a}} \psi \frac{(t-b)}{a} dt .$$ (1)

The indexes $c(a, b)$ are called wavelet coefficients of signal $x(t)$, a is dilation and b is translation, $\Psi(t)$ is the transforming function, the mother wavelet. It is so called because the wavelet derived from it analyzes signal at different resolutions $(1/a)$. Low frequencies are examined with low temporal resolution while high frequencies with more temporal resolution. A wavelet transform combines both low pass and high pass filtering in Spectral decomposition of signals [9].

978-1-4673-5289-5/12 $31.00 © 2012 IEEE

III. SUPPORT VECTOR MACHINE

The Support Vector Machine (SVM) [12] is a novel machine learning technique based on a statistical learning theory proposed by the Vapnik group in 1995. It has gained increasing attention in areas that range from pattern recognition to regression estimation, due to its perfect learning performance. Based on the structural risk minimization principle [13], it can perfectly improve the generalization ability of a learning machine. At the same time, an optimization problem can be transformed into a convex quadratic programming problem. The solution of a quadratic programming problem is the unique optimization solution of the whole. Hence, SVM does not have the problems with local extrema that are present for traditional neural networks, which require large numbers of training samples.

The SVM performs pattern recognition for two-classes problems by determining the hyperplane of separation with the maximum distance to the narrowest points of the positioning of formation. These points are called the vectors of support [14].

SVM was originally designed for linear binary classification. In practice, many applications of SVM are nonlinear classification problems. For nonlinear classification problems, we can use nonlinear transforms. A transformation, $\Phi(x)$, maps the data from the input space to a feature space that allows linear separation. We then seek the optimization separation plane in feature space. One only needs an inner product operation in feature space. The inner product operation may be implemented by a certain function (called a kernel function). In SVM, we introduce a kernel function $K(x_i, x_j) = \Phi(x_i) \cdot \Phi(x_j)$ to perform the transformation [15]. Then the basic form of SVM can be obtained [16]:

$$f(x) = \mathrm{sgn}\left(\sum_{i=1}^{l} \alpha_i y_i K(x_i, x_j) + b \right). \qquad (2)$$

In our case, radial basis kernel was used to differentiate between gaseous and solid embolus. More details about recent developments of SVM can be found in [17].

IV. EXPERIMENTAL SET-UP

The experimental set-up consists of a nonrecirculating flow phantom containing a 0.8mm diameter vessel, see figure (1). A continuous flow carries the Sonovue microbubbles through the insonified vessel. The concentration of contrast agent microbubbles is chosen to obtain comparable fundamental scattering amplitude than non perfused tissue used to simulate the response of a solid embolus.

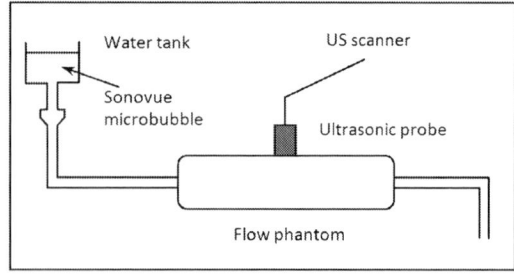

Fig. 1. Experimental set-up

The ultrasound waves were generated by a VF13-5 probe connected to a Siemens Antares scanner. The acquisitions were performed at 1.82MHz transmitting frequency in Tissue Harmonic Imaging (THI) mode. Two concentrations of the contrast agent were used, 5µl and 10µl. The microbubbles were administered into a 200 ml volume of Isoton [2].

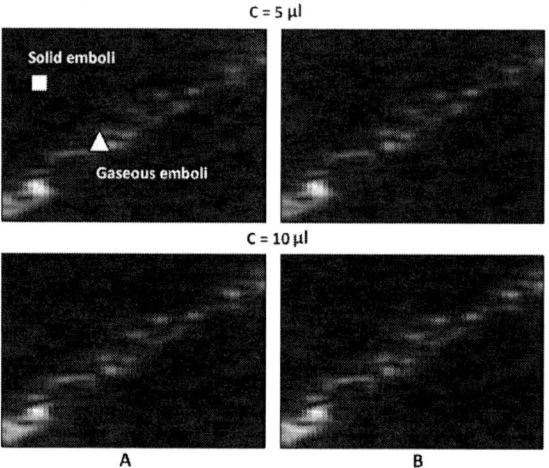

Fig. 2. Examples of grey scale images acquired: A. MI= 0.2, B. MI= 0.6 for two microbubbles concentrations

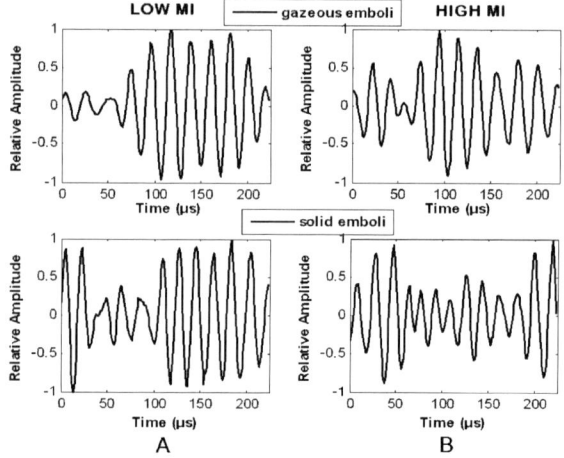

Fig. 3. Examples of RF signals A. MI= 0.2, B. MI= 0.6

Figure 2 shows the positions where RF signals corresponding to a solid embolus and gaseous embolus are extracted. Figure 3 displays two examples of RF signals extracted from the obtained experimental grayscale images. Panel A presents a RF signal backscattered by a solid and a gaseous embolus at low MI (0.2). We note that the acoustic pressure is not sufficient to start nonlinear microbubbles oscillations. Panel B displays the RF signal of each type of embolus at a higher MI (0.6).

V. FEATURE EXTRACTION BASED DWT

With the DWT, the original signal can be transformed into different time–frequency scales through the wavelet analysis, it uses two functions as high-pass filters and low pass filters. The high-frequency filter generates a detailed version of the signal (*d*), while the low-frequency filter produces its approximate version (*a*).

In our study, we have decomposed the signal *x(t)* into two levels, so we can write that:

$$x(t) = a2 + d2 + d1. \qquad (3)$$

Wavelet functions used for this study were standard DWT functions available in Matlab Wavelet toolbox [18], namely Daubechies (1 to 32), Coiflet (1 to 5), Symlet (2 to 8), and Biorthogonal (1.1, 1.3, 1.5, 2.2, 2.4, 2.6, 2.8, 3.1, 3.3, 3.5, 3.7, 3.9, 4.4, 5.5, 6,8). However, Matlab wavelet toolbox was not used in the classification. Instead, these functions were integrated into the automatic classification system.

The detail and approximation coefficients are not directly used as the classifier inputs. In this study, all of the method of standard deviation, root mean square, energy, and Shannon-entropy given in Table 1 [19] were used as the features extractors. All methods were individually applied to the detail coefficients of each level and the approximation coefficients at 2nd level.

TABLE I. FORMULATIONS OF FEATURE EXTRACTION TECHNIQUES

Feature extraction techniques	Formulations of feature extraction techniques
Standard deviation	$\sigma^2 = \dfrac{1}{N}\sum_{j=1}^{N}\left(x_{ij} - \mu\right)$ μ: mean of *x*
Root mean square	$rms = \sqrt{\dfrac{1}{N}\sum_{j=1}^{N}x_{ij}^{\;2}}$
Energy	$E = \sum_{j=1}^{N}\left|x_{ij}\right|^2$
Shannon-entropy	$H = \sum_{j=1}^{N}x_{ij}^{\;2}\log\left(x_{ij}\right)^2$

So, we obtain 12 features for each RF signal (solid or gaseous embolus). These features are used as input to the SVM classifier which provides in its output a value of 1 or -1 for gaseous or solid emboli respectively.

VI. RESULTS AND DISCUSSION

Classification results for four data sets with each of the wavelet functions used in the classification algorithm are summarized in Tables 2, 3, 4, and 5. Since no other parameters than the wavelet filter type changed, the results give a good indication on the suitability of the wavelet function for the particular acquisition. In Table 2, detection results are given for different Biorthogonal types of the wavelet function. Best detection was achieved by Biorthogonal3.1. For the Coiflet type wavelet, Coiflet3 achieved the best results (Table 3). For the Daubechies type wavelet, the best result was obtained by Daubechies 6 (Table 4). For the Symlet type wavelet, the best result was obtained by Symlet 7 (Table 5).

TABLE II. DETECTION WITH BIORTHOGONAL-WAVELET FUNCTIONS

Wavelet type	Classification rate (%)			
	C=5µl		C=10µl	
	Low MI	High MI	Low MI	High MI
Bior1.1	79,17	75,00	70,83	79,17
Bior1.3	79,17	79,17	79,17	79,17
Bior1.5	75,00	83,33	83,33	79,17
Bior2.2	83,33	87,50	83,33	79,17
Bior2.4	87,50	83,33	83,33	83,33
Bior2.6	83,33	87,50	83,33	83,33
Bior2.8	83,33	79,17	83,33	91,67
Bior3.1	**87,50**	**91,67**	**91,67**	**95,83**
Bior3.3	83,33	87,50	87,50	87,50
Bior3.5	83,33	91,67	87,50	95,83
Bior3.7	83,33	83,33	87,50	91,67
Bior3.9	83,33	91,67	87,50	95,83
Bior4.4	83,33	83,33	87,50	91,67
Bior5.5	79,17	87,50	87,50	91,67
Bior6.8	83,33	87,50	87,50	91,67

TABLE III. DETECTION WITH COIFLET-WAVELET FUNCTIONS

Wavelet type	Classification rate (%)			
	C=5µl		C=10µl	
	Low MI	High MI	Low MI	High MI
Coif1	83,33	83,33	83,33	83,33
Coif2	87,50	87,50	83,33	83,33
Coif3	**87,50**	**91,67**	**87,50**	**91,67**
Coif4	87,50	75,00	87,50	87,50
Coif5	83,33	75,00	87,50	87,50

TABLE IV. DETECTION WITH DAUBECHIES-WAVELET FUNCTIONS

Wavelet type	Classification rate (%)			
	C=5µl		C=10µl	
	Low MI	High MI	Low MI	High MI
db 1	79,17	79,17	83,33	75,00
db 2	62,50	83,33	91,67	83,33
db 3	75,00	79,17	91,67	79,17
db 4	83,33	83,33	79,17	87,50
db 5	87,50	91,67	79,17	91,67
db 6	**91,67**	**91,67**	**91,67**	**95,83**
db 7	79,17	83,33	83,33	87,50
db 8	83,33	79,17	79,17	91,67
db 9	70,83	83,33	83,33	91,67
db 10	70,83	70,83	87,50	91,67
db 11	83,33	79,17	75,00	87,50
db 12	70,83	70,83	83,33	91,67
db 13	75,00	91,67	83,33	91,67
db 14	79,17	75,00	83,33	87,50
db 15	75,00	70,83	83,33	95,83
db 16	75,00	70,83	83,33	91,67
db 17	62,50	75,00	79,17	87,50
db 18	75,00	70,83	83,33	87,50
db 19	75,00	75,00	70,83	91,67
db 20	75,00	75,00	79,17	83,33
db 21	75,00	79,17	83,33	83,33
db 22	70,83	75,00	83,33	87,50
db 23	75,00	75,00	83,33	83,33
db 24	75,00	75,00	75,00	87,50
db 25	70,83	75,00	79,17	91,67
db 26	75,00	75,00	83,33	83,33
db 27	75,00	79,17	79,17	83,33
db 28	70,83	75,00	79,17	87,50
db 29	70,83	75,00	83,33	87,50
db 30	75,00	79,17	83,33	87,50
db 31	70,83	75,00	79,17	87,50
db 32	70,83	75,00	70,83	83,33

TABLE V.   DETECTION WITH SYMLET-WAVELET FUNCTIONS

| Wavelet type | Classification rate (%) | | | |
| | C=5µl | | C=10µl | |
	Low MI	High MI	Low MI	High MI
sym2	75,00	75,00	87,50	83,33
Sym3	75,00	79,17	87,50	87,50
Sym4	79,17	87,50	83,33	83,33
Sym5	79,17	87,50	79,17	79,17
Sym6	87,50	87,50	87,50	83,33
Sym7	**87,50**	**91,67**	**91,67**	**87,50**
Sym8	83,33	83,33	83,33	83,33

TABLE VI.   BEST CLASSIFICATION RATES OF THE WAVELET FUNCTIONS

| Wavelet type | Classification rate (%) | | | |
| | C=5µl | | C=10µl | |
	Low MI	High MI	Low MI	High MI
Bior3.1	87,50	91,67	91,67	95,83
Coif3	87,50	91,67	87,50	91,67
db 6	**91,67**	**91,67**	**91,67**	**95,83**
Sym7	87,50	91,67	91,67	87,50

Classification results for the wavelet functions given in the tables (2, 3, 4 and 5) indicate that there is no analytical justification for the choice of a particular wavelet function for a particular signal, so the required wavelet filter should be determined experimentally. The performance of the classification system greatly depends on the selection of the mother wavelet. From table 6, the wavelet corresponding to the highest classification rates is db6, so we can say that db6 is the most appropriate wavelet function for this application.

VII.   CONCLUSION

The DWT offers a method by which RF signals from the experiment can be analyzed for the classification of the microemboli within the flow phantom. DWT is used to decompose the input signal into approximation and detail coefficients. The method offers the ability to provide an automatic classification system.

In this work, a comparative study of different wavelet family for microemboli classification has been done using SVM and RF signals. This study gives the choice of a suitable wavelet mother for this application. The effects of Biorthogonal, Coiflets, Daubechies, and Symlets wavelet functions on four different acquisitions have been examined. The classification rates for each wavelet family are also presented. The wavelet corresponding to the highest classification rate was selected as the suitable mother wavelet. Therefore, conclusively we can say that, among 59 mother wavelet functions, db6 appears to be the most appropriate wavelet function for this application.

The technique proves effective improving classification; it still needs a clinical trial and verification to demonstrate the additional benefit.

The main message of this study is to validate this strategy in a simple and a controlled experimental environment before further pre-clinical and clinical validations are undertaken.

ACKNOWLEDGMENT

The authors would like to acknowledge the support (datasets) of UMR INSERM U 930, Tours (France).

REFERENCES

[1] J. Molloy, H.S. Markus, "Asymptomatic embolization predicts stroke and TIA risk in patients with carotid artery disease," Stroke 30 (7) pp. 1440–1443, 1999.

[2] N. Benoudjit 1, K. Ferroudji 1, M. Bahaz 1, A. Bouakaz 2 ; " In vitro microemboli classification using neural network models and RF signals ". Ultrasonics 51. pp. 247–252. 2011.

[3] K. Ferroudji, N. Benoudjit, M. Bahaz, A. Bouakaz; "Feature Selection Based on Rf Signals and Knn Rule: Application to Microemboli Classification" 7th International Workshop on Systems, Signal Processing and their Applications (WOSSPA), Algeria. pp. 251 – 254. 2011.

[4] Guangmin Sun, Xiaoying Dong, Guandong Xu, « Tumor tissue identification based on gene expression data using DWT feature extraction and PNN classifier» Neurocomputing, Volume 69, Issues 4–6, Pages 387-402, January 2006.

[5] Wu JD, Kuo JM. An automotive generator fault diagnosis system using discrete wavelet transform and artificial neural network. Expert Syst Appl;36:9776–83, 2006.

[6] Dattatray V. Jadhav, Raghunath S. Holambe, « Feature extraction using Radon and wavelet transforms with application to face recognition» Neurocomputing, Volume 72, Issues 7–9, Pages 1951-1959, March 2009.

[7] N. Saravanan, K.I. Ramachandran, « Incipient gear box fault diagnosis using discrete wavelet transform (DWT) for feature extraction and classification using artificial neural network (ANN)» Expert Systems with Applications, Volume 37, Issue 6, Pages 4168-4181, June 2010.

[8] Wei-Yen Hsu, Chao-Hung Lin, Hsien-Jen Hsu, Po-Hsun Chen, I-Ru Chen; « Wavelet-based envelope features with automatic EOG artifact removal: Application to single-trial EEG data », Expert Systems with Applications, Volume 39, Issue 3, Pages 2743-2749, 15 February 2012.

[9] G.K. Kharate, A.A. Ghatol and P.P. Rege, « Selection of Mother Wavelet for Image Compression on Basis of Image », IEEE - ICSCN 2007, MIT Campus, Anna University, Chennai, India. pp.281-285, Feb. 22-24, 2007.

[10] Ernest Nlandu Kamavuako, Winnie Jensen, Ken Yoshida, Mathijs Kurstjens, Dario Farina, « Acriterion for signal-basedselection of wavelets for denoisingintrafascicularnerverecordings », Journal of Neuroscience Methods, Volume 186, Issue 2, Pages 274-280, 15 February 2010.

[11] Loris Nanni *, Alessandra Lumini, « Wavelet selection for disease classification by DNA microarray data », Expert Systems with Applications 38,Pages 990–995, 2011.

[12] Cortes, C., & Vapnik, V. "Support vector networks". Machine Learning, 20, pp. 273–297. 1995.

[13] Bilge Karaçalı, Rajeev Ramanath, Wesley E. Snyder, "A comparative analysis of structural risk minimization by support vector machines and nearest neighbor rule", Pattern Recognition Letters, Volume 25, Issue 1, Pages 63-71,5 January 2004.

[14] S.S. Mehta, N.S. Lingayat, Development of entropy based algorithm for cardiac beat detection in 12-lead electrocardiogram, Signal Process. 87, Pages 3190–3201, 2007.

[15] John Shawe-Taylor, Nello Cristianini, "Support Vector Machines and other kernel-based learning methods", Cambridge University Press, 2000.

[16] Olivier bousquet "Introduction aux Support Vector machine (SVM)". Orsay, 15 Novembre 2001.

[17] Wang, L.P. (Ed.): "Support Vector Machines: Theory and Application". Springer, Berlin Heidelberg New York,2005.

[18] Wavelet Toolbox (2.0) User´s Guide, The Math Works, Natick, MA, 1998.

[19] Hüseyin Erişti, Ayşegül Uçar, Yakup Demir, « Wavelet-based feature extraction and selection for classification of power system disturbances using support vector machines », Electric Power Systems Research 80, 743–752, 2010.

INFLUENCE OF G722.2 SPEECH CODING ON TEXT-INDEPENDENT SPEAKER VERIFICATION

Meriem FEDILA

Speech Communication & Signal Processing Lab.
Faculty of Electronics and Computer Science, USTHB
P.O. Box 32, El-Alia, Bab Ezzouar, 16111,
Algiers, Algeria.
fedila_m@yahoo.fr

Abderrahmane AMROUCHE

Speech Communication & Signal Processing Lab.
Faculty of Electronics and Computer Science, USTHB
P.O. Box 32, El-Alia, Bab Ezzouar, 16111,
Algiers, Algeria.
namrouche@usthb.dz

Abstract— This paper deals with the text-independent speaker recognition over wireless telephone network. The performed speaker verification system, based on Gaussian Mixture Model (GMM), was designed to use the information extracted directly from the coded parameters embedded in the ITU-T G.722.2 bitstream. Experiments were performed over the ARADIGIT database. The robustness in verification accuracy is also studied. The obtained results show that the use of Immittance Spectral Frequency (ISF) features extracted from ITU-T G.722.2 bitstream would improve significantly the recognition performance compared with the Mel Frequency Cepstral Coefficients (MFCC) features extracted from transcoded speech.

Keywords— ITU-T G722.2; Speaker Verification; GMM; ISF; MFCC.

I. INTRODUCTION

Speaker verification interest has increased considerably in the last decade. Compared to others biometrics modalities (face recognition and fingerprints), the speech allows identity verification to be performed remotely, because the speech signal can easily be transmitted over communication channels.

Speaker verification is the process of making a decision about acceptance or rejection of a specified identity claimed from a speaker. There are two types of speaker verification systems: "text-dependent" speaker verification, where the system requires the user to utter words that has been previously pronounced, and "text-independent" speaker verification, where the user can utter any speech segment.

Today, speaker verification systems are in the way for their inclusion in mobile communications [1], [2]. This has led to the following objectives: a) to ensure robustness of the system against additive noise as well as channel and session variabilities. b) to improve the performance of the automatic speaker recognition when using speech codecs (i.e. GSM, AMR-NB, AMR-WB).

Speech codecs were mostly designed for narrowband speech of telephone bandwidth (300 to 3400 Hz), but the evolution of broadband multimedia services has spawned an increased interest in wideband speech. Thus, this paper proposes an in-depth look at the influence of Adaptative MultiRate WideBand (AMR-WB) speech coding on text independent speaker verification performance. Few contributions [3] were made on this subject, whereas the effect of narrowband speech codecs (GSM, GSM-HFR, AMR-NB, etc.) on automatic speaker recognition has been more extensively studied [4],[5].

The aim of this paper is to investigate the efficiency of the automatic speaker verification using ITU-T G722.2 speech coding. We are particularly focused on the performance recognition obtained with the encoded bit stream using ISF features.

In this work, the ITU-T G.222.2 [6] speech coding is used to encode and decode the speech. Experiments were performed over the ARADIGIT corpus [7] using GMM speaker modeling technique.

The organization of this paper is as follows. The speaker recognition system used in all the experiments is presented in section 2. Section 3 describes the ITU-T G722.2 speech coding. The performance evaluation of this work is described in section 4. Finally, the paper is concluded in section 5.

II. SPEAKER VERIFICATION SYSTEM

A. Speaker Verification via Likelihood Ratio detection

The approach to speaker verification is to apply a likelihood ratio test to the input utterance.

Given an observation X, and a hypothesized speaker S, the task of speaker verification can be stated as a hypothesis test between:

978-1-4673-5289-5/12 $31.00 © 2012 IEEE

H0: X is from the hypothesized speaker S.
H1: X is not from the hypothesized speaker S.

The optimum test to decide between these two hypotheses is a likelihood ratio (LR) test given by:

$$\frac{p(X|H0)}{p(X|H1)} \begin{cases} \geq \theta \; accept \; H_0 \\ < \theta \; reject \; H_0 \end{cases} \quad (1)$$

where $p(X|H0)$ is the Likelihood that the utterance was produced by λ_S and $p(X|H1)$ is the Likelihood that the utterance was not produced by the claimed speaker.

Figure 1 shows the basic components found in speaker verification systems based on LRs. The role of the front-end processing is to extract from the speech signal features that convey speaker-dependent information. The output of this stage is typically a sequence of feature vectors representing the test segment, X= {x₁,..., x_T}, where x_t is a feature vector indexed at discrete time t ϵ{1,2,...,T}. These feature vectors are then used to compute the likelihoods of H0 and H1. Mathematically, a model denoted by λ_S represents H0, that characterizes the hypothesized speaker S. For example, one could assume that a Gaussian distribution best represents the distribution of feature vectors for H0, so that λ_S would contain the mean vector and covariance matrix parameters of the Gaussian distribution (GMM). The model $\lambda_{\bar{S}}$ represents the alternative hypothesis, H1. The likelihood ratio statistic is then $p(X|\lambda_S) / p(X|\lambda_{\bar{S}})$. Often, the logarithm of this statistic is used giving the log likelihood ratio

$$\Lambda(X) = \log p(X|\lambda_S) - \log p(X|\lambda_{\bar{S}}) \quad (2)$$

Generally, the likelihood that the utterance was not produced by the speaker is determined from a collection of background speaker models.

B. Gaussian Mixture Model

Gaussian Mixture Model (GMM) is a well-known statistical approach for speaker recognition [9]. A GMM is used to estimate the Probability Density Function (PDF) of a sequence of feature vectors.

$$p(\bar{x}|\theta) = \sum_{i=1}^{M} \pi_i b_i(\bar{x}) \quad (3)$$

Where: π_i, i =1,2,..., M, are the weight, and M is the total number of Gaussian components.
$b_i(\bar{x})$, i =1,2,...,M are the probability density function (PDF) given by:

$$b_i(\bar{x}) = \frac{1}{(2\pi)^{d/2}|\Sigma_i|^{1/2}} exp\left\{-\frac{1}{2}(\bar{x}-\mu_i)'\Sigma_i^{-1}(\bar{x}-\mu_i)^T\right\} \quad (4)$$

The complete Gaussian mixture PDF is represented by the mean vector μ_i, covariance matrices Σ_i and mixture weights π_i of all the component densities. These parameters are collectively represented by the notation:

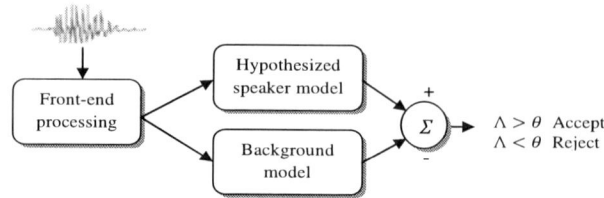

Figure 1. Block diagram of a speaker verification system [8].

$$\lambda = \{\pi_i, \mu_i, \Sigma_i\} \quad (5)$$

Where λ is the GMM for each speaker. It may be estimated using the iterative Expectation-Maximization algorithm [10].

C. Decision

Once a verification score has been obtained, a threshold value is needed to decide if the observation was produced by an impostor or by a registered speaker. The threshold is set to create a tradeoff between rejecting the true claimants, which is also known as the false rejection error (FR) and accepting false claimants, known as the false acceptance error (FA). If the match score is higher than the set threshold, the claimed speaker will be accepted as a genuine or else it will be treated as an impostor.

III. G722.2 SPEECH CODING

G722.2 is a ITU-T standard 7 kHz wideband speech codec. It also known as AMR-WB, based on ACELP (Algebraic Codebook Excited Linear Prediction) codec and operates on 20ms frames. The G722.2 consists of nine source codec modes with bit-rates of 23.85, 23.05, 19.85, 18.25, 15.85, 14.25, 12.65, 8.85 and 6.60 Kbit/s. All modes are sampled at 16 KHz and processed at 12.8 kHz.

The encoder performs the analysis of the LPC, LTP (Long Term Prediction) and fixed codebook parameters at 12.8 kHz sampling rate. At each frame of 20ms, the speech signal is analyzed to extract the parameters of the CELP model: LP filter coefficients, adaptive and fixed codebooks indices and gains. In addition, to these parameters, high-band gain indices are computed in 23.85 Kbit/s mode.

These parameters are encoded and transmitted via transmission channel. At the receiver side, the transmitted indices are extracted from the received bit-stream with the decoder, and then, decoded to obtain, for each received frame, the coder parameters: the ISP (Immittance Spectral Pair) vector, the 4 fractional pitch lags, the 4 LTP filtering parameters, the 4 innovative codevectors, and the 4 sets of vector quantized pitch and innovative gains. In 23.85 Kbit/s mode, also high-band gain index is decoded. Table I shows the bit allocation for each quantized speech parameter at 12.65 Kbits/s (20 ms speech frame) [6].

978-1-4673-5289-5/12 $31.00 © 2012 IEEE

TABLE I. BIT ALLOCATION OF THE AMR-WB CODING ALGORITHM FOR 20-MS FRAME

mode	parameter	1st subframe	2nd subframe	3rd subframe	4th subframe	Total per frame
12.65 Kbit/s	VAD flag					1
	ISP					46
	LTP filtering	1	1	1	1	4
	Pitch delay	9	6	9	6	30
	Algebraic code	36	36	36	36	144
	Gains	7	7	7	7	28
	Total					**253**

IV. RESULTS AND DISCUSSION

A. Speech database

ARADIGITS [7] consists of a set of 10 digits of the Arabic language (zero to nine) spoken by 110 speakers of both genders with three repetitions for each digit. This database was recorded by Algerians speakers from different regions aged between 18 and 50 years in a quiet environment with an ambient noise level below 35 dB, in WAV format. Files were acquired at a sampling rate of 22,050 kHz, and then were downsampled to 16 KHz.

In our work, a total of 60 speakers were used. Out of 60 speakers, 16 speakers were used for training and client evaluation, and other 24 speakers (different from training) for impostor evaluation. 20 different speakers with 10 different samples (zeros to nine) were used for a background model. We only used 8 (zeros to seven) samples for client speaker training. For client and impostor evaluation, 2 samples (eight to nine) for each speaker were used.

B. Features extraction

In this work we adopted three different experimental setups as show in Fig.2. A baseline experimental "setup 1" consists of a conventional Speaker Verification system applied to the original set of waveforms in ARADIGIT. The front end processing for this system is as follows. A 12-dimensional Mel-cepstral vector MFCC is extracted from the speech signal every 10 ms using a 20 ms window. Similarly, experimental "setup 2" applies the same scheme as in "setup 1" to a transcoded version of the database. Experimental "setup 3" uses 16 ISF (Immittance Spectral Frequency) features derived from the G722.2 bitstream, This approach is investigated in order to reduce the computational load related to the synthesis process. Note that the G722.2 codec uses ISF as representation for the LP coefficients to describe the spectral envelope of a speech frame. In both setup2 and setup3, we adopted the mode of bit-rate 12.65Kbit/s.

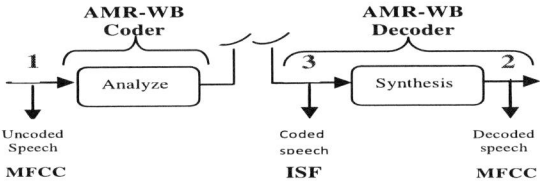

Figure 2. Features extraction methods.

C. Experimental Results

1) Performance measure of the speaker verification system

In this work, the performances of the speaker verification system are evaluated in terms of two criterions [11] as follows:

- Receiving operating characteristic (ROC) curve, and,
- Equal error rate (EER).

EER corresponds to the operating point where FAR=FRR. Graphically, it corresponds to the intersection of the ROC curve with the first bisector curve.

2) A baseline system

As shown in Fig.3, the baseline system performs excellent with only 16 Gaussian mixtures and 12 MFCC features compared with other numbers of Gaussian mixtures.

A conventional Speaker Verification system applied to the original set of waveforms in ARADIGIT (12 MFCC features) is able of achieving an equal error rate of 4.16% with 16 components (see Fig.4).

3) Experiments on G722.2 coded speech

Table II summarizes the equal error rates (EERs) from different experiments. In the first experiment, we obtained a best performance with uncoded ARADIGIT (ERR=4.16%) using 12 MFCC features and GMM with 16 components. In the second experiment, the results show performance degradation when using G722.2 decoded database at 12.65Kbit/s compared to the initial version of ARADIGIT.

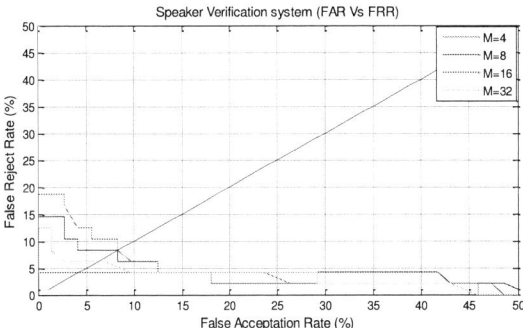

Figure 3. ROC curve with different number of components.

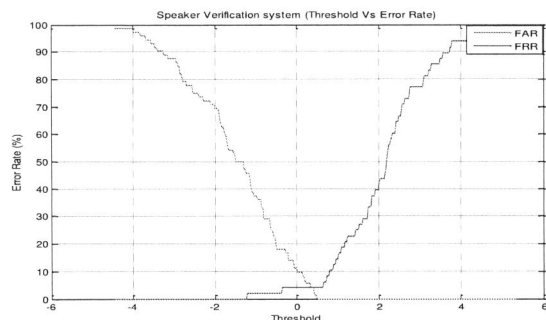

Figure 4. ROC curve of clean speech (uncoded speech) with 16 Gaussian mixtures.

TABLE II. ERROR EQUAL RATE (%) USING MFCC EXTRACTED FROM UNCODED SPEECH, ISF EXTRACTED FROM CODED SPEECH AND MFCC EXTRACTED FROM DECODED SPEECH (AT 12.65KBIT/S).

Features representation	EER (%)
setup 1 **MFCC**	4.16%
setup 2 **MFCC**	8.33%
setup 3 **ISF**	4.16%

Experiment three shows that the use of the ISF parameters, extracted from G722.2 bitstream, can significantly improve performance in a speaker verification system, compared to decoded speech. Thus, the EER decreases from 8.33% to 4.16%.

4) Influence of transmission channel

Figure 5.a presents the ROC curves obtained from speech degraded by Binary Symmetric Channel BSC at p of 0.1 (error probability) input into G722.2 at 12.65kb/s, while Fig.5.b and fig.5.c present the same for Additive White Gaussian Noise (AWGN) and Rayleigh noise respectively at an SNR of 15 dB. The advantages of using ISF features extracted from G722.2 bitstream can be clearly observed from these figures. Obviously, substantial reduction was achieved in EER from 26.4% to 8.33% for AWGN channel, from 45.83% to 42.756% for Rayleigh channel and from 40.27% to 27.78% for BSC channel. The robust feature set of ISF provides significant performance under degraded noise conditions compared to MFCC features extracted from G722.2 decoded speech.

V. CONCLUSION

In this paper we have investigated the influence of G722.2 speech coding on a text independent speaker verification system based on GMM classifier. The speaker verification performance using MFCC features extracted from G722.2 transcoded speech was measured, and compared with MFCC extracted from original speech. We have also explored the possibility of working directly in the compressed domain, so that no decoding is needed.

The obtained results show that the use of the ISF parameters extracted from G722.2 bitstream provides significant performance in clean conditions and under severely degraded noise conditions compared to MFCC features.

REFERENCES

[1] M. Chowdhury, S. Selouani and D. O'shaughnessy, "Text-independent distributed speaker identification and verification using GMM-UBM speaker models for mobile communications," International conference on Information Science, Signal Processing and their Applications (ISSPA), pp. 57- 60, 2010.

[2] S. Marcel, C. McCool, P. Matejka, T. Ahonen, and J. Cernocky. Mobile biometry (MOBIO) face and speaker verification evaluation. Idiap Research Report Idiap-RR-09-2010, Idiap, May 2010.

[3] M. Fedila, A. Amrouche, "Automatic speaker recognition for mobile communications using AMR-WB speech coding", International Conference on Information Sciences, Signal Processing and their Applications (ISSPA), pp. 1034-1038, Montreal, Canada, 3-5 July 2012.

[4] A. Krobba, M. Deybeche, A. Amrouche. "Evaluation of Speaker Identification System Using GSMEFR Speech Data", in Proc. International Conference on Design & Technology of Integrated Systems (DTIS), pp. 1-5, Tunis, March, 2010.

[5] S. Grassi, M. Ansorge, F. Pellandini, P.-A. Farine, "Distributed Speaker Recognition using the ETSI Aurora Standard", in Proc. on the 3rd COST 276 Workshop on Information and Knowledge Management for Integrated Media Communication, pp. 120-125, 2002.

[6] ITU-T Recommendation G.722.2, "Wideband Coding of Speech at around 16 kbit/s using Adaptive Multi-Rate Wideband (AMR-WB)", 2003.

[7] A. Amrouche, "Automatic speech recognition using connectionist models (Reconnaissance automatique de la parole par les methods connexionnistes)", Doctoral thesis, Faculty of Electronics and Computer Science, USTHB. 2007.

[8] F. Bimbot and Al, "A Tutorial on Text-Independent Speaker Verification," EURASIP Journal on Applied Signal Processing, 430–451, 2004.

[9] D.A. Reynolds, "Speaker Identification and Verification using Gaussian Mixture Speaker Models", Speech Communication, Vol. 17, pp. 91–108, 1995.

[10] J. Bilmes, "A Gentle Tutorial of the EM Algorithm and its Application to Parameter Estimation for Gaussian Mixture and Hidden Markov Models," Report, University of Berkeley, ICSI-TR-97-021, 1997.

[11] A. Mayoue, "Biosecure Tool: Performance Evaluation of a Biometric Verification System, " GET-INT, Ver. 1.0, Dec., 2007.

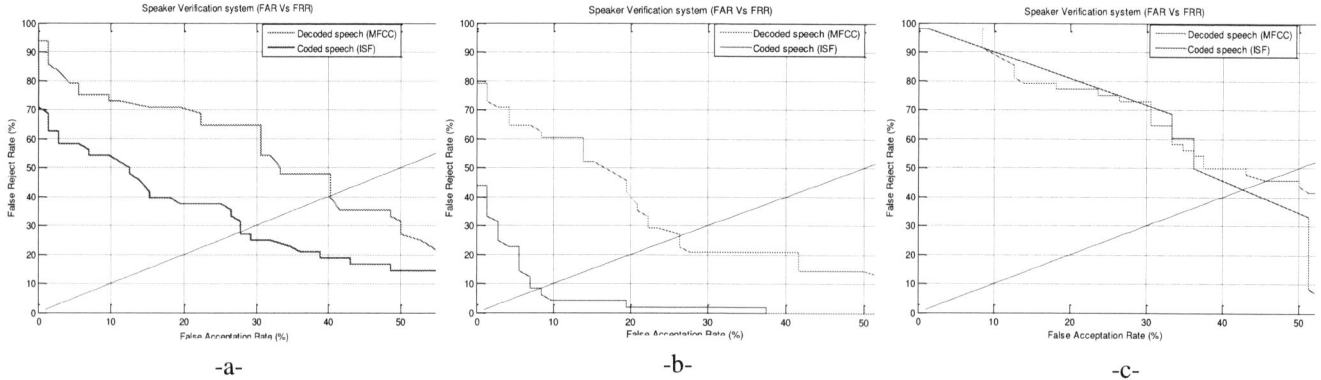

Figure 5. ROC curve of coded speech (ISF) and decoded speech (MFCC) with -a- BSC channel (p=0.1), -b- AWGN channel (SNR=15dB) and -c- Rayleigh channel (SNR=15dB).

Evaluation of the Performance of Block Matching Algorithms in Wavelet-based t+2D Video Coding

M. Marzougui, A. Zoghlami, M. Atri and R. Tourki
University of Monastir, Faculty of Sciences at Monastir
Laboratory of Electronic and Microelectronic, 5000 Monastir, Tunisia
Email: mehrez.marzougui@fsm.rnu.tn

Abstract—Motion estimation is the process of determining the movement of blocks between adjacent video frames by which image information is assessed for similarities that can be reused in subsequent frames. However, computational complexity and resource sharing of the motion estimation algorithm poses great challenges for real time applications. Fast search algorithms emerged as important search technique to achieve real time tracking results. Among the different motion estimation algorithms block matching is the most common used technique. This paper presents a system-level implementation of a wavelet-based t+2D video coding. Three block matching motion estimation algorithms are implemented in order to evaluate the overall system performance. In fact, since the chip design and layout process is time consuming and expensive, it is very important to be able to predict the overall system performance in a high-level implementation before its circuit layout is deployed. The aim is to discuss the impact of different block matching motion estimation algorithms on video coding performance. Because of the complexity of the entire motion estimation system, decision in choosing one algorithm versus the other algorithms is often empirical and crucial.

Index Terms—motion estimation, block matching algorithm, t+2D wavelet transform, video compression, computation, PSNR.

I. INTRODUCTION

With the advance of multimedia systems and the internet, digital networking and video storage systems have been gaining a lot of popularity. Video compression becomes necessary for an efficient data storage as well as transmission of internet video through a limited bandwidth channel. Compression is useful because it helps reduce the consumption of resources such as data space or transmission capacity. The design of data compression schemes involve trade-offs among various factors, including the degree of compression, the amount of distortion introduced when using lossy data compression, and the computational resources required to compress and uncompress the data. Recent video coding are based on predicting a new frame from a previous frame and then predicting the current block from previous block in the same frame. This form of prediction is known as motion estimation.

Video coding standards such as H.261, H.262, H.263, MPEG-1, MPEG-2 and MPEG-4 utilize motion estimation to reduce the temporal redundancy in the video sequence and a 2D transform, such as 2D DCT or 2D wavelet transform, is performed in the spatial domain. Finally, entropy coding is realized. In this paper, we focus on the t+2D analysis

where temporal redundancy is first exploited through a motion compensated multiresolution decomposition and the resulting temporal subband frames are then spatially decomposed with a wavelet transform. Thus, temporal and spatial scalability are achieved, in particular when also the motion information is properly managed between resolution levels. The Block Matching Algorithm (BMA) is the most popular used motion estimation method for its simplicity and regularity. The basic operation of a block matching algorithm is picking up the best candidate image block in the reference image frame by calculating and comparing the matching functions between the current image block and all the candidate blocks inside a confined area in the reference frame [1]. The acceptance criterion is based on minimizing the mean square error or mean absolute difference between the two sets of pixels, and the relative displacement between the two blocks is taken to be the motion vector.

A straightforward block matching algorithm is the full search in which a measure of the difference between every block in a search window from the previous frame and the current block is calculated. This algorithm search all locations in a specific search range and selects the position where the cost function of block matching is minimized. However, the heavy computational load of the full search can be significant problem in real time video coding. To reduce the computation time needed for an exhaustive search, many fast algorithms have been proposed such as the Three Step Search (TSS) [2], [3], Four Step Search (4SS) [4], New Three Step Search (NTSS) [3], [5] and Diamond Search [6]. However, many of them decrease the coding time at the expense of coding quality.

This paper is organized as follows. Section II presents block matching algorithms in general, and the simulation methodology. We choose three well known algorithms and analyze them in depth in this paper. They are the three step search, the new three step search and the four step search. Section III presents an overview about a system-level implementation of a wavelet-based t+2D scheme used in most powerful video coding. In section IV, simulation results for the different block matching algorithms in terms of computational complexity (number of computations) and PSNR are compared and presented. The last section concludes our work based on the results obtained in section IV.

II. BLOCK MATCHING ALGORITHMS

To reduce the computational complexity of full search algorithm, the number of search location or the number of operations should be decreased. This is the motivation for developing fast search algorithms over full search method that restricts the search to a few points. In this section we present Three Step Search algorithm (TSS), New Three Step Search algorithm (NTSS) and Four Step Search algorithm (4SS). All the mentioned block matching algorithms use Mean Absolute Difference (MAD) as cost function.

The three step search algorithm (TSS) has been widely used in block matching motion estimation due to its simplicity and effectiveness [2], [3], [5], [7]. It searches for the best motion vectors in a coarse to fine search pattern. It starts with selecting initial step size in the search area of current frame. Furthermore, it sets the step size S=4 and then eight blocks at a distance of step size from the center location (0,0) (around the block center) are considered for comparison. At this level, cost function is calculated at each block and new origin is assigned to the pixel giving the minimal cost function. The step size is then halved and the same process is repeated until the step size becomes equal to 1. At that stage, it finds the location with the least cost function and the macro block at that location is the best match. One problem that occurs with the three step search is that it uses a uniformly allocated checking point pattern in the first step, which becomes inefficient for small motion estimation.

The three step search algorithm has been used in some low bit-rate video compression applications due to its simplicity and effectiveness. However, it is inefficient in terms of computation for the image with small motions since it uses a uniformly allocated checking point pattern in its first step. That is, TSS has unnecessarily many checking points when the motion is small for the block. New Three Step Search (NTSS) algorithm was proposed for small motioned images by R. Li et. all [8], [9]. The NTSS uses a center biased searching scheme and having provisions for half way stop to reduce computational cost. Unlike TSS algorithm, NTSS employs 17 checking points in the first step for lowest weight using a cost function, and makes the search adaptive to the distribution of the motion vector. This adaptive search is based on using 8 search locations at a distance of S=4 similar to TSS and 8 other at S=1 away from search origin. If the lowest cost function is at the origin then the search is stopped right here and the motion vector is set as (0, 0). If the lowest cost function is at any one of the 8 locations at S = 1, then we change the origin of the search to that point and check for weights adjacent to it.

Four step search algorithm (4SS) based on the center biased motion vector distribution characteristic is proposed in [4]. Similar to NTSS algorithm, 4SS has a half-way stop provision. It starts with selecting initial step in the 15x15 searching area. Furthermore, it sets the step size S=2 and then eight blocks on a 5x5 window located at the center of the searching area are considered for comparison. The minimum cost function

is calculated at each block. If the minimum is found at the center of the search window, the search jumps to the fourth step, otherwise, the search jumps to the second step. In the second step, the search window is sill maintained as 5x5 pixels wide. Depending on the calculated minimum cost function, we might end up checking at 3 locations or 5 locations. If the previous minimum cost function is located at the corner of the previous search window, 5 additional checking points are used. If the previous minimum cost function is located at the middle of horizontal or vertical axis of the previous search window, 3 additional checking points are used. If the minimum cost function is found at the center of the search window, the search jumps to the fourth step; otherwise to the third step. The third step uses the same searching pattern strategy as the second step but finally it will go to the fourth step. In the fourth step the window size is dropped to 3x3 (step size S = 1) and the direction of the overall motion vector is considered as the minimum cost function among these 9 searching points.

III. SYSTEM LEVEL IMPLEMENTATION OF A WAVELET-BASED T+2D VIDEO CODING

The discrete wavelet transform (DWT) has gained widespread acceptance in signal processing and image compression. Because of their inherent multiresolution nature, wavelet coding schemes are especially suitable for applications where scalability and tolerable degradation are important. It converts an input signal x_i into one high-pass wavelet coefficient series and one low-pass wavelet coefficients series. In practice, such transformation will be applied recursively on the low-pass series until the desired number of iterations is reached.

Lifting scheme of DWT has been recognized as a faster approach compared to classical wavelet transform with mirror filters. Lifting decomposition is easily invertible, any type of operation, linear or non-linear, can be incorporated into the prediction and update steps. This lifting scheme consists of three basic steps: polyphase operation, prediction and update. Input samples are first split into odd (x_{2i+1}) and even (x_{2i}) samples. Motion estimation is performed between the input frames and generated motion vectors are used for motion compensation operation in the prediction and update steps as shown in Fig. 1. In the prediction step, odd frames are predicted by even frames and replaced by the prediction errors to form the corresponding high-pass temporal subband. Then, in the update step, the low-pass temporal subband is obtained by updating even frames with the normalized high-pass subband samples. The lifting implementation of the wavelet transform allows for a motion compensated temporal transform, based on any wavelet kernel and any motion model, without sacrificing the perfect reconstruction property.

In a lifting-based MCTF (Motion Compensated Temporal Filtering) coder, a wavelet transform is applied along the motion trajectories with an open loop structure [10] to take advantage from temporal redundancy between neighboring frames as shown in Fig. 2.

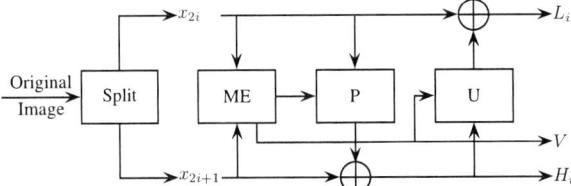

Fig. 1. Lifting-based Motion Compensated Temporal Filtering analysis.

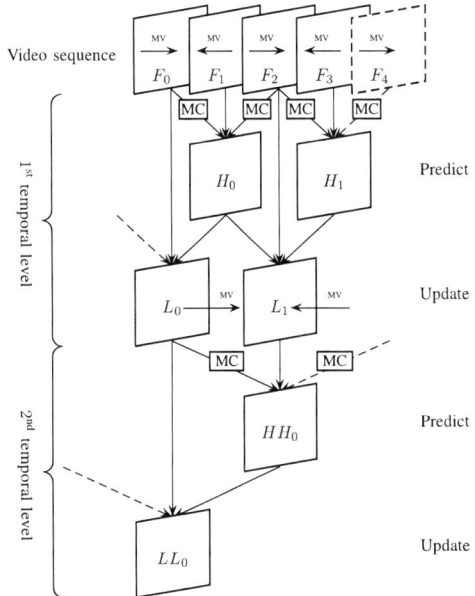

Fig. 2. Motion Compensated Temporal Filtering with lifting.

The MCTF schemes mainly used are Haar-MCTF and 5/3-MCTF. Haar-MCTF performs unidirectional motion estimation and produces one L-frame and one H-frame from each pair of input frames. The 5/3-MCTF employs bidirectional motion estimation and thus produces one H-frame from three input frames and one L-frame from five input frames [11]. However, longer filters such as 9/7 [12] take better use of the temporal redundancies, but the quality on the lower bit-rates is influenced by the number of motion vectors which is transmitted negatively.

Motion Compensated Temporal Filtering is a revolution in the context of video coding standards since it provides a filtering scheme t+2D. It not only allow to consider motion in video sequence but also an efficient temporal decorrelation in wavelet based video coding. The low-pass and high-pass resultant frames are transformed using 2D spatial wavelet transform for the removal of spatial redundancy and then quantized and coded using embedded coding, such as Motion-Compensated Embedded Zero blocks Coder (MC-EZBC) [13]. Motion vectors are also coded using lossless entropy coding and integrated throughout the bitstream.

In spatial redundancies reduction, the wavelet transform has been shown to give better results than the DCT based schemes. The wavelet based scheme not only eliminates the visual artifacts due to DCT coding, but also allows for a multiresolution approach. In other words, the bit stream can be partially decoded in order to reconstruct a low resolution image. The DWT has several advantages of multi-resolution analysis and subband decomposition, which has been successfully used in image processing.

Figure 3 shows the block diagram of our wavelet based video compression system, in which sequence of frames is given as input to the DWT and motion estimation and compensation is performed on the low frequency coefficients obtained from DWT.

The motion estimation and compensation directly applied on the coefficients resulting from the discrete wavelet transform, rather than on natural images. It will reduce the number of computations for the motion estimation. We use the GCC compiler with the standard SystemC library as a modeling language for the design.

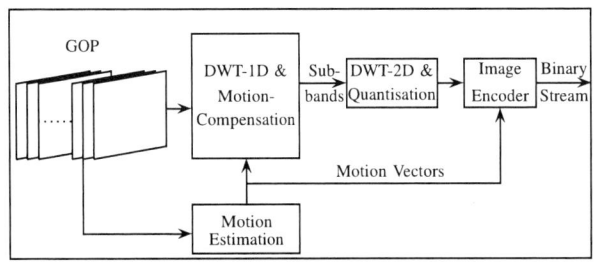

Fig. 3. t+2D video coding block diagram.

IV. PERFORMANCE EVALUATION OF MOTION ESTIMATION ALGORITHMS: SIMULATION AND DISCUSSION

Since different block matching algorithms are used, their image qualities are not identical. Peak signal-to-noise ratio (PSNR) is used as an indicator for quality comparison. Another measure of the effectiveness of a motion estimation algorithm is the number of tested blocks to provide motion vectors also called computation.

To choose the adequate block matching algorithm for the proposed video coding, we used 31 frames of "Tennis Player", "Susie", "Claire", "Call Train" and "Garden" video sequences. Experiments were performed with the Tree Step Search (TSS), the New Three Step Search (NTSS) and the Four Step Search (4SS) algorithms. We note that the search parameter P influences the coder performances. Indeed, the more the search zone the number of compared blocks increase. The macro block was a square of side 16 by 16 pixels and the search parameter was ±7 pixels. Image size was SIF (352 by 240) for each sequence. MAD (Mean Absolute Difference) as error criterion for finding optimal motion vector was used. The simulation results are presented in terms of PSNR, and also

TABLE I
AVERAGE PSNR AND COMPUTATIONS FOR DIFFERENT SEARCH
ALGORITHMS AND DIFFERENT SEQUENCES.

Sequence		NTSS	TSS	4SS
Tennis Player	PSNR	28,727	28,514	28,519
	Computations	17,610	23,088	16,759
Call Train	PSNR	30,269	29,426	29,804
	Computations	19,783	23,637	18.765
Claire	PSNR	41,755	41,644	41,673
	Computations	16,325	23,239	16,291
Garden	PSNR	23,552	23,056	23,310
	Computations	22,067	23,165	19,034
Susie	PSNR	36,242	35,621	35,744
	Computations	17,260	23,063	16,176

MSE (Mean Squared Error), and the number of compared macro blocks for each frame. In fact, the higher the PSNR, the closer the distorted image is to the original. In general, a higher PSNR value should correlate to a higher quality image.

These criteria for error and performance are described by the equations (1), (2) and (3). N is the size parameter of the matching block and i, j are related to position information in the matching block. $C_{i,j}$ points to original pixel of current frame and $R_{i,j}$ shows predicted pixel from past frame.

$$\text{MAD} = \frac{1}{N^2} \sum_{i=0}^{N-1} \sum_{j=0}^{N-1} |C_{ij} - R_{ij}| \qquad (1)$$

$$\text{MSE} = \frac{1}{N^2} \sum_{i=0}^{N-1} \sum_{j=0}^{N-1} (C_{ij} - R_{ij})^2 \qquad (2)$$

$$\text{PSNR} = 10 \log_{10}(\frac{255^2}{\frac{1}{N^2} \sum_{i=0}^{N-1} \sum_{j=0}^{N-1} (C_{ij} - R_{ij})^2}) \qquad (3)$$

Different experimentations, for the cited above conventional algorithms, have been done and discussed. Table 1 shows average PSNR performance and computation for previously mentioned algorithms with listed sequences above.

The aim of this study is to choose the adequate search algorithm which reduce the amount of computations without serious degradation of predicted image. According to our study, their PSNR values in all five sequences are very close to each other. In general, all the three search algorithms have a better PSNR performance on slowly moving pictures (such as Tennis Player and Claire) but have a poorer PSNR performance on fast moving pictures (such as Garden). As shown in Table 1, the average PSNR value is still quite high at the acceptable quality of 27dB expect for the fast moving pictures where the average PSNR is about 23dB. On the other hand, it is clear that the four step search algorithm outperforms all the other algorithms in terms of computation. The average computation of the four step search algorithm is about 16 compared blocks while it is about 23 compared blocks for the three step search algorithm. Overall the four step search algorithm is giving the acceptable PSNR with reduced number of computations for slow and fast video sequences.

If the picture quality is our major concern, the new three step search may be the best choice . In fact, The new three step search is higher by roughly 1dB in PSNR. However, for large size pictures and/or large search ranges, the three step search and the new three step search algorithms can be costly and time-consuming.

V. CONCLUSION

The Motion estimation is a process which determines the motion between two or more frames of video sequence. Block matching algorithms are the most popular and efficient of the various motion estimation techniques. Comparative study on results, in a high level implementation, in terms of reconstructed image quality and search speed of block matching algorithms may result in an optimized hardware implementation. In this paper, a brief description and a system-level implementation of motion compensation based video compression are presented. Three block matching motion estimation algorithms, namely Three Step Search (TSS), New Three Step Search (NTSS), and Four Step Search (4SS) algorithms are compared and implemented. Experimental studies have proved that Four Step Search algorithm is the best matching motion estimation algorithm that achieve best tradeoff between search speed (number of computations) and reconstructed picture.

REFERENCES

[1] A. Ahmadi and M. M. Azadfar, *Implementation of fast motion estimation algorithms & comparison with full search method in H.264*, IJCSNS International Journal of Computer Science & Network Security. vol. 8, no. 3, pp. 139-143, March 2008.

[2] Z. Chen, *Efficient Block Matching Algorithm for Motion Estimation*, International Journal of Signal Processing. 5:2 2009.

[3] L. Tao, Y. S. Ying, S. Z. Feng and G. Peng, *A improved three-step search algorithm with Zero Detection and vector filter for motion estimation*, International Conference on Computer Science and Software Engineering. 2008.

[4] L. M. Po and W. C. Ma, *A Novel Four-Step Search Algorithm for Fast Block Motion Estimation*, IEEE Trans. Circuits Syst. Video Technol. vol. 6, no. 3, pp. 313-317, Jun. 1996.

[5] R. Li, B. Zeng and M. L. Liou, *A new three-step search algorithm for block motion estimation*, Circuits and systems for video Technology, IEEE Transaction. August 2002.

[6] A. Barjatya, *Block Matching Algorithms for Motion Estimation*, spring 2004.

[7] T. Koga, K. Inuma, A. Hirano, Y. Ijima and T. Ishiguro, *Motion Compensated Interframe coding for video conferencing*, Proc. Nat. Telecommun. Conf. New Orleans, LA, pp. G.5.3.1-G.5.5.5, 1981.

[8] T. Liang and P. Kuo, *A novel fast block-matching algorithm for motion estimation using adaptively asymmetric patterns*, International Journal of Innovative Computing, Information and Control. vol. 4, no. 8, August 2008.

[9] R. Li, B. Zeng and M. L. Liou, *A new three-step search algorithm for block motion estimation*, IEEE Trans. Circuits Syst. Video Technol. vol. 7, pp. 833-844, Dec. 1997.

[10] J. R. Ohm, *Multimedia Communication Technology*, Berlin, Germany: Springer-Verlag, 2005.

[11] M. Flierl and B. Girod, *Investigation of motion-compensated lifted wavelet transforms*, Proc. Picture Coding Symp. Apr. 2003, pp. 59-62.

[12] Dr.B Eswara Reddy and K Venkata Narayana, *A Lossless Image Compression using Traditional and Lifting based Wavelets*, Signal & Image Processing : An International Journal (SIPIJ), Vol.3, No.2, April 2012.

[13] P. Chen and J. W. Woods, *Bidirectional MC-EZBC with Lifting Implementation*, IEEE Trans. Circuits and System for video technology. vol. 14, pp. 982-993, oct. 2004.

978-1-4673-5289-5/12 $31.00 © 2012 IEEE

Automated Design Technique for Constant-g_m Rail-to-Rail for OTA Input Stage

Ahmed Reda Mohamed, Mohamed F Ibrahim, and Fathi Farag
Electronics and Communication. Dept., Faculty of Engineering,
Zagazig University, Egypt.
ffarag@zu.edu.eg

Abstract— **This paper presents a novel technique for an automated design to produce a constant transconductance (g_m). This constant g_m is valide over the entire common-mode input range for input stage of Low Voltage (LV) operational transconductance amplifier (OTA) based on DC level shifter. The issue lies at the input sage so, this automated design technique is created to rail-to-rail input stage of OTA. The key parameter is the bias current in DC level shifter (I_{sh}). The proposed technique is responsible for finding the optimal I_{sh} to obtain minimum variation on the g_m of the OTA input stage. This technique is based on a script written on Linux, operating system, to be connected to the BSIM MOST model and netlist of the desired topology. Utility of the physics-based g_m/I_D characteristics, this technique is more suitable for short channel transistors in sub-micron processes. This work allows the design problem to be cast as a program. So, it offers an efficient, reliable, and fast way to implement high-performance of analog integrated circuits. The results demonstrate that g_m variation can be restricted within ±1.25 %. The circuit is simulated in IBM 0.13μ CMOS technology with a single power supply 1.5-V.**

Keywords- Constant transconductance g_m; Rail-to-rail input stage; DC level shifter; LV; g_m/I_D characteristics.

I. INTRODUCTION

Recently, due to market requirements and technology constrains, the research in analog integrated circuits has been become in the direction of Low-Voltage Low-Power (LV LP) design. The environment of portable systems is the most requirements in side of market and the dramatically growth of submicron technology has been forced the researchers to work at low voltage supply. These LV circuits have to show also a reduction of power consumption to maintain a longer battery lifetime[1].

The reduction of supply voltage results in reduction in the input common mode range (V_{icmr}). The key problems are how to achieve a high rang of input common mode signal (V_{icm}) and a constant g_m with minimum variation.

Rail-to-rail input stage allows input common mode signal to move from negative and positive supply rails [2]. In order to obtain an input stage with rail-to-rail input common mode signal, a complementary (N - P MOS) differential pairs have to be driven in parallel as shown in Fig. 1(a). The total small signal transconductance (g_{mt}) of such a combination depends on the common mode input voltage as show in Fig. 1(b). Assume both pairs have the same g_m value ($g_m = g_{mn} = g_{mp}$).

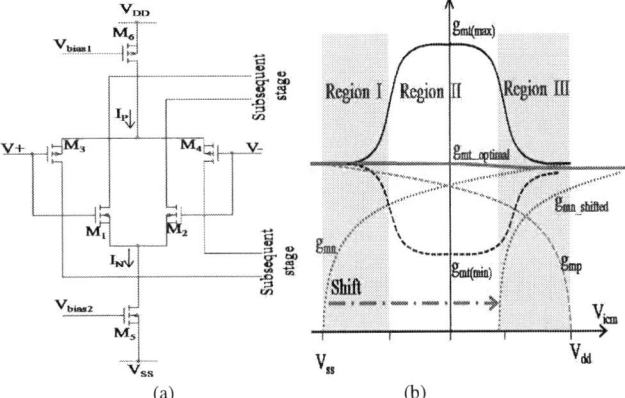

Fig. 1 (a) Complementary input stage, (b) g_{mt} vs V_{icm}

In region I; when V_{icm} is near to the negative supply rail, only P-differential pair is on and N-differential pair is cut off because $|V_{GSn}| < |V_{thn}|$ (Threshold voltage).The total transconductance (g_{mt}) of complementary differential pair equals to g_{mp} ($g_{mt} = g_{mp} = g_m$). In region II; g_{mt} may have a two distinct values shown in Fig. 1(b). One of these values is maximum transconductance ($g_{mt(max)}$), where $g_{mt(max)}$ has twice of g_m ($g_{mt} = 2g_m$) because N-channel and P-channel input differential pairs are turned on at the same time in the middle of V_{icmr}. And, the another value is minimum transconductance ($g_{mt(min)}$), where $g_{mt(min)}$ equals to zero, because both N and P differential pairs are off, which is the conduction gap phenomena [3]. In region III; the value of g_{mt} returns again to have g_{mn} ($g_{mt} = g_{mn}$) because, N-differential pair is on and P-differential pair is cut off because $|V_{GSp}| < |V_{tn}|$. Unfortunately, this variation in g_{mt} according to V_{icm} is unwanted phenomena because that affects on the Gain Bandwidth Product (GBW), stability performance, non-constant Slew Rate (SR), and power consumption. So, the output will introduce some nonlinear distortion [4].

In the past years, a number of constant g_m techniques have been reported and these can be categorized by scheme or operating region [5-6]. Many of efforts are based on weak inversion region and how to keep the sum of tail current of N(P) - differential pair ($I_{N(P)}$) constant . But, working in weak inversion region generates a small g_m, small GBW, and large area [7]. Others are based on square law characteristic of MOS transistor in strong inversion to keep ($\sqrt{I_N} + \sqrt{I_P}$) constant. But, using square law is not exactly for short channel

978-1-4673-5289-5/12 $31.00 © 2012 IEEE

transistors in sub-micron processes [8]. The limitations of these trends produce a large variation in g_{mt}.

Nowadays, several works have been done in the field of analog-circuit design automation. Analog-circuit design automation is one of the important obstacles especially for System On Chip (SOC). One reason for this is the nonlinear relations between components' sizing and required specifications. Moreover, it is hard for ordinary designer to make good trade-offs among these relations. For these reasons, some endeavors have been reported aiming to the development CAD tools for reducing the time spent through design cycle [9].

In this work, a novel technique for automated design is discussed. This technique is applied on rail-to-rail OTA input stage for constant g_m with DC level shifter circuit

This article is organized as following: DC level shifter circuit is described in section II. The proposed technique is developed in section III. In section IV, simulation results are presented . Finally, conclusion is presented in section V.

II. DC LEVEL SHIFTER CIRCUIT

The proposed technique is applied to a CMOS input stage consists of N and P MOS differential pairs connected in parallel (M_1 and M_2, M_3 and M_4) with auxiliary DC level shifters (M_7, M_8) shown in Fig. 2. The total small signal trans-conductance of the complementary (N-P) differential pairs is g_{mt}. Essentially, two N-MOS source followers (M_7, M_8) with active load (M_9, M_{10}) are used for DC level shifter[10]. The input signal is directly connected to N-MOS source follower and the P-channel input differential pair. The output of N-MOS source follower is connected to the input of N-channel differential pair. Thus, the shifted input signal is fed to the input of N-differential pair. The auxiliary circuit shifts the g_{mn} curve of N-differential pair to right or left. Thus, the transition region of N-MOS is shifted. Moreover, the transition regions of N channel and P channel input differential pairs are overlapped and total transconductance is maintained constant.

$$V_{in,min1} = V_{GSn} + V_{Dsat} \qquad (1)$$
$$V_{in,min2} = V_{GS-sh} + V_{GSn} + V_{Dsat} \qquad (2)$$
$$V_{GS-sh} \; \alpha \; \sqrt{\left(\frac{I_{sh}}{(W/L)_{sh}}\right)} \qquad (3)$$

where; V_{GSn} :is the gate-source voltage of $M_{(1,2)}$.

V_{GS-sh} :is the shift voltage of $M_{(7,8)}$ in V_{icm}.

V_{Dsat} :is the drain-source saturation voltage of $M_{(5)}$.

I_{sh} :is the DC current in $M_{(7,8)}$,key parameter.

$(W/L)_{sh}$:is the dimension of $M_{(7,8)}$.

The minimum input voltage ($V_{in,min1}$) for N differential pair shown in Fig. 1(a) is given by (1). After addition DC level shifter shown in Fig. 2, the minimum input voltage ($V_{in,min2}$) is given by (2).

Both M_5 and M_6 are used to provide the tail currents (I_N, I_P) of (N,P) differential pairs from biasing voltages.

Using (2) and (3), both I_{sh} and $(W/L)_{sh}$ are used to improve and compensate any error occurred in threshold voltage (V_{th}) variation, temperature dependant, or mathematical calculations. Thus, I_{sh} is selected to be the key parameter to control the input shift voltage. Finally the

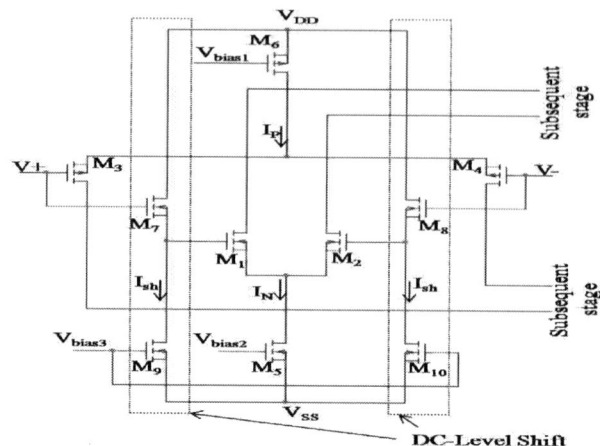

Fig. 2 Rail-to-rail for constant g_m using DC level shift circuit.

subsequent stage circuits can be used to add the currents from constant g_m of N- and P channel differential pairs together and converts to a single-ended output. These subsequent stages are edited for testing later.

III. PROPOSED TECHNIQUE

The flow chart shown in Fig. 3 describes the proposed technique.

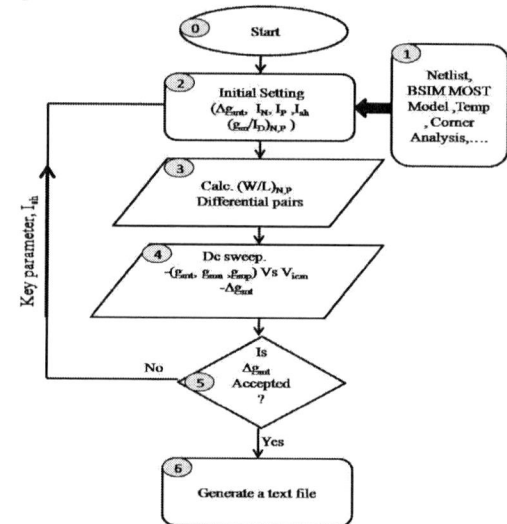

Fig. 3 Flow chart of the proposed technique.

First block is responsible for attaching the netlist of desired topology, BSIM MOST model [11], the temperature (27), and the MOST parameters.

$$\Delta g_{mt} = \frac{g_{mt(max)} - g_{mt(min)}}{g_{mt(max)}} \qquad (4)$$

The initial setting for tail currents of N and P channel differential pairs (I_N, I_P) , I_{sh} , and acceptable limit variation in the total transconductance (Δg_{mt}) ,given by (4), are set in second block. I_N and I_P can be estimated according to available power consumption. Within third block, g_m/I_D technique is used. This technique is based on a unified synthesis methodology in all the regions of operation of MOS

978-1-4673-5289-5/12 $31.00 © 2012 IEEE   58

transistor [9]. The choice of g_m/I_D is based on its relevance for the three following reasons: It is strongly related to the performances of analog circuits. It gives an indication of device operating region. It provides a tool for calculating the transistors dimensions. This technique exploits the transconductance over DC drain current ratio (g_m/I_D) relationship versus the normalized current, $I_n \equiv I_D/(W/L)$. To obtain dimensions (W/L) of differential pairs (N-P), this steps are followed:

1- Assume that $(g_m/I_D)_N = (g_m/I_D)_P$ and $L > L_{min}$ [12].

2- The normalized current $I_n \equiv I_D/(W/L)$ of N-differential pair can be obtained from Fig. 4 which is based on Semi-empirical approach [13]. By the same way, normalized current of P-differential pair can be obtained.

3- According to the available power consumption, I_N and I_P are estimated. Thus, it is easy to find dimensions (W/L) of differential pairs (N-P).

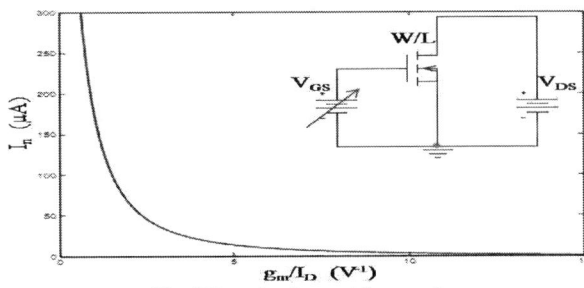

Fig. 4 Normalized current I_n vs g_m/I_D.

In fourth block, variable shifts according to (3) based on I_{sh} are done. According to variation on I_{sh}, g_{mn} curve moves to right or left until the proposed technique to find the best value of I_{sh} whose minimum variation in g_{mt}. In fifth block, the proposed technique terminates its work if the variation on g_{mt} is acceptable or returns to second block. Finally, a text file is generated in the sixth stage. Entire the generated text file lies a summary which indicates the proper I_{sh}, minimum g_{mt}, maximum g_{mt}, relative error in g_{mt} and dimensions (W/L).

IV. SIMULATION RESULTS

The proposed technique in Fig. 3 is applied in the circuit in Fig. 2. The number of curves, limitation of sweep (begin and end) of I_{sh}, allowable variation on g_{mt} can be adjusted in the proposed technique. According to I_{sh}, the g_{mn} curve is shifted as shown in Fig. 5. The proposed technique is smart enough to determine the value of I_{sh} for the required minimum error occurred in g_{mt} (Δg_{mt}). The deviation of g_{mt} versus I_{sh} is plotted in Fig 6. I_{sh} is selected to compensate any variation in g_{mt}.

To verify the proposed script, it is necessary to simulate an OTA by the designed rail-to-rail input stage with a summing current circuit [14-15], as shown in Fig. 7. This OTA is loaded by a capacitor whose value larger than the parasitic capacitance. The simulation result is shown in Fig. 8. There is a slightly variation of ±1.3 % in the unity gain frequency (GBW) due to the variation in g_{mt}.

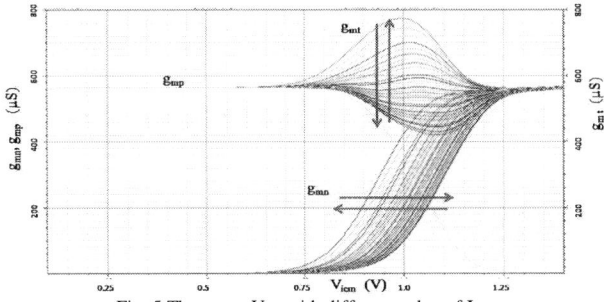

Fig. 5 The g_{mt} vs V_{icm} with different value of I_{sh}

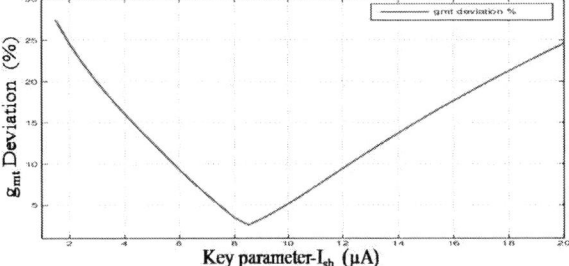

Fig.6 The g_{mt} deviation vs I_{sh}

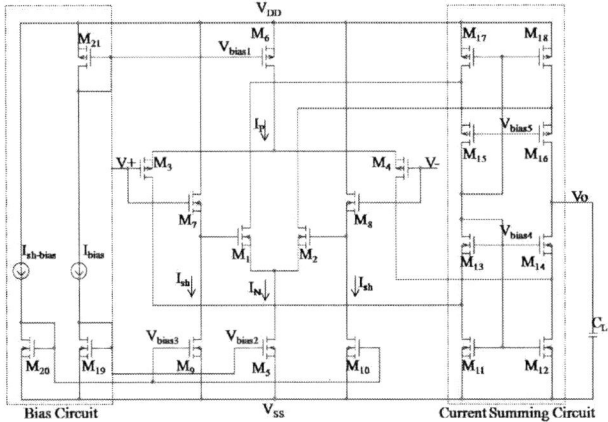

Fig. 7 Rail-to-rail input stage with current summing circuit

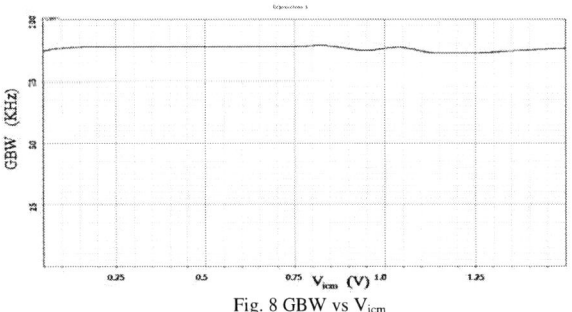

Fig. 8 GBW vs V_{icm}

$$\Delta\% = \frac{g_{mt-Script} - g_{mt-sim}}{g_{mt-Script}} \quad (5)$$

Moreover, a comparison between g_{mt} generated from the script ($g_{mt-Script}$) and g_{mt} generated from simulation ($g_{mt-sim} = \frac{GBW}{2\pi\,C_L}$) is done, and it illustrates a good result with relative error ($\Delta\%$) ± 0.5 % given by (5), as shown in Fig. 9.

978-1-4673-5289-5/12 $31.00 © 2012 IEEE

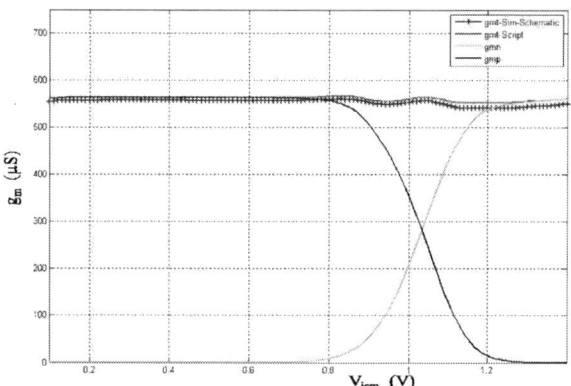

Fig.9 The $g_{mt-Script}$ vs g_{mt-sim}

Summary of the generated text file can be concluded in table I. The performance of the proposed technique compared to previous work is shown in table II.

TABLE I
SUMMARY OF THE GENERATED TEXT FILE FOR L=1 μm

Transistor	W(μm)
M_1, M_2	20
M_3, M_4	70
M_5	100
M_6	200
M_7, M_8	100
M_9, M_{10}	100
I_{sh}	85 μA

TABLE II
PERFORMANCE OF THE PROPOSED TECHNIQUE

Technique	$\Delta g_{mt}(\%)$	Universality
This work	±1.25	Universal
Ref[5]	3	Universal
Ref[14]	4.65	Universal
Ref[2]	5	Square law
Ref[16]	5.4	Weak inversion
Ref[17]	±3.4	Weak inversion
Ref[4]	±4	Square law
Ref[18]	14.5	Square law

V. CONCLUSION

A novel automated design technique has been presented to produce a constant g_m over the entire common-mode input range. This is done for complementary N-P differential pairs based on DC level shifter. The design issue and key parameter have been cast in a program written by script. This script is able to be connected to simulation, netlist, and BSIM MOST model. This technique offered an efficient, reliable, and fast way to get a constant trans-conductance within ±1.25 %. The proposed technology has been designed to be universal due to using the physics-based g_m/I_D characteristics. The final results are completely independent of initial values. The simulation results illustrated a good agreement to the theoretical with relative error Δ%= ±0.5 %. IBM 0.13μm CMOS technology parameter has been used for circuit simulation.

ACKNOWLEDGMENT

The authors would like to thank the MOSIS Company and IBM foundry for providing them with the model's parameters used in this work.

REFERENCES

[1] G. Ferri and N. C. Guerrini, *Low-Voltage, Low-Power CMOS Current Conveyors*, Kluwer Academic Publishers, 2003.

[2] J. H. Botma, R. F. Wassenaar, and R. J. Wiegerink, "A Low Voltage CMOS Opamp with a Rail-to-Rail Constant-g_m Input Stage and a Class AB Rail-to-Rail Output Stage," *IEEE International Symposium on Circuits and Systems*, vol. 2, pp. 1314-1317, May 1993.

[3] F. A. Farag, C. Galup-Montoro, and M. C. Schneider, "Digitally Programmable Switched Current FIR Filter for Low-Voltage Applications," *IEEE Journal of Solid-State Circuits*, vol. 35, pp. 637-641, April 2000.

[4] M. Wang, T. L. Mayhugh, Jr., S. H. K. Embabi, and E. Sanchez-Sinencio, "Constant-g_m Rail-to-Rail CMOS Opamp Input Stage with Overlapped Transition Region," *IEEE Journal of Solid-State Circuits*, vol. 34, no.2, pp.148-156, Feb. 1999.

[5] S. Yan, J. Hu, T. Song, and E. Sanchez-Sinencio, "Constant-g_m Techniques for Rail-to-Rail CMOS Amplifier Input Stages: A Comparative Study," *IEEE International Symposium on Circuits and Systems*, vol. 3, pp. 2571-2574, May 2005.

[6] R. Hogervost, J. P. Tero, and J. H. Huijsing, "Compact CMOS Constant-g_m Rail-to-Rail Input Stage with g_m-Control by an Electronic Zener Diode," *IEEE Journal of Solid-State Circuits*, vol. 31, no. 7, pp. 1035-1040, July 1996.

[7] J. H. Huijsing, R. Hogervorst, and K. J. de Langen, "Low-Power Low-Voltage VLSI Operational Amplifier Cells," *IEEE Trans. Circuits and Systems-I*, vol. 42, no.11, pp. 841-852, Nov. 1995.

[8] R. Hogervorst, R. J. Wiegerink, P. A. L. de Jong, J. Fonderie, R.F. Wassenaar, and J. H. Huijsing, "CMOS Low-Voltage Operational Amplifiers with Constant-g_m Rail-to-Rail Input Stage," *IEEE International Symposium on Circuits and Systems*, vol. 6, pp. 2876-2879, May 1992.

[9] A. Girardi, F. P. Cortes, and S. Bampi, "A Tool for Automatic Design of Analog Circuits Based on g_m/I_D Methodology," *IEEE International Symposium on Circuits and Systems*, pp. 4643-4646, May 2006.

[10] Z. Wu1 and F. Rui1, "Design of a Rail-to-Rail Constant-g_m CMOS Operational Amplifier," *Computer Science and Information Engineering*, Vol. 6, pp.198-201, April 2009.

[11] http://www-device.eecs.berkeley.edu/bsim/

[12] C. saint and J. saint, *Ic Mask Design Essential Layout Techniques*, McGraw Hill,2002.

[13] P. Jespers, *The g_m/I_D Methodology, A Sizing Tool for Low-voltage Analog CMOS Circuits*, Springer ,2010.

[14] G. Dai, P. Huang, L. Yang, and B. Wang, "A Constant g_m CMOS Opamp with Rail-to-Rail Input/Output Stage," *Solid-State and Integrated Circuit Technology* ,pp. 123 – 125, Nov. 2010.

[15] H. Chow and P. Weng, "A Low Voltage Rail-to-Rail Opamp Design for Biomedical Signal Filtering Applications," *IEEE International Symposium on Electronic Design*, pp. 232 – 235, Jan. 2008.

[16] Y. Zhang, Q. Meng, Z. Wang, and S. Chen, "Constant-g_m Low-Power Rail-to-Rail Operational Amplifier," *Wireless Communications and Signal Processing* , pp.1- 4,Nov. 2009.

[17] A. Masoom and Kh. Hadidi, "A 1.5-V, Constant-g_m, Rail-to-Rail Input Stage Operational Amplifier," *International Conference on Electronics, Circuits and Systems*,pp. 632 – 635, Dec. 2006 .

[18] E. Lee, A. Lam, and T. Li, "A 0.65V Rail-to-Rail Constant g_m Opamp for Biomedical Applications," *IEEE International Symposium on Circuits and Systems*, pp. 2721-2724, May 2008.

Inductor Implementation Using CMOS Current Conveyor Integrator for Low Voltage Low Power Applications

Fathi A. Farag

Electronics and Communications Dept., Zagazig University,
Zagazig, Egypt
ffarag@zu.edu.eg

Abstract— A new topology of CMOS floating and ground passive coil simulation are proposed in this work. The proposed active circuits are realized using the second-generation current conveyor integrator (CCII-Int). The ground coil is modeled only using one current mode integrator. Moreover, the floating inductor is simulated by two current-mode integrators. The proposed inductor circuits are considered as low voltage and low power (LVLP) since it's used CMOS inverter as class AB transconductance circuit. The high input impedance of the CMOS inverter simplifies the op-amp design since it's loaded only by capacitive load. An LC passive butter worth filter is designed and realized using the proposed circuits. The proposed circuit is designed using the MOST parameters of the IBM 0.13μ CMOS technology.

Keyword : CMOS Current conveyor, LVLP, CMOS inductor, current-mode integrator

I. INTRODUCTION

It is well-known; replacement of conventional inductors by synthetic ones in passive LC ladder filters belongs to high-order low-sensitivity filter design. Therefore, the ground and floating coils are needed for complete passive LC circuit simulation ability in system-on-chip. The impedance converter has been used for this reason [1] using the current conveyor. The second generation current conveyor (CCII) has proved to be a functionally flexible building block for active only filter design and signal processing applications [2-5]. It satisfies higher signal bandwidth, greater linearity and dynamic range. Moreover, the trend to low voltage – low power (LVLP) circuit design has enormous challenged due to the dramatically growth of submicron technologies and battery life time of the portable devices.

The circuit shown in Fig. 1, is consider as two stages amplifier with class AB output section [5,6]. The circuit is connected as unity gain amplifier (the differential stage and the output section -M_6/M_7). The input current translates to voltage in the internal output of the first section (V_{DS4}). The transfer function is square root relation and the rail to rail output ability shown in Fig. 1(c), which makes this cell to work at low voltage supply with accepted dynamic range. Moreover, the proposed cell considers as low power operation

due to the class AB operation ability. The CCII input/output characteristic can be concluded as in equation 1.

$$\begin{bmatrix} Iy \\ Vx \\ Iz \end{bmatrix} = \begin{bmatrix} 0 & 0 & 0 \\ 1 & 0 & 0 \\ 0 & 1 & 0 \end{bmatrix} \begin{bmatrix} Vy \\ Ix \\ Vz \end{bmatrix} \quad (1)$$

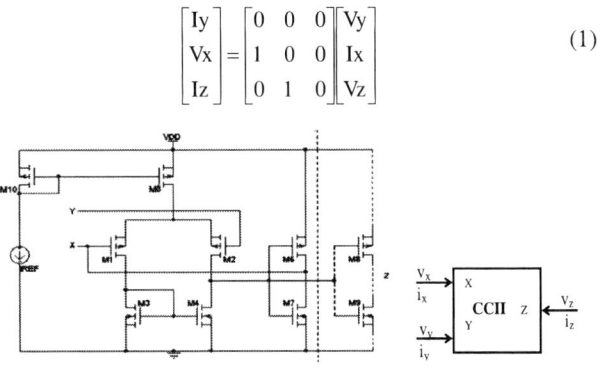

(a). Circuit schmatic.   **(b).** Circuit symbol.

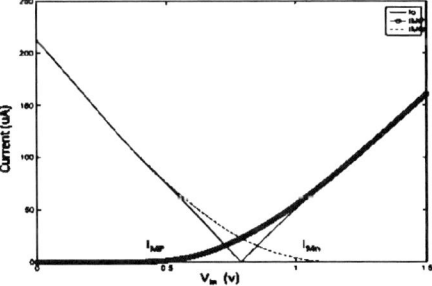

(c). The I-V Characteristic of the CMOS inverter.

Fig. 1, The basic CCII cell proposed in [5, 6].

This paper is organized as following: the current-mode integrator [7] for LVLP applications is reviewed in section II. The proposed inductors (ground/floating) realization and implementation are presented in section III. Moreover, illustrated passive filter is designed and tested for verification in section IV. Some practical parameter effects are studied in section V. Finally, our proposal is concluded in section VI.

978-1-4673-5289-5/12 $31.00 © 2012 IEEE

II. THE CURRENT-MODE INTEGRATOR

In this section, the current mode-integrator in [7], will be reviewed. The current mode integrator is depicted in Fig. 2. In this circuit; the current is convoyed to output terminal with integrated function. The proposed circuits is based on the second-generation current conveyor circuit (CCII) published in [5]. In this circuit, the input current is processed as following:

- The input current is integrated into the feedback capacitor which generates an output voltage lagging by 90° phase shift.
- The CMOS inverter (M_9 & M_{10}) as transconductance "OTA" cell is used to obtain the output current.
- The terminal voltages V_x, V_y, and V_z must be equal which is minmized the error due to output impedance of the CMOS inverter.

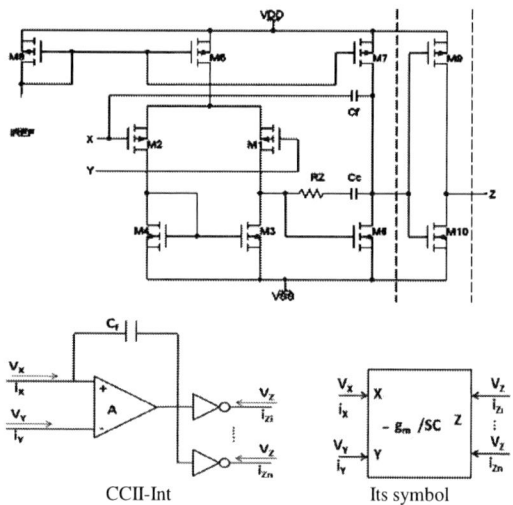

Fig. 2, The second-generation current-mode conveyor integrator [7].

Ideally, the CCII integrator characteristic has been summarized in the following equation

$$\begin{bmatrix} I_z \\ V_x \\ I_y \end{bmatrix} = \begin{bmatrix} 0 & -\frac{g_m}{S.C} & 0 \\ 1 & 0 & 0 \\ 0 & 0 & 0 \end{bmatrix} \begin{bmatrix} V_y \\ I_x \\ V_z \end{bmatrix} \quad (2)$$

The CCII integrator is simulated using IBM 0.13μ CMOS technology transistor parameters. Fig. 3, depicts the time response simulation of the proposed CCII integrator which shows the good integration relation between the input current and the convoyed current.

III. THE PROPOSED INDUCTOR CIRCUITS

III-a) Ground Coil

In this section, the active only ground inductor is realized using the current-mode integrator as shown in Fig. 4. The inductor implementation is based on the CCII current

conveyor [5,7] for low voltage and low power applications. The circuit is implemented using the current–mode integrator with $i_z/i_x = -g_m/SC$ relation.

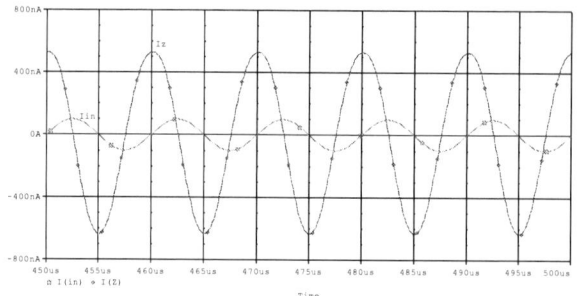

Fig. 3. The simulation time response of the integrator.

The input voltage is converted as current using the linear element "R". The creation current go back to input terminal through the z terminal. Finally the voltage to current relation achieves the coil ($j\omega L$) characteristic. Therefore, the proposed circuit in Fig. 4, is considered as simulated ground coil. The theoretical analysis of the circuit shown in Fig. 4.(a), leads to :

$$L = \frac{RC}{g_m} \quad (3)$$

Where, C is integrated capacitor and g_m is total transconductance of the CMOS inverter.

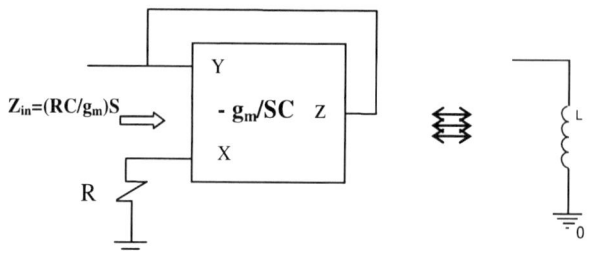

(a). Coil realization.   **(b).** Equivalent circuit.

Fig. 4, The proposed ground coil.

III-b) Floating Coil Realization

In this subsection, the floating coil realization is proposed as shown in Fig. 5(a). The circuit is constructed using two current-mode integrators for I and -I creation. In this figure, voltage terminal is transferred across the resistor R terminals thanks to current conveyor characteristic ($V_x = V_y$). The generated current is conveyor to output terminals (z) with integration relation " $i_z \alpha (i_x/s)$ ". Then, the output currents "i_z" is forced to follow at the output terminals with coil behavior characteristic. Moreover, the coil equivalent value can be approximated as in equation 3.

978-1-4673-5289-5/12 $31.00 © 2012 IEEE   62

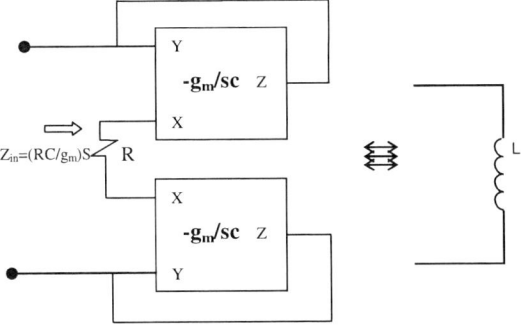

(a). Realizetion. (b). Equavelant coil.

Fig. 5, The proposed floating coil.

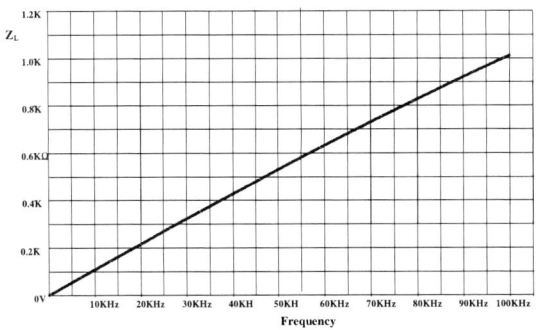

Fig. 7, The floating coil simulation.

The proposed circuits are constructed only using two analog cells which are opamp circuit and the conventional CMOS inverter. Therefore, the proposed circuits are very suitable for Field Programmable Analog Array (FPAA) [8] for system-on-chip applications. The proposed circuits are designed using IBM 0.13μ CMOS technology transistor parameters. The g_m/I_D approach [9] has been employed for circuit design, which is based on the current mode MOST model [10]. The designed circuits " opeational amplifier and CMOS inverter), "transistors sizes and current biasing" are shown in Fig. 6. The amplifier is designed for 5pF capacitive load.

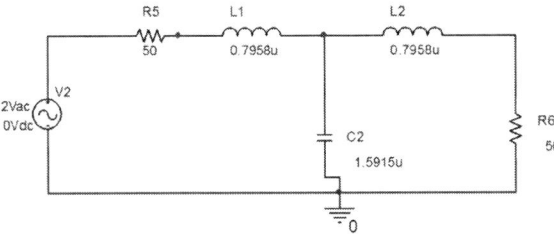

Fig. 8, Third order maximium flat low pass audio filter.

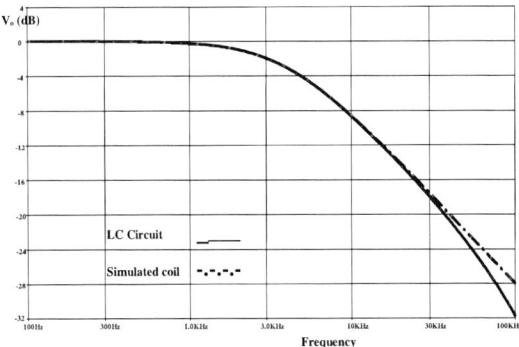

Fig. 9, The Simualtion result.

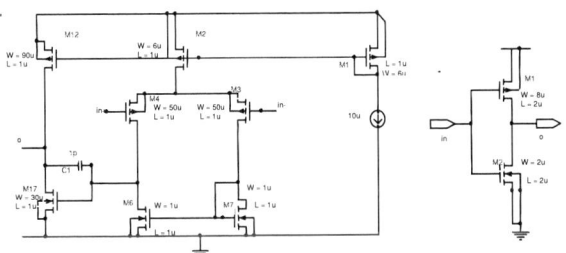

Fig. 6, The Opamp and CMOS inverter schematics.

IV. VERIFICATION AND SIMULATION RESULTS

In this section, the desinged circuits are simulated for functionality testing. The simulation result of the proposed inductor is shown in Fig. 7, which illustrates good mactching to the passive coil charactarestic. Also, a 3ed order Butter worth low pass filter has been considered for 4 KHz cut off frequency and 50 ohm load /source impedances. The passive low pass filter is shown in Fig. 8. The circuits are simulated with 1.5V supply voltage and the biased voltage becomes 685mV. The integrated capacitor was 4pF. The coil resistance (R) is calculated using Eq. 3 (273 ohm). The frequency response of the LC passive filter and the proposed one is shown in Fig. 9. The simulation result manifests a good comparison between the passive filter and simulated one.

V. PRACTICAL CONSIDERATION

In practical, some considerations should be highlighted due to its significant effect on the systematic error and finite limitation of circuit parameters. One of the major sources of errors is the finite open loop gain of the op-amp. Fig. 10 shows the coil in general connection schematic.

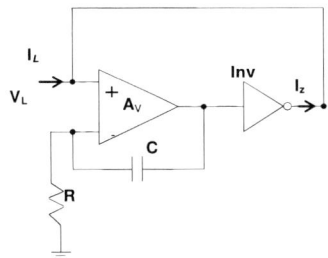

Fig. 10, The general connection of the ground coil.

978-1-4673-5289-5/12 $31.00 © 2012 IEEE

The input impedance can be approximated as

$$V_L = \left(\frac{1}{g_m A_V} + \frac{R}{g_m A_V r_o} + \frac{RC}{g_m} S \right) I_L \qquad (4)$$

Which means that, the equivalent coil Z_{in} is :

$$Z_{in}(s) = r_L + sL \qquad (5)$$

From Equations 4 and 5, the internal resistance coil "r_L" and the coil value (L) are:

$$r_L = \frac{1}{g_m A_V} + \frac{R}{g_m A_V r_o} \qquad \text{and} \qquad L = \frac{RC}{g_m} \qquad (6)$$

where;

A_V: is the open loop gain of the op-amp.
g_m ($g_{mn}+g_{mp}$) is transconductance of each inverter.
g_{mp}: is the inverter PMOS transconductance.
g_{mn}: is the inverter NMOS transconductance.
$r_o = r_{on} // r_{op}$ output impedance of the ith's inverter
r_{op}: is the output resistance of PMOS.
And, r_{on}: is the output resistance of NMOS.

So, the coil resistor is very small since it is function of the closed loop gain ($g_m r_o * A_v$). Therefore, the proposed circuit is presented a high quality factor coil since the closed loop gain ($g_m r_o * A_v$) is very large. The last analysis shows the importance of having a large gain op-amp for highly accurate current mirror and circuit performance.

VI. CONCLUSION

In this paper, inductor characteristic (ground and floating coil) has been simulated. The current-mode integrator has been employed for the circuit realization. Therefore, the proposed coil (floating or ground) are suitable for LVLP applications. The proposed circuits have been designed and tested using IBM 0.13µ CMOS technology transistor parameters. The simulation results show good agreement with the expected behavior. Many studies such as noise, IP3,…etc, must be done for full circuit characterization. Third order LC Low pass filter has been implemented using the proposed circuits. The simulation result has been illustrated high matching degree between the ideal LC filter and the active

simulations. A 2.0mw has been consumed from 1.5V supply voltage.

ACKNOWLEDGMENT

The author would like to thank the MOSIS Company and IBM foundry for providing him with the model's parameters used in this work.

REFERENCE

[1] I. A. Khana and M. H. Zaidi "A Novel Generalized Impedance Converter Using Single Second Generation Current Conveyor" Active and Passive Elec. Comp., Vol. 26(2), pp. 91–94, 2003

[2] J. M.Islamia, J. Nagar, C. Tomazou, F. J. Lidgey, and D. G. Haigh, "Analogue IC Design: the Current-Mode Approach" Stevenage, Herts., UK: Peregrinus, 1990.

[3] C. Bartolozzi, S. Mitra, and G. Indiveri "An Ultra-Low Power Current-Mode Filter For Euromorphic Systems and Biomedical Signal Processing" ISCAS , pp. 130-133, 2006

[4] Y.S. Hwang, D.S.Wu, J.J. Chen, C.C. Shih, and W.S. Chou, "Design of current mode MOSFET-C Filters using OTRAs" ,International Journal of Circuit Theory and Applications, Vol. 37, pp. 397-411, 2009.

[5] F. A. Farag, and H. F. A. Hamed, "Low Voltage Low Power CMOS Current Conveyor Cells" International Computer Engineering Conference (ICENCO2004), p.p642-645, Dec.2004 Cairo, Egypt.

[6] F. A. Farag and Y. A. Khalef " Digitally Controlled Variable Gain Amplifier Based on Opamp-Inverter Current Conveyor" International Conference of Microelectronics (ICM) 2010, pp 224-227 Cairo ,Egypt

[7] F. A. Farag " CMOS Current-Mode Integrator for Low Voltage and Low Power Applications" " sent to publish in ISCAS 2013.

[8] E. Oruklu1, R. Hanley, S. Aslan, C. Desmouliers, F. M. Vallina, and J. Saniie "System-on-Chip Design Using High-Level Synthesis Tools" Circuits and Systems, 2012, 3, 1-9, Published Online January 2012 (http://www.SciRP.org/journal/cs).

[9] A. H. Ghariani, and M. Samet, "Design and Optimization of CMOS OTA with gm/Id Methodology using EKV Model for RF Frequency Synthesizer Application" 12th IEEE International Conference on Electronics, Circuits and Systems(ICECS 2005), pp. 1-5, 2005

[10] A. I. A. Cunha, M.C. Schneider, and C. Galup-Montoro, "An MOS Transistor Model for Analog Circuit Design," IEEE Journal of Solid-State Circuits, vol. 33, pp. 1510-1519, October 1998.

978-1-4673-5289-5/12 $31.00 © 2012 IEEE

Power Recovery From Data Line in Avionic Applications

Jinying Zhang, Saeid Hashemi, Masood Karimian, Zied Koubaa, and Mohamad Sawan

Polystim Neurotechnologies Laboratory

Electrical Engineering Department, Polytechnique Montreal, QC, Canada

jinying.zhang@polymtl.ca

Abstract—In this paper, we propose a novel power recovery approach to reduce the mass of the cables installed in aircrafts. The power recovery unit harvests energy from data field bus which is recently presented to modify classical Avionics Full Duplex Ethernet (AFDX) networks, to power up the Smart Sensor Interface (SSI) module entirely. A top-down modeling approach in Verilog-A is used to build complex modules in the Power Conversion Chain (PCC) which makes power transfer and distribution possible. The effects of frequency and the size of capacitor reservoir on settling time and output voltage ripple are studied. Simulation results in system level prove that the proposed power recovery scheme could procure and deliver significant amount of power to SSI which makes the structure self-powered and therefore additional power lines are saved.

I. INTRODUCTION

Bulky cabling is one of the major challenges in current as well as future aircrafts. Each application device requires a power as well as typically a data connection. More cabling eventually leads to more weight, space consumption, fuel wasting and thus more CO_2 emission. Recent researches in power/data transmission have led to introducing new schemes where less number of cables are required in avionic application. These approaches include wireless communication [1], Power Line Communication (PLC) [2-3] and Power over Ethernet (data) (PoE) [4-5] techniques. They however suffer from limitations which partially outweigh their advantages and limit their application. Compared to wired communication, wireless communication is less popular and more power hungry.

Power lines are usually not designed to carry data, and actually behave as a very harsh and noisy transmission medium, and are impaired by colored background noise and impulsive noise [6]. Thus maintaining signal integrity over power lines requires robust signaling techniques and additional hardware. To the best of our knowledge, the researchers only reported the use of PoE technique to power up End Systems in conventional Ethernet/AFDX networks, not to the devices connected to the field bus, which are numerous and widespread in aircrafts.

Ethernet is about to become the new standard physical layer solution for networks in many fields of industry [5]. This trend also applies to be the aeronautic industry, where AFDX as an Ethernet based protocol is already widely used in modern avionic systems [7]. Here, controller Area Network (CAN) bus is introduced to interface AFDX through End Systems and Gateways so as to make the network much more scalable. The CAN bus is meant to interconnect numerous sensors and actuators.

In this paper, we propose a novel approach for power recovery from data lines. This technique aims basically at harvesting power from CAN bus which transmits data signals and powering up our designated Smart Sensor Interface (SSI) module. This SSI module is meant to interface the L/RVDT, MEMS and optical position sensors to the avionic network in aircrafts. The proposed method offers the potential to significantly reduce the mass of the cables installed, and correspondingly decrease the cost, wiring complexity and weight, especially aiming at energy management and greener avionics for future aircrafts.

The Power Conversion Chain (PCC) of the SSI is presented and its complex components are modeled in Verilog-A using a top-down modeling approach. The models make it possible to study over power transfer and distribution throughout the SSI. Simulation results in system level prove that the proposed power recovery scheme could procure and deliver significant amount of power to SSI which makes the structure self-powered and, correspondingly, additional power lines are saved.

II. PROPOSED POWER RECOVERY SCHEME

We propose the architecture for a power recovery unit which harvests power from CAN bus connected to recently modified AFDX networks, where a local field bus is introduced to interface SSI and the CAN bus. The proposed scheme harvests power from data line to power up the SSI module entirely.

A. Idle State in CAN Bus (ARINC 825)

The proposed architecture uses the idle state of the data line to perform power harvesting process so as to maintain data integrity. According to ARINC 825 Standard - which

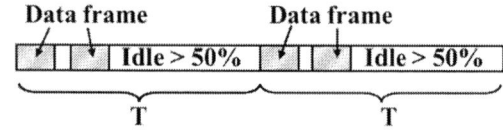

Fig. 1. Schematic of idle state in CAN bus (ARINC 825)

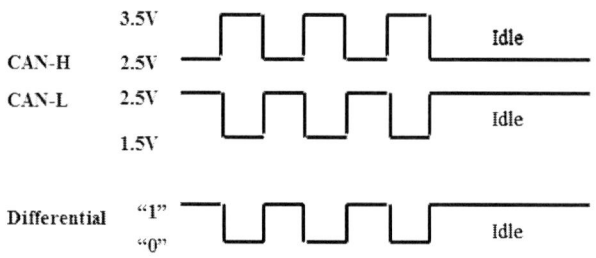

Fig. 2. Voltage levels of idle state in CAN bus (ARINC 825)

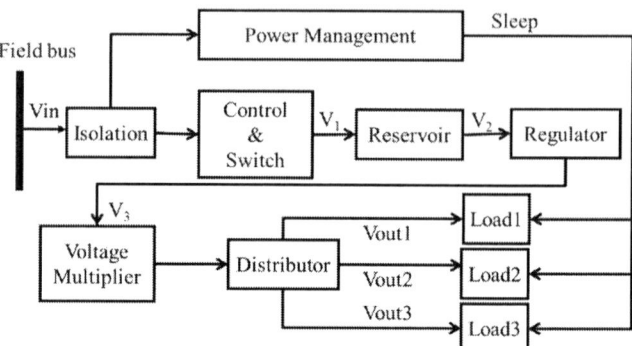

Fig. 3. Proposed architecture for power recovery path

provides data transmission protocol in up-to-date avionic applications - and taking the error handling procedure into account, the transmission time is meant to be not longer than 50% during each period T (Major Time Frame) [8], as illustrated in Fig. 1, in order to provide enough time for repeating data frames when faulty transmission occurs. In order to spare as much room as possible to maintain high data transmission efficiency, we suggest utilizing the latter half of each period T for energy harvesting.

According to ARINC 825 Standard [8], the voltage levels are illustrated in Fig. 2. During the idle state the voltage is 2.5 V both in CAN-H and CAN-L cables.

B. Power Recovery Path

The required power is scavenged through a power recovery path whose architecture is shown in Fig. 3. The power recovery unit consists of a high-efficiency low-voltage switch module managed by a control block, a reservoir to store energy, a low drop-off voltage regulator, and a multi-stage voltage multiplier (VM) in a cascade configuration. The control block is used to detect the start-up of idle state in CAN bus and to enforce the switch to be open when the harvesting process begins. The VM module is inserted to generate voltages higher than the regulated voltage for exciting the specific loads. It also includes voltage distribution to supply different voltage levels required by different loads. An isolation module is needed to be embedded in this distribution block to isolate voltage supply for different loads, e.g., analog one and digital one. Moreover, a power management block is used to decide the working mode of each load. An isolation block is inserted between the field bus and power recovery unit in order to eliminate the undesired influence from power recovery unit to the field bus.

When the power budget is not enough to support all loads at the same time, the power management module would allocate the available power to the loads with higher priority and lower consumption. Among various possible priority agendas, three kinds of them are considered. Firstly, priority is offered to some loads and all loads are working in a given sequence. Secondly, some loads would be enforced to move to the sleep mode to leave power to other ones which accomplish more important tasks at that time. Thirdly, auxiliary power lines are launched to compensate for power shortage. Moreover, three approaches could be applied to detect the power hungry state. Firstly, the bus current exceeds the standing ability of the bus. Secondly, the working voltage on the load drops. Thirdly, the overall output power exceeds the maximum supply value.

C. Power Budget Calculations

In our design, the period T is chosen as 2 ms according to the design experience from our industrial partner, that is Thales Canada Inc. Based on our choices of architectures for various modules in the power recovery scheme and their circuit components, we obtained from simulation that the power conversion efficiency of the unit combined from isolation, control and switch blocks will be up to 90%. The voltage regulator block may give high power efficiency around 90% when applying the structure reported in [9]. The power efficiency of voltage multiplier is as high as 75% if the circuit in [10] is applied. A high accuracy distributor is required to supply accurate supply voltages to different digital blocks and an efficiency of more than 50% is expected in future circuit design. To store more energy and maintain the power recovery unit more useful, we need a large volume reservoir and an off-chip capacitor may serve for this application. Considering all above mentioned assumption, the overall power conversion efficiency of the power recovery path is calculated to be around 30%. On one hand, the voltage level of CAN bus is 2.5 V during the idle state and the bus can maintain a maximum current of 50 mA. As mentioned above, we utilize the latter half part of each signal period T for energy harvesting. Thus, the maximum power, with no influence in data integrity, the bus can deliver is (2.5 V * 50 mA / 2) * 30% = 18.75 mW. On the other hand, the designed SSI module may consume power of around 13 mW. Therefore, the assumption results in system level indicate that the proposed power recovery approach may realize the idea towards building a self-powered SSI module.

III. SIMULATION RESULTS AND DISCUSSIONS

A top-down modeling approach based on Verilog-A is used to build complex modules of this power recovery path in Cadence 6.1. This helps to investigate some critical

978-1-4673-5289-5/12 $31.00 © 2012 IEEE

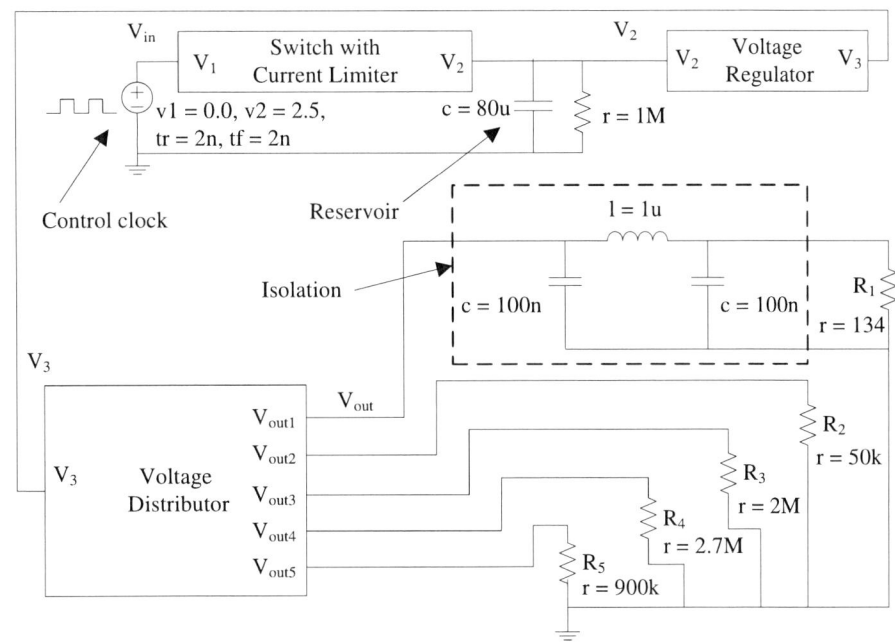

Fig. 4. Schematic of power conversion path to power up analog module and digital module in SSI

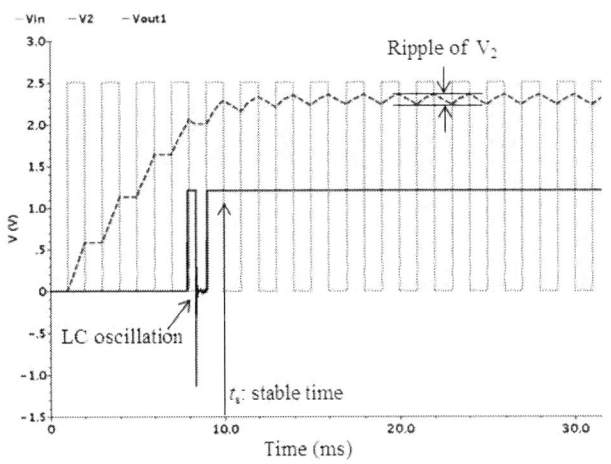

Fig. 5. Transient response of voltages in power conversion path

parameters and detailed performances in the above architecture. Based on the power assumption in Section II, the proposed power recovery approach may procure power for SSI module. Thus, the power management block is not considered within the simulated scheme for power recovery path. Future work will include powering up SSI module as well as the position sensors, so we need to improve the power conversion efficiency of the proposed power recovery unit, and the power management block should be considered when the power budget is not enough to support all loads.

A. Schematic of PCC to power up SSI

Fig. 4 presents the schematic of power conversion path to power up SSI, including analog and digital modules. A control clock is used to control the switch with a current limiter. The clock has amplitude of 2.5 V which stands for the

state in the field bus during its idle time. This clock should be synchronized with the internal clock of the data line and the switch is enforced to get open from the middle to the end of each message period T, during which the power harvesting process is performed. Then the control clock should have the same period, T. When T is 2 ms, which is currently practiced in our developed AFDX network, the transient simulation result of voltages is shown in Fig. 5, in which the ripple of V_{in} is defined and settling time, t_s, is illustrated. The width of the reset signal should be larger than this t_s (10 ms) in order that the chip would perform properly.

The SSI modules, developed in our laboratory, have critical requirement on supplied voltages to work desirably, so the outputs of the power conversion path should be very accurate. In Fig. 4, the regulator is designed to decrease the output voltage ripple to 1/5 in amplitude compared to that of the input one. The distributor will continue to reduce its input voltage ripple to 1/2 when the input voltage has a ripple of less than 0.04 V. Consequently, the output voltage ripple would be maximally 0.02 V which meets the requirement of our designed SSI module.

LC oscillation peak appears in the V_{out1} as presented in Fig. 5 due to the insertion of π–type isolation block. The output voltage oscillates because the regulator is set that the output signal is zero when its input voltage is below 2.0 V. In the stable period, this isolation block helps to maintain a desired output for the analog load in order to isolate from the digital loads which may generate glitches.

B. Influnces of frequency and reservoir

Different data transmission frequencies are considered and various capacitor values of reservoir are chosen to investigate their influences on the performances of the power conversion path. As mentioned previously, the message period is designated as T, then the frequency is f = 1/T. Table 1 summarizes the simulation results of the power conversion path in terms of settling time and ripple voltage. Generally, the settling time, t_s, will arise with the increase of reservoir volume meanwhile the voltage ripple, V_{ripple}, on the output decreases. These are expected as the larger capacitor, with the given load, contributes in longer time constant. A special case is the settling time at 2.0 kHz when capacitor reservoir is 20 µF. This value is large because the ripple of V_2 is somehow big while our designed distributor is set to response properly only to the input voltage with a ripple of less than 0.04 V.

Moreover, the voltage ripple, V_{ripple}, on the output reduces with the increase of frequency while the settling time, t_s, keeps almost the same. This is predictable as increases in input frequency contributed in more frequent charging of the reservoir and reducing the ripples accordingly. At a specific frequency, the system could not generate a proper output when the capacitor reservoir is too small. Thus, an off-chip large capacitor needs to be considered to store more energy and maintain the power recovery unit more useful.

TABLE 1. SIMULATION RESULTS OF POWER CONVERSION PATH

T (ms)	f (kHz)	C (µF)	t_s (ms)	V_{ripple} (mV)
2.0	0.5	20	-	-
		40	-	-
		80	11.5	9.5
		160	20.0	6.4
1.0	1.0	20	-	-
		40	5.9	12.5
		80	9.6	6.2
		160	22.0	3.2
0.5	2.0	20	17.5	13.0
		40	5.5	6.1
		80	10.8	3.0
		160	20.4	2.1

IV. CONCLUSIONS

A novel approach of power recovery from data line is proposed which tackles the problems associated with bulky wiring issue in aircrafts. The technique aims at harvesting power from field bus through a designed power conversion path. All modules of power recovery unit are modeled in Verilog-A, making it easy to replace them with real designs as needed and accelerates simulation time. The effects of frequency and capacitor reservoir on settling time and output voltage ripple are studied, which gives instructions for the optimization in future design. Simulation results prove that the presented method helps to build a self-powered SSI module and it offers the potential to significantly reduce the mass of the cables installed in aircrafts. Future work will include powering up SSI module as well as the position sensors, so we need to improve the power conversion

efficiency of the proposed power recovery unit, and the power management block should be considered when the power budget is not enough to support all loads. These modules will be eventually realized using CMOS 130 nm technology. An off-chip large capacitor (80 µF) is also considered to store more energy and maintain the power recovery unit more useful.

ACKNOWLEDGMENT

The authors would like to thank Natural Sciences and Engineering Research Council of Canada (NSERC), Bombardier aerospace, Thales Canada Inc., MITACS, Consortium for Research and Innovation in Aerospace in Quebec (CRIAQ), CMC Microsystems, and Polystim neurotechnologies for financial support and design tools.

REFERENCES

[1] J. H. Liu, "Communication schemes for aerospace wireless sensors", IEEE/AIAA 27th Digital Avionics Systems Conference, pp. 5.D.4-1 - 5.D.4-9, 2008.

[2] K. Kilani, V. Degardin, P. Laly and M. Lienard, "Transmission on aircraft power line between an inverter and a motor: impulsive noise characterization", IEEE International Symposium on Power Line Communication and Its Applications, pp. 301-304, 2011.

[3] C. H. Jones, "Communication over aircraft power lines", IEEE International Symposium on Power Line Communication and Its Applications, pp. 149-154, 2006.

[4] B. Zhang, L. Zhang, K. Liu and Y. Ma, "Aircraft cockpit distributed simulation based on embedded Ethernet and PoE", Applied Mechanics and Materials, Vol. 58-60 (2011), pp. 1402-1407.

[5] C. Heller, E. Heidinger, S. Schneele et al., "Power-over-Ethernet for avionic networks", IEEE/AIAA 29th Digital Avionics Systems Conference, pp. 5.A.2-1-5.A. 2-11, 2010.

[6] http://www.eetimes.com/design/industrial-control/4218852/What-is-Power-Line-Communication-?pageNumber=2

[7] "664P2-2 Aircraft Data Network, Part 2-Etehrnet Physical and Data Link Layer Specification", Aeronautical Radio Incorporated (ARINC) 2009, Annapolis, MD.

[8] "General Standardization of CAN (Controller Area Network) Bus Protocol for Airborne Use ARINC SPECIFICATION 825-2", published by AERONAUTICAL RADIO. INC., pp. 54-56, 2011.

[9] Y. Hong, C. F. Chan, J. Guo et al,. "Design of passive UHF RFID tag in 130nm CMOS technology", IEEE Asia Pacific Conference on Circuits and Systems, pp. 1371-1374, 2008.

[10] S. Hashemi, "Characterization of integrated MOS circuits under voltage stress and application to power conversion chains of electronic implants", Master Thesis of Ecole Polytechnique Montreal, pp. 77, 2004.

Control of Chaotic Behaviour in Buck-Boost DC-DC Converters

A.N. Natsheh*, J.G. Kettleborough**

*Faculty of Engineering, Al-Ahliyya Amman University, Post Code 19328 Amman, Jordan
** Department of Electronic and Electrical Engineering
Loughborough University, Loughborough, Leicestershire, LE11 3TU, UK
ammar_natsheh@yahoo.com

Abstract -- **Chaos control *is* used to design a controller that is able to eliminate the chaotic behaviour of nonlinear systems that experience such phenomena. The paper describes the control of the bifurcation behaviour of a DC-DC buck-boost converter used to provide an interface between energy storage batteries and photovoltaic (PV) arrays for renewable energy sources. The paper presents a delayed feedback control scheme in a peak current-mode controlled DC-DC buck-boost converter operating in the continuous current conduction mode. MATLAB/SIMULINK simulation results show the effectiveness and robustness of the scheme.**

I. INTRODUCTION

DC-DC converters contain a switch, inductor, diode and capacitor. The buck converter (step down) reduces the output voltage from the input voltage and the boost converter (step up) increases the output voltage. A Buck-Boost converter is capable of increasing, maintaining or decreasing the output voltage with respect to the input. Buck-Boost DC-DC switch-mode converters are non-linear devices and recently it has been observed that they may behave in a chaotic manner, through period doubling. A survey of the literature established considerable investigations into chaos in buck and boost converters [1-8], but little relating to the buck-boost converter.

DC-DC converters are common and they deliver power at different voltage levels. They are used in many applications, such as renewable energy systems. In reality the converter input might be a photovoltaic (PV) array that charges a battery, the load. Fixed switching rate buck and boost converters have all been shown to exhibit chaotic behavior within certain operating conditions [1-8]. Converters are usually designed so that in its normal operating region it functions in a stable manner. Outside of these conditions the converter can experience period doubling bifurcations (forking) that can develop into chaos (unpredictability). When chaotic behavior occurs, it can cause additional losses, noise and even destroy the electronics. Chaotic behaviour is therefore extremely undesirable and a requirement exists to understand this behaviour and control it.

This paper investigates chaotic behaviour in a buck-boost converter operating under a peak current-mode control scheme and then investigates the control of chaotic behaviour using a delayed feedback control strategy, which is based on the idea of stabilization of unstable periodic orbits that exist in the chaotic attractor [9].

II. BIFURCATION and CHAOS THEORY

Bifurcation theory, originally developed by Poincare, is used to indicate the qualitative change in behavior of a system in terms of the number and the type of solutions, under the variation of one or more parameters on which the system depends.

Bifurcation is a route to chaos for many non-linear systems. The chaos can appear quickly or after several bifurcations, depending on the system and its parameters. Bifurcations are when the expected result forks and has two possibilities. Through period doubling the expected result can have several possibilities.

In bifurcation problems, in addition to state variables, there are control parameters. The relationship between any of these control parameters and any state variable is called the state-control space. In this space, locations at which bifurcations occur are called bifurcation points. Bifurcations of an equilibrium or fixed-point solution are classified as either static bifurcations, such as saddle-node, pitch fork, or transcritical bifurcations; or as dynamic bifurcations which are also known as the Hopf bifurcation that exhibits periodic solutions. For the fixed-point solutions, the local stability of the system is determined from the eigenvalues of the Jacobian matrix of the linearized system. On the other hand, with periodic-solutions, the system stability depends on what is known as the Floquet theory and the eigenvalues of the Monodromy matrix that are known in the literature as Floquet or characteristic multipliers. The types of bifurcation are determined from the manner in which the Floquet multipliers leave the unit circle.

Chaos is the unpredictability in a system. The theory was first developed in 1960 by Lorenz. It is possible to experience chaos in many systems and it can be investigated in converters by adjusting the converter pulse width modulation (PWM). Chaos is an undesirable effect and can cause additional losses, noise, other unwanted outputs and possibly the catastrophic failure of the converter or its connected electronics.

III. MATHEMATICAL MODELLING AND MATLAB/SIMILINK SIMULATION

A simplified current-mode controlled buck-boost converter is shown in Fig. 1, with a feedback path comprising a comparator and a flip-flop. The comparator compares the inductor current with a reference value, to control the state of switch S. When the switch is closed, the diode is reverse biased and the input provides energy to the inductor. When the switch is open, the inductor's energy is transferred to the output. During this interval, no energy is supplied by the input.

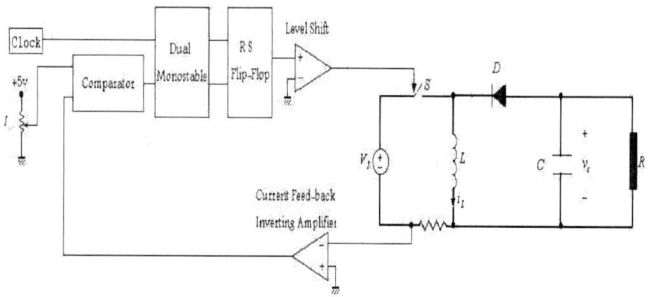

Fig. 1. Current-mode controlled DC-DC buck-boost converter circuit

Based on the parameters of Table 1 and according to the block diagram of Fig. 1, a MATLAB/SIMULINK model was developed based on the system equations:

$$\frac{di_L}{dt} = \frac{1}{L}[v_{in} - v_C(u_s - 1)]$$

$$\frac{dv_C}{dt} = \frac{1}{C}[-\frac{v_C}{R} + i_L(1 - u_s)]$$

(1)

Where u_s takes the value 0 or 1 depending on whether the switches 1 or 2 are closed or open. The structure of the SIMULINK model is given in Fig. 2.

Circuit Component	Value
Switching period	μs100
Input voltage V_i	V12
Inductance L	mH0.94
Capacitance C	μF470
Load resistance R	Ω22

Table 1 Circuit Parameters

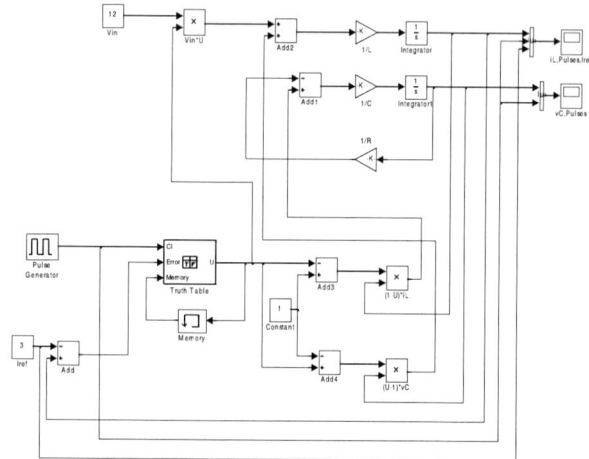

Fig. 2. MATLAB/SIMULINK Model for a Buck-Boost Converter

IV. OPEN LOOP PERFORMANCE ANALYSIS

The buck-boost converter experiences bifurcations that are seen through period doubling and quadrupling of the inductor current waveform. Period-1 operating waveforms are shown in Figs. 3 and 4.

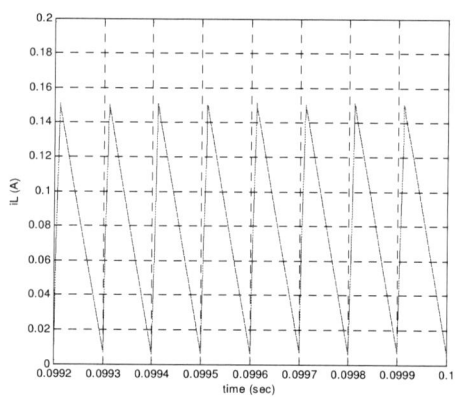

Fig. 3. "Period-1" regime at $I_{ref} = 0.15A$

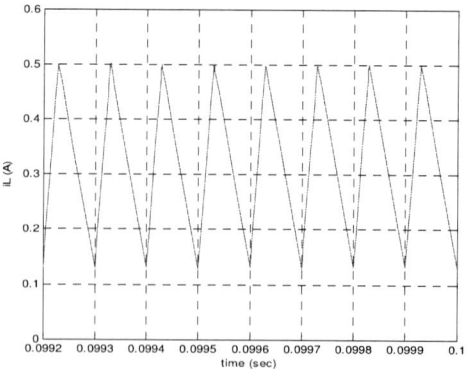

Fig. 4. "Period-1" regime at $I_{ref} = 0.5A$

An unstable period-2 waveform is shown in Fig. 5. It is noticed that the circuit misses a reset as the inductor current is still increasing linearly to I_{ref}. At I_{ref} the switch is reset (turned-off) and the inductor then starts to transfer its energy to the load. Before it can transfer all its energy the clock triggers the switch and the inductor current starts to rise again. On reaching I_{ref} the inductor current is able to transfer all of its energy to the load before the clock turns the circuit on. This process occurs in all the waveforms shown but will not be explained any further. A Fourier transform of the waveform would show the dominant frequency to be 5kHz not the 10kHz required. As I_{ref} is increased it causes the converter to bifurcate more and this rapidly progresses into chaos when the circuit demand is high.

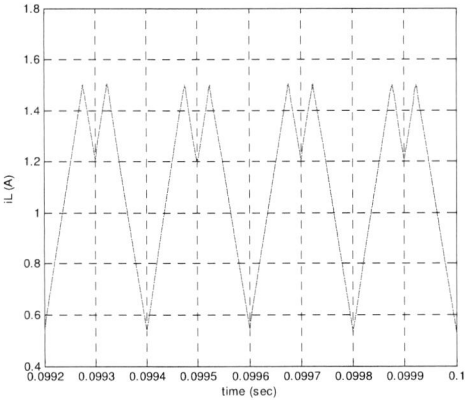

Fig. 5. "Period-2" unstable regime at $I_{ref} = 1.5A$

Fig. 6 shows the stable operation of the converter in period-2. The demand is 1.8A and the inductor waveform, due to period doubling, is now 5kHz. This mode of operation is stable but could be dangerous in some situations.

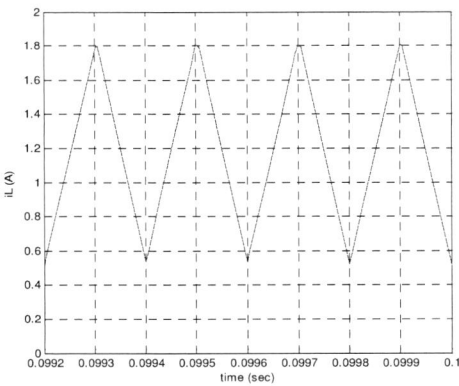

Fig. 6. "Period-2" stable regime at $I_{ref} = 1.8A$

Fig. 7 shows the converter operating in unstable period-8. The demand is 2.2A and the period has multiplied by 8. The converter does not yet exhibit chaotic behaviour.

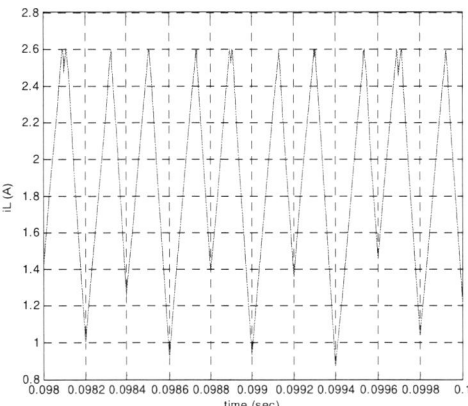

Fig. 7. "Period-8" unstable regime at $I_{ref} = 2.6A$

In Fig. 8 the converter is chaotic. The converter becomes chaotic at $I_{ref} >= 3A$ with a 22 ohm load resistance and an input voltage of 12V.

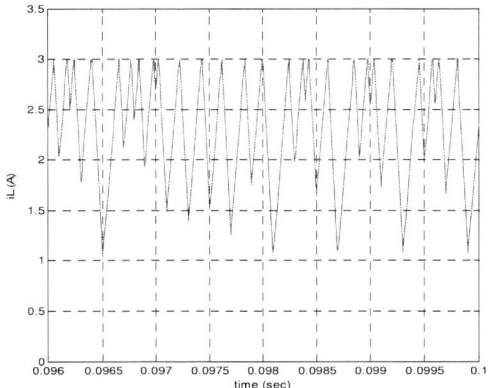

Fig. 8. "Chaotic" regime at $I_{ref} = 3A$

Chaos that occurs in a system can have dramatic effects. It can reduce the efficiency, causes the system to be unreliable, be the source of a failing system or produce unexpected dangerous events in linked systems. Load changes, noise and increasing the switching frequency can all cause the PWM controlled converter to exhibit bifurcations and ultimately chaos. Some converters, depending on component selection, produce chaos on bifurcation from the period-2 region. In this paper the converter only produced chaotic behaviour after the period-8 region.

V. CONTROL OF CHAOS

Pyragas [9] has suggested that chaotic behaviour may be eliminated from the system if one applies the delayed feedback control scheme shown in Fig. 9. The feedback control force *F(t)*, applied to the system is the difference between the current value of some system variable *y(t)*, and its value τ seconds previously, multiplied by a constant *K*, where K is the feedback strength. The idea behind the scheme relies on the fact that a skeleton of a chaotic attractor is formed by an infinite set of unstable periodic orbits with different periods. If the value of the time delay τ is exactly equal to the period T of one of the orbits, then at the appropriate values of K, the orbit can become stable and chaos will be eliminated. In simple terms, an unstable periodic orbits embedded in the chaotic attractor is selected then control is used to fix this orbit through small changes in the parameters. When the chaos passes sufficiently close to the orbit it is made linear. As the controller *is* connected to the open-loop model, the inductor current at switch turn-on now does not have many values and the periodic regime is stable. Once control is achieved, i.e. the phase trajectory reaches the periodic orbit; the control force $F(t)$ is zero at any instant. This is called non-invasive control and implies that virtually no power is spent in the control loop to support the desired behaviour of the system.

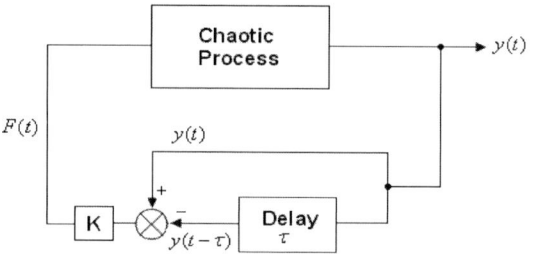

Fig. 9. Delayed current feedback control system

The closed-loop performance of a converter with the specification of Table 1 was simulated using a MATLAB/SIMULINK model. The structure of the SIMULINK model is given in Fig. 10. The simulations were obtained by solving numerically Eqns. (1), setting u_S to 1 at every clock pulse and u_S to 0 when:

$$i_L(t) = I_{ref} - K[i_L(t-\tau) - i_L(t)] \tag{2}$$

It is clear that the periodic regime observed is stable. The waveforms of Fig. 11 have the desired frequency of 10 kHz ("period-1" cycles) at $I_{ref} = 3A$.

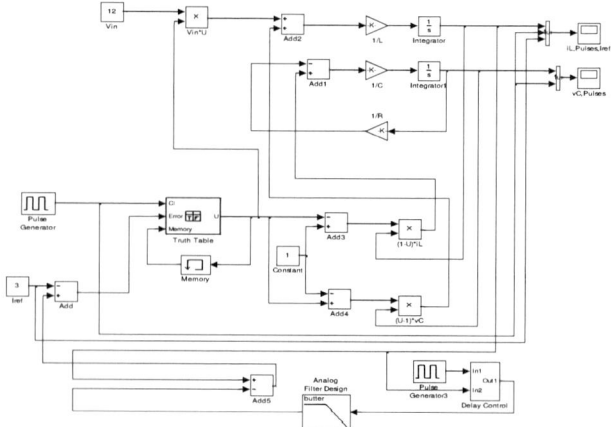

Fig. 10. MATLAB/SIMULINK Model for a Buck-Boost Converter with a Feedback Control Scheme

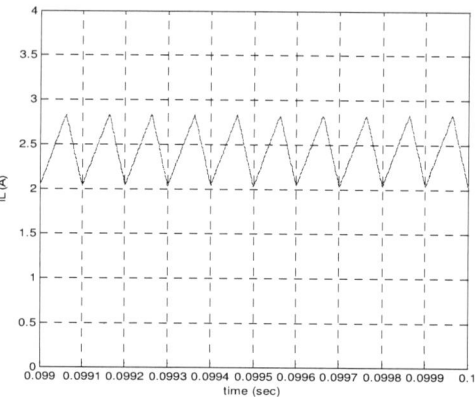

Fig. 11. "Controlled Chaotic" regime at $I_{ref} = 3A$ using delayed feedback controller

CONCLUSION

The simulated time waveforms for $i_L(t)$ at $I_{ref} = 3A$ with the converter connected to the delayed current feedback (the control circuit) showed the converter is operating in a "period-1" mode as required. The study shows the effectiveness of the designed delayed feedback control mode. The results indicated that delayed feedback control mode results in a greater range of stable period-1 operation, compared with the traditional method.

REFERENCES

1. Deane, J. H. B. and Hamill, D. C., "Instability, Subharmonic and Chaos in Power Electronic Circuits," IEEE Transactions on Power Electronics, Vol. 5, 1990, pp. 260-268.
2. Deane, J. H. B., "Chaos in a Current-Mode Controlled Boost DC-DC Converter," IEEE Transactions on Circuits and systems-1: Fundamental Theory and Applications, Vol. 39. No. 8, August 1992, pp. 680-683.
3. Hamill, D. C., Deane, J. H. B., and Jefferies D. J., "Modeling of Chaotic DC-DC Converters by Iterated Nonlinear Mapping," IEEE Transactions on Power Electronics, Vol. 7, No. 1, 1992, pp. 25 – 36.
4. Tse, C. K.," Flip Bifurcation and Chaos in Three-State Boost Switching Regulators," IEEE Transactions on Circuits and systems-1: Fundamental Theory and Applications, Vol. 41, No. 1, January 1994, pp. 16-21.
5. Tse, C. K., and Chan W. C. Y., "Chaos from a current programmed Cuk converter, " Int. J. Circuit Theory Applications, Vol. 23, May-June 1995, 217-225.
6. Marrero, J. L. R., Font, J. M., and Verghese, G. C., "Analysis of the chaotic regime for DC-DC converters under current-mode control," Power Electronics Specialists Conference, PESC 1996, pp. 1477-1483.
7. Banerjee, S., "Nonlinear Modeling and Bifurcation in Boost Converter," IEEE Transactions on Power Electronics, Vol. 13, No. 2, 1998, pp. 253 – 260.
8. Natsheh, Ammar, N., Janson, Natalia B., and Kettleborough, J. Gordon; "Control of Chaos in a DC-DC Boost Converter" *ISIE 2008. IEEE International Symposium on Industrial Electronics, 2008, Cambridge University-UK, June 30 2008-July 2 2008 Pages: 317 – 322.*
9. K. Pyragas, Continuous control of chaos by self-controlling feedback, *Physics Letters, 170,* 1992, 421-428.

978-1-4673-5289-5/12 $31.00 © 2012 IEEE

Modeling of Op-Amp Nonlinearities in Pipelined ADC

[1]Samir Barra, [3]Souhil Kouda, [1,2]Abdelghani Dendouga, and [1]Nour-Eddine Bouguechal

[1]Advanced Electronics Laboratory -LEA
Department of Electronics
University of Batna
Chahid Boukhlouf M. El-Hadi Avenue
Batna, Algeria
barrasamir@hotmail.com

[2]Microelectronics and Nanotechnology Division
Center for Development of Advanced
Technologies - (CDTA)
August 20 1956 City, BP 17, Baba Hassen
Algiers, Algeria

[3]Department of Electronics
University of M'sila
B.P 166 Ichbelia
M'sila, Algeria

Abstract—in this paper, we present a behavioral SIMULINK model for the simulation of operational amplifiers (op-amp). The model includes most of the nonlinearities which affect the performance of these circuits, such parameters (noise, finite gain, finite bandwidth, and slew-rate).With the proposed models it is possible to accurately predict with fast simulations the linearity of sample and hold (S/H) circuit and Multiply digital-to-analog (MDAC) circuit that constructed the pipelined ADC

Keywords— behavioral modeling, nonlinearities, Operational amplifier circuit, S/H, MDAC, Pipelined ADCs

I. INTRODUCTION

Analog-to-Digital Converters (ADCs) transfer the analog information to digital data. ADC is a bridge from the real world to the digital world. As the digital technique develops, many digital applications such as PDA, cell phone with camera, Digital TV, 4G communication and WLAN etc., are very popular, in which ADC plays an important role of converting analog audio, image and video signal into digital processing. Pipelined ADC is the most popular ADC architecture for sampling rates from a few MSPS up to 100MSPS, with resolutions from 8 to 14 bits. The operational amplifier (op-amp) is an essential analog building block of Pipelined ADCs [1-2-3-4]. It often limits the performance- of pipelined ADC, the gain of the op-amp will limit the accuracy and the bandwidth of the op-amp will limit the speed of the pipelined ADC. The design of the op-amp becomes increasingly more difficult as device dimensions and the supply voltage Scale down. The modeling of the op-amp for understanding the functioning and predicting the source of error that affects performance of MDAC and S/H. Behavioral description can be achieved by the use of different high level languages like MATLAB SIMULINK [5-6-7-8-9], VHDL-AMS [10-11], and Spice [12]. This paper introduces a methodology for modeling the non-linearity of op-amp in pipelined ADC environment.

II. PIPELINE ADC

The pipeline ADC is constructed by using Switched Capacitor (SC) circuit, which exploits the charge storing abilities of CMOS to achieve precision in signal processing and is preferred in mixed signal A/D interfaces [6]. The conceptual block diagram of a generic pipeline ADC is consisting of an arbitrary cascade of k stages and a Sample- and-hold (S/H) circuit at the front [7]. Each stage re- solves partial code words of length n_i, i=1,..., k , which are all re-ordered and combined at the digital correction block to obtain the output of the converter. The inner structure of a pipeline stage comprises

four blocks, as illustrated in Fig.1. (b): a flash sub-ADC with $N_i \leq 2^{n_i}$, output codes, and a sub-DAC with N_i output levels, a substractor, and an S/H residue amplifier with gain G_i. The latter three blocks are implemented in practice by a single subcircuit, which is often referred to as Multiplying DAC (MDAC).

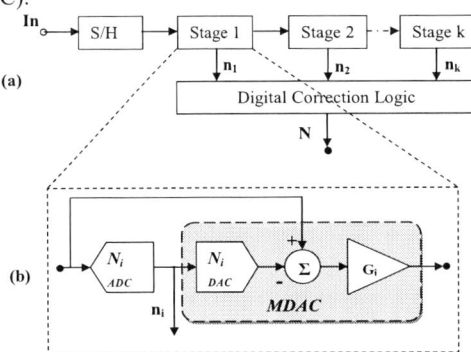

Fig. 1. Block diagram of pipelined ADC (a), and conceptual block diagram structure of a single stage (b).

A. Sample-and-hold (S/H) circuit

The function of the S/H circuit is to track/sample the analog input signal and to hold that value while subsequent circuitry digitizes it. The basic operations include sampling the signal on the sampling capacitor(s) and transferring the signal charge onto the feedback capacitor by using an op amp in the feedback configuration. For simplicity, single-ended configurations are chosen Fig. 2 (a) [13]. The output voltage of the S/H is given by:

$$V_{out} = 1.V_{in} \qquad (1)$$

The ratio of the feedback capacitor to the total capacitance at the summing node can be much closer to one ($C_s/C_f=1$).

B. Multiply digital-to-analog converter circuit

The multiplying digital to analog converter (MDAC), is a single switched capacitor circuit that can be implemented the function of S/H operation, D/A conversion, subtraction, and amplification of the remainder. With the charge conservation concept, the output in hold phase is given by [14]:

$$V_{out} = \left(\frac{C_f + C_s}{C_f} \right).V_{in} - \left(\frac{C_s}{C_f} \right).V_{DAC} \qquad (2)$$

978-1-4673-5289-5/12 $31.00 © 2012 IEEE

where C_s is the sampling capacitor, C_f is the feedback capacitor, and V_{DAC} is the output voltage of the DAC circuit in the MDAC circuit.

The MDAC circuit in the 1.5-bit/stage architecture is very simple as shown in Fig. 2 (b).

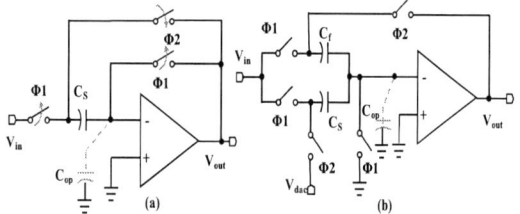

Fig. 2. Block diagram of S/H (a), Block diagram of MDAC (b), where Φ1 is the sampling phase, and Φ2 is the hold phase.

III. THE NON-IDEALITIES OF OP-AMP

In this section, main nonlinearities of the operational amplifier (op-amp) and their effects in the pipelined ADC are introduced.

A. Non-linear op-amp DC gain

Ideally, the DC gain of the op-amp is infinite, theoretically its transfer function is:

$$V_{out} \cong A.V_{in} \qquad (3)$$

A is the amplification factor.

In practice, the actual gain is limited by circuit constraints and in particular by the operational amplifier open-loop gain A0, that the gain of the operational amplifier depends on its output, in such a way that the transfer function is approximated given by (4) [15]:

$$A \cong A_0 \left(1 + \alpha_1 |V_0| + \alpha_2 |V_0|^2 + \alpha_3 |V_0|^3 + \ldots\right) \qquad (4)$$

V_0 represents the SC am-op output, and (α_1, α_2, α_3...) are the amplification factors parasites.

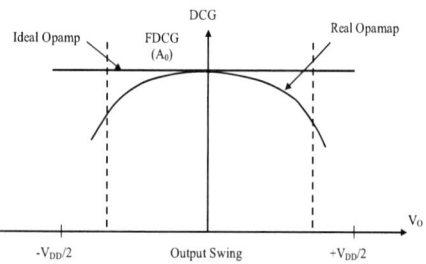

Fig. 3. DC gain (DCG) of an amplifier as function of output voltage.

The nonlinearity of DC gain of the op-amp resulting from its dependency on the output voltage as shown in Fig. 3. The actual gain is usually finite and nonlinear; which induces nonlinearities into the pipelined ADC. The transfer function of the SC circuit with the finite DC gain effect of op-amp becomes [13]:

$$V_{out} = G.V_{in} \left(\dfrac{1}{1 + \dfrac{1}{A.\beta}} \right) \qquad (5)$$

where G is the ideal gain of SC circuits, and the term in parentheses represents the finite gain error of SC circuits, A is the DC gain of the op-amp and β represents the feedback factor.

B. Bandwidth and slew rate

The effect of the finite bandwidth (GBW) and the slew-rate (SR) are related to each other and may be interpreted as a non-linear gain [16].

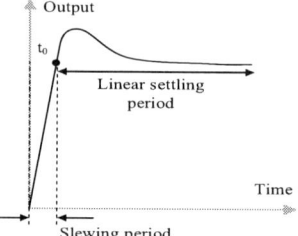

Fig. 4. Typical settling behavioral of the step response.

During the initial settling period, the slope of the output signal is probably limited by the SR of op-amp, which proceeds the nonlinear settling. Once the slope of output signal is less than the SR, The response goes into the linear response. The transition timing point between them is at t_0. We add the settling error behavior into (3). This response is given [17]:

$$V_{out} = G.V_{in} \left(\dfrac{1}{1 + \dfrac{1}{A.\beta}} \right) \left(1 - e^{\frac{-t}{\tau}}\right) \qquad (6)$$

The exponential term in the second bracket represents the settling error of the single pole op-amp, and $\tau = 1/(2\pi.\beta.GBW)$ is the time constant of the SC circuit, where GBW and β are respectively the unity gain frequency and feedback factor of the SC circuit. The slope of this curve reaches its maximum value when t=0, resulting in [17]:

$$\frac{d}{dt} v_{out}(t) \Big|_{\substack{max \\ t=0}} = G.V_{in} \left(\dfrac{1}{1 + \dfrac{1}{A.\beta}} \right) / \tau \qquad (7)$$

We must consider now two separate cases [18]:

1-The value specified by (7) is lower than the operational amplifier slew-rate, SR. In this case, no SR limitation appears and the evolution of Vout is described by (6).

2-The value specified by (7) is larger than SR. In this case, the operational amplifier is in slewing. Therefore, the first part of the transient response of V_{out} is limited by SR ($t<t_0$), and the other part recovers to the linear settling behavior ($t>t_0$). The following equations hold (assuming $t_0 <T_s/2$):

978-1-4673-5289-5/12 $31.00 © 2012 IEEE   74

$$t < t_0 \quad V_{out} = SR.t \tag{8}$$

$$t > t_0 \quad V_{out} = V_{out}(t_0) + (V_{in}G - SR.t_0).\left(1 - e^{-\frac{t-t_0}{\tau}}\right) \tag{9}$$

Imposing the condition for the continuity of the derivatives of (8) and (9) in t_0, we obtain:

$$t_0 = \frac{V_{in}.G}{SR} - \tau \tag{10}$$

C. Op-amp thermal noise

The calculation of op-amp thermal noise is dependent on the op-amp architectures. We first assume that the noise sources only include the noises produced by input transistors in the differential pair. Fig.5 is the small signal model for op-amp used for noise calculation [18].

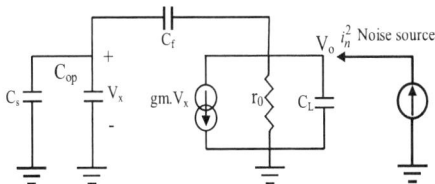

Fig. 5. Small signal model for op-amp.

The transfer function of this model is [18]:

$$H(s) = \frac{V_0}{\overline{i_n}} = \frac{r_0}{(1 + gm.r_0.f).\left(1 + \frac{s.C_{LT}.r_0}{1 + gm.r_0.f}\right)} \tag{11}$$

where I_n is the noise current source which can be seen at the most right side from Fig.5, $C_{LT}=C_L+ \beta(C_S+C_{op})$, is the total load capacitance, β is the feedback factor, C_L is load capacitor, C_{op} is the input capacitor of the op-amp, gm is transconductance of MOSFET transistor. The noise current source can be written as:

$$\overline{i_n}^2 = \gamma.4KT.gm.\Delta f \tag{12}$$

where γ is coefficient equal to 2/3 for long channel transistors and it can be higher than unity in submicron MOSFET, γ is approximately 1 in 0.18μm technology.

So the input-referred noise power can be expressed as:

$$\overline{V_{in}^2} = \frac{\overline{V_0^2}}{G^2} = \frac{\int_0^\infty \left(\left|H(s|_{j\omega})\right|^2.\overline{i_n}^2\right)}{G^2} = \frac{2}{3}.KT.\frac{1}{\beta}.\frac{1}{C_{LT}}.\left(\frac{C_F}{C_S + C_F}\right) \tag{13}$$

where G is the gain of MDAC.
Then the input-referred noise of MDAC can be written as:

$$\sigma_{mdac}^2 = \frac{2}{3}.KT.\frac{1}{\beta}.\frac{1}{C_{LT}}.\left(\frac{C_F}{C_S + C_F}\right) \tag{14}$$

where $\beta = C_S/(2^B.C_S+C_{op})$, and $C_{LT} = CL+ \beta (C_S+C_{op})$, and B is the stage resolution.

The op-amps in S/H and MDAC are the same. The input-referred noise of S/H can be written as:

$$\sigma_{s/h}^2 = \frac{2}{3}.KT.\frac{1}{\beta}.\frac{1}{C_{LT}} \tag{15}$$

where $\beta = C_f/(C_f+C_{op})$

D. Flicker noise

Flicker noise or 1/f noise is present due to trapping and de-trapping effects at the silicon-oxide interface. Since flicker noise is inversely proportional to the transistor gate area, this noise component typically increases with technology scaling [19]. The noise voltage can be roughly expressed as:

$$V^2 = \frac{K_f}{C_{ox}.WL}.\frac{1}{f} \tag{1}$$

where K_f and C_{ox} are technology parameters, and W and L are the transistor dimensions.

E. Op-amp input offset

There are two basic forms of offset, which have different effects on the ADC transfer. Firstly, there is input offset, which adds up with the input signal to the stage. This offset is due mainly to the amplifier and to a lesser extent to the switches [20]. The transfer function in this case is of the form:

$$V_{out_i} = G_i\left(V_{in_i} + V_{off_i}\right) - D_i.V_{ref} \tag{1}$$

where the offset gets multiplied up by the stage gain.

The second form of offset is that due to the comparators. This has the effect of shifting either one or both of the decision levels of the sub-ADC. The total offset from all sources must remain within the bounds of $\pm V_{ref}/4$ [20].

IV. SIMULATION RESULTS

To validate the proposed models of the various nonidealities affecting the operation of op-amp in pipelined ADC, we performed several simulations with SIMULINK on the ideal and the non-ideal model of pipelined. The design of the model was validated by reconstructing the digital output into its original form, the ideal and non-ideal model was tested with sinusoidal waveform inputs.

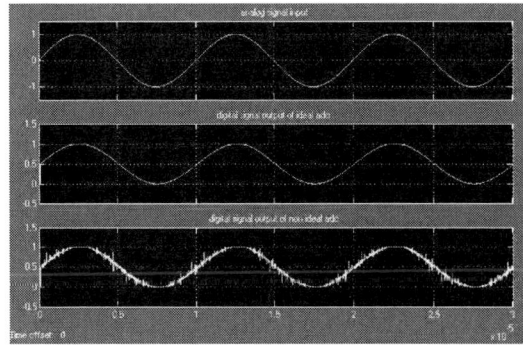

Fig. 6. Analog input and reconstructed outputs.

The original input and the two digital output waveform reconstructed are shown in Fig. 6. As we expected, the significant noise sources make an impact on the output signal of non-ideal pipelined ADC. Fig. 7 show the FFT output for a non-ideal model of pipelined ADC.

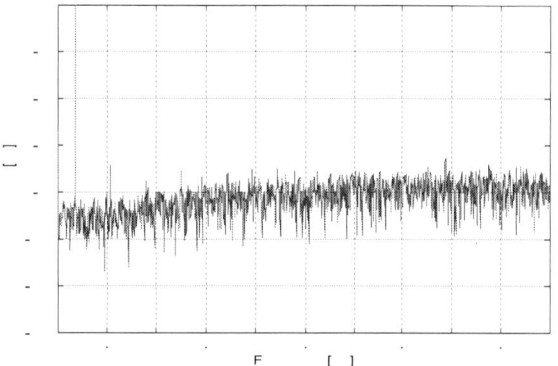

Fig. 7. FFT of the ADC output showing the effect of errors on the fundamental component and on other frequencies with F_{in}=1,78MHz, F_s=100MHz.

We assessed the potential possibilities of the proposed approach and the SIMULINK model comparing the characteristics of the modeled ADC with real ADCs in the other work such as [1]. The comparison shows a good agreement between the behaviorally modeled ADC and the real ADC [1]. Table I summarizes the simulated ADC performance.

TABLE I.   PERFORMANCE COMPARISON.

Parameters	This work Ideal Model	This work Non-ideal Model	[1]
INL (LSB)	-0,3/+0,3	-0.3/+0.3	-0,6/+0,8
DNL (LSB)	-0,3/+0,2	-0.4/+0.2	-1/+0,7
SNR(dB)	66,86	61,52	/
SFDR(dB)	62,95	57,52	65
SNDR(dB)	66,87	61,55	54
Resolution (bit)	10	10	10
Technology	0.18μm	0.18μm	0.18μm

V.   CONCLUSION

In this work we have proposed of models of two key building blocks of pipelined ADC (MDAC and S/H), the model of both circuit takes into account at the behavioral level most of the op-amp nonidealities, such as white noise, finite DC-gain, finite BW, and SR. The model proposed of pipelined ADC with nonidealities has been validated by comparing the results obtained of the non-ideal model and the ideal model of pipelined ADC with real pipelined ADCs.

REFERENCES

[1] Kickback Naga Sasidhar, Youn-Jae Kook, Seiji Takeuchi, Koichi Hamashita, Kaoru Takasuka, Pavan Kumar Hanumolu, and Un-Ku Moon. "Low Power Pipelined ADC Using Capacitor and Opamp Sharing Technique With a Scheme to Cancel the Effect of Signal". IEEE journal of solid-state circuits, vol. 44, no. 9, pp. 2392-2401, Sep. 2009.

[2] S.-W. Sin Seng-Pan U, and R.P. Martins, "1.2-V, 10-bit, 60–360 MS/s time-interleaved pipelined analog-to-digital converter in 0.18 mm CMOS with minimised supply headroom", IET Circuits Devices Syst., Vol. 4, Iss. 1, pp. 1-3, 2010.

[3] Charles T. Peach, Un-Ku Moon, and David J. Allstot, "An 11.1 mW 42 MS/s 10 b ADC With Two-Step Settling in 0.18 μm CMOS", IEEE Journal of solid-state circuits, Vol. 45, No. 2,pp. 391-400, Feb. 2010.

[4] Jin-Fu Lin, Student Member, Soon-Jyh Chang, Chun-Cheng Liu, and Chih-Hao Huang, "A 10-bit 60-MS/s Low-Power Pipelined ADC With Split-Capacitor CDS Technique", IEEE transactions on circuits and systems, vol. 57, no. 3, pp.163-167, Mar 2010.

[5] Erkan Bilhan, Pedro C. Estrada -Gutierrez, Ari Y. Valero-Lopez, and Franco Maloberti, "Behavioral Model of Pipeline ADC by using SIMULINK", In the Proc. of the Southwest Symposium on Mixed-Signal Design, SSMSD, Feb. 2001, pp.147-151.

[6] M.Ramalatha, A.P.Karthick, and S.Karthick, "A High Speed 12-Bit Pipelined ADC Using Switched Capacitor and Fat Tree Encoder". In the Proc. of the International Conference on Advanced in Computational Tools for Engineering Applications, ACTEA, Jul. 2009, pp.391-395.

[7] Jesús Ruiz-Amaya, and José M. de la Rosa, Manuel Delgado-Restituto and Angel Rodríguez-Vázquez "Behavioral Modeling, Simulation and High-Level Synthesis of Pipeline A/D converters", In the Proc. of IEEE international symposium on circuits and systems", ISCAS, May. 2005, vol. 6, pp.5609 -5612.

[8] E. Mancini, S. Rapuano, and D. Dallet, "A Distributed Test System for Pipelined ADCs", Measurement, vol. 42, Issue 1, pp.38–43, Jan.2009.

[9] Samer Medawar, Peter Händel, Senior, Niclas Björsell, and Magnus Jansson," Input-Dependent Integral Nonlinearity Modeling for Pipelined Analog-Digital Converters", IEEE Transactions on Instrumentation and Measurement, vol. 59, Iss. 10, pp: 2609 – 2620, Oct. 2010.

[10] Antonio J. Acosta, Eduardo J. Peral´ıas, Adoracio´n Rueda, and Jose´ L. Huertas, "VHDL Behavioural Modelling of Pipeline Analog To Digital Converters", Measurement, Vol.31, Iss. 1, pp 47–60, Jan. 2002.

[11] J. A. Diaz-Madrid , G. Doménech-Asensi, J. A. Lpez-Alcantud, and H. Neubauer, "VHDL-AMS Model of a 40m/S 12 bits Pipeline ADC", in the Proc. of the International Conference on Mixed Design of Integrated Circuits and Systems, MIXDES, Jun. 2006, pp.555-560.

[12] Grzegorz Zareba, and Olgierd A. Palusinski, "Behavioral Simulator of Analog-to-Digital Converters for Telecommunication Applications", In the Proc. Of IEEE International Conference on Behavioral Modeling and Simulation, BMAS, Oct. 2004, pp.135-140.

[13] Jin-Fu Lin, Soon-Jyh Chang, Te-Chieh Kung, Hsin-Wen Ting, and Chih-Hao Huang "Transition-Code Based Linearity Test Method for Pipelined ADCs with Digital Error Correction", IEEE Transactions On Very Large Scale Integration (VLSI) Systems, Vol.18,Issue 12, pp.1-12, Nov. 2010.

[14] Jin-Fu Lin; and Soon-Jyh Chang," A High Speed Pipelined Analog-to-Digital Converter Using Modified Time-Shifted Correlated", In the Proc. of IEEE International symposium on circuits and systems, ISCAS, Sept. 2006, pp.5367-3570.

[15] Bibhu Datta Sahoo; and Behzad Razavi, "U-PAS: A User-Friendly ADC Simulator for Courses on Analog Design", In the Proc. of the IEEE International Conference on Microelectronics Systems Education, MSE '09, Jul.2009, pp.77-80.

[16] Piero Malcovati, Simona Brigati, Fabrizio Francesconi, Franco Maloberti, , Paolo Cusinato, and Andrea Baschirotto, ,"Behavioral Modeling of Switched-Capacitor Sigma–Delta Modulators", IEEE Transactions on Circuits And Systems-Fundamental Theory And Applications, vol. 50, No. 3,pp.352-364, Mar. 2003.

[17] Samir Barra, Souhil Kouda, Abdelghani Dendouga, and Nour-Eddine Bouguechal, "Simulink Behavioral Modeling of a 10-bit Pipelined ADC", International Journal of Automation and Computing", Vol 10 No 2, Apr. 2013.

[18] Samir Barra, Abdelghani Dendouga, Souhil Kouda, and Nour-Eddine Bouguechal, "Contribution to The Analysis and Modeling of The Non-Ideal Effects of Pipelined ADCs using MATLAB", Journal of Circuits Systems and Computers, Vol. 22, No. 2, Feb. 2013.

[19] Yannis Tsividis, "Operation and Modeling of the MOS transistor", Second Edition, Boston, McGraw-Hill, pp.422, 1999.

[20] Patrick J. Quinn, and Arthur H.M. Van Roermund, "Accuracy Limitations of Pipelined ADC", In the Proc. of IEEE International Symposium on Circuits and Systems, ISCAS, Feb.2005, vol. 3, pp.1956-1959

High Bandwidth $0.18\mu m$ CMOS Transimpedance Amplifier for Photoreceiver Circuit

Escid* .H, Gachi .N, Sebti .A

USTHB - Faculty of Electronics and Computer Science
Instrumentation Laboratory, BP. 32, Bab Ezzouar 16111,
Algiers, Algeria
hescid@yahoo.fr*, moi-fedoua@hotmail.com,
miminaeln@hotmail.fr

Slimane .A

Advanced Techniques Development Center - CDTA
Microelectronics laboratory, Cité 20 aout 1956 Baba Hassen,
Algiers, Algeria
aslimane@cdta.dz

Abstract—**This paper presents a transimpedance amplifier for photoreceiver circuit. The proposed structure operates at a data rate of 10 Gb/s at a BER of 10^{-12} and was implemented in a 0.18 µm CMOS process. The structure achieves a wide bandwidth (6.3 GHz). We used NMOS transistors as active resistor to increase bandwidth and to reduce noise level. With a photodiode capacitance of 0.25 pF, the proposed TIA has a gain of 60 dBΩ, a phase margin of 56°, and an input courant noise level of about 23 pA/Hz$^{0.5}$. It consumes a DC power of 21.2 mW from 1.8 V supply voltage.**

Keywords-CMOS Technology; Transimpedance; bandwidth enhancement;

I. INTRODUCTION

The recent rapid progress of data transportation volume and speed has brought great demands of high bandwidth, low cost and low power integrated optical communication circuits [1]. A high speed optical receiver generally consists of a photodiode, a transimpedance amplifier (TIA) and a limiting amplifier (LA). The TIA, which converts and amplifies the induced photo current into voltage for following signal processing, must has a large bandwidth to support high-bit rate applications. In the design of a low cost and low power TIA for optical communication system, deep sub-micron CMOS technology is the primary candidate for its low cost and easy of fully-integration [1-4].

The major difficulty in designing the wideband TIA lies in the large natural photodiode capacitance at the input node. Therefore, different techniques are introduced to extend bandwidth.

This paper describes the design of an integrated TIA that operates at a data rate of 10 Gb/s. The next section focuses on implementation of CMOS transimpedance amplifier and conclude this paper with the simulation results.

II. TRANSIMPEDANCE AMPLIFIER

The purpose of the preamplifier stage is to convert a miniscule photocurrent into a usable voltage, and several solutions exist for achieving this goal. Prior to the transimpedance amplifier, common solutions were to use either a low impedance or high impedance open-loop voltage amplifier, and convert the photocurrent to an input voltage

using a resistor to ground. There are pros and cons for each topology, but the transimpedance amplifier suits us well due to its high bandwidth and high transimpedance gain.

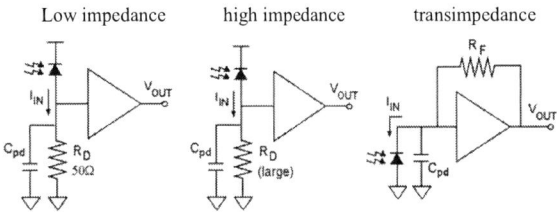

Figure 1. Preamplifier configuration

The TIA given in Fig. 3 is the most common implementation of the preamplifier circuit for high-speed optical receivers. C_T is the total capacitance which includes photodiode(C_{PD}), and parasitic(C_S) capacitances. Merits for using transimpedance amplifiers include its ability to couple a relatively large transimpedance gain with high bandwidth. These two factors directly relate to sensitivity of the detection circuit and its speed respectively. The TIA can be viewed as a single ended voltage amplifier with a single feedback resistor providing current feedback. The high input resistance of the voltage amplifier is now replaced with the feedback resistance, which is lowered by A, the open-loop voltage gain.

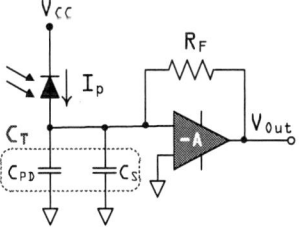

Figure 2. Closed loop amplifier used in an optical receiver

The overall transimpedance gain and bandwidth are:

$$\frac{V_{Out}}{I_{in}} = -R_F \frac{A}{(1+A)+sR_FC_T} \approx -R_F \frac{1}{1+s\left[\frac{R_FC_T}{1+A}\right]} \qquad (1)$$

$$f_{-3dB} = \frac{1+A}{2\pi R_FC_T} \qquad (2)$$

For sufficiently high A, the transimpedance gain is simply the feedback resistance. The cutoff frequency of the closed-loop response is directly related to the open-loop voltage gain of the system. Therefore, we want to maximize the open-loop voltage gain of the system. The dominant pole of the system is due to the diode capacitance, and the feedback resistance, which must be balanced to simultaneously maximize the system transimpedance gain and bandwidth.

The value of the photodiode capacitance must be low to increase the bandwidth of the amplifier. However, the low input impedance of the topology reduces this capacitive effect and allows for enlargement of the transmission capability.

III. DESIGN AND IMPLEMENTATION OF A TIA

Careful selection of the open-loop voltage amplifier topology is necessary for maximizing the Transimpedance Bandwidth product. We compared several open-loop topologies before selecting the Push-Pull inverter. This configuration looks like a digital inverter, but the transistor gates are biased so that the transistors are in saturation. At this bias point the inverter is in its high-gain region. An additional advantage to this topology is the self-biasing nature of the circuit. The feedback resistor couples the DC voltage at the output back to the input, and the inverter automatically sets itself up to be in the high-gain region. The Push-Pull inverter offers a high gain, high bandwidth, and medium noise performance when compared to other topologies.

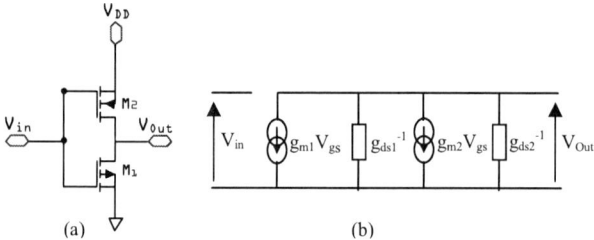

Figure 3. (a) Push-Pull inverter configuration, (b) Small signal equivalent circuit

The gain of the Push-Pull inverter will be given by :

$$A = \frac{V_{Out}}{V_{in}} = \frac{g_{m1} + g_{m2}}{g_{ds1} + g_{ds2}} \qquad (3)$$

A. Basic structure design

An improved CMOS implementation of the transimpedance amplifier is presented in Fig. 3. The TIA takes a current from the input and converts it into a voltage signal. The transistors M_1 and M_2 form the push-pull inverter which is used to maximize the transconductance of the amplifier and increases its gain bandwidth product (GBP) [4],[5].

By choosing a functional point where the two transistors are in saturation mode, and considering the current through the PMOS transistor equal to that which passes through the NMOS transistor, we can obtain the ratio of dimensions of the transistors and it given by the following expression:

$$\Gamma = \frac{\left(W/L \right)_p}{\left(W/L \right)_n} = \frac{k_n}{k_p} = \frac{\mu_n}{\mu_p} \qquad (4)$$

where μ_n and μ_p are electrons and hole mobility. This ratio will be taken equal to 4.5 for 0.18μm technology.

If the feedback resistor is implemented with a MOS transistor in the triode region, its value can be continuously adjusted via the gate voltage of the device. A PMOS device takes up less area for the same resistance value. However, in an n-well process, the capacitance associated to the well limits the maximum attainable bandwidth. For this reason, n-channel transistors were used in the proposed configurations.

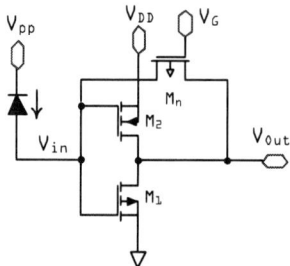

Figure 4. Original 3-Transistors design for preamplifier stage

To obtain a high open-loop gain, the amplifier in the present paper is implemented with five cascaded stages, as depicted in Fig. 5.

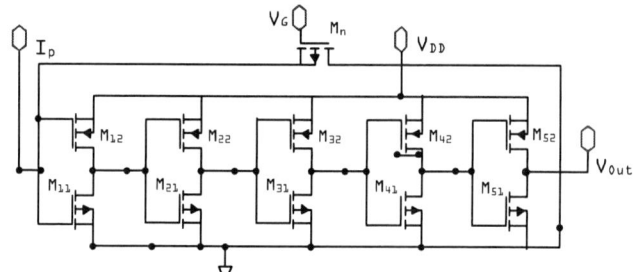

Figure 5. The CMOS transimpedance preamplifier

B. Noise optimization

The equivalent thermal noise power at the input node for the circuit given in Fig .6 is given by:

$$<I_{tot}^2> = \frac{4kT}{R_F} B + \frac{8}{3} \frac{kT}{g_m} \left[\frac{\left(1 + A^5\right)^2}{R_F^2} B + \left(2\pi C_T\right)^2 B^3 \right] \qquad (5)$$

where k is Boltzmann's constant, T is the absolute temperature, C_T is the total input capacitance, and B indicates the useful bandwidth. The use of a NMOS transistor as a feedback resistor eliminates the factor $4KTB/R_F$. In addition to this advantage it reduces considerably the surface during the integration of knowing that such a resistance value, carried out in poly, take a place in the chip. Thus, this structure keeps the same performance as with resistance. The feedback resistor will be determined using:

$$R_F = \frac{1}{\dfrac{W}{L}\mu C_{ox}(V_{GS}-V_T)} \tag{6}$$

The last term in (4) is the more prominent then the noise power, depends mainly on B^3 and C_T^2/g_m. This term gives the absolute minimum noise that can occur at the input, assuming R_F extremely high. However, the noise performance is close to the optimum when [6]:

$$0.2\,(C_{PD}+C_S)<(C_{gs}+C_{gd})<2(C_{PD}+C_S) \tag{7}$$

To optimize the noise, we can therefore increase the transconductance of attack transistors. Thus, instead of large transistors, we connect several identical transistors in parallel, whose number is defined by the following expression [3]:

$$M = \frac{\delta}{\delta+1}\frac{C_T+C_1}{C_1} \tag{8}$$

where C_1 is the input capacitance of the preamplifier, δ is a factor which reflects the inequality given in (7), it varies between 0.1 and 2.0. We took the value of M equal to 5.

But this increases the capacitance. These capacitances degrade gain and cause an increase in the dissipation in one hand and congestion in the other.

C. Optimized structure

To increase the bandwidth of an amplifier, it is important to identify the key component that is responsible for limiting the bandwidth. To have a broadband characteristic for the given amplifier and to improve the bandwidth, first a low feedback resistance is required, which requires the use of a wide transistor or increase the number of the feedback NMOS transistors, this provides a high bandwidth, low noise, improving the phase margin affecting the stability of the structure, and a wide dynamic range.

IV. SIMULATION RESULTS

All post-layout simulation results have been performed using the RF transistor models of BSIM3v3.3, based on TSMC 0.18μm CMOS process, and by using *Cadence* software tools.

The improved structure exhibits a transimpedance gain (see Fig. 6) of 1kΩ (60 dBΩ).

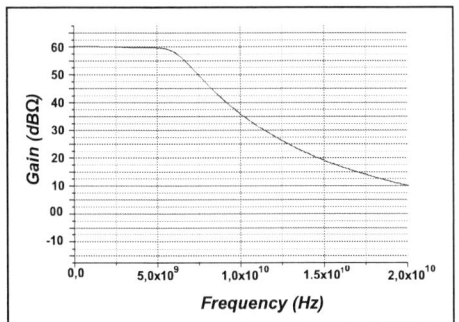

Figure 6. The transimpedance gain of the preamplifier.

The simulated noise at the input is depicted in Fig. 7, which shows a maximum at low frequencies, then it falls and tends to zero at high frequencies along the desired frequency range. The TIA has an input current noise equal to $23\ pA/\sqrt{Hz}$.

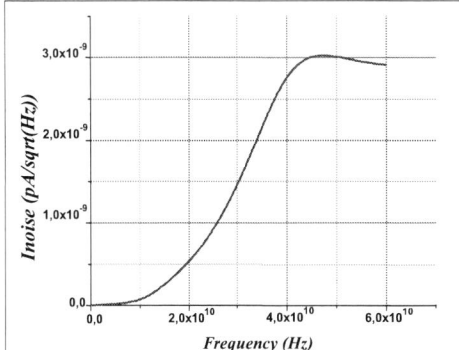

Figure .7 Input noise current of the TIA

A DC analysis giving the output voltage swing as a function of the photocurrent is shown in Fig. 8. This characteristic shows a good linearity of the amplifier as well as a high dynamic range along a high input current range. We easily see that optimized structure gives a good dynamic range.

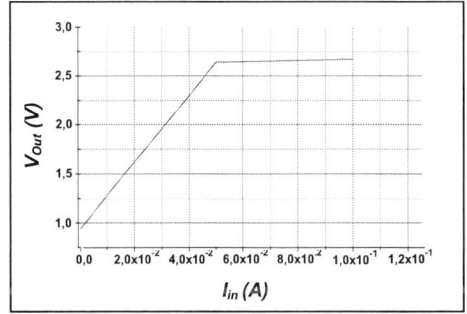

Figure 8. Output voltage vs the input photo current

A layout of the proposed TIA (Fig. 9) has been drawn using 0.18μm CMOS technology. The area of the circuit is 11539 μm^2

Figure 9. Layout of the optimized TIA structure

TABLE I. PERFORMS SUMMARY OF THE PROPOSED TIA AND COMPARISON WITH RECENT PUBLICATIONS.

Ref	Tech CMOS	C_{PD} (pF)	BP (GHz)	Gain (dBΩ)	Noise ($pA/\sqrt{Hz}$)	Power (mW)
[7]	0.35μm	1.5	1.9	65	9.7	17 under 3.3V
[8]	0.5μm	0.6	1.2	64	0,6	115 under 3.3V
[9]	0.18μm	0.2	4.36	85	13	150 under 1.8V
[10]	0.35μm	0.4	2.75	54.5	12,76	53.5 under 3.3V
[11]	0.35μm	0.1	5.75	19.7	25.8	87.4 under 3.3V
TW*	0.18μm	0.25	6.3	60	23	21.2 under 1.8V

Performances summary

V. CONCLUSION

This paper has presented the preamplifier stage of the optical receiver in a high speed optical communication system. A TIA consisting of a self-biasing inverter with NMOS feedback resistance was proposed to transform small input currents produced by a photodiode to usable voltage levels. It was designed in a low-cost 0.18μm CMOS process and is suitable for low voltage Gigabit Ethernet applications. The structure gives better performances; namely a good transimpedance gain (60 dBΩ), a high gain bandwidth product enough, good dynamic range for a photocurrent ranging up 48.5 mA and especially a relatively low noise level (23 $pA/\sqrt{Hz}$) giving better reception sensitivity combined with a wide bandwidth (6.3 GHz) to achieve a high transmission speed (10 Gb/s).

REFERENCES

[1] M. Li, B. Hayes-Gill, I. Harrison, "6 GHz transimpedance amplifier for optical sensing system in low cost 0.35 μm CMOS", Electronics letters, 42, issue 22, pp. 1278-1279, 2006.

[2] D. Coppée, J. Genoe, J.H. Stiens, R.A. Vounckx, M. Kuijk, "Calculation of the Current Response of the Spatially Modulated Light CMOS Detector", IEEE Transactions on Electron Devices, Vol.48, pp. 1892-1902, September 2001.

[3] T. K. Woodward and Ashok V. Krishnamoorthy, "1-Gb/s Integrated Optical Detectors and Receivers in Commercial CMOS Technologies", IEEE Journal of Selected Topics in Quantum Electronics, Vol. 5, No. 2, March/April 1999.

[4] Sunderarajan S. Mohan, Maria del Mar Hershenson, Stephen P. Boyd, and Thomas H. Lee, "Bandwidth Extension in CMOS with Optimized On-chip Inductors", IEEE J. Solid-State Circuits, Vol. 35, No. 3, pp. 346-355, Marsh 2000.

[5] S. J. Sim, J. M. Park, S. M. Park, "A 1.8 V, 60 dB Ohm 11 GHz transimpedance amplifier with strong immunity to input", IEEE International Symposium on Circuits and Systems (ISCAS), May 2006.

[6] H. Escid, M. Attari, "Low Noise and High bandwidth 0.35 μm CMOS Transimpedance Amplifier", Proceedings of International Conference on Microelectronics (ICM'09), Marrakech Morocco, pp. 26-29, December 2009.

[7] C. D. Motchenbacher and J. A. Connelly, "Low Noise Electronic System Design", (John Wiley and Sons, Inc., New York, 1993.

[8] M. Ingels, M. Steyaert, " A 1Gb/s, 0.5 μm CMOS Optical Receivers With Full Rail to Rail Output Swing", IEEE Journal of Solid State Circuits, Vol. 34, pp. 971-977, July 1999.

[9] F. Beaudoin, M. N. El-Gamal, "A 5-Gbit/s CMOS optical receiver frontend", Circuits and Systems, MWSCAS-2002. , Vol. 3, pp. 168-171, 2002.

[10] H. Escid, M. Attari, "5 Gb/s Low Noise and High Bandwidth 0.35 μm CMOS Transimpedance Amplifier", AMSE Journal ' Best of Book', Vol. 11, issue 2, No 1-2, pp. 15–24, 2010.

[11] H. Escid, M. Attari, M. Ait Idir, W. Mechti, "0.35 μm CMOS Optical Sensor for an Integrated Transimpedance Circuit", International Journal On Smart Sensing & Intelligent Systems, Vol. 4, No. 3, pp. 467-481, September 2011.

Power Electronics Circuit for Speed Control of Experimental Wind Turbine

Adel Merabet[1], John Kerr[1], Vigneshwaran Rajasekaran[1] and Derek Wight[2]

[1]Division of Engineering, Saint Mary's University, Halifax, NS, Canada

[2]Quanser inc., Markham, ON, Canada

adel.merabet@smu.ca, john.kerr@smu.ca, vigneshwaran.rajasekaran@smu.ca, derek.wight@quanser.com

Abstract— In this paper, a power electronics circuit is proposed to control the rotational speed of a wind turbine. This interface is based on two DC-DC converters, where a buck converter is used to control the speed and a boost converter is used to maintain a constant voltage for the load. The power electronics circuit is implemented on a small laboratory size wind turbine and experimentation is carried out to show its capability of speed control.

Keywords– wind turbine; DC generator; DC-DC converter; buck, boost converters.

I. Introduction

Wind energy conversion systems in small scale capacity are prominently viewed as a viable option of renewable energies for stand-alone load applications [1]. However, their efficiency in all wind speed regimes is related to the way of regulating speed and power. Several configurations of such systems are available such as fixed pitch, variable pitch and power electronics interfaced wind turbines, and depending on the configuration, speed and power control can be performed for high efficiency operation [1], [2].

In below rated wind speed, maximum power point tracking control scheme can be achieved by controlling the rotational speed to follow a reference at optimum value of tip-speed ratio, where it is common to keep the pitch angle constant [3], [4]. In above rated wind speed, where the nominal speed/power regulation is achieved by changing pitch angle, such control cannot be performed on fixed pitch wind turbines. Therefore, power electronics can be used for speed control applications in both regimes [5].

In DC generator based wind turbine, DC-DC converters interface can be used to operate the turbine at optimum operating point below rated wind speed or at nominal speed above rated wind speed [5]-[7].

In this paper, a power electronics circuit is designed and integrated into a laboratory scale wind turbine experimental system to control its speed under constant pitch angle. The circuit is based on DC-DC converters and the gate's switching will allow changing the electromagnetic torque of the generator, which in turn varies the rotational speed. The state of switching is a PWM (Pulse-Width Modulation) signal generated from a PI controller.

This work was supported by The Natural Sciences and Engineering Research Council of Canada (NSERC) under the Engage Grant.

II. Wind Turbine-Generator Experimental System

1. Wind Turbine

The wind turbine, manufactured by Quanser inc., is installed in a tunnel as shown in Figure 2. It has five blades and is connected to the DC generator through a gearbox of ratio 1:1 to change the vertical rotation to a horizontal one as the generator is mounted vertically. The generator is directly connected to the load and the shaft speed can only be controlled by changing the pitch angle through the pitch actuator as no power electronics for control is included in this system [8].

The torque at the shaft of the turbine, produced by the wind, is given by the following expression

$$T_t = 0.5 \, \pi \, \rho \, C_t \, r^3 \, v_w^2 \qquad (1)$$

where, ρ is the air density, C_t is the torque coefficient, r is the radius of the turbine blade, v_w is the wind speed.

2. DC Generator

The armature of the DC generator is modeled as a circuit with resistance R connected in series with and inductance L, a voltage source E representing the back emf (electromotive force) in the armature when the rotor rotates due to the wind turbine, and the generated voltage is V.

The electrical equations of the armature circuit are

$$\begin{cases} V(t) = E(t) - Ri(t) - L\dfrac{di(t)}{dt} \\ E(t) = K_b \omega(t) \end{cases} \qquad (2)$$

where, i is armature current, and K_b is back-emf constant.

The mechanical dynamic response of the rotor driven at a speed ω by the turbine torque T_t is expressed as

$$\frac{d\omega(t)}{dt} = \frac{1}{J}(T_{em} - T_t) - \frac{B}{J}\omega(t) \qquad (3)$$

where, T_{em} is the torque developed by the generator, J is the rotor inertia, and B is the viscous-friction coefficient.

The developed electromagnetic torque T_{em} is given by

$$T_{em}(t) = K_i i(t) \qquad (4)$$

where, K_i is the torque constant.

978-1-4673-5289-5/12 $31.00 © 2012 IEEE

III. Power Electronics Circuit

The power electronics interface consists of two DC-DC converters as shown in Figure 1. The DC-DC buck converter is used to control the rotational speed of the generator shaft and the DC-DC boost converter is used to provide the required voltage to the load.

1. DC-DC buck converter

The DC-DC buck converter is based on a MOSFET connected in series with the armature of the generator. This MOSFET allows the flow of current to the load by switching its gate ON, which will produce an electromagnetic torque T_{em} as expressed in Eq. (4). This torque will act against the turbine torque T_t to reduce the generator speed following the dynamic (3). In case of speed increase, the gate of the MOSFET is switched OFF and the generator operates under no load with only the turbine torque affecting its speed variation. When the speed reaches the reference and in order to keep it around it, the gate keeps switching ON-OFF depending on the speed error e_ω as follows

$$e_\omega = \omega_{ref} - \omega < 0 \rightarrow \text{gate: ON} \rightarrow \omega\downarrow \quad (5.a)$$
$$e_\omega = \omega_{ref} - \omega > 0 \rightarrow \text{gate: OFF} \rightarrow \omega\uparrow \quad (5.b)$$

where, ω_{ref} is the speed reference

The switching ON-OFF of the gate is achieved by a PWM signal generated from a PI controller for speed tracking.

The ratio of voltage between the input (generator voltage) and the output (intermediate voltage) depends on the duty cycle D of the PWM signal to be sent to the MOSFET's gate such that

$$V_{out} = V_{in} \times D \quad (6)$$

The duty cycle D is carried out using the expression

$$D = \frac{t_{ON}}{T} \quad (7)$$

where, t_{ON} is the time when the switch is ON and T is the period.

A low pass filter is included to smooth the output voltage of the converter by eliminating high frequencies above the cut-off frequency

$$f = \frac{1}{2\pi\sqrt{L_1 C_3}} \quad (8)$$

where, C_3 is capacitance and L_1 is inductance.

During the state OFF of the switch, the generator is disconnected from the circuit and the inductance L_1 plays the role of powering the load, through the diode D_1 to have a closed circuit, until the gate switches back ON.

The output voltage is reduced due to the voltage drop across the inductance L_1, which justifies the need of a boost converter to provide an adequate voltage to the load.

2. DC-DC boost converter

This converter is used to increase the intermediate voltage (output voltage of the buck converter) and maintain a constant voltage for the resistive load. It is based on the LX1741 voltage boost controller, where the output voltage V_{out} depends on the adjustable voltage by the expression

$$V_{out} = V_{adj}\left(1 + \frac{R_1}{R_2}\right) \quad (9)$$

where, R_1, R_2 are resistors to define the voltage scale, and V_{adj} has a range of 0.9 to 1.5 V and given from the computer through the Data Acquisition Board Q8 [8].

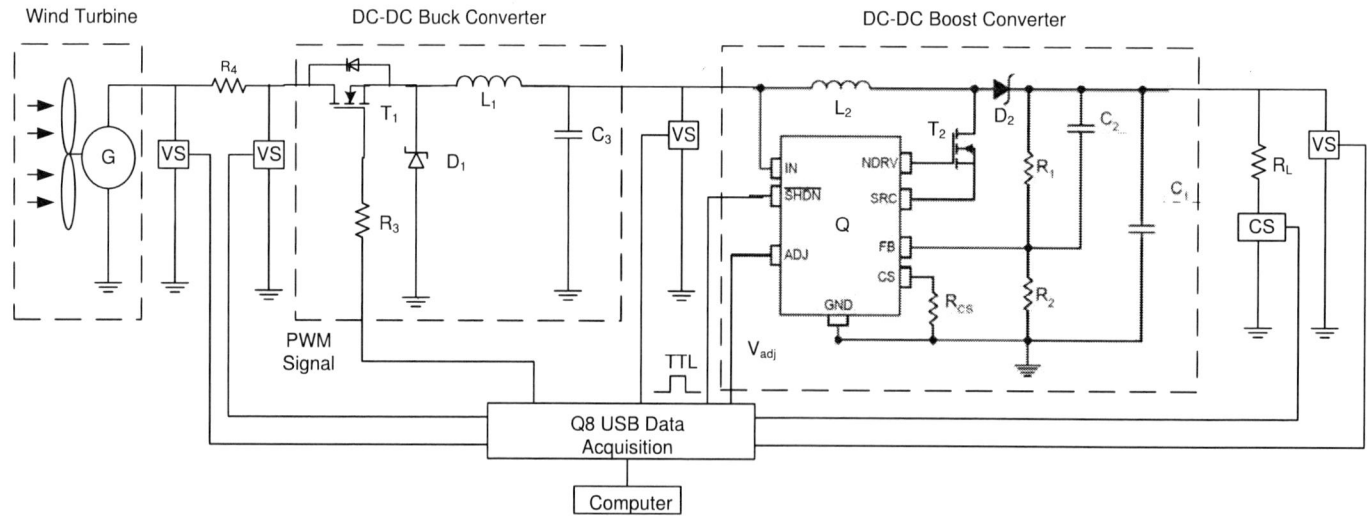

Figure 1. Schematic of the power electronics interface for speed control of the wind turbine

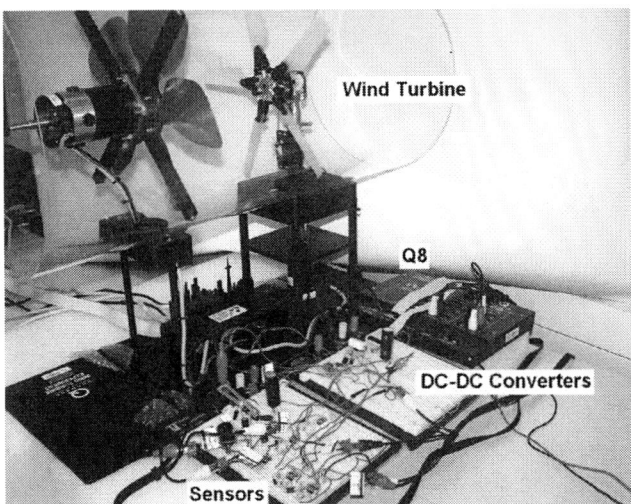

Figure 2. Experimental set-up of the power electronics interface for the wind turbine

IV. SPEED CONTROL

The objective of the power electronics interface is to control the rotational speed of the turbine-generator shaft by applying an electromagnetic torque T_{em} to the generator through sending a PWM signal to the gate of the buck converter's MOSFET. In this work, the signal is generated from a PI controller to allow tracking a speed reference

$$S = K_p (\omega - \omega_{ref}) + K_i \int (\omega - \omega_{ref}) dt \qquad (10)$$

where, S is the controller output, ω_{ref} is the speed reference, K_p is the proportional gain, and K_i is the integral gain.

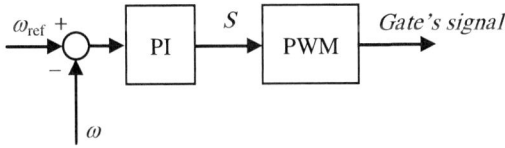

Figure 2. Speed control strategy

The PWM signal is generated following the Figure 3, where the output of the PI controller is compared to a saw signal of frequency f_r. Then, the binary signal is amplified to a voltage signal and sent to the gate through the resistance R_3 as shown in Figure 1.

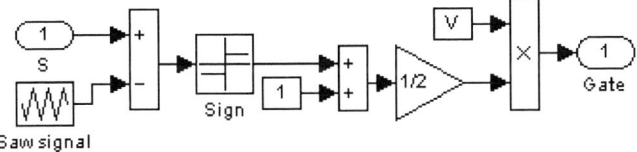

Figure 3. PWM generation

V. EXPERIMENTAL RESULTS AND DISCUSSION

Experimentation is carried out to validate the performance of the power electronics interface for speed control. The choice of proportional gain K_p and integral gain K_i of the PI controller is determined by trial and error method, so that the controller is tuned to achieve good performance. The wind speed is artificially generated by a DC blower motor.

Different voltages (generator, output of the buck converter and load) and currents (load) are measured by voltage and current sensors (VS and CS) as shown in Figure 1. The measurements are sent to the computer (PC), through a real-time data acquisition board Q8-USB, to be analysed by using the software package QUARC with MATLAB/Simulink. QUARC is a powerful rapid control prototyping tool that significantly accelerates control system design and implementation [8].

The speed control algorithm is implemented in Simulink and the PWM signal is sent as an analog signal to the gate of the MOSFET. The gains of the controller are chosen by trial and error in order to achieve a good performance.

Figures 4, 5 and 6 show the speed tracking performance of PI controller for step changes profile, the intermediate and load voltage, and the PWM signal, respectively.

Initially when the speed goes from zero to the reference, the generator is not connected to the load as the switch of the MOSFET gate is OFF. The only torque affecting the machine is the turbine torque as the electromagnetic torque is zero. When reaching the reference (slightly above), the switch turns ON and the load is connected, this creates an electromagnetic torque T_{em} to reduce the speed of the shaft following the dynamic equations (3) and (4). Then, the buck converter switches OFF and ON to track the constant reference. Similar behavior occurs in case of increase in the reference step. The performance of the speed tracking is successfully achieved as shown in Figure 4.

In Figure 5, the intermediate voltage and load voltage are null until the speed reaches the reference as the switch is OFF and the converters and load are not connected to the generator. Also, peaks of voltage occur at 60 s and 120 s due to the change in the reference profile. At 60 s, the reference has increased and the PI controller takes into consideration this change by adjusting the PWM signal (Figure 6) to switch OFF the MOSFT gate, so the electromagnetic torque is null, and the speed increases, due to the turbine torque, in order to reach the new reference step. At 120 s, the reference has decreased and the PI controller acts by generating a PWM signal (Figure 6) to switch ON the MOSFET gate so the electromagnetic torque is produced to reduce the speed of the generator to the new reference step.

978-1-4673-5289-5/12 $31.00 © 2012 IEEE

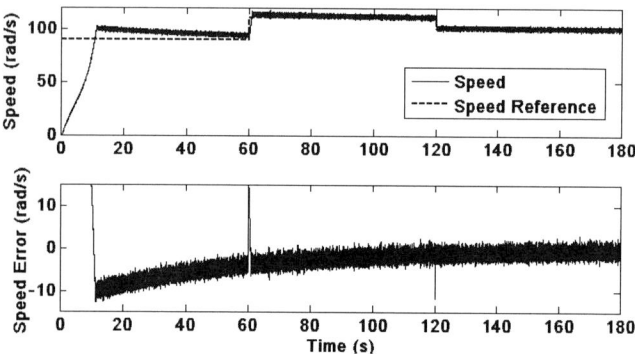

Figure 4. Speed tracking and error

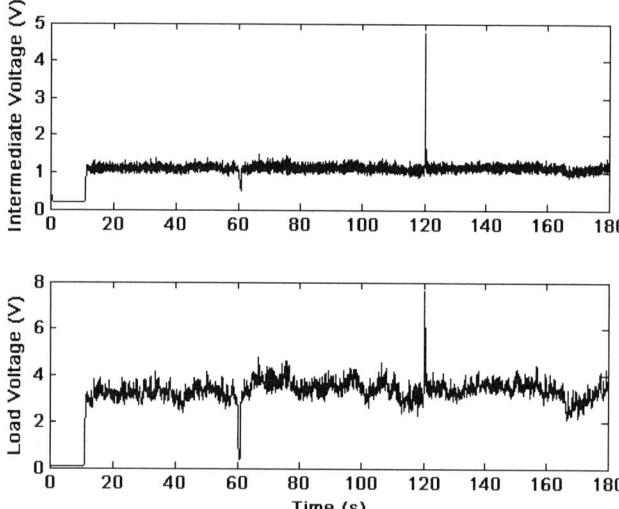

Figure 5. Intermediate and load voltages

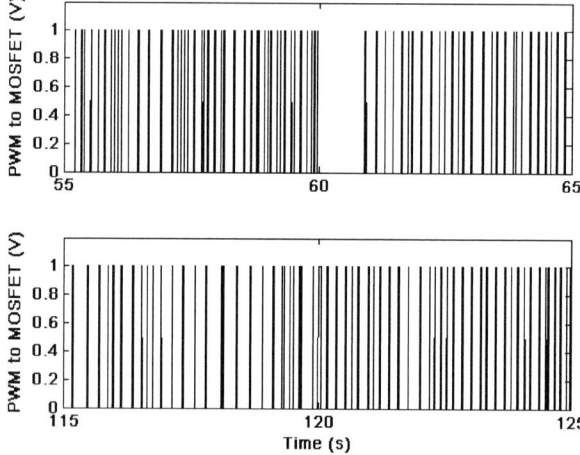

Figure 6. PWM signal generated by the PI controller

VI. CONCLUSIONS

A power electronics circuit, based on two DC-DC converters, is implemented to control the rotational speed of a wind turbine. The buck converter is used to control the speed by switching ON-OFF the MOSFET gate to produce an electromagnetic torque. This torque decelerates the generator to reduce its speed and no electromagnetic torque allows increasing the speed due to the turbine torque. At the output of this converter, the voltage drops. Therefore, a boost converter is used to increase and maintain a required voltage for the load. The power electronics circuit is tested on a small laboratory size wind turbine and experimental results have shown good speed tracking performance.

APPENDIX

Wind turbine: r= 14cm, ρ=1.14 kg/m^3.
DC generator: R=2.47Ω, L=500μH, K_b=7.05mV/rpm, K_t=0.015 A/mN , J=110 g/cm^2
DC-DC buck converter: T_1=IRLU014; R_3 =1 kΩ (current limiting resistor); D_1 = 1N5822; L_1 = 330 μH; C_3 = 1.5 mF
DC-DC boost converter: T_2=FDV303N; R_1=280 Ω; R_2=49.9 Ω; R_{CS}=1 kΩ; D_2=1N5822; L_2=47 μH; C_1=4.7 μF; C_2=1.2 nF; V_{adj}=1.29 V; Q=LX1741; R_4= 0.5 Ω; TTL= 5V (digital signal).
PI controller: K_p= 5, K_i= 0.1.
PWM: f_r= 10 kHz, V=10V
Load: R_L= 500Ω

REFERENCES

[1]. S. Drouilhet, E. Muljadi, R. Holz, and V. Gevorgian, "Optimizing small wind turbine performance in battery charging applications," in *NREL/TP-441-7808*. Golden, CO: National Renewable Energy Laboratory, 1995.

[2]. N. Yamamura, M. Ishida, and T. Hori, "A simple wind power generating system with permanent magnet type synchronous generator," IEEE Int. Conf. Power Electronics Drive Systems, vol. 2, Hong Kong, Jul. 27–29,1999, pp. 849–854.

[3]. Md. Arifujjaman, M.T. Iqbal, J.E. Quaicoe, "Maximum power extraction from a small wind turbine emulator using a DC-DC converter controlled by a microcontroller," Int. Conf. of Electrical and Computer Engineering, 2006, pp. 213 – 216.

[4]. R. Esmaili, L. Xu, and D.K. Nichols, "A new control method of permanent magnet generator for maximum power tracking in wind turbine application," IEEE Power Engineering Society General Meeting, June 12-16, 2005, vol.3, pp. 2090–2095.

[5]. V. Lazarov, D. Roye, D. Spirov, Z. Zarkov, "New control strategy for variable speed wind turbine with DC-DC converters," 14th International Power Electronics and Motion Control Conference, Sofia, Bulgaria, Sept. 6-8, 2010, pp. T12-120 - T12-124.

[6]. M. Bottu, M.L. Crow, A.C. Elmore, "Design of a conditioner for smoothing wind turbine output power," North American Power Symposium (NAPS), Sept. 1 – 6, 2010, pp. 26–28.

[7]. J. C. Mayo-Maldonado, R. Salas-Cabrera, H. Cisneros-Villegas, R. Castillo-Ibarra, J. Roman-Flores, M. A. Hernandez-Colin, "Maximum power point tracking control for a DC-generator/multiplier-converter combination for wind energy applications," World Congress on Engineering and Computer Science, October 19-21, 2011, San Francisco, USA 2011, vol. I.

[8]. "Wind Turbine Experiment," Quick-Start Guide, Quanser inc. 2010.

Energy Monitoring System for Security and Power Management Applications

Sepideh Shariati, Radu Muresan, Anthony Vannelli

School of Engineering
University of Guelph
Guelph, Ontario, Canada
{sshariat, rmuresan, vannelli} @uoguelph.ca

Abstract—**In this paper, we introduce an energy monitoring system composed of a mixed-signal field-programmable gate array (FPGA) device and a custom designed energy measurement circuit. The energy measurement circuit is designed as a current integrator over a fixed interval using the switched-capacitor (SC) technique. The circuit is implemented using the 65 nm CMOS technology. Simulation results show that our circuit provides energy measurement results with a precision of up to 3% for irregular input waveforms. The circuit is designed to accommodate the low sampling rate of mixed-signal FPGA devices in order to support both power management and security applications for embedded devices.**

Keywords—**energy measurement circuit, mixed-signal FPGA, switched-capacitor integrator**

I. INTRODUCTION

With the increased use of portable devices, such as notebook computers, personal digital assistants and cellular phones, security and power management are essential [1]. Security is important to protect personal information and secure communications while power management aims to increase the battery lifetime. The so-called power analysis attacks (PAA) compromise the security of any embedded device by retrieving the stored secret key of the device [2]. There exist various countermeasures against PAA at the hardware and software level [2]. One of the countermeasures relevant to this paper is the current flattening technique implemented at the software level [3]. However, implementing effective current flattening in software requires energy consumption information at the instruction level [3]. Furthermore, in order to optimize the battery lifetime, a number of techniques have been proven to be effective, such as dynamic power management (DPM) [4] and dynamic voltage scaling (DVS) [5]. Key to these techniques is the ability to profile the energy consumed by the device. The goal of this paper is to present an energy monitoring system that can generate an energy profile for security and power management purposes. The concept of integrating power management and security at system-on-chip (SoC) level has

been introduced in [1] where a capacitor bank controlled by a capacitor switch-box is integrated within the power management system. This concept can be extended to include mixed-signal field-programmable gate arrays (FPGAs). The energy monitoring system we propose in this paper is composed of two main blocks (i.e., mixed-signal FPGA and energy measurement circuit) as shown in Fig. 1. The FPGA device used in our design is the Actel Fusion FPGA. The Fusion device was selected for its low-power characteristics, its analog interface programmable capabilities and its suitability for applications targeting power management [6]. As shown in Fig. 1, the Fusion device samples the target device voltage as well as its current through the Actel external resistor (R_{Actel}) through channels A_{V3} and A_{C3}, respectively. Sampling is at a fixed interval; however, since the frequency of the analog interface of the Fusion device is considerably low (limited up to 10 MHz), the information of the current of the target device will be lost between sampling intervals. The energy measurement circuit is therefore used to compensate for this low sampling rate and to prevent current information loss of the target device between sampling intervals. The output voltage of the energy measurement circuit sampled from channel A_{V2} is proportional to the measured currents of the target device provided by the current sensor (R_{energy}). After each monitoring interval, the Fusion device sends the Reset Pulse signal and a new monitoring cycle is started. The connection to high-level software is established through a universal asynchronous receiver/transmitter intellectual property (UART IP) core.

Other types of energy monitoring FPGA-based systems were also investigated in the literature. In [7], Remscrim et al. used an Altera Cyclone I FPGA-based system to measure the energy of a spectral envelope application. In [8], authors developed a digital circuit as an energy meter to estimate active and reactive energy consumption. The meter was implemented with an Actel FPGA device.

In this paper, we present the complete IP core architecture of the FPGA device, we propose an energy measurement circuit as a custom chip using the 65 nm CMOS technology

978-1-4673-5289-5/12 $31.00 © 2012 IEEE

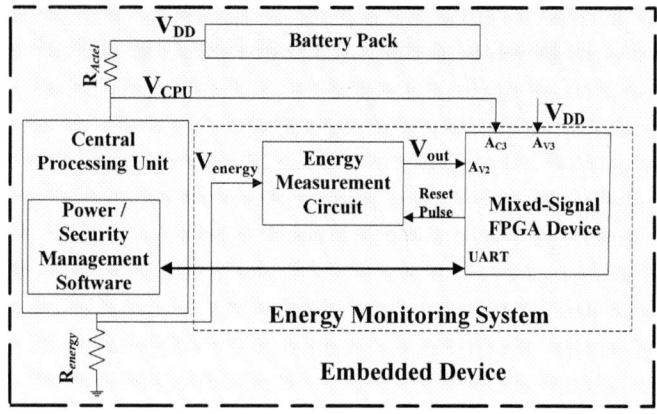

Figure 1 : Block diagram of energy monitoring system for security and power management.

and we investigate the relation between our circuit's results and mathematical model results.

The rest of this paper is organized as follows: Section II presents the IP architecture of the FPGA device; Section III describes the theory of operation of the energy measurement circuit and the design constraints; Section IV shows our experimental results; and Section V concludes the paper.

II. IP CORE ARCHITECTURE OF THE FPGA DEVICE

The Fusion FPGA device can integrate a number of IP cores that can be used to configure various SoC architectures. Fig. 2 presents the IP configuration of our architecture. We developed the IP architecture of the Fusion device around the Cortex-M1 IP soft-core in order to support all our interfaces. Specifically, the designed architecture needs to interface with the custom energy measurement chip to collect measurements, and with the power and security management high-level software applications, to transmit the measurements (See Fig. 1).

As shown in Fig. 2, Cortex-M1 shares the advanced high-performance bus (CoreAHBLite) with memory interfaces, such as nonvolatile memory (NVM) and static random access memory (SRAM), and AHB-APB (advanced peripheral bus) bridge. The APB interfaces low-speed IP core devices, specifically the core timer, analog interface, UART and the general purpose input/output (GPIO). The phase-locked loop (PLL) and 100 MHz RC oscillator are used to generate the system and peripheral clocks. The key feature of our IP architecture is the integration of the IP core timer to design a hardware-based interrupt system. The hardware interrupt can be programmed to interrupt Cortex-M1 at fixed intervals and initiate the analog interface of the Fusion device to perform the required measurements. The combination of the electrical characteristics of the Fusion's analog interface and the application program allows our Fusion device to measure the voltage, the current and the energy of the target device at a minimum rate of 15 μs. As a result, energy information will be lost within the sampling interval. Depending on the clock frequency of the targeted embedded device, a large number of instructions can be executed during this interval.

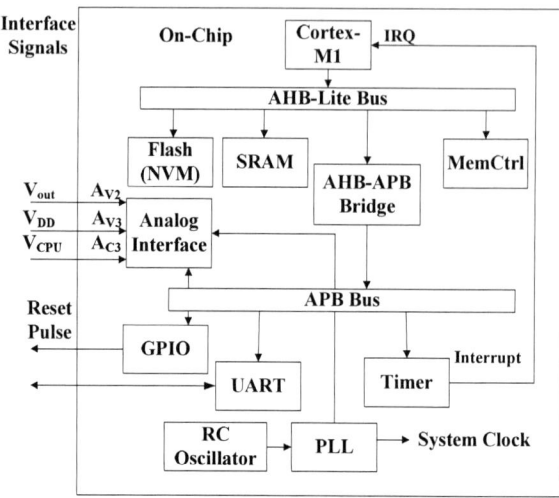

Figure 2: Actel FPGA device on-chip IP core architecture.

However, this is unacceptable for the software current flattening countermeasure technique. Also, sampling the power consumption at fixed intervals will reduce the overall precision of the energy measurement. In order to address these issues, we implemented an energy measurement circuit capable of integrating the instantaneous current of the device and storing the integration information for each interval.

III. ENERGY MEASUREMENT CIRCUIT

Fig. 3 depicts the proposed energy measurement circuit composed of a conventional switched-capacitor (SC) integrator, a non-overlapping clock circuit, a reset block, a differential amplifier and a low-pass filter [9], [10].

A. Theory of Operation

The conventional SC integrator circuit provides an accurate integration by switching a sampling capacitor (C_1) into an amplifier with a feedback integration capacitor (C_2). The circuit uses MOS transistors as switches (M_1-M_4) for the control of the sampling and integration operations. These switches are operated by two non-overlapping clocks (CLK_1 and CLK_2) to ensure that sampling and integration operations do not overlap. The schematic of the non-overlapping clocks can be seen as part of Fig. 3.

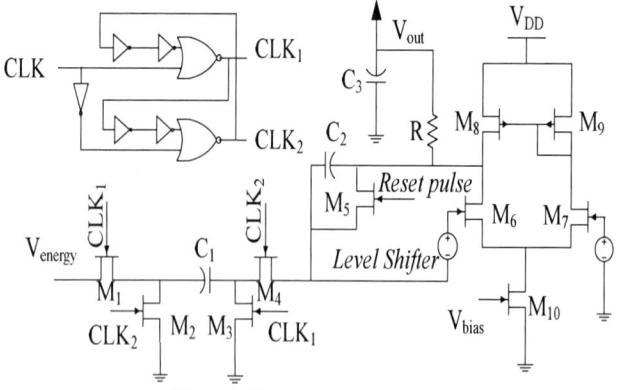

Figure 3: Energy measurement circuit.

This circuit comprises two main NOR gates and two cross-coupled inverter chains for creating non-overlap delay. The single-stage differential amplifier is sized to bring the voltage output (V_{out}) to appropriate input levels required by the FPGA device. This amplifier uses the n-channel transistors as differential pairs, and p-channel transistors as a current mirror load. A voltage generator in series with the gate of M_6 acts as a level shifter to stabilize the DC bias voltage at differential pairs. The advantage of our chosen amplifier is that it uses less silicon circuit area and provides negligible input load [9].

The SC integrator circuit performs the mathematical operation of integration [9].The output voltage is equal to Eq. 1 where f is the frequency of the non-overlapping clocks, and V_{energy} is the amplitude of the sensed voltage.

$$V_{out}(t) = f \cdot (C_1/C_2) \, \square \, V_{energy}(t) \, dt \qquad (1)$$

B. Design Constraints

The design of the energy measurement circuit is associated with following constraints:

1) General design specification: The maximum input voltage is a key design parameter. The maximum input voltage is proportional to the maximum measurable current of the target device and R_{energy}. In this design, the maximum measurable current was chosen as 400 mA. Additionally, R_{energy} was set to 0.1 Ω in order to reduce power dissipation and supply voltage fluctuation. The maximum input voltage therefore was 40 mV. The supply voltage should be compatible with the voltage reference of the Fusion device (V_{ref} = 2.56 V). For this reason, high voltage MOS switches are used with a 2.5 V supply voltage. All transistors are set at minimized size (w_{NMOS} = 400 nm, w_{PMOS} = 800 nm, l = 280 nm) as optimization of silicon area is important. The sampling capacitance, C_1, should be small enough to charge up rapidly. On the other hand, C_1 should not be too small such that the parasitic capacitances will dominate and influence the circuit performance [11]. Moreover, we determined through simulations that the integration capacitance, C_2, should be at least two times greater than the sampling capacitance in order to ensure that the SC integrator does not saturate over the sampling interval. Based on the V_{energy} chosen range, the saturation conditions and the capacitor integration area trade off, we determined C_1 and C_2 to be 1 pF and 2 pF, respectively. The clock period of 1 μs was chosen based on the condition that $R_{ON} \cdot C \ll T$.

2) Circuit saturation: The integration is accurate up to 70 μs, after which the circuit is fully saturated. Therefore, the circuit needs to reset before 70 μs in order to maintain integration accuracy. In our design, we selected a reset time of 15 μs both to prevent saturation impact and to be consistent with the sampling rate of the Fusion device. As it is shown in Fig. 3, we added a reset block to our circuit which is composed of a reset switch (M_5) and a reset pulse source.

3) Noise reduction: During the simulations, we observed some noise with the frequency of 2 MHz; a frequency two times greater than the frequency of the clock. In order to reduce this noise, we introduced a simple RC low–pass filter with a frequency two times greater than the clock frequency (See Fig. 3).

IV. EXPERIMENTAL RESULTS

The proposed energy measurement circuit was simulated in Cadence environment using 65 nm CMOS technology and 2.5 V supply voltage. First, the performance of the differential amplifier that impacts the precision of the energy measurement circuit was optimized using AC analysis. Fig. 4 shows that the dc gain of the amplifier is 20 dB, the unity-gain frequency is 5.37 GHz and the phase margin is 87°.

Second, the behaviour of the circuit was investigated by applying three different types of inputs (sine, ramp and irregular) to the circuit. Of these inputs, the irregular waveform is our primary interest as it best models the characteristics of the target device. Fig. 5 illustrates the irregular waveform input and its integration output results. As seen from Fig. 5, the energy measurement circuit goes through cycles of reset and integration. The integration result is sampled every 15 μs by the Fusion device. After the sampling process, the Fusion device sends the Reset Pulse signal and a new integration cycle is initiated.

Based on our experiments, we formulated two matrices to compare the circuit's results (simulation results) with the ideal results (mathematical model results). The integration results obtained from the circuit were collected in a circuit output matrix, while the ideal results calculated by the Cadence built-in calculator were stored in a calculator output matrix. The contents of these matrices were then analyzed through MATLAB. The analysis indicated that there is a linear relationship between the circuit's output and the ideal calculator's output. Therefore, we derived Eq. 2 from Eq. 1, where $V_{out}(t)$ is the circuit output and $\square V_{energy}(t)$ dt is the calculated mathematical integration output.

$$K \cdot V_{out}(t) + A = f \cdot (C_1/C_2) \, \square \, V_{energy}(t) \, dt \qquad (2)$$

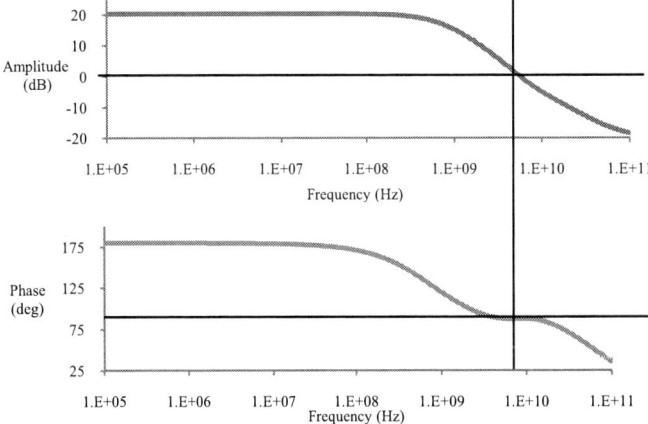

Figure 4: AC response and phase margin of the differential amplifier.

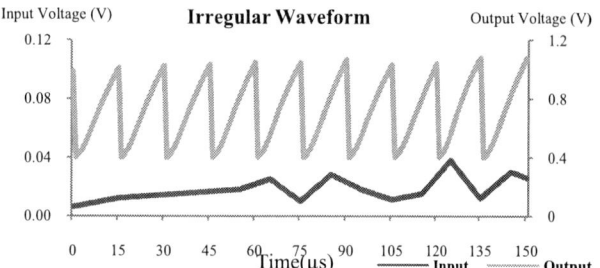

Figure 5: Irregular waveform input and its integration output results.

We calculated the values of constants K and A (energy conversion constants) for all three inputs. These values for different inputs show minimal variation. As a result, K and A can be assumed to be independent from the input signals and the average values of K and A, equal to 1.5001 and –1.4324 respectively, can be considered an accurate estimation of these constants. In order to assess the overall precision of the energy measurements, we applied these final values to the circuit's output matrix. Fig. 6 plots the comparisons between the updated circuit's output and the calculator's output for 10 sampling points. Table 1 summarizes K and A for each input, and displays the average error percentages after applying the final values of K and A to each input. According to Table 1, a maximum error of 6% ± 3 is obtained for the irregular input while the regular sine input gives much smaller error. It can be seen that the error in measurement is inversely proportional to the regularity of the input signal.

V. CONCLUSIONS AND FUTURE WORK

In this paper, we presented an energy monitoring system. We designed the FPGA IP architecture that interfaces a 65 nm CMOS energy measurement circuit and a power and security management application software. The energy measurement circuit outputs a voltage that is proportional to the integral of the instantaneous current consumption of an embedded device. This voltage can be sampled at fixed intervals no larger than 70 µs. We calculated the associated constants, K and A, for energy conversion applications. By applying these constants to the circuit output matrix, we determined that our circuit results have at most 6% ± 3 error relative to the ideal results when irregular input signal is applied. We further identified that the amount of error decreases with the input wave regularity. Our simulation results indicate that the energy measurement circuit can accurately accumulate the energy information within each 15 µs intervals. Further research can explore the possibility of integrating the energy measurement circuit as an IP core within any FPGA mixed-signal device that allows the integration of analog components.

TABLE 1. ENERGY CONVERSION CONSTANTS AND THE AVERAGE ERROR VALUES FOR EACH INPUT

Input Wave	K	A	Error Percentage
Sine	1.0194	-0.9363	3% ± 1
Ramp	1.7806	-1.7218	6% ± 2
Irregular	1.7004	-1.6392	6% ± 3

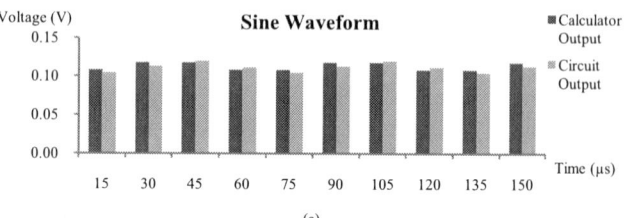

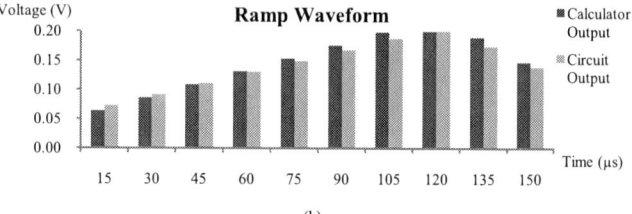

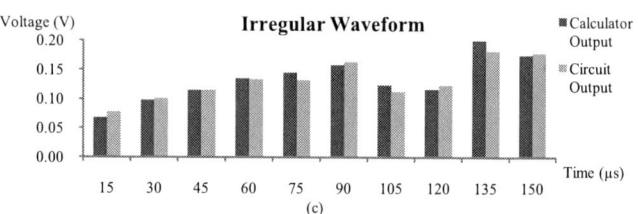

Figure 6: (a) Sine wave circuit and calculator results. (b) Ramp wave circuit and calculator results. (c) Irregular wave circuit and calculator results.

REFERENCES

[1] M. Mayhew and R. Muresan,"Integrated capacitor switchbox for security protection," in IEEE ISCAS 2012, pp. 1452 - 1455.

[2] S. Mangard, E. Oswald and T. Popp, Power Analysis Attacks: Revealing the Secrets of Smart Cards. Springer-Verlag New York Inc, 2007.

[3] R. Muresan and C. Gebotys, "Current flattening in software and hardware for security applications," in Proc. CODES+ISSS, 2004, pp. 218-223.

[4] L. Benini, D. Bruni, A. Mach, E. Macii and M. Poncino, "Discharge current steering for battery lifetime optimization," IEEE Trans. Computers, vol. 52, pp. 985-995, 2003.

[5] H. Visairo and P. Kumar, "A reconfigurable battery pack for improving power conversion efficiency in portable devices," in Proc. 7th ICCDCS, 2008, pp. 1-6.

[6] Microsemi Corporation. Fusion Family of Mixed-signal FPGAs, August 2012. [Online]. Available at: http://www.actel.com/documents/Fusion_DS.pdf

[7] Z. Remscrim, J. Paris, S. B. Leeb, S. R. Shaw, S. Neuman, C. Schantz, S. Muller, and S. Page, "FPGA-based spectral envelope preprocessor for power monitoring and control," in 25th Annual IEEE APEC, 2010, pp. 2194-2201.

[8] S. L. Toral, J. M. Quero, and L. G. Franquelo, "Power energy metering based on random signal processing (EC-RPS)", in Proc. IEEE ISCAS, 1998, pp.435-438.

[9] D. Johns and K. Martin, Analog Integrated Circuit Design, John Wiley and Sons, United States, 1997.

[10] B. J. Hosticka, R. Brodersen and P. Gray, "MOS sampled data recursive filters using switched capacitor integrators," IEEE Journal of Solid-State Circuits, vol. 12, pp. 600-608, 1977.

[11] S. Korkmaz and G. Dundar, "Quasi settled switched capacitor integrator," in Proc. 18th Int. MIXDES, 2011, pp. 362-367.

978-1-4673-5289-5/12 $31.00 © 2012 IEEE

Studies on the influence of corona on overvoltage surges by simulation using the ATP/EMTP

Zahira Anane
Department of Electrical engineering
Automatic Laboratory of Setif
Faculty of technology, UFAS, Setif, Algeria
zahiraelect@yahoo.fr

Abdelhafid Bayadi
Department of Electrical engineering
Automatic Laboratory of Setif
Faculty of technology, UFAS, Setif, Algeria
a_bayadi@yahoo.fr

Abstract—Electrical systems are subjected to constraints more severe continuation with overvoltage's which can be due to lightning and switching strikes. Under the influence of the intense corona effect which accompanies of atmospheric overvoltage, these overvoltages undergo deformations at the same time as an attenuation of their amplitude, this phenomenon of distortion, which is superimposed on the distortion by skin effect, is due to the dissipation of energy by injection of space charges around the conductor, this process with place as soon as the instantaneous voltage exceeds the threshold voltage of the corona effect conductors. In this paper, an analogical model of the corona effect has been implemented in the Alternative Transients Program/ Electromagnetic Transients Program (ATP/EMTP). This model was incorporated into a transmission lines, the line is divided on a number of the short sections. This study is to have the attenuation and the distortion of the overvoltage waves due to the corona effect on this transmission line.

Keywords- Corona effect; Overvoltage surge;, Critical field; ATP/EMT;, Attenuation and deformation of Transmission lines.

I. INTRODUCTION

Corona effect begins appearance when the value of the electric field around the conductor is higher than the value of the critical field on the conductor surface, when it can be calculated by Waters empirical formula [1], [2].

$$E_0 = 23,8m\left[1 + \frac{0,67}{r_0^{0,4}}\right] \tag{1}$$

The effect of the air density on the field critical values can be taken into account using:

$$E_0(\delta) = E_0\,\delta^b \tag{2}$$

.

Therefore the basic concept, adopted by all researchers in the modeling of the corona effect, is the increase in the conductor capacitance when the applied voltage exceeds the appearance threshold of the corona effect.

In this paper, Kudyan and Schih [4] analogical model of corona effect has been incorporated, in the ATP/EMTP simulation software.

Corona effect plays an important role in the determination of attenuation and distortion of the overvoltage waves which are propagate along of the transmission lines. For this purpose, we have implanted this model in overhead transmission lines;

the line is divided into short segments, with the corona model connected between them.

II. MODELING OF THE CORONA EFFECT

Like cities in the references [3-7] Kudyan and Shih in 1981 had represented a corona model shown in figure 1. In this model, the resistance and capacitance represent the process of the energy losses and the change in the line capacitance. Source DC in this circuit represents the corona appearance voltage. It is noted that this model was proposed at the beginning by Wagner [5, 7].

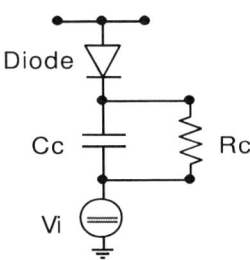

Figure 1. Equivalent Corona Circuit proposed by Kudyan and Shih.

III. RESULTS AND DISCUSSIONS

The selected line is subdivided in a number of identical elementary sections of length rather short. Each section to constitute a portion without losses connected in cascade with a branch of two elements C_c and G_c assembled in parallel to represent the corona losses, and another branch Re and L_e to represent the Joule losses and the return by the ground (Fig 2).

The characteristic impedance of the line Z_c in $[\Omega]$ is calculated by the following formula:

$$Z_c = 60\log\left(\left(H + \sqrt{H^2 + R_0^2}\right)/R_0\right) \tag{3}$$

978-1-4673-5289-5/12 $31.00 © 2012 IEEE

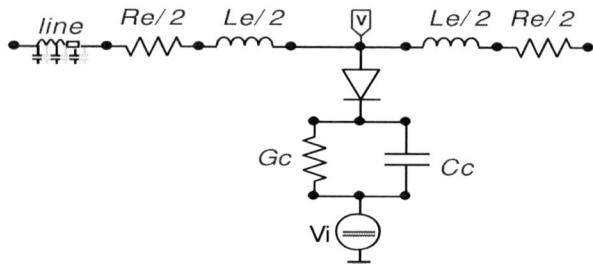

Figure 2. Structure of a typical section

A parametric study of the influence of the various factors on the results was carried out. In first place, the line consists of a radius of $R_0 = 1,18$cm placed at a height of 20m, carried to an impulse of fast voltage of the lightning type (1/5µs) of amplitude 1800kV.The line is divided into 450 sections, each one with length of 5 m.

A. Effect Of The Joule Losses And The Return By The Ground Without Corona Discharge

For the distances measuring points of: 0, 625 and 1280m, the figure 3 shows the waveforms for a combination of the two effects, Joule losses and return by the ground, by two Re values and two other values of Le.

Where we have an attenuation of wave due to the increase and we have a delay in the wave caused by the inductance.

B. Influence Of The Corona Capacitance

Various values of the corona capacitance were selected to have its effect on the wave propagation:

$$C_c = 0,25 C_g, C_c = 0,5 C_g \ et \ C_c = C_g$$

The simulation results at a distance measuring point of 625m are shown in figure 4, where it is noted that with the increase in corona capacitance the deformation is more marked and the wave damping is more significant. This results compared with results in the literature, and with results obtained with other corona models in [2, 3, 6, 10, 11].

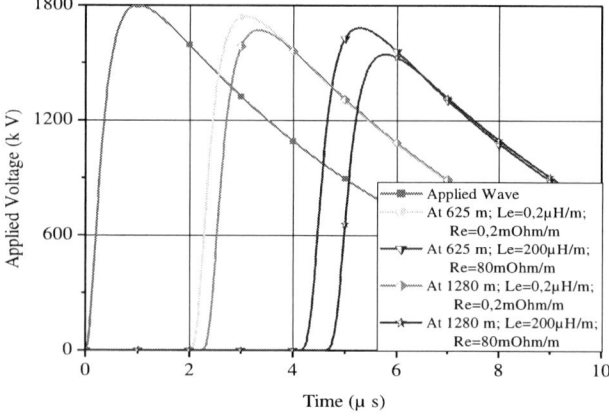

Figure 3. Combined Influence of the Joule losses and the inductance.

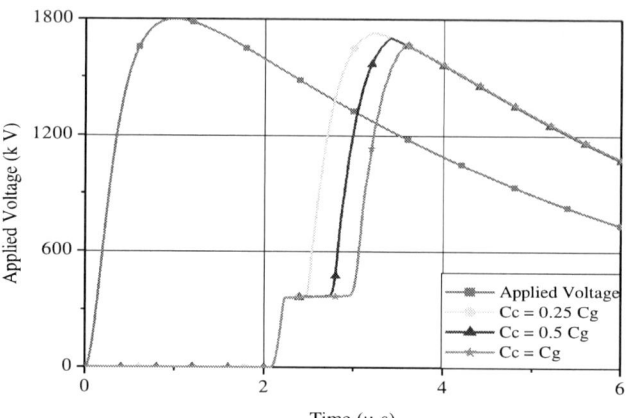

Figure 4. Effect of the corona capacitance

C. Influence Of The Corona Conductance

In [5] and [8], the plotted results are identical for the two cases, for two values of 60M Ω and of 600MΩ . From these results, it was concluded that the shunt resistors doesn't any effect on the model response.

By adopting these two values the results obtained by adopted approach EMTP are shown in figure 6 from a distance 625 m. It is clear that these results confirm this conclusion.

D. Influence Of Integration Step

To see the influence of the integration step on the model response , we applied various values of ΔT which are values as low as possible, the line is divided with short sections of 5m.

Figure 6 shown this influence for ΔT equal: 0,01;0,0025;0,005 and 0,0075 µs for the voltage wave taken at a distance of 625 m of the transmission line origin. These results are also obtained by [6].

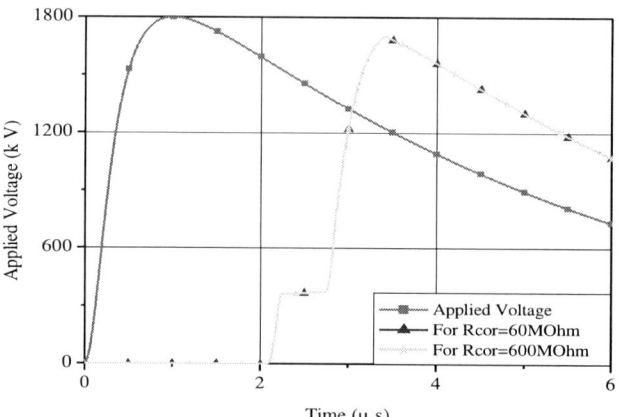

Figure 5. Effect of the corona resistance-simulation results

978-1-4673-5289-5/12 $31.00 © 2012 IEEE

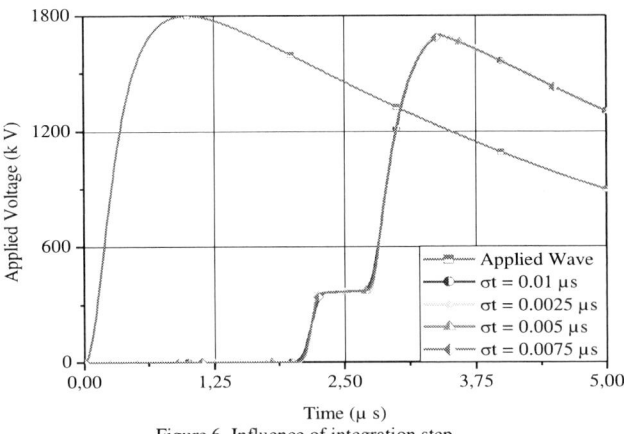

Figure 6. Influence of integration step

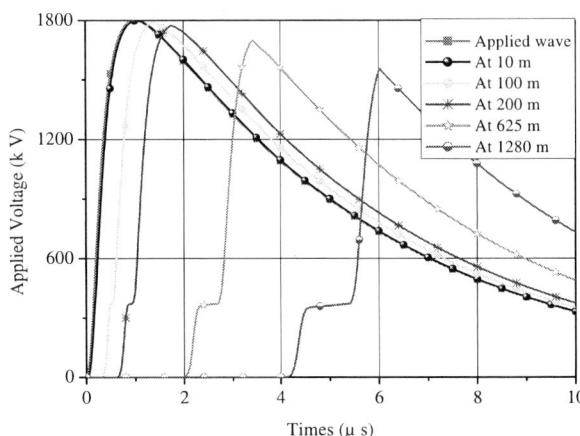

Figure 8 . Effect of measured point of: 0, 10, 100, 200, 625 and 1280m.

E. The Effect Of Section Length

To evaluate the effect of the section length we use of a short section length 10m and 20m, and we have plotted at a distance measuring ponit of 600m. The result also shows that there is a small reduction in magnitude of the peak voltage and steepness of wave front when a 10m section length was used (Figure 7).

These results are obtained and confirmed by the results of [5]; and by [8.

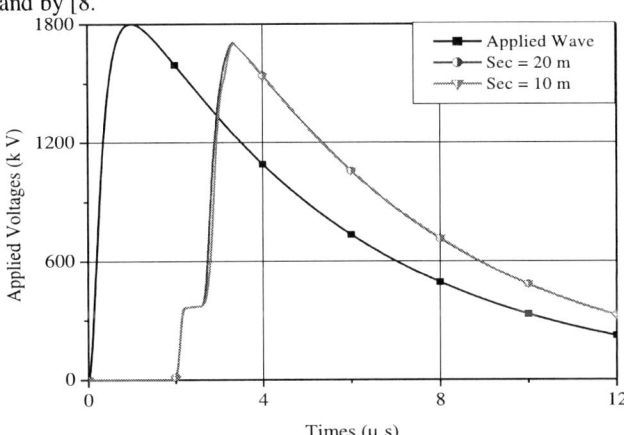

Figure 7. Influence of section length- simulation results.

F. Effect Of The Measurement Points

We considered it useful to note the importance of the measurement points.

The cases considered are shown in figure 9 when we choose the distance measuring points of 10, 100, 200, 625, and 180m. Of these figures we note that the effect of damping and deformation of the wave shock becomes sensitive only starting from a few tens of meters.

G. Influence Of Return By The Ground And The Joule Losses

The effect combined of the two parameters with corona discharge was presented in figure 9 with a distance of 625 m where Re = 80MΩ/m and Le= 200μH/m and Cc = 0, 625μF.

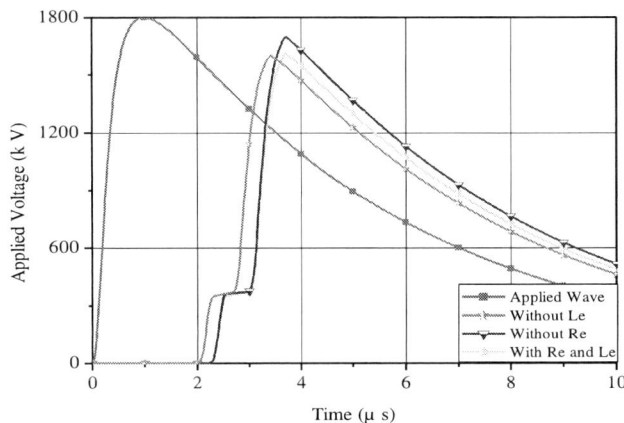

Figure 9. Combined Effect of R_c, L_e and C_c

IV. NOT-STANDARDS WAVE PROPAGATION

We can replace in the preceding line the shock wave of the Heidler type by an oscillatory shock wave of 1800kV of amplitude whose equation used is:

$$V(t) = Vmax \left(1 - e^{-\alpha t} \cos \omega t\right) \qquad (4)$$

The figure 10 shows that the deformation appears only on the first cycle of this wave. This remark was also noted by the authors of the reference [6], [10], [11].

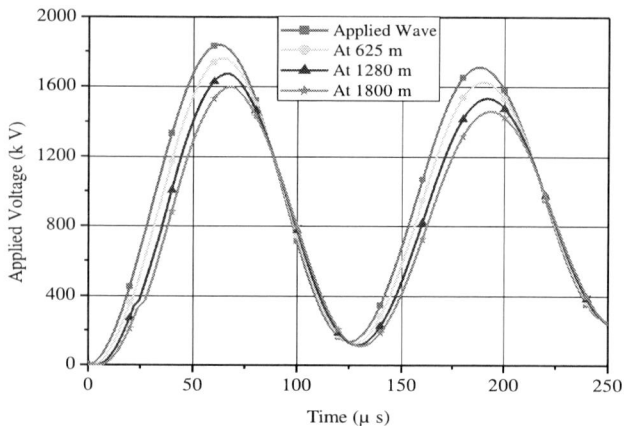

Figure 10. Calculation of overvoltage's on various points.

V. CONCLUSION

The propagation of the shock waves at summer analyzed with the injection of a corona model on a transmission line represented by a series of the identical sections assembled in cascade.

The results of simulation show on the one hand that the corona resistance does not have any effect on the wave form, on the other hand, the principal causes of the deformation and of the damping of shock wave are due to the various losses of energy such as the losses by Joule effect, of the return by the ground and the losses by corona effect.

This problem has been posed for one half-century already. However work was completed without computers, on weak radius conductor and small lengths of propagation (3 km). A new support adapted to the lines of today, as well as a refinement of the means of simulation became today necessary.

I. NOMENCLATURE

r_0 :	Is the interior ray of conductor, in [cm].
m :	Coefficient of the conductor surface state.
B :	Empirical constant between 0.5 and 0.67.
δ:	relative density of the air.
H :	Height of the conductor in [m].
C_g :	Geometrical capacity of line in [µF].
C_c :	Corona capacitance in [µF].
ε_0:	Permittivity of the air.
E_0 :	Electric field ahead streamer system in [kV/cm].
R_e :	Joule losses in [Ω].
L_e :	The return by the ground in [H].
Z_c :	Characteristic impedance in [Ω]

REFERENCES

[1] M. Al-Tai, H. S. B. Elyyan, D. M. German, A. Haddad, N. Harid, R. T. Waters, 'The simulation of surge corona on transmission lines', IEEE Transactions on Power Delivery, Vol. 4, No.2, pp. 1360-1368, April 1989.

[2] N. Harid, R. T. Waters, 'Statistical study of impulse corona inception parameters on line conductors', IEE PROCEEDINGS-A, Vol. 138, No. 3, pp. 161-168, May1991.

[3] D.A. Rickard, N. Harid, R. T. Waters, 'Modelling of corona at a high-voltage conductor under double exponential and oscillatory impulses', IEE Proc-Sci. Meus. Teclinol, Vol. 143, No. 5, pp. 277-284, September. 1996

[4] H. M. Kudyan, C. H- Shih, 'A nonlinear circuit model for transmission lines in corona', IEEE Transactions on Power Apparatus and Systems, Vol. PAS-l00, No. 3, pp.1420-1430, March 1981.

[5] S. Carneiro, J. R. Marti, 'Evaluation of corona and line models in electromagnetic transients simulations', IEEE Transactions on Power Delivery. Vol. 6, No. 1, pp. 334-342, January 1991.

[6] A. Gherbi, 'Influence de l'effet couronne sur les surtensions de choc dans les lignes de transmission de l'énergie électrique', thèse de magister, Université de sétif, Décembre 1995.

[7] F. Castellanos, 'Full frequency dependent phase-domain modeling of transmission lines and corona phenomena', PHD thesis, University of British Columbia, February 1997.

[8] M. Z. A. Kadir, W. F. W. Ahmed, J. Jasni, H. Hizam, 'The importance of corona effect in lighting surge propagation studies', Journal Of Applied Science 8(19), Asian Network For Scientific Information, pp. 3446-3452, 2008.

[9] M. Afghahi, R.J. Harrington, 'Charge model for studying corona during surges on overhead transmission lines '', IEEE Proc, Vol. 130, Pt. C, No. 1, pp. 16-21, January 1983

[10] Z. Anane, A. Bayadi, 'Implantation Of A Static Model Of The Corona Effect In The ATP-EMTP Software', 2011 7th International Workshop on Systems, Signal Processing and their Applications (WOSSPA), PP. 39-42, Algeria.

[11] Z. Anane, 'Modélisation de l'effet couronne pour l'analyse de la propagation des ondes de chocs sur les lignes de transmission aériennes' These of Magister, University of Setif. Algeria, February. 2011.

A High-resolution DCO with MOS Capacitors

Zixuan. Wang, Cheng. Huang, Jianhui. Wu
National ASIC system engineering research centre
Southeast University
Nanjing, 210096 P. R. China
wangzx_inf@hotmail.com

Abstract—**A digitally controlled oscillator (DCO) that achieves a minimum frequency tuning step of 20 kHz without any dithering is presented. Three tuning stages are employed to obtain a wide frequency range of 1 GHz in the classical LC tank. The fine tuning bank is realized by inverse connection of two pairs of pMOS transistors and a tiny unit capacitance of 0.47 fF is achieved. A prototype integrated in 130nm CMOS technology exhibits a phase noise of -118.7 dBc/Hz @1MHz offset and a power dissipation of 2 mW under a supply of 1.2 V. The core area size is 330um×480um.**

Keywords- DCO; ADPLL; hign resolution; phase noise;

I. INTRODUCTION

With the explosive growth of the wireless communication industry, the research on transceiver becomes hot spot rapidly. As a key part of transceiver, phase locked loops (PLLs) provide pure local oscillator (LO) signal for the system [1]. During last decade, the conventional PLLs have been replaced by all digital PLLs (ADPLLs) gradually. Since the inherent noise immunity of digital circuit, ADPLLs have better tolerance to process variations and ambient conditions than analog implementation of PLLs. Besides, ADPLLs have advantage of easy scalability with process shrink.

The most important performance of PLL is phase noise. The DCOs domain the outband noise of ADPLLs. High frequency resolution of DCOs is benefit to decreasing the quantization noise and depressing the outband phase noise. Therefore, the design of DCOs with fine frequency tuning step becomes a big challenge for most researchers. In 2009, a DCO, which has been proposed in [2], achieves frequency resolution of 250 kHz by using MOS varactor. In order to suit digital communication with other modules in the ADPLL, a digital-analog converter (DAC) has to be added at the front of DCO. The additional DAC circuit inevitably increases the core area and power dissipation. Furthermore, this circuit structure does not accord with the concept of DCO. Therefore, it is still sensitive to process voltage supply and temperature (PVT) variations like VCOs. A fully digital implementation of oscillator, which avoids any analog tuning controls, has been presented in [3] for bluetooth radio. For achieving fine tuning step, a digital $\Sigma \Delta$ modulator is utilized. Although the resolution of DCO is enhanced through high-speed $\Sigma \Delta$ dithering, more core area and power dissipation have to be costed simultaneously. Recently, some implementations of DCOs utilize tiny capacitors based on transmission lines, as described in [4, 5]. Since additional assistant circuit is unnecessary, the DCOs consume lower power dissipation. However, the DCOs lose the advantage of easy scalability due to the use of passive device. Moreover, the length of transmission lines also dominates the chip size of DCO.

In this paper, we propose an LC-DCO composed of three tuning stages, which realizes wide tuning range and high frequency resolution without any analog controls. MOS capacitors are used to realize minimum tuning step of 20 kHz and oscillating frequency from 3.20GHz to 4.28GHz by utilizing LC tank that contains 14 control bits. The structure of the proposed DCO is introduced in section II. Section III presents the circuit details of the DCO implementation, followed by the simulate results and layout in section IV. In the conclusion, we discuss the features of the proposed DCO in summary.

II. THE PROPOSED DCO

To achieve wide frequency range, the DCO utilizes three tuning stages that contain a five bits coarse array, a five bits medium array and a four bits fine array, as shown in Fig. 1. The tank uses a differential inductor of 1.968nH with a simulated quality factor of 15. The fixed capacitor C_{fix} of 50fF is implemented in MIM technology. The transistors of cross-coupled pair are biased to work always in the saturation region.

In the coarse tuning stage, an array of MIM capacitors are employed to achieve a frequency range of about 1GHz, as shown in Fig. 2. The estimated values of the capacitors vary from 476 fF to 29.77 fF for Most Significant Bit (MSB) to Least Significant Bit (LSB), respectively. The MIM capacitors are chosen to work through turning on the corresponding nMOS transistors. So there are 32 separate frequency channels in the coarse tuning bank.

The medium tuning bank consists of five pairs of pMOS capacitors that employ conventional method of MOS capacitor [4, 5]. As shown in Fig. 3, the sources and drains of transistors in every capacitor unit are connected to control voltage Mn. The capacitor unit works in region A and region B when Mn connects vdd and gnd respectively, as shown in Fig. 5. Therefore, the capacitance realized by every capacitor unit of medium tuning bank is determined by

$$\Delta C_{Medium} = C_B - C_A \qquad (1)$$

Each pair of pMOS capacitors can realize capacitance of 1.948 fF with good linearity. The frequency range tuned by medium tuning bank varies from 37 MHz to 89 MHz within the effective frequency range of the DCO (i.e., 3.2GHz to

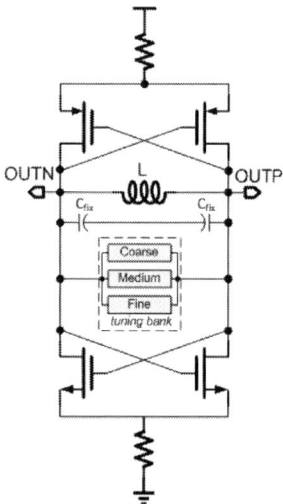

4.2GHz).

Figure 1.   Structure of the proposed DCO

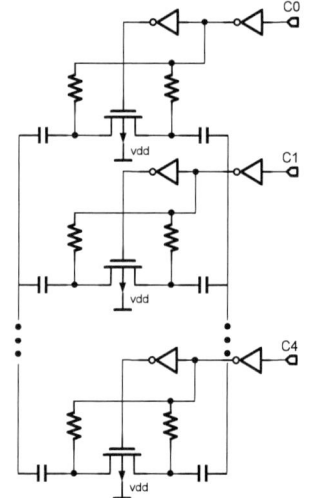

Figure 2.   Schematic of the coarse tuning bank

An improved MOS capacitor presented in this paper has much finer unit capacitance compared to the conventional MOS capacitor in Fig. 3. As shown in Fig. 4, the proposed capacitor unit comprises two pMOS pairs that are inversely connected in parallel. The bulks of pMOS transistors are connected to sources and drains so that the transistors have approximate capacitance in the region of accumulation and inversion. Fig. 5 shows the capacitance-voltage curve for PAIR1, PAIR2 and conventional MOS capacitor. For PAIR1, high control voltage makes PM1 and PM2 work in region F while low control voltage implies transistors work in region E. On the contrary, high and low control voltage imply region C and D respectively for PAIR2 (capacitance in region D approximates that in region B). Therefore, the capacitance

realized by every capacitor unit of fine tuning bank is determined by

$$\Delta C_{Fine} = (C_F + C_C) - (C_E + C_D) \qquad (2)$$

Form (2), we can see that the unit capacitance depends on the capacitance difference between accumulation region and inversion region. By the means of reducing the size difference between PAIR1 and PAIR2, a tiny unit capacitance and an accurate tuning step can be achieved. In this work, the sizes of transistors in PAIR1 and PAIR2 are 40/13 and 90/13, respectively. The minimum unit capacitance of fine tuning bank is 0.47 fF and the minimum tuning step is 20 kHz. It is noted that extreme attention should be paid to design such tiny capacitance in order to avoid parasitic capacitance from surpassing the required value of the capacitors.

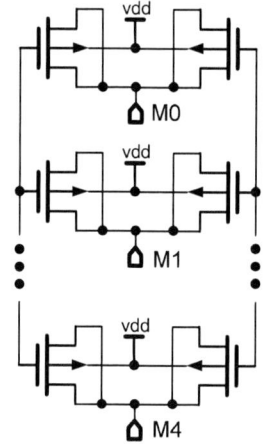

Figure 3.   Schematic of the medium tuning bank

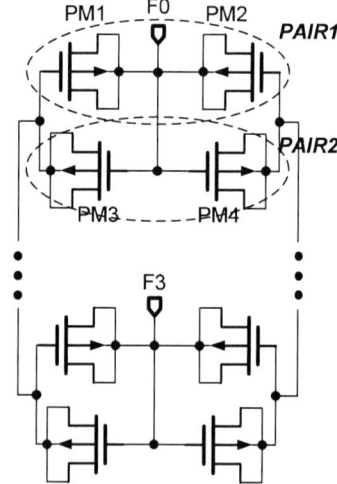

Figure 4.   Schematic of the fine tuning bank

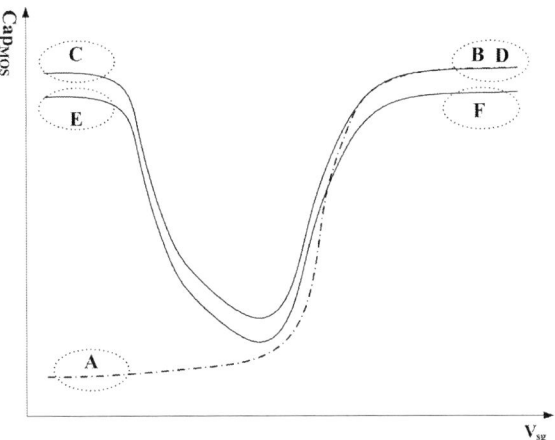

Figure 5. Curve of Capacitance-Voltage

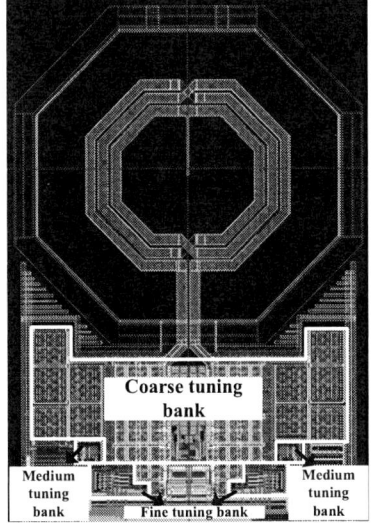

Figure 6. Layout of proposed DCO

III. LAYOUT AND SIMULATION

The proposed DCO has been designed in TSMC 130nm CMOS process. Fig. 6 shows the full layout diagram. The core area size is 330um×480um. The output of DCO is designed with buffer circuits to amplify the amplitude of signals and isolate DCO from ambient conditions. The simulated DC power dissipation consumed by the whole circuit is 2 mW at 1.2 V supply, of which about 40% is consumed by the buffer circuit.

The frequency tuning range achieved by the DCO is 1 GHz, and the minimum frequency step is 20 kHz. Fig. 7 shows the graph of output frequency versus the control words, which implies the gain of the DCO. It can be seen that the DCO has a good linearity of tuned frequency within the effective region. In the low frequency region, the gain of DCO is slightly

smaller than that of high frequency region. Moreover, the linearity becomes worse at highest frequency, and this is mainly due to more significant parasitic problem.

The curves of phase noise at 3.2 GHz, 3.6 GHz and 4.2 GHz are shown in Fig. 8. The phase noise has a difference of 3.8 dBc between the lowest frequency and highest frequency and the value at 3.6 GHz is -118.7 dBc/Hz @1MHz offset.

The comparison between the performance of the proposed DCO and other implementations is reported in Table 1. The advantages of the proposed DCO are high frequency resolution, low power dissipation and small area. Reference [5] has slight superiorities in frequency resolution and Phase noise but the power dissipation consumed is over 15 times higher than the DCO consumed in this paper.

Figure 7. Transfer characteristic of the proposed DCO

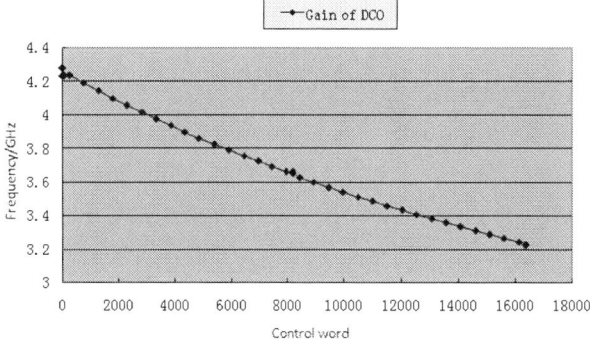

TABLE I. SUMMARY RESULTS AND COMPARISON WITH OTHER DCOs

	This work	[2]	[3]	[4]	[5]
Centre frequency (GHz)	3.6	1.2	2.4	9	5.1
Tuning range (MHz)	1000	1000	600	380	510
Frequency resolution (kHz)	20	250*	23*	100	18
Phase noise (dBc/Hz)	-118.7 @1MHz	-97 @1MHz	-110 @500kHz	-105 @1MHz	-118 @1MHz
Power dissipation (mW)	2	N/A	N/A	9	32
Area (mm²)	0.16	N/A	N/A	0.49	0.2
Technology (nm)	130	130	130	180	180

* without dithering

IV. CONCLUSION

We have proposed a 14-bit DCO for wireless communication. The fully digitally controlled oscillator does not employ any DAC or delta-sigma modulator. Three tuning

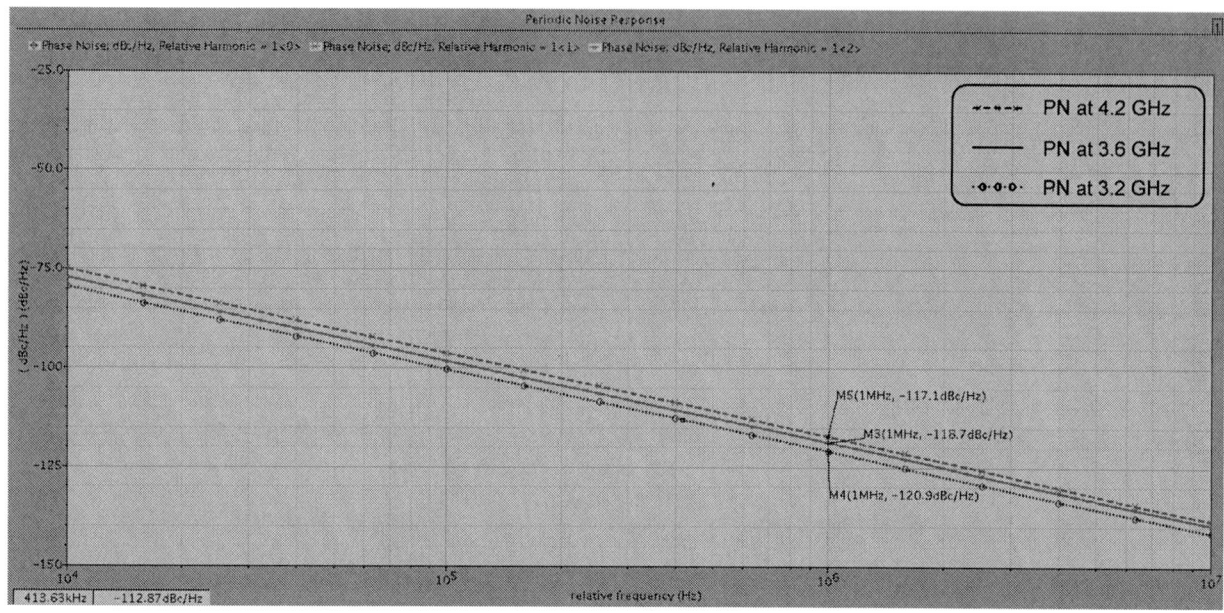

Figure 8.   Phase noise of the proposed DCO

stages are used to achieve a tuning range of 1GHz. PMOS capacitors are employed in the fine tuning stage and a minimum frequency step of 20 kHz is realized without any analog control or dithering. Moreover, the DCO has advantages of low power dissipation of 2mW and small area of 0.16 mm². Because the finest tuning bank is realized by all MOS transistors, the DCO also has good scalability and promotion of performance with the development of technology.

ACKNOWLEDGMENT

This work was supported by the National Natural Science Foundation of China (No. 60871079 and No. 61176031) and Natural Science Foundation of Jiangsu (BK2011334).

REFERENCES

[1] R. B. Staszewski, H. Chih-Ming, N. Barton, L. Meng-Chang, and D. Leipold, "A digitally controlled oscillator in a 90 nm digital CMOS process for mobile phones," Solid-State Circuits, IEEE Journal of, vol. 40, pp. 2203-2211, 2005.

[2] V. Kratyuk, P. K. Hanumolu, K. Ok, M. Un-Ku, and K. Mayaram, "A Digital PLL With a Stochastic Time-to-Digital Converter," Circuits and Systems I: Regular Papers, IEEE Transactions on, vol. 56, pp. 1612-1621, 2009.

[3] R. B. Staszewski, K. Muhammad, D. Leipold, H. Chih-Ming, H. Yo-Chuol, J. L. Wallberg, C. Fernando, K. Maggio, R. Staszewski, T. Jung, K. Jinseok, S. John, D. Irene Yuanying, V. Sarda, O. Moreira-Tamayo, V. Mayega, R. Katz, O. Friedman, O. E. Eliezer, E. de-Obaldia, and P. T. Balsara, "All-digital TX frequency synthesizer and discrete-time receiver for Bluetooth radio in 130-nm CMOS," Solid-State Circuits, IEEE Journal of, vol. 39, pp. 2278-2291, 2004.

[4] W. Chaivipas, T. Ito, T. Kurashina, K. Okada, and A. Matsuzawa, "Fine and wide frequency tuning digital controlled oscillators utilizing capacitance position sensitivity in distributed resonators," in Solid-State Circuits Conference, 2007. ASSCC '07. IEEE Asian, 2007, pp. 424-427.

[5] R. K. Pokharel, K. Uchida, A. Tomar, H. Kanaya, and K. Yoshida, "Low phase noise 18 kHz frequency tuning step 5 GHz DCO using tiny

capacitors based on transmissi on lines," in Silicon Monolithic Integrated Circuits in RF Systems (SiRF), 2010 Topical Meeting on, 2010, pp. 8-11.

A microcontroller-based pulse generator for isothermal I-V measurements

Maurizio Costagliola, Vincenzo d'Alessandro, Grazia Sasso, Niccolò Rinaldi

Department of Biomedical, Electronics and Telecommunications Engineering
University of Naples Federico II
Naples, Italy
Mail: maurizio.costagliola@unina.it

Abstract — **An *in-house* general-purpose microcontroller-based pulse generator is presented and described in detail. Besides producing a pulse train with assigned specifications, the system is provided with an auxiliary unit devoted to measure *on-board* the I-V characteristics of various device categories. The pulser is successfully adopted for the isothermal characterization of packaged power BJTs and MOS transistors, as well as of on-wafer SiGe:C HBTs.**

Keywords – Isothermal characterization, pulsed measurements, self-heating, thermal resistance.

I. INTRODUCTION

A reliable experimental characterization of semiconductor devices is steadily requested for several purposes, e.g., model parameters extraction, process diagnostics, and safe operating area definition. However, measurements are often performed under dc bias conditions, which may give rise to a significant self-heating and a consequent distortion of the I–V curves in comparison to the isothermal case. Besides, electrothermal effects – traditionally associated to high-power devices – are nowadays exacerbated also in state-of-the-art HF transistors, like SiGe HBTs and silicon-on-glass BJTs, which is ascribable to a twofold reason: (i) the self-heating thermal resistances of these devices have grown to several thousands of K/W due to the scaling process and to the low thermal conductivity of the materials surrounding the active device region [1], [2]; (ii) the frequency performance is improved by shifting the operating point to higher current densities, which increases the power density consumption.

Advanced equipments suited to bias transistors with short pulse widths (even lower than 1 μs) are therefore needed to annihilate self-heating during the measurement process, thus ensuring isothermal conditions [3], [4]. Unfortunately, the cost of commercial curve tracers including this feature may exceed 100 k\$. This has motivated a relevant effort to develop cheaper, yet reliable, systems [5]-[7].

In this contribution, an *in-house* general-purpose pulse generator is proposed, which can be exploited for the isothermal I–V characterization of a large variety of semiconductor devices, both in package and on wafer. The circuit is based on a microcontroller unit (MCU) and allows the setting (via firmware) of period and duty-cycle of the voltage

pulse train; in addition, it is equipped with a section based on an instrumentation amplifier for *on-board* measurements.

The system is described in detail in Section II, and successfully applied to various transistor typologies in Section III. Conclusions are then given in Section IV.

II. THE PULSE SYSTEM

The proposed circuit can be described as follows. The block diagram and the PCB prototype are shown in Figs. 1 and 2, respectively.

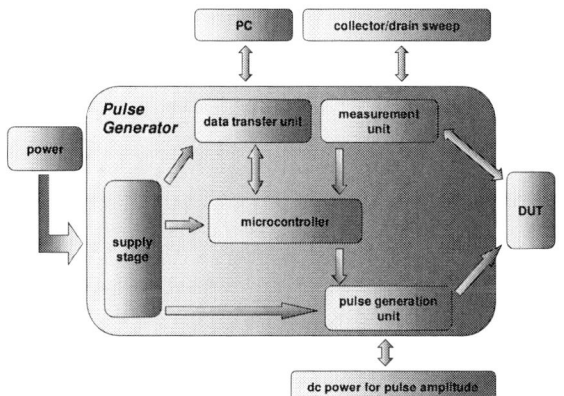

Figure 1. Schematic block diagram of the proposed circuit.

Figure 2. Picture of the realized prototype.

A. The supply stage

The system requires different voltages for the dc-supply of the *on-board* integrated circuits, namely,

- 3.3 V and 2.5 V for the MCU;
- 5 V for the Max232 integrated circuit;
- 15 V for the half-bridge driver;
- +/-15 V for the instrumentation amplifier of the measurement unit.

The supply stage is based on an external 12 V power supply. Various regulators are used to generate the requested voltages.

B. The measurement strategy and the control unit

The control unit is based on the Microchip 28-pin MCU dsPIC33FJ16GS402, which represents the core of the system and includes: a high-speed PWM module, a high-speed 10-bit ADC, an UART serial peripheral, and a 2-Kbytes data SRAM. The developed firmware of the MCU coordinates the measurement flow by means of the following steps: (1) generation of the PWM signal to be sent to the half-bridge driver (this signal has the same width T_{on} and period T as the desired pulse); (2) acquisition – via the 10-bit ADC – of a prescribed number of samples (programmed by firmware), which are stored in the internal SRAM (see Section II.D); (3) data transfer through the UART peripheral.

C. The pulse generation unit

The pulse generation is demanded to a power MOS with high breakdown voltage and low on-resistance so as to satisfy the desired pulse specifications; in particular, the STP22NF03L transistor was selected and mounted on the prototype. The drain of this device is connected to an external voltage supply, while the gate-source voltage V_{GS} is provided by the half-bridge driver. As a result, the MOS is switched on and off accordingly to the period and duty-cycle of the PWM signal. In particular, the positive V_{GS} value is high enough to push the transistor into deep linear mode, thereby ensuring a low voltage drop between drain and source. This gives rise to a train of pulses on the source terminal, with amplitude slightly lower than the supplied drain voltage (due to the V_{DS} drop), and desired period and duty-cycle. The amplitude can be measured *on-board* by the 10-bit ADC of the MCU.

D. The measurement unit

The proposed system, besides generating a pulse train with assigned features, is also equipped with an *ad-hoc* unit – based on the instrumentation amplifier INA110KP – which can be enabled for *on-board* measurements. The operating principle of this block can be described as follows. A known resistance R_C is connected between an external voltage supply and the collector/drain of the DUT, whose base/gate is biased with the pulse train. The input pins of the INA100KP are connected in parallel to R_C to sense and amplify the voltage drop due to the collector/drain current as the device is turned on by the pulses; the output pin is connected to a channel of the 10-bit ADC of the MCU (a 3.3 V Zener diode is used to protect the analog part of the ADC channel). Moreover, another ADC channel is devised to measure the supply voltage. In conclusion, both the voltage and current corresponding to each point of an I–V characteristic are determined and subsequently stored in the internal SRAM.

The measurement unit is conceived to sense the current within a wide range of values (from 10 nA to 1 A), which is accomplished as follows. A group of 10 different resistors are available, which have a terminal tied to the external voltage supply and the other floating; R_C is selected among them consistently with the current span to be investigated by connecting the floating terminal to the collector/drain of the DUT through a jumper. In principle, the resistance R_C can be also given by a chosen combination of paralleled resistors belonging to the aforementioned group.

The trigger for the ADC acquisition is the rising edge of the PWM generated by the MCU with a delay programmed via firmware. This allows safely performing the measurement during the pulse width T_{on}.

Lastly, it must be remarked that the measurement unit can be kept fully deactivated as the system is exploited only to generate the pulse train while the data acquisition is demanded to external instruments (*off-board* mode). However, in this case the synchronization between pulse train and data acquisition becomes quite cumbersome.

E. The data transfer unit

At the end of the programmed number of acquisitions, the measurement task terminates, the ADC is turned off, and a firmware routine coordinates the data transfer via the UART peripheral. The Max232 is employed to adapt the voltage levels to the RS-232 protocol. The I–V data buffered in the internal SRAM are then post-processed off-line.

F. Pulser specifications

The proposed circuit is suited to generate a minimum pulse width T_{on} of 400 ns, as shown in Fig. 3. The main pulser specifications are reported in Table I.

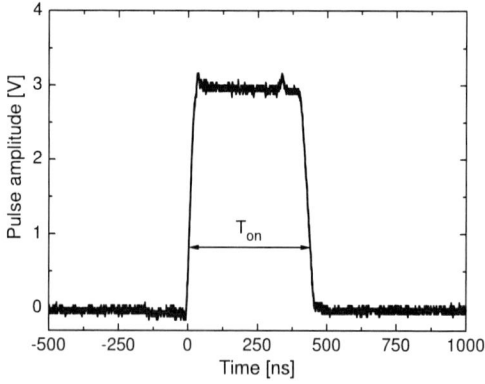

Figure 3. Generated pulse with amplitude of 3 V and width T_{on} of 400 ns.

TABLE I. SYSTEM SPECIFICATIONS

Specification	Value
Maximum pulse amplitude	15 V
Current range	10^{-8} A ÷ 1 A
Minimum pulse width	400 ns
Minimum duty cycle	0.074%
Maximum duty cycle	100%
Start-up time	10 μs
Maximum number of samples	480
Maximum number of sample averages	120

III. EXPERIMENTAL RESULTS

The proposed pulse generator was used to experimentally characterize various device categories, i.e., in-package power Si BJTs and MOS transistors, as well as on-wafer SiGe HBTs for RF applications. A comparison with the corresponding dc data – measured through an HP4142B parameter analyzer – was also carried out in order to highlight the reduction in self-heating effects achieved with the pulsed bias. The applied T_{on} was in all cases longer than the settling time of the DUTs.

First, the constant-V_{BE} I_C–V_{CE} curves of the commercial packaged 2N3415 NPN power BJT (featuring BV_{CEO}=25 V, I_{Cmax}=500 mA, and junction-to-ambient thermal resistance R_{TH}=200°C/W) were measured by applying the pulse train to the base terminal and sweeping the collector voltage by means of an external dc voltage source. In particular, the pulsed characteristics were obtained by varying the width T_{on} of the applied pulse for a period T=100 μs (i.e., by varying the duty-cycle). Fig. 4 clarifies that a reduction in T_{on} allows increasingly countering the self-heating impact on the curve slope induced by the well-known positive temperature coefficient (TC) of the collector current [8]. In particular, T_{on}<10 μs (i.e., duty-cycle <10%) was found to be suited for achieving isothermal conditions. For the sake of clarity, the common-emitter configuration adopted for this analysis is depicted in the figure inset. An isothermal curve family obtained by varying V_{BE} is shown in Fig. 5; all the characteristics were measured by applying a pulse train with T_{on}=5 μs and T=100 μs.

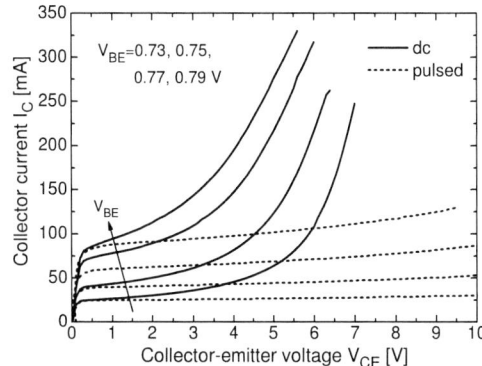

Figure 5. Experimental dc (solid lines) and pulsed (dashed) output characteristics of a 2N3415 BJT for various V_{BE} values. The pulsed curves were obtained with T_{on}=5 μs and T=100 μs.

The proposed system allows *on-board* measurements also for on-wafer transistors. As an illustrative case-study, state-of-the-art SiGe:C HBTs featuring f_T/f_{max}=260/350 GHz [9] were experimentally analyzed. Fig. 6 reports both the dc and pulsed I_C–V_{CE} characteristics of a device with high R_{TH} (≈5000 K/W). Again, the pulsed curves were obtained by considering a pulse train with T_{on}=5 μs and T=100 μs, for which self-heating is nearly eliminated; it should be noted that in this case the current increase observed at high V_{CE} values is induced by weak avalanche effects (BV_{CBO} was measured to be about 5 V).

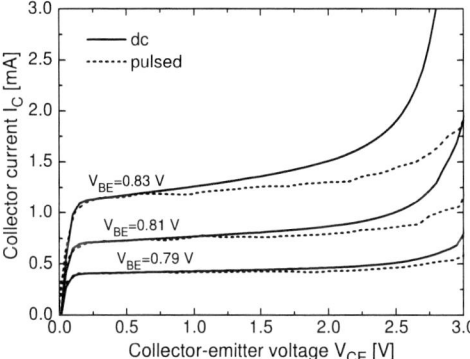

Figure 6. Experimental dc (solid lines) and pulsed (dashed) I_C–V_{CE} characteristics of a SiGe:C HBT for various V_{BE} values. The pulsed curves were obtained with T_{on}=5 μs and T=100 μs.

Fig. 7 reports the I_D–V_{DS} characteristics measured for the commercial packaged BS170 NMOS transistor (characterized by V_{TH}=2.1 V, BV_{DS}=60 V, I_{Dmax}=500 mA, R_{TH}=150°C/W) at different gate-source voltages V_{GS}. A pulse train with T_{on}=10 μs and T=100 μs was found to guarantee isothermal conditions. A close inspection of the curves plainly confirms that the TC of the drain current I_D reverses its sign by increasing V_{GS}, which can be explained as follows. The thermal behavior of a MOS transistor is related to the temperature dependence of threshold voltage V_{TH} and channel mobility $μ_n$. Both these key parameters decrease with increasing temperature. However, the influence on the drain current is different: the reduction in V_{TH} leads to a positive TC,

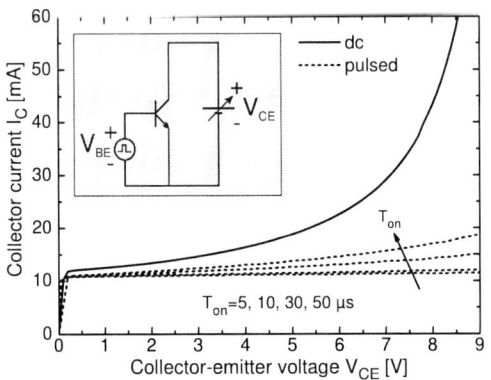

Figure 4. Experimental I_C–V_{CE} characteristics of a 2N3415 BJT measured at V_{BE}=0.7 V under pulsed conditions for different pulse widths (dashed lines), along with the corresponding dc curve (solid).

978-1-4673-5289-5/12 $31.00 © 2012 IEEE

whereas the μ_n lowering entails a negative TC. At low V_{GS} (low I_D) the first effect prevails, while the second dominates at high V_{GS} (high I_D) [10], [11]. The isothermal measurements reveal that the device is weakly subject to the (merely electrical) channel modulation effect, as can be evinced by the low slope of the characteristics.

It is noteworthy that the I–V characteristics measured for the packaged devices shown in Figs. 5 (BJT) and 7 (MOS) were found to be comparable to those reported in the corresponding datasheets.

The proposed system was also employed as a mere pulse generator for the *off-board* monitoring of the transient thermal impedance evolution of on-wafer multifinger GaN HEMTs [12].

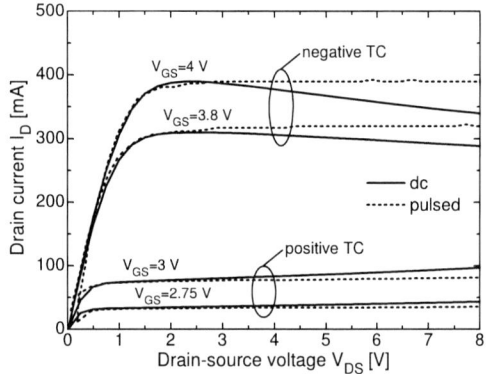

Figure 7. Experimental dc (solid lines) and pulsed (dashed) I_D–V_{DS} characteristics of a BS170 NMOS for various V_{GS} values. The pulsed curves were obtained with T_{on}=10 μs and T=100 μs.

IV. CONCLUSIONS

A novel general-purpose pulse generator has been designed, realized and characterized. The system is provided with a section devised for *on-board* measurements, which allows a reliable synchronism between pulse train and data acquisition. The versatility of the pulser has been demonstrated by obtaining isothermal I–V characteristics for different types of transistors, including on-wafer SiGe:C HBTs with very high thermal resistances.

This allows concluding that – in spite of the much lower cost – the proposed system can be considered as a trustworthy alternative to advanced commercial curve tracers.

REFERENCES

[1] N. Nenadović, V. d'Alessandro, L. K. Nanver, F. Tamigi, N. Rinaldi, and J. W. Slotboom, "A back-wafer contacted silicon-on-glass integrated bipolar process – Part II: A novel analysis of thermal breakdown," IEEE Transactions on Electron Devices, vol. 51, no. 1, pp. 51-62, 2004.

[2] V. d'Alessandro, I. Marano, S. Russo, D. Céli, A. Chantre, P. Chevalier, F. Pourchon, and N. Rinaldi, "Impact of layout and technology parameters on the thermal resistance of SiGe:C HBTs," in Proc. IEEE Bipolar/BiCMOS Circuits and Technology Meeting, 2010, pp. 137-140.

[3] A. Hammache, G. Brassard, M. Bouchard, F. Beauregard, C. Akyel, and F. M. Ghannouchi, "Thermal characterization of MESFETs using I-V pulsed and dc measurements," in Proc. IEEE Instrumentation and Measurement Technology Conference, 1997, pp. 664-667.

[4] A. K. Sahoo, M. Weiß, S. Fregonese, N. Malbert, and T. Zimmer, "Transient electro-thermal characterization of Si-Ge heterojunction bipolar transistos," Solid-State Electronics, vol. 74, pp. 77-84, 2012.

[5] A. Platzker, A. Palevsky, S. Nash, W. Struble, and Y. Tajima, "Characterization of GaAs devices by a versatile pulsed I-V measurement system," in IEEE International Microwave Theory and Techniques Symposium Digest, 1990, vol. 3, pp. 1137-1140.

[6] J. P. Teyssier, P. Bouysse, Z. Ouarch, D. Barataud, T. Peyretaillade, and R. Quéré, "40-GHz/150-ns versatile pulsed measurement system for microwave transistor isothermal characterization," IEEE Transactions on Microwave Theory and Techniques, vol. 46, no. 12, pp. 2043-2052, 1998.

[7] M. Marchetti, K. Buisman, M. Pelk, L. Smith, and L. C. N. de Vreede, "A low-cost pulsed RF & I-V measurement setup for isothermal device characterization," in 70th ARFTG Microwave Measurement Symposium – High Power RF Measurement Techniques Digest, 2007.

[8] N. Rinaldi and V. d'Alessandro, "Theory of electrothermal behavior of bipolar transistors: Part I – Single-finger devices," IEEE Transactions on Electron Devices, vol. 52, no. 9, pp. 2009-2021, 2005.

[9] P. Chevalier, F. Pourchon, T. Lacave, G. Avenier, Y. Campidelli, L. Depoyan, G. Troillard, M. Buczko, D. Gloria, D. Céli, C. Gaquière, and A. Chantre, "A conventional double-polysilicon FSA-SEG Si/SiGe:C HBT reaching 400 GHz f_{MAX}," in Proc. Bipolar/BiCMOS Circuits and Technology Meeting, 2009, pp. 1-4.

[10] Z. Prijić, Z. Pavlović, S. Ristić, and N. Stojadinović, "Zero-temperature-coefficient (ZTC) biasing of power VDMOS transistors," Electronics Letters, vol. 29, no. 5, pp. 435-437, 1993.

[11] A. Castellazzi, M. Honsberg-Riedl, and G. Wachutka, "Thermal characterisation of power devices during transient operation," Microelectronics Journal, vol. 37, no. 2, pp. 145-151, 2006.

[12] S. Russo, V. d'Alessandro, M. Costagliola, G. Sasso, and N. Rinaldi, "Analysis of the thermal behavior of AlGaN/GaN HEMTs," Material Science and Engineering B, vol. 177, no. 15, pp. 1343-1351, 2012.

978-1-4673-5289-5/12 $31.00 © 2012 IEEE

RADFET dosimeter design for environment monitoring applications

M. Meguellati[1], F. Djeffal[*,1] D. Arar[1], T. Bendib[1,2] and L. Khettache[1]

[1] LEA, Department of Electronics, University of Batna, Batna 05000, Algeria
[2] Department of Electronics, University of M'sila, M'sila 28000, Algeria
[*] E-mail: faycaldzdz@hotmail.com, faycal.djeffal@univ-batna.dz
Tel/Fax: 0021333805494

Abstract— In this paper, a radiation sensitive FET (RADFET) dosimeter design (called the Dual-Dielectric Gate All Around DDGAA RADFET dosimeter) to improve the radiation sensitivity performance and its analytical analysis have been proposed for RADFET dosimeter-based applications (monitoring, robotics, medical sciences,…). The proposed device has been implemented in SIMULINK tool to show the impact of the proposed dosimeter on the environment monitoring applications. The obtained results make the DDGAA RADFET dosimeter a promising candidate for environment monitoring applications.

Index Terms—dosimeter, RADFET, traps, irradiation, sensitivity, Genetic Algorithm.

I. INTRODUCTION

The Gate All Around GAA MOSFETs have emerged as excellent devices to provide the electrostatic integrity needed to scale down transistors to minimal channel lengths, and allowing a continuous progress in digital and analog applications. In addition to a better electrostatics than the conventional bulk MOSFET, the use of these devices have advantages relative to the electronic transport, mainly due to (i) the reduced surface roughness scattering because the lower vertical electric field and (ii) the reduction of the Coulomb scattering because the film is made of undoped/low-doped silicon [1-5]. Design and modeling guidelines of GAA MOSFETs have been discussed in previous work [2-5]. Employing this design for environment monitoring applications (irradiation measurement) becomes more beneficial if the device is made in vertical cylindrical recrystallized silicon due to highly flexible process integration options. There have been several reports of MOSFETs fabricated in recrystallized silicon for high-density digital integrated circuits [5].

Radiation sensitive MOSFETs (RADFETs) have been focus of interest both from applications and fundamental research point of views. In electronic industry these devices are considered as attractive alternatives for nuclear industry, space, radiotherapy and environment monitoring applications due to their reliability, low power consumption, non-destructive read-out of dosimetric information, high dose range, and compatibility to standard CMOS technology and on-chip signal processing [6-8]. The main RADFET disadvantage is the relatively low sensitivity. In this context, the submicron multi-gate design may be considered as attractive alternative to overcome this disadvantage because of the high electrical performance and reliability provided by the

multi-gate structure in comparison with single-gate one. However, as semiconductor devices are scaled into the deep submicron domain, short-channel effects (SCEs) begin to plague conventional planar CMOS-based devices. To avoid the electrical constraints and improve the sensitivity performance, a new design and enhancement of conventional (bulk) RADFET become important. In this work, a new design of RADFET called the Dual-Dielectric Gate All Around (DDGAA) RADFET dosimeter, in which the manufacturing processes and sensitivity performances will be greatly improved, is proposed for deep submicron CMOS-based dosimeter applications. The (DDGAA) RADFET dosimeter design presented in this paper is basically surrounded dual-dielectric layers (SiO_2 and Si_3N_4) with low p-channel (Si) doping concentration. The results showed that the analytical model is in agreement with the 2-D numerical simulation over a wide range of device parameters. The proposed structure has been analyzed and validated by the good sensitivity and electrical performance obtained in deep submicron regime in comparison with the conventional (bulk) design. Finally, the proposed dosimeter model was used as a subcircuit in the SIMULINK software for the simulation of the environment monitoring applications. In addition, in this work, we present the applicability of genetic algorithm optimization (GA) approach to optimize the radiation sensitivity of DDGAA RADFET for integrated CMOS-based dosimeters.

II. THEORY DEVELOPMENT AND MODEL DERIVATION

A. Interface potential analysis

Schematic cross-sectional view of the proposed (DDGAA) RADFET dosimeter is presented in Fig.1. The insulator consists of a thermal oxide (SiO_2) grown on a (100) n on n+ expitaxial silicon substrate (channel), and a low pressure CVD silicon nitride layer(Si_3N_4) deposited on top of the oxide. ND/S represents the doping level of the drain/source region, respectively. The channel region is bounded by source and drain spacing at $x= 0$ and L, respectively, where L is the gate length. With a negatively applied gate bias, holes generated in the SiO_2 layer are transported and trapped at the SiO_2 /Si_3N_4 interface producing a measurable threshold-voltage shift as it is shown in Fig. 1. The investigation reported in this work for gamma radiation sources can also be applied qualitatively to other radiation sources (protons, electrons, …).

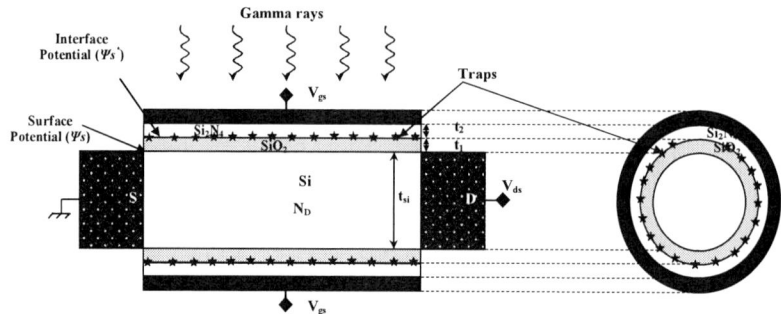

Figure 1. Cross-sectional view of the proposed DDGAA RADFET design

For deep submicron devices, the solution of 2D Poisson's equation satisfying suitable boundary conditions is required to model the interface potential. Refer to Fig. 1, the 2D Poisson's equation for the channel region is given by

$$\frac{1}{r}\frac{\partial}{\partial r}\left(r\frac{\partial}{\partial r}\psi(r,x)\right)+\frac{\partial^2}{\partial x^2}\psi(r,x)=\frac{q\cdot N_D}{\varepsilon_{si}} \qquad (1)$$

The boundary conditions for $\psi(x,r)$ are found by satisfying continuity of both the normal component of the electric displacement at the (Si/SiO$_2$) interfaces, and the potential at the source/drain sides. Using the same parabolic potential profile in vertical direction [2] and applying the symmetry condition of $\partial\psi/\partial r=0$ for r=0, we obtained the following expressions of 2-D channel potential as $\psi(r,x)=\frac{C_{ox}}{\varepsilon_{si}\cdot t_{si}}\left[V_g^*-\psi_s(x)\right]r^2+\left(1+\frac{C_{ox}t_{si}}{4\varepsilon_{si}}\right)\psi_s(x)-\frac{C_{ox}t_{si}}{4\varepsilon_{si}}V_g^*$ (2)

where $\psi_s(x)$ represents the surface potential, C_{ox} represents the insulator capacitance ($C_{ox}=2\pi\varepsilon_1 L/\ln(1+2t_1/t_{si})$), t_{si} is the silicon thickness, the effective oxide and silicon nitride layer is defined as $t_{oxeff}=t_1+t_2(\varepsilon_1/\varepsilon_2)$ with t_1 is the thickness of the SiO$_2$ ($\varepsilon_1=\varepsilon_{ox}$) layer and t_2 is the thickness of the Si3N4 layer (ε_2), V_{bi} is the junction voltage between the source/drain and intrinsic silicon, $V_{bi}=(kT/q)\ln(N_{D/S}/n_i)$, n_i is the intrinsic silicon density, Vds represents the drain-to-source voltage and k is the Boltzmann constant. V_g^* represents the effective voltage at the gate which is introduced to simplify notations and alleviate derivations for symmetric structure as $V_g^*=V_{gs}-V_{fb}$, with V_{fb} is the flat-band voltage. Substituting (2) in (1), we obtain the differential equation that deals only with surface potential as

$$\frac{d^2\psi_s(x)}{dx^2}-\frac{1}{\lambda^2}\psi_s(x)=D_1 \qquad (3)$$

with $\lambda=\sqrt{\dfrac{\varepsilon_{si}\cdot t_{oxeff}\cdot t_{si}}{4\cdot\varepsilon_{ox}}}$ and $D_1=\dfrac{q\cdot N_D}{\varepsilon_{si}}-\dfrac{1}{\lambda^2}\cdot V_g^*$

where λ represents the natural length of the analyzed (DDGAA) RADFET dosimeter. This parameter gives the scaling capability (downscaling ability) of the device. D_1 is a factor which represents the impact of the applied gate voltage and channel doping on the surface potential.

The differential equation that deals only with interface potential is given by

$$\frac{d^2\psi_s^*(x)}{dx^2}-\frac{1}{\lambda^2}\psi_s^*(x)=D_2 \qquad (4)$$

with $D_2=\alpha-\beta V_{gs}^*$ and $\alpha=\dfrac{qN_D\varepsilon_2 t_1}{\varepsilon_{si}(\varepsilon_2 t_1+\varepsilon_1 t_2)}$, $\beta=\dfrac{\varepsilon_2 t_1}{\lambda^2(\varepsilon_2 t_1+\varepsilon_1 t_2)}$

where ψ_s^* represents the interface potential at SiO$_2$ /Si$_3$N$_4$ interface which satisfies the continuity of the normal component of the electric displacement at the interface.

This resolution of this Equation allows us the calculation of the interface potential without (before) irradiation.

In the case of RADFET under irradiation new term should be introduced in order to include the radiation-induced interface-traps effect [2,12]. So, the parameter D_2 can be written, in this case, as, $D_2=\alpha-\beta V_{gs}^*-qN_f/\varepsilon_2 t_2$, with N_f represents the irradiation induced localized interface charge density per square area. The second term in this expression represents the impact of the irradiation induced localized interface charge density on the interface potential.

The surface and interface potentials can be, respectively, expressed as

$$\psi_S(x)=-\lambda^2 D_2+\frac{\phi_D\sinh\left(\dfrac{x}{\lambda}\right)-\phi_S\sinh\left(\dfrac{x-L}{\lambda}\right)}{\sinh\left(\dfrac{L}{\lambda}\right)} \qquad (5)$$

With $\phi_D=V_{ds}+\lambda^2 D_2$ and $\phi_S=V_{bi}+\lambda^2 D_2$

$$\psi_S^*(x)=\frac{\varepsilon_1 t_2}{\varepsilon_2 t_1+\varepsilon_1 t_2}V_{gs}^*+\frac{\varepsilon_2 t_1 x}{\varepsilon_2 t_1+\varepsilon_1 t_2}\psi_S(x) \qquad (6)$$

B. Threshold voltage shift model

Schematic cross-sectional view of the proposed (DDGAA) RADFET The basic concept of RADFET dosimeter is to convert the threshold voltage shift, ΔV_{th}, induced by radiation, into absorbed radiation dose, where $\Delta V_{th} = V_{th} - V_{th0}$ with V_{th} and V_{th0} represent the threshold voltage after and before irradiation, respectively. Based on the surface potential model given by Eq.(5), the threshold voltage can be derived using the condition of the minimum channel potential $\psi_{s\min}\big|_{V_{gs}=V_{th}} = 2.\phi_B$, with $\psi_{s\min} = \psi_s(x_{\min})$, V_{th} is the threshold voltage value, and ϕ_B represents the bulk potential of silicon body given as $\phi_B = (K_B T/q).\ln(N_D/n_i)$. The location of the minimum surface potential can be obtained analytically by solving $\dfrac{d\psi_s(x)}{dx} = 0$ [2].

The solution of the equation $\psi_{s\min}\big|_{V_{gs}=V_{th}} = 2.\phi_B$ at low drain-source voltage for long channel lengths ($L \gg \lambda$) can be given as

$$V_{th} = \frac{\left(2A\phi_B + \lambda^2\alpha + \dfrac{qN_f}{\varepsilon_2 t_2}\right)\sinh\left(\dfrac{L}{\lambda}\right) + (V_{bi} - V_{ds})\sinh\left(\dfrac{L}{2\lambda}\right)}{\left(\beta\lambda^2 - \dfrac{B}{A}\right)\sinh\left(\dfrac{L}{\lambda}\right) - 2\sinh\left(\dfrac{L}{2\lambda}\right)} \quad (7a)$$

with: $A = \dfrac{\varepsilon_1 t_2 - \varepsilon_2 t_1}{\varepsilon_1 t_2}$, $B = \dfrac{\varepsilon_2 t_1}{\varepsilon_1 t_2}$

$$V_{th0} = V_{th}\big|_{N_f=0} = \frac{\left(2A\phi_B + \lambda^2\alpha\right)\sinh\left(\dfrac{L}{\lambda}\right) + (V_{bi} - V_{ds})\sinh\left(\dfrac{L}{2\lambda}\right)}{\left(\beta\lambda^2 - \dfrac{B}{A}\right)\sinh\left(\dfrac{L}{\lambda}\right) - 2\sinh\left(\dfrac{L}{2\lambda}\right)} \quad (7b)$$

From (7a) and (7b), the threshold voltage shift can be given as

$$\Delta V_{th} = \frac{\dfrac{qN_f}{\varepsilon_2 t_2}\sinh\left(\dfrac{L}{\lambda}\right)}{\left(\beta\lambda^2 - \dfrac{B}{A}\right)\sinh\left(\dfrac{L}{\lambda}\right) - 2\sinh\left(\dfrac{L}{2\lambda}\right)} \quad (7c)$$

The RADFET radiation sensitivity S, given by [8,9,11]:

$$S = \frac{\Delta V_{th}}{D} \quad (9)$$

where D represents the absorbed radiation dose.

III. RESULTS AND DISCUSSION

In Figure 2, the variation of DDGAA RADFET sensitivity versus the absorbed radiation dose, D, has been compared with conventional (bulk) RADFET. For both designs, the output response of the RADFETs is linear with absorbed radiation dose.

It is clearly shown that DDGAA RADFET has higher sensitivity, $S = 95.45\mu V/Gy$, in comparison with conventional RADFET design, $S = 30.68\mu V/Gy$. This means that DDGAA RADFET has better electrical and scaling performances in comparison with the conventional design. So, our design provides a high sensitivity, better electrical and technological performances in comparison with the conventional structure. These results make the proposed design as a promising candidate for CMOS-based dosimeters.

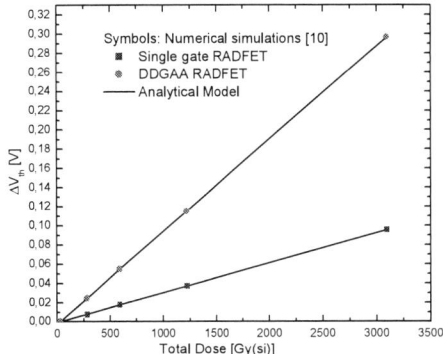

Figure 2. Variation of threshold voltage shift in function of the absorbed radiation dose for the conventional and DDGAA RADFET designs.

A. GA-based sensitivety optimisation

GA optimization has been defined as finding a vector of decision variables satisfying constraints to give acceptable values to objective function. It has recently been introduced to study the complex and nonlinear systems and has found useful applications in engineering fields. Due to the simple mechanism and high performance provided by GA for global optimization, GA can be applied to find the best design of DDGAA RADFET in order to improve the radiation sensitivity by satisfying of the following objective function:

- Maximization of the RADFET radiation sensitivity $S(X)$

where X represents the input normalized variables vector which is given as $X = (t_{si}, t_1, t_2, L)$.

For the purpose of GA-based optimization of the radiation sensitivity of DDGAA RADFET, routines and programs for GA computation were developed using MATLAB 7.2 and all simulations are carried out on a Pentium IV, 3GHz, 1GB RAM computer. For the implementation of the GA, tournament selection is employed which selects each parent by choosing individuals at random, and then choosing the best individual out of that set to be a parent. Scattered crossover creates a random binary vector. It then selects the genes where the vector is unity from the first parent, and the genes where the vector is zero from the second parent, and combines the genes to form the child. An optimization process was performed for 20 population size and maximum number of generations equal to 200, for which stabilization of the fitness function was obtained.

The steady decrease in objective function in each generation until it reaches a best possible value can be attributed to the selection procedure used namely Roulette wheel selection.

The use of GA in this work is strongly motivated by some advantages such as their simplicity and flexibility of modeling and optimization. In addition, the obtained results show that the GA is a powerful tool to solve of the global optimization problem without significant cost.

The radiation sensitivity values of the DDGAA RADFET with and without optimization are shown in Table. 1. It is clearly shows that the radiation sensitivity, for optimized design ($S=162.22\ \mu V/Gy$) is better than the both conventional RADFET ($S = 30.68 \mu V / Gy$) and DDGAA RADFET without optimization ($S = 95.45 \mu V / Gy$).

TABLE I.   DDGAA RADFET DESIGN PARAMETERS

Symbol	Optimized design	Design without optimization	Conventional design
L(nm)	100	100	100
t_{si}(nm)	50	20	20
t_1(nm)	5	5	5
t_2(nm)	15	5	-
S(μV/Gy)	162.22	95.45	30.68

B. Implementation into SIMULINK Software

In order to show the impact of our approach on the environment monitoring applications, we propose the implementation of the developed dosimeter into SIMULINK Software as it is shown in Fig. 3. Figure.3 represents the SIMULINK-dosimeter- block, which is the device that determines the absorbed radiation dose measured in the analyzed environment.

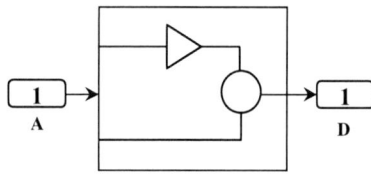

(DDGAA)RADFET dosimeter
Figure3. Dosimeter- SIMULINK-block

Connection A is absorbed radiation dose conserving port that connect to the where radiation is being monitored. Port D is a physical signal port that outputs the absorbed radiation dose value. It is to note that the developed SIMULINK-dosimeter-block can be used for engineering applications to control the environment.

IV.   CONCLUSION

In this paper, we compared new sensor design, DDGAA RADFET, with conventional planar RADFET through 2-D

analytical investigation. A two-dimensional analytical analysis comprising radiation-induced interface-traps effect, 2D surface and interface potentials, threshold voltage shift and sensitivity model for DDGAA RADFET has been developed. The threshold voltage shift behavior of the proposed design was more effectively improved than those of the conventional planar RADFET. Also, we confirmed that DDGAA RADFET had advantages in CMOS scaling in comparison with planar RADFET. Application of the GA-based design approach to DDGAA RADFET has also been discussed. It can be concluded that proposed GA-based approach is efficient and gives the promising results. The optimized design is used to elaborate a new SIMULINK- dosimeter- block, which could be used for environment monitoring applications.

REFERENCES

[1] The international technology roadmap for semiconductors (ITRS); 2007. <http:// public.itrs.net>.

[2] Z. Ghoggali, F. Djeffal, N. Lakhdar, "Analytical analysis of nanoscale Double-Gate MOSFETs including the hot-carrier degradation effects," Int. J. Electronics, pp. 119-127, 97(2), 2010.

[3] D. Jiménez, B. Iñiguez, J. Suñé, L.F. Marsal, J. Pallarès, J. Roig and D. Flores, "Continuous Analytic I-V Model for Surrounding-Gate MOSFETs," IEEE Electron Devices Letters, Vol.25, pp. 571-573, 2004.

[4] H. Kaur, S. Kabra, S. Haldar and R.S. Gupta, "An analytical drain current model for graded channel cylindrical/surrounding gate MOSFET," Microelectronics Journal, vol. 38, pp. 352-359, 2007.

[5] M. Meguellati, F. Djeffal, "New Dual-Dielectric Gate All Around (DDGAA) RADFET dosimeter design to improve the radiation sensitivity," Nuclear Instruments and Methods in Physics Research Section A, Vol. 683, pp. 24–28, 2012.

[6] A. G. Holmes-Siedle, "The space charge dosimeter-General principles of a new method of radiation dosimetry," Nuclear Instruments and Methods, vol. 121, pp. 169–179, 1974.

[7] A. Kelleher, W. Lane, L. Adams, "A design solution to increasing the sensitivity of pMOS dosimeters: The stacked RADFET approach," IEEE Trans. Nuclear Science, vol. 42, pp.48–51, 1995.

[8] A. Jaksic, G. Ristic, M. Pejovic, A. Mohammadzadeh, C. Sudre, W. Lane, "Gamma-ray irradiation and post-irradiation responses of high dose range RADFETs," IEEE Trans. Nuclear. Science, vol. 49, pp. 1356–1363, 2002.

[9] J.R. Schwank, S.B. Roeske, D.E. Beutler, D. J. Moreno, M.R. Shaneyfelt, IEEE Trans. Nuclear Science, vol. 43, pp. 2671–2678, 1996.

[10] Atlas User's manual, SILVACO TCAD, 2008.

[11] F. Djeffal, T. Bendib, M. Meguellati, D. Arar and M.A Abdi, "New Dual-Dielectric Gate All Around (DDGAA) RADFET Dosimeter Design to Improve the Radiation Sensitivity," Proceedings of the World Congress on Engineering 2012 Vol II, WCE 2012, July 4 - 6, 2012, London, U.K.

[12] M. Meguellati, F. Djeffal, D. Arar, F. Douak and L. Khettache, "New RADFET Dosimeter Design For Radioactive Environment Monitoring Applications," Engineering Letters, vol. 20, pp. 20_4_06, 2012.

978-1-4673-5289-5/12 $31.00 © 2012 IEEE

The thermal effect on the output conductance in AlGaN/GaN HEMT's

A. Bellakhdar [1], A. Telia [1], L. Semra [1],

[1] Laboratoire des Microsystèmes et Instrumentation (LMI), Département d'Electronique Faculté des Sciences de l'Ingénieur, Université Mentouri Constantine, Algeria
e mail : as.bellakhdar@mail.lagh-univ.dz

A. Soltani [2]

[2] Institut d'Electronique, de Microélectronique et de Nanotechnologie (IEMN-CNRS 8520), Université des Sciences et Technologie de Lille, 59655 Villeneuve d'Ascq, France

Abstract—**The aim of this work is to study the potential offered by microwave power in the device AlGaN/GaN HEMT by studying the thermal effect and self heating on the output conductance taking into account the effects of spontaneous and piezoelectric polarization.**

We presented an analytical model for the concentration of electrons "n_s" in the 2DEG and the current I_{ds} in the channel of HEMT for strong inversion regime by solving the Poisson equation and Schrödinger self-consistent calculation. Thanks to the analytical description of the current I_{ds} we are able to demonstrate changes in the output conductance g_d according to the voltage V_{ds}, including the spontaneous and piezoelectric polarization, self heating and thermal effects on the HEMT drain conductance g_d.

We noted that when the temperature increases, the electron mobility decreases, this is due to a deterioration of transport properties which is dominated by the scattering with phonons. This decrease in mobility leads to a drop in drain current therefore a degradation of electrical characteristics that determine performance of the transistor.

Keywords—*GaN, HEMT, AlGaN/GaN, drain conductance, self heating.*

I- INTRODUCTION

High-electron-mobility transistors (HEMTs) based on $Al_mGa_{(1-m)}N$/GaN heterostructures are very promising for applications in high power and high temperature electronics operating at high frequencies owing to their outstanding electrical properties such as high breakdown voltage, high peak electron velocity and high sheet electron concentration in two-dimensional electron gas (2DEG) structures [1–2]. Performance of HEMTs can be estimated from some important electrical quantities that characterize them and which are good indicators for the performance of high frequency component. They must therefore have a high transconductance, low output conductance and a high cutoff frequency.

The output conductance characterizes a transistor in the saturated operation. It measures changes in the drain current relative to changes in drain voltage, with constant V_{gs}. Calculating the power gain of the component requires knowledge of the output conductance, which allows among other to account for the saturation drain current. From a physical standpoint g_d characterizes the confinement of electrons in the channel.

In this paper, we presented an analytical model for the concentration of electrons "n_s" in the 2DEG and the current I_{ds} in the channel of HEMT for strong inversion regime by solving the Poisson equation and Schrödinger self-consistent calculation. Thanks to the analytical description of the current I_{ds} we are able to demonstrate changes in the output conductance g_d according to the voltage V_{ds}, including the spontaneous and piezoelectric polarization, self-heating and thermal effects on the $Al_mGa_{(1-m)}N$/GaN HEMT drain conductance g_d.

II- THEORETICAL ANALYSIS

Fig. 1 shows the cross-sectional schematic of the AlGaN/GaN 2DEG HEMT structure.

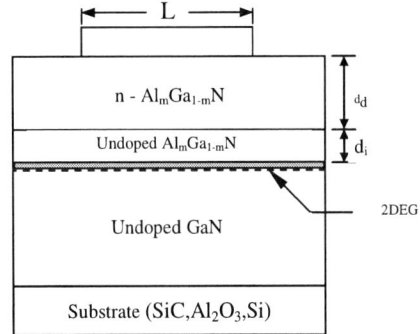

Figure 1. Schematic AlGaN/GaN HEMT structure used

The device structure consists of a thick GaN quantum well channel separated by an undoped $Al_mGa_{(1-m)}N$ spacer layer (d_i) n-type $Al_mGa_{(1-m)}N$ supply layer (d_d). The function of the undoped spacer layer is to reduce impurity scattering. The undoped GaN layer can be grown on a (SiC, Al_2O_3) substrate. When gate voltage V_{gs} is applied through a Schottky contact, the sheet carrier concentration of 2-DEG formed at the AlGaN/ GaN heterojunction is given by [3].

$$n_s\left(x,T,m\right)=\frac{\varepsilon(m)}{q\left(d_d+d_i+\Delta d\right)}\left(V_{gs}-V_{th}\left(T,m\right)-V_c\left(x\right)\right) \quad (1)$$

Where q is the electron charge, m is the Al mole fraction in $Al_mGa_{1-m}N$, $\varepsilon(m)$ is the AlGaN dielectric constant, d_d is the AlGaN barrier thickness, d_i is the spacer layer thickness, Δd is the effective thickness of 2-DEG, V_{gs} is the applied gate–source voltage, and $V_c(x)$ is the channel potential at x . The temperature and mole fraction dependent threshold voltage $V_{th}(T,m)$ of the AlGaN/GaN HEMT is given by the following expression:

$$V_{th}\left(T,m\right)=\phi_m(m)-\Delta E_C\left(T,m\right)-\frac{qN_dd_d^2}{2\varepsilon(m)}-\frac{\sigma(m)}{\varepsilon(m)}\left(d_d+d_i\right) \quad (2)$$

With N_d the doping density of AlGaN layer, σ (m); the total polarization induced charge density and $\varepsilon(m)$ the Schottky barrier height [4].

$$\varepsilon(m) = 0.84+1.3m \text{ [eV]}; \quad (3)$$

N_d is the doping concentration of the n-AlGaN layer and $\Delta E_c(T,m)$ is the conduction band discontinuity at the $Al_mGa_{1-m}N$/ GaN interface. $\Delta E_c(T,m)$ is defined as [5]:

$$\Delta E_C\left(T,m\right)=0.70\left[E^{AlGaN}_g\left(T,m\right)-E^{GaN}_g\left(T\right)\right]\text{[eV]} \quad (4)$$

The major factor which controls 2-DEG is the Al composition of AlGaN. When Al composition becomes high, it also directly affects the polarization induced sheet charge density. The polarization induced charge density σ_{pz} (m), is given by [6].

$$\left|\sigma_{pz}\right|=\left|P_{sp_{Al_mGa_{(1-m)}N}}\left(m\right)+P_{pz_{Al_mGa_{(1-m)}N}}\left(m\right)-P_{sp_{GaN}}\right| \quad (5)$$

The drain current in the 2-DEG channel can be obtained from the current density equation as [7]:

$$I_{ds}\left(T,m,x\right)=wq\mu(T,x)\left(\begin{array}{c}n_s\left(T,m,x\right)\dfrac{dV_c\left(x\right)}{dx}\\+\dfrac{K_BT}{q}\dfrac{dn_s\left(T,m,x\right)}{dx}\end{array}\right) \quad (6)$$

Where w is channel width, K_B is Boltzmann constant, $V_c(x)$ is a channel potential.

$\mu\left(T,x\right)$ is the doping and temperature dependent mobility which is given by [3] as:

$$\mu\left(T,x\right)=\frac{\mu_0(T)}{1+\dfrac{1}{E_1}\dfrac{dV_c\left(x\right)}{dx}}$$
$$E_1=\frac{E_cv_{sat}}{\mu_0(T)E_c-v_{sat}} \quad (7)$$

$\mu_0(T)$ is the low field mobility, E_c is the critical electric field and v_{sat} is the saturation velocity.
Using equation (1) and equation (7) in equation (6), and integrating along the channel we can obtain:

$$I_{ds}\left(T,m\right)=\frac{-A_2\left(T,m\right)+\sqrt{A_2\left(T,m\right)^2-4A_1\left(T,m\right)A_3\left(T,m\right)}}{2A_1\left(T,m\right)} \quad (8)$$

With

$$A_1\left(T,m\right)=\left(\frac{\mu_0\left(T\right)E_c-v_{sat}}{E_cv_{sat}}\right)\left(2R_s+R_d\right)$$
$$-\left(\frac{w\mu_0\left(T\right)q^2D\varepsilon\left(m\right)}{\left(\varepsilon\left(m\right)+2q^2D\left(d_d+d_i\right)\right)}\right)\left(R_d^2+2R_sR_d\right) \quad (9)$$

$$A_2\left(T,m\right)=\left(\frac{2w\mu_0(T)q^2D\varepsilon(m)}{\left(\varepsilon(m)+2q^2D(d_d+d_i)\right)}\right)\left(V_{ds}\left(R_s+R_d\right)-V'_{gs}\left(T,m\right)\left(2R_s+R_d\right)\right)$$
$$-L-\left(\frac{\mu_0(T)E_c-v_{sat}}{E_cv_{sat}}\right)V_{ds} \quad (10)$$

$$A_3\left(T,m\right)=\left(\frac{2w\mu_0(T)q^2D\varepsilon(m)}{\left(\varepsilon(m)+2q^2D(d_d+d_i)\right)}\right)\left(V'_{gs}\left(T,m\right)V_{ds}-\frac{V_{ds}^2}{2}\right) \quad (11)$$

$$V'_{gs}\left(T,m\right)=V_{gs}-V_{th}\left(T,m\right)-\frac{K_BT}{q} \quad (12)$$

Where R_s and R_d are the source and drain resistances respectively and V_{ds} is drain to source voltage.

$D=\dfrac{4\pi m^*}{h^2}$ is conduction band density of state of 2DEG system, $m*$ is effective mass of electron and h is Planck constant. The drain conductance is an important microwave parameter that determines the maximum voltage gain attainable from a device. The drain conductance of the $Al_mGa_{1-m}N$/GaN HEMT is evaluated as [3]:

$$g_d\left(T,m\right)=\frac{\partial I_{ds}\left(T,m\right)}{\partial V_{ds}}\bigg|_{V_{gs}} \quad (13)$$

$$g_d(T,m) = \frac{1}{2A_1(T,m)} \left(\partial A_2(T,m)/\partial V_{ds} \left(\frac{A_2(T,m)}{\sqrt{A_2(T,m)^2 - 4A_1(T,m)A_3(T,m)}} - 1 \right) - \frac{2A_1(T,m)\partial A_3(T,m)/\partial V_{ds}}{\sqrt{A_2(T,m)^2 - 4A_1(T,m)A_3(T,m)}} \right) \quad (14)$$

With:

$$\frac{\partial A_1(T,m)}{\partial V_{ds}} = 0$$

$$\frac{\partial A_2(T,m)}{\partial V_{ds}} = A(R_s + R_d) - \left(\frac{\mu_0(T)E_C - v_{sat}}{E_C v_{sat}} \right)$$

$$\frac{\partial A_3(T,m)}{\partial V_{ds}} = A \left(V'_{gs}(m) - V_{ds} \right)$$

$$A = \frac{2 w \mu_0(T) q^2 D \varepsilon(m)}{\left(\varepsilon(m) + 2q^2 D (d_d + d_i) \right)}$$

The mobility of electrons in semiconductor depends strongly on temperature and electric field present. This mobility tends to decrease with increasing temperature. Fig.2 shows the evolution of the electron mobility in wurtzite GaN at low fields as a function of temperature for a doping concentration $N_d = 1 \times 10^{17}$ cm^{-3}.

The mobility model used in our simulation is an analytical formulation proposed by Caughey and Thomas [8] adapted to silicon carbide. It allows describing the variation of mobility on the basis of the total concentration of ionized dopant and temperature. The first step of this technique consists in an adequate approximation of the doping level dependence of the mobility at room temperature on the base of well known Caughey-Thomas approximation:

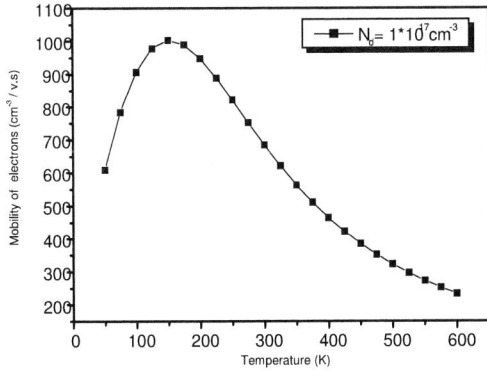

Figure 2. the evolution of the electron mobility as a function of temperature

$$\mu_i(N) = \mu_{min,i} + \frac{\mu_{max,i} - \mu_{min,i}}{1 + \left(\dfrac{N}{N_{g,i}} \right)^{\gamma_i}} \quad (15)$$

Where i = n, p for electrons and holes respectively for GaN , model parameters $\mu_{max,i}$ $\mu_{min,i}$, $N_{g,i}$, and

γ_i depend on the type of semiconductor material, and N is the doping concentration[8].

III- RESULTS AND DISCUSSION

Fig.3 shows the transfer characteristics of a HEMT based heterostructure AlGaN/GaN on the gate-source voltage when considering the thermal effect and the effect of self-heating. A degradation of drain-source current with increasing temperature of 300 K to 475 K was observed, this degradation becomes important when one considers the effect of self-heating.

Fig.4 shows the current-voltage characteristics of a HEMT based heterostructure AlGaN/GaN when considering the thermal effect and the self-heating (dashed lines). These characteristics show a negative resistance at higher drain voltage level. It is observed that, with increasing temperature from 300K to 425K, explicitly from an external source or by injecting more current in the transistor, a drain current decay. This is due to a decrease in the low field mobility of electrons where the electron transport in 2DEG channel is dominated by scattering with phonons. The self-heating is a local increase of crystal temperature due to dissipated Joule electric power. It is seen that the saturation drain current drops significantly when the temperature rises above room temperature from 300K to 425K. Also the saturation drain current does not vary much with the higher drain voltages. Our results can be compared with those obtained by Gangwani and al. [3]. The striking constant behaviour of drain current suggests the usefulness of these devices for high power applications [9]. Drain conductance g_d is deduced from the current I_{ds}. Fig 5 shows the drain conductance versus the drain-source voltage when considering the thermal effect and the self-heating (dashed lines). Then a strong increase in temperature significantly alters the performance of transistors. View the decrease in mobility by increasing the temperature from 300K to 425K, the saturation current decreases then increases the drain conductance.

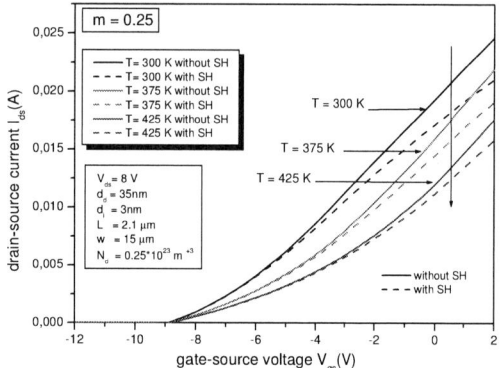

Figure 3. the transfer characteristics of HEMT with and without the effect of self-heating (SH) and thermal effect for different temperature.

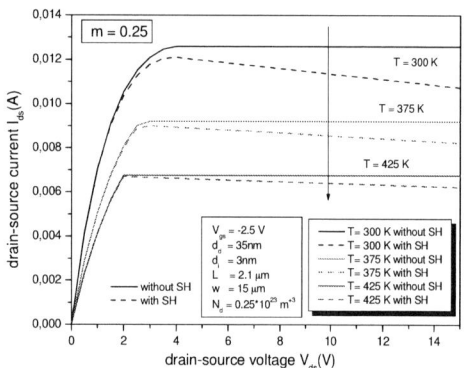

Figure 4. Dependence of drain current I_{ds} on drain voltage V_{ds} for various values of temperature for V_{gs}= -2.5V and m= 0.25. When self-heating is not considered (solid lines) and when self-heating (SH) is considered (dashed lines).

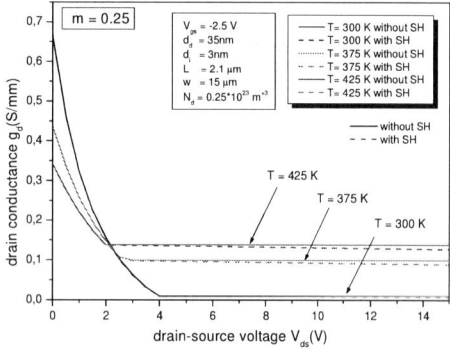

Figure 5. Dependence of drain conductance g_d on drain voltage V_{ds} for various values of temperature for V_{gs}= -2.5V and m= 0.25. When self-heating is not considered (solid lines) and when self-heating (SH) is considered (dashed lines).

IV-   CONCLUSION

In this work, we presented the simulation results obtained and interpreted by studying the influence of the thermal effect and self-heating performance of the HEMT based heterostructure AlGaN/GaN. We noted that increasing temperature, the electron mobility decreases, this is due to a deterioration of transport properties which is dominated by the scattering with phonons. This decrease in mobility leads to a drop in drain current therefore a degradation of electrical characteristics that determine performance of the transistor.

V-   REFERENCES

[1]Wu Y F, Keller B P, Keller S, et al. Very high breakdown voltage and large transconductance realized on GaN heterojunction field effect transistors. Appl Phys Lett, 1996, 69: 1438–1440

[2] Miyoshi M, Ishikawa H, Egawa T, et al. High-electron-mobility AlGaN/AlN/GaN heterostructures grown on 100-mm-diam epitaxial AlN/sapphire templates by metalorganic vapor phase epitaxy. Appl Phys Lett, 2004, 85: 1710–1712

[3]Parvesh Gangwania, Ravneet Kaur a, Sujata Pandeyb, Subhasis Haldar c, Mridula Guptaa, R.S. Guptaa,_ Modeling and analysis of fully strained and partially relaxed lattice mismatched AlGaN/GaN HEMT for high temperature applications, Superlattices and Microstructures 44 (2008) 781-793

[4] Ambacher O, Foutz B, Smart J, Shealy JR, Weimann NG, Chu K, et al. Two dimensional electron gases induced by spontaneous and piezoelectric polarization undoped and doped AlGaN/GaN heterostuctures. J Appl Phys 2000;87:334–44.

[5] Manju K. Chattopadhyay , Sanjiv Tokekar, Temperature and polarization dependent polynomial based non-linear analytical model for gate capacitance of AlmGa1_mN/GaN MODFET, Solid-State Electronics 50 (2006) 220–227

[6] Ambacher O, Smart J, Shealy JR, Weimann NG, Chu K, Murphy M, et al. Two-dimensional electron gases induced by spontaneous and piezoelectric polarization charges in N- and Ga-face AlGaN/GaN heterostuctures.
J Appl Phys 1999;85:3222–33.

[7] S.M. Sze, Physics of Semiconductor Devices, second ed., Willey, New York, 1981.

[8] T. T. Mnatsakanov , M. E. Levinshtein , L. I. Pomortseva , S. N. Yurkov , G. S. Simin , M. A. Khan « Carrier mobility model for GaN». Solid-State Electronics 47 (2003), pp 111–115.

[9] Parvesh,Ravneet kaur,Sujata Pandey,Subhasis Haldar,Mridula Gupta and R S Gupta, "High Temperature Performance of AlGaN/GaN HEMT". 2nd National Conference Mathematical Techniques: Emerging Paradigms for Electronics   and IT Industries, September 26-28, 2008. MATEIT-2008.

Compact modeling of long channel Double Gate MOSFET transistor

Billel Smali, Saida Latreche and Samir Labiod

Laboratory of Hyperfrequency and Semiconductor (LHS), Electronic department, faculty of sciences engineering

Mentouri Constantine University, 25000, Algeria

E-mail: billel1248@yahoo.fr, latreche.saida@gmail.com, samir.labiod@gmail.com

Abstract—In this work, we present a compact modeling of long channel Double Gate MOSFET transistor with an efficient procedure to compute the mobile charge density for this model. In the first time, the static behavior of the symmetrical DG MOSFET is obtained using a relationship between charges and voltages. The model is based on the formalism EKV developed for the MOSFET bulk. In second time, to define the explicit solution of the gate charge density in weak and strong inversion, we use the Taylor series development. From that, we get an efficient algorithm that computes the gate/mobile charge density of the model with a faster computation time and without any iterative calculation. Our results are compared with the iterative calculation using the Newton-Raphson method, especially compared with 2-D numerical simulations using ATLAS-TCAD software.

Keywords: Double Gate MOSFET transistor ; EKV model; Compute the mobile charge density.

I. INTRODUCTION

In the last few years, the CMOS technology approaching the limit caused by the quantum and physical effects appears in nanoscale devices [1]. As a solution to these problem different architectures was developed, the undoped DG MOSFET architecture is one of the best candidates for the future integrated circuits. This structure can offer many advantages such as: an ideal 60 mV/decade subthreshold slope, reduced short channel effects (SCEs), free dopant associated fluctuation effects [2, 3]. Compact modeling of nanoscale devices is important to describe the relation between the physical process, geometry and the electrical behavior, especially important for circuits design applications [4].

Some compact models of the DG MOSFET structure have been presented in the literature [5, 6]. However, most of them are good for the physical device but less useful for circuit simulation, because these models are built on complicated analytical expressions or implicit solutions that need to be numerically solved, generally an iterative calculation.

In this work, we present a compact modeling of symmetrical long channel DG MOSFET transistor.

In the first part, from 1-D Poisson's equation of the long channel undoped DG MOSFET we get a relationship between charges and voltages. However, this nonlinear equation needs an iterative procedure to compute the mobile charge density, but this is not suitable for circuit simulation.

In the second part, by applying the Taylor series development in the fundamental equation of the model in weak and strong inversion, we find a polynomial equation. From the solution of this polynomial equation, the explicit solution of the gate charge density in weak and strong inversion is obtained. An efficient algorithm that computes the mobile/gate charge density without any iterative calculation and without any problem of time convergence is defined, and which gives the advantage for the model to be used in circuit simulation. Finally, our results obtained using MATLAB are compared with 2-D numerical simulation obtained by Silvaco-ATLAS software, we also made a comparison with the Newton-Raphson method, applying to the fundamental equation of the model.

II. COMPACT MODEL DESCRIPTION

In this work, a long channel DG *n*MOSFET operate in symmetrical mode is considered. The studied structure is presented as follows:

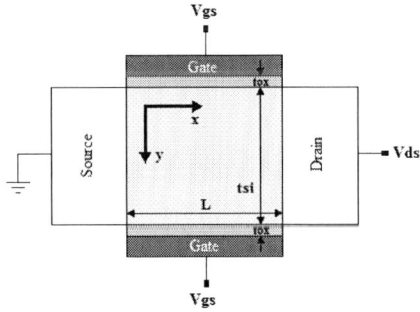

Fig 1. Structue of the DG MOSFET where L, tsi, and tox are the channel length, channel thickness and oxide thickness,recpectivy.

By considering a DG MOSFET with n type, we can ignore the charge density of acceptors and the 1-D Poisson's equation can be written as follows [7]:

$$\frac{d^2\psi(y)}{dy^2} = \frac{d^2(\psi(y)-V_{ch})}{dy^2} = \frac{e.n_i}{\varepsilon_{si}}\exp\left(\frac{\psi(y)-V_{ch}}{U_T}\right) \quad (1)$$

Where $\Psi(y)$ is the electrostatic potential, e is the elementary charge, ε_{si} is the permittivity of silicon, n_i is the intrinsic carrier density, U_T is the thermal voltage and V_{ch} is the quasi Fermi potential in the channel.

After a first integration of Poisson's equation and applying the Gauss law to the interface of silicon and gate oxide, we get [8]:

$$V_g - V_{ch} = \frac{Q_g}{C_{ox}} + U_T . \ln\left(\frac{Q_g^2}{2.\varepsilon_{si}.e.n_i.U_T} + \frac{Q_g}{e.n_i.d}\right) \quad (2)$$

Where V_g is the gate voltage, Q_g is the gate charge density per unit surface and $d = t_{Si}/2$.

By applying the normalization of charge and voltages as EKV MOSFET model in (2), we get the fundamental relation of this model which relates the gate charge density and potentials [9]:

$$v_g - v_{ch} - v_{to} = 4.q_g + \ln q_g + \ln\left(1 + q_g . \frac{C_{ox}}{C_{si}}\right) \quad (3)$$

Where C_{ox}, C_{si} are the silicon layer capacitance, oxide layer capacitance, respectively.

To define the normalized mobile charge density, (3) need to be numerically solved.

The normalized threshold voltage of long channel DG MOSFET can be defined as the gate voltage when the mobile charge density vanishes, we get [8]:

$$v_{to} = -\ln\left(\frac{e.n_i.t_{si}}{8.C_{ox}.U_T}\right) \quad (4)$$

Considering the drift diffusion transport model, the drain current is obtained by integrating the mobile charge density in the x direction of the channel (from source to drain) [10]:

$$I_d = -\mu.\frac{W}{L}\int_{V_s}^{V_d} Q_m.dV_{ch} \quad (5)$$

Where μ is the constant mobility in channel, W is the channel width.

The mobile charge density per unit surface is defined as follows:

$$Q_m = -2.Q_g$$

After integrate (5) from source to drain and using the normalization of variable, we get:

$$i = -q_m^2 + 2.q_m + 2.\frac{C_{si}}{C_{ox}}.\ln\left(1 - q_m.\frac{C_{ox}}{2.C_{si}}\right)\Big|_{q_{ms}}^{q_{md}} \quad (6)$$

Finally, the drain current is obtained after denormalization of (6) as follows:

$$I_{ds} = i.4.\mu.C_{ox}.U_T^2.\frac{W}{L} \quad (7)$$

III. NUMERICAL PROCEDURE TO COMPUTE THE MOBILE CHARGE DENSITY

To compute the mobile charge density without any iterative calculation, the explicit solution of the gate charge density in a weak and a strong inversion is defined. We rewrite the fundamental equation of the model (3) with $v (= v_g - v_{ch} - v_{to})$, $\alpha (= C_{ox}/C_{si})$ and $q(= q_g)$, we obtain [11]:

$$v = 4.q + \ln[q.(1 + \alpha.q)] \quad (8)$$

A. Explicit solution of the gate charge density in strong inversion

In this case, the inversion charge density is important $(q \gg 1)$. By applying the Taylor series development to the first order of the term $ln[q.(1 + \alpha.q)]$ around $q = q_t$, we get the following expression:

$$v = 4.q + \ln\left[q_t\left(1 + \alpha.q_t\right)\right] + 2.\frac{q - q_t}{q + q_t}\left(\frac{1 + 2.\alpha.q_t}{1 + \alpha.q_t}\right) \quad (9)$$

This expression (9) can be rewrite as the following polynomial equation:

$$4.q^2 + \left(4.q_t + a + b - v\right)q + [q_t(b - a - v)] = 0 \quad (10)$$

Where $a = \frac{2.(1 + 2.\alpha.q_t)}{1 + \alpha.q_t}$ and $b = \ln(q_t(1 + \alpha.q_t))$

Therefore, the expression (10) represents a second order polynomial, where its solution is given by:

$$q_0 = 1/2.\left[\left(\frac{v - a - b}{4} - q_t\right) + \sqrt{\left(\frac{v - a - b}{4} - q_t\right)^2 + 2a.q_t}\right] \quad (11)$$

By substituting the solution of q_0 in $ln[q.(1 + \alpha.q)]$ of the expression (8), we get the explicit solution of the gate charge density in the strong inversion:

$$q_s = 1/4.\{v - \ln[q_0.(1 + \alpha.q_0)]\} \quad (12)$$

B. Explicit solution of the gate charge density in weak inversion

In this case the inversion charge density is very small $(q<<1)$, which implies that the logarithmic term is dominant compared with the first term in the expression (8). The expression (8) can be rewritten as follows [11]:

$$v = \ln q + 1 / F_W (\ln q) \qquad (13)$$

Where $F_W (\ln q) = \dfrac{1}{4.q_t \exp(\Delta \ln q) + \ln(1 + \alpha.q.\exp(\Delta \ln q))}$ and

$\Delta \ln q = \ln q - \ln q_t$

By applying the Taylor series development to the first order of the term $F_W(ln\ q)$ around $\Delta ln\ q=0$, we can get:

$$F_W (\ln q) = \frac{1}{4.q_t + \ln(1 + \alpha.q_t)} \cdot \left[1 - \frac{q_t \left[4 + \alpha(1 + 4.q_t) \right]}{(1 + \alpha.q_t).(4.q_t + \ln(1 + \alpha.q_t))} \Delta \ln q \right] \qquad (14)$$

The substituting of the solution of (14) into (13) let us obtain a new polynomial expression which is defined as:

$$\frac{1}{b}.\ln^2 q + \left(1 + \frac{\ln q_t - v}{b} \right).\ln q + \left(a - v + \frac{v.\ln q_t}{b} \right) = 0 \qquad (15)$$

Where $a = 4.q_t + \ln(1 + \alpha.q_t)$ and $b = \dfrac{a(1 + \alpha.q_t)}{q_t \left[4.(1 + \alpha) + \alpha \right]}$

The solution of the expression (15) is given by:

$$q_0 = \exp \left\{ 1/2 . \left[v + (b + \ln q_t) - \sqrt{[v - (b + \ln q_t)]^2 + 4ab} \right] \right\} \qquad (16)$$

After substituting the solution of (16) into the expression $v = 4.q_0 + \ln[q.(1 + \alpha.q)]$, the explicit solution of the gate charge density in weak inversion can be written as:

$$q_w = \frac{\exp(v - 4q_0)}{1/2 + \sqrt{1/4 + \alpha.\exp(v - 4q_0)}} \qquad (17)$$

After defining the gate charge density in a weak and a strong inversion, a linearization of the gate charge density and its approximate value is done in both cases, from that we can get:

$$q = q_0 [1 + \beta(1 + K.\beta)] \qquad (18)$$

Where $\beta = \dfrac{v - 4q_0 - \ln[q_0(1 + \alpha q_0)]}{4q_0 + \dfrac{1 + 2\alpha q_0}{1 + \alpha q_0}}$

With k is a parameter which depends on the operating region of the DG MOSFET transistor (fixed after numerical experiments, $kw = 0.35$ for the weak inversion and $ks = 0.13$ for the strong inversion).

In order to separate the operating regions of the DG MOSFET transistor (weak and strong inversion), we use the voltage transition which is defined as following:

$$v_t = 4q_t + \ln[q_t(1 + \alpha q_t)] \qquad (19)$$

We consider q_t as the transition charge (from weak to strong inversion of the operating regions of the device), which is fixed after some numerical simulation ($q_t=0.01$, 0.02, 0.1, $0.3... 0.4$) at 0.3, worked well in our case.

Finally, the mobile charge density is defined as a function of the gate charge density ($=-2q_g$).

IV. RESULTS AND DESCUSIONS

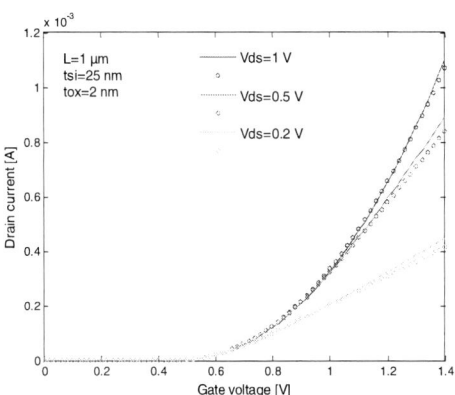

Fig 2. Drain current versus gate voltage at different drain voltage
Lines: Compact Model; Symbols: Silvaco Atlas-TCAD.

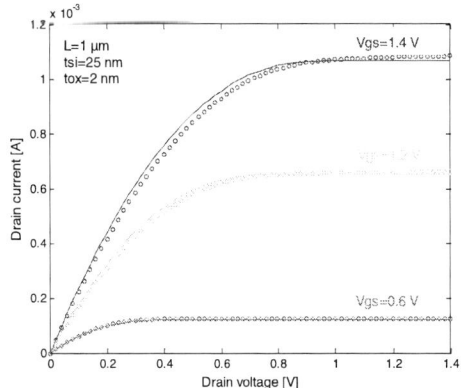

Fig 3. Drain current versus drain voltage at different gate voltage
Lines: Compact Model; Symbols: Silvaco Atlas-TCAD.

978-1-4673-5289-5/12 $31.00 © 2012 IEEE

As shown in Fig. 2 and Fig. 3, for different values of the gate voltage (0.2, 0.5 and 1V) and the drain voltage (0.6, 1.2 and 1.4V), the drain current of long channel (1μm) model of symmetrical DG MOSFET have a good behavior from a weak to a strong inversion compared with 2-D numerical simulation.

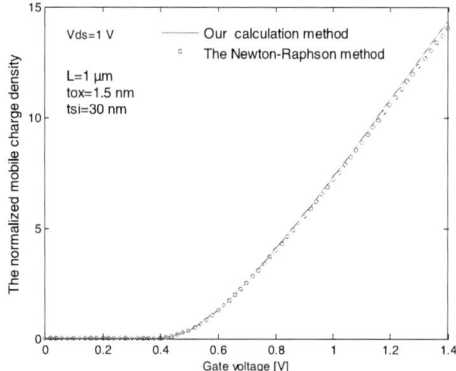

Fig 4. Comparing the computing mobile charge density of our method calculation with the Newton- Raphson method.

As shown in Fig. 4, that we also compare our method with the iterative calculation using the Newton-Raphson method. We can observe a good agreement with this method.

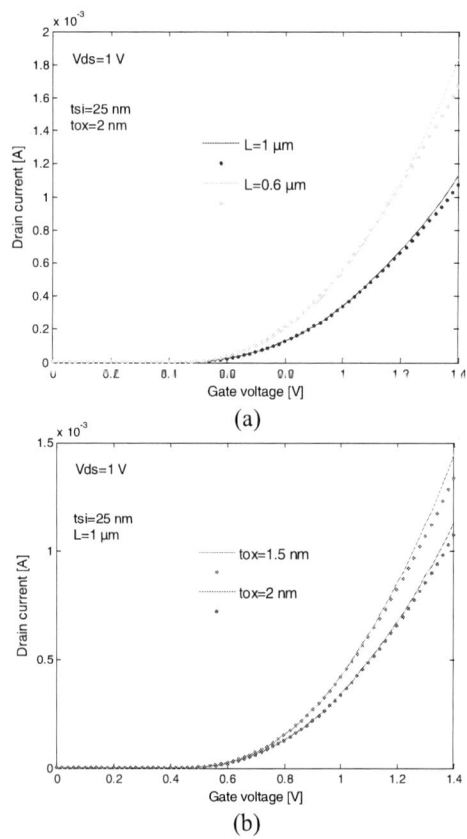

Fig 5. The Transfer characteristic: (a) For different channel length, (b) For different oxide thickness. Lines: Our solution; Symbols: Atlas

Fig. 5 show the influence of the channel length Silicon (0.6, 1μm) and the oxide thickness (1.5, 2 nm) on the drain current model as a function of the gate voltage. We can see, that the decreases in the technological parameters of the device such as: the channel length (L) and the oxide thickness (t_{ox}) are accompanied with a significant increase in the drain current (I_{ds}).

V. CONCLUSION

A compact modeling of long channel DG MOSFET with an efficient method to calculate the gate/mobile charge density for the model has been presented. The model describes the static behavior of long channel device and it is continue from weak to strong inversion. The proposed method can compute the mobile charge density without any iterative procedure and with a faster computation time (time simulation).

Our results show a good agreement compared with the iterative calculation using the Newton-Raphson method, especially, with 2-D numerical simulations from SILVACO-ATLAS software.

The normalized equations of the presented model associated with the proposed method give the opportunity for them to be used in circuit simulation.

REFERENCES

[1] L Y. Li, H.-M. Chou, "A comparative study of electrical characteristic on sub-10-nm double-Gate MOSFETs," IEEE Trans. Nanotechnology, vol. 4,no. 5, pp. 645–647, 2005.

[2] F. Balestra, S. Cristoloveanu, M. Benachir, and T. Elewa, "Double-gate silicon-on-insulator transistor with volume inversion: A new device with greatly enhanced performance," IEEE Electron Device Lett, vol. EDL-8, no. 9, pp. 410–412, 1987.

[3] F. J. Garcia-Sánchez, A. Ortiz-Conde, and J. Muci, "Subthreshold behavior of undoped DG MOSFETs," in Proc. IEEE Conf. Electron Devices, Solid-State Circuits, Dec. 19–21, 2005, pp. 75–80.

[4] N. Arora, "MOSFET Modeling for VLSl Circuit Simulation: Theory and Practice", World Scientific, ISBN-13 978-981-256-862-5, 1993.

[5] G. Baccarani, S. Reggiani, "A Compact Double-Gate MOSFET Model Comprising Quantum-Mechanical and Non-Static Effects," IEEE Transactions on Electron Devices,vol. 6, no. 8, pp. 1656 1666, 1999.

[6] Y. Yorozu, M. Hirano, K. Oka, and Y. Tagawa, "Electron spectroscopy studies on magneto-optical media and plastic substrate interface," IEEE Transl. J. Magn. Japan, vol. 2, pp. 740–741, August 1987 [Digests 9th Annual Conf. Magnetics Japan, p. 301, 1982].

[7] Y. Taur, X. Liang, "A continuous, analytic drain-current model for DG MOSFETs," IEEE Electron device Letters, vol. 25, no. 2, pp. 107-109, 2004.

[8] J.M.Sallese, F. Krummenacher, "A design oriented charge-based current model for symmetric DG MOSFET and its correlation with the EKV formalism," Solid-State Electronics, vol. 49, no. 3, pp. 485-489, 2005.

[9] C.Enz, F.Krummenacher, A.Vittoz, "An analytical MOS Transistor Model Valid in All Regions of Operation Dedicated to low voltage and low current applications," Analog and integrated Circuits and Signal Processing, vol. 8, pp. 83-114, 1995.

[10] A. Amara, "Planar Double-Gate Transistor from technology to circuit,", ISBN-978-1-4020-9341-8, 2009.

[11] F. Pregaldiny, F. Krummenacher, "An explicit quasi-static charge-based compact model for symmetric DG MOSFET,"NSTI-Nanotech, ISB-0-97 67985-8-1, 2006.

978-1-4673-5289-5/12 $31.00 © 2012 IEEE

The Electro-Thermal Sub Circuit Model for Power MOSFETs

Messaadi Lotfi
Department of Electronics, University of Batna
Advanced Electronic Laboratory (LEA)
Rue Mohammed Elhadi Boukhlouf 05000, Batna, Algeria
lotfi.messaadi@gmail.com

Smail Toufik
Department of Electronics, University of Batna
Advanced Electronic Laboratory (LEA)
Rue Mohammed Elhadi Boukhlouf 05000, Batna, Algeria

Abstract— **an empirical self-heating SPICE MOSFET model which accurately portrays the vertical DMOS power MOSFET electrical and thermal responses is presented. This macro-model implementation is the culmination of years of evolution in MOSFET modeling. This new version brings together the thermal and the electrical models of a VDMOS MOSFET. The existing electrical model [2], [3] is highly accurate and is recognized in the industry. The sequence of the model calibration procedure using parametric data is described. Simulation response of the new self-heating MOSFET model track the dynamic thermal response and is independent of SPICE's global temperature definition.**

Keywords- Device characterization; device modeling; high power discrete devices; modeling; MOS device; power semiconductor devices; semiconductor devices; simulation; thermal design; Spice.

I. INTRODUCTION

Many power MOSFET models available today are based on an ideal lateral MOSFET device. They offer poor correlation between simulated and actual circuit performance in several areas. They have low and high current inaccuracies that could mislead power circuit designers. This situation is further complicated by the dynamic performance of the models. The ideal low power SPICE level-1 NMOS MOSFET model does not account for the nonlinear capacitive characteristics Ciss, Coss, Crss of a power MOSFET. Higher level SPICE MOSFET models may be used to implement the non-linear capacitance with mixed results. The need for this higher level modeling accuracy becomes apparent in high frequency applications where gate charge losses as a proportion of overall losses become significant. The inherent inaccuracies of modeling a power VDMOS with the SPICE MOSFET model dictated need for an alternative approach; a macro-model. A macro-model such as the one defined by Wheatley and Hepp [1] can address the short comings of the ideal low power SPICE MOSFET model. Highly accurate results are possible by surrounding a temperature independent gain block (implemented using three level-1 MOSFET models) with resistive, capacitive, inductive and other SPICE circuit elements. It is possible to develop a model from parametric measurements in a single iteration. The model extraction procedure from parametric data must follow a given sequence. Many of the changes to the model affect different behavior. Failure to follow this sequence will result in repeated model

calibration iterations. The MOSFET model reference on which this work is based has been explained in [1], [2], [3], [10].

II. SELF-HEATING SPICE MOS MODEL

The self-heating macro-model from Figure 1 is the evolution of years of work and improvements from numerous authors [1-7]. A significant advantage of this model is that knowledge of device physics or process details are not necessary to implement the parametric data within the model.

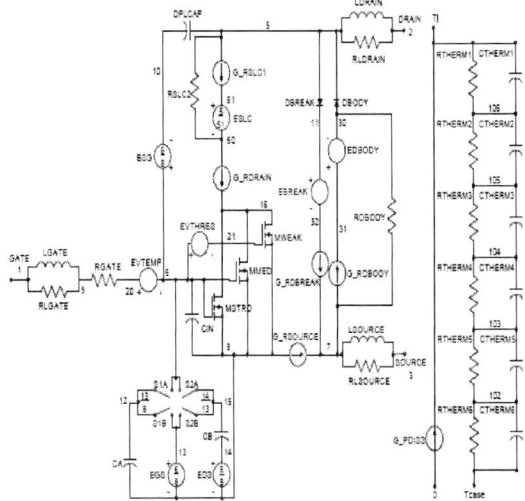

Figure 1.   Self-heating MOSFET macro-model independent of global temperature definition

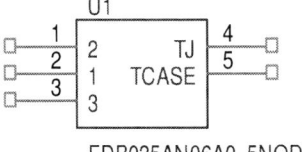

FDB035AN06A0_5NODE

Figure 2.   Self-heating MOSFET symbol

III. ELECTRO THERMAL SPICE MOS MODEL

Ability to describe the value of a resistor and its temperature coefficients as a behavioural model referenced to a voltage node is necessary to express dependence on junction temperature. PSPICE resistor ABMs do not permit voltage node references. Dynamic temperature dependence of the MOSFET's resistive element (expressed as separate lumped elements) and of the diode's resistive component cannot be implemented without a resistor ABM. This limitation is overcome with a voltage-controlled current source ABM expression (Figure 3). By using the nodes of the current source for voltage control, resistor behaviour may be expressed as I = V/R(Tj). The resistance R(Tj) is replaced by a behavioural model expression dependent on the voltage node Tj representation of junction temperature. This voltage-controlled current source ABM model was used to implement voltage dependent expressions of RDRAIN, RSOURCE, and RSLC1. Temperature dependent resistive elements of diodes DBODY and DBREAK were separated from the diode model, and expressed as voltage-controlled current source ABM models G_RDBODY and G_RDBREAK. A very large value resistor RDBODY was added to improve convergence. EDBODY is added in series with DBODY to incorporate the temperature dependency of the intrinsic body diode forward conduction drop. Junction temperature information is implemented by the inclusion of the MOSFET's thermal network ZθJC and current source G_PDISS. The thermal network parameters are supplied in Fairchild Semiconductor data sheets. G_PDISS calculates the MOSFET instantaneous operating loss, and expresses the result in the form of a current. This is a circuit form implementation of the junction temperature from expression (1)

$$T_J = P_{DISSIPATION} \cdot Z_{\theta JC} + T_{CASE} \qquad (1)$$

Where Tj = junction temperature, Pdissipation = instantaneous power loss, $Z_{\theta JC}$ = thermal impedance junction- to-case and Tcase = case temperature. The unit conversion for the electrical analogy of the thermal system is listed in Table 1.

Table 1 Electrical/thermal analogy

Electrical	Thermal
Ohm	°C/Watt
Farad	Joules/°C
Amp	Watt
Volt	°C

IV. PARAMETER EXTRACTION METHODOLOGY

The sequence of the parameter extraction procedure is very important since many of the changes to the library affect different behaviour. For instance, changing parameters in the transfer curve affect the saturation curves. The recommended methodology is shown below.

A. *The transfer curve,* **B.** *The saturation curve,* **C.** *The body diode forward conduction,* **D.** *Breakdown voltage,* **F.** *Capacitance (Crss, Coss, Ciss)* **G.** *Gate charge* **H.** *Temperature coefficients,* **I.** *Thermal model*

Extraction is achieved more rapidly if data is plotted log-log, semilog, versus t, etc. First extraction may take days. It becomes a rapidly learned process with repeated usage.

B. Transfer Curve

Three level-1 MOSFET transistors are used to model the gain block for the full current range from the sub-threshold region through high current. The three transistor models are MweakMOD, MmedMOD and MstroMOD. The parameters VTO and KP of each transistor are used for alignment of the model with measured data.

.MODEL MmedMOD NMOS (VTO=3.3 KP=9 IS=1e-30 N=10 TOX=1 L=1u W=1u RG=1.36 T_ABS=25)

.MODEL MstroMOD NMOS (VTO=4.0 KP=275 IS=1e-30 N=10 TOX=1 L=1u W=1u T_ABS=25)

.MODEL MweakMOD NMOS (VTO=2.72 KP=0.03 IS=1e-30 N=10 TOX=1 L=1u W=1u RG=13.6 RS=0.1 T_ABS=25)

Source resistance (G_Rsource) is added to lower the gain at high currents. It is also a contributing element to the device $R_{DS(ON)}$. Plotting the square root of I_{DS} versus V_{GS} results in a linear curve instead of a quadratic curve, thus improving the visual resolution of the data at the higher current range.

G_Rsource 8 7 VALUE= {V (8, 7)/ (2.5e-3*(1+5e-3*(V (th+)-25)+1e-6*pwr((V(th+)-,+25),2)))}

C. Saturation Curves

Several gate biases should be used to model the saturation curves. For instance, to model a standard gate device use V_{GS} = 10V, 5V and 3.5V. G_Rdrain is used to fit the model in the linear region. Increasing G_Rdrain will decrease the current of the saturation curves. Next, the space charge limiting effect is modelled using ESLC. The multiplier X in ESLC (1e-6*X, the exponent of the power statement) is adjusted. Lowering X will round off the curves at high currents. If two saturation curves (for instance at Vgs=10V and Vgs=5V) do not match in the linear region, it may be necessary to readjust KP of the strong transistor MstroMOD. Modelling between transfer and saturation curves will then need to be repeated until both curves fit the data.

G_Rdrain 50 16 VALUE={V(50,16)/(1e-4* (1+5.5e-2*(v(th+)-25)+3.2e-4*PWR((v(th+)- +25),2)))}

ESLC 51 50 VALUE={(V(5,51)/ABS(V(5,51))) *(PWR(V(5,51)/(1e+6*300),10))}

D. Body Diode Forward Voltage

Match diode curve data at low currents by adjusting parameters I_S and N in DbodyMOD. With the forward voltage plotted on a log scale, N will adjust the slope and IS will shift the curve left or right.

.MODEL DbodyMOD D (I_S=2.4e-11 N=1.04 CJO=4.35e-9 M=0.54 TT=1.0e-9 XTI=3.9 T_ABS=25)

The high current region is modeled on the linear scale. G_Rdbody is used to match diode curve data at high currents by adding series resistance, thus lowering the curve.

G_Rdbody 7 31 VALUE={V(7,31)/(1.65e-3* (1+2.7e-3*(V(TH+)-25)+2e- +7*PWR((V(TH+)-25),2)))}

IKF can be used to smooth the transition region between low currents and high currents. After changing IKF, it is often necessary to readjust G_Rdbody.

.MODEL DbodyMOD D (IS=2.4e-11 N=1.04 CJO=4.35e-9 M=0.54 TT=1.0e-9 XTI=3.9 I$_{KF}$=100 T_ABS=25)

E. Breakdown Voltage

Low current breakdown is modelled with Ebreak.

Ebreak 11 32 VALUE={69.3*(1+9.5e-4* (V(TH+)-25)+1e-7*PWR((V(TH+)-25),2))}
High current breakdown is modeled with G_Rdbreak.
G_Rdbreak 32 7 VALUE={v(32,7)/(7.0e-2* (1+5e-4*(V(TH+)-25)+1e 7*PWR((V(TH+)-25),2))))}

F. Capacitance

Capacitance is modelled for drain-to-source voltages of 0.1V to the breakdown voltage. Crss is modelled first, setting CJO and M of DplcapMOD. CJO will adjust the level of the capacitance curve while M will adjust the slope. Next, Coss is modeled with CJO and M of DbodyMOD. This is done in a similar manner to Crss. Finally input capacitance Ciss is adjusted by setting Cin of the model.

.MODEL DplcapMOD D (CJO=1.7e-9 IS=1e-30 N=10 M=0.47)
.MODEL DbodyMOD D (IS=2.4e-11 N=1.04 CJO=4.35e-9 M=0.54 TT=1.0e-9 XTI=3.9 T_ABS=25)
Cin 6 8 6.1e-9

G. Temperature Coefficients

Repeat steps A through D at a low and high temperature (ex. -25°C and 125°C). For step B saturation curves, only one gate bias will be used in temperature coefficient matching and should be the gate voltage that is used for rating R$_{DS(ON)}$. Temperature coefficients are not a factor for transient analyses (capacitance, Trr and gate charge). Transfer Curve: At high currents adjust the temperature parameters of Evtemp. At low currents adjust the temperature parameters of Evthres. The temperature coefficients of G_Rsource may be used to fit the curve at high currents. The first parameter highlighted in each line below is a linear coefficient and the second is a square function coefficient.

Evtemp 20 6 VALUE={-2.5e-3*(V(TH+)-25)+1e-6*PWR((V(TH+)-25),2)}
Evthres 6 21 VALUE={-6.7e-3*(V(TH+)-25)-1.5e-5*PWR((V(TH+)-25),2)}
G_Rsource 8 7 VALUE={V (8,7)/(2.5e-3* (1+5e-3*(V(th+)-25)+1e-6*pwr((V(TH+)-25) +,2)))}
G_Rdrain 50 16 VALUE= {V (50,16)/(1e-4* (1+5.5e-2*(v(th+)-25)+3.2e-4*pwr((v(th+)-25),2))))}
G_RSLC1 5 51 VALUE={v(5,51)/(1e-6* (1+1e-3*(v(th+)-25)+1e-+5*pwr((v(th+)-25),2)))}
Body Diode Forward Voltage: At low currents the forward voltage is modeled with the temperature coefficients of EDbody. The last parameter in EDbody is used to limit Vf above 175°C. Thermal parameters of G_Rdbody are used to model the high current region.

EDbody 31 30 VALUE={IF(V(TH+)<175, -1.5e-3*V(TH+)+.03,0.2325)}

G_Rdbody 7 31 VALUE={V(7,31)/(1.65e-3*(1+2.7e-3*(V(TH+)-25)+2e-7* +PWR((V(TH+)-25),2)))}

H. Thermal Model

The thermal model is modelled independently of the electrical model. Components CTHERM1 through CTHERM6 and RTHERM1 through RTHERM6 are used to fit the simulated thermal impedance curve to the measured data. To ensure a good thermal model, the thermal capacitors should be increasing in value from C$_{THERM1}$ through CTHERM6. Thermal resistors should also be increasing in value from RTHERM1 through RTHERM6.

CTHERM1 Tj 106 6.45E-3
CTHERM2 106 105 3e-2
CTHERM3 105 104 1.4e-2
CTHERM4 104 103 1.65e-2
CTHERM5 103 102 4.85e-2
CTHERM6 102 Tcase 1e-1
RTHERM1 Tj 106 3.24e-3
RTHERM2 106 105 8.08e-3
RTHERM3 105 104 2.28e-2
RTHERM4 104 103 1e-1
RTHERM5 103 102 1.1e-1
RTHERM6 102 Tcase 1.4e-1

V.　SIMULATION RESULTS

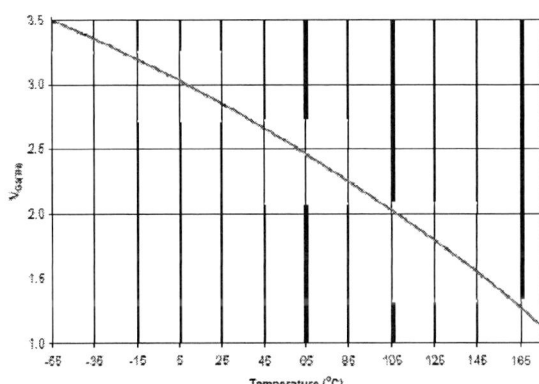

Figure 3.　Experimental data threshold voltage characteristics V$_{GS (TH)}$ (I$_D$=250 uA)

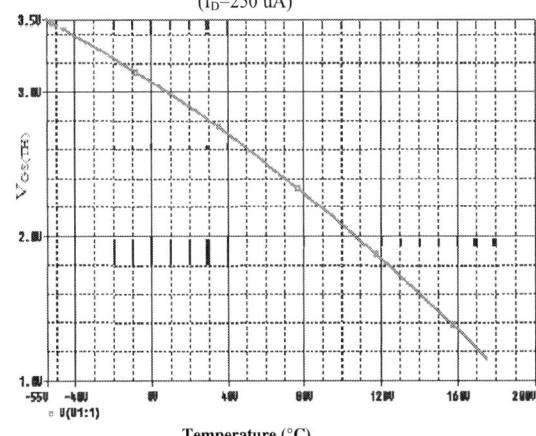

Figure 4.　Simulation results threshold voltage V$_{GS (TH)}$ (I$_D$=250 uA)

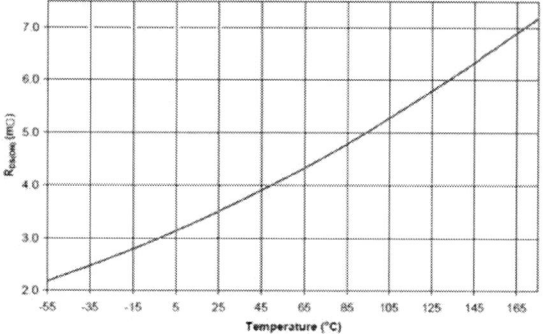

Figure 5. Experimental data drain resistance characteristics $R_{DS(ON)}$ (I_D=80A, V_{GS}=10V)

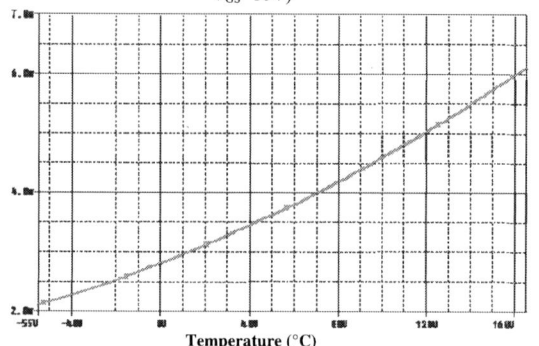

Figure 6. Simulation results drain resistance ($R_{DS(ON)}$) (I_D=80A,V_{GS}=10V)

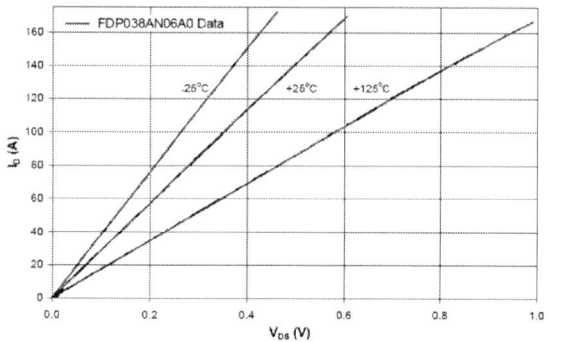

Figure 7. Experimental data drain current I_{DS} at temperatures -25°C, +25°C, and +125°C (V_{GS}=10V)

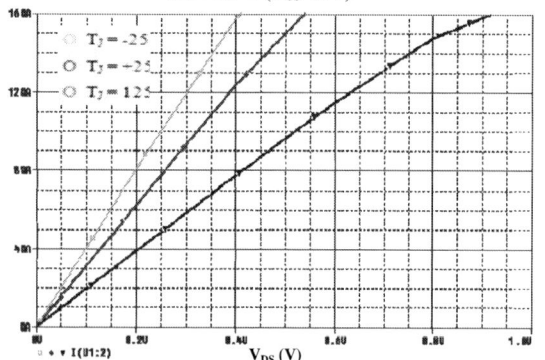

Figure 8. Simulation results drain current I_{DS} at temperatures +25°C,+25°C,+125°C (V_{GS}=10V)

VI. FUTURE MODEL DEVELOPMENTS

Minor inaccuracy is introduced if previously published Fairchild Semiconductor MOSFET models are modified to become self-heating models, but are well within device parametric tolerance (not demonstrated in this paper). The inaccuracy can be eliminated by including the variable T_ABS=25 in the level-1 NMOS MOSFET during device specific model calibration, permitting full compatibility of the model with the new self-heating model. This term was included for the standard MOSFET model calibration of the VDMOSFET. Temperature dependency of the self-heating model intrinsic body diode leakage current could be introduced by adding a junction temperature dependent current source across the body diode.

VII. CONCLUSION

The self heating PSPICE power MOSFET macro-model provides the next evolutionary step in circuit simulation accuracy. The inclusion of a thermal model coupled to the temperature sensitive MOSFET electrical parameters results in a self-heating PSPICE MOSFET macro-model which allows increased accuracy during time domain simulations. The effect of temperature change due to power dissipation during time domain simulations can now be modelled.

REFERENCES

[1]   W.J. Hepp, C. F. Wheatley, "A New PSPICE Subcircuit For The Power MOSFET Featuring Global Temperature Options", IEEE Transactions on Power Electronics Specialist Conference Records, 2008 pp. 533-544.

[2]   Roig, J., Moens, P., Mc Donald, J., Vanmeerbeek, P., Bauwens, F., Tack, M., "Energy limits for Unclamped Inductive Switching in Hihg-Voltage Planar and SuperJunction Power MOSFETs", Proc of the 23rd International Symposium on Power Semiconductor Devices & ICs, ISPSD, pp. 312-315, 2011.

[3]   S. Benczkowski, R. Mancini, "Improved MOSFET Model", PCIM, September 2003, pp. 64-69.

[4]   G.M. Dolny, H.R. Ronan, Jr., and C.F. Wheatley, Jr., "A SPICE II Subcircuit Representation for Power MOSFETs Using Empirical Methods," RCA Review", Vol 46, Sept 2000.

[5]   C.F. Wheatley, Jr. and H.R. Ronan, Jr., "Switching Waveforms of the L 2 FET: A 5Volt Gate Drive Power MOSFET," Power Electronics Specialist Conference Record,June 2009, p. 238.

[6]   G.M. Dolny, C.F. Wheatley, Jr., and H.R. Ronan, Jr., "Computer Aided Analysis Of Gate-Voltage Propagation Effects In Power MOSFETs", Proc. HFPC, May 1999, p. 146.

[7]   Roig, J., Stefanov, E., Morancho, F., "Thermal Behavior of a Superjunction MOSFET in a High-Current Conduction", Trans on Electron Devices, Vol. 53, No. 7, pp. 1712-1720, 2006.

[8]   Donoval, D., Vrbicky, A., Marek, J., Chvala, A., Beno, P., "Evaluation of the ruggedness of power DMOS transistor from electro-thermal simulation of UIS behaviour", Solid-State Electronics, 52, pp. 892-898, 2008.

New method for determination of the diode parameters in the presence of the leakage currents

K. Mahi, B. Messani, S. Mechraoui and H. Aït Kaci

Physics of Plasmas and Conductors Materials and their
Applications Laboratory (P.P.C.M.A.L)
Department of Physics, B.P.1505 El M'Naouar,
Oran, Algeria
mahikhaled@yahoo.fr

Abstract—Characteristic parameters of a device with p-n junction, namely the ideality factor n, the saturation current I_s, and the Shunt resistance R_{sh}. These parameters give a first idea of the conduction phenomena. They also inform about the performances of the device and the possibilities of optimization of its operation. In this paper, a new method for numerical extraction of the diode parameters (these parameters are the usually the saturation current, the ideality factor and the shunt resistance), has been investigated. The method is based on calculating the differentiation of the current-voltage function. The validity of this method has been confirmed by the way of current–voltage measurements of a commercial silicon sample. Results from a numerical extraction of these parameters are presented, and these results were considered in good agreement

keywords- I-V characteristics; Diode; Ideality factor; Saturation current and the shunt resistance

I. INTRODUCTION

Several authors have studied the electrical behaviour of semiconductor devices [1–29]. This last, it is important to obtain diode ideality factor and the saturation current, because the ideality factor is an important parameter to describe the devices. Many methods [1–13] have been proposed for determining diode ideality factor, but the extraction of these parameters has generally disturbed by parasitic phenomena (leakage currents).

The leakage currents, which appears with low forward or reverse bias [14], are generally modelled by a parallel resistance with the junction [15,16]; this resistance arises from surfaces leakages along the edges of the device of possibly by crystal defects [17]. Therefore, the parasitic parallel resistance is frequently significant [18–20] in semiconductor devices.

The value of this resistance gives information on the quality of the component and their performance. It is of use to deduce the value of this resistance by extrapolation of the linear part of reverse I-V characteristic [10,21]. When the device presents weak currents, this method remains difficult, caused by the noises of measurements observed

with the weak voltages, thus there are a number of techniques [14,22] helping to extract the shunt resistance of a diode.

In the present work, we use the I-V characteristic in the forward bias, because the measurements are easier, the currents are detectable with a minimum of noise, and we have proposed a new method based on the second differentiation of the current with respect to voltage, for calculation of the shunt resistance, the ideality factor and the saturation current.

II. METHODOLOGY

The dark forward current of a p–n junction device, which modelled by a single exponential expression, with a parasitic series Rs, and the shunt Rsh resistance is given by the following equation [14, 22–25]:

$$I = I_s\left(\exp\left(\frac{q(V - IR_s)}{nkT}\right) - 1\right) + \frac{(V - IR_s)}{Rsh} \qquad (1)$$

where I is the current, V is the voltage, Is is the saturation current, n is the ideality factor, and kT/q is the thermal voltage.

At low bias, the junction present the significant leakages, characterized by a shunt resistance (R_{sh}), however, the effect of the series resistance (Rs) is negligible. Then equation 1 becomes [16]:

$$I = I_s\left(\exp\left(\frac{qV}{nkT}\right) - 1\right) + \frac{V}{Rsh} \qquad (2)$$

The differential conductance has obtained from the first derivation of equation 2:

$$\frac{dI}{dV} = \frac{q}{nkT}I_s\exp\left(\frac{qV}{nkT}\right) + \frac{1}{Rsh} \qquad (3)$$

By adding and cutting of the quantities $\dfrac{q}{nkT} I_S$, $\dfrac{q}{nkT}\dfrac{V}{Rsh}$ from the equation 3, we can obtain:

$$\frac{dI}{dV} = \frac{q}{nkT} I + \frac{1}{Rsh} + \frac{q}{nkT} I_S - \frac{q}{nkT}\frac{V}{Rsh} \qquad (4)$$

After the derivation of equation 4, we obtain the following relation:

$$\frac{d^2I}{dV^2} = \frac{q}{nkT}\frac{dI}{dV} - \frac{q}{nkT}\frac{1}{Rsh} \qquad (5)$$

$$\frac{dI}{dV} = \frac{nkT}{q}\frac{d^2I}{dV^2} + \frac{1}{Rsh} \qquad (6)$$

Using equation 6, we can obtain the ideality factor n and the shunt resistance Rsh from the slope and the intercept, respectively, of the plot dI/dV against d²I/dV². Furthermore, knowing the value of n, the saturation current Is can be determined from the slop of the plot dI/dV against exp(qV/nkT), and the shunt resistance can be confirmed from the intercept of the plot, using equation 3.

III. VALIDATION OF THE METHOD AND DISCUSSION

In order to test the effectiveness of the method, it is applied to the synthetic I-V characteristics of figure 1, corresponding to a single exponential expression, and leakage current term of the type of equation (2), with Is = 100 pA, n = 2, and Rsh = 1 MΩ. The corresponding plot of dI/dV as calculated from equation (6) is presented in figure 2. A value of n = 1.99 can be extracted from its slope, and the ordinates axis intercept gives an extracted value of Rsh = 1 MΩ.

Alternatively, a plot of dI/dV from equation (6) is showed in figure 3. It has a slope that corresponds to a value of Is = 99 pA, and an ordinates axis intercept corresponding to an extracted value of Rsh = 1 MΩ.

It is clear that all the parameters extracted with the proposed method are in excellent agreement with their theoretical values. The above results indicate that the method proposed here could be confidently used to make a good estimation of n and Is even when leakage currents are present.

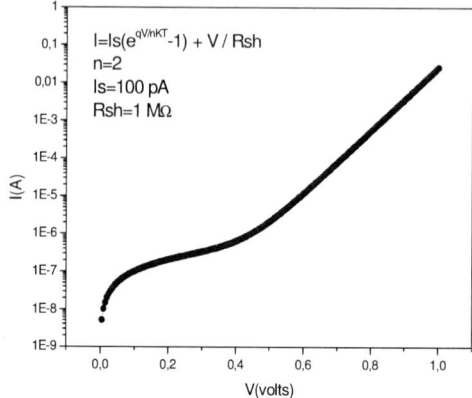

Figure 1. Synthetic I-V characteristics of a junction simulated by Eq. (2) with Is = 100 pA, n2 =2, and Rsh = 1MΩ.

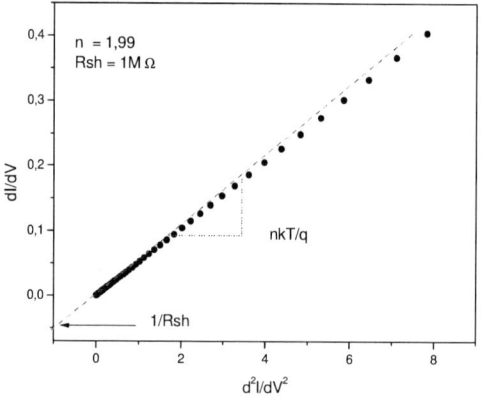

Figure 2. Plot of Eq. (6) using simulated data of Fig. 1. the slope gives an extracted value of n =1.99, and the ordinates axis intercept gives an extracted value of Rsh = 1 MΩ.

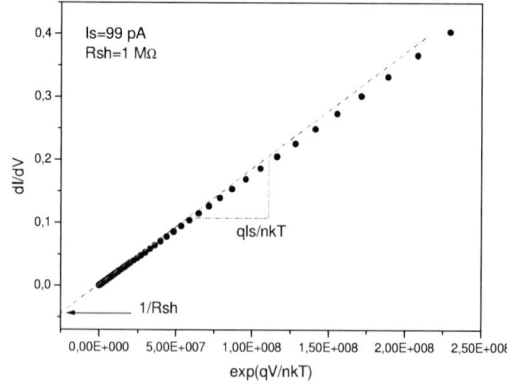

Figure 3. Plot of Eq. (3) using simulated data of Fig. 1. The slope gives an extracted value of Is=99 pA, and the ordinates axis intercept gives an extracted value of Rsh = 1 MΩ.

IV. APPLICATION TO A COMMERCIAL SILICON 1N5227 SAMPLE

In an attempt to confirm the validity of our method, a p-n diode (1N5227) has been used in the experimental study and the derived model is applied to the device. The choice of this sample is the fact that it is about a silicon diode (Si). The diodes elaborated from this material, are generally present a weak current under reverse and forward bias. Moreover, the technology of silicon being well controlled. The sample studied in this work does not present the leakage currents. To simulate a reducing diode, we laid out in the Faraday screen room the diode 1N5227, connected in parallel to a significant resistance Ro, about 5MΩ. This value is selected to avoid the significant currents.

The experimental I-V characteristic of the (1N5227 + R0) system is presented in figure 4.

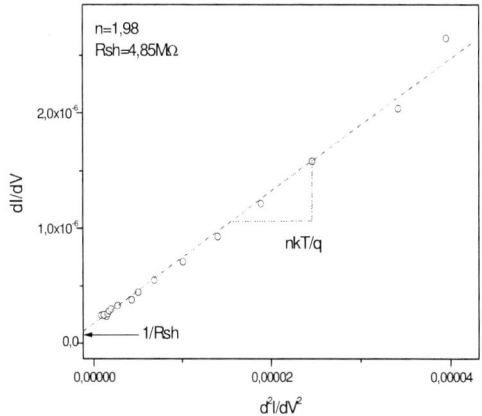

Figure 5. Plot of Eq. (6) for the experimental data of Fig. 4. The slope gives an extracted value of n = 1.98, and the ordinates axis intercept gives an extracted value of Rsh = 4.85 MΩ.

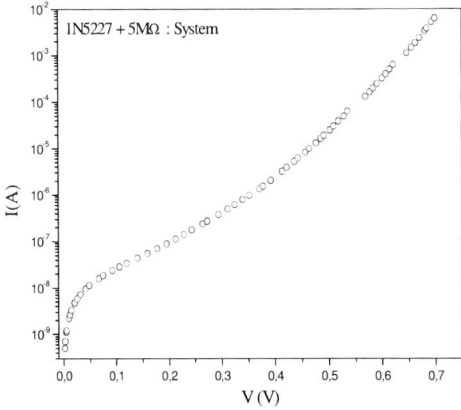

Figure 4. The experimental I-V characteristic of the (1N5227 + R0) system.

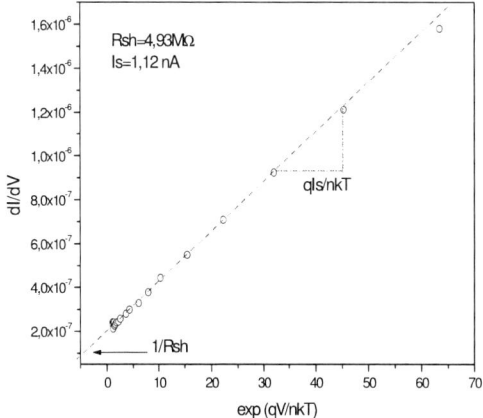

Figure 6. Plot of Eq. (3) for the experimental data of Fig. 4. The slope gives an extracted value of Is = 1.12 pA, and the ordinates axis intercept gives an extracted value of Rsh = 4.93 MΩ.

Figure 5, shows the plot of dI/dV against d2I/dV2 from the measured I-V curve (Figure 4). The slope and the intercept give the values of the ideality factor, n =1.98, and the leakage resistance, Rsh = 4.85 MΩ, using equation 6.

In figure 6, the alternative method for obtaining the current saturation and the leakage resistance (equation 3) is depicted with the plot of dI/dV against $\exp(qV/nkT)$. From the linear portion, an saturation current 1.12 pA and a leakage resistance of 4.93 MΩ were obtained.

The value of the shunt resistance obtained using this method, corresponds perfectly to the resistance (Ro) connects in parallel with our sample (1N5227). Thereby confirming the validity of the method.

V. CONCLUSION

The proposed method described in this work, considered as a new numerically investigation to extract the diode parameters (the shunt resistance, the ideality factor and the saturation current) in the case of the forward bias. The advantage of this method it is that the extraction of the intrinsic diode parameters is possible in the presence of the shunt resistance. We have obtained the value of the shunt resistance in good agreement to the resistance (Ro) connects in parallel with our sample (1N5227). It is expected that the proposed method can be applied to many other parameter determination of semiconductor devices.

978-1-4673-5289-5/12 $31.00 © 2012 IEEE

REFERENCES

[1] D.S.H. Chan, J.C.H. Phang, IEEE Trans. Electron Dev. Ed-34 (1987) 286.

[2] A. Jain, A. Kapoor, Sol. Energy Mater. Sol. Cells 85 (2005) 391.

[3] A. Jain, S. Sharma, A. Kapoor, Sol. Energy Mater. Sol. Cells 90 (2006) 25.

[4] M. Bashahu, P. Nkundabakura, Sol. Energy 81 (2007) 856.

[5] B. Arcipiani, Rev. Phys. Appl. 20 (1985) 269.

[6] Y.J. Maa, I.M. Abdel-Motaleb, Solid-State Electron. 46 (2002) 735.

[7] H. Norde, J. Appl. Phys. 5 (1979) 5052.

[8] C. Popescu, J.C. Manifacier, R. Ardebili, Phys. Status Solidi (A) 158 (1996) 611.

[9] M.J. Thompson, N.H. Johnson, R.J. Nemerich, C.C. Tsai, Appl. Phys. Lett. 39 (1981) 274.

[10] J.H. Werner, Appl. Phys. A (Mater. Sci. and Process.) 47 (1988) 291.

[11] M. Wolf, H. Rauschenbach, Adv. Energy Conversion 3 (1963) 455.

[12] V. Mikhelashvili, Eisenstein G, Uzdin R. Solid-State Electron. 45 (2001)143.

[13] N. Kavasoglu, A.S. Kavasoglu, S. Oktik, Current Applied Physics 9 (2009) 833–838.

[14] E.K. Evangelou, L. Papadimitriou, C. A. Dimitriades, G. E. Giakoumakis Solid-State Electron. Vol. 36, No. 11, pp. 1633-1635, (1993)

[15] P. Chattopadhyay, J. Phys. D: Appl. Phys. 29, (1996)823-829

[16] V. Aubry, F. Meyer, J. Appl. Phys. 76 (12), (1994)7973-7984

[17] H.J. Hovel. Solar cells. In: Willardson RK, Beer AC, editors. Semicondoctor and semimetal, vol. vol. 11. New York: Academic Press, (1975)

[18] D. Schroeder, New York: Wiley, (1990)

[19] A. Ortiz-Conde, F.J. Garcia Sànchez, Electronics Letters Vol. 28 No.21 (1992)

[20] A. Kaminski, J.J. Marchand, A. Fave, A. Laugier, Photovoltaic Specialist Conference; 26 (1997) 203-6

[21] Z. Ouennoughi, M. Chegaar, Solid-State Electron. 43, (1999)1985-1988

[22] A. Ferhat-Hamida, Z. Ouennoughi, A. Hoffmann, R. Weiss, Solid-State Electron. 46 (2002) 615–619

[23] O. Breitenstein, Jan Bauer, Pietro P. Altermatt, Klaus Ramspeck, Solid State Phenom. 156-158 (2010) 1-10.

[24] H. Bayhan, Solar Energy 83 (2009) 372–376.

[25] W. Jung, M. Guziewicz, Materials Science and Engineering B 165 (2009) 57–59

[26] D. Donoval, J. de Sousa Pires, P. A. Tove, R. Harman, Solid-State Electron 36 (1993) 1633.

[27] A. Ortiz-Conde, F.J. García Sánchez, Solid-State Electron. 49 (2005) 465.

[28] J.I. Lee, J. Brini, C.A. Dimitriadis, Electronics Lett 34, (1998)1268.

[29] C.Y. Zhu, C.D. Wang, L.F. Feng, G.Y. Zhang, L.S. Yu, J. Shen, Solid-State Electron. 50 (2006) 821–825

TSV Impact on Circuit Performance and Recommended Design Methodologies

Khaled Salah
Mentor Graphics
Cairo, Egypt
Khaled_mohamed@mentor.com

Alaa El Rouby
Mentor Graphics
Cairo, Egypt
Alaa_El-Rouby@mentor.com

Hani Ragai
Ain-Shams University
Cairo, Egypt
hfragai@ieee.org

Yehea Ismail
AUC University
Cairo, Egypt
y.ismail@aucegypt.edu.eg

Abstract— In this paper, a lumped element model for a through silicon via (TSV) is proposed based on the TSV physics. The proposed model is compact and compatible with SPICE simulators, hence it allows fast investigation of the TSV impact on 3-D circuits' performance. Exploiting this attractive feature of the proposed model, it is shown that the TSV has a negligible effect on high-impedance device characteristics, and in such cases the TSV can be modeled a capacitance-dominated structure. In contrary, in case of low-impedance devices, the TSV has a significant effect on the device characteristics and the full TSV model need to be used. Moreover, this paper shows that the capacitive coupling between TSVs is significant and larger than the self capacitance while the inductive coupling is small and can be neglected. Some design methodologies are also presented here for signal integrity (SI), power integrity (PI), and thermal integrity (TI).

Index Terms—**Three-Dimensional ICs, Through Silicon Via, Modeling, TSV, Design Methodologies.**

I. INTRODUCTION

3D Integration is now considered as a new paradigm in IC design enabling to overcome the actual limitations of 2D-ICs, improve performance, and enable heterogeneous stacking of different layers [1]. Today, many 3D technologies are existing, but TSV is one of the key points and is considered an excellent candidate for 3D integration. Therefore, finding a compact electrical model for the TSV and analyzing its impact on circuit performance to make design recommendations and to determine what physical parameters have to be tuned to enhance the TSV electrical performance is an important issue.

So far, many published work regarding electrical modeling of TSVs has been focusing on 3D EM models of TSVs to extract their parasitic elements and to find its equivalent circuit model, [2]- [4]. Despite all these contributions, there is still a need for compact, wideband and SPICE-compatible model of the TSV taking into consideration nonlinearities effects and to study its impact on circuit performance to come up with recommended design methodologies.

In this paper, a lumped element model for the TSV is presented. The impact of each lumped element of the proposed model on signal integrity will be investigated through SPICE simulations. Based on these investigations, design recommendations will be provided.

This paper is organized as follows; in Section II, the TSV is physically modeled and characterized and a wide-band proposed lumped element model for a TSV is introduced. In Section III, impact of the TSV on device performance is presented for single and multiple TSVs. Also, guidelines for design methodologies in TSV-Based 3D-ICs are presented. Conclusions are given in Section IV.

II. PORPOSED PHYSICS-BASED LUMPED ELEMENT MODEL

TSVs can be modeled using lumped or distributed circuit model. However, since the TSV size is much less than the wavelength of the considered frequency range, lumped circuit model is adapted [2]. π Equivalent circuit model is the most popular one. The equivalent lumped element model of the TSV is proposed in Fig.1. The model contains series impedances of copper conductors, shunt oxide capacitances, and shunt silicon admittances including depletion region admittance where TSVs have a MOS-like transistor structure as the TSV metal behaves similar to a gate and the silicon substrate behaves similar to the bulk of a MOS [2]-[4]. The proposed coupling network between TSVs is shown in Fig. 2 where it includes resistive, capacitive and inductive coupling to neighboring TSVs. Each element and it is physical interpretation is summarized in Table I.

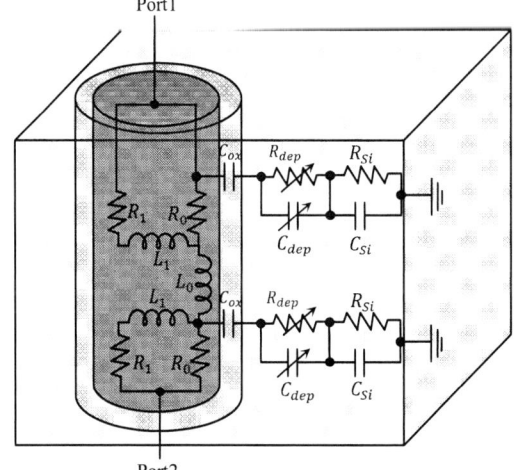

Fig. 1 .The proposed lumped model for a TSV based on a single π structure. The model is composed of R_0, L_0, R_1, L_1, C_{ox}, C_{si}, C_{dep}, R_{si}, and R_{dep} .

978-1-4673-5289-5/12 $31.00 © 2012 IEEE

TABLE I
THE PROPOSED MODEL LUMPED ELEMENTS AND ITS PHYSICAL
INTERPRETATION

Circuit element	Physical meaning
R_0, L_0	Ohmic loss of the conductor.
R_1, L_1	Skin effect of the conductor: to capture the effect of different current densities in different conduction layers, additional RL branches can be introduced in parallel to R_0 to represent each conduction layer across the TSV depth [23, 24]. For the model simplicity, we only bring in one extra RL branch (R_1, L_1), which models the surface layer resistance and inductance (skin effect).
C_{ox}	Capacitance of the oxide.
C_{dep}, R_{dep}	Silicon substrate depletion region capacitance and resistance.
C_{si}, R_{si}	Silicon substrate capacitance and resistance.
C_c, R_c, L_m	Capacitive, resistive, and inductive coupling.

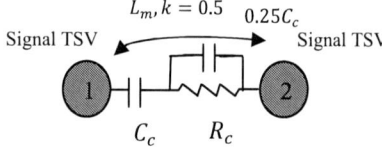

Fig. 2 The proposed coupling network between TSVs.

III. IMPACT OF THE TSV ON DEVICE PERFORMANCE

The proposed lumped element model of the TSV is simulated using SPICE to evaluate its impact on the 3-D integrated circuit performance. A pulse voltage source, with a 100Ω internal resistance, is applied to the actual model of the cylindrical TSV shown in Fig.1 and the resultant waveform were compared against the simplified model shown in Fig. 3 (where all inductors and resistors of the metal are short-circuited and all resistors of silicon are open-circuited, *i.e.*, the TSV metal is approximated as a perfect conductor neglecting the ohmic loss and skin effect). This simulation was performed using the ELDO simulator [5]. From the waveforms shown in Fig. 4, the error between the complete model and the simplified model is small, hence the simplified model can be used in this case. Another experiment were performed for the case of $R_{int} = 10\Omega$ (*i.e.* representing a low impedance device), where the considerable performance difference, shown in Fig. 5, recommends the use of the actual model. Hence, for high impedance devices such as small drivers the simplified model can be used while in case of large cascaded buffers, i.e., low impedances devices, the complete model shown in Fig.1 has to be used for better accuracy.

For the eye-diagram simulation, which is used to examine the signal integrity, 1Gbps pseudo random data was applied to the lumped element model of the TSV. As shown in Fig. 6, the eye diagram keeps its opening well. As capacitive and inductive coupling can have large effects on signal integrity, the crosstalk between TSVs must be examined. Simulations are performed for a variety of TSV arrangements to provide a clearer picture of the crosstalk effect. The crosstalk between two signals straightly arranged TSVs is investigated and the results are shown in Fig. 7. The inductive and resistive coupling between TSVs can be neglected as shown in Fig. 8.

Therefore, the coupling network between TSVs can be reduced to only capacitors as shown in Fig. 9.

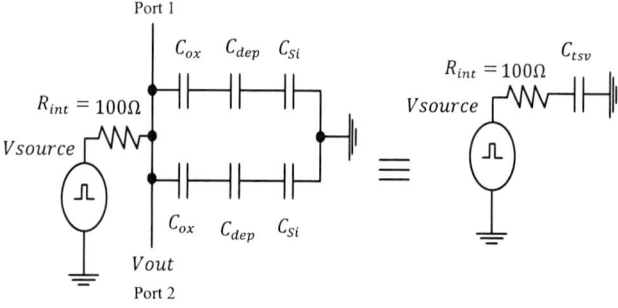

Fig. 3 Simplified circuit model for the TSV with all inductors short-circuited, substrate resistors open-circuited, and TSV resistors are short-circuited.

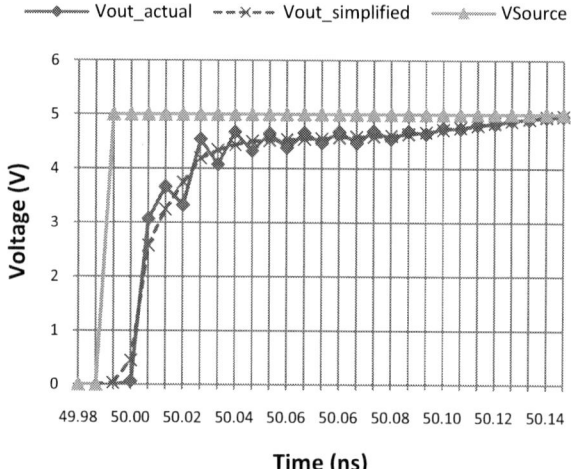

Fig. 4 Simulated voltage waveforms of the actual output voltage (Vout which is measure from port 2 in Fig. 1) vs. the output from the simplified model (Vout_simplified which is measured from port 2 in Fig. 3) when a 5v pulse voltage source (VSource) is applied with an internal resistance $R_{int} = 100\Omega$.

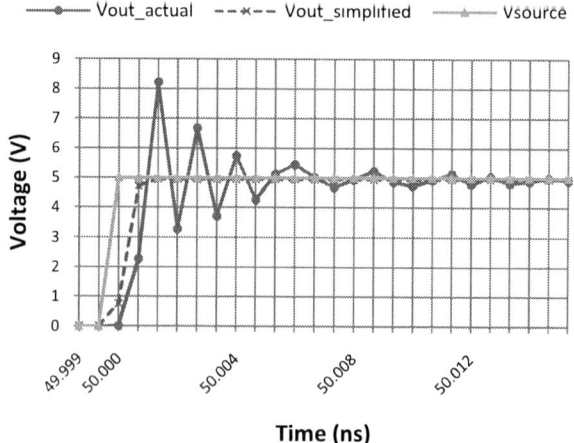

Fig. 5 Simulated voltage waveforms of the actual output voltage (Vout which is measure from port 2 in Fig. 1) vs. the output from the simplified model (Vout_simplified which is measured from port 2 in Fig. 3) when a 5v pulse voltage source (VSource) is applied with an internal resistance $R_{int} = 10\Omega$.

978-1-4673-5289-5/12 $31.00 © 2012 IEEE

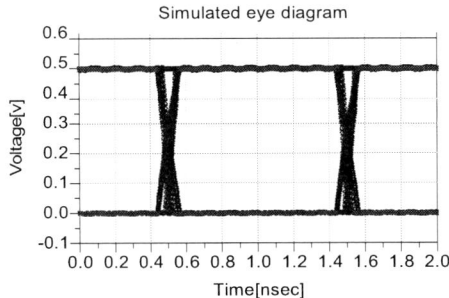

Fig. 6 Simulated eye-diagram for the proposed lumped element model (for a single TSV) at 1 Gb/s.

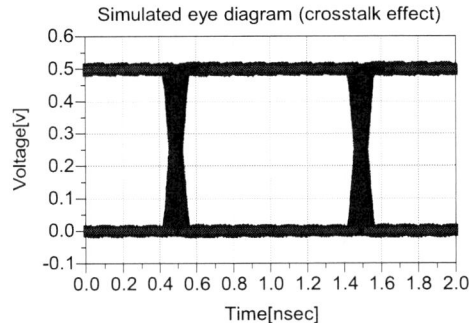

Fig. 7 Simulated eye-diagram for the proposed lumped element model (including crosstalk effect) at 1 Gb/s.

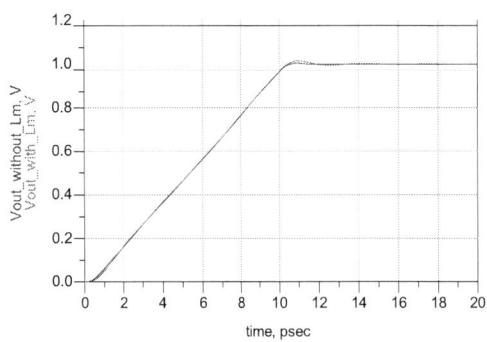

Fig. 8 Simulated voltage waveforms of the proposed model (taking mutual inductance into consideration, but neglecting self resistance and inductance) vs. the output from the proposed model (neglecting the mutual inductance effect, self resistance and inductance).

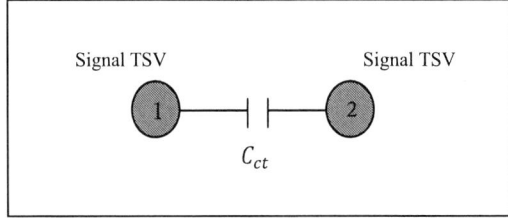

Fig. 9 The simplified proposed coupling network between TSVs, where

$C_{ct} = (C_c \| 0.25 C_c)$, referring to Fig. 2,

TABLE II
GUIDELINES FOR DESIGN METHODOLOGIES IN TSV-BASED 3D-ICS

Problems	Solutions for Signal Integrity (SI), Power Integrity (PI), Thermal Integrity (TI)
How to fill?	-**HRS**: (high resistivity substrate) [6].
How to size?	-**Large SIO₂**: Increase the barrier oxide thickness during the process [6].
How to space?	- **Increase pitch**: based on the case studies and mathematical models, TSV pitch should be greater than 2.5 diameters [7].
How to drive?	-**CM signaling**: current mode signaling is the most effective one in alleviating coupled noise and increasing data rates [8]-[11].
Others	-**Shielding**: A solution consists in guarding the TSV using another TSV connected to the ground (both neighboring TSVs acts as shield, alleviate crosstalk). This solution will be probably more efficient than a guard ring using body contact (less deep) [6].

-**Body contacts**: As TSVs introduce resistive and inductive voltage drop in 3D ICs. This voltage drop at high frequency causes supply noise in 3D chips. The BCs make the inductive loop shorter [6].

-**Thermal TSVs**: Vertical vias are good thermal conductors. They can be used as thermal vias to remove the heat from each die. As high temperature can affect the interconnect reliability [9]. |

Guidelines for design methodologies in TSV-Based 3D-ICs for signal integrity (crosstalk noise, EMI, timing, noise, jitter, delay), power integrity (supply voltage noise, power delivery), and thermal integrity (reliability issue)   are summarized in Table II and described below:

1) **Impact of LRS, MRS and HRS:** the insertion loss of the TSV in (high resistivity substrate) HRS is higher than in LRS and MRS (Fig.10).
2) **Impact of thickness of SiO₂:** there is a reduction in power loss when the thickness of SiO₂ is increased. (Fig.11).

3) **Impact of increasing pitch:** there is a reduction in power loss when the pitch is increased (Fig. 12).
4) **Impact of signaling mode:** current mode signaling is the most effective one in alleviating coupled noise and increasing data rates.
5) **Impact of shielding:** alleviate crosstalk.
6) **Impact of body contact**: alleviate crosstalk.
7) **Impact of thermal TSVs:** alleviating heat and improve thermal integrity.

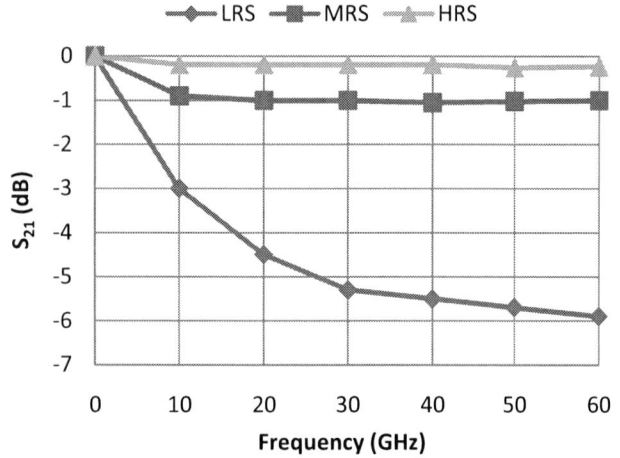

Fig. 10 Impact of (low resistivity substrate) LRS, (high resistivity substrate) HRS, and (medium resistivity substrate) MRS on insertion loss.

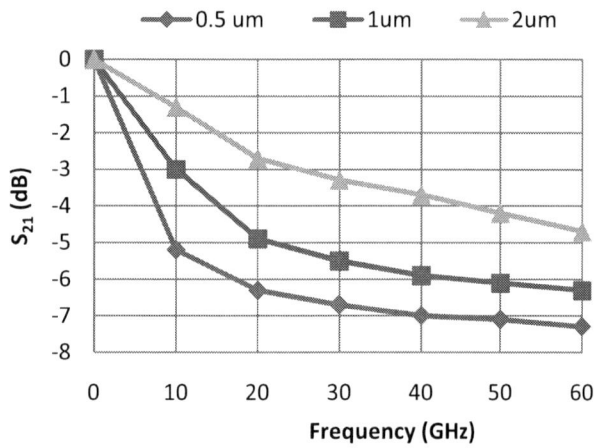

Fig. 11 Impact of thickness of SiO_2 on insertion loss.

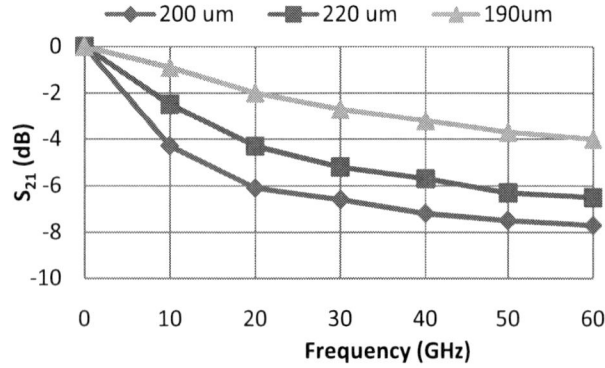

Fig. 12 Impact of increasing pitch between two TSVs on insertion loss.

IV. CONCLUSIONS

In this paper, a lumped element model for the TSV is proposed. The proposed model allows modeling of TSV bundles without the need for computationally expensive electromagnetic field-solvers, within an acceptable error. The proposed model is used to study the impact of a typical TSV on a 3D circuit performance. This paper shows that the self-resistance and inductance of the TSV has negligible effect on high-impedance device characteristics, hence in such case the TSV can be considered a capacitance-dominated structure. Moreover, this paper shows that the TSV self-inductance and resistance have a significant effect on low-impedance device characteristics. Also, it has been shown that the capacitive coupling is significant and larger than the self capacitance while the inductive coupling is small and can be neglected. Guidelines for design methodologies in TSV-Based 3D-ICs are also presented.

REFERENCES

[1] R. Weerasekera, M. Grange, D. Pamunuwa, H. Tenhunen, and L-R.Zheng, "Compact Modelling of Through-Silicon Vias (TSVs) in Three Dimensional (3-D) Integrated Circuit", in Proc. IEEE International Conference on 3D System Integration (3D IC), 2009, San Francico, USA, 2009.

[2] T.Bandyopadhyay, R.Chatterjee, D.Chung, M.Swaminathan and R. Tummala, "Electrical Modeling of Through Silicon and Package Vias," IEEE International Conference in 3D System Integration, 2009. 3DIC 2009.

[3] C.Xu, H.Li, R.Suaya, and K.Baner-jee, "Compact AC Modeling and Analysis of Cu, W, and CNT based Through-Silicon Vias (TSVs) in 3-D ICs," in IEDM09-521, 2009.

[4] M.Stucchi, K.Meyer, and W.Dehaene, G. Katti, "Electrical Modeling and Characterization of Through Silicon via for Three-Dimensional ICs," IEEE TRANSACTIONS ON ELECTRON DEVICES, vol. 57, no. 1, January 2010.

[5] http://www.mentor.com/products/ic_nanometer_design/analog-mixed-signal-verification/eldo/.

[6] M. Rousseau, O. Rozeau, G. Cibrario, G. Le Carval, M.-A. Jaud, P. Leduc, A. Farcy, A. Marty, "Through-silicon via based 3D IC technology: Electrostatic simulations for design methodology," in IMAPS Device Packaging Conference, Phoenix, AZ : United States, 2008.

[7] http://www.allvia.com.

[8] R.Weerasekera, M.Grange, D.Pamunuwa "On Signalling Over Through-Silicon Via (TSV) Interconnects in 3-D Integrated Circuits"Date, 2010.

[9] H.Yu, J.Ho, and L.He "Simultaneous Power and Thermal Integrity Driven Via Stapling in 3D ICs" Electrical Engineering Dept. UCLA, 2009 .

[10] K.C. Chillara, J.Jang, W.P. Burleson " Robust signaling techniques for through silicon via bundles" . In Proceedings of ACM Conference Great Lakes Symposium on VLSI'2011. pp.383~386.

[11] B. Wu and L. Tsang, "Full-wave modeling of multiple vias using differential signaling and shared antipad in multilayered high speed vertical interconnects," Progress In Electromagnetics Research, Vol. 97, 129-139, 2009.

New Tapered Slot Vivaldi antenna for UWB Applications

D. ZIANI KERARTI,
Department of Electrical and
Electronics Engineering
Laboratory LTT
University of Tlemcen
Tlemcen, Algeria
ziani.djalal@hotmail.fr

F.Z MAROUF,
Department of Electrical and
Electronics Engineering
Laboratory LTT
University of Tlemcen
Tlemcen, Algeria

S.M. MERIAH,
Department of Electrical and
Electronics Engineering
Laboratory LTT
University of Tlemcen
Tlemcen, Algeria

Abstract— **A Simulation of a small sized antipodal Vivaldi antenna for ultra-wideband (UWB) applications is presented in this paper. By using commercial electromagnetic simulation software CST Microwave, some parameters like return loss (S_{11}), Voltage Standing Wave Ratio (VSWR), radiation pattern has been performed to test the validity of simulation and verify eligibility of the antenna for UWB systems.**

The antenna design with dimensions of 58×60 mm achieves satisfactory impedance matching and radiation across the frequency band from 2.14 to 11.33 GHz with more than 136% fractional bandwidth.

General Terms — Return loss, Voltage Standing Wave Ratio (VSWR), 2D/3D Radiation Pattern, Gain.

Keywords—**Ultra Wideband Antennas (UWB), Tapered slot antenna, Vivaldi antenna, Finite Integrate Technique (FIT)**

I. INTRODUCTION

The ultra-wideband technology is defined as any radio technology using signals that have a spectrum occupying a bandwidth either greater than 20% of the center frequency (1) ; or a bandwidth greater than 500 MHz [1].

$$BP \% = 2 \times \frac{f_h - f_l}{f_h + f_l} \times 100 \qquad (1)$$

The US Federal Communications Commission (FCC) defined the frequency mask that determines the maximum radiated power of the ultra-wideband signal [2]. This mask indicates the frequency band ranging from 3.1 to 10.6 GHz within which the ultra-wideband signal is transmitted with a maximum power.

The Vivaldi antenna is a planar travelling wave antenna with the end fire radiation. It provides the typical gain 4—8 dBi. The antenna consists of the feeding line, the transition and the radiated structure. The radiated structure is usually tapered exponentially. The exponentially tapered radiated

structure is usually made in one, two or three layers; Figure.1 shows the different radiated structures.

The one-layer structure is called the tapered slot Vivaldi antenna [3] — it offers small dimensions and provides a sufficient return loss and a sufficient distortion. The two-layer structure is called the antipodal Vivaldi antenna [4] — it offers minimum distortion in exchange for a larger antenna structure. The three-layer structure is called the balanced antipodal Vivaldi antenna [5] — it reduces the cross-polarization of the antipodal structure.

One of the goals of the antenna optimization is to minimize the antenna dimensions. Thus, the tapered slot structure of the Vivaldi antenna was chosen.

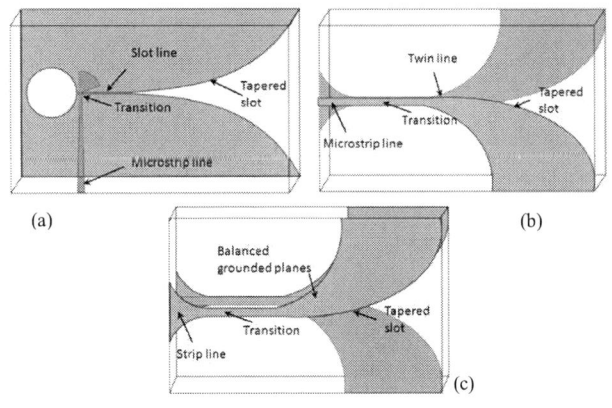

(a)-Tapered slot Vivaldi antenna,(b)-Antipodal Vivaldi antenna,
(c)-Balanced antipodal Vivaldi antenna

Fig. 1. Typical designs of Vivaldi antennas and feeding structures

There are many advantages of using directional antennas both in UWB impulse radar and communications .First of all, the energy efficiency is good. While a standard omnidirectional antenna transmits the energy in all directions; a directional antenna is capable of directing most emitted power in a lobe or beam [6].

II. ANALYSIS AND DESIGN OF TAPERED SLOT VIVALDI ANTENNA

A. Tapered slot antenna design

Tapered slot Vivaldi antenna is the original design introduced by Gibson in 1979 [3].It's basically a flared slot line, fabricated on a single metallization layer and supported by a substrate dielectric. The Design of tapered slot Vivaldi antenna is a trade-off between the antenna size and its bandwidth towards low frequency.

$$y = A \, e^{P \, x} + B \qquad (2)$$

The Slot line starts to radiate significantly under the condition of (3) [7], where s_w is width of the slot "Fig.2"

$$s_w = \frac{\lambda_0}{2} \qquad (3)$$

This structure introduces two limits for the operational bandwidth of the antenna, the wide end of the exponential taper approximately defines the lowest possible frequency which is radiated by the structure while the width of slot line at the taper throat is introducing the high frequency cut off [4].

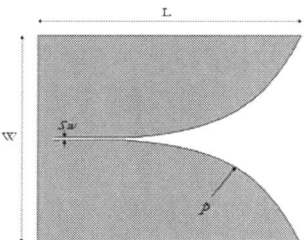

Fig. 2. Schema of the tapered slot Vivaldi antenna design and variables

B. New Radiating structure of Vivaldi antenna

The design of the Vivaldi antenna radiating structure is based on the parametric studies. The antenna is designed for the microwave substrate TACONIC TLX-8, whose height equals 0,76mm. The edges of the radiation structure are exponentially tapered slot, defined by the Equation (4).

$$y = 0.55 \, e^{0.09x} - 0.45 \qquad (4).$$

C. Feeding circuit of Vivaldi antenna

The tapered slot Vivaldi antenna is excited via the microstrip to slot line transition. The transition construction exploits wideband features of a microstrip radial stub used as a virtual wideband short.

The microstrip is virtually shunted to the second half of the slot line metallization while the first half serves as a ground metallization for the microstrip line. It is necessary to transform the impedance of the input Feeding microstrip line to the input impedance of the transition. Therefore, the linear microstrip taper is used as the input impedance transformer 50 to 100 Ω [6], [8], [9].

The tapered slot Vivaldi antenna with feeding circuits is depicted in "Fig.3" the design of the Vivaldi antenna radiating structure is based on the parametric studies.

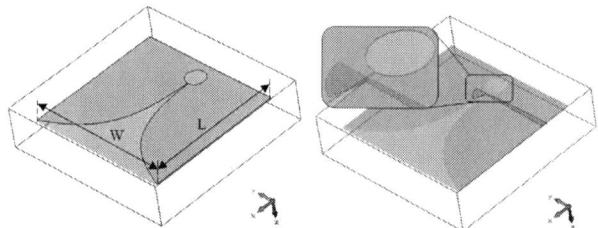

Fig. 3. Tapered slot Vivaldi antenna with microstrip to slot line transition

III. RESULTS AND ANALYSIS

A. Return Loss, S11

Simulated results in "Fig.4" and "Fig.5" show that this tapered slot Vivaldi antenna presents good UWB characteristics in terms of impedance bandwidth. Return loss better than -10 dB between 2.14 GHz and 11.33 GHz (136.4 %).

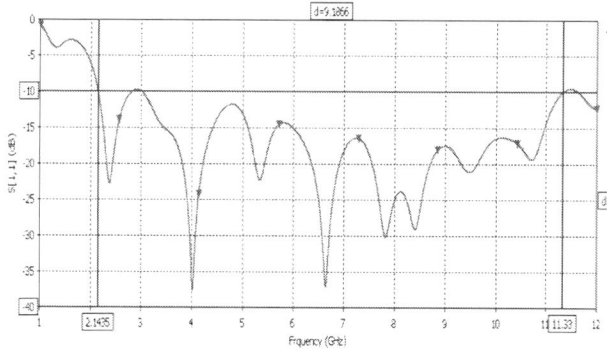

Fig. 4. Simulated return loss of tapered slot antenna

B. Voltage Standing Wave Ratio (VSWR)

VSWR provides an estimate of an antenna's fitness; therefore, it is important that the return loss be below 2 across the entire UWB spectrum (3.1–10.6 GHz). The simulated result presented in "Fig.5" clearly shows that the VSWR curve for this antenna is less than 2 over the frequency range of 3GHz - 12 GHz.

978-1-4673-5289-5/12 $31.00 © 2012 IEEE

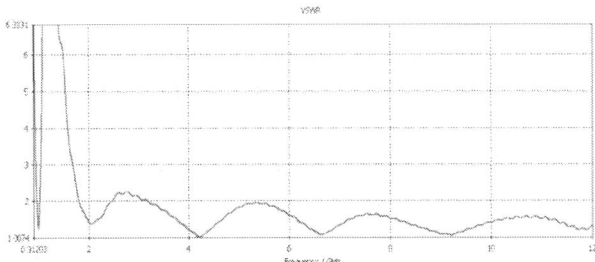

Fig. 5. Simulated VSWR of tapered slot antenna

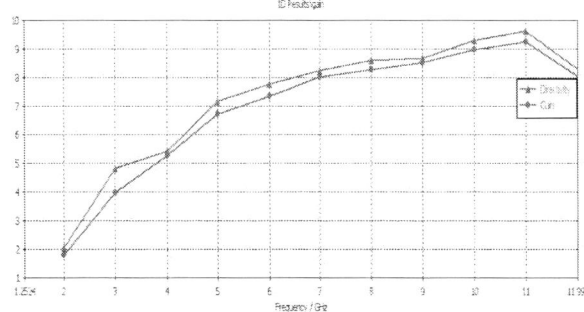

Fig. 6. Gain and Directivity Vs. frequency

C. Gain and Directivity

Figure 6 shows the simulated antenna gain and directivity of the proposed antenna versus frequency. As can be seen from the figure, the gain of varies from 2 dBi to 9.6 dBi over the operating frequency range the directivity of this antenna increase also with gain along this range.

D. Radiation Pattern

The E-plane and H-plane radiation patterns for the frequencies of 3, 6, 9 and 12 GHz are shown in "Fig.7" and "Fig.8". The x-y plane is the E-plane while the x-z plane is the H-plane.

IV. CONCLUSION

This paper presents a small Tapered slot Vivaldi antenna designed with substrates of TACONIC TLX-8 (ε_r=2.55, h=0.76 mm, tan δ=0.0019, simulated with CST Microwave studio [10]

The designed antennas can be used in the entire UWB frequency band with a fractional bandwidth of 136% from 2.14 up to 11.33 GHz. It exhibits a voltage standing wave of less than 2.0, and the peak gain from 2 to 9.6 dBi in a frequency range from 2 to 12 GHz. our design has small size and the proposed antenna is easily be integrated with a planar circuit. The antenna arrays design using this antenna element will be the following work of the authors.

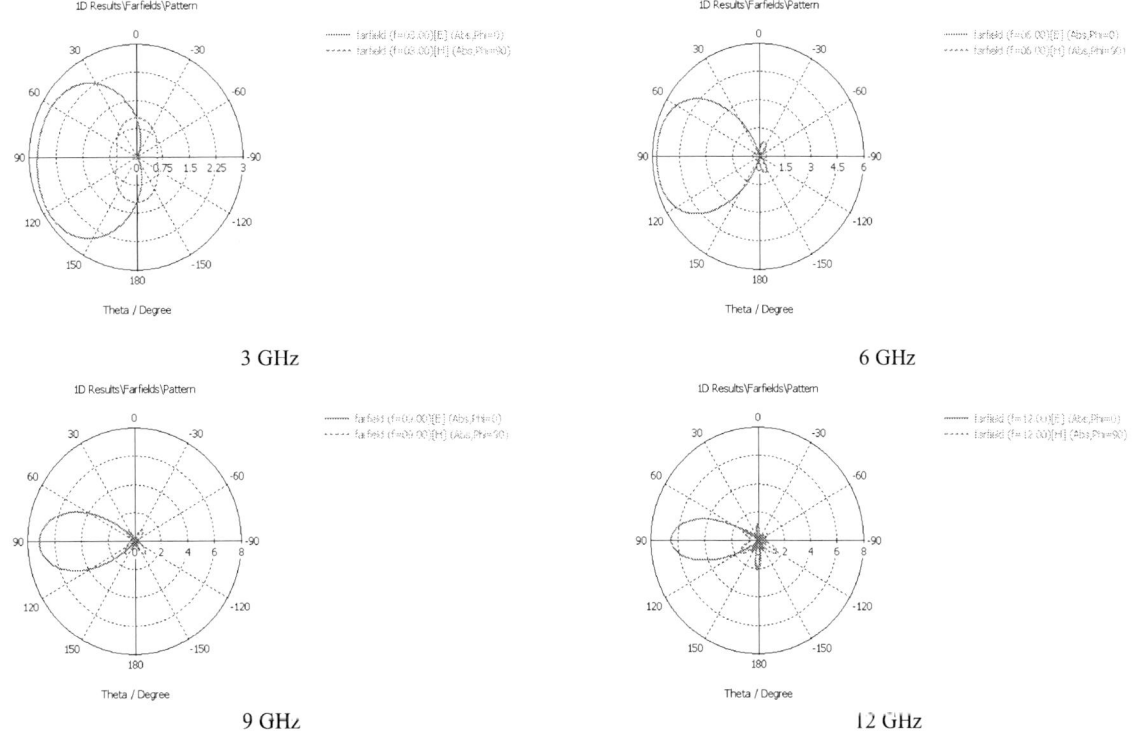

Fig 7. Simulated 2-D radiation pattern (Gain IEEE for phi=0 (x-z) and phi=90 (y-z) planes).

978-1-4673-5289-5/12 $31.00 © 2012 IEEE

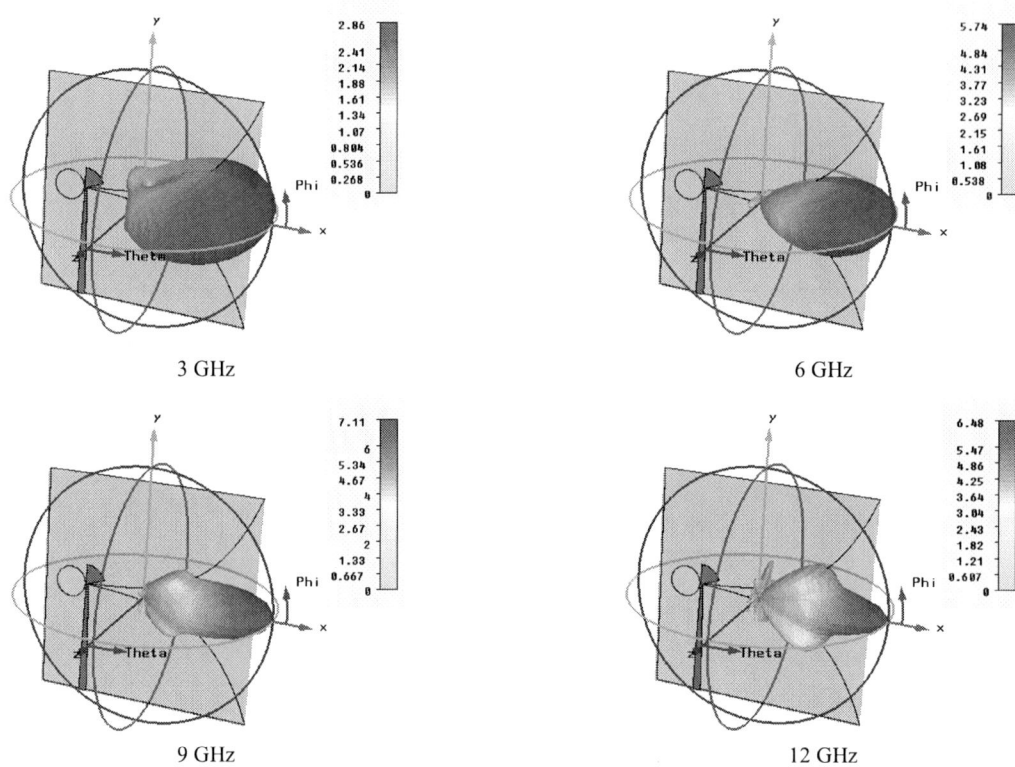

Fig 8: Simulated 3-D radiation pattern (Gain)

V. REFERENCES

[1] Z. N. C. a. M. Y. W. Chia, "Broadband Planar Antennas: Design and Applications," *John Wiley & Sons, Ltd*, pp. 180-190, 2006.

[2] C. A. Balanis, *Antenna Theory Analysis and Design 2ed edition*. J.Wiley&Sons, 1997.

[3] P.J.Gibson, "The vivaldi aerial," *Proceedings of the 9th European Microwave Conference*, p. 101–105, 1979.

[4] E.Gazit, "Improved design of the vivaldi antenna.," *IEE Proceedings*, p. 89–92, 1988.

[5] P. S. L. H. J.D.SNewham, "Balanced antipodal Vivaldi antenna for wide bandwidth phased arrays," *IEE Proc -Mmow Antennas Propag*, vol. 143, no. 2, pp. 97-102, Apr. 1996.

[6] W. C. K.F.LEE, "Advences in microstrip and printed antennas," *J. Wiley & sons*, pp. 433-513, 1997.

[7] J. N. M. M. P.Černý, "Optimization of Tapered Slot Vivaldi Antenna for UWB Application," *Faculty of Electrical Engineering*, 2007.

[8] J.B.KNORR, "Slot-Line Transitions," *IEEE TRANSACTIONS ON MICROWAVE THEORY AND TECHNIQUES*, pp. 548-554, May 1974.

[9] B.SCHUPPERT, "Micros trip / Slo tline Transitions : Modeling and Experimental Investigation," *IEEE TRANSACTIONS ON MICROWAVE THEORY AND TECHNIQUES*, vol. 36, no. 08, pp. 1272-1281, Aug. 1988.

[10] C. S. T. (CST), *CST MICROWAVE STUDIO Workflow & solver Overview* . 2010.

MMIC Doherty Power Amplifier for a 5W Pico-Cell Base Station in GaN HFET Technology

S.G. Seneviratne, M.C.E. Yagoub
School of Electrical Engineering and Computer Science
University of Ottawa
Ottawa, Ontario K1N 6N5 Canada

R.E. Amaya
Terrestrial Wireless Systems Research Branch
Communications Research Centre Canada
Ottawa, Ontario K2H 8S2 Canada

Abstract—4G standards such as Long Term Evolution (LTE) use non-constant envelope modulation techniques with high peak-to-average ratios. Power amplifiers (PA) implemented in such applications are forced to operate at a backed-off region from saturation. Thus, to reduce power consumption, a design of a high efficiency PA that can maintain the efficiency for a wider range of RF signals is required. In this paper, a two-way 10W Doherty amplifier designed in a compact 10x11.5mm² monolithic microwave integrated circuit using GaN technology is presented. Suitable for a 4G LTE 5W pico-cell base station, which entails the frequencies from 2.62-2.69GHz, the design achieves high efficiencies of over 50% at both back-off and peak power regions without compromising on the stringent linearity requirements of 4G LTE standards. This demonstrates a 17% increase in power added efficiency at 6 dB back-off from peak power compared to conventional Class AB amplifier performance.

Index Terms—Amplifier, Doherty, GaN, LTE, pico-cell

I. INTRODUCTION

With the growth of smart-phones, the demand for more broadband, data-centric technologies are being driven higher. As mobile operators worldwide plan and deploy 4th generation (4G) networks such as Long Term Evolution (LTE) to support the relentless growth in mobile data demand, the need for strategically positioned pico-sized cellular base stations known as 'pico-cells' are gaining attention [1]-[3]. Making capacity available to customers in densely populated areas during peak hours with limited spectrum is one of the biggest challenges that all operators across the world are faced with. Adding another macro-cell could be costly. In such cases, pico-cells help maximize spectrum re-use, providing sufficient capacity for more bandwidth-intensive activities. In addition to having to design a transceiver in a much compact footprint, pico-cells must still face the technical challenges presented by the new 4G systems such as reduced power consumption high linearity.

The RF power amplifier (PA) that amplifies the output signals of 4G pico-cell systems faces challenges to minimize size, achieve high average efficiencies and broader band-widths while maintaining linearity and operating at higher frequencies [4]. 4G standards such as LTE use non-constant envelope modulation techniques with high peak to average ratios. Power amplifiers implemented in such applications are forced to operate at a backed-off region from saturation. Therefore, to reduce power consumption, a design of a high

efficiency PA that can maintain the efficiency for a wider range of radio frequency signals is required.

In this paper, a 2.62-2.69GHz 2-way 10W Doherty power amplifier, suitable for a 4G LTE band 7 pico-cell base station, is presented. Designed in a compact 10x11.5mm² monolithic microwave integrated circuit (MMIC) using GaN technology, the power amplifier achieves high efficiencies of over 50% at both back-off and peak power regions without compromising on the stringent linearity requirements of 4G LTE standards. This demonstrates a 17% increase in power-added efficiency at 6 dB back-off from peak power compared to conventional Class AB amplifier performance.

II. DESIGN CONSIDERATIONS

The RF power amplifiers (PA) that amplify the output signals of 4G pico-cell systems face challenges to minimize size, achieve high average efficiency and broad bandwidths while maintaining linearity and operating at higher frequencies. Amongst these many requirements of a base station amplifier, linearity and efficiency are the most crucial. To meet linearity requirements, power amplifiers can be operated at a back-off power level from the peak output power in a linear and efficient class of operation, such as class AB. But due to the high peak-to-average power ratio (PAPR) of the signal, this can lead to very low efficiency at the signals average output power. On the other hand more efficient classes of operation such as class B are extremely non-linear, even with linearization techniques such as digital pre-distortion this would not be a suitable solution. Therefore, the design of power amplifiers generally forces trade-offs between linearity and efficiency.

To facilitate the PA design to focus more on the efficiency aspect of the design by essentially minimizing linearity requirements from the design goals, several linearization techniques such as Feed Forward Linearization, Cartesian Feedback, LINC (Linear amplification using Nonlinear components) and Digital pre-distortion [5]-[8] have been developed and implemented. However, power amplifiers also possess memory effects that contribute to the distortion of the input signal when the signal has a broad envelope bandwidth, thus making most standard linearization inefficient. Although some Digital pre-distortion algorithms do correct for some memory effects [9], it is fundamentally important to minimize the memory effects at the circuit design level. Linearity can

978-1-4673-5289-5/12 $31.00 © 2012 IEEE

also be optimized at the device level through the use of improved device processes, such as the use of field plated HEMT structures in GaN devices [10].

Since a pico-cell base station is much smaller compared to a macro-cell, an additional challenge is the implementation of a highly-efficient highly-linear power amplifier in a compact area. This is where new wide bandgap device technologies such as GaN HFETs present architectural benefits. Recent developments have shown that GaN fabrication has made it possible to reach power densities up to 30W/mm [11].

Achieving high efficiency for a single power level can be attained through harmonic tuning [12], switch mode amplifier designs [13] and single stage class B or class AB designs. The problems with such techniques are that either they are extremely non-linear or they have poor average efficiency when a non-constant envelope signal is being amplified. To address these issues, several efficiency enhancement techniques such as Chireix Outphasing [14], Envelope elimination and restoration (EER) [15], Doherty [16] and Envelope tracking (ET) [17] have been suggested and studied to date. Among these techniques, the Doherty technique is the most feasible to be implemented on an IC design.

The Doherty technique has already been implemented using a variety of device technologies, predominantly using LDMOS and GaAs. With these technologies, the reported efficiency numbers have been lower than the theoretical efficiencies, especially at frequencies higher than 2.5GHz. Device technologies such as the GaN present an attractive option for Doherty design due its low output capacitance and high output impedance. This aspect of GaN devices simplifies significantly the matching networks. Therefore, in this paper GaN power cells are used to implement the Doherty design.

A Cree's CGH25120F GaN HEMT transistor [18] was then selected and, as preliminary step, evaluated to asses GaN device technology on a PCB level.

III. GaN Transistor Characterization

To highlight GaN device technology performance, CGH25120F transistor was tested via a classical printed circuit board (PCB) design of a single stage class AB amplifier. Using the CGH25120F large signal model, gain, drain efficiency and compression points have been characterized with the measured and simulated test setups. Tables I and II summarize the results showing close agreement between simulated and measured data collected at the centre band of 2.65GHz.

Note that single-tone Load-pull simulations were performed with fundamental source impedance of 4.32-10.9j, drain voltage of 28V and gate voltage of -3.02V to achieve a drain current of 500mA at 2.650GHz. At the given bias point, input power and frequency, output power contours elliptically in a 0.6dB step and power added efficiency (PAE) in a 5% step. From Fig. 1-a, we can see that although the highest achievable efficiency for this device is 55.97%, the evaluation board has been designed at an impedance where the efficiency is around 40% at an output power of 46dBm. This indicates a good compromise between linearity and efficiency. This

assumption was further confirmed by observing the 3rd-order intermodulation (IMD3) contours extracted from the two-tone Load-pull simulations (Fig. 1-b). In fact, although the PAE and the delivered power contours rotate in the same direction, the IMD3 contours rotate differently, indicating that a higher efficiency impedance point could yield lower linearity.

TABLE I. CGH25120F: Measured and Simulated Impedances.

	Measured	Simulated with Momentum	Simulated with ADS
Impedance seen by the device gate	4.39-10.3j	4.32-10.9j	4.04-12.8j
Impedance seen by the device drain	3.98-1.37j	3.35-2.87j	4.22-2.53j

TABLE II. Measured and Simulated Performance Data of the CGH25120F Evaluation Board (Pout is the Output Power and P-3dB the 3dB Compression Point)

	P-3dB (dBm)	Gain (dB) for Pout of 46dBm	Drain efficiency for Pout of 46dBm
Simulated	50.4	14.03	39.90 %
Measured	51.2	13.68	45.03 %

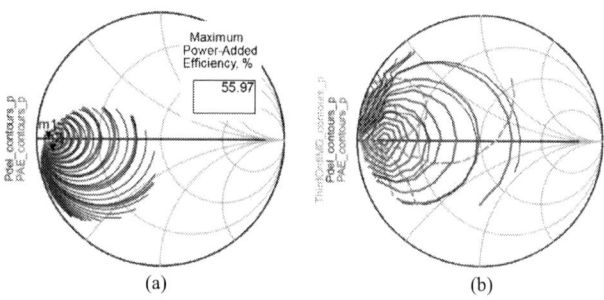

Fig. 1. CGH25120F evaluation board: a) PAE (red) and power delivered (blue) Load-pull contours at the load impedance, b) IMD3 (green), PAE (red) and power delivered (Blue) Load-pull contours.

IV. Doherty Amplifier Design for a 5W Pico-Cell Base Station

In this section, the proposed two-way MMIC symmetrical Doherty amplifier for a 5W base station and its performance evaluation in LTE band 7 are presented. Two 8 x 200µm GaN HFETs have been used for each main and peaking amplifiers in Doherty configuration (Fig. 2) on a 10x11.5mm² chip to meet design requirements (Table III). To assess the large-signal performance of the Doherty amplifier, single-tone and two-tone large signal simulations have been performed with a fixed drain bias of 20V. Figure 3 shows the single-tone PAE response of the Doherty amplifier, clearly depicting a higher efficiency over a wider range of output power while compared to a Class AB power amplifier of the same power sizing.

TABLE III. Design Requirements (for a 5W Base station)

Parameter	Design specifications
Frequency (LTE Downlink operating band 7)	2.62-2.69GHz
Power average output power (PA-output)	40dBm
Gain at PA-output	>10dB
Average power added efficiency (PAE)	>35%

The key point to note is the 17% increase in efficiency at the back off power level of the Doherty power amplifier. Although a relatively constant efficiency is expected in the full 6dB back off range, a slight dip in the PAE between the two peak points can be noticed. This is due to the lower gain of the peaking amplifier from Class C operation. This situation could be overcome by having an unequal power division at the input, delivering higher power to the peaking amplifier than the main amplifier [19].

With nominal bias points for main and peaking amplifiers of Vd=20V, Vgs_main=-3.0V, Vgs_peak=-4.9V at 2.655GHz, we observed some distortion caused by the low biasing of the peaking amplifier. Then, by altering the bias conditions of the peaking amplifier, proper turn on characteristics can be achieved to suppress these distortion levels, leading to results within required specifications (Fig. 4).

V. CONCLUSION

The 10W MMIC Doherty amplifier reported in this paper has demonstrated the feasibility of achieving high efficiency at the back off power levels while maintaining reasonable linearity compared to conventional PA designs. Compared to existing MMIC PAs, the proposed amplifier achieved excellent efficiency at back off power levels, demonstrating a PAE of 50.5% in the 6dB back off power from a peak power of 46dBm, for the full LTE band 7 which entails the frequency band from 2.62-2.69GHz.

REFERENCES

[1] M. Steer, "Beyond 3G," IEEE Microwave Magazine, Vol. 8, pp. 76-82, 2007.

[2] S. Sesia, I. Toufik, M. Baker, LTE-The UMTS long term evolution-from theory to practice, J. Wiley & Sons, 2011.

[3] S. Barbera, P.H. Michaelsen, M. Saily, K. Pedersen, "Mobility performance of LTE co-channel deployment of macro and pico cells," IEEE Wireless Communications and Networking Conf., pp. 2863-2868, 2012

[4] S. Ramanath, E. Altman, V. Kumar, M. Debbah, "Optimizing cell size in Pico-cell networks," Int. Symp. Modeling and Optimization in Mobile, Ad Hoc, and Wireless Networks, 2009.

[5] M.K. Nezami, "Feedforward power amplifier linearization for 4G base station," IEEE Malaysia Int. Conf. on Communications, pp. 880-884, 2009.

[6] K.K. Mohammed, R.B. Mohammed, "Linearization of power amplifier class AB using Cartesian feedback," Int. Multi-Conf. on Systems Signals and Devices, 2010.

[7] P. Garcia, A. Ortega, J. de Mingo, A. Valdovinos, "Nonlinear distortion cancellation in OFDM systems using an adaptive LINC structure," IEEE Int. Symp. on Personal, Indoor and Mobile Radio Communications, pp. 1506-1510, 2004.

[8] J.A. Sills, R. Sperlich, "Adaptive power amplifier linearization by digital pre-distortion using genetic algorithms," Radio and Wireless Conf., pp. 229-232, 2002.

[9] R. Sperlich, J.A. Sills, J.S. Kenney, "Power amplifier linearization with memory effects using digital pre-distortion and genetic algorithms," Radio and Wireless Conf., pp. 355-358, 2004

[10] R. Borges, "Gallium nitride electronic devices for high power wireless applications" RF design, pp-72-82, 2001

[11] A.Z. Markos, "Efficiency enhancement of linear GaN RF power amplifiers using the Doherty technique," University of Kassel, Germany, 2008

[12] B. Kopp, D.D. Heston, "High-efficiency 5-watt power amplifier with harmonic tuning," IEEE MTT-S Int. Microwave Symp., pp. 839-842, 1988

[13] R. Meshkin, A. Saberkari, M. Niaboli-Guilani, "A novel 2.4 GHz CMOS class-E power amplifier with efficient power control for wireless communications," IEEE Int. Conf. on Electronics, Circuits, and Systems, pp. 599-602, 2010

[14] W. Gerhard, R. Knoechel, "Improvement of power amplifier efficiency by reactive Chireix combining, power back-off and differential phase adjustment," IEEE MTT-S Int. Microwave Symp., pp. 1887-1890, 2006

[15] S. Hong et al., "High efficiency GaN HEMT power amplifier optimized for OFDM EER transmitter," IEEE MTT-S Int. Microwave Symp., pp. 1247-1250, 2007

[16] W.H. Doherty, "A new high efficiency power amplifier for modulated waves," Proc. IRE, Vol. 24, pp-1163-1182, 1936

[17] H. Harju, T. Rautio, S. Hietakangas, T. Rahkonen, "Envelope tracking power amplifier with static predistortion linearization," 1European Conf. Circuit Theory and Design, pp. 388-391, 2007

[18] Cree, Inc. (2012) "CGH25120F Datasheet". Retrieved from http://www.cree.com/products/pdf/CGH25120F.pdf

[19] D. Kang et al., "30.3% PAE HBT Doherty power amplifier for 2.5~2.7 GHz mobile WiMAX," IEEE MTT-S Int. Microwave Symp. , pp. 796-799, 2010.

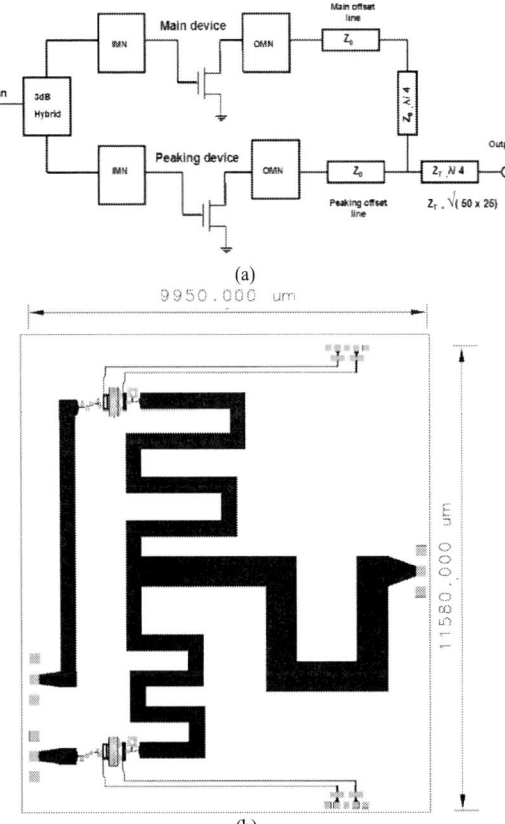

Fig. 2. Doherty Amplifier: a) design blocks, b) Layout.

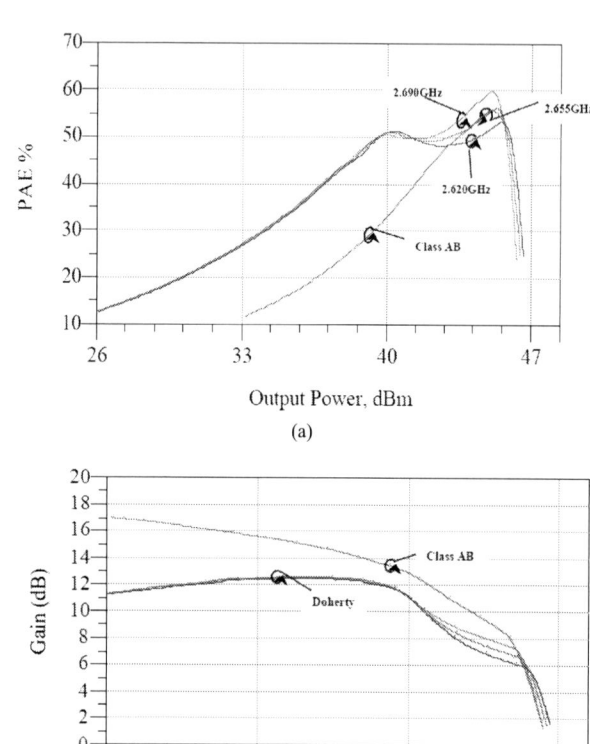

(a)

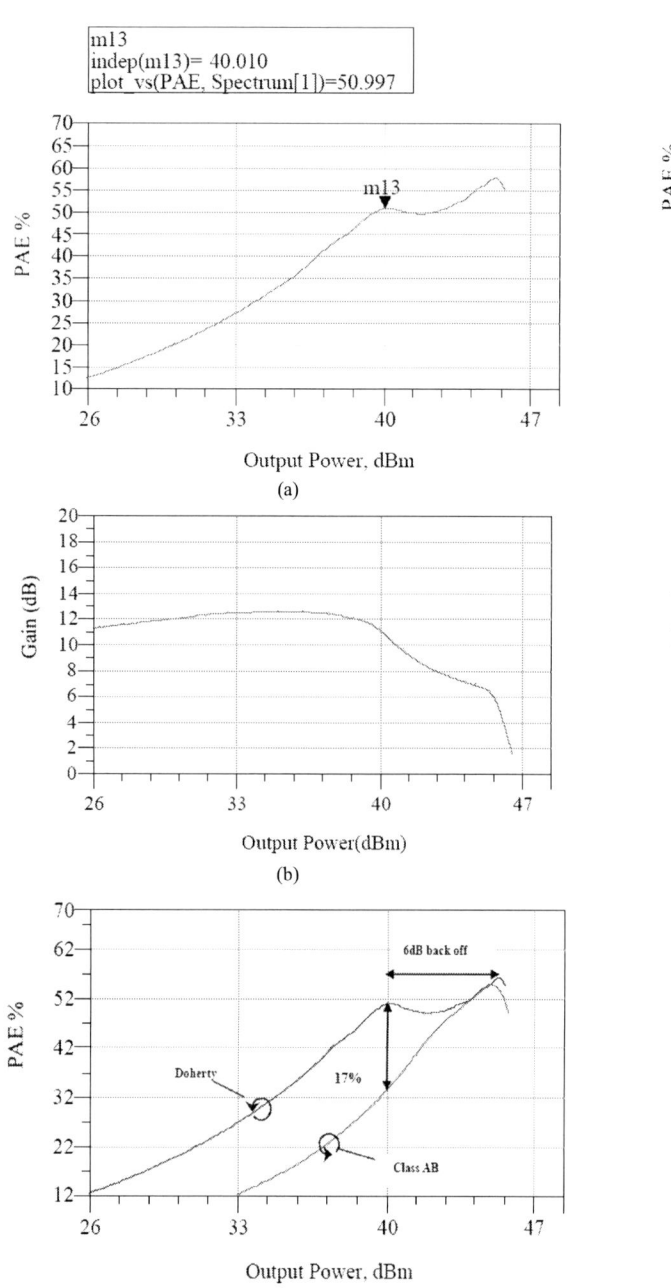

Fig. 4. Frequency performance of the Doherty Amplifier.

Fig. 3. Single-tone simulations with Vds=20V, Vgs_main=-3.0V,Vgs_peak=-4.9V, a) PAE of the Doherty amplifier, b) Gain of the Doherty amplifier, c) Comparison of PAE of Class AB and Doherty amplifiers

A Novel Ultra-Wide Stopband Microstrip Low-Pass Filter for Rejecting High Order Harmonics and Spurious Response

Applications in Wideband Microstrip Circuits and Systems

M. CHALLAL[1,2], A. AZRAR[1] and D. VANHOENACKER JANVIER[2]

[1] Dept. of Electronic, IGEE
University of Boumerdes
Boumerdes, Algeria
mchallal@umbb.dz

[2] ICTEAM, Electrical Engineering
Université catholique de Louvain
Louvain-La-Neuve, Belgium

Abstract— **In this paper, an ultra-wide stopband microstrip low-pass filter (LPF) to reject higher harmonics and spurious response for wideband microwave applications is proposed. It is based on quasi-triangular (QT) defected ground structure (DGS) resonators and open stubs. An equivalent circuit model is also presented. The introduced LPF has small size, a low insertion loss and a return loss less than -20 dB. In addition, a round -20 dB suppression level ranging from 4 GHz to more than 20 GHz is achieved. The simulated results obtained by circuit model and full-wave EM show good agreement with the measured ones.**

Keywords-Quasi-Triangular (QT); Defected ground structure (DGS); Ultra-wide stopband; Low-pass filter (LPF).

I. INTRODUCTION

A microstrip low-pass filter (LPF) is one of the fundamental components in RF/Microwave wireless communication systems. Low cost, low insertion loss, ultra-wide stopband and compact size are necessary to meet modern RF/Microwave communication systems requirements. Due to these features and, convenient integration with other microwave circuits, planar resonators have progressively been taken into consideration to be employed in microwave filter design. One of the techniques that can be applied in microstrip LPF is using a defected ground structure (DGS) technique instead of cascading several resonator cells. Many types of LPF for performance improvement have been introduced [1-5]. However, their performances such as stopband bandwidth, insertion loss and filter area do not completely achieve the communication systems requirements.

In this paper, we propose a novel ultra-wide stopband LPF using only two quasi-triangular (QT) DGS along with open stubs for rejecting higher harmonics and spurious response in wideband microstrip circuits and systems applications. Its equivalent circuit model is also analyzed and discussed. This type of structure avoids employment of cascaded LPF units and allows achievement of an ultra-wide stopband with very good insertion and return losses in the LPF passband. The simulation

results achieved by circuit model and full-wave EM show a good agreement to the measurement ones.

II. DGS-LPF DESIGN CONCEPT AND CIRCUIT MODELING

Figure 1 shows the proposed DGS-LPF with its equivalent circuit model. It is composed of a QT defected areas etched in the ground plane below a 50 Ω microstrip line and H shape open stubs [5]. The filter is designed on a RO4350B Rogers material with a permittivity of the dielectric (εr) of 3.63 and a thickness (h) of 0.254 mm. The conductor strip of the microstrip line (50 Ω) on the top plane has a calculated width w of 0.52 mm.

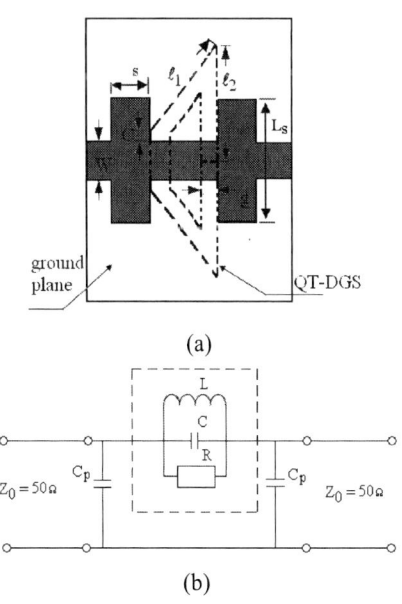

(a)

(b)

Figure 1. The proposed of DGS-LPF (a) Geometry, and (b) Equivalent circuit model

The physical parameters of this DGS-LPF structure, ℓ_1 = 8.14 mm, ℓ_2 = 5.5 mm, c = 0.24 mm, g = 1, L_S = 6 mm and W_S = 1.3 mm, are considered.

The proposed DGS-LPF can be modelled as one resonator along with two shunt capacitors Cp which correspond to the open stubs as shown in Figure 1.b. The circuit elements are extracted using the following expressions [5]:

$$C = \frac{\omega_0}{2Z_0(\omega_0^2 - \omega_c^2)} \quad (1)$$

$$L = \frac{1}{\omega_0^2 C} \quad (2)$$

$$R = \frac{2Z_0}{\sqrt{\frac{1}{|S_{11}(\omega_0)|^2} - (2Z_0(\omega_0 C - \frac{1}{\omega_0 L}))^2 - 1}} \quad (3)$$

where $\omega_0 (= 2\pi f_0)$ and $\omega_c (= 2\pi f_c)$ are respectively the angular resonant and 3-dB cutoff frequencies of the DGS pattern.

According to the basic theories of transmission lines, an open stub is modeled as an equivalent capacitor. The equivalent capacitance with the characteristic impedance (Z_s) and length (ℓ_s) can be obtained from [6] as:

$$C_p = \frac{1}{\omega Z_s} tan\left(\frac{2\pi \ell_s}{\lambda_g}\right) \quad (4)$$

where λ_g represents the guided wavelength.

For the assumed circuit model, the parameters L, C, R and C_p are respectively 4.70 nH, 3.89 pF, 3.50 kΩ and 1.09 pF. The structure is investigated using the full-wave EM IE3D simulator. Circuit model and EM simulations results are illustrated in Figure 2 which shows the characteristics of an LPF with 3-dB cutoff frequency (fc) is equal to 3.1 GHz. It can be observed from Figure 2 that the insertion loss is equal to 0.10 dB and the return loss is better than 26 dB in the whole passband. Furthermore, a large suppression band at attenuation level of -20 dB within 06-20 GHz is achieved in the stopband.

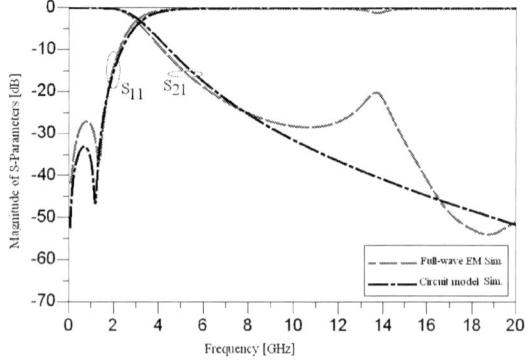

Figure 2.   Circuit model and EM-Simulations of the proposed DGS-LPF

III.   ULTRA-WIDE STOPBAND LPF DESIGN, CIRCUIT MODELING, IMPLEMENTATION AND MEASUREMENT

The proposed compact ultra-wide stopband LPF is shown in Figure 3. It is composed of two identical QT-DGS units and H- open stubs. This structure avoids employment of LPF units and allows significant enhancement of the characteristics shown in Figure 2 of the considered structure in the previous section.

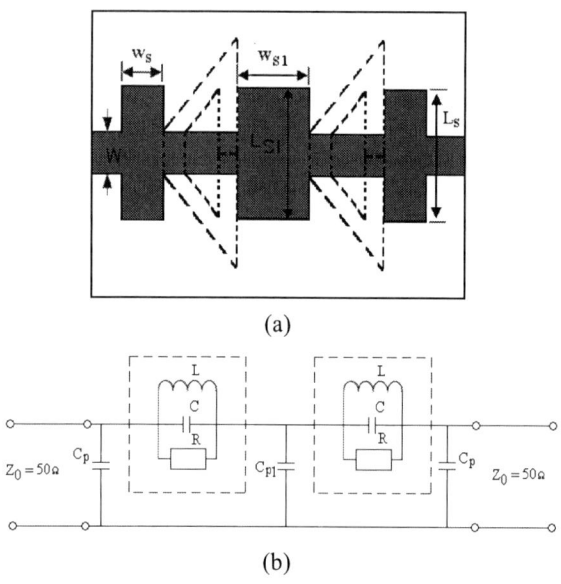

(a)

(b)

Figure 3.   Proposed ultra-wide stopband DGS-LPF (a) Geometry, and (b) Circuit model

The circuit model and full-wave EM simulations results are shown in Figure 4.

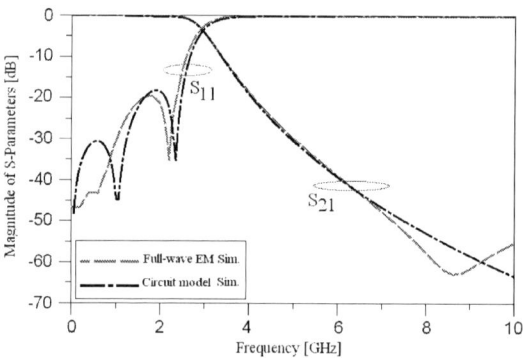

Figure 4.   Circuit model and EM-Simulations of the proposed ultra-wide stopband DGS-LPF

From Figure 4, it is clear that the proposed LPF behaves well in both passband and stopband. It is observed from Figure 4 that the LPF has an insertion loss about 0.10 dB and a return loss better than 20 dB in the whole passband. Besides, reasonably good agreement between circuit model and full-wave EM simulations can be seen except some difference appears at more than 7 GHz for insertion loss and at less than 1.5 GHz for return loss. It could be resulted from the simplicity of the lumped circuit model that the distributed effects are not included in this model. This result shows that the circuit model provides quite good performances and confirms its validity.

The proposed LPF with two DGSs in the metallic ground plane and H-open stubs on the top layer with size of 25 x 11 mm^2 is fabricated as shown in Figure 4.

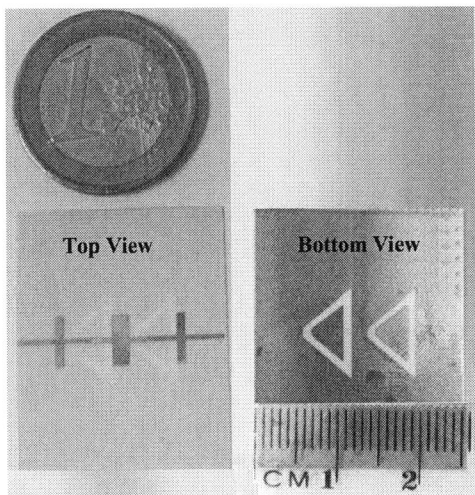

Figure 5.   Photography of the proposed ultra-wide stopband DGS-LPF

Figure 6 shows the measured and the simulated results. It is observed from Figure 6 that the measured results agree with the simulated ones. From the measured results, it is seen that the fabricated LPF has an insertion loss lower than 0.1 dB in the filter pass-band and stopband suppression at a level lower than -20 dB from 4 GHz to more than 20 GHz. The small deviations between the simulated and measured results may most probably be caused by the usual connectors and manufacturing errors.

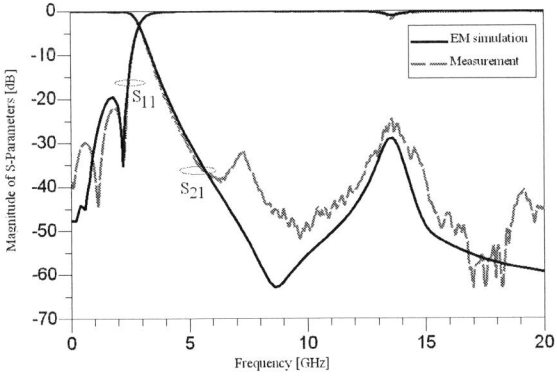

Figure 6.   Measured and Simulated S-parameters of the proposed ultra-wide stopband DGS-LPF

The performance of the proposed filter is summarized in Table I with other reported filters for comparison. It can be observed from Table I that the proposed filter provides good performances in stopband rejection and passband insertion loss and smaller in size than those reported in literature.

TABLE I.   COMPARISON OF THE PROPOSED DGS-LPF WITH OTHER RELATED LPF

	Substrate dielectric constant/ height (mm)	Size (mm^2) x X y	fc (GHz)	Stopband (dB) with -20 dB rejection	Passband insertion loss (dB)	Passband return loss (dB)
Ref. [2]	4.4/0.8	21 x 20	03.5	4.3 - 15.8	< 2	-
Ref. [3]	3.38/1.524	71 x 13	02.4	3.26 - 10	< 2.26	> 5
Ref. [4]	4.4/0.8	27 x 23	03.7	3.75 - 20	< 1	-
This work	3.63 / 0.254	25 x 11	3.1	4 - >20	0.1	> 20

IV.   CONCLUSION

In this paper, a novel ultra-wide stopband microstrip low-pass filter (LPF) based on defected ground structure (DGS) technique has been introduced and investigated. The proposed LPF provides a low insertion loss of 0.1 dB, a return loss much lower than -20 dB, suppression levels approximately -20 dB from 4 GHz to more than 20 GHz and has small size. It has been shown that the simulations results achieved by circuit model and full-wave EM were in excellent agreement with the measurement ones. The proposed compact and high performance LPF can be broadly used to reject higher harmonics and spurious response for wideband microstrip circuits and systems applications.

ACKNOWLEDGMENT

The authors are grateful to D. Spote and P. Simon technical staff at the Electromagnetics Microwave Communication Laboratory, ICTEAM, UCLouvain.

REFERENCES

[1] D. Ahn, J. S. Park, C. S. Kim, Y. Qian, and T. Itoh, "A Design of the low-pass filter using the novel microstrip defected ground structure, " *IEEE Trans. Microw. Theory Tech.,* vol. 49, pp.86-91, 2001.

[2] H. J. Chen, T. H. Huang, C. S. Chang, L. S. Chen, N. F. Wang, Y. H Wang, M. P. Houng, "A novel cross-shaped DGS applied to design ultra-wide stop-band low-pass filters," *IEEE Microw. Wireless Comp. Lett.,* vol. 16, no. 5, pp. 252–254, May 2006.

[3] S. W. Ting, K.W. Tam, and R.P. Martins, "Miniaturized microstrip lowpass filter with wide stop-band using double equilateral u-shaped defected ground structure, " *IEEE Microw. Wireless Compon Lett.,* vol. 16, pp. 240–242, May 2006.

[4] P. Y. Hsiao, R. M. Weng, "An ultra-wide stop-band low-pass filter using dual reverse U-shaped DGS", *Microw. and Optical Technology Lett.,* vol. 50, no. 11, pp. 2783-2780, November 2008.

[5] X. Chen, L. Wang, L. Weng and X. Shi, "Compact low pass filter using novel elliptic shape DGS, " *Microw. and Optical Technology Lett,* vol. 51, no. 4, pp. 1088-1091, Apr. 2009.

[6] D. M. Pozar, Microwave Engineering, 3rd edition, John Wiley & Sons, Inc. 2005.

Wireless Microelectronic Nose Network for Gas Classification

Muhammad Hassan[1], Amine Bermak[1,2]
[1]Hong Kong University of Science and Technology
[2]American University of Ras Al Khaimah
Emails: mhassan@ust.hk, amin.bermak@aurak.ae

Abstract—Electronic nose systems, containing array of gas sensors, are emerging platforms in numerous disciplines of life for gas classification. Repeated performance of these systems is not guaranteed because of their drift due to temperature and humidity and hence make them inefficient to distinguish gases correctly over time with stand-alone unit. Wireless sensors network is the promising solution to improve the performance of gas identification over single unit. This paper, adopts spike rank order encoding scheme due to its concentration independent feature and proposes wireless microelectronic nose network to get improved performance over single unit. In our scheme, weights are assigned to relative position of spikes in each training gas and these weights play a decisive role about decision of testing gas when received ranks from the nodes are not fully matched with any of the training gas signature. In the network, nodes send their ranks sequence to base station on gas exposure. Base station identifies testing gas as one of the training gas if its reference ranks signature is received from the nodes. If no training gas rank signature is received from any node then the node with maximum ranks matching sequence to any training gas is selected and mismatch pattern is searched in others node. On failure to find a matched sequence, decision is based on the inter-spikes weights. Experimental results validate the improved performance of this approach over single electronic nose system.

Index Terms—Gas sensors, Gas classification, Electronic nose, Sensor network, Rank order matching

I. INTRODUCTION

The human sense of smell is one of the primary senses and used to distinguish different odors. Over the past decades, it is highly realized to develop system that could mimic human nose and can be used for industrial applications. In recent years, advancement in semiconductor technology and manufacturing process attracted many researchers to model biological olfactory mechanism attempting to mimic olfactory system in human nose and develop electric nose system targeting for efficient odor discrimination. Electronic nose system targeted many real life applications such as odor tracking using mobile robot [1], water quality monitoring [2], disease diagnosis and food spoilage detection [3], odor tracking around livestock [4], air quality monitoring [5] [6], agricultural environment monitoring [7] [8] and toxic gases detection [9] in industrial and residential area. Various types of gas sensors are used in electronic noses including electrochemical, fiber optic, metal-oxide, field-effect and conducting polymer gas sensors [10].

Array of gas sensors, pre-processing unit and gas identification engine are three core blocks of electronic nose system.

Gas sensors transform gas signature to electrical signals. Pre-processing unit performs signal conditioning on transformed electrical signal and forwards it to gas identification unit. Gas identification unit matches signature of the sensed gas with training gas. Many electronic nose systems are developed to classify different gases. In [11], an electronic nose, containing thick film doped tin-oxide gas sensors, is developed and artificial neural network (ANN) is used to classify car exhaust gases. In [12], an electronic nose system is developed by integrating ferric oxide gas sensor array to analyze hydrogen, methane and their mixtures. A principal component regression method is used to distinguish gases. An electronic nose based on committee machine is proposed in [13]. The committee machine integrates five different classifiers namely gaussian mixture model, multilayered perception, radial basis function, k-nearest neighbor and probabilistic principal component analysis. FPGA is used to implement these classifiers and dynamic reconfiguration is used to time multiplex these classifiers. An analog VLSI implementation is proposed in [14] for an olfaction chip by integrating gas identification circuit. A microelectronic nose is developed in [15] by integrating tin-oxide gas sensor array and rank order encoding scheme is proposed using biologically inspired time of first spike technique. The algorithm is simple in implementation as compared to aforementioned work and gas concentration independent formulation is the most distinguished characteristic of this algorithm.

On the other hand, wireless sensors network technology is targeting many daily life application because of compact size and low cost. In [17], wireless electronic nose systems are deployed to monitor real time gas mixtures. Commercial gas sensors with ZigBee wireless interface are used to monitor air quality [18]. In [19], wireless sensor network is proposed for hazardous gas detection.

In this paper we are proposing wireless microelectronic nose network to identify testing gas as one of the trained gas or a new gas. Our work is the extension of the work proposed in [15] because in real life it is not easy for a single node to identify testing gas because of uneven distribution of gases and sensor drift due to aging, temperature and humidity. In our technique, multiple nodes acquire testing gas signature and transmit it to base station. Base station analyzes the ranks of all nodes to distinguish gas as one of the training gas in reference gas library or a new unknown gas. Experiments are

978-1-4673-5289-5/12 $31.00 © 2012 IEEE

performed to validate this concept.

The remainder of the paper is organized as follows. In section II, we describe our gas classification technique. Performance evaluation of our technique is described in section III. Finally, conclusion and future work is outlined in section IV.

II. GAS CLASSIFICATION TECHNIQUE

Metal-oxide semiconductor (MOS) sensors use metal oxide based thick sensing film for sensing gases. The MOS sensors may contain a single sensing element or array of elements and the tin-oxide is the most commonly used sensing element in these sensors. The sensors can detect a wide range of gases and volatile organic compounds with change in resistance and their sensitivity follow power law relation [20] as shown in following equation.

$$S_{x,g} = \frac{(R_s)_{x,g}}{(R_s)_{x,air}} = \lambda_{x,g} C_g^{\zeta_{x,g}}; \tag{1}$$

Where x indicates the xth sensor in the sensor array, g represents the exposed gas to sensors, C is the gas concentration, S is the sensitivity of sensor and ζ and λ are sensor and gas dependent parameters.

A training procedure is required for gas sensor in order to distinguish gases. As shown in Eq. 1, sensitivity depends upon gas concentration. The training system based on the knowledge of concentration is not feasible for field applications. In [15] and [16], concentration independent rank order gas identification algorithm based on relative position of neighboring spikes is proposed and results in following equations.

$$\log S_{x,g} = p_{x,g} \log S_{1,g} + q_{x,g}; \tag{2}$$

$$t_{x,g} = \frac{\log S_{x,g}}{p_{x,g}} \tag{3}$$

$$t_{x,g} - t_{y,g} = \frac{\log \lambda_{x,g}}{p_{x,g}} - \frac{\log \lambda_{y,g}}{p_{y,g}} \tag{4}$$

where $p_{x,g} = \frac{\zeta_{x,g}}{\zeta_{1,g}}$ and $q_{x,g} = \log \lambda_{x,g} - p_{x,g} \log \lambda_{1,g}$. Where $t_{x,g}$ corresponds to the time-to-first spike of the sensor in response to a certain gas g and $p_{x,g}$ parameter is computed during training phase of gases. Eq. 4 represents concentration independent and unique relative timing between neighboring spikes and hence results in a unique set of ranks for each gas.

A stand-alone sensor is not sufficient for gas classification because of sensor drift with time and dynamic environmental changes. We are proposing microelectronic nose sensor network and suggesting a technique by adopting formulation for concentration independent relative spike timing [15] to classify gases. All the nodes containing electronic nose in the network send their firing ranks to base station. On receiving data, base station firstly decides about the ranks of sensed gas and then it matches the ranks with training gases available in the reference library. The library is build during training phase and signature of every gas is stored in the library. we assign weight to each inter-spike position of training gas through following equation.

$$w_{x,y} = \frac{|t_{x,r} - t_{y,r}|}{\max |t_{i,r} - t_{j,r}|} \tag{5}$$

Where $w_{x,y}$ is the weight assigned to spikes relative position of sensor x and y of the node. $t_{x,r}$ and $t_{y,r}$ correspond to time of first spike of sensor x and y for training gas r. $t_{i,r}$ and $t_{j,r}$ represent time of two neighboring spike of the same training gas. Hence, Eq. 5 assigns weight to each neighboring spike position of training gas by normalizing it with maximum neighboring spike relative position. The lower the inter-spike weight means that the corresponding spikes swapping probability is high as compared to the higher inter-spike weights.

In case of receiving similar full ranks from majority of nodes, base station considers it as ranks of testing gas. Full ranks of node indicate sequence of time of first spikes of all sensing elements of gas sensor array. Upon fully matching of these ranks with any training gas from reference gas library, testing gas is identified as corresponding training gas otherwise it is included in reference library as a new gas.

In case of no similar response from the nodes, base station selects the node with maximum matching ranks with any of the training gas ranks and exceed some threshold value e.g 8 out of 16 ranks. For the unmatched ranks, similar sequence as of corresponding training gas is searched at the mismatched positions in the other node's ranks. If it is found then the testing gas is recognized as corresponding training gas otherwise the base station computes the inter-spikes weights of matching sequences and if exceed some threshold e.g. 80% of the total weight then the testing gas is declared as the corresponding training gas.

III. EXPERIMENTAL SETUP AND PERFORMANCE EVALUATION

We use 4 x 4 tin-oxide gas sensor in our experiment that is fabricated using in-house 5 μm CMOS process [21]. Fig. 1 shows the microphotograph of this gas sensor array. Different post treatment scheme is used for every row and column in the sensor array to get different responses among each other when exposed to certain gas. There is no post treatment scheme on the sensor in the first row and column. Three gases namely ethanol, hydrogen (H_2) and carbon mono-oxide (CO) are used in the training phase so that spike ranks from the gas sensor array can be stored in memory as a reference for these gases. The trained gases target numerous applications of daily life. Ethanol or alcohol is a flammable liquid and it is the main cause behind many accidents. Hence it is necessary to identify a drunk driver by sensing alcohol. Hydrogen is highly flammable and it is important to detect its presence for safety. Continuous exposure of CO may severely affect health system.

In order to acquire a reference rank pattern for aforementioned 3 gases, electronic nose [15] is placed in gas chamber and exposed to these gases at concentrations 20 ppm to

978-1-4673-5289-5/12 $31.00 © 2012 IEEE

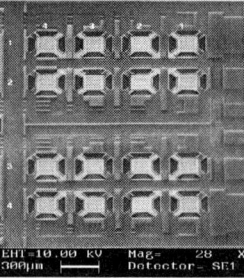

Fig. 1. Microphotograph of the fabricated 4 x 4 tin oxide gas sensor array. [21]

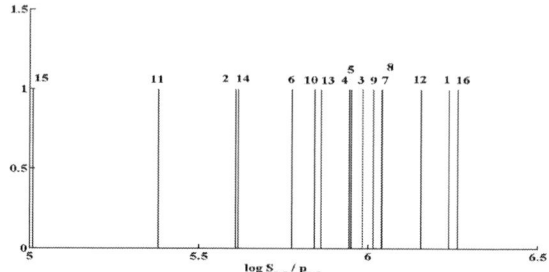

Fig. 4. Carbon mono-oxide reference rank order

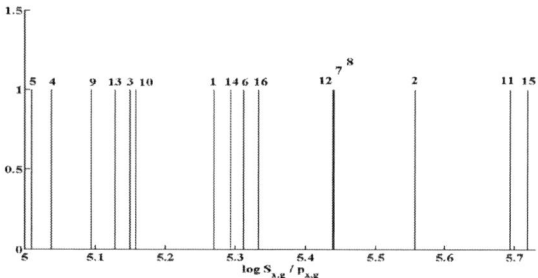

Fig. 2. Ethanol reference rank order

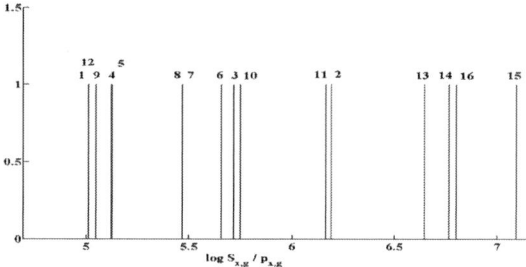

Fig. 3. Hydrogen reference rank order

200 ppm and the concentrations are controlled through mass flow controller. Acquired ranks order through electronic nose system are transmitted to base station through wireless link. For reference neighboring spike timing, relative spike timing is averaged for mostly repeated ranks order and hence we get a reference signature of ranks for corresponding trained gas. Parameter $p_{x,g}$ is calculated using Eq. 2. Sequence of ranks for ethanol, H_2 and CO are shown in Fig. 2, Fig. 3 and Fig. 4 respectively. The numbers written over spikes represent corresponding sensor numbers of the node. These sequence of ranks are independent of concentration and supposed to be repeated for corresponding gas exposure every time.

To evaluate the effectiveness of our gas classification technique, testing data sets, containing responses of 5 nodes in each set, are acquired and analyzed. Fig. 5, Fig. 6 and Fig. 7 show one set of spike ranks on exposure to ethanol, hydrogen and carbon mono-oxide respectively. In these figures, radius indicates the ID of sensor in the node that generates spike and

the number written on the outside of outer circle denotes the rank of the sensor. In case of no full match of test gas ranks with trained gas ranks, threshold of 8 ranks is set for further evaluation. It implies that if 8 spikes of testing gas are matched with 8 spikes in any trained gas then for non matching spikes, others nodes are searched to find matched ranks as of the trained gas at the same position of unmatched spikes ranks. If these are found then the testing gas is considered as the same training gas otherwise weights of matching neighboring spikes is computed and if exceeds 80 % of total weight of corresponding training gas then the testing gas is declared as the same training gas.

In Fig. 5, no node response is exactly matched with any trained gas pattern but node 2 and node 5 have 14 matching ranks with ethanol. Non matching ranks in node 2 are at position 12 and 13. The sequence should be 7 and 8 at these positions to fully match with training signature of ethanol. These sequence are found in node 3, node 4 and node 5 at the same positions. So the sensed gas is considered as ethanol. Similarly, if we take node 5 then its non matching ranks at positions 4 and 5 are found in node 1 and node 2 at the same position. So, in both ways this gas is identified as ethanol. In Fig. 6, node 1 response is exactly matched with trained pattern of hydrogen and ranks of other nodes do not match with any other trained gas in the library. Hence the testing gas is recognized as hydrogen. In Fig. 7, 13 ranks of node 3 are matched with signature of CO while other nodes' matching ranks are less than 13. For unmatched ranks of node 3, there is no CO reference signature in others nodes at the unmatched position. Inter-spike weight is computed for matched ranks of node 3 and it is 86.3% of total inter-spike weight of trained CO. Therefore, we identify this testing gas as CO gas.

It is obvious from the results that sensor ranks varies on exposure to same gas and 100% gas classification performance can not be achieved with single node. Our network approach is more realistic in field applications and shows significantly higher performance.

IV. CONCLUSIONS AND FUTURE WORK

In this paper, microelectronic nose sensor, containing tin-oxide gas sensor array, is used as a platform to distinguish different gases. Spike rank order encoding scheme is adopted due to its simplicity and concentration independent character-

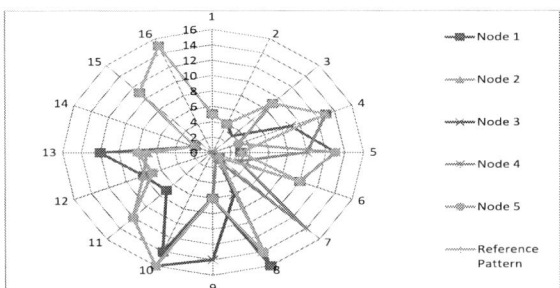

Fig. 5. Nodes spike ranks on exposure to Ethanol

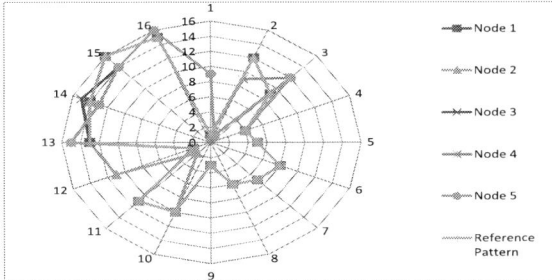

Fig. 6. Nodes spike ranks on exposure to Hydrogen

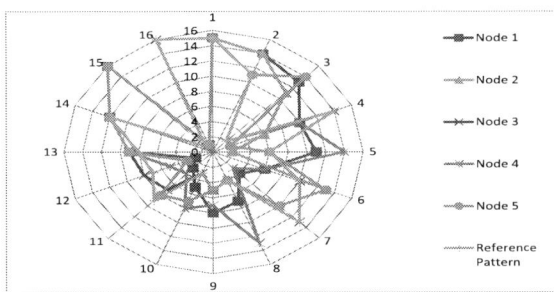

Fig. 7. Nodes spike ranks on exposure to Carbon mono-oxide

istics. The performance of the scheme is limited because of sensor drift with time and environmental changes and hence similar responses might not be repeated for same gas exposure over time with single node. In this paper, microelectronic nose sensor network is proposed to get improved performance with spike rank order encoding scheme. Responses from nodes with in the network are collected at base station and then base station decides about the status of the testing gas through analyzing received ranks from the nodes. Experimental results validate the effectiveness of this approach over testing gases. Moreover, single unit of microelectronic nose can not identify unknown gases and training of all the gases and their mixture is not possible due to limited experimental setup. Our network approach is useful in the aforementioned scenario to identify unknown gases by receiving same signatures from multiple nodes. In future, we will extend this work to train more gases in lab and sensors will also be deployed in the field to analyze real time performance of our scheme on exposure to various gases.

REFERENCES

[1] D Martinez, O Rochel and E. Hugues "A biomimetic robot for tracking specific odors in turbulent plumes," Autonomous Robots, Volume 20, Number 3, 2006, pp. 185-195.

[2] J. W. Gardner, H. W. Shin, E. L. Hines, C. S. Dow, "An electronic nose system for monitoring the quality of potable water," Sensors and Actuators B-Chem, 2000, 69, pp. 336-341.

[3] I. A. Casalinuovo, D. D. Pierro, M. Coletta and P. D. Francesco, "Application of electronic noses for disease diagnosis and food spoilage detection," Sensors 2006, 6, pp. 1428-1439.

[4] L. Pan, R. Liu, Peng, S. Peng, S. X. Yang and S. Gregori, "Real-time monitoring system for odours around livestock farms," Proceedings of the IEEE International Conference on Networking, Sensing and Control, London, 2007, pp. 883-888.

[5] A. C. Romain, D. Godefroid, M. Kuske, J. Nicolas, "Monitoring the exhaust air of a compost pile as a process variable with an E-nose," Sensors and Actuators B-Chem, 2005, 106, pp. 29-35.

[6] S. Zampolli, I. Elmi, F. Ahmed, M. Passini, G. C. Cardinali, S. Nicoletti, L. Dori, "An electronic nose based on solid state sensor arrays for low-cost indoor air quality monitoring applications," Sensors and Actuators B-Chem, 2004, 101, pp. 39-46.

[7] G. Qu, J.J.R. Feddes, W.W. Armstrong, R.N. Coleman, and J.J. Leonard, "Measuring odour concentration with an electronic nose," Transactions of the ASAE, 2001, Vol. 44, No. 6, pp. 1807-1812.

[8] G. Hudon, C. Guy, and J. Hermia, "Measurement of odour intensity by an electronic nose," Journal of Air and Wastes Management Association, 2000, Vol. 50, pp. 1750-1758

[9] X. Zhang, M. Zhang, J. Sun, C. He, "Design of a bionic electronic nose for robot," ISECS International Colloquium on Computing, Communication, Control, and Management, 2008, pp. 18-23.

[10] J. W. Gardner and P. N. Bartlett, "Electronic noses, principles and applications." Oxford University Press, 1999.

[11] H. K. Hong, C. H. Kwon, S. R. Kim, D. H. Yun, K. Lee and Y. K. Sung, "Portable electronic nose system with gas sensor array and artificial neural network," Sensors and Actuators B, vol. 66, no. 1-3, Jul 2000, pp. 49-52.

[12] K. Song, Q. Wang, H. Zhang and Y. Cheng, "Design and implementation a real-time electronic nose system," International Instrumentation and Measurement Technology Conference, Singapore, May 2009.

[13] M. Shi, A. Bermak, S. Chandrasekaran, A. Amira, and S. B. Belhouari, "A committee machine gas identification system based on dynamically reconfigurable FPGA," IEEE Sensors Journal, vol.8, no.4, April 2008, pp.403-414.

[14] T. J. Koickal, A. Hamilton, S. L. Tan, J. A. Covington, J. W. Gardner and T. C. Pearce, "Analog VLSI circuit implementation of an adaptive neuromorphic olfaction chip," IEEE Transactions on Circuits and Systems, vol.54, no.1, Jan 2007, pp.60-73.

[15] H. T. Chen, K. T. Ng, A. Bermak, M. K. Law, D. Martinez, " Spike latency coding in biologically inspired microelectronic nose," IEEE Transactions on Biomedical Circuits and Systems, vol.5, no.2, April 2011, pp.160-168.

[16] H. T. Chen, "A portable electronic nose micro-system based on bio-inspired log-spike processing," M. Phil. Thesis, Department of Electronics and Computer Engineering, Hong Kong University of Science and Technology, Hong Kong, July 2009.

[17] Y. W. Kim, S. J. Lee, G. H. Kim, G. J. Jeon, "Wireless Electronic Nose Network for real-time Gas Monitoring System," International Workshop on Robotic and Sensors Environments, Nov. 2009, pp. 169-172.

[18] V. D. Lecce, R. Daria, J. Uva, "A Wireless Electronic Nose for Emergency Indoor Monitoring," SENSORCOMM 2011, pp. 274-279.

[19] A. Somov, A. Baranov, A. Savkin, M. Ivanov,L. Calliari, R. Passerone, A. Karpov and A. Suchkov, "Energy-aware gas sensing using wireless sensor networks," EWSN 2012, pp. 245260.

[20] N. Yamazoe and K. Shimanoe, "Theory of power laws for semiconductor gas sensors," Sensors and Actuators B-Chem, vol. 128, 2008, pp. 566573.

[21] B. Guo, A. Bermak, P. C. H. Chan and G. Yan, "An Integrated Surface Micro-machined Convex Micro-hotplate Structure for Tin Oxide Gas Sensor Array,", IEEE Sensors Journal, Vol. 7, No. 12, Dec. 2007 pp. 1720-1726.

A 0.9 V High Gain and High Linear Bleeding CMOS Mixer for Wireless Applications

Sid Ahmed Tedjini, Abdelhalim Slimane,
Mohand Tahar Belaroussi,
Centre de Développement des Technologies Avancées
Cité 20 aout 1956 Baba Hassen, Alger, Algérie.
Tel :+213 (0) 21 35 10 18 / 40 / 75

Mohamed Trabelsi
École Nationale Polytechnique d'Alger
Avenue Hacen Badi El harrach, 16200 Alger, Algérie
Telephone: +213 (0) 21 52 53 01/03

Abstract—In this paper, a double balanced bleeding CMOS mixer topology, appropriate for low voltage applications, is presented. A low voltage mixer is designed with high gain and good linearity. The bleeding technique and derivative technique are used to make this trade off. The mixer is designed in 0.18 μm CMOS technology and simulated with Spectre RF at 2 GHz input frequency which can work with supplies as low as 0.9 V. The simulation results show the conversion gain of 18 dB, an input referred third order (IIP3) of 9 dBm, input-referred second-order intercept point (IIP2) of +64 dBm, DSB noise figure of 11 dB and 7 mW power dissipation.

I. INTRODUCTION

Mixers are key building blocks for RF transceivers in wireless communications. In the receiver front end, mixers link together the low-noise-amplifier, local oscillator and IF stage of which the performances are interrelated. Their nonlinear behavior due to the fundamental nonlinear characteristic of the MOS transistors makes analysis and design difficult. This behavior can be the source of noise and troublesome spurious signals at the various frequencies, especially in the coexistence of modern wireless communication standards. Considerable effort has been expended by designers in the design and optimization of the current commuting mixer based on Gilbert cell topology. This choice entails advantages such as large gain conversion, low supply voltage, relatively good linearity and good port to port isolation.

Gain conversion and linearity are the key mixer's design, while the noise is less important because the gain provided by the low noise amplifier makes the subsequent stages noise less significant as suggested by Friis formula. Although, high LNA gain makes the linearity of the mixer challenging. The total third order input intercept point IIP3 of the the direct conversion receiver is given by [1]:

$$\frac{1}{IIP_{3,TOT}^2} \approx \frac{G_{LNA}^2}{IIP_{3,MIXER}^2} + \frac{1}{IIP_{3,LNA}^2} \qquad (1)$$

The above equation shows that the total IIP3 of the receiver is dominated by the first term of the second side when the LNA provides a high gain. Therefore, the linearity of the mixer must be high enough to improve the total linearity. The problem of linearity in mixers is largely addressed in literature to improve the third order input intercept point. Initially used for low

noise amplifiers and recently extended to mixers, the derivative superposition method is undoubtedly one of the most efficient linearization techniques used in RF circuits [2].

In this work, a double bleeding CMOS mixer using derivative superposition linearization technique is presented. The mixer was simulated with Spectre RF at 2 GHz and implemented using 0.18 μm CMOS technology. The designed mixer operates at supply voltages down to 0.9 V. That would make it suitable for low voltage wireless applications.

This paper is organized as follows. In section II, the mixer design is presented, a short description of the bleeding and superposition techniques are also given. Section III presents the simulation results. Finally, a conclusion is provided in section IV.

II. MIXER DESIGN

A. Basic mixer

Gilbert cell mixer is a very challenging design for low voltage applications. However, the pseudo differential topology depicted in Fig. 1 uses less stacked stages which allows to lower the supply voltage. Consequently, this configuration is good candidate for low voltage design.

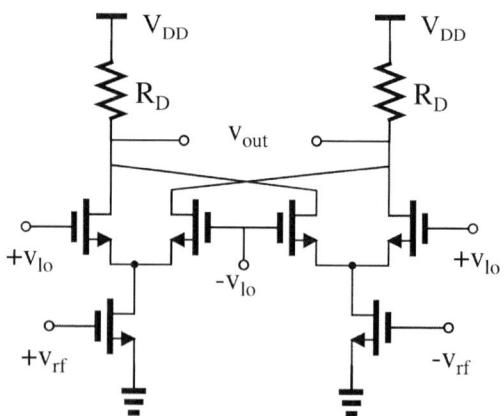

Fig. 1. Pseudo differential mixer

In terms of linearity, pseudo differential pair achieves better IIP3 than a fully differential one. While its IIP2 is worse due to the poor common rejection ratio[3].

978-1-4673-5289-5/12 $31.00 © 2012 IEEE   140

B. Bleeding technique

For ideal commutation stage, the non linearity of the whole mixer is dominated by the non linearity of the transconductance stage.

The fully differential stage IIP3 is given by well known formula [4]:

$$V_{IIP3} = 4\sqrt{\frac{2}{3}}(V_{GS} - V_{TH}) = 8\sqrt{\frac{1}{3}\frac{I_D}{\mu_n C_{ox}}} \qquad (2)$$

and its gain conversion is given by:

$$G_v = \frac{2}{\pi}g_m R_D = \frac{2}{\pi}\sqrt{2\mu_n C_{ox}\frac{W}{L}I_D}R_D \qquad (3)$$

Obviously, both linearity and gain conversion can be improved by increasing the DC current. However, a large DC current provokes significant voltage drop through the load resistors leading to head room problem and decreasing the linearity. To overcome this problem, the current bleeding technique is used. As depicted in Fig. 2, a DC current is injected in the common source commuting pair to increase the gain of the input transconductor and to avoid a large current through the loads. Therefore, large load resistors can be used for further amplification without head room problem.

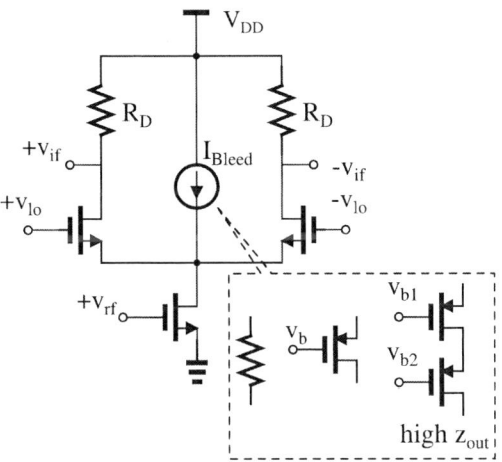

Fig. 2.   Bleeding technique

The current source must be much ideal as possible meaning a high output resistance, otherwise the AC current leaks to the bleeding current source causing gain conversion drop. The bleeding technique can be implemented with small value resistor regarding to the load resistors. Therefore, a large current flows to the input stage. For such values, polysilicon resistors are used, however they experience more then 20% of error due to process variation in CMOS technology[5], leading to DC operation point variation in derivative superposition technique. In the proposed design, a cascode current mirror is used instead of resistors to increase the output resistance and reduce AC current leakage (Fig. 2).

C. Derivative superposition technique

The current voltage relationship for a MOSFET device is given by:

$$i_d = g_m v_{gs} + g_2 v_{gs}^2 + g_3 v_{gs}^3 + \cdots \qquad (4)$$

The most linearization techniques attempt to cancel the third order trasnconductance g_3 to increase the linearity of a given circuit. The Fig. 3 shows the g_m, g_2 and g_3 of the NMOS. Depending on bias conditions, the g_3 is positive in a weak inversion and becomes more negative as V_{GS} increases.

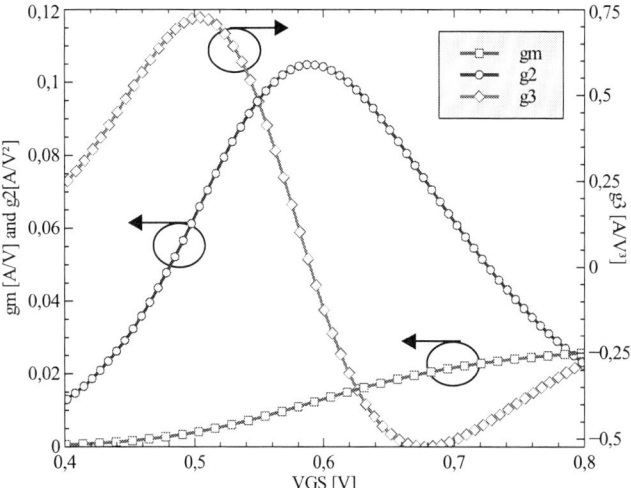

Fig. 3.   Variation of NMOS transconducance against V_{GS}

The derivative superposition technique, which is widely used to improve the linearity of the low noise amplifier [6] and the mixer as well [2], explores this behavior to cancel out the g_3 [6].

As depicted in Fig. 4, the derivative superposition technique, called also multiple gated transistor technique, employs an auxiliary NMOS device M_B biased in weak inversion to operate with the main NMOS device M_A biased in strong inversion [6].

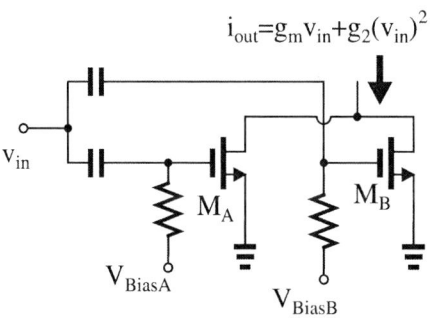

Fig. 4.   Derivative superposition technique with NMOS

The g_3 of M_B is shifted thereby to cancel the g_3 of the main device, as depicted in Fig. 5.

978-1-4673-5289-5/12 $31.00 © 2012 IEEE   141

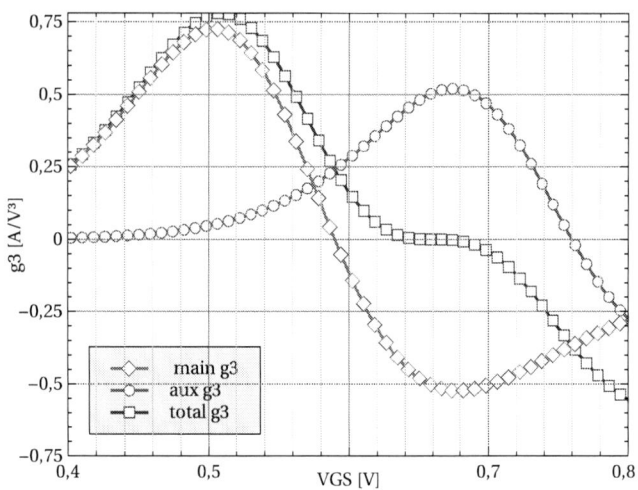

Fig. 5. g_3 cancellation with derivative superposition technique

Fig. 6 shows the complete mixer (the biasing block is not shown for simplicity). The commuting stage is biased in subthreshold region, driving a small amount of DC current to avoid head room problem and to reduce flicker noise which is proportional to DC current. In other hand, a small DC current degrades the IIP2 [3].

III. SIMULATION AND RESULTS

To demonstrate the proposed idea, the mixer is implemented in 1P6M RF TSMC 0.18 μm CMOS technology and simulated with Spectre RF for 2 GHz RF frequency. All mixer's parameters are summarized in table I

Device	Design value
M_A: RF main	75 μm/0.18 μm
M_B: RF aux	52 μm/0.18 μm
M_C: LO switch	21μm/0.18 μm
R_D	1.5kΩ

TABLE I
MIXER'S PARAMETERS

Fig. 7 shows 18 dB of gain conversion over 100 MHz, while the input frequency varies from 2140 MHz to 2170 MHz making this mixer desirable for WCMDA applications. The mixer requires only 150 mV of LO magnitude.

Performed by a quasi periodic steady state analysis [7], the third order intercept point reaches +9 dBm as depicted in Fig. 8. The mixer achieves +64 dBm of second order intercept point and -16 dBm of compression point. under 0.9 V of supply voltage, and consumes only 7.8 mA including the bias circuit.

Table II shows the comparison with the stat of the art of mixers. The bleeding technique is used in [8][9] and [10], while a folded switching mixer and a conventional one are used respectively in [11] and [12]. The proposed mixer achieves better performance in terms of gain conversion and linearity with moderate noise figure in comparison with

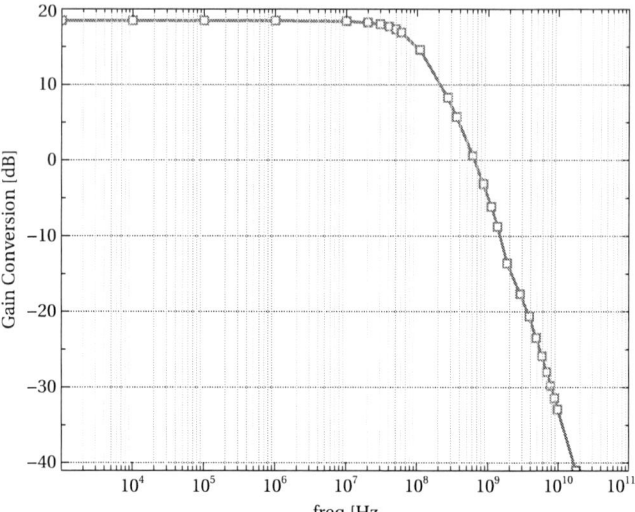

Fig. 7. Gain conversion

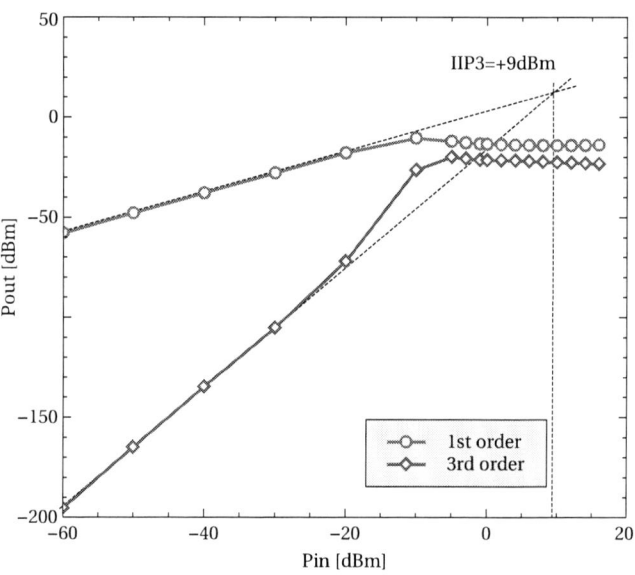

Fig. 8. Simulated IIP3

other works. The supply voltage is also lower as the current consumption as well.

IV. CONCLUSION

In this work, a high linearity and high gain mixer is presented. Combining derivative superposition with bleeding technique, the proposed mixer is successfully designed and implemented in 0.18 μm CMOS technology. It achieves good performances in terms of linearity (IIP3) and gain conversion, moderate DSB noise figure and low supply voltage compared to other CMOS mixers that use the bleeding technique. One can conclude that the proposed mixer is suitable to low voltage wireless applications.

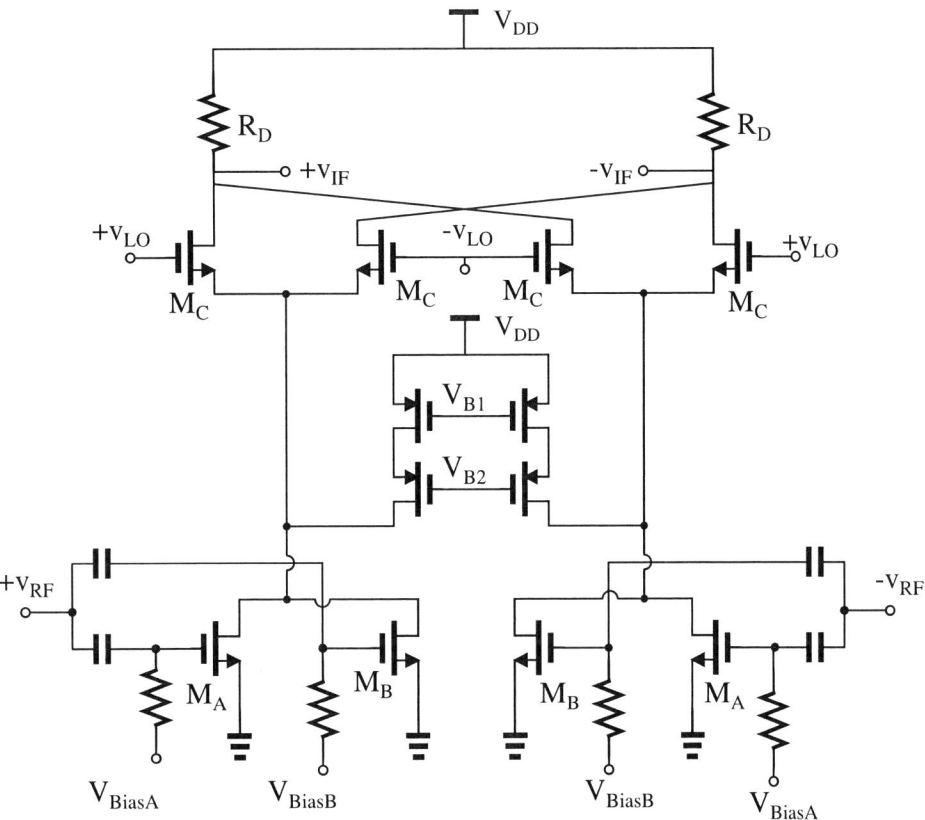

Fig. 6.   Proposed mixer with combined bleeding and derivative superposition techniques

ref	tech CMOS [μm]	freq [GHz]	V_{DD} [V]	idc [mA]	G_v [dB]	IIP3 [dBm]	NF [dB]	Pdc [mW]
[2]	0.18	3.4-6.8	1	2.9	7.2	+3	14.4	2.9
[8]	-	0.9	-	-	4	-1.6	11	-
[9]	0.18	2.4	-	-	17	-0.7	17	-
[10]	0.35	2.1	3	6	12	2.2	9.3	18
[11]	0.18	2.5	1	-	9	-1	12	-
[12]	0.8	1.9	1.8	4.8	0.5	-6	10.2	8.64
this work	0.18	2.1	0.9	7.8	18	+9	11	7

TABLE II
COMPARISON WITH STATE OF THE ART MIXERS

REFERENCES

[1] B. Razavi, *RF Microelectronic.*   Prentice Hall, 1999.

[2] J. D. Chen, "A low voltage high linearity ultra wideband down conversion mixer in 0.18 μm CMOS technology," *Microelectronics Journal,* vol. 1, no. 14, pp. 1–9, 14 Nov 2009.

[3] D. Manstretta, M. Brandolini, and F. Svelto, "Second-order intermodulation mechanisms in CMOS downconverter," *IEEE J. Solide-State Circuits,* vol. 38, no. 3, pp. 394–406, March 2003.

[4] B. Lueng, *VLSI for Wireless Communication.*   Printice Hall, 2002.

[5] T. Wu, S. Tseng, C. Meng, and G. Huang, "GaInP/GaAs HBT subharmonic gilbert mixers using stacked LO and leveled LO topologies," *IEEE Transcactions O, Microvave Theorie and Techniques,* vol. 55, no. 5, pp. 880–888, May 2007.

[6] D. Webster, J. Scott, and D. Haigh, "Control of circuit distortion by the derivative superposition method," *IEEE Microwave Guided Wave Lett,* vol. 6, pp. 123–125, Mars 1996.

[7] K. Mayaram and al, "Computed aided circuit analysis tools for RFIC simulation: Algorithms, features and limitations," *IEEE Trans. Circuits Syst II,* vol. 47, no. 4, pp. 274–285, 2000.

[8] S. Lee and J. Choi, "Current-reuse bleeding mixer," *Electronics Letters,* vol. 36, no. 8, pp. 696–697, Apr 13 2000.

[9] G. Z. Fatin, Oskooei, M. Savadi, Kanani, and Z. D. Koozeh, "A technique to improve noise figure and conversion gain of CMOS mixers," in *50th Midwest Symposium on Circuits and Systems, Montreal, CANADA, Sep 05-Aug 08,* pp. 379–382.

[10] S. Douss, F. Touati, and M. Loulou, "Design Optimization methodology of CMOS Active Mixers for Multi Standard Receivers," *International joural of Electronics; circuits and systems,* vol. 1, no. 1, pp. 1–9, Apr 13 2008.

[11] V. Vidojkovic, J. V. D. Tang, A. Leeuwenburgh, and A. V. Roermund, "A high gain, low voltage folded-switching mixer with current-reuse in 0.18 μ m CMOS," in *IEEE Radio Frequency Integrated Circuits Symposium (RFIC), Ft Worth, TX, Jun 06-08, 2004.*

[12] P. Sullivan, B. Xavier, and W. Ku, "Low voltage performance of a microwave CMOS Gilbert cell mixer," *IEEE J. Solid-State Circuits,* vol. 32, no. 7, pp. 1151–1155, Jul 1997.

Improved antenna diversity system by low correlation between elements exploiting Single-Negative Metamaterials

Belgacem Aouadi and Jamel Belhadj Tahar

Innov'COM, Higher School of Communications of Tunis, University of Carthage, Tunisia

b_aouadi@yahoo.fr

Abstract— In this paper, we propose a two-element diversity-antenna array at the worldwide interoperability for microwave access WiMAX band 3.5 GHz (3.4-3.6GHz) for the Multiple-Input Multiple-Output (MIMO) applications. Each element is a coplanar waveguide (CPW) fed wide rectangular slot antenna with a widened tuning stub. An insulator which consists of Single-Negative Metamaterials (SNG) is added between the two elements (back to back) in order to enhance the isolation characteristic for the designed MIMO antennas. Antenna characteristics such as, scattering parameters, envelope correlation and diversity gain with and without the presence of SNG are provided. The proposed technique achieved a 17.09 dB reduction in mutual coupling at 3.5 GHz.

Index Terms—Correlation, mutual coupling, multiple-input multiple-output (MIMO) systems, single-negative metamaterials (SNG).

I. INTRODUCTION

A wireless communication system with a diversity-antenna provides a much higher channel capacity compared with a system using a single antenna. In conventional Single-Input Single-Out (SISO) communication systems, channel capacity is limited with the theoretical channel capacity set forth by Shannon's theorem. In Multi-Input Multi-Output (MIMO) systems, channel capacity can be proportionally increased by employing multiple antennas in both the transmitter and receiver ends [1]. The advantages of MIMO systems are well known in the meantime, and have led to a large number of publications, as well as the emergence of commercial systems based on this technology. Distance between antenna elements has to be smaller for limited space as a mobile handset. Therefore antenna performance become worse ought to big mutual effect caused by radiated electromagnetic wave of antenna elements [2],[3]. High coupling coefficient between two antennas introduces higher correlation thus lower data rate. Usually, reducing coupling influence can be achieved by separating two antennas at a distance of a half wavelength or more, but this method is not suitable for practical portable devices [4]. In order to overcome this problem, we propose to use Single-Negative Metamaterials (SNG) to reduce the coupling and therefore the envelope correlation of MIMO antennas. The SNG is inserted between the two antennas (back to back) to decrease the correlation between them as will be shown in a later section. One efficient way of realizing SNG metamaterials is to use Split Ring Resonators (SRRs) which provide negative permeability when excited with a specific polarization [5], [6].

Following this introduction, we discuss the design of each radiating element and the SNG structure in section II. The simulations results of the MIMO antennas with and without SNG metamaterials are compared in Section III. Finally, we conclude the paper in Section IV.

II. PROPOSED STRUCTURE GEOMETRY

A. Configuration of the Antenna

CPW-fed antennas for Wireless Local Area Network (WLAN) applications have been increasingly investigated in recent years because a CPW structure has many interesting advantages such as wide operating bandwidth, single metallic layer, and easy integration with active devices or Microwave/Millimeter-Wave Monolithic Integrated Circuits (MMICs) [7].Therefore, the designs of CPW-fed antennas have recently become more and more attractive.

In this paper, we suggest a coplanar waveguide (CPW) fed wide rectangular slot antenna with a widened tuning stub to operate at 3.5 GHz and its configuration is shown in Fig. 1. The proposed radiating element is etched on an inexpensive FR4 substrate with dimensions of 48*64.26 mm^2, thickness of 0.6 mm and relatively permittivity of 4.4. The printed wide rectangular radiating slot has dimensions of L1, L2, L3 and L4 as illustrated in Table 1.

A 50 Ω CPW has a signal strip of length Lf , width Wf, and a gap of spacing g between the signal strip and the coplanar ground plane. The widened tuning stub with a length of L and a width of W is connected to the end of the CPW feed line. The spacing between the tuning stub and edge of the ground plane is S. Ansoft High-frequency structure simulator (HFSS V.13) [8] was used to optimize the necessary parameters to achieve the operating frequency. The simulated S11 parameter for the proposed antenna is shown in Fig. 2. The antenna resonates at 3.5 GHz with a return loss of -19.95dB and a bandwidth of 500MHz which satisfy the requirements of WLAN applications.

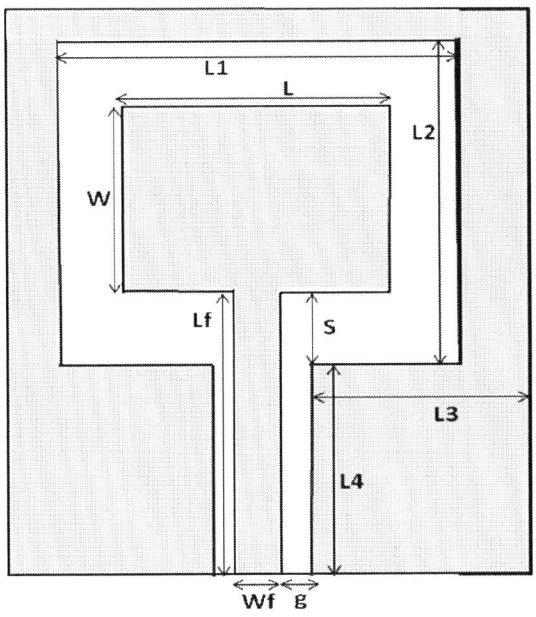

Fig. 1. Geometry of the proposed CPW-fed wide rectangular slot antenna with a widened tuning stub.

TABLE I. DIMENSIONS OF THE PROPOSED ANTENNA

Dimensions	Value (mm)
L1	40
L2	30.26
L3	21.235
L4	30
L	26
W	20
Lf	33
Wf	1.53
g	2
s	3

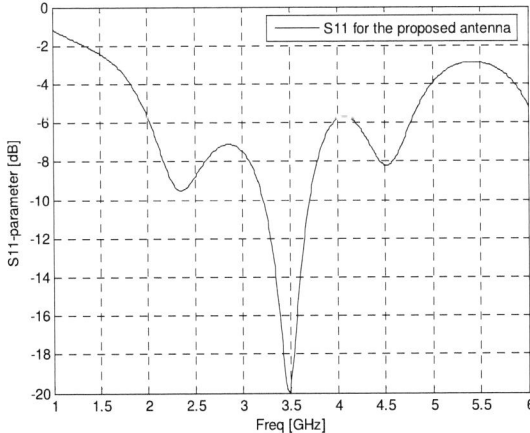

Fig. 2. Simulated return loss for the proposed antenna.

B. Configuration of the SNG metamaterials

Metamaterial is an artificial structure, whose properties may not be found in nature. When its geometrical structure is

much smaller than its operating wavelength, some special properties, such as negative permittivity and negative permeability, can be obtained [9]. The SNG metamaterials considered here are Split Ring Resonators (SRRs) to have negative real permeability values around the same operational frequency as the antenna elements [10].The SRR structure stores magnetic energy so that the magnetic field perpendicular to the surface of the ring, as acting as an inductor and the gap introduces a capacitor to store electric energy. SRR can be equivalent to a LC resonate circuit. The resonance frequency is determined by the capacitance and inductance of its structure [11]. The proposed SNG structure is printed on a Rogers RO3003 substrate with 7.14×7.14 mm^2 surface area, 0.8mm thickness with relative permittivity of 3.38 and loss tangent of 0.0027. Then, it is excited by an electromagnetic wave with propagation vector (k) along the z-axis, electric field vector (E) along the x-axis and magnetic field vector (H) along the y-axis, as shown in Fig.3. The retrieved value of the permeability was -5.215 at the operating frequency as shown in Fig.4. The permeability retrieval method is based on a procedure reported in [12].

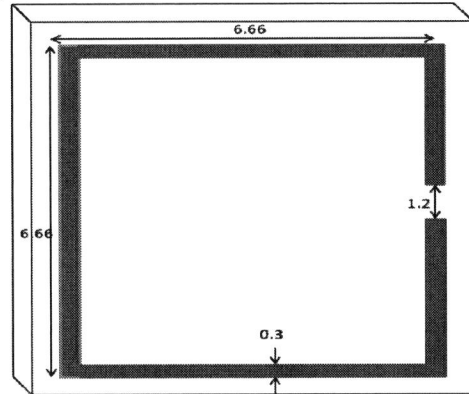

Fig. 3. Dimensions (in millimeter) of the proposed SNG metamaterials.

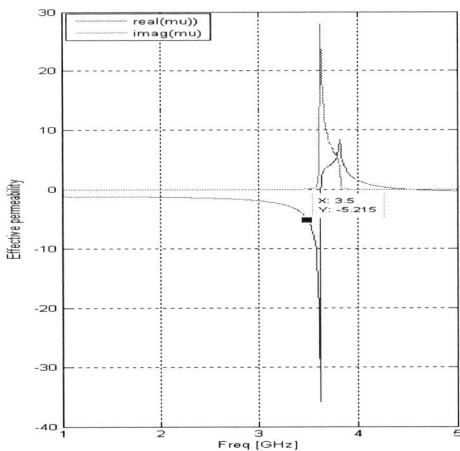

Fig. 4. The extracted effective magnetic permeability of the proposed SNG metamaterials.

978-1-4673-5289-5/12 $31.00 © 2012 IEEE

In this work, a group of 6 sets of SNG slabs were inserted between the antennas. Each set consists of 9 unit cells. The optimized separation distance between the SNG slabs is 8.64 mm. The two identical antennas were separated (back to back) by 7.74 mm ($\lambda//11$ where λ is the free space wavelength). The MIMO system with SNG metamaterials is presented in Fig.5.

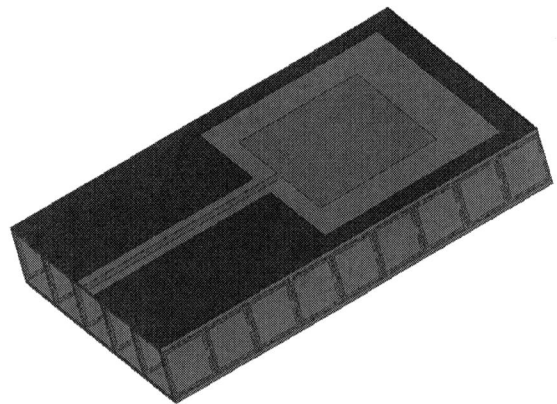

Fig. 5. Configuration of the MIMO system with SNG metamaterials.

III. PERFORMANCES RESULTS

A. Mutual coupling

Fig. 6 depicts the isolation of MIMO antenna with and without SNG metamaterials. By placing the insulator between the antennas, the mutual coupling, S21 has been reduced at the resonance frequency to become -26.33 dB despite of -9.24 initially (Air case).

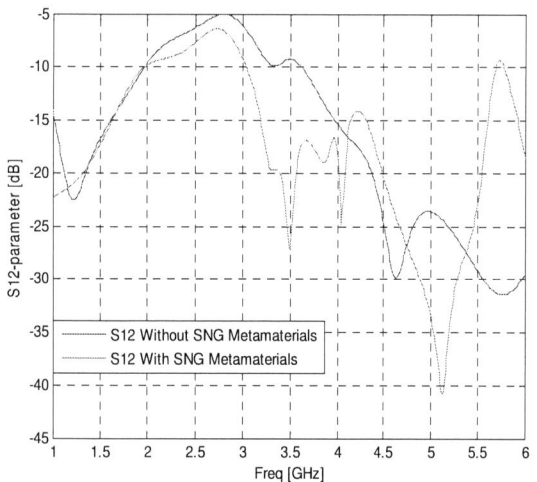

Fig. 6. Mutual coupling of the proposed MIMO antennas with and without SNG metamaterials.

B. Analysis of correlation characteristics

The performance of MIMO antenna elements can be evaluated using the envelope correlation parameter. In fact, the correlation between received signals is a basic parameter for the use of an antenna array or an antenna pair in diversity systems. In many systems the signal can arrive from different directions. In the case of uniform distribution of the angle of arrival the correlation matrix between element port responses can be simply calculated from the scattering matrix measured by a vector network analyzer [13]. For multiple antennas, the envelope correlation is defined as [14]:

$$\rho = \frac{\left| \sum_{n=1}^{N} S_{i,n} * S_{n,j} \right|^2}{\prod_{k=i,j} [1 - \sum_{n=1}^{N} S_{k,n} * S_{n,k}]} \quad (1)$$

Using the equation (1), we can compute the envelope correlation between antennas i and j in an N * N MIMO antennas system. Results, as shown in Fig. 7, have been compared with and without insulator between the antennas. The envelope correlation has been reduced to achieve 0.081 when using the magnetic resonators instead of 0.139 without the added SNG metamaterials.

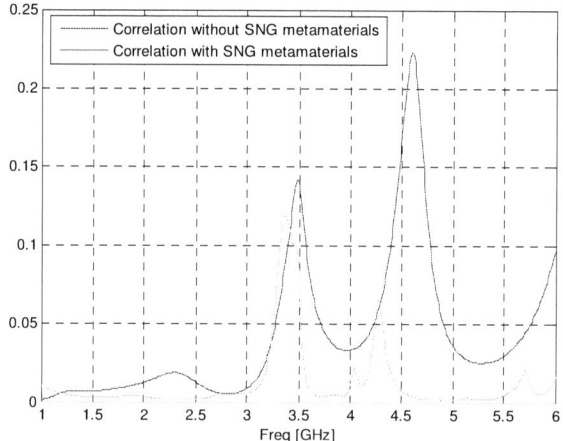

Fig. 7. Enveloppe correlation of the proposed MIMO antennas with and without SNG metamaterials.

C. MIMO Diversity Gain

The concept of diversity gain is that more than two antennas are used to receive a signal and those are combined the replicas of received signal in a desirable way to improve the communication link performance. The diversity gain depends on the number of antenna elements [15]. The diversity gain can be calculated from envelope correlation of the MIMO antennas using the expression in [16].

$$Gapp = 10 * \sqrt{1 - |\rho|} \quad (2)$$

As shown in Fig.8, it's obviously that the diversity gain becomes ameliorated thanks to the SNG metamaterials to get the value of 6.13dB compared to 2.33dB without the added insulator.

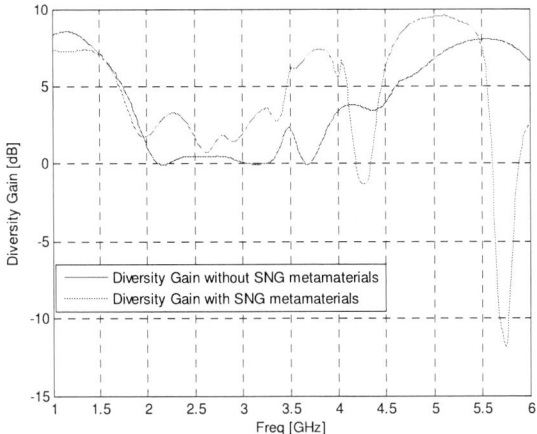

Fig. 8. Diversity Gain of the proposed MIMO antennas with and without SNG metamaterials.

IV. CONCLUSION

In this paper, we have compared the correlation characteristic, mutual coupling and diversity gain of a 2* 2 MIMO communication system with and without employing SNG metamaterials, as an insulator between the two radiating elements at 3.5 GHz. The simulation results show that better isolation can be achieved when using the magnetic resonators, as compared to the air case. Furthermore, the correlation and diversity gain have been significantly ameliorated. Therefore SNG metamaterials can be used in MIMO antennas system to reduce correlation and it is suitable for future WLAN applications.

REFERENCES

[1] G. 1. Foschini and M. 1. Gans, "On limits of wireless communications in a fading environment when using multiple antennas," Wireless Personal Comm., vol. 6, no. 3, pp. 311-335, Mar. 1998.

[2] M. Manteghi and Y. RaI1mat-Samii, "A novel miniaturized triband PIF A for MIMO applications," Microwave Optical Technol. Lett., vol.49, no. 3, pp. 724-731, Mar. 2007.

[3] R. A. Bhatti, S. Vi, and S. O. Park, "Compact antenna array with port decoupling for L TE-standarlized mobile phones," IEEE Antennas and Wireless Prop. Lett., vol. 8, pp. 1430-1433,2009.

[4] H. Shin and J. H. Lee, "Capacity of multiple-antenna fading channels: spatial fading correlation, double scattering, and keyhole," IEEE Transactions on Information Theory, vol. 49, no. 10, pp. 2636–2647, 2003.

[5] J. Pendry, A. Holden, D. Robbins, and W. Stewart, "Magnetism from conductors and enhanced nonlinear phenomena," IEEE Trans. Microw. Theory Tech., vol. 47, no. 11, pp. 2075–2084, Nov 1999.

[6] R. Marques, F. Medina, and R. Rafii-El-Idrissi, "Role of bianisotropy in negative permeability and left-handed metamaterials," Phys. Rev. B, vol.65, pp. 144 440–6, 2002.

[7] S. Kim, J.Choi, and Y.Kim, "CPW-fed broadband G-shaped monopole antenna for WLAN applications," Microwave and optical technologu letters, vol.48, pp.2310-2311, 2006.

[8] Ansoft HFSS, http://www.ansoft.com.

[9] G. Ajay, C. J. Lee and M. Achour, "Compact metamaterial quad-band antenna for mobile application," Antennas and Propagation Society International Symposium, 2008. IEEE AP-S 2008, pp.1–4, 2008.

[10] J. B. Pendry, A. J. Holden, D. J. Robbins, and W. J. Stewart, "Magnetism fromconductors and enhanced nonlinear phenomena," IEEE Trans. on Microwave Theory Tech., vol. 47, pp. 2075-2083, Nov. 1999.

[11] Ricardo Marqués, Ferran Martín and Mario Sorolla, Metamaterials with Negative Parameters: Theory, Design and Microwave Applications, Wiley Inter-Science, 2008.

[12] X. Chen, T. M. Grzegorczyk, B. Wu, J. Pacheco, and J. A. Kong, "Robust method to retrieve the constitutive effective parameters of metamaterials," Physical review. E, Vol. 70, No.1, pp. 016608.1-016608.7, Jan. 2004.

[13] S. Blanch, J. Romeu and I. Corbella," Exact representation of antenna system diversity performance from input parameter description," Electronics Letters, Vol. 39, No., May 2003.

[14] P. Ferrer, J. Arbesu, and J. Romeu, "Decorrelation of two closely spacede antennas with a metamaterial AMC surface," Microwave and Optic. Tech. Lett., Vol. 50, No. 5, May 2008.

[15] D. Gesbert, M. Shafi, D.-S. Shiu, P. J. Smith, and A. Naguib, " From theory to practice: An overview of MIMO space time coded wireless systems,"IEEE JSAC, 21(3), 2003.

[16] Rosengren and P.-S. Kildal. "Radiation efficiency, correlation, diversity gain and capacity of a six monopole antenna array for a MIMO system: Theory, simulation and measurement in reverberation chamber," IEEE Microwave, Optics and Antennas, vol. 152, no. 1, pp. 7–16, Feb. 2005.

978-1-4673-5289-5/12 $31.00 © 2012 IEEE

EC-CPW Fed Elevated Patch Antenna

Adel Saad Emhemmed[1], Nuredin Ali Ahmed[2], Khaled Elgaid[3]

[1] Electrical and Electronic Engineering Department, University of Tripoli, Tripoli, Libya, adel@ee.edu.ly
[2] Department of Computers and System Engineering, Azzaytuna University, Tarhona, Libya, nuredin@mtit.com.ly.
[3] Department of Electronics and Electrical Engineering, University of Glasgow, Glasgow,UK, kelgaid@elec.gla.ac.uk.

Abstract— This paper presents a novel G-band (140GHz - 220GHz) integrated antenna. The antenna incorporates a newly proposed elevated center coplanar waveguide (EC-CPW) fed elevated patch antenna on a high dielectric substrate (GaAs). The proposed antenna consists of an (EC-CPW) feed line, λ/4 transformer used for matching, supporting posts, and a 5.5μm height elevated radiating patch. The antenna topology effectively creates a low-substrate dielectric constant and hence eliminates undesired substrate effects, since the antenna substrate is essentially air, the lowest possible dielectric constant. This increases the radiation efficiency, gain, and the radiation bandwidth. Close agreement between simulated and measured data have shown a good match of -20dB at 172GHz and bandwidth of 7.9GHz from 169.2GHz to 177.1GHz (return loss<-10 dB). Furthermore, the presented antenna exhibits an excellent constant broadside radiation pattern which makes it ideal for array antennas applications where directionality is an important factor avoiding pixels interferences. In addition the proposed antenna fabrication process is compatible with III-V MMICs process technology. This makes it ideal for fully integrated millimeter wave focal plane arrays applications.

Index Terms— *G-band antenna, integrated millimeter-wave antenna, integrated sub millimeter-wave antenna, elevated patch antenna, elevated coplanar waveguide, surface micromachining.*

I. INTRODUCTION

The demand for broadband wireless communication services is growing at an extensive rate all over the world. The intensive development and wide applications of new generations of communication systems using sub millimeter-wave have increased the need for new antenna designs. The most common requirements these systems pose on antennas are broad bandwidth, high gain, a high performance radiation pattern, and fabrication compatibility with integrated circuits and MMICs. Considering these requirements, printed mm-wave antennas appear to be a suitable choice of antenna technology for wireless communication systems, avoiding the use of horn antennas which are bulky, impossible to integrate and avoiding the loss of the transition required between the MMICs and the horn antenna [1, 2].

Printed mm-wave antennas are used in broad range of wireless short range communication applications in several different bands. This is due to the natural absorption of sub millimeter-wave spectrum band in Earth's atmosphere at different rates. These attenuation properties by atmospheric constituents and gases at different rates for different frequencies allow the reuse of the frequencies bands, reduce interferences and multi path issues, and enhance network security [2, 3].

There is a significant trend which is appearing in the new generation of wireless communication systems to integrate RF, analog and digital parts on single substrate at operating frequencies beyond 100GHz. This trend is very challenging for antenna design due to different substrate property requirements. For example, the signal in the analog and digital circuitry often carried through high dielectric constant substrate. However, printed antennas on high dielectric substrates show significant performance degradation. This is due to the excitation of surface waves which leads to lower efficiency, reduced bandwidth, degraded radiation patterns and undesired coupling between the various elements in array configurations [4-8].

Recently, micromachining technology has been applied to antennas to improve the performance. In order to reduce the substrate effect, several techniques have been reported such as making a cavity around and underneath the patch antenna using bulk micromachining [4], electronic bandgap (EBG) structures [5], stacking substrates and coupling through apertures [6], and elevating the antenna patch over an air cavity using a membrane or by using posts to lift the patch into the air [7]. The elevated patch antenna approach can be considered as an alternative to a conventional printed antenna approach, with concomitant advantages of ultra wide bandwidth, low loss, and reduced dependence on substrate effects. Further the elevated structure is part of the MMICs process flow used to realize airbridges and some interconnect, hence avoiding the addition of extra lithography steps. However, feed network loss has prevented them from being efficiently implemented in integrated form [9]. Therefore, using elevated centre coplanar wave guide (EC-CPW) to feed an elevated patch has the advantage of less loss at high frequencies compared with the conventional transmission line structures, hence avoiding electromagnetic coupling between the feed and the substrate as the signal wavelength becomes comparable to the substrate thickness. [10].

In this paper, an EC-CPW fed elevated patch antenna scheme is proposed for the first time. This scheme is realized by elevating both the centre CPW line and patch by gold posts and using λ/4 transformer to match the patch to the feed line.

978-1-4673-5289-5/12 $31.00 © 2012 IEEE

II.   ANTENNA TOPOLOGY

The aim is to integrate patch antennas into circuit designs on high dielectric substrates such as GaAs without losing the advantages of low dielectric materials. This is achieved by elevating both the patch and centre conductor of the CPW feed line. The elevated rectangular patch and EC-CPW feed are supported using a number of gold posts which offer a mechanically strong and rigid solution. In order to match the elevated patch, EC-CPW feed line is inset into the patch where the resonant resistance is equal to feed line impedance, and also EC-CPW feed is matched with feeding post by an elevated λ/4 transformer. The input impedance of the inset fed patch antenna mainly depends on the inset distance and to some extent on the inset width (spacing between feeding line and patch conductor) [11]. Therefore, the spacing between the patch conductor and feed line in this design is kept constant, 15µm, and inset length is optimized using the simulation software HFSS and final length is 50µm. Fig. 1 shows the proposed antenna structure. This topology effectively creates a low-substrate dielectric constant and undesired substrate effects can be eliminated, since the antenna substrate is essentially air (the lowest possible dielectric constant), and most of the electric field of EC-CPW is confined in the air region between the centre line and the ground, not in the substrate. This will help increase the radiation efficiency, gain, and the radiation bandwidth. In addition, the type, thickness and stack of the substrates can be varied without significant change of antenna performance. This will allow the use of high dielectric substrates with different thicknesses.

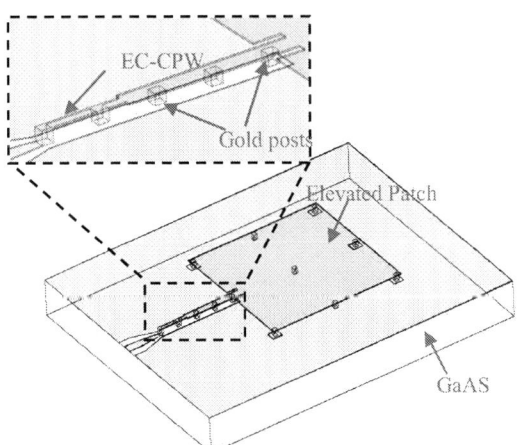

Fig. 1. EC-CPW fed elevated patch antenna.

For operation at G-band simulation suggested the geometry of the rectangular patch antenna to be W × L of 806µm × 738µm. Also, the antenna ground plane extents Ws × Ls of 1.5 mm × 1.65 mm which is sufficiently large to reduce diffraction from the edges and reduces ripples in the main radiation pattern and minimizes backward radiation. The elevated central conductor CPW feed and patch were built on a 630µm thickness GaAs substrate with dielectric constant of 12.9.

III.   FABRICATION

The antenna fabrication was based on surface micromachining technology. A 5.5µm height elevated CPW feed and patch antennas were fabricated on a 630µm thickness GaAS substrate with dielectric constant of 12.9. The height of the structure can be varied as required for optimum antenna performance. Two steps were required to realize the proposed structures. First the ground plans was realized by e-beam lithography techniques. This was done by exposing a spun PMMA on the GaAS followed by development, metallization of 1.2µm thickness and liftoff. The second step uses photolithography techniques, incorporating different two different photoresists thicknesses followed by electroplating 2µm thick gold. The fabrication process is compatible with both III-V and Si MMICs process technology. Fig. 2 shows the SEM image of the fabricated antenna.

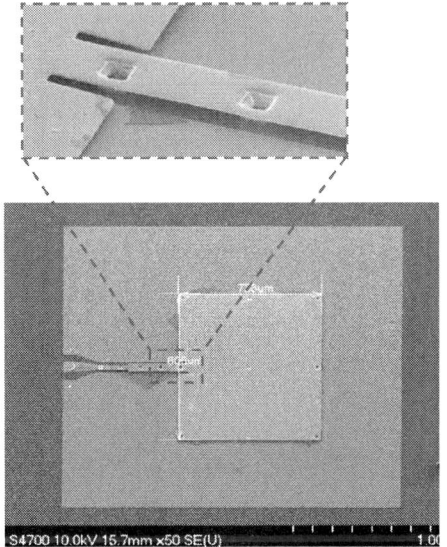

Fig. 2. SEM photo of the fabricated antenna

IV.   SIMULATION AND MEASURMENTS

The proposed antenna is designed using an Ansoft HFSS simulator which is based on the finite element technique. Waveport at the antenna input terminal was used for the simulation with meshing at λ/4 to obtain higher simulation resolution. The fabricated antennas were characterized using an Agilent PNA Vector Network Analyzer with 140GHz to 220GHz OML Heads. On wafer LRRM calibration techniques was used on ISS calibration standard to calibrate the VNA.

Fig. 3 shows the measured and the simulated input match (return loss) data of the antennas. A very good agreement between simulation and measurement was obtained, validating the electromagnetic simulation. The experimental results show a good match of -20dB at 172GHz and bandwidth of 7.9GHz from 169.2GHz to 177.1GHz (return loss<-10 dB). However, there is quite high loss at higher frequency more, this probable because of discontinuity of electrical field due to the inset feed and λ/4 transformer.

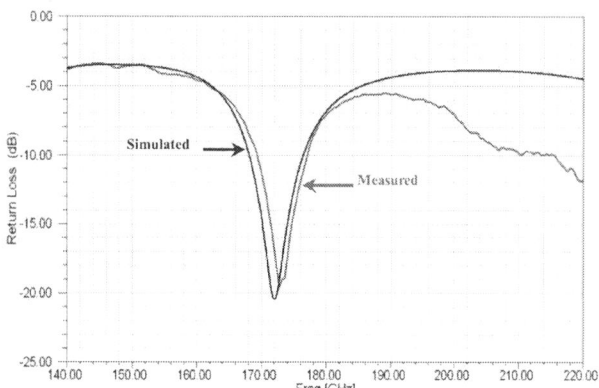

Fig.3. Simulated and measured results of return loss.

Simulation predicts a nearly constant broadside radiation pattern in both E and H plane cross the designed bandwidth with beam-width of *64°* in the E-plane and *46°* in the H-plane at resonant frequency as show in Fig. 4 The simulated antenna gain in the vertical axis at resonant frequency is about *6.7dB* with the front-to-back ratio (worst case) of about *7.8dB*, showing high radiation efficiency of the antenna structure. Fig. 5 shows the simulated 3-D gain radiation pattern of the antenna.

V. COMPARISON WITH CONVENTIONAL PATCH ANTENNA

A conventional edge-fed microstrip patch antenna directly printed on top of GaAs substrate with the same elevated patch dimensions was simulated using HFSS simulation software to compare its performance with the elevated patch antenna. Fig. 6 shows the return loss of the patch antenna directly printed on the substrate. It can be clearly observed that the antenna loss is very high and increases with frequency and there is no clear resonant frequency. This because at G-band frequencies there is possibility to excited seven surface wave modes in the GaAs substrate, the first mode TM0 has zero cut-off frequency and last mode TM6 has cut-off frequency of 207.06GHz.

Also, the elevated patch antenna has demonstrated a significantly improved radiation performance than the conventional patch antenna directly printed on top of the same type of substrate. A non-elevated patch on the same substrate gives a maximum predicted directivity of 4.8dB with a bad front-to-back ratio of 0.203dB, but no longer at the broadside direction as a consequence of surface wave triggered in this electrically thick dielectric substrate, as shown in Fig. 7. It can be clearly observed that there is a deep drop in the antenna radiation performance without elevation due to the diffraction of surface waves at the edge of the substrate. In contrast, the elevated patch has a clear main beam in the z-direction.

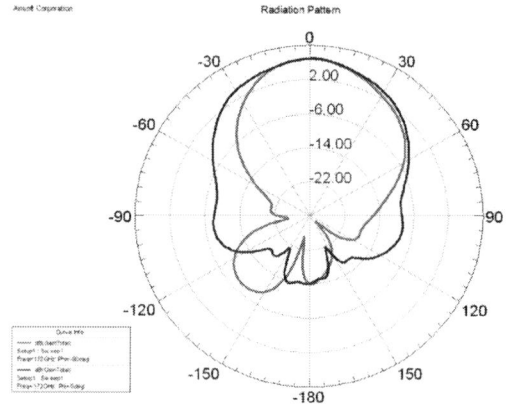

Fig. 4. 2-D radiation pattern in E and H plane at resonant frequency (gain)

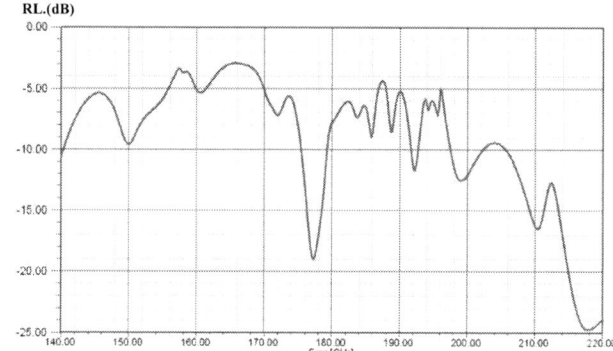

Fig. 6 Return loss of the patch antenna directly printed on the substrate.

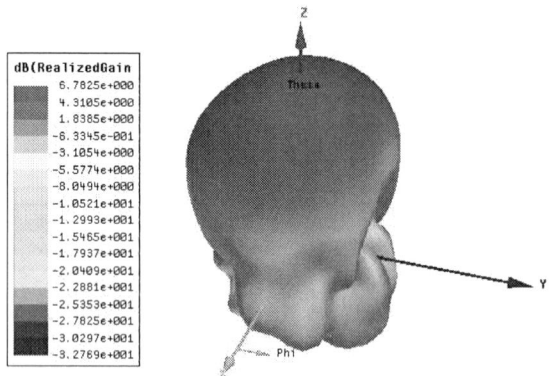

Fig. 5. Simulated 3-D radiation pattern.

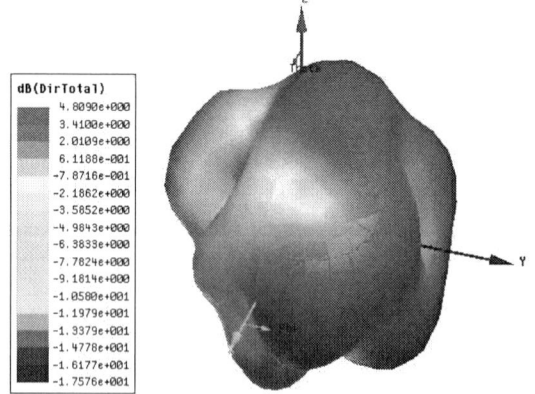

Fig. 7 3-D radiation pattern of the patch antenna directly printed on the substrate.

978-1-4673-5289-5/12 $31.00 © 2012 IEEE

VI. CONCLUSION

It can be concluded that the proposed antenna configuration in this work offers significant improvements in performance compared with those previously reported. The antenna scheme offers an easy method to integrate antenna with other MMICs, eliminate the undesired substrate effects, gives good matching with the feed line and maximizes antenna performance on high dielectric substrates. The structure also demonstrates broadband impedance matching and symmetry radiation pattern in G-band frequencies.

REFERENCES

[1] D. M. Pozer, ''Consideration for Millimetre Wave Printed Antennas'', IEEE Transactions on Antennas And Propagation, Vol. AP-31, NO. 5, pp. 740-747, Sep 1983.

[2] H. LiQuan, ''Some Advances In Millimeter Wave Application Systems'', IEEE Asia Pacific Microwave Conference, pp. 749-752, Dec 1997.

[3] S. Gunnarsson, N. Wadefalk, J. Svedin, S. Cherednichenko, I. Angelov, H. Zirath, I. Kallfass, and A. Leuther, '' A 220 GHz Single-Chip Receiver MMIC With Integrated Antenna'', IEEE Microwave and Wireless Components Letters, Vol. 18, NO. 4, pp. 284-286, Apr 2008.

[4] I. Papapolymerou, R. Drayton, and L. Katehi, ''Micromachined Patch Antennas'', IEEE Transactions On Antennas And Propagation, Vol. 46, NO. 2, pp. 275-283, Feb 1998.

[5] Y. Lee, X. Lu, Y. Hao, S. Yang, R. Ubic, J. Evans and C. Parini, ''Directive Millimetre-wave Antenna Based on Free Formed Woodpile EBG Structure'', IEEE Electronic Letters, Vol. 43, No. 4, pp. 195-196, Feb 2007.

[6] G. Gauthier and L. Katehi, ''A 94-GHz Aperture-Coupled Micromachined Microstrip Antenna'', IEEE Transactions On Antennas And Propagation, Vol. 47, No.12, pp.1761-1766, Dec 1999.

[7] H. Lee, J. Kim, S. Hong, and J. Yoon, ''Micromachined CPW-Fed Suspended Patch Antenna For 77 GHz Automotive Radar Applications'', IEEE European Microwave Conference, Vol. 3, pp. 249-252, Oct 2005.

[8] Bo Pan, Y. Yoon, J. Papapolymerou, M. Tentzeris and M. Allen, ''Design and Fabrication of Substrate-Independent Integrated Antennas utilizing Surface Micromachining Technology'', IEEE Microwave Conference Proceedings, Vol. 2, Dec 2005.

[9] Y. Yoon, Bo Pan, J. Papapolymerou, M. Tentzeris, and M. Allen, '' Surface-micromachined millimeter-wave antennas'', IEEE The 13th International Conference on Solid-State Sensors, Actuators and Microsystems, Vol. 2, pp. 1986-1989, Jun 2005.

[10] A. S. Emhemmed, I. McGregor, K. Elgaid, "Elevated Conductor Coplanar Waveguide-fed Three-level Proximity-coupled Antenna for G-band Applications," Microwaves, Antennas & Propagation, IET, vol 4, pp. 1910-1915, Nov. 2010.

[11] M. Ramesh and Y. KB, "Design Formula for Inset Fed Microstrip Patch Antenna," Journal of Microwaves and Optoelectronics, vol. 3, pp. 5-7, December 2003.

Multi-Band Meta-material Structures Based On Hexagonal Shaped Magnetic Resonators

H. Benosman

Laboratory of telecommunication, Faculty of Technology
University of Abou-Bekr Belkaïd
Tlemcen, Algeria

N.Boukli Hacene

Laboratory of telecommunication, Faculty of Technology
University of Abou-Bekr Belkaïd
Tlemcen, Algeria

Abstract— In the present paper, we propose a new multi band metamaterial structure composed of a unit cell that has multiple hexagonal-shaped metal rings printed one inside the other. The introduced metamaterial composition exhibits multi-resonant behavior due to distinct magnetic resonances caused by the self and mutual couplings of each inclusion. The number of distinct magnetic resonances is determined by the number of hexagonal-shaped rings placed in the unit cell. Using three rings in this paper, three distinct resonance frequencies can be shifted to desired values simply by changing the design parameters. Transmission and reflection spectra of the proposed structure are obtained using the CST Microwave Studio to observe its resonance frequencies. Then, the effective permeability and effective permittivity are retrieved from the already computed complex S-parameters to demonstrate the presence of three distinct negative permeability bandwidths.

Keywords: left handed material, split ring resonator, negative permeability.

I. INTRODUCTION

There is much interest in meta-materials for the microwave and optical applications because their electromagnetic properties are vastly different from ordinary materials. A metamaterial is a structural composite with unique electromagnetic properties due to the interaction of waves with the finer-scale periodicity of conventional materials [1]. The person who is responsible in discovering the concept of metamaterials is Veselago in 1967 [2]. He assumes an unknown material that has a negative permeability and permittivity in the same frequency range and it shows the abnormal of electromagnetic properties when the uniform plane-wave propagation was studied [2-4]. As a result, Veselago referred to the materials as left-handed materials (LHM), which mean the electric field, magnetic field, and propagation vector are related by a left-handed rule which results in the phase and group velocities of an electromagnetic wave being anti-parallel.

The first structure to prove the existence of metamaterials was a split ring resonator structure that has a magnetic response without magnetic materials [5]. Magnetic properties of the SRR are considered as the substrate of the microstrip patch antenna instead of high permittivity materials because we can, not only reduce the size of the patch, but also improve the bandwidth of

the antenna using high permeability materials for substrates of microstrip patch antennas [6] [7]. The SRR structure has a ring with a gap, which makes the inductance and capacitance. The transmission coefficient of the SRRs is minimum at the magnetic resonance frequency, which is made from the inductance and capacitance. It is important to design SRR structures in the substrate of the microstrip components because the shape, the orientation and the arrangement of SRR structures change the inductance and capacitance. This means we can control and optimize the design of microstrip components by designing structures in the substrate. There are several structures that have been introduced by the previous researchers starting with a symmetrical ring structure, then an omega, and finally an S structure [8]. The SRR which has hexagonal cross section, different from the circular or square cross section of the SRRs in previous studies has been studied experimentally by Zhang fu-li for X band [10].

In this paper, a new three-ring SRR unit cell design is introduced for multi-band metamaterial applications. For a given substrate material, design parameters are the side lengths and widths of metal strips, gap distances for each ring, and the separation distances between the rings. As a proof of concept, several multi-band SRR arrays are designed and simulated in this paper for three different cases (for N = 1, 2 and 3). Complex S parameters transmission and reflection characteristics of the proposed hexagonal SRR are obtained by CST microwave studio, these parameters are used for estimating the resonance frequencies of this novel structure and also for the retrieval of the associated effective medium parameters ε and μ. In this work, Nicolson-Ross-Weir technique was chosen to attain the permittivity and the permeability since the technique is widely used to convert the S-parameters. Moreover, this approach provides an easy as well as simple calculation method. The detail equations were presented in [9]

II. DESIGN AND SIMULATION SETUP

The schematic view of the proposed unit cell is given in figure 1 for the case of N = 3. Metallic inclusions are made of copper with the thickness of 0.035 mm laid out on the face of a dielectric substrate (RO4003) who has a permittivity ε=4.3, and a thickness of 0.81 mm. Two waveguide ports were set at the top and bottom of the X-

978-1-4673-5289-5/12 $31.00 © 2012 IEEE

axis, the substrate with multi-ring are centered in the waveguide and it is excited by an electromagnetic wave with the propagation vector k in x-direction, the electric field vector E in y-direction, and the magnetic field H in z-direction. Perfect electric conductor (PEC) boundary conditions were implemented on the left and the right of the Y-axis, and perfect magnetic conductor (PMC) boundary conditions were placed in front and back of the Z-axis. A frequency domain solver was used to simulate the SRR-material construction in the CST MWS. Using this setup, single-ring, two-ring and three-ring SRR arrays are designed and simulated.

Each one of the SRR arrays simulated in the first step of this work are composed of only one type of single-ring hexagonal-shaped unit cells shown in figure 2, where the side lengths (L) of the SRR metal rings are chosen to be 4 mm, 3 mm and 2 mm for the SRR-A, SRR-B and SRR-C type unit cells respectively. The same gap distance g=0.33 mm and the same metal strip width w=0.33 mm are used in these three different unit cells.

In the next step, a two-ring SRR unit cell is designed by combining the SRR-A and SRR-B unit cell topologies which are aligned in the gap-to-gap configuration as shown in Figure 3(a). Similarly, a three-ring SRR unit cell is also designed by combining all three types of unit cells SRR-A, SRR-B and SRR-C as given in figure 3(b).

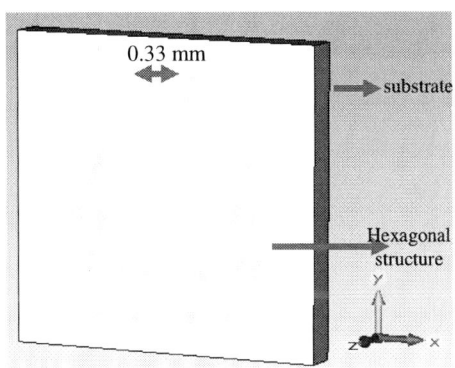

Fig.1: CST MWS simulation setup.

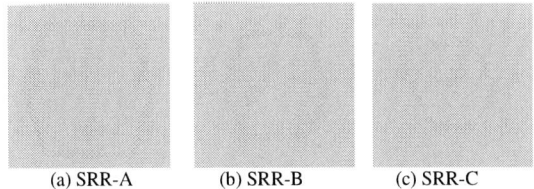

(a) SRR-A   (b) SRR-B   (c) SRR-C

Fig. 2: Front view of the single-ring hexagonal-shaped SRR unit cells with different parameters.

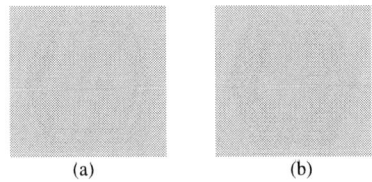

(a)   (b)

Fig.3: Multi-ring Hexagonal SRR unit cell structures with two and three-rings.

III. RESULLTS

Transmission spectra calculated by CST MWS for the single-ring array topologies SRR-A, SRR-B and SRR-C are shown in figure 4. As it is seen, these parameters are frequency dependent complex functions which satisfy certain requirements of causality. The dip in the phase of the transmission indicates the presence of negative region (or the location of the resonance in terms of the frequency) which can easily be observed from the figure. In figure 4, the simulation results are obtained for different values of side lengths (L) of the SRR metal rings 4, 3, and 2 mm. The magnetic resonance frequencies associated occurs respectively at 5.19 GHz, 6.56 GHz, and 8.37 GHz. As the side lengths decreases from 4 mm to 2 mm, the minimum of the transmission and the dip of its phase shift from the lower to higher frequencies. Thus, these kinds of structures can be considered as a resonant LC circuit with a resonance frequency which depends only of the inductance and capacitance of the structure $\omega = \dfrac{1}{\sqrt{LC}}$. The capacitance in the hexagonal split ring resonator is formed by edge coupling between the inner and outer ring. When the side length is decreased, the capacitance will decrease, and thus the resonance frequency will increase. Then, the minimum of the transmission and the dip of its phase will shift to the higher frequencies.

The magnitude and phase spectra of the S21 and S11 parameters are next computed for the two-ring SRR topology of figure 3 (a). Resulting plots are given in figure 5 together with the plots for the real and imaginary parts of the effective permittivity, and effective permeability. It is clear from figure 5 (a) that the structure resonates at two different frequencies, which are 5.08 GHz, and 6.28 GHz. Similarly figure 5 (b) shows the phase spectra of S11 and S21. The permeability is resonant in real part and takes negative values in the frequency range between 4.97 GHz and 5.35 GHz, 6.22 GHz and 6.4 GHz as it appears in the figure 5 (c), Whereas, the real part of effective permittivity never goes below zero figure 5 (d). On the other hand, the three-ring SRR array has three distinct magnetic resonances at 5.06 GHz, 6.17 GHz, and 8 GHz respectively. Figure 6 (c) shows that this structure has three bands with negative real effective permeability, which are from 4.95 to 5.3 GHz, from 6.15 to 6.26 GHz and from 7.98 to 8.03 GHz. Computed values of resonance frequencies of all structures under investigation are summarized in Table 1.

These results demonstrate that a desired number of magnetic resonances can be realized by selecting the number of SRR rings within the limits of geometrical constraints.

TABLE I. RESONANCE FREQUENCIES FOR THE SIMULATED SRR TOPOLOGIES

Structure	Resonance Freq .(GHz)
SRR-A	5.19
SRR-B	6.56
SRR- C	8.37

two-ring SRR	5.08, 6.28
three-ring SRR	5.06, 6.17, 8

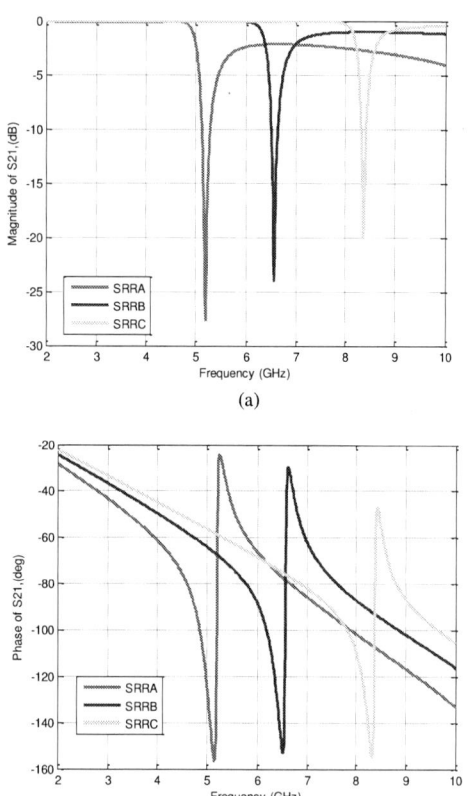

(a)

(b)

Fig.4: Magnitude and phase of S_{21} for the single-ring hexagonal SRR.

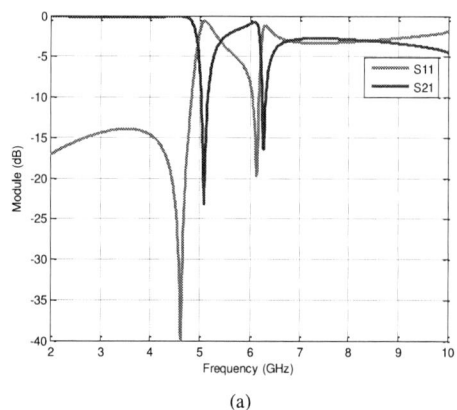

(a)

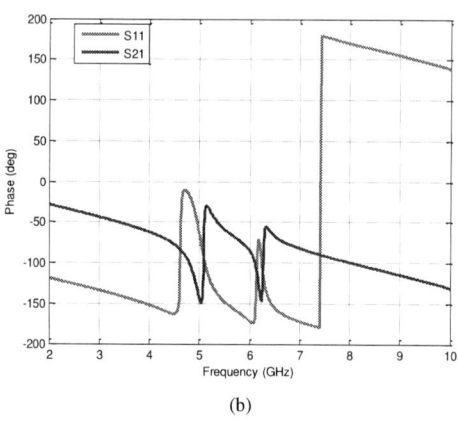

(b)

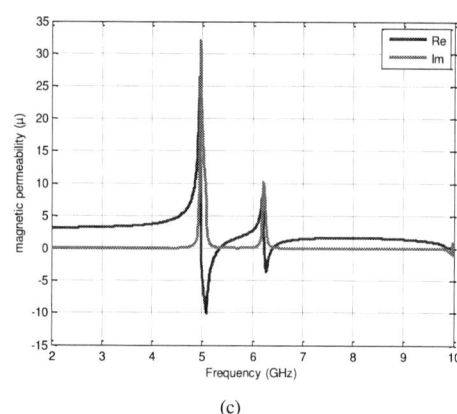

(c)

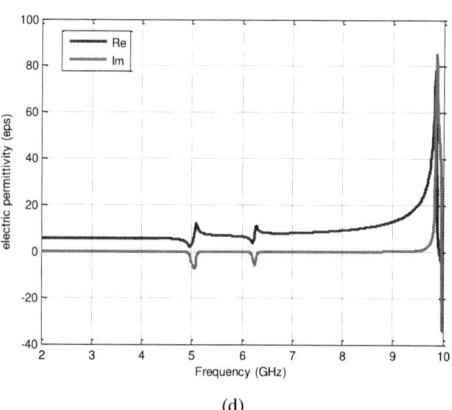

(d)

Fig.5: Magnitude and phase of S parameters, real and imaginary parts of the permittivity, and permeability for the two-ring SRR.

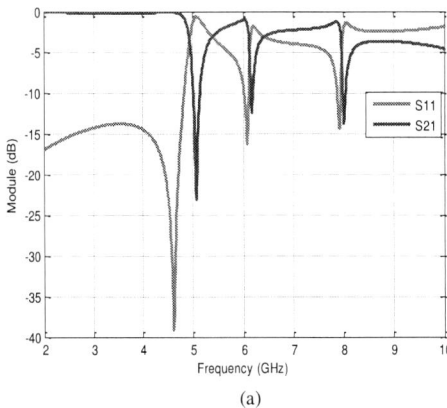

(a)

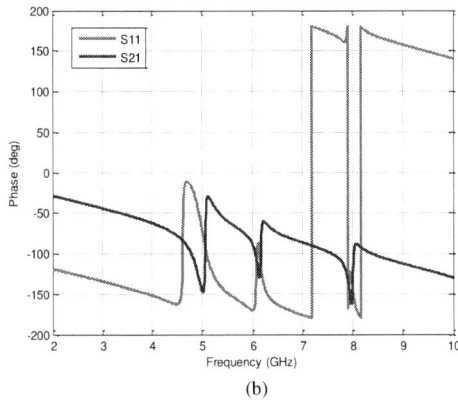

(b)

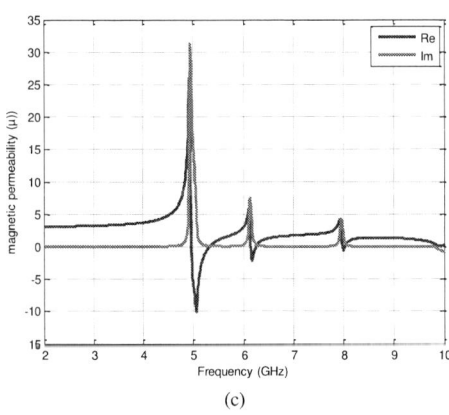

(c)

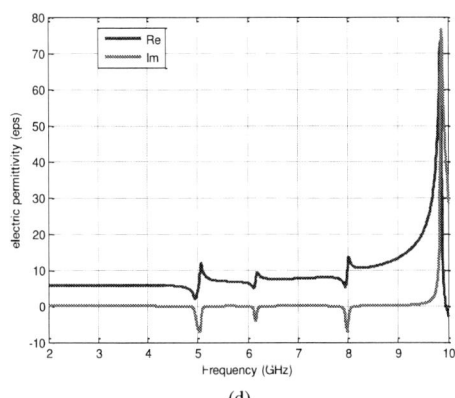

(d)

Fig.6: Magnitude and phase of S_{11} and S_{21} for the three-ring SRR, real and imaginary parts of the permeability, and permittivity.

IV. CONCLUSION

In this paper, three-ring hexagonal SRR structure for a new artificial magnetic inclusion has been introduced and the possibility of multiple magnetic resonances with negative permeability bands are demonstrated for this SRR topologies where the number of resonances is determined by the number of concentric rings. These results clearly show that the structure can be an alternative to the well known ordinary SRR structure to be used in the design of composite left-handed metamaterials especially when a multi-band magnetic resonator operation is needed. The proposed resonant structure can be a good candidate for various metamaterial applications in different portions of the electromagnetic spectrum.

REFERENCES

[1] Semichaevsky and A. Akyurtlu; "Homogenization of metamaterial-loaded substrates and superstrates for antennas" Progress In Electromagnetics Research. Vol 71, 129-147,2007.

[2] Veselago, V. G., "The electrodynamics of substances with simultaneously negative values of ε and μ," Sov. Phys. Uspekhi, Vol. 10, 509-514, 1968.

[3] L. Le-wei; Y. Hai-ying; W. Qun, and C. Zhi-ning: "Broad-bandwidth and low-loss metamaterials: theory, design and realization". Journal of Zhejiang University SCIENCE A. Vol. 7, 5-23,2006.

[4] E. Nader and R. W. Ziolkowski. "A positive future for double-negative metamaterials". IEEE Transactions on Microwave Theory and Techniques. Vol. 53, 1535-1556, 2005.

[5] Pendry, J. B., A. J. Holden, D. J. Robbins, and W. J. Stewart, "Magnetism from conductors and enhanced nonlinear phenomena," IEEE Transactions on Microwave Theory and Techniques, Vol. 47, 2075-2084, 1999.

[6] P. M. T. Ikonen; S. I. Maslovski; C. R. Simovski; S. A. Tretykov. "On Artificial magnetodielectric Loading for Improving the Impedance Bandwidth Properties of Microstrip Antennas". IEEE transactions on antennas and propagation. Vol. 54, 1654-1662,2006.

[7] H. Mosallaei; K. Sarabandi." Antenna Miniaturization and Bandwidth Enhancement Using a Reactive Impedance Substrate". IEEE transactions on antennas and propagation. Vol. 52, 2403-2414, 2004.

[8] L. Ran; J Huangfu; H. Chen; X. Zhang; K. Cheng; T. M. Grzegorczyk, and J. A. Kong. "Experimental study on several left-handed metamaterials". Progress In Electromagnetics Research , Vol 51, 249–279, 2005.

[9] Rhode&Schwarz, "Measurement of dielectric material properties" 2006.

[10] F.Zhang; Q.Zhao; Y.Liu; C.Luo. "Behaviour of Split Ring Resonators and Left Handed Metamaterials ". CHIN. PHIS. LET, Vol. 21, 1330, 2004.

Design and Implementation of passive UHF RFID system

Rafik Khelladi, Mustapha Djeddou and Mustapha Benssalah
Communication Systems Laboratory
Military Polytechnic School
Algiers, Algeria
Emails: (r.khelladi, djeddou.mustapha, bensmusta)@gmail.com

Abstract—In this paper, the implementation of passive UHF RFID, using Agilent Advanced Design System, is constructed. The implementation of this system is divided into two main parts, the reader part and the tag part. The architecture of each part is described in details and is implemented. During the design process, we propose models for building blocks of different encoding types used in RFID system. A dipole reader antenna, with a center frequency of 915 MHz is also proposed and designed. Finally, we present some implementation results.

Index Terms—UHF passive RFID system, RFID reader, RFID tag, Agilent ADS.

I. INTRODUCTION

Radio Frequency IDentification (RFID) is a generic term used to describe a system that transmits the identity of an object or person wirelessly using radio waves. Nowadays, RFID technologies are applied to various fields such as transportation, medical treatment, smart card, attendance system, item tracking, supply-chain management, etc. [1].

Generally, the RFID system consists of two basic components, a reader and a tag. The reader is composed of a radio-frequency module, a control unit and an antenna. The tag attached to an object, consists of an antenna and a silicon chip containing a small amount of information [2]. According to the operating frequency, RFID systems can be distinguished into four frequency bands: the low frequency (LF) band around 125 kHz, the high frequency (HF) band at 13.56 MHz, the ultra high frequency (UHF) band at 860-960 MHz, and the microwaves band at 2.4 GHz. For LF and HF RFID systems, the read range is less than 1m; for UHF RFID system, the read range generally reaches around 5m [3]. The RFID system can be also classified into passive tags which have no independent electrical power source and have no radio transmitter. Semi passive tags have a local battery only to power the tag circuitry, and active tags with a battery for both a local power source and a radio transmitter [4]. In this paper, we are interested in the design of the passive UHF RFID system. This operation is achieved through modeling and implementation using the rapid prototyping environment of Agilent ADS (Advanced Design System) [7]. The latter environment provides a block diagram interface, an integrated design environment to designers of RF electronic products, graphic and programming functionalities.

Agilent ADS supports many steps of the design process: schematic capture, layout, frequency-domain and time-domain circuit simulation, and electromagnetic field simulation.

The paper is organized as follows: in section 2, we build and describe different blocks of RFID system implementation, section 3 deals with the results of the implementation. Finally, concluding remarks are drawn in section 4.

II. RFID SYSTEM IMPLEMENTATION

A RFID reader is a radio transceiver, meaning that it consists of a transmitter and receiver working together to communicate with the tag. In this section, we describe models of the reader and the tag parts.

A. Reader transmitter model

The reader transmitter, shown in Fig.1, is implemented using the co-simulation between the numeric front end and the radio front end. The transmitter is mainly made up of a source encoder, raised cosine filter that removes the Inter Symbol Interference to satisfy Nyquist criteria. In other hand, in the radio front end, we find a non-ideal mixer for up-conversion, a high-power amplifier and band pass filter with bandwidth lying (860-960) MHz, used to remove the out- of-band spectrum. The implementation of the radio front end is depicted in Fig.2.

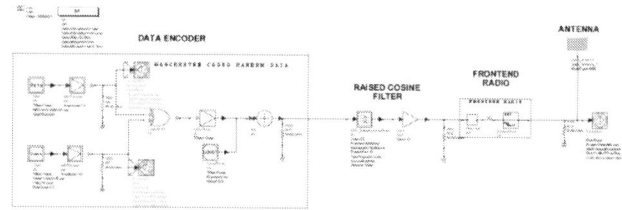

Fig. 1. RFID transmitter implementation

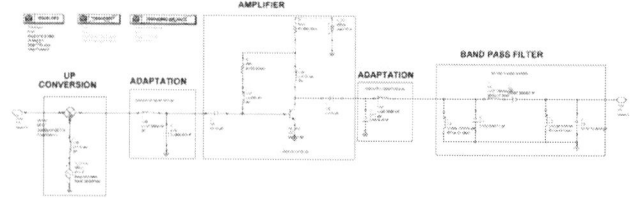

Fig. 2. Transmitter radio frontend

In Gen2 (second RFID generation), the reader transmitter uses DSB-AM (Double Side Band-Amplitude Modulation) or SSB-AM (Single Side Band-Amplitude Modulation) [3]. According to the architecture of the reader transmitter shown in Fig.1, only the DSB modulation is presented.

1) Transmitter data encoding: In Gen2 protocols, the generated binary data by the reader is coded using Manchester encoding [5]. In Fig.3(a) we propose schematic blocks to build the Manchester encoder. The obtained Manchester encoding result for the sequence "011001001100" is depicted in Fig.3(b).

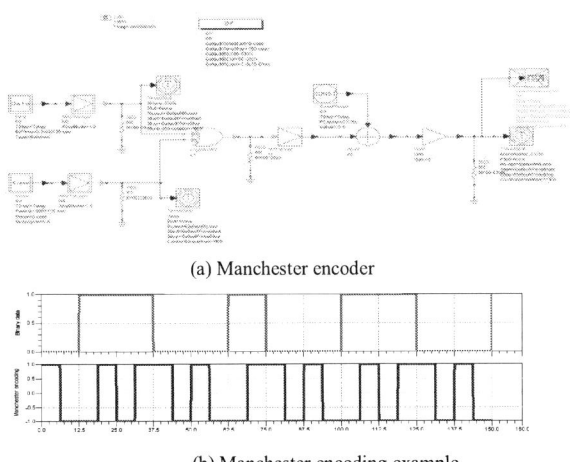

(a) Manchester encoder

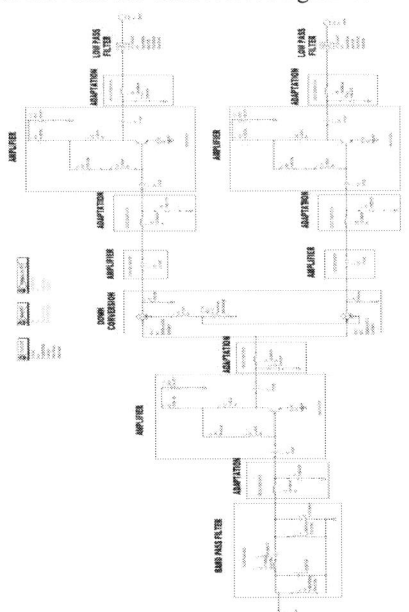

(b) Manchester encoding example

Fig. 3. Manchester encoder & encoding

B. The reader receiver

The basic receiver architecture for a RFID receiver is based on a direct-conversion I/Q demodulator. Fig. 4 sketches the radio frontend of the receiver. The received signal from the tag

Fig. 4. Receiver radio frontend

is directed first to a band filter, to remove signals laying outside the frequency band of interest. Then, the resulting signal is split and directed to two branches of demodulators. One branch has a mixer excited with the local oscillator signal and the other has a mixer excited with the local oscillator signal shifted by 90 degrees. The demodulated signal is then filtered to remove the carrier frequency and harmonics.

To explain how the receiver works, we present in Fig.5 the return link communication implementation (from the tag to the reader).

While the reader is listening for the tag response (return link), the reader transmitter must send the continuous wave (CW) to power the tag, and to backscatter the tag data encoded as FM0. It means that the tag will respond to reader, by sending over the CW, the data stored on it [6]. The communication channel is modeled as an additive white Gaussian noise channel to introduce noise sources from environment.

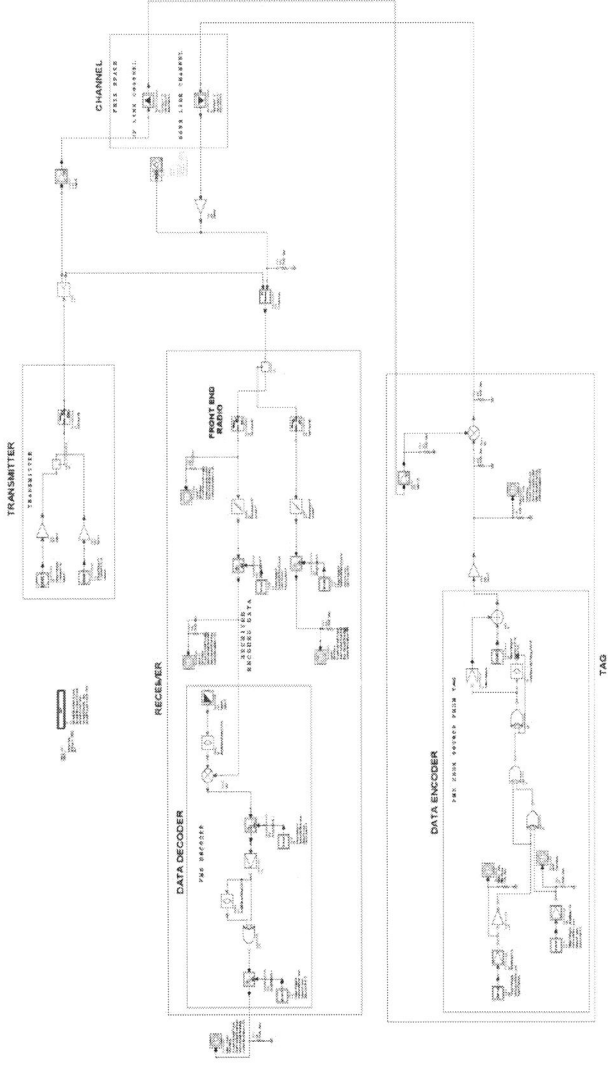

Fig. 5. Return link implementation

978-1-4673-5289-5/12 $31.00 © 2012 IEEE   157

C. The tag model

1) Signal detection: Generally, The passive tag is composed of two envelope detectors; each one includes a diode used to rectify the voltage and a capacitor used for signal smoothing. The first envelope detector has a big capacitance value that creates a constant voltage used to power the tag. The second one uses a smaller capacitance value to recover the information from the reader [4].

In Fig.6, we propose an implementation of the envelope detectors, the top envelope detector has a capacitance value of 1 *nF*, is used to recover the modulated transmitted signal. The second one on the bottom, has about 4.6 *μF* as capacitance value, and is used to recover the power source.

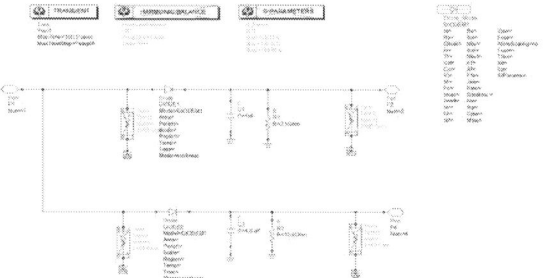

Fig. 6. Envelope detectors

2) Binary data encoding: For the return link, tags should encode the backscattered data as either FM0 baseband or Miller encoding [3]. Only FM0 encoding is presented in this paper.

In Fig.7(a), we propose a model of blocks for generating the FM0 encoding. The data-0 or data-1 should be encoded like the form in Fig.7(b). FM0 inverts the phase at every symbol boundary; a data-0 has an additional mid-symbol phase inversion [3].

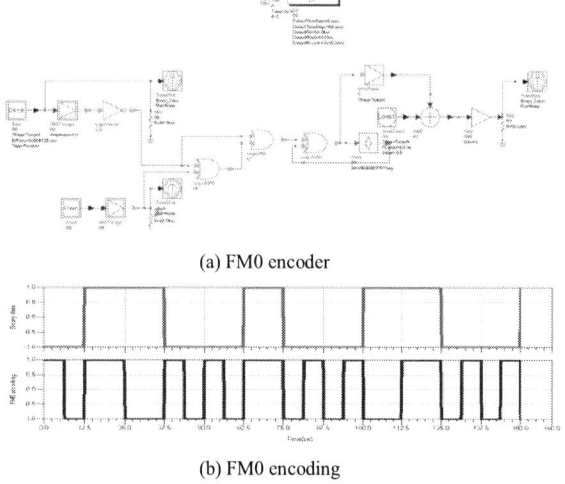

(a) FM0 encoder

(b) FM0 encoding

Fig. 7. FM0 encoder and encoding data

Every communication process between readers and tags are based on identification or authentication protocols. This part is

not included in this work, but we are interested on RFID tag signal in its general form.

D. Reader antenna

In this section, we propose the design of a compact printed dipole antenna for UHF RFID reader. Fig.8 gives the overall dimension of the antenna. Our dipole antenna mainly consists of two radiating arms separated by a 6 mm width slot. At the end of the slot, there is a 1 mm shorting strip, which connects two radiating arms. The antenna is a simple meandered dipole which occupies the volume of ($78 \times 30 \times 1.6$ mm^3). The width of the radiating arms is fixed to 4 mm as an optimized result.
The feeding of the antenna is done in A and B. The value of the distance between A and B is fixed to 6 mm to achieve a good impedance matching.

It is worth to mention that the length of the conventional dipole antenna is about half the resonant wave length that is 164 mm at 915MHz. Tthe size of the antenna is reduced by 54% compared to the size of a conventional dipole antenna.

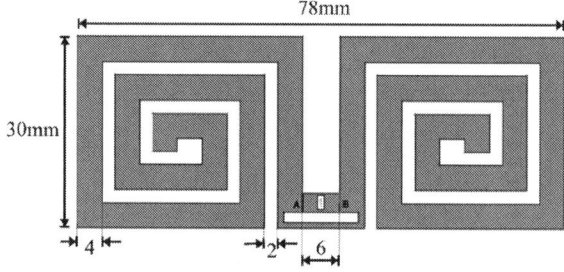

Fig. 8. Reader dipole antenna

For the demonstration purpose, the proposed antenna was designed on a FR4 epoxy substrate with a dielectric constant 4.32 and loss tangent of 0.017. In this case, the final antenna parameters are optimized using the commercial electromagnetic (EM) IE3D simulator. Fig. 9 shows the antenna response behavior at 915 MHz with return loss about -27 dB.

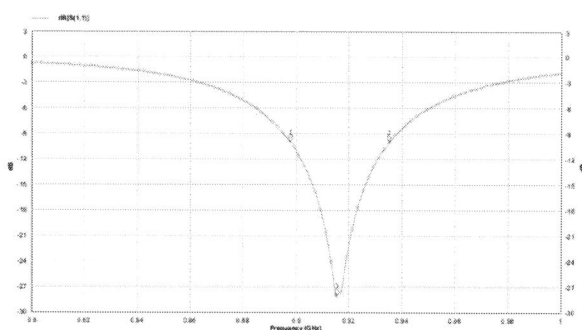

Fig. 9. Reader dipole antenna return loss

Finally, based on the antenna's results, it can be stated that the designed model of antenna is suitable for the RFID application.

978-1-4673-5289-5/12 $31.00 © 2012 IEEE

III. RFID IMPLEMENTATION RESULTS

In this section, some implementation results are presented for each part of UHF RFID system.

A. Implementation results of the transmitter

Figure 10 shows the DSB transmission of Manchester encoding with 80 kbits/s data rate for 30% modulation depth.

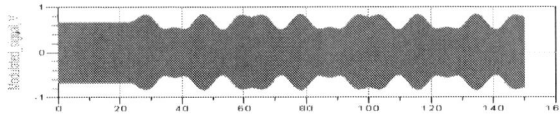

Fig. 10. DSB transmission of Manchester encoding (30% modulation depth)

The Gen 2 modulated signal shown in Fig.10 will be transmitted and will be recovered by the tag.

B. Implementation results of the reader receiver

Figure 11 shows the FM0 encoded data from the tag on the top, the demodulated signal just after the low-pass filter in the receiver, and in the bottom, the encoded received data.

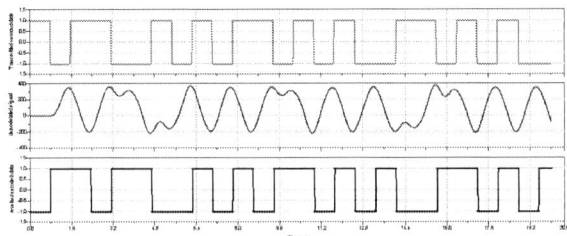

Fig. 11. FM0 demodulated signal and encoded received data

Bit Error Rate (BER) is used to evaluate the performance of the communication from the tag to the reader (return link communication). Figure 12 depicts the BER vs. SNR of the return link communication of FM0 encoded data, where the BER is inversely proportional to the SNR. From the Fig.12, the BER reaches 10^{-3} for an SNR of 9 dB.

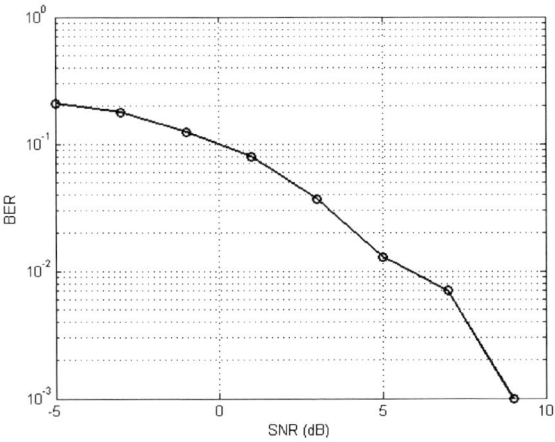

Fig. 12. BER performance of the return link communication

C. Implementation results of the tag

Figure 13 shows the envelope detection of the Manchester DSB modulated signal from the reader transmitter.

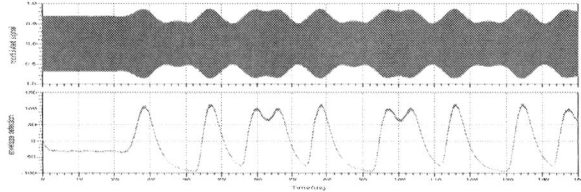

Fig. 13. Envelope detection

The detected envelope is then rectified by a limiter in order to get the received encoded data as shown in Fig.14. We present, on the top of Fig.14, the transmitted encoded data from the reader transmitter, and the bottom received encoded data by the tag. We can see that the tag can recover the same data as transmitted with a time delay.

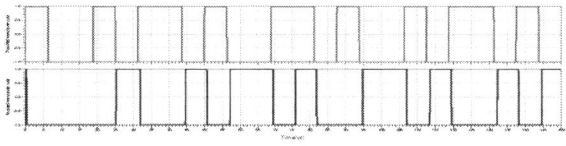

Fig. 14. Received encoded data

IV. CONCLUSION

In this paper, both RFID reader and tag is modeled and implemented using Agilent ADS environment. During the implementation, we proposed to use the co-simulation between the numeric front end and the RF front end for each part of the RFID system (reader and tag). Different blocks for encoding types are proposed. A low-cost reader antenna is also designed. From the obtained results, the received data by either the receiver or the tag are successfully and entirely demodulated. Finally, the implemented RFID system allows us to state that the passive UHF RFID system can be easily manufactured using low-cost components.

REFERENCES

[1] L. Junhuai, L. Hongying, J. Zhang, "Design and Implementation of a RFID-based Exercise Information System", Second International Symposium on Intelligent Information Technology Application, 2008.

[2] V. D. Hunt, A. Puglia and M. Puglia, RFID-A Guide to Radio Frequency IDentification, Wiley-Interscience, 2007.

[3] L. Jin, T. Cheng, "Analysis and Simulation of UHF RFID System", IEEE ICSP2006 Proceedings, 2006.

[4] D. Dobkin, "The RF in RFID Passive UHF RFID in practice", 2008.

[5] H. Yifeng, Hao Min, "System Modeling and Simulation of RFID", AUTO-ID LABS, white paper series/version1, September 2005

[6] "RFID: Radio Frequency IDentification: Applications and Implications for Consumers: A Workshop Report From the Staff of the Federal Trade Commission", DIANE Publishing, 2005

[7] http://www.agilent.com/find/eesof.

Pattern Nulling of Linear Array Antennas with Mutual Coupling Effects using Taguchi Method

A. Recioui and H. Bentarzi

Laboratory signals and systems, Institute of Electrical and electronic Engineering
University M'hamed Bougara
Boumerdes, Algeria.
a_recioui@umbb.dz

Abstract— **The purpose of the present work is to employ the Taguchi's optimization technique, which is a relatively new optimization technique, for null steering in the antenna radiation pattern by controlling only the element positions of a nonuniform linear array for a specified beamwidth and minimum achievable sidelobe level. As a practical part, an array of parallel dipoles is designed to account for the mutual coupling effects on the performance of the array in terms of directivity, sidelobe level and null placement. Illustrative examples are considered to impose single or multiple nulls in the antenna pattern at specific interference directions with the maximum side lobe level reduction of array antenna design.**

Keywords_ Antenna Arrays, Mutual Coupling, Optimization, Sidelobe Level, Taguchi method.

I. INTRODUCTION

An antenna array may be regarded as a spatial filter i.e. it allows signal from a certain direction to pass through while rejecting or stopping all other signal sources (from other directions) impinging on the array. To achieve this filtering objective, one would like to design an array with high gain at the desired signal direction (look direction) and low gain at the non look directions. Placing nulls in the array far field pattern at specified directions, is becoming important in communication systems to maximize signal to interference ratio. Another vital issue in modern wireless communication systems would be to reduce the amount of interference over the array aperture. This is because capacity of the system is directly proportional to the amount of interference that it can tolerate [2]. One characteristic of the radiation pattern that can be directly related to interference reduction is the level of sidelobes which determines the amount of interference the array picks up. Hence, this property is chosen as a performance indicator whenever interference reduction is a concern.

The control of the radiation pattern (Array synthesis) has its importance. To improve the radiation efficiency, this pattern must be oriented to the desired directions, while nulls (zero radiation energy) directed towards the interferers. Antenna array synthesis can be classified into several categories according to the desired radiation pattern characteristics. The very common required characteristics are: the null steering and direction control in the antenna radiation pattern, a desired distribution in the entire visible region, narrow beam width and low side lobes, decaying minor lobes [3]. Methods on null

steering include controlling the amplitude-only [4], the phase-only [5], the complex weights (both the amplitude and phase) [6, 7] and position-only [8] of the array elements have been dealt with in literature. The amplitude-only control approach suffers from the fact that it results in a complex feeding network that involves amplifiers and attenuators which complicates the design of the network. Furthermore, power management is another issue to consider. The phase-only control (phased arrays) has its power advantages, but use phase shifters in the feeding network which is hard to tune at the exact element phase. The complex weight approach suffers from both amplitude-only and phase-only problems. The position-only control; (considered in the present work), uses mechanical shifters to place the elements at their corresponding positions. It has been considered here for it allows the investigation of the mutual coupling effect.

In array design and synthesis, ideal isotropic elements are usually considered for they simplify the procedures. It turns out also that the losses are generally ignored. However, in practical use, different loss mechanisms are present and in more realistic models they need to be included. One common problem is the mutual coupling between the real elements of the array. Mutual coupling is a common problem in the applications of antenna arrays. It significantly affects the operation of almost all types of antenna arrays as the element patterns are perturbed, which causes difficulties in conventional array pattern synthesis and design methods. Compared with a single antenna, an antenna array is able to provide spatial information of the signal distributions. However, this function critically relies on the independence or distinctiveness of the signals received or transmitted from different antenna elements in the array. In reality, this is simply impossible because antenna elements will interact with each other, i.e., they will mutually couple with each other. Hence, this effect has to be taken into account in the optimization procedure for the designed array to be realistic and practical [3, 13].

Thanks to the rapid development of computer technology, many optimization techniques such as genetic algorithms, particle swarm optimization, simulated annealing, artificial neural networks and gradient-based techniques have been implemented in the form of computer codes [9]. Compared to other optimization techniques, Taguchi's optimization method is easy to implement and very efficient in reaching optimum solutions. In addition, it reduces the number of tests required in the optimization process compared to GA or PSO [9].

II. PROBLEM FORMULATION

A linear array with an even number of isotropic elements ($2N$). If the array elements are symmetrically situated and excited around the center of the linear array, the far field array factor of such an array is real, and can be written as [1]:

$$AF_a(\theta) = 2\sum_{k=1}^{N} a_k \cos[\frac{2\pi}{\lambda} x_k(\sin(\theta) - \sin(\theta_0)] \quad (1)$$

Where λ is the wavelength, θ is the scanning angle from broadside, θ_0 is the direction of the main beam ($0°$ for broadside), a_k is the amplitude of the kth element constrained to be 1 for all elements, and x_k is the position of the kth element with respect to the array center.

This array factor has to be normalized with respect to the highest value. To perform null placement in conjunction with the lowest peak SLL, the optimization objective function is chosen to be to contain two terms to account for null placement and sidelobe level reduction, respectively. The term that accounts for null placement is defined as:

$$F_1 = \int_0^\pi |AF_d(\theta) - AF_p(\theta)| d\theta \quad (2)$$

To achieve the lowest peak SLL, the second term in the optimization fitness function is chosen to be merely the sidelobe level defined as:

$$F_2 = SLL = \max[20 Log(AF(\theta))] \quad \textit{In the sidelobe region} \quad (3)$$

Then the fitness function to minimize is:

$$f = F_1 + F_2 \quad (4)$$

The symmetry dictates that the optimization can be done on half the array with the other half symmetrically constructed which reduces the problem of optimizing a $2N$ element array to N dimensions.

The positions are constrained to be such that the minimum distance between the elements is never less than 0.25λ with a maximum allowable spacing of λ. Hence, the problem can be stated otherwise as:

$$\text{Minimize } f \qquad \text{Subject to}$$
$$x_i \in \{0, x_{max}\} \text{ and } 0.25 < |x_i - x_j| < 1 \quad (5)$$
$$\min(x_i) = 0.125 \text{ for } i, j = 1..N \quad i \neq j$$

The mutual impedance mat3rix will have entries that are computed based on the induced EMF method presented in [3] and are given as:

$$Z_{mn} = \begin{cases} 30(0.5772 + j30 S_i(2kl)] & \text{if } m=n \\ 30[2 C_i(u_0) - C_i(u_1) - C_i(u_2)] - j30[2 S_i(u_0) - S_i(u_1) - S_i(u_2)] & \text{if } m \neq n \end{cases} \quad (6)$$

with
k: the wave number;
if d_h is the horizontal distance between the elements, then:

$$u_0 = kd_h \quad (7)$$

$$u_1 = k(\sqrt{d_h^2 + l^2} + l) \quad (8)$$
$$u_2 = k(\sqrt{d_h^2 + l^2} - l) \quad (9)$$

And $C_i(u)$ and $S_i(u)$ are cosine and sine integrals defined as [3]:

$$C_i(u) = \int_\infty^u \frac{\cos(x)}{x} dx \quad (10)$$

$$S_i(u) = \int_0^u \frac{\sin(x)}{x} dx \quad (11)$$

This mutual impedance matrix will alter the element excitations and hence changes the radiation pattern characteristics which affect the performance of the array. The change in excitation can be written as:

$$V = ZI \quad (12)$$

Where V is the voltage excitation vector; I is the original excitation vector (constrained to be 1 for all elements in this case). Z is the impedance matrix whose entries are defined by equation (6).

Mutual coupling effect implies that the element excitation currents will not be solely dependent on the element self impedance, but on the other elements' impedances as well.

Notice, the elements excitations are the same as the ones considered for isotropic radiators. Furthermore, the consideration of mutual coupling introduces a phase in the element's excitation (initially zero). This is taken into account by considering another formula for the array factor that is:

$$AF(\theta) = 2\sum_{k=1}^{N} (a_{kR} \cos(\varphi_k) - a_{kL} \sin(\varphi_k)) \quad (13)$$

With
$$\varphi_k = 2\pi x_k [\cos(\theta) - \cos(\theta_0)] \quad (14)$$

With the different terms defined as in (1). Notice that if the phase is zero ($a_{kL}=0$), equation (13) reduces to equation (1).

III. THE TAGUCHI METHOD

Taguchi's method was developed based on the concept of the orthogonal array (OA), which can effectively reduce the number of tests required in a design process [9]. It provides an efficient way to choose the design parameters in an optimization procedure.

Before presenting the Taguchi procedure, it is worth understanding what OAs are and how are they generated [9, 18, 19, 20]. Let S be a set of s symbols or levels (the simplest symbols are integers 1, 2, 3…). A matrix A of N rows and k columns with entries from S is said to be an OA *with s levels and strength t* ($0 < t < k$) if in every N×t subarray of A, each t-tuple based on S appears exactly the same times as a row. The notation OA(N, k, s, t) is used to represent an OA.

A Initialization procedure

The optimization procedure starts with the problem initialization, which includes the *selection of a proper OA and the design of a suitable fitness function*. The selection of an OA(N, k, s, t) mainly depends on the number of optimization parameters. In general, to characterize the nonlinear effect, three levels ($s = 3$) are found sufficient for each input parameter. Usually, an OA with a strength of 2 ($t = 2$) is efficient for most problems because it results in a small number of rows in the array [7].

B Design of input parameters

The input parameters need to be selected to conduct the experiments. When the OA is used, the corresponding numerical values for the three levels of each input parameter should be determined.

In the first iteration, the value for level 2 is selected at the center of the optimization range. Values of levels 1 and 3 are calculated by subtracting/adding the value of level 2 with a variable called *level difference* (LD). The *level difference* in the first iteration (LD1) is determined by the following equation:

$$LD_1 = \frac{Max - Min}{Number \ of \ levels + 1} \qquad (15)$$

Where *Max* and *Min* represent the upper and lower bounds of the optimization range, respectively.

C Conduct Experiments and Build a Response Table

After determining the input parameters, the fitness function for each experiment can be calculated. These results are then used to build a response table for the first iteration by averaging the fitness values for each parameter n and each level m using the following equation:

$$F_{av} = \frac{s}{N} \sum_{i, OA(i,n)=m} f_i \qquad (16)$$

D Identify Optimal Level Values and Conduct Confirmation Experiment

Finding the largest fitness value ratio in each column can identify the optimal level for that parameter. When the optimal levels are identified, a confirmation experiment is performed using the combination of the optimal levels identified in the response table. This confirmation test is not repetitious because the OA-based experiment is a fractional factorial experiment, and the optimal combination may not be included in the experiment table. The fitness value obtained from the optimal combination is regarded as the fitness value of the current iteration.

E Reduce the Optimization Range

If the results of the current iteration do not meet the termination criteria, the process is repeated in the next iteration. The optimal level values of the current iteration are used as central values (values of level 2) for the next iteration. To reduce the optimization range for a converged result, the LDi is multiplied with a reduced rate (*rr*) to obtain LD_{i+1} for the $(i + 1)^{th}$ iteration:

$$LD_{i+1} = rr \times LD_i = RR(i) \times LD_i \qquad (17)$$

Where $RR(i)$ is called *reduced function*. When a constant rr is used, $RR(i) = rr^i$. The value of rr can be set between 0.5 and 1 depending on the problem. The larger rr is, the slower the convergence rate.

If LD_i is a large value, and the central level value is located near the upper bound or lower bound of the optimization range, the corresponding value of level 1 or 3 may reside outside the optimization range. Therefore, a process of checking the level values is necessary to guarantee that all level values are located within the optimization range. If an excessive situation happens, reassigning the level value for the parameter will be performed. A simple way is to use the boundary values directly.

F Check the Termination Criteria

When the number of iterations is large, the level difference of each element becomes small from equation (17). Hence, the level values are close to each other and the fitness value of the next iteration is close to the fitness value of the current iteration. The following equation may be used as a termination criterion for the optimization procedure:

$$\frac{LD_i}{LD_1} < \text{converged value} \qquad (18)$$

Usually, the converged value can be set between 0.001 and 0.01 depending on the problem. The iterative optimization process will be terminated if the design goal is achieved or if equation (18) is satisfied.

IV. RESULTS AND DISCUSSIONS

Taguchi's method is used to optimize the positions of a 20-element array with the inter-element spacings restricted to be in the range $[\frac{\lambda}{4}, \lambda]$. Some examples illustrate the capability of the Taguchi's method for null control and side-lobe level reduction. The converged values (LD) and the reduced rate (rr) vary from one example to the other.

In the first example; it is attempted to place a single null in the radiation pattern in the direction 50°.

Fig. 1 shows the optimum designed non-uniformly spaced linear array's pattern with the null direction at 50° considering and ignoring mutual coupling effect. In this case, the converged value and the (rr) rate are set to be 0.0002 and 0.87, respectively. Ignoring mutual coupling, the null is placed exactly in the desired direction at a depth of –65 dB with a minimum achievable sidelobe level being –24.38 dB.

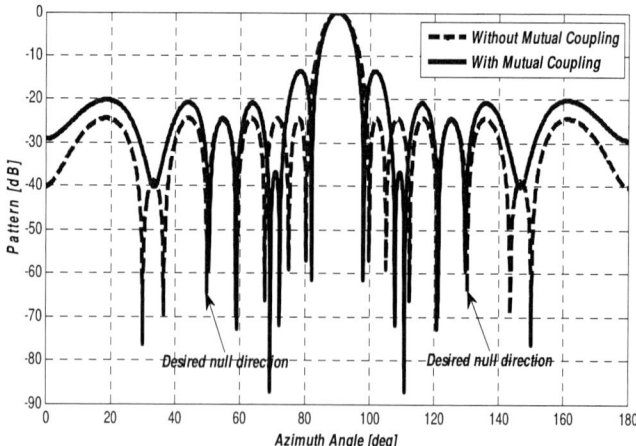

Fig. 1 Normalized pattern of 20 element linear array of half wave dipoles with a desired symmetrical null direction at θ =50° with and without mutual coupling

Considering the mutual coupling effect, it is seen that the null did not undergo any change in location. However, the depth has been reduced from –65 dB to –55 dB.

The sidelobe level also underwent a dramatic degradation and it jumped from –24.38 to –15.13 dB in the neighborhood of the main lobe and –20.15 starting from around 20° farther from the main lobe. This is a clear proof that the ignorance of mutual coupling is not acceptable for a better realistic linear array design.

In the next example, it is attempted to place two distinct nulls in the radiation pattern in the directions 33° and 55°. The converged value and the (rr) rate are set to 0.002 and 0.85, respectively. The algorithm converged and the optimum pattern is shown in fig. 2. It is noticed that the nulls have been placed exactly at the desired directions with a depth of –75 dB for the 33° null and –60 dB for the 55° null. The sidelobe level achieved in this case is at –20.67 dB down the main lobe.

The inclusion of mutual coupling into the array pattern calculation showed a similar behavior to the first example. The null locations have been slightly altered with their depths reduced from –75 dB to –63 dB for the 33° null and from –65 dB to –55 dB for the 55° null. The overall sidelobe level increased from –20.67 dB to –14.05 dB which is not convenient for communication systems.

V. CONCLUSION

In this piece of work, an attempt to demonstrate how important it is to consider one effect; usually ignored in literature, in any design of linear array antennas. The mutual coupling effect has been shown to dramatically alter the far field radiation pattern as it changes the element excitations due to neighboring elements.

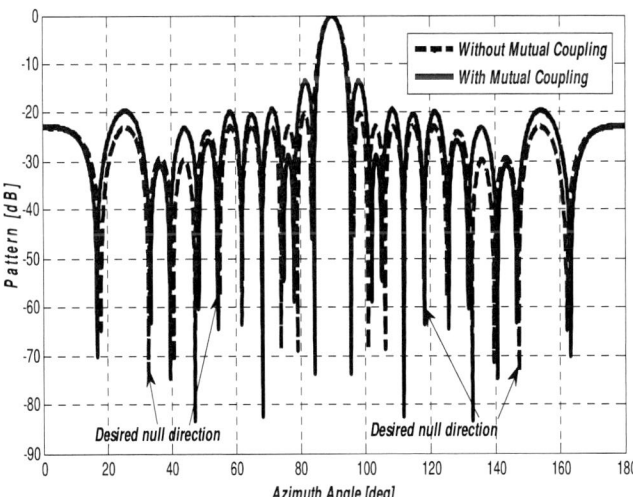

Fig. 2 Normalized pattern of 20 element linear array of half wave dipoles with two desired symmetrical nulls located at θ =33°and θ =55 with and without mutual coupling

The use of the Taguchi method has been to prove how powerful is this technique of optimization in reaching the global optimum solution of any design that can be modeled as a constrained optimization problem. The investigation carried out in this piece of work can be conducted to any design including antenna arrays with any elementary antenna type. The presence of analytic expressions for a side-by-side array of half wave dipoles made it easier to assess the effects of mutual coupling on the performance of the array.

REFERENCES

[1] B. Babayigit, K. Guney and A. Akdagli, 2008 "A Clonal Selection Algorithm for Array Pattern Nulling by Controlling the Positions of Selected Elements", Progress In Electromagnetics Research B, Vol. 6, 257–266.

[2] Guney K. and Akdagli A, 2001"Null steering of linear antenna arrays using a modified Tabu search algorithm", Progress in Electromagnetics Research (PIER), Vol 33.

[3] Constantine A. BALANIS, 2005 "Antenna theory: analysis and design" Third edition, John Wiley & sons Inc.,.

[4] Dale J. Shpak, 1996 "A method of optimal pattern synthesis of linear arrays with prescribed nulls", IEEE transactions on antenna and propagation, Vol. 44, No.3.

[5] M. Mouhamadou and P. Vaudon, 2006, "Smart Antenna Array Patterns Synthesis: Null Steering and Multi-user Beamforming by Phase Control", Progress In Electromagnetics Research, PIER 60, 95–106.

[6] A. Recioui, A. Azrar, 2007, "Use of Genetic Algorithms in Linear and Planar array synthesis based on Schelkunoff method", Microwave and optical technology letters, Vol. 49, Issue 7.

[7] Abdelmadjid Recioui, Arab Azrar, Hamid Bentarzi, Mokrane Dehmas & Mouloud Challal, "Synthesis of Linear Arrays with Sidelobe Level Reduction Constraint using Genetic Algorithms", 2008, international journal of microwave and optical technology, Vol. 3, No. 5.

[8] Khodier MM and Christodoulou CG, 2005, "Linear array geometry synthesis with minimum sidelobe level and null control using particle swarm optimization", IEEE Trans Antennas Propag. Vol. 53, pp 2674–2679.

[9] Wei-Chung Weng, Fan Yang and Atef Elsherbeni, 2007, "Electromagnetics and Antenna Optimization Using Taguchi's Method", Morgan and Claypool Publishers.

[10] Er MH, "Linear antenna array pattern synthesis with prescribed broad nulls", 1990, IEEE Trans Antennas Propag. Vol. 38,. pp. 1496–1498

[11] Thomas A. Milligan, "Modern Antenna Design", 2005, John Wiley & Sons Inc,.

[12] Marco A. Panduro, David H. Covarrubias, Carlos A. Brizuela and Francisco R. Marante, "A multi-objective approach in the linear antenna array design", Int. J. Electr. Comm., (AEU) 59 205-212.

[13] M. Rattan, M. S. Patterh and B. S. Sohi, 2008, " Design of a Linear Array of Half Wave Parallel Dipoles using Particle Swarm Optimization", Progress In Electromagnetics Research M, Vol. 2, 131–139,.

[14] Hon Tat Hui, 2007, " Decoupling Methods for the Mutual Coupling Effect in Antenna Arrays: A Review", Recent Patents on Engineering, 1, 187-193, Bentham Science Publishers Ltd.

[15] S. Durrani and M. E. Bialkowski, 2004, "Effect of mutual coupling on the interference rejection capabilities of linear and circular arrays in CDMA systems", IEEE Trans. Antennas Propagat., Vol. 52, no. 4, pp. 1130– 1134, Apr..

[16] Weng, W., F. Yang, and A. Elsherbeni, 2007, "Linear antenna array synthesis using Taguchi's method: A novel optimization technique in electromagnetic", IEEE Trans on Antennas and Propagation, Vol. 55, 723-730.

[17] Wei-Chung Weng and Charles T. M. Choi, 2009; "Optimal Design of CPW Slot Antennas Using Taguchi's Method", IEEE Transactions on magnetic, Vol. 45, NO. 3.

[18] A. S. Hedayat, N. J. A. Sloane, and J. Stufken, 1999, "Orthogonal Arrays: Theory and Applications", Springer-Verlag: New York.

[19] N. Dib, S. K. Goudo, H. Muhsen, 2010, "Application of Taguchi's optimization method and self adaptive diffrrential evolution to the synthesis of linear antenna arrays", Progress In Electromagnetics Research, PIER 102, 159-180.

978-1-4673-5289-5/12 $31.00 © 2012 IEEE

Fast and Accurate Analysis Method of a Circular Patch Antennas using Neurospectral Method

N. Hamdiken, T. Fortaki
Advanced Electronics Laboratory
University of Batna
05000, Batna, ALGERIA
hamdikenazih@yahoo.fr

R. Addaci
LEAT, University of Nice-Sophia Antipolis-CNRS,
Bât. Forum, Campus Sophi@Tech,
930 Route des Colles, 06903 Sophia Antipolis, France
rafik.addaci@unice.fr

Abstract—**In this paper, fast and accurate analysis method is used to calculate the real and imaginary part of the resonant frequency of a circular patch antenna is presented. This method is based on artificial neural network (ANN), that is offered a very small computation time compared with other methods used to model this antenna such as Chew, Howell, Wolff, Demeryd, Nirun Kumprasert. The results of the real and imaginary part of the resonant frequency obtained by the ANN method for the circular antenna are in very good agreement with the experimental results available in the literature.**

Keywords-component; Artificial neural networks, Circular patch antenna, Microstrip antenna, Modeling method, Neurospectral technique.

I. INTRODUCTION

During the last decade, statistical learning algorithms have generated much interest in academic and industrials domain. They have been successfully implemented as powerful tools for the accomplishment of modeling and prediction tasks. The artificial neural network is one of these statistical learning algorithms which are firmly implanted in various industries domain to solve a variety of problems such as models identification (pattern recognition), prediction, optimization and control [1]-[2].

Considered as an effective tool for nonlinear approximation, the ANN technique is used in this work, to calculate the real and imaginary part of the resonant frequency of a circular microstrip antenna, depending on the radius of the patch, the high and the relative permittivity of the used substrate.

Microstrip antennas have been extensively studied experimentally, analytically and numerically. Many numerical methods have been served engineers and researchers in the analysis and design of these antennas for several years [3-7]. These methods, with different levels of complexity, require different computational efforts, and can generally be divided into two groups: simple analytical methods and rigorous numerical methods. The simple analytical methods can give a good intuitive explanation of the radiations properties of the antenna. The exact mathematical formulations in rigorous methods involve extensive (complex) numerical procedures.

They are also long and not easily included in a computer aided design system (CAD).

To overcome these limitations and thereby reduce significantly the computation time, we present a new approach which used neural networks in conjunction with the spectral technique for the analysis of microstrip antennas. This method named: neurospectrale method, allows precise determination of the dyadic Green's function associated with the integrated structure [8].

To be validated, the results obtained by this new approach are compared with those of the spectral approach (SDA) and other theoretical and experimental results of the literature.

II. IMPLEMENTATION NEUROSPECTRAL TECHNIQUE STRATEGY

A. Circular microstrip antenna parameters

Microstrip antennas have drawn considerable attention of researchers in last decades because their many unique and attractive properties, such as low weight, small size, low cost manufacturing, polarization diversity, compatibility with hybrid circuits and MMIC (Microwave and Monolithic Integrated Circuits). Fig. 1 shows the geometrical configuration of a circular microstrip disk antenna.

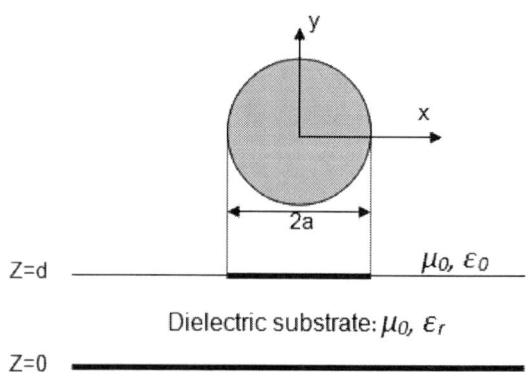

Figure 1. Geometry of a circular microstrip antenna.

A perfectly conducting circular disk is placed on the top of a dielectric substrate backed by a perfectly conducting ground plane. The disk has a physical radius "a". The dielectric substrate has a relative dielectric constant of "ε_r" and thickness "d". This shape of microstrip antenna is chosen in this work because of its inherent importance and because it is largely used in antenna arrays.

B. Neurospectral model development

Artificial neural networks (ANN) play an important role in optimization technique of a large electromagnetic structure [1]-[2], and multilayer perceptrons (MLP) network model, which are among the simplest architectures and then the most used, have been adapted for the calculation of the complex resonance frequency of a circular microstrip antenna. The MLP can be trained by the use of multiple algorithms. In this work, the back propagation algorithm is used for training the MLP [9].

The network model used for calculating the complex resonance frequency of the proposed antenna is shown in Fig.2. The neurons "a", "d", "ε_r" in the input layer distribute the signals to the hidden layer neurons. The transfer functions of the hidden layers used is the sigmoid function (SF), by cons the output layer is based on a linear transfer function (LF).

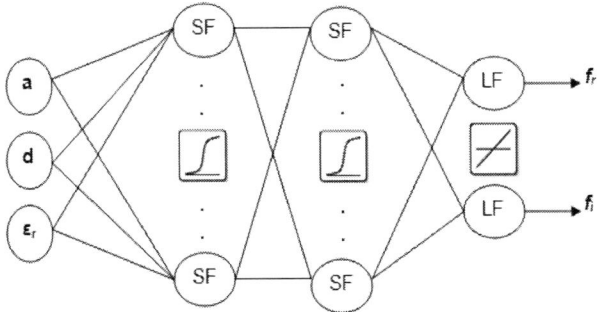

Figure 2.   The Shock and the transfer functions of each layer.

In order to generate network learning, it is necessary to create a training database, which must contain both the network input and desired output.

The databases must undergo a preprocessing to be adapted to the input neurons and make more efficient the training of the ANN, and this by removing the artificial discontinuities in the input function space. Then we must make an appropriate normalization, which takes into account the vales of the magnitude accepted by the network.

In our case, the database contains 931 examples obtained from the spectral domain approach (SDA) method. This database is divided into two sets; 810 example for training and 121 examples for the test which are reserved for the evaluation of the final network performance.

Once the two databases are created, the learning of the network is done using MATLAB software [10]. The possibilities, offered by this software, on learning methods are very extensive. In this work, the used method is the Bayesian regularization back-propagation algorithm.

III.   RESULTS AND DISCUSSION

The optimized model has two hidden layers of 5 and 7 neurons respectively. This configuration gives a best performance while maintaining a fast calculation times. This model is verified for the values of "d", "a" and "ε_r" comprised in the data space reported into the Table I.

TABLE I.   OPTIMAL PARAMETERS OF THE ANN MODEL

Number of neurons	Input		3	
	Hidden 1		5	
	Hidden 2		7	
	Output		2	
Data limits		d(cm)	a(cm)	ε_r
	Min	0.02	0.5	1
	Max	1.2	5	10
Tolerance		<10-3		

By using this method we have calculated the complex resonant frequency in function of the ratio (d/a). Table II summarizes the results of the real part of resonant frequency obtained by the different modeling techniques [3-7] and by the proposed method for different combinations of "a", "d", "ε_r". The results obtained by the neurospectrale method are in very good agreement with those obtained by the existing methods [3-7].

However, the method of Chew and those with which we have compared the results of the neurospectrale technique, require a ratio of (d/a) not exceeding a certain value; 0.24 for Chew for example. This limitation is not presents on the technique proposed in this work. This is because the neural model creates a cartography showing the relation between inputs and outputs of the network based on the database used during the learning and thus can give the corresponding output for any d/a and "ε_r" combination.

TABLE II.   THE RESONANT FREQUENCY: COMPARISON OF THE NS RESULTS WITH THOSE OBTAINED USING [3], [6] AND SDA METHODS FOR DIFFERENT CONFIGURATIONS

N° of example	a (cm)	d (cm)	ε_r	f_r(SDA) GHz	f_r(Ch) GHz [3]	f_r(De) GHz [6]	f_r(NS) GHz
1	3.493	0.1588	2.5	1.559	1.557	1.537	1.554
2	1.27	0.0794	2.59	4.187	4.183	4.159	4.185
3	3.493	0.3175	2.5	1.528	1.527	1.478	1.521
4	2.99	0.235	4.55	1.36	1.35	1.332	1.355
5	2	0.235	4.55	2.012	2.01	1.965	2.013
6	1.04	0.235	4.55	3.737	3.733	3.661	3.734

N° of example	a (cm)	d (cm)	ε_r	f_r(Ho) GHz [4]	f_r(Wo) GHz [5]	f_r(Ni) GHz [7]
1	3.493	0.1588	2.5	1.58	1.569	1.555
2	1.27	0.0794	2.59	4.29	4.267	4.175
3	3.493	0.3175	2.5	1.58	1.526	1.522
4	2.99	0.235	4.55	1.379	1.384	1.358
5	2	0.235	4.55	2.061	2.067	2.009
6	1.04	0.235	4.55	3.963	3.95	3.744

To show the validity of our method, we have traced in Figure 3, the relative error obtained by the "NS", "Ch" and "De" methods calculated by taking the "SDA" as reference method ("Ho", "Wo" and "Ni" methods are not taken into account in order to not overload the figure).

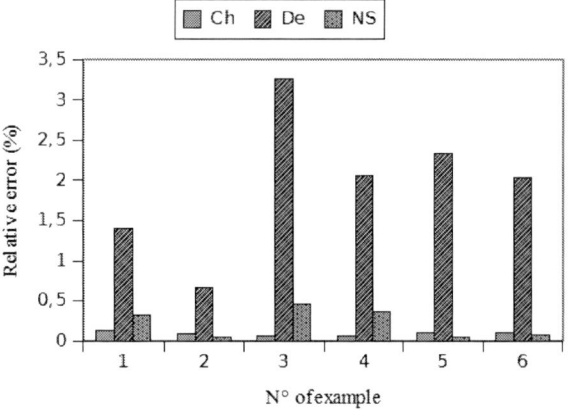

Figure 3. Comparison of the relative error between Ch, De and NS methods.

IV. CONCLUSION

The neuronal method presented in this study has a high accuracy and requires no complicated mathematical function. While the learning takes approximately five minutes, after this, the proposed method is capable to predict and provide promptly, in few microseconds, the desired outputs. The comparison of the obtained results with those obtained by measurements or by others simulated methods prove that this method can be very useful for the development of fast algorithms for computer aided designs (CAD).

REFERENCES

[1] R.E. Uhrig, "Introduction to artificial neural networks", Proceedings of the 21st international conference on industrial electronics, control, and instrumentation, vol. 1, pp. 33-37. 6-10 November 1995.

[2] G. Antonini and A. Orlandi, "Gradient evaluation for neural-networks-based electromagnetic optimization procedures", IEEE Transactions on microwave theory and techniques, vol. 48, no. 5, pp. 874-876, may 2000.

[3] W.C. Chew and Q. Liu, "Curve-fitting formulas for fast determination of accurate resonant frequency of circular microstrip patches", IEE Proceedings H Microwaves Antennas and Propagation, vol.135, no.5, pp. 289-292, October 1988.

[4] J. Q. Howell, "Microstrip antennas", IEEE Trans. Antennas Propagat. vol. 23, no. 1, pp. 90-93, January 1975.

[5] I. Wolff and N. Knoppik, "Rectangular and circular microstrip disk capacitors and resonators", IEEE Trans. on Microwave Theory and Tech, vol.22, no.10, pp. 857-864, October 1974.

[6] A. G. Demeryd, "Microstrip disc antenna covers multiple frequencies", Microwave Journal. vol. 21, pp. 77-79, May 1978.

[7] N. Kumprasert and W. Kiranon, "Simple and Accurate Formula for the Resonant Frequency of the Circular Microstrip Disk Antenna", IEEE Transactions on Antennas and Propagation, vol. 43, no. 11, pp. 1331-1333, November 1995.

[8] R.K. Mishra, A. Patnaik, "Designing Rectangular patch antenna using the neurospectral method", IEEE Transactions on Antennas and Propagation, vol. 51, no. 8, pp. 1914-1921, August 2003.

[9] M.H. Hassoun, Fundamentals of artificial neural networks, New Delhi, Prentice Hall of India, 1999, Chapter 8.

[10] Matlab 7.9.0.529.

Efficient ECC Implementation Architecture Suitable for RFID Technology

Mustapha Benssalah
Ecole Militaire Polytechnique
Communication System Laboratory
BP 17 Bordj El Bahri, Algiers, Algeria
Email: bensmusta@gmail.com

Mustapha Djeddou
Ecole Militaire Polytechnique
Communication System Laboratory
BP 17 Bordj El Bahri, Algiers, Algeria
Email:djeddou.mustapha@gmail.com

Karim Drouiche
LPTM, Cergy Pontoise University
2,av. Adolphe Chauvin BP 222 Pontoise
Cergy Pontoise CEDEX, France
kdrouiche@free.fr

Abstract—In this paper, we provide an investigation on the feasibility of embedding the ECC authentication service in RFID chip-set. Thus, we deal with the implementation of the most costly operation, which is the scalar point multiplication. It is also considered as the core of the security of the ECC protocols. Two architectures are studied and implemented using the Montgomery algorithm adopting the affine and projective coordinates. According to the obtained results, we show that the scalar point multiplication implementation using affine coordinates is efficient in terms of speed and area than the projective coordinates architecture.

I. INTRODUCTION

Most of the practical authentication protocols, used in the Radio Frequency Identification (RFID) systems, are lightweight protocols that use simple functions like PRNG, CRC and XOR. This is the case of some standards developed by the EPC-global such as EPC Class 1 Generation 2, which meet the RFID requirements but unfortunately provides little security [1]. An example of an RFID system is illustrated in Figure 1.

On the other hand, with the increasing of counterfeiting problem and attacks in this filed, high security level is an urgent need. Recently, many attempts to introduce the Elliptic Curves Cryptography (ECC) authentication process in this technology have been proposed [2],[3]. The choice of ECC is dictated by their short key sizes compared to other cryptographic existing system like RSA [2]. In fact, implementing ECC on such a small device with limited power budget is a challenging task. In this paper, we provide a survey of the implementation of Montgomery scalar point multiplication algorithm using affine and projective coordinates representations, where we show that the affine coordinates implementation is advisable for the miniaturized devices than projective coordinates, although researchers in literature encourage the use of projective coordinates. This paper is organized as follows: In section 2, the RFID and ECC protocols requirements are described. Section 3 describes the used algorithm and the system coordinates representations. Section 4 is devoted to the implementation of scalar point multiplication using Montgomery algorithm In section 5, some concluding remarks are drawn.

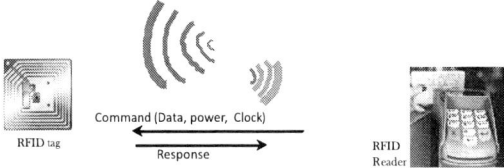

Fig. 1. RFID system principle

II. RFID & ECC PROTOCOLS REQUIREMENTS

In this section, we summarize the RFID tags challenging implementation constraints and some computational problems over the elliptic curve group. Our attention is directed to those are frequently used to design secure cryptographic protocols, for which there is no algorithm to be able to solve any of them up to date [4].

1- The RFID tag constraints are as follows: the storage area (the chip presents a limited memory, around of 12 k equivalents gates), computational resources (computational power, energy power), communication cost (bandwidth).

2- The used protocol must be scalable so that the server must be able to identify a large number of tags using the same reader, on the other hand, the protocol should include a robust algorithm to fight against all possible attacks using one of the following ECC problems [2]:

a- Elliptic curve discrete logarithm problem (ECDLP): Given two points P and Q over an elliptic curve $E_P(a,b)$ that, the ECDLP is to find an integer k given $Q = kP$, which is difficult.

b- Computational Diffie-Hellman problem (CDHP): Given three points P, sP and mP over $E_P(a,b)$, the CDHP is to find the point smP over $E_P(a,b)$, which is hard.

c- Decisional Diffie-Hellman problem (DDHP): Given four points $(P, sP, mP$ and $tP)$, decide whether or not $tP = smP$, i.e. decide $t = sm$ mod$p$ or not.

d- Elliptic curve factorization problem (ECFP): Given two points $(P, Q) \in E_P(a,b)$ where $Q = sP + tP$. The ECFP is to find the point sP and tP over $E_P(a,b)$ which is impossible.

978-1-4673-5289-5/12 $31.00 © 2012 IEEE

III. ALGORITHMS AND SYSTEM COORDINATES REPRESENTATIONS

As seen in the previous section, the ECC protocols are based on one or more of the fourth cited ECC problems ECDLP, ECDH, ECFP and ECHP which are based on the scalar point multiplication operation. Therefore, there exist different algorithms to implement the scalar point multiplication. We may cite: binary method, Frobenuis map method, sliding window methods, signed digit representation and combination of these methods as described in [4]. The different methods are summarized in the Figure 2. On the other hand, these algorithms can be coded using different coordinates representations, we can cite: affine and projective coordinates. The table I summarized the most used projective coordinates representations.

Representation	Mapping to affine coordinates
Projective	$x = X/Z,\ y = Y/Z$
Jacobian	$x = X/Z^2\ y = Y/Z_3$
Lopez-Dahab	$x = X/Z,\ y = Y/Z^2$

TABLE I
TYPES OF PROJECTIVE MAPPINGS [4].

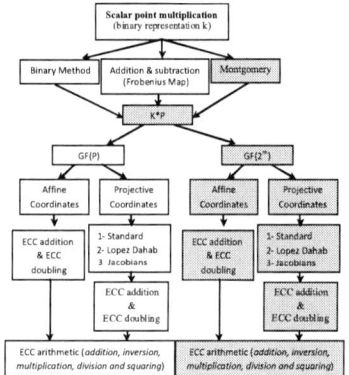

Fig. 2. Point multiplication computation methods

For our implementations, we adopt Montgomery algorithm because it is an efficient algorithm which offers a competitive performance in comparison to non-window methods for scalar multiplication [4]. Montgomery algorithm has the added advantage that it computes only over the x-coordinates in each iteration and hence requires lesser storage area. Another advantage, is that Montgomery algorithm is secure against timing and power attacks due to the regular structure of the instructions used, which is independent of the bit of the scalar.

IV. SCALAR POINT MULTIPLICATION IMPLEMENTATION

The efficiency of the elliptic curve cryptosystem implementation is determined by the efficiency of the most costly operation which is the scalar point multiplication. In this paper, we provide an investigation into the feasibility of introducing the ECC authentication in RFID systems. Therefore, we studied and compared the implementation performances of two

scalar point multiplication architectures using Montgomery algorithm for affine and projective coordinates representations. The main aim is to define the best architecture that satisfies the RFID tag requirements i.e, a very small chip, a few clock cycles and a low power consumption. For the two considered architectures, we do not convert back results to affine representation in the tag, since this conversion can be done by the terminal (reader) having more computing power compared to the tag.

A. Montgomery implementation using projective coordinates

The first architecture design uses projective coordinates representation of points on the elliptic curve, and it adopts Montgomery Lopez-Dahab algorithm (M-LD in short) [4]. Thus, the projective coordinates' representation is adopted to avoid inversion operation in each iteration of the Montgomery algorithm.

Knowing that the Montgomery algorithm structure in both coordinates representations consists of a succession of point addition and doubling, which are performed by the arithmetic operation such as multiplication, division and squaring. We have implemented these operations in the most optimized structures given in the literature combined with optimal FSM (Finite State Machine) over $GF(2^m)$.

For this architecture, we have considered five (05) temporary registers (X_A, Z_A, X_B, Z_B, T_1, and T_2) for the execution of the M-LD algorithm. The FSM structure for the execution of this algorithm is illustrated in Figure 3.

B. Montgomery implementation using affine coordinates

The second architecture uses affine coordinates representation which allows the use of less storage of points and temporary registers (area), at the expense of an additional inverter (or divider), which is a costly operation. For this second architecture, in order to optimize the different operations and the VHDL code, we have implemented one component for the two mains operations: point addition and doubling selected by a switch "operation". The second FSM for the execution of the point addition and doubling is nested in the first one. Their schemes are given in the Figure 4 and 5. (dashed steps and arrows are eliminated in our implementation).

The different architectures are coded using VHDL language and are synthesized using ISE Xilinx 10.1 software, using a Virtex 4 (XC4VLX200 speed-5). The main arithmetic units used in these implementations are: the bit serial multiplier, binary algorithm polynomial and squarer units over $GF(2^m)$.

The bit-serial multiplier is considered as the simplest solution which requires the least area comparing to the digit-serial multiplier. For more details of the multiplier core see [3]. The architecture of this multiplier has a smaller critical path delay, and his complexity is: $mAND + (m + t - 1)$ XOR gates and $3m$ Flip-flops (where $t = 3$, for a trinomial reduction polynomial and $t = 5$, for a pentanomial reduction polynomial). The latency for the multiplier output is m clock cycles [3]. For the squaring operation, we have considered the squarer proposed in [3] and [5], which is a fast squarer with

978-1-4673-5289-5/12 $31.00 © 2012 IEEE

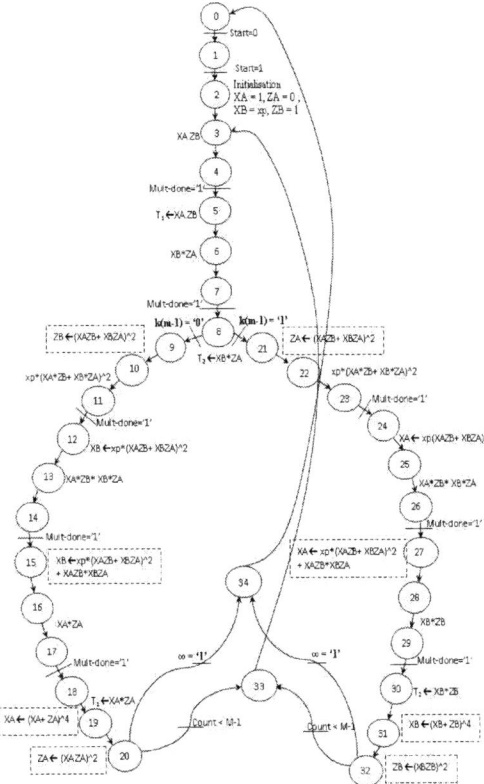

Fig. 3. Montgomery algorithm projective coordinates FSM

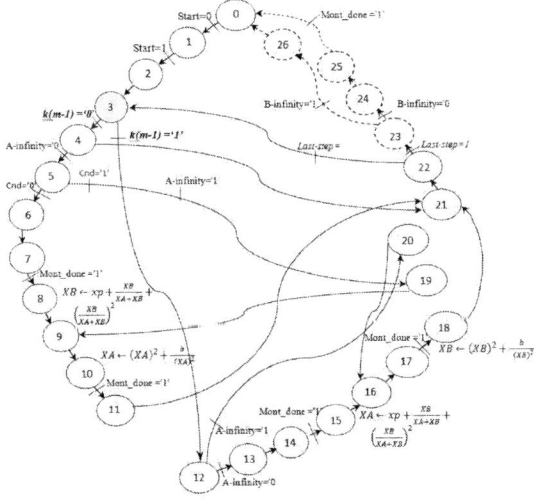

Fig. 4. Montgomery algorithm affine coordinates FSM

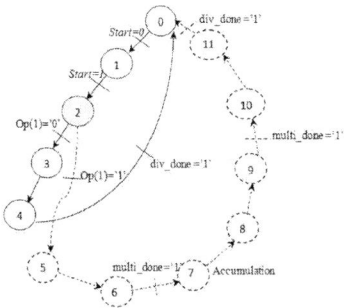

Fig. 5. Point addition and doubling shared FSM

fields that provide short-term security to a high security level. We consider the binary fields recommended in the standards SECG and NIST [6]) which are: $GF(2^{113})$, $GF(2^{131})$ and $GF(2^{163})$. Their corresponding reduction polynomials are presented below:

- SECG, filed size (113 bits), $f(x) = x^{113} + x^9 + 1$;
- SECG, filed size (131 bits), $f(x) = x^{131} + x^8 + x^3 + x^2 + 1$;
- SECG, filed size (163 bits), $f(x) = x^{163} + x^7 + x^6 + x^3 + 1$.

The obtained implementation results for the two considered architectures are summarized in Tables II and III.

Resources	GF(113)	GF(131)	GF(163)
Slices	773	960	1195
Flip Flops	846	972	1198
LUTs	1414	1727	2185
Frequency Max(MHz)	160	157	167
Execution time(μs)	319.3	437.2	636
Equiv.Gates (k)	8.484	10.362	13.110

TABLE II
IMPLEMENTATION RESULTS FOR THE THREE CONSIDERED BINARY FIELDS
(AFFINE COORDINATES)

Resources	GF(113)	GF(131)	GF(163)
Slices	1209	1495	1708
Flip Flops	1183	1365	1522
LUTs	2235	2720	3101
Frequency Max(MHz)	157	160	172
Execution time(μs)	406.7	536.3	772
Equiv.Gates (k)	13.410	16.320	18.606

TABLE III
IMPLEMENTATION RESULTS FOR THE THREE CONSIDERED BINARY FIELDS
(PROJECTIVE COORDINATES)

By analyzing the obtained results (Tables II and III, figures 6 and 7), we note that:

-Although the literature states that affine coordinates implementations are not usually suggested for constrained devices due to the need for an inverter or divider, we show that, through the use of optimized arithmetic components (fast squarer, divider) and efficient FSM, the architecture using affine coordinates outperforms the second architecture (using

a single clock cycle latency. This squarer is implemented as a non-clocked component with a hard wired XOR circuit, see [5] for more details. For the divider, we have considered the binary algorithm polynomial, which is based on polynomial division in binary arithmetic, for more details, see [4].

Now, we test our designs with some standardized binary

Architecture	FPGA Device	Area (Slices)	Frequency (MHz)	Cycles	Time (μs)
Chelton [7]	XC4VLX200	16209	153.9	3010	19.55
Machhout [8]	Virtex II pro	12861	167.83	-	2070
Jarvinen [9]	XC2V8000-5	18079	90.2	9561	106
Bencherif [10]	XC5VSX95T-1136	-	561	3468	6.1799
Montgomery Projective-Coord	XC4VLX200	1708	172	132850	772
Montgomery Affine-Coord	XC4VLX200	1195	167	106280	636

TABLE IV

COMPARISON OF PROPOSED APPROACH WITH ALTERNATIVE ARCHITECTURES IN THE LITERATURE.

projective coordinates) in terms of speed and area requirements, for all considered Galois field (113, 131 and 163 bits),
knowing that, the projective coordinates representation is the
most recommended in literature. This important result can
be explained by the fact that, in affine coordinates, we use
less storage of points and temporary registers (02 registers to
store the point coordinates (x,y) and 02 registers for temporary
variables). Comparing to the projective coordinates, we use not
less than 05 registers for temporary variables and 03 registers
for used points, knowing from previous implementations ([7],
[2]), that the memory requirements for storage of points and
temporary variables can contribute substantially more than
50% to the overall size of the RFID tag.

As illustrated in the Figure 6, the first architecture (affine) is
faster than the second, which can be explained by the fact that
in the affine representation, in each iteration of the algorithm,
we execute 02 divisions, 02 squaring and 03 additions. On
the other side for projective representation, we execute 06
multiplications, 04 squaring and 03 additions in each iteration.

For the impact of GF size on the system, clearly, when
we reduce the field sizes, we reduce the processing workload
(chip size and speed). As a result, we reduce the security level
of the ECC system (as illustrated in Figures 6 and 7). We
conclude that, if we have serious constraints, we can use, for
example, $GF(2^{113})$ at the expense of a short term security for
the application. This latter choice use about $8.488K$ equivalent
gates and a speed of $319\mu s$ (case of affine coordinates).

Table IV collects some FPGA based architectures available
in the literature for $GF(2^{163})$. A strict comparison is not
entirely straightforward because of the large existing variety
of FPGAs, curves, and implemented features. We can compare
our results to [7] which uses the same hardware target.
Despite that, the implementation strategy is different (speed
or speed/area); our architecture outperforms this architecture
in terms of slices (equiv. gates).

V. CONCLUSION

In this paper, two efficient architectures for the scalar
point multiplication using affine and projective coordinates
are presented. We have shown that the affine coordinates
implementation is recommended for constrained devices like
RFID tags, through the use of optimized arithmetic components and efficient FSM. The obtained results for the affine
coordinates representation can fulfill easily the RFID challenging constraints in terms of memory requirement (storage

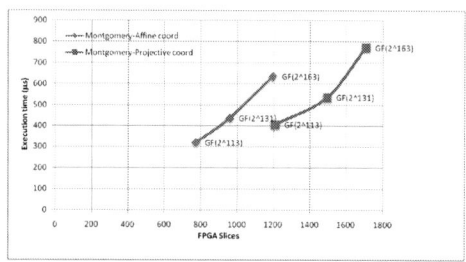

Fig. 6. Execution time versus FPGA versus memory requirement

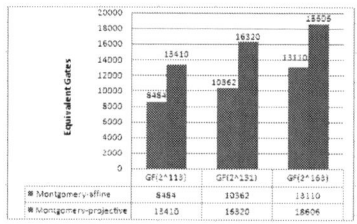

Fig. 7. Memory requirement (affine and projective coordinates)

space) and execution time. Consequently, the ECC using affine
coordinates is the most suitable for the authentication of the
RFID Technology.

REFERENCES

[1] X. Yi, L. Wang & D. Mao and Y. Zhan. *An Gen2 Based Security Authentication Protocol for RFID System*, Procedia 24, 2012 13851391.
[2] M. Hutter, *RFID Authentication Protocols Based on Elliptic Curves*, SECRYPT- International Conference on Security and Cryptography, 2009.
[3] S. Kumar, *Elliptic curve cryptography for constrained devices*, PhD thesis, Ruhr-University Bochum, Germany, June 2006.
[4] J.P. Deschamps, *Hardware Implementation of Finite-Field Arithmetic*, McGraw-Hill, Inc., New York, NY, USA, 2009.
[5] J. P. Deschamps, *Squaring over $GF(2^{163})$*, Tech. Rep. DESS21, University Rovira Virgili, 2006.
[6] National Institute of Standards and Technology, 2000. FIPS PUB 186-2: Digital Signature Standard. http://xml.coverpages.org/FIPS-186-2.pdf
[7] W. N. Chelton and M. Benaissa, *Fast elliptic curve cryptography on FPGA*", IEEE Trans. Very Large Scale Integr. Syst. 16, 198-205, 2008. URL http://dl.acm.org/citation.cfm?id=1378394.1378404.
[8] M. Machhout, Z. Guitouni, K. Torki, L. Khriji and R. Tourki, *Coupled FPGA/ASIC implementation of elliptic curve cryptoprocessor*, International Journal of Network Security & Its Applications (IJNSA) 2, 2010.
[9] K. Jarvinen, M. Tommiska, J. Skytta. *A scalable architecture for elliptic curve point multiplication*, Proceedings of the IEEE International Conference on Field-Programmable Technology, FPT 2004. IEEE, pp. 303-306.
[10] M. A. Bencherif, H. Bessalah and A. Guessoum, *Reconfigurable Elliptic Curve Crypto-Hardware Over the Galois Field $GF(2^{163})$*, American Journal of Applied Sciences 6 (8): 1596-1603, 2009.

R F MEMS CIRCUIT FOR SPACE COMMUNICATIONS SYSTEMS

Souad. OUKIL [1]
Faculté de Génie Electrique
USTOMB, BP 1505 El M'Naouar
Oran, ALGERIA
e-mail: sadoukil@yahoo.fr

Abdelmadjid. BOUDJEMAI [2]
Space Mechanics Division
National Center of Space
Techniques
P.O. Box 13 Arzew 31200,
ALGERIA
e-mail: a_boudjemai@yahoo.fr

Nabil. BOUGHENMI [3]
Faculté de Génie Electrique
USTOMB, BP 1505 El M'Naouar
Oran, ALGERIA
e-mail:
boughanminabil@yahoo.com

Abstract— One of the most important considerations in designing a spacecraft is weight. By reducing the weight of a spacecraft, it is possible to increase the payload, which improves agility and also reduces the launch cost [1]. Missions costs are directly proportional to its total weight, thus, the trend will be to replace bulky and heavy components of space carriers, communication and navigation platforms and of scientific payloads. RF-MEMS technology has the potential of replacing many of the mechanical and semiconductor switches used in mobile and satellite communication systems. In many cases, such RF-MEMS switches would not only reduce substantially the size and power consumption, but also promise superior performance. The paper reviews the recent development of RF MEMS switches and switch matrices. Several configurations are presented for multi-port RF-MEMS switches.

Key words — **RF, MEMS Switch, Finite element, Communications systems, Space.**

I. INTRODUCTION

Satellite subsystems that are currently being examined in the MEMS community include attitude determination sensors and control and actuators, propulsion, communications, power, flight computers, mission instrumentation, satellite scale thermal management, and the spacecraft chassis. Within this list, the most mature systems are those that offer the greatest benefit from implementation in MEMS or those with terrestrial market pressure. The current pressure within the personal communication industry is driving a number of technologies that can directly cross over into the space community. Two of the more dramatic devices that will both soon be ready for satellite integration and flight testing are MEMS gyroscopes and RF switching components [2, 3, 4, 5].

Wireless satellite communications and space-based sensors are the major beneficiaries of the MEMS technology. Minimum size, power consumption, and high reliability are the critical requirements for both the applications, which can be satisfied by deploying MEMS-based devices and sensors. MEMS technology conserves

energy and minimizes power consumption, weight, and size. RF-MEMS devices such as switches, phase shifters, switched antennas, filters, and amplifiers have become critical to increasing communications satellite capability while realizing significant reduction in launch weight and cost. This is important not only in large satellites employing phased array antennas and switch matrices that are deployed in constellations to downlink data but also in the case of nano satellites, which are very small, lightweight spacecraft containing microelectronic equipment, RF-MEMS components, and payloads.

Space communications systems are "ripe" for the insertion of MEMS-based technologies, in part due to the growth in commercial communication developments. One of the most exciting applications of MEMS for microwave communications in spacecraft concerns the implementation of "active aperture phase array antennas." These systems consist of groups of antennas phase-shifted from each other to take advantage of constructive and destructive interference in order to achieve high directionality. Such systems allow for electronically steered radiated and received beams, which have greater agility and will not interfere with the satellite's position. Optical communications could also play an important role in low-power, low mass, and long-distance missions such as the Realistic Interstellar Explorer (RISE) mission, which seeks to send an explorer beyond the solar system, which requires traveling a distance of 200 to 1000 AU from the Sun within a timeframe of about 10 to 50 years. The primary downlink for such a satellite would need to be optical because of the distances and weight limits involved. It has been proposed that a MEMS implementation of the beam-steering mechanism may be necessary to achieve the desired directional accuracy with a sufficiently low mass. MEMS in space communication may well fall under the trendy term "disruptive technology" for their potential to redefine whole systems [5,6,7].

MEMS are unique from a radiation-effects point of view because they contain electronic control circuits coupled with mechanical structures, both of which are potentially sensitive to radiation damage. The electronic circuits in MEMS are either CMOS or bipolar technologies that are known potentially to exhibit great sensitivity to radiation damage. It is not at all obvious that radiation doses that produce measurable changes in performance in electronic circuits will have any effect on mechanical structures; however, they can. The performance of MEMS devices in space depends, critically, on the characteristics of the radiation environment. Before a device can be used in space it must be qualified to ensure that it will survive the rigors of the space environment. Radiation qualification is one of many different qualification procedures that must be performed. Others include temperature, pressure, and vibration. In the absence of specific guidelines for qualifying MEMS devices for a radiation environment, radiation test engineers make use of standard radiation qualification procedures that have been developed for microelectronics.

II. RADIO FREQUENCY MEMS SWITCH FOR SPACE SYSTEMS

RF MEMS switch performances and reliability; have been largely improving during the last years. Many examples of successful concepts for RF switches can be found in academia and companies. Fig.1 reports a summary of state-of-the-art MEMS switches.

MEMS-based RF switches are an attractive option for space-based applications. In comparison with mechanical coaxial or waveguide devices and PIN diodes, MEMS switches provide excellent RF performance with low insertion loss, high isolation, and high linearity, while also offering low power consumption, small size and light weight.

Satellite payloads typically have hundreds of switches integrated in the form of switch matrices to provide system redundancy. In case of any failure in the amplifiers, these switch matrices reroute the signal to a spare amplifier and maintain the full functionality of the system.

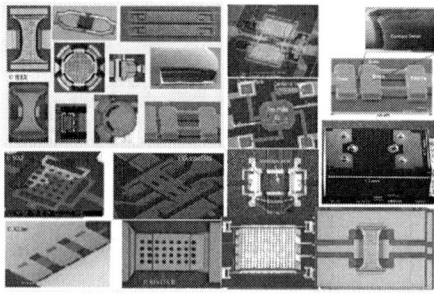

Figure 1: Pictures of state-of-the-art RF MEMS devices [8], [5], [4].

For the communications field MEMS currently offers solutions to filtering, switching local oscillation [9], [10]. Radios are currently adapting an increasingly integrated process as semiconductor materials are able to handle higher frequencies. By using MEMS wherever possible not only is there a weight and volume benefit from the device itself but

from the lack of associated connectors, cabling, casework and packaging.

The MEMS weight and volume remains constant because for MEMS RF switches the size is frequency independent and for mechanical filtering the carrier package dominates the weight and volume figure over the die mass [11, 12]. For the table 1 the numbers reflect a full duplex communications link.

TABLE I.   TABLE 1: SUMMARY OF MEMS COMMUNICATION COMPONENTS ADVANTAGES.

Frequency Range	Macro sized mass [24]	Macro sized volume [24]	MEMS mass	MEMS volume
X-Band	1.5 kg	880 cm³	100 g	0.5 cm³
S-Band	2.0 kg	2700 cm³	100 g	0.5 cm³
Ku-Band	1.2 kg	608 cm³	100 g	0.5 cm³

III. MODELING OF RF MEMS SWITCHES

A. Electrostatic model

The movable part of an electrostatic RF MEMS switch, for both the cases of a clamped-clamped bridge or a cantilever beam, can be described in a compact form by using a 2-D model of the traditional parallel plates capacitor, as depicted in Fig. 2. In this case, the anchors of the membrane are modelled by a linear spring k, attached to the top electrode of the capacitor, which is free to move along the z-direction.

A thin dielectric layer of thickness t_d and dielectric constant ε_r lays over the bottom electrode, which is fixed to the substrate, whereas the rest of the capacitor is filled with air. Note that such a capacitor works as the series of two capacitances, one filled only with air and one filled with the dielectric layer.

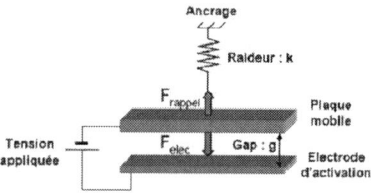

Figure 2: Spring-capacitance system modeling the electro-mechanical coupling in a MEMS switch.

The threshold voltage (or the actuation voltage) is an important index of the RF MEMS switch's function, in wireless communication system, the magnitude of RF MEMS switches' actuation voltage will directly affect the electric circuit design and the performance of the whole system such as the system size, consumption etc. In structure design, the basic mechanics model of different cantilever beam MEMS switches looks like a cantilever beam with a force in one end, as shown in Fig.3, L is the length of the cantilever beam; δ is the flexibility degree; F is a force; g0 is the original space between the parallel plank capacitor.

978-1-4673-5289-5/12 $31.00 © 2012 IEEE

Because the width of the cantilever beam is far larger than the thickness and the direction of both the force and the movement are in one plane as shown in Fig.3, the movement of the cantilever beam can be analyzed as a flat problem.

According to simple beam theory, we can get the relation between the voltage and the flexibility degree of the cantilever beam:

$$F_{rappel} = k(g_0 - g) \qquad (1)$$

$$F_{elec} = \frac{1}{2} \frac{\varepsilon_0 \times S \times V^2}{g^2} \qquad (2)$$

$$V = \sqrt{\frac{2k}{\varepsilon_0 \times S} g^2 \times (g_0 - g)} \qquad (3)$$

For a cantilever beam RF switch, the dynam
and frequency responsibility of the cantilever beam affect the executive function of the switch greatly.

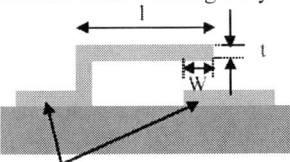

Coplanar of line Conductor

Figure 3 : Structure micromécanique à géométrie simple

The displacement characteristic of the top electrode describes a hysteresis as a function of bias voltage, as illustrated in Fig. 4. The pull-out voltage is always smaller than the pull-in, since the electrostatic force, acting after the actuation and then depending on the down-state capacitance, is higher then the one acting before the snap-down.

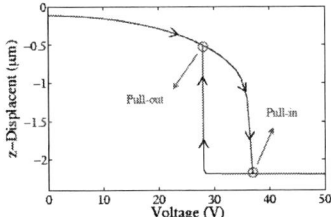

Figure 4: Displacement along z-axis of a clamped-free bridge as a function of bias voltage. Note the hysteresis loop that characterizes any kind of RF MEMS switch.

B. Mechanical model

A MEMS switch is made of a suspended conductive membrane constrained to a fixed support (substrate). The membrane can be a fixed-fixed beam or a cantilever beam according to the type and the number of anchor points. A fixed-fixed beam or air bridge is usually anchored to the substrate by means of at least two anchor points or anchor springs. A cantilever beam or free-clamped beam is anchored at one or more points at the same side.

In electrostatic MEMS switches, the suspended membrane is flexible enough to be considered elastically

movable. The movement is driven by an electrostatic force provided by applying a difference of potential between the membrane and a fixed electrode. The reaction force depends on the mechanical behaviour of the membrane, which in most of the cases can be easily modelled by a linear spring constant. The MEMS switch used in this paper is shown in the figure 5.

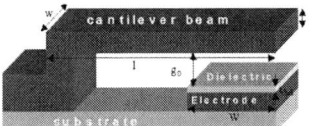

Figure 5: MEMS switch geometry

The RF switch characteristics are given in table 3.

TABLE II.   TABLE 2: THE RF SWITCH CHARACTERISTICS

cantilever beam Length	l	235 µm
width	W	100µm
Tope electrode length	ω	70µm
Thickness of the beam	t	3µm
Gap of the capacitive system	g0	1.8µm
Thickness of the dielectric one	tdiel	0.2µm
Permittivity of the dielectric one	εr diel	9.6

IV.   FINITE ELEMENT ANALYSIS OF THE RF MEMS SWITCH

Specifically, the RF MEMS switches can be classified according to the following three factors: A. the structure of the radio frequency electric circuit (series and parallel etc.); B. the mechanical structure (cantilever beam and film structure etc.); C. the contact method (capacitance type and resistance type). Current research of the RF MEMS switch is concentrated on cantilever beam structure and film structure switches. Cantilever beam structure is generally the series resistance type. Then according to the signal connections, the cantilever beam MEMS switches can be classified into inline MEMS-series switches and broadside MEMS-series switches [13, 14, and 15]. The basic structure of the switch is shown in Fig.5.

The finite element model (FEM) of a cantilever beam switch is given by figure 6.

The boundary condition in the FEM simulation concerns the one edge of the short side which is constrained (displacement of x, y and z are zero, and rotation of x, y and z are zero) in the cantilevered MEMS switch.

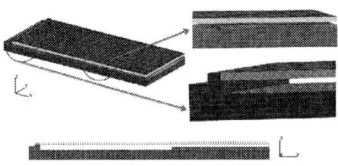

Figure 6: FEM model

The mechanical properties of materials used in the switch are given in the table 3.

TABLE III.   TABLE 3: MATERIALS MECHANICAL PROPERTIES

	Young modulus E(Gpa)	Poisson ration ν	Density ρ (kg/m³)
polysilicon	170	0.22	2300
Silicon	169	0.36	2392
SiN	1100	0.07	3500
Al₂O₃	125	0.26	3900

V.   RESULTS AND DISCUSSION

Figure 7 shows the frequencies modes of a cantilever beam switch obtained using Msc. Patran/Nastran software

Figure 7 shows the results of the structure modal forms, and which makes it possible to see where are made the most deformations and which elements.

The coloured fringes give the amplitude of the displacement vector describing the shape of each mode. The black colour corresponds to null displacement and the red one presents the maximum amplitude.

Figure 7 shows the displacement of the cantilever beam MEMS switch in Z direction, the maximum is about 5 mm witch is enough for the MEMS switch.

The lowest frequency was in 1st mode (2460.7 Hz), which gives the better the vibration direction

The frequency was increasing with each subsequent mode of vibration.

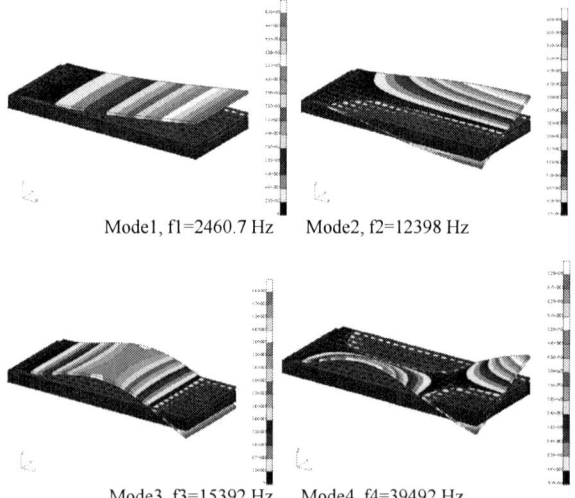

Mode1, f1=2460.7 Hz   Mode2, f2=12398 Hz

Mode3, f3=15392 Hz   Mode4, f4=39492 Hz

Figure 7: Various modes of the cantilever beam MEMS switch

VI.   CONCLUSION

This paper takes the cantilever beam MEMS switch as an example to discuss mechanical principle and function simulation of the RF MEMS switch in satellite communication system, whose main structure is a cantilever beam controlled by electrostatic force.

This paper discusses the significant structural elements in RF MEMS switches

The first mode of natural frequency of the switch, which is about 2460.7 Hz, gives the better the vibration direction

With all these technologies there is no quick and easy solution. Thus, the challenges facing the MEMS community are great when it comes to the development of an integrated MEMS spacecraft.

REFERENCES

[1]   A.Boudjemai, R. Amri, A. Mankour, H. Salem, M. H. Bouanane, D. Boutchicha, Modal Analysis and Testing of Hexagonal Honeycomb Plates Used for Satellite Structural Design, Materials & Design (September 2011) doi:10.1016.

[2]   G.M. REBEIZ "RF MEMS theory, design and technology"John Wiley & Sons, 2003.

[3]   G.M. REBEIZ, J.B. MALDAVIN "RF MEMS switches and switch circuits" IEEE Microwave Magazine, Dec. 2001, pp.59-71

[4]   Robert Osiander, M. Ann Garrison Darrin, John L. Champion, MEMS and Microstructures in Aerospace Applications, © 2006 by Taylor & Francis Group, LLC.

[5]   H. Helvajian, ed, "Microengineering Aerospace Systems", The Aerospace Press, 1999

[6]   S. Cass, "MEMS in Space", IEEE Spectrum, p.56, July 2001

[7]   N.F. de Rooij, S.Gautsch, D. Briand, C. Marxer, G. Mileti,W. Noell, H. Shea5, U. Staufer and B. van der Schoot, MEMS FOR SPACE,978-1-4244-4193-8/09/$25.00 ©2009 IEEE

[8]   [8] G. Rebeiz, K. Entesari, I. Reines, S. J. Park, M. El-tanani, A. Grichener, and A. Brown, \Tuning in to RF MEMS," Microwave Magazine, IEEE, vol. 10, no. 6, pp. 55 (72, 2009).

[9]   Nguyen, C., Howe, R. An integrated CMOS Micromechanical Resonator High-Q

[10] Oscillator, IEEE Journal of Solid-State Circuits, Col. 34, No. 4, April 1999, pp. 440-455.

[11] A surface micromachined miniature switch for telecommunication applications with signal frequencies from DC up to 4 GHz, 8th International Conference on Solid-State Sensors and Actuators and Eurosensors IX. Digest of Technical Papers (IEEE Cat. No.95TH8173). Stockholm, Sweden. pp. 384-7 vol.2. 25-29 June 1995.

[12] Micromachined Transducers Sourcebook, Gregory Kovacs, New York, NY: McGraw-Hill, 1998.

[13] Space Mission analysis and Design. ed. Larson W. and Wertz J. Torrance, CA: Microcosm, 1992

[14] R.R.Mansour, "RF MEMS for space applications", IEEE Proc. of the 2005 International Conference on MEMS, NANO and Smart Systems.

[15] Y. Yashchyshyn, "Reconfigurable Antennas: the State of the Art", Intl Journal of Electronics & Telecommunications, 2010, vol 56, No3, pp319-326.

A Highly Efficient Substitution Matrix Loader for Pairwise Sequence Alignment

M.Nazrin Md.Isa, Khaled Benkrid, Thomas Clayton
System Level Integration Group,
School of Engineering,
University of Edinburgh,
EH9 3JL, Edinburgh
{m.n.isa,k.benkrid,t.clayton}@ed.ac.uk

Abstract—This paper presents a novel substitution matrix loader architecture for pairwise sequence alignment. The search for sequence homology using DP-based alignment matrix computation is an important tool in molecular biology. It can be implemented either by optimal or sub-optimal approaches. Both of these methods require frequent and rapid access to the amino acids probability scores for PE (Processing Element) configuration especially in a folded systolic array. Typical FPGA implementations configure look-up tables in the pipeline PEs either by using a serial configuration chain with different look-up tables or by run time reconfiguration of the same look-up table. In the former case, configuration time increases proportionally to the number of look-up tables, while the latter case suffers from the limited reconfiguration bandwidth. Therefore, in this paper, we propose a highly efficient parallel loader to optimize both time and space complexities of protein sequence alignment in folded systolic arrays, using only two configuration elements (CEs). In addition, the proposed loader enables PEs to be updated with substitution matrix scores concurrently, with the worst case configuration time of 2 x the depth of the PE's look-up table (in clock cycles). This allows for further optimization of the most time consuming alignment matrix computation through efficient scheduling of alignment matrix computation and PE configuration. Implementation results show that the proposed architecture achieves $k.N_{PE}$ speed-up in configuration time (where k is the folding factor and N_{PE} is the number of PEs) compared to classical approaches, at virtually no area overhead.

Keywords: Field Programmable Gate Arrays, Sequence Alignment, Folded Systolic Array, Substitution Matrix, Smith Waterman, Needleman-Wunsch .

I. INTRODUCTION

Sequence alignment is a fundamental tool in molecular biology with a plethora of applications including in drug engineering, forensics and early disease diagnosis. Searching for sequence homology could be done either by optimal or sub-optimal search techniques. The former searches for optimal scores between database sequences and query sequences e.g. a newly discovered biological sequence. The search technique is usually based on dynamic programming (DP) e.g. the Smith Waterman and the Needleman-Wunsch algorithms. On the other hand, sub-optimal methods use heuristic approaches as a tradeoff between speed and sensitivity. Given a threshold score T, sequences with scores higher than T will be selected for full

dynamic programming based alignment. In either case, aligning sequences in a realistic time becomes an issue due to the exponential growth of database sequences. Advancement in computing technologies has seen the use of parallel architectures such as FPGAs in [1], [2], [3], [4],[5] and GPUs in [6], [7], [8], [9] to accelerate the time consuming DP-based algorithm. In hardware, the algorithm is usually accelerated using a linear systolic array. The latter consists of an array of processing elements (PEs) with one PE holding one amino acid residue. For sequences longer than the maximum possible number of PEs on a particular chip (N_{PE}), the computation is performed by a folded systolic architecture with k-folds, whereby k is equal to the sequence length divided by N_{PE}. This way, PEs are reused between passes to complete a whole pairwise sequence alignment. Typical FPGA implementations configure the pipeline PEs with k fold factors by using a serial configuration chain, which updates probability scores in the PEs sequentially includes in [1], [2]. This requires each PE to have k different look-up tables and consequently, both configuration time and space complexities increase proportionally to $k.N_{PE}$. Another method uses run time reconfiguration (RTR) to configure a look-up table in the PE during alignment matrix computation [3]. Although, this approach optimizes space complexity, it suffers from the limited bandwidth of the configuration mechanism (e.g. the maximum bandwidth of the Internal Configuration Access Port, ICAP). Therefore, in this paper, we present an alternative way to optimize the configuration and computation architecture both in time and space complexities. A highly efficient parallel loader is proposed which updates all PEs simultaneously using only two configuration elements (CE) per PE.

The remainder of this paper is organized as follows. In the following section, important background on biological sequence alignment and a brief overview of an optimal algorithm are presented. Then, section III details the internal architecture and operation of the proposed loader. In section IV, the performance of the loader in terms of speed, area and power is discussed before conclusions are laid out.

II. BACKGROUND

Biological sequences diverge from their common ancestors due to the process of mutation, selection and random genetic drift [10]. For instance, mutation involves three main processes: substitution of residues, insertion of new residues and deletion of existing residues. Both insertion and deletion are referred to as gaps. Gaps in alignments are penalized when scoring. The cost of gaps depends on its length and generally, there are two different ways to penalize gaps: linear and affine gap penalties. Equation (1) shows the Needleman-Wunsch algorithm with a linear gap penalty [10] as an example of optimal alignment algorithms which is used to compute the best global alignment (and score). In it, given a query sequence, $X = x_1,x_2,x_3..x_i...x_M$ (of length M) and $Y = y_1,y_2,y_3..y_j...y_N$ (of length N), this DP-based alignment algorithm searches for the best alignment between subsequences of x and y using an alignment matrix $F(i,j)$. This M by N matrix calculates the largest score among three alternatives in a recursive manner (see (1)). Here $s(x_i,y_j)$ represents the probability of substituting residue x_i for residue y_j according to a probability model stored in the form of a *substitution matrix* (see Fig. 1 for instance).

$$F(i,j) = \max \begin{cases} F(i-1, j-1) + s(x_i, y_j) \\ F(i-1, j) - d \\ F(i, j-1) - d \end{cases} \qquad (1)$$

BLOSUM50, *BLOSUM62* and *PAM* are examples of substitution matrices [10], [11] and [12]. Each column and row in a substitution matrix represent individual amino acid residues with the intersection representing a score denoting the likelihood of substituting one residue for the other. Note that entries on the main diagonal are highlighted in bold, and represent scores of identical residue pairs.

	A	C	D	E	F	G	H	I	K	L	M	N	P	Q	R	S	T	V	W	Y
A	**5**	-1	-2	-1	-3	0	-2	-1	-1	-2	-1	-1	-1	-1	-2	1	0	0	-3	-2
C	-1	**13**	-4	-3	-2	-3	-3	-2	-3	-2	-2	-2	-4	-3	-4	-1	-1	-1	-5	-3
D	-2	-4	**8**	2	-5	-1	-1	-4	-1	-4	-4	2	-1	0	-2	0	-1	-4	-5	-3
E	-1	-3	2	**6**	-3	-3	0	-4	1	-3	-2	0	-1	2	0	-1	-1	-3	-3	-2
F	-3	-2	-5	-3	**8**	-4	-1	0	-4	1	0	-4	-4	-4	-3	-3	-2	-1	1	4
G	0	-3	-1	-3	-4	**8**	-2	-4	-2	-4	-3	0	-2	-2	-3	0	-2	-4	-3	-3
H	-2	-3	-1	0	-1	-2	**10**	-4	0	-3	-1	1	-2	1	0	-1	-2	-4	-3	2
I	-1	-2	-4	-4	0	-4	-4	**5**	-3	2	2	-3	-3	-3	-4	-3	-1	4	-3	-1
K	-1	-3	-1	1	-4	-2	0	-3	**6**	-3	-2	0	-1	2	3	0	-1	-3	-3	-2
L	-2	-2	-4	-3	1	-4	-3	2	-3	**5**	3	-4	-4	-2	-3	-3	-1	1	-2	-1
M	-1	-2	-4	-2	0	-3	-1	2	-2	3	**7**	-2	-3	0	-2	-2	-1	1	-1	0
N	-1	-2	2	0	-4	0	1	-3	0	-4	-2	**7**	-2	0	-1	1	0	-3	-4	-2
P	-1	-4	-1	-1	-4	-2	-2	-3	-1	-4	-3	-2	**10**	-1	-3	-1	-1	-3	-4	-3
Q	-1	-3	0	2	-4	-2	1	-3	2	-2	0	0	-1	**7**	1	0	-1	-3	-1	-1
R	-2	-4	-2	0	-3	-3	0	-4	3	-3	-2	-1	-3	1	**7**	-1	-1	-3	-3	-1
S	1	-1	0	-1	-3	0	-1	-3	0	-3	-2	1	-1	0	-1	**5**	2	-2	-4	-2
T	0	-1	-1	-1	-2	-2	-2	-1	-1	-1	-1	0	-1	-1	-1	2	**5**	0	-3	-2
V	0	-1	-4	-3	-1	-4	-4	4	-3	1	1	-3	-3	-3	-3	-2	0	**5**	-3	-1
W	-3	-5	-5	-3	1	-3	-3	-3	-3	-2	-1	-4	-4	-1	-3	-4	-3	-3	**15**	2
Y	-2	-3	-3	-2	4	-3	2	-1	-2	-1	0	-2	-3	-1	-1	-2	-2	-1	2	**8**

Figure 1. Blosum 50 (rearranged to alphabetical order), with 20 columns by 20 rows of different amino acid residues.

III. THE PROPOSED EFFICIENT LOADER

The main function of our proposed novel loader is to load all PEs of a pairwise sequence alignment array with their corresponding substitution matrix columns simultaneously. This allows for efficient data transfer as the configuration time significantly reduced to $1/k.N_{PE}$ compared to conventional

serial configuration techniques. Consequently, a PE with only two configuration elements could be used for any folding factor, whereby as one CE is being used for alignment matrix computation, the other CE is being updated with probability scores for subsequent fold computation. This optimizes both time and space complexities of the DP-based algorithms in hardware. Fig. 2 illustrates the loader and its parallel configuration chain connected to a PE. The port description of the loader is given in Table I.

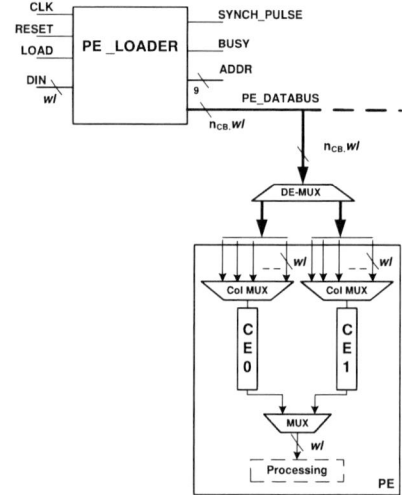

Figure 2. The parallel loader and the PE with two configuration elements (CE$_0$ and CE$_1$).

TABLE I.   PORT DESCRIPTIONS OF THE EFFICIENT LOADER

Port	Description
CLK	Clock signal for synchronize the loader operations
RESET	When HIGH, it resets the loader to load different substitution matrix (if required)
LOAD	When HIGH, it triggers the loader to start shifting data into circular buffers, i.e. when substitution matrix memory is ready with probability scores
ADDR	Supply address to the substitution matrix memory to fetch its contents into the loader
DIN	Read out substitution matrix scores from the memory
PE_DATABUS	Data bus to update PEs with probably scores
BUSY	When HIGH, it indicates that the loader is shifting in substitution matrix scores into its circular buffers
SYNCH_PULSE	A pulse signal whereby each pulse interval represents valid scores to configure the PEs

Fig. 3 illustrates the internal architecture of the loader with n_{CB} circular buffers to hold the columns of the substitution matrix e.g. $n_{CB} = 20$ for the substitution matrix of Fig. 1. The buffer has n_{row} shift registers, with each shifting one element of a particular substitution matrix row into the buffer, in turn, every clock cycle. The wordlength of the substitution matrix elements wl is parameterizable. For the case of Blosum50, 5-bit two's complement is enough to represent its elements, in which case the loader operates as 5-bit serial-in-serial-out shift register during initial configuration mode, and once in running mode, it operates as 5-bit serial in $n_{col} \times wl$-bit parallel out circular shift register. Details of these operations are presented in subsections A and B.

A. Initial Configuration Mode

Right shift operation is the fundamental operation of the loader during initial configuration mode. The operation starts by shifting elements of a given substitution matrix serially column by column into the corresponding buffers which are pipelined together in a long chain. Each element is shifted into the buffer chain every clock cycle. Consequently, all substitution matrix elements of one column are completely loaded into the buffer chain within n_{row} clock cycles. Note that, the thick broken line arrow in Fig. 3 depicts the flow of the shift operation during configuration mode. It begins to fill the last buffer i.e CB_{n-1} with the first n_{row} elements in the last column of the substitution matrix and continues with the following n_{row} elements to buffer CB_{n-2}. This sequential shift operation continues until CB_0. This way, all scores will be loaded into the buffer chain according to their corresponding column. Once scores are completely loaded i.e. the memory read is finished, the loader is ready to configure the PEs. The initial configuration time (in clock cycles) to read an entire substitution matrix into the loader depends on the size of the substitution matrix and is mathematically expressed in (3).

$$t_{initload} = n_{col} \times n_{row} \qquad (3)$$

Where, n_{col} is the number of columns and n_{row} is the number of rows of the substitution matrix. Details of the configuration mode function are described by the pseudo code in Fig. 4.

B. Running Mode

During this mode, all elements in a buffer are circulated every clock cycle following the direction of the arrow with the thin dotted line as shown in Fig. 3. Data circulation within the circular buffers ensures that valid scores are available for PE configuration within a maximum duration of $2.n_{row}$, as expressed in (4).

$$t_{config} \leq 2 \times n_{row} \qquad (4)$$

This way, all PEs will be configured with probability scores concurrently with the worst case configuration time of $2 \times n_{row}$.

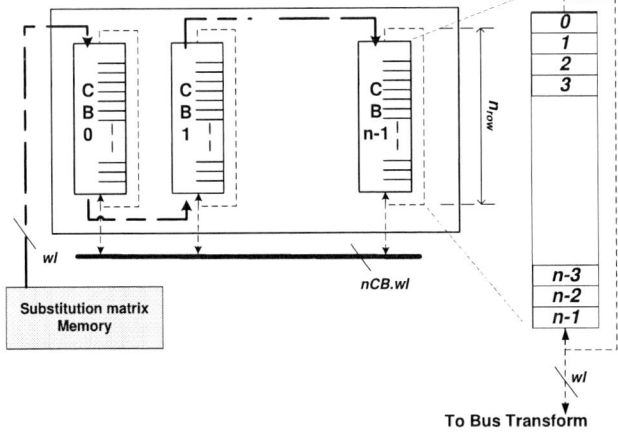

Figure 3.   The Parallel Loader with the circular buffers

The loader architecture offers efficient way of fast configuration of the probability scores instead of updating PEs using a serial configuration chain as outlined in section I. Note that, the initial configuration mode runs only once whereas the loader remains in the running mode until it is reset to load other amino acid models.

Input: Substitution matrix probability scores, n_{CB}, n_{row}
CBcounter $\leftarrow n_{CB}$ -1
n $\leftarrow n_{row}$-1
For every element of a given substitution matrix
  If (CBcounter!=0)
    Shift element into the buffer chain
      Decrement CBcounter by one
      If (n !=0)
        Shift score into the corresponding row of the buffer
        Decrement n by one
      Else reset n to n_{row}-1
      End if
  Else
    Circulate scores inside their circular buffer
    Generate synch_pulses at each start of circulation
  End if
End for
Output: valid substitution matrix scores at synch_pulse intervals

Figure 4.   Pseudo code for initial configuration mode and running mode.

C. Bus transformation

For the purpose of data transfer to the PE, a large bus is designed (see the bold horizontal line in Fig.3) to broadcast probability scores to all systolic array PEs. Indeed, all output ports of the circular buffers (*wl*-bit each) are joint together to form a large bus of width $n_{CB} \times wl$. The bus (*PE_DATABUS*) creation is shown in the Verilog pseudo code of Fig. 5.

For every circular buffer beginning from n=20 until 0
  Generate
    PE_DATABUS [n × wl: (n-1) × wl] =CB$_{n-1}$
  Endgenerate
Endfor

Figure 5.   Pseudo code for a large bus creation

In the timing diagram shown in Fig. 6, the substitution matrix memory is assumed to be already filled with probability scores. During initial configuration mode, the loader operation is marked by the *BUSY* signal being HIGH. Once all scores are entirely loaded, the loader is ready for PE configuration, thus synchronization pulses (*SYNCH_PULSE*) are emitted every n_{row} clock cycles, whereby at each pulse interval, valid scores are available on the *PE_DATABUS* for PE configuration.

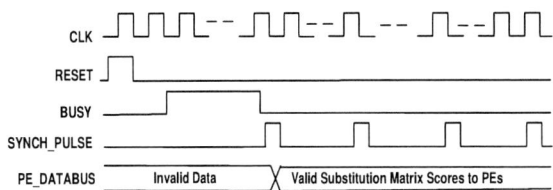

Figure 6.   Valid Substitution matrix scores available to the PE during SYNCH_PULSE intervals.

PE configuration happens at any stage during this interval i.e. when its own probability scores are output by the circular buffers. Indeed, the query residue inside each PE selects its corresponding substitution matrix column (see multiplexer *Col MUX* in Fig 2).

IV. IMPLEMENTATION RESULTS

This section discusses the loader's implementation efficiency in terms of area, speed and power performance. The loader design was captured using Verilog HDL in a parameterizable manner. Two different CAD tools are used in this evaluation: Xilinx ISE 13.1 targeting a Xilinx XC5VLX110-3 FPGA and Cadence Build Gates version 2005 with 0.18um UMC process technology for ASIC implementation.

The operating frequency of the loader when synthesized on the XC5VLX110-3 FPGA (65nm based CMOS technology) is 396 MHz. Note that the loader's circular buffers are implemented efficiently on Xilinx FPGAs using the slices' LUT configuration referred to as SRL32[13]. Note also that typical systolic array FPGA implementations have PEs operating at a lower frequency e.g. 100-200 MHz. This means that the configuration time could be reduced by a factor of ~4 for higher performance if the loader is clocked separately. In terms of area utilization, the Xilinx ISE reported that the design utilizes 87 logic slices. We also synthesized the design using Cadence Build Gates with 0.18um UMC process technology. From the area report, we found that, the total area occupied by the loader was 142,980.92 um^2. The speed reported was 1GHz.

In an attempt to measure the power consumption of the design, we used a file generated from a map report (*sml.ncd*) to measure its static power and a simulation activity file (*sml.vcd*) to estimate the dynamic power of the loader. These files are then used as input parameters to the Xilinx ISE Xpower Analyzer. The power consumption of the loader was estimated to ~1.269W, with 1.144W as static power and 0.126W as dynamic power. The implementation results are summarized in Table II.

TABLE II.   AREA, SPEED AND POWER PERFORMANCE

ASIC Implementation frequency (GHz)	1.0
ASIC Implementation Area (um^2)	142,980.92
FPGA logic slices (#slices)	87
FPGA Implementation Frequency (MHz)	396.1
Worst case Configuration time (ns)	101.0
Static Power (W)	1.14
Dynamic Power(W)	0.13

V. CONCLUSION

An efficient loader targeting the configuration of folded systolic arrays for pairwise sequence alignment was presented in this paper. The loader has an ability to configure the systolic array's processing elements in parallel with a worst case configuration time $2.n_{row}$ clock cycles where n_{row} is the number of rows of the substitution matrix. Unlike conventional substitution matrix loaders, which configure processing elements through a serial configuration chain, this loader enables concurrent PEs configuration regardless of the number of processing elements, using a fixed number of configuration elements (equal to 2). Implementation results show that the loader occupies a very small footprint (87 slices on a Virtex-5 FPGA) with a typical maximum clock frequency ~4 times faster than a typical systolic array operating frequency. As a result, this loader is able to optimize the time consuming configuration operation of optimal sequence alignment algorithms, especially in folded architectures, through the concurrent scheduling of alignment matrix computation and processing element configuration.

REFERENCE

[1] K. Benkrid, L. Ying, and A. Benkrid, "A Highly Parameterized and Efficient FPGA-Based Skeleton for Pairwise Biological Sequence Alignment," *IEEE Transactions on Very Large Scale Integration (VLSI) Systems*, vol. 17, pp. 561-570, 2009.

[2] M. N. Isa, K. Benkrid, T. Clayton, C. Ling, and A. T. Erdogan, "An FPGA-based parameterised and scalable optimal solutions for pairwise biological sequence analysis," presented at 2011 NASA/ESA Conference on Adaptive Hardware and Systems (AHS), 2011.

[3] T. F. Oliver, B. Schmidt, and D. L. Maskell, "Reconfigurable architectures for bio-sequence database scanning on FPGAs," *IEEE Transactions on Circuits and Systems II*, vol. 52, pp. 851-855, 2005.

[4] Y. Yoshiki, M. Yosuke, M. Tsutomu, and K. Akihiko, "High Speed Homology Search Using Run-Time Reconfiguration," in *Proceedings of the Reconfigurable Computing Is Going Mainstream, 12th International Conference on Field-Programmable Logic and Applications*: Springer-Verlag, 2002.

[5] Y. Yamaguchi, et al., "FPGA-Based Smith-Waterman Algorithm: Analysis and Novel Design Reconfigurable Computing: Architectures, Tools and Applications," vol. 6578, *Lecture Notes in Computer Science*: Springer Berlin / Heidelberg, 2011, pp. 181-192.

[6] K. Dohi, K. Benkrid, C. Ling, T. Hamada, and Y. Shibata, "Highly efficient mapping of the Smith-Waterman algorithm on CUDA-compatible GPUs," presented at 21st IEEE International Conference on Application-specific Systems Architectures and Processors (ASAP), 2010.

[7] Y. Liu, D. L. Maskell, and B. Schmidt, "CUDASW++: optimizing Smith-Waterman sequence database searches for CUDA-enabled graphics processing units," *BMC Res Notes*, vol. 2, pp. 73, 2009.

[8] Y. Liu, B. Schmidt, and D. L. Maskell, "CUDASW++2.0: enhanced Smith-Waterman protein database search on CUDA_enabled GPUs based on SIMT and virtualized SIMD abstractions," *BMC Res Notes*, vol. 3, pp. 93.

[9] Y. Munekawa, F. Ino, and K. Hagihara, "Design and implementation of the Smith-Waterman algorithm on the CUDA-compatible GPU," presented at BioInformatics and BioEngineering, 2008. BIBE 2008. 8th IEEE International Conference on, 2008.

[10] R. Durbin, Eddy, S., Krogh, A., Mitchison, G, *Biological Sequence Analysis: Probabilistic Models for Proteins and Nucleic Acids*: Cambridge University Press, Cambridge UK, 1998.

[11] S. Henikoff and J. G. Henikoff, "Amino acid substitution matrices from protein blocks," *Proceedings of the National Academy of Sciences*, vol. 89, pp. 10915-10919, 1992.

[12] M. O. Dayhoff, R. M. Schwartz, and B. C. Orcutt, "{A model of evolutionary change in proteins}," *Atlas of protein sequence and structure*, vol. 5, pp. 345-351, 1978.

[13] "Virtex-5 Family User Guide," Xilinx, Inc., San Jose, CA 2009.

Implementation of Spectral Subtraction Method on FPGA using High-Level Programming Tool

Mohammed Bahoura
Department of Engineering,
Université du Québec à Rimouski,
300, allée des Ursulines, Rimouski, Qc, Canada.

Hassan Ezzaidi
Department of Applied Sciences,
Université du Québec à Chicoutimi,
555, boul. de l'Université, Chicoutimi, Qc, Canada.

Abstract—**This paper presents a real-time architecture of spectral subtraction technique applied to speech enhancement. The proposed architecture is easily and quickly implemented on Field Programmable Gate Array (FPGA) using high-level programming tool in MATLAB/SIMULINK environment. Speech enhancement results obtained with fixed-point format implementation are compared to those obtained with the floating-point format one. The maximum operating frequency and resource utilization are presented for a Virtex-6 FPGA chip.**

I. INTRODUCTION

Spectral subtraction method was originally introduced by Boll [1] for speech enhancement. An improved version was proposed by Berouti et al. [2] to reduce the residual musical noise. Other improvements have been suggested in the last three decades. The spectral subtraction approach has been also applied to other kinds of sounds like underwater sounds [3], respiratory sounds [4], etc. These methods are firstly evaluated using software platform such as MATLAB. However, real-world applications such as hands-free devices, cellular phones or hearing aids require real-time implementation of these speech enhancement algorithms. The hardware implementation of these algorithms is a great challenge that requires a tradeoff between complexity, computation speed and efficiency of these algorithms. In the last two decades, several spectral subtraction based architectures are implemented on digital signal processor (DSP) for real-time speech enhancement. However, at the best of our knowledge, only few ones are implemented on Field Programmable Gate Array (FPGA) [5], [6] the last years. Implementation on FPGA can be done using hardware description language (HDL) such as VHDL and Verilog, or using high-level programming tool such as Xilinx system generator (XSG) [7]. In fact, XSG enables the use of MATLAB/SIMULINK environment to create and verify hardware designs for Xilinx FPGAs quickly and easily. It provides a library of SIMULINK blocks bit and cycle accurate modeling for arithmetic and logic functions, memories, and DSP functions. It also includes a code generator that automatically generates HDL code from the created model [8].

In this paper, we propose an FPGA implementation architecture of the spectral subtraction method for speech enhancement using XSG. Not like the previously published architectures, the proposed one very simple and allows one to implement this algorithm easily and quickly on FPGA.

II. SPECTRAL SUBTRACTION

The noisy signal $y(t)$ is assumed to be composed of the clean signal $s(t)$ and the uncorrelated additive noise $d(t)$.

$$y(n) = s(t) + d(t) \tag{1}$$

In the frequency domain, equation (1) becomes

$$Y(\omega) = \mathcal{F}\{y(n)\} = S(\omega) + D(\omega) \tag{2}$$

where $Y(\omega)$, $S(\omega)$ and $D(\omega)$ are the Fourier transform of the corresponding signals. The Fourier transform of the corrupted signal can be expressed in polar form $Y(\omega) = |Y(\omega)|e^{j\varphi_y(\omega)}$, where $|Y(\omega)|$ and $\varphi_y(\omega)$ are the magnitude and the phase of $Y(\omega)$, respectively.

The spectral subtraction techniques are based on the principle that an estimate of the clean signal spectrum can be obtained by subtracting an estimate of the noise spectrum from the noisy signal spectrum. To ensure non negative magnitude spectrum, the first approach used a simple half-wave rectifier [1].

$$|\widehat{S}(\omega)| = \max\left\{|Y(\omega)| - |\widehat{D}(\omega)|,\ 0\right\} \tag{3}$$

where the noise spectrum $|\widehat{D}(\omega)|$ is estimated by its average value taken during non-signal periods. However, this process is accompanied by perceptually annoying noise, namely musical noise. Berouti et al. [2] proposed a modified spectral subtraction method to reduce the residual musical noise. It consists in subtracting an overestimate of the noise power spectrum and limiting the resultant spectrum from going below a preset minimum level (spectral floor).

$$|\widehat{S}(\omega)|^\gamma = \max\left\{|Y(\omega)|^\gamma - \alpha|\widehat{D}(\omega)|^\gamma,\ \beta|\widehat{D}(\omega)|^\gamma\right\} \tag{4}$$

where $\alpha \geq 1$ is the over subtraction factor and $0 < \beta << 1$ is the spectral flooring parameter. γ is the exponent determining the transition sharpness, with $\gamma = 1$ providing magnitude spectral subtraction [1] or $\gamma = 2$ for power spectral subtraction [2].

The enhanced signal spectrum is obtained using the magnitude estimate $|\widehat{S}(\omega)|$ of the enhanced signal and the phase $\varphi_y(\omega)$ of the corrupted input signal.

$$\widehat{S}(\omega) = |\widehat{S}(\omega)|e^{j\varphi_y(\omega)} \tag{5}$$

978-1-4673-5289-5/12 $31.00 © 2012 IEEE

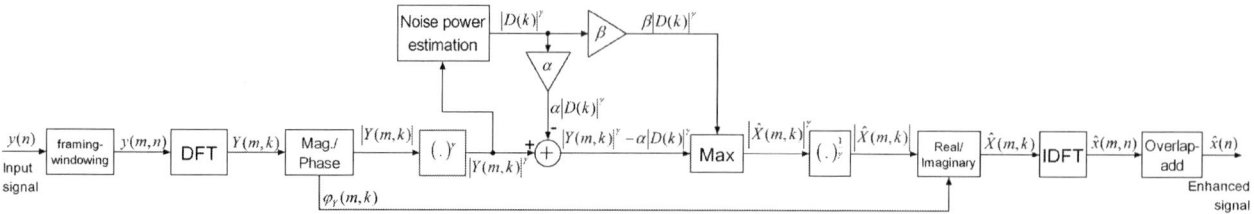

Fig. 1. Block diagram of spectral subtraction technique. Parameter values used in this paper are: $\gamma = 2$, $\alpha = 6$, and $\beta = 0.001$.

Finally, the enhanced signal, $\widehat{s}(n)$, is reconstructed by inverse Fourier transform.

$$\widehat{s}(n) = \mathcal{F}^{-1}\{\widehat{S}(\omega)\} \tag{6}$$

Fig. 1 shows a practical signal enhancement system based on the spectral subtraction technique. Discrete-time input signal, $y(n)$, is split into overlapped frames of N samples, $y(m, n)$, where m is the frame time index and n is the sample time index within the given frame. Then, a discrete Fourier transform (DFT) is applied to each frame to perform spectral subtraction before reconstructing the enhanced frame $\widehat{s}(m, n)$ by applying the inverse discrete Fourier transform (IDFT).

III. FPGA IMPLEMENTATION

The spectral subtraction technique is implemented on FPGA using Xilinx System Generator (XSG) [7] and the Virtex-6 FPGA ML605 Evaluation Kit [9]. Fig. 2 presents the proposed architecture that is mainly based on the block diagram described in Fig. 1 by taking into account the latencies (delays) introduced by some XILINX blocks. Table I presents delays inserted by different computational unites.

A. Discrete Fourier transform

The DFT of a given input frame, $y(m, n)$, is computed using the Xilinx FFT block [10].

$$Y(m - m_1, k) = \sum_{n=0}^{N-1} y(m, n)e^{-jnk2\pi/N} \tag{7}$$

where N is the transform size, k is the frequency bin index ($0 \leq k \leq N - 1$), and $j = \sqrt{-1}$. Pipelined streaming input/output option has been chosen to achieve continuous computation of the short-time Fourier transform. Signal is implicitly segmented on non-overlapping rectangular windows of $N = 256$ samples. The Xilinx FFT block provides two outputs corresponding to real, $Y_r(m - m_1, k)$, and imaginary, $Y_i(m - m_1, k)$, parts of the transformed signal. In fact, this block presents a latency of $m_1 = 611$ samples (see Table I).

B. Magnitude and phase

The Xilinx CORDIC block [11] is used to compute the magnitude, $|Y(m - m_2, k)|$, and phase, $\varphi_y(m - m_2, k)$, of the inputs $Y_r(m - m_1, k)$ and $Y_i(m - m_1, k)$.

$$|Y(m - m_2, k)| = \sqrt{Y_r^2(m - m_1, k) + Y_i^2(m - m_1, k)} \tag{8}$$

TABLE I
DELAYS INSERTED BY DIFFERENT COMPUTATIONAL UNITES USING
4-MULTIPLIER STRUCTURE FOR FFT BLOCK [10] AND $N = 256$.

XSG block	Latency (samples)
Forward Fast Fourier Transform	611
Magnitude and Phase computation	23
Cosine and Sine Computation	20
Inverse Fast Fourier Transform	611

$$\varphi_y(m - m_2, k) = \arctan\left(\frac{Y_r(m - m_1, k)}{Y_i(m - m_1, k)}\right) \tag{9}$$

As shown in Table I, this block presents a latency of 23 samples. So, the cumulative delay is now $m_2 = m_1 + 23 = 634$ samples.

C. Noise power estimation

The noise power spectrum $|\widehat{D}(k)|^2$ is estimated by its average value taken during the first $L = 5$ frames.

$$|\widehat{D}(k)|^2 = \frac{1}{L}\sum_{l=0}^{L-1}|\widehat{D}(lN, k)|^2 \tag{10}$$

where $|\widehat{D}(m - m_2, k)| = |Y(m - m_2, k)|$ because of the signal absence. The first frame ($l = 0$) corresponds to $m = 0$ for the $y(m, n)$ signal and $m = m_2$ for the $|Y(m - m_2, k)|$ signal. The average value of the noise power spectrum is computed using an RAM-based accumulator that is incremented by $0.2|\widehat{D}(lN, k)|^2$ at each frame ($l < L$). Using the RAM controller subsystem described in Fig. 2, this RAM is enabled on write and read mode during the first L frames, and on the read mode for the others frames. This step is accomplished without inserted delays.

D. Magnitude Subtraction

Based on comparator and multiplexer component, the magnitude subtraction subsystem (Fig. 2) implements the following expression:

$$|\widehat{S}(m - m_2, k)|^2 = \max\{|Y(m - m_2, k)|^2 - \alpha|\widehat{D}(k)|^2, \ \beta|\widehat{D}(k)|^2\} \tag{11}$$

where $\alpha = 6$ and $\beta = 0.001$. This step is also accomplished without inserted delays.

978-1-4673-5289-5/12 $31.00 © 2012 IEEE   180

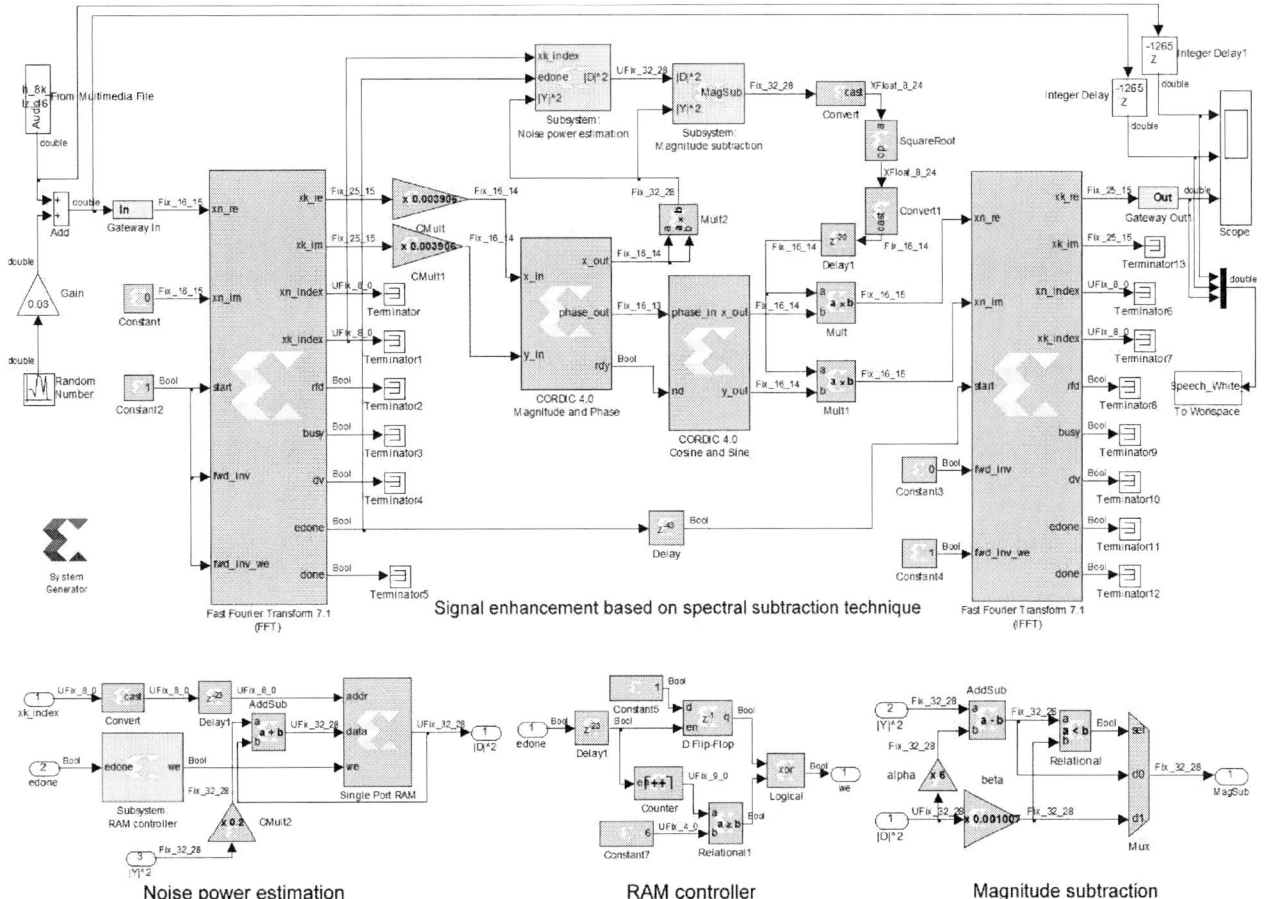

Fig. 2. Spectral subtraction architecture based on Xilinx system generator blockset for speech enhancement. Architectures of the Noise power estimation, the RAM controller, and the magnitude subtraction subsystems are also detailed.

E. Real and imaginary parts

A second Xilinx CORDIC block is used to compute the cosine and sine of the input phase $\varphi_y(m-m_2,k)$. As shown in Table I, this block presents a latency of 20 samples. A square root block is used to extract the magnitude $A_s = |\widehat{S}(m-m_2,k)|$ from its input $|\widehat{S}(m-m_2,k)|^2$. An equivalent delay (20 samples) is inserted in the magnitude line to obtain $A_s = |\widehat{S}(m-m_3,k)|$, where $m_3 = m_2 + 20 = 654$. The Real and imaginary parts are computed using two multiplier blocks:

$$\widehat{S}_r(m-m_3,k) = A_s \cos\left(\varphi_y(m-m_3,k)\right) \quad (12)$$

$$\widehat{S}_i(m-m_3,k) = A_s \sin\left(\varphi_y(m-m_3,k)\right) \quad (13)$$

F. Inverse Discrete Fourier transform

Configured as IDFT, a second Xilinx FFT block is used to recover the enhanced signal. Taking into account the latency of this block (Table I), the enhanced signal is obtained by:

$$\widehat{s}(m-m_4,n) = \frac{1}{N}\sum_{k=0}^{N-1}\widehat{S}(m-m_3,k)e^{jnk2\pi/N} \quad (14)$$

TABLE II

RESOURCE UTILIZATION AND MAXIMUM OPERATING FREQUENCY OF THE VIRTEX-6 XC6VLX240T CHIP USING 4-MULTIPLIER STRUCTURE [10].

Resource utilization	
Flip Flops (301,440)	8,870 (2.9%)
LUTs (150,720)	7,762 (5.1%)
Bonded IOBs (600)	42 (7.0%)
RAMB18E1s (832)	6 (0.7%)
DSP48E1s (768)	43 (5.6%)
Maximum Operating Frequency	16.560 MHz

where $\widehat{S}(m-m_3,k) = \widehat{S}_r(m-m_3,k) + j\widehat{S}_i(m-m_3,k)$. Then, the global delay of this architecture is $m_4 = m_3 + 611 = 1265$.

Table II gives the resource requirement and the maximum operating frequency using 4-multiplier structure for complex multiplications in the FFT/IFFT blocks. These values are obtained for the fixed-point format shown in Fig. 2. The required resources are negligible for this kind of FPGA.

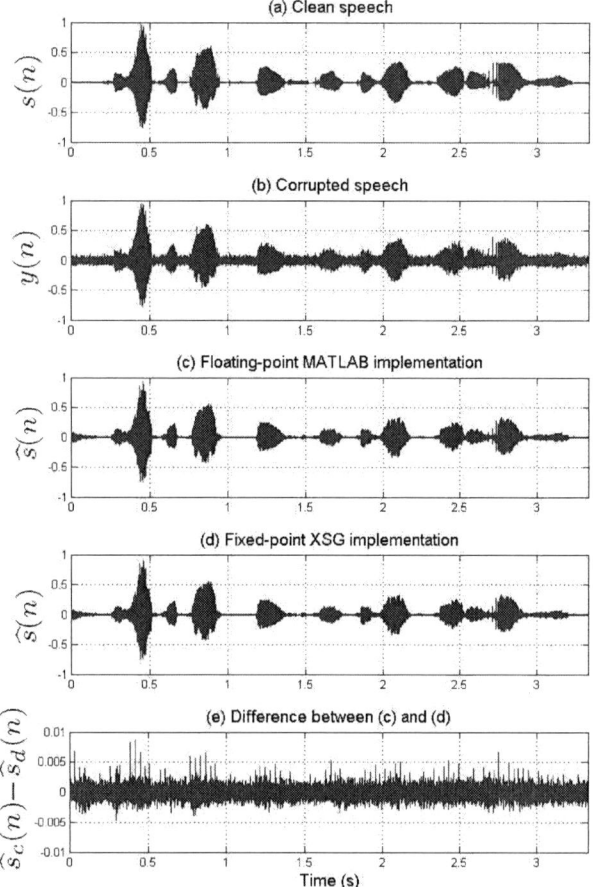

Fig. 3. Speech enhancement results: (a) clean speech signal, (b) speech signal corrupted with artificial white noise, (c) enhanced speech with floating-point MATLAB implementation, (d) enhanced with fixed-point XSG implementation, and (e) Difference between MATLAB and XSG implementations.

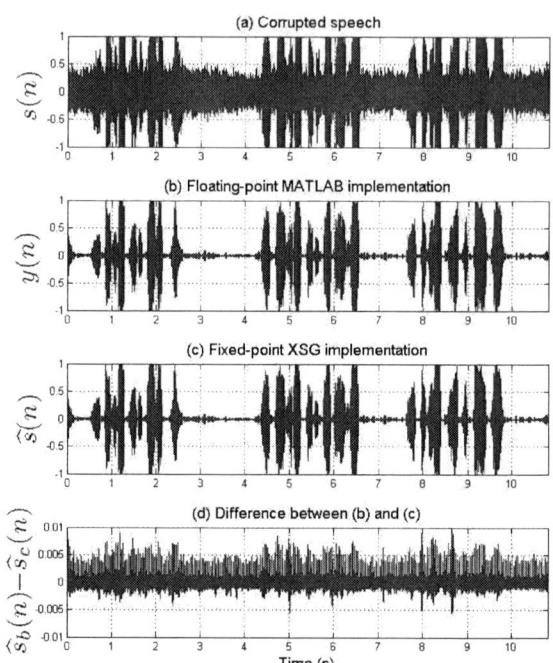

Fig. 4. Speech enhancement results: (a) noisy speech recorded in an aircraft cockpit, (b) enhanced speech with floating-point MATLAB implementation, (c) enhanced with fixed-point XSG implementation, and (d) Difference between MATLAB and XSG implementations.

IV. RESULTS AND DISCUSSION

The implemented spectral subtraction technique has been evaluated using natural speech corrupted by white noise and speech recorded in real environment. The speech signals are sampled at 8000 Hz. As shown in Figs. 3 and 4, the fixed-point XSG implementation of the spectral subtraction technique gives the same performances as the floating-point MATLAB implementation. It can be seen that noise was progressively estimated during the first five frames. Then, it was considerably removed from the noisy speech.

V. CONCLUSION

In this paper, spectral subtraction method for speech enhancement has been successfully implemented on FPGA. The noise spectrum is firstly estimated in the silence period and subtracted later from noisy signal. As a future work, a voice activity detector (VAD) will be added to this architecture to update continuously the noise power spectrum.

REFERENCES

[1] S. F. Boll, "Suppression of acoustic noise in speech using spectral subtraction," *IEEE Transactions on Acoustics, Speech, and Signal Processing*, vol. 27, no. 2, pp. 113–120, 1979.

[2] M. Berouti, R. Schwartz, and J. Makhoul, "Enhancement of speech corrupted by acoustic noise," in *IEEE International Conference on Acoustics, Speech, and Signal Processing, ICASSP'79.*, vol. 4, 1979, pp. 208–211.

[3] Y. Simard, M. Bahoura, and N. Roy, "Acoustic Detection and Localization of whales in Bay of Fundy and St. Lawrence Estuary Critical Habitats," *Canadian Acoustics*, vol. 32, no. 2, pp. 107–116, June 2004.

[4] G.-C. Chang and Y.-F. Lai, "Performance evaluation and enhancement of lung sound recognition system in two real noisy environments," *Computer Methods and Programs in Biomedicine*, vol. 97, no. 2, pp. 141–150, 2010.

[5] J. Whittington, K. Deo, T. Kleinschmidt, and M. Mason, "FPGA implementation of spectral subtraction for in-car speech enhancement and recognition," in *2nd International Conference on Signal Processing and Communication Systems, ICSPCS 2008*, 2008.

[6] U. Mahbub, T. Rahman, and A. B. M. H. Rashid, "FPGA implementation of real time acoustic noise suppression by spectral subtraction using dynamic moving average method," in *IEEE Symposium on Industrial Electronics and Applications, ISIEA 2009*, vol. 1, 2009, pp. 365–370.

[7] Xilinx, "Xilinx System Generator for DSP," www.xilinx.com/tools/sysgen.htm, Xilinx Inc.

[8] M. Bahoura and H. Ezzaidi, "FPGA-Implementation of a Sequential Adaptive Noise Canceller using Xilinx System Generator," in *21th IEEE International Conference on Microelectronics (ICM'09)*, 19-22 Dec. 2009, pp. 213–216.

[9] Xilinx, "Virtex-6 FPGA ML605 Evaluation Kit," www.xilinx.com/products/devkits/EK-V6-ML605-G.htm, Xilinx Inc.

[10] ——, *LogiCORE IP Fast Fourier Transform v7.1, Product Specification*, www.xilinx.com/support/documentation/ip_documentation/xfft_ds260.pdf, DS260 March 1, 2011.

[11] ——, *LogiCORE IP CORDIC v4.0, Product Specification*, w3.xilinx.com/support/documentation/ip_documentation/cordic_ds249.pdf, DS249 March 1, 2011.

FPGA Design of a Real-Time Obstacle Detection System Using Stereovision

Bendaoudi Hamza
Laboratoire systèmes numériques
Ecole Militaire Polytechnique
Algiers, Algeria
bendaoudi_hamza@yahoo.fr

Khouas Abdelhakim
Faculté des sciences
UMBB
Boumerdès, Algeria
akhouas@umbb.dz

Cherki Brahim
LAT Laboratory
Abou Bekr Belkaid University
Tlemcen, Algeria
b_cherki@yahoo.fr

Abstract—**Obstacle detection using stereovision is an important issue in intelligent vehicle and robot navigation, especially for the Advanced Driver Assistance Systems. This paper presents real-time obstacle detection system designed and implemented on single Field Programmable Gate Array (FPGA). The proposed hardware architecture combines stereo vision algorithms to compute the disparity map, V-disparity image, and Hough transform for obstacle detection. Considering the particular aspect of the V-disparity image and the real-time constraint, the Hough transform is only applied to detect the obstacles corresponding lines. The proposed system was tested in indoor environment using Virtex-II FPGA based prototyping board. For 640x480 pixels images, the proposed system can treat up to 180 frames/s when running at full rate, with a minimum detection time of 5.5 ms.**

I. INTRODUCTION

Obstacle detection system is very crucial to the safety of mobile robots in autonomous robot navigation. Among the various sensors used for obstacle detection, stereovision allows extracting the scene depth. Stereovision sensors are very promising and have two main advantages: they are passive and provide the highest volume of information about the scene compared to radar and laser sensors [1]. Stereovision algorithms extract 3-dimensional information, namely the disparity map (DM). DM is derived from the point-by-point matching between two images acquired from a stereo setup. Several stereovision algorithms are proposed and implemented in hardware platforms [2-4]. Given their advantages, many obstacle detection algorithms based on stereovision have been also developed. Researchers working in the field of the road safety and advanced driver assistance systems, proposed an approach for obstacle detection based on the construction of the V-disparity image (VDI). This image, which is obtained from DM, provides an efficient and simple geometrical representation of the road scene. The analysis of VDI makes it possible to extract the obstacle positions and the road profile. Real-time stereovision applications requires huge calculations, thereby adequate hardware solutions are needed. Ventroux *et al.* developed an expensive obstacle detection system using 5 FPGAs, 4 DSPs,

an ARM processor and 6 banks of RAM [3]. The proposed solution uses V-disparity algorithm and achieves a rate of 22 frames/s (f/s) on 640x480 images size, and dissipates about 100 W when running the application at full rate. A software application based on the V-disparity algorithm that turns in multiprocessor system-on-chip (MP-SoC) was developed by Greiner *et al.* [4]. The application is able to process 40 stereoscopic pairs per second with 256 lines of 512 pixels images. An implementation in Pentium 4 2.4 GHz for a mobile robot was done by Bai *et al.* [5]. For a resolution of the input image pairs of 320x240 pixels, this solution achieves a rate of 4.4 f/s. Cong *et al.* developed a software implementation for Unmanned Ground Vehicle [6]. For images of 640x480 pixels, the implementation achieved a frame rate of 5 f/s. The previously developed systems have various disadvantages in regard to the implementation platform and stereo matching method such as frame rate, scalability, area efficiency and cost. To overcome these disadvantages, we propose dedicated hardware architecture for a real-time obstacle detection system. The entire stereovision based obstacle detection system, including stereo matching, VDI construction, Hough Transform (HT), and distance calculation is designed and implemented on a single FPGA. The detection system is pipelined and synchronized with a pixel sampling clock. Synchronizing the entire system with individual data element rather than control flow eliminates the sequential bottleneck of the image data treatment and improves the system performances. Compared to the previously developed systems, our obstacle detection system is real time, compact, low cost and very suitable for mobile robots applications.

The remainder of this paper is organized as follows: Section II presents common background information about the stereovision process for obstacle detection, Section III gives a detailed description about the hardware architecture and implementation of the proposed system, Section IV discusses and evaluates the experimental results, and Section V draws the conclusion.

II. OBSTACLE DETECTION BACKGROUND

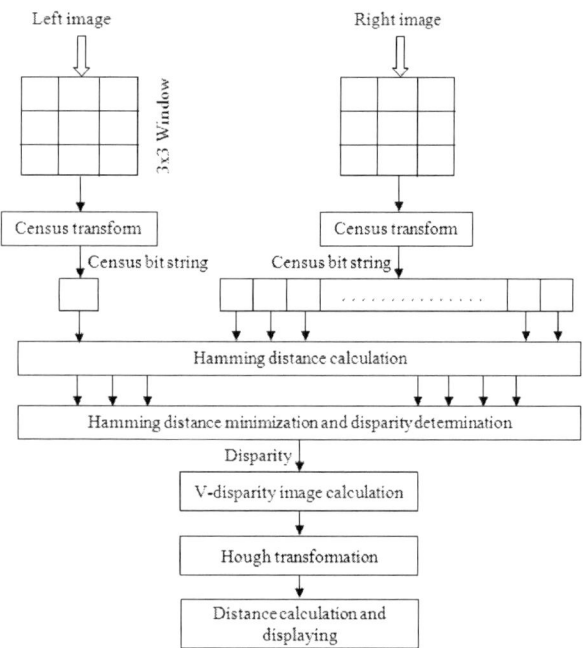

Figure 1. Obstacle detection system overview.

Figure 2. Binarized VDI for three different thresholds.

The stereovision approach used in this work is based on Census algorithm for DM and VDI calculation. Census algorithm is an area based approach, which is used for its robustness to the random noise within a window, and its bit-oriented cost computation [2]. The detection method is generic and can detect all kinds of obstacles, even in case of partial occlusion. The Census algorithm is robust with respect to noise or bad conditions (e.g. weather, night, low lighting, …etc.) [6], thus, it is suitable for obstacle detection in road context. As shown in Fig 1, our obstacle detection system includes the following main modules:

- Right and left images acquisition ;
- Disparity map calculation ;
- V-disparity image construction ;
- Global surfaces extracting using HT ;
- Obstacle distance calculation.

A. Disparity Map Calculation

Stereovision algorithms intend to recover depth information using two images of the same scene taken from distinct viewpoints. The displacement of the objects in both images is counted in pixels and called disparity. The disparity is inversely proportional to the pixel depth. DM contains the disparity of each pixel of the original image, it is computed by stereovision algorithms and used with triangulation for object distance calculation [7]. In this work, we use Census area-based approach for stereovision computation because it's hardware-friendly, simple to implement and mostly used for real-time disparity calculation. The Census transform introduced in [8] uses non-parametric local transforms that tolerate factionalism.

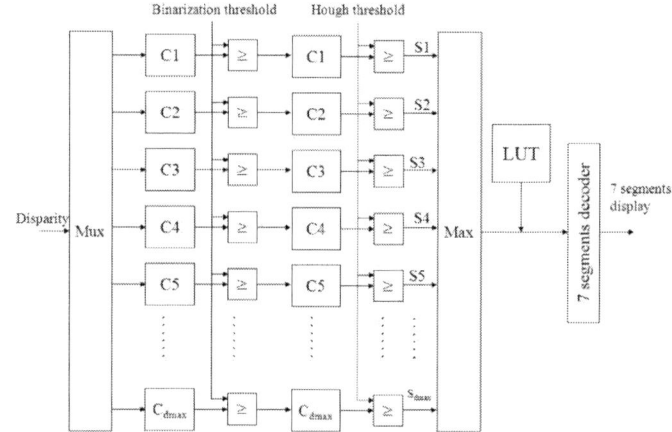

Figure 3. Hardware architecture of VDI construction, HT computation and distance calculation.

The census transform maps the window surrounding the center pixel to a bit vector representing the local information about the center pixel and its neighboring pixels. If the intensity value of a neighboring pixel is less than center pixel's intensity value, the corresponding bit is set to 1, otherwise it is set to 0 [9]. The center pixel's intensity value is replaced by the bit string composed of set of comparison results. The matching between two pixels is calculated using a similarity function, in case of the Census algorithm. As a similarity function, we use the Hamming distance which is based on binary comparison to determine the number of bits that differ between two bit strings. The comparison is made between one pixel of the left image and several pixels candidates from the right image. The number of these pixels is defined as the disparity range. The Hamming distance is calculated and compared for the disparity range pixels. The order of the pixel that minimise the Hamming distance in disparity range is taken as disparity.

B. V-Disparity Image Calculation

VDI was first introduced by Labayrade *et al.* for obstacle detection using DM produced by stereovision process [10]. VDI can be considered as a 3D graphical representation of DM. After DM calculation, VDI is constructed by accumulating the number of pixels with the same disparity that occur on a given image line. The number of lines of the VDI is equal to DM lines number, and the columns number is equal to the disparity range. The obstacle is represented by a vertical line in the VDI, and its abscissa along the

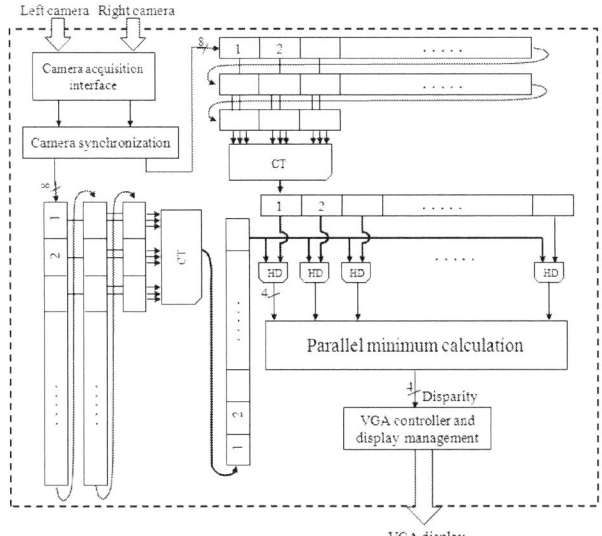

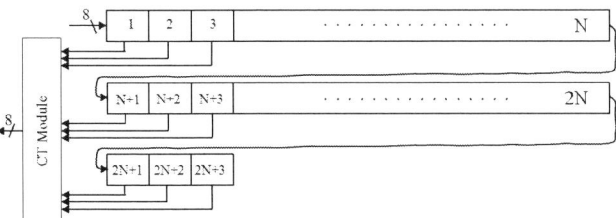

Figure 4. Hardware architecture of the real-time obstacle detection system using a 3x3 pixels window size.

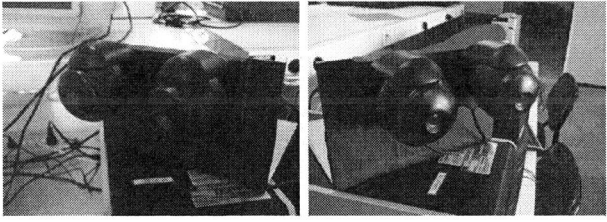

Figure 5. 3x3 pixels selection using shift registers.

Figure 6. Stereoscopic bench used.

horizontal axis represents the distance between the obstacle and the camera.

C. Obstacle Detection and Distance Calculation

Hough Transform is a tool for image parametric curves detection proposed by Hough in [11]. In our application, the obstacles are represented by vertical lines in VDI. An approach of HT is used to detect only vertical lines. Equation (1) is used for obstacle distance calculation, where d is the calculated disparity, f is camera lens focal length, and b is the distance between the two cameras.

$$distance = \frac{f \cdot b}{d} \qquad (1)$$

Before applying HT, VDI is binarized with a carefully selected threshold in order to improve the detection process.

Fig. 2 shows example of image results of binarized VDI for three different thresholds.

III. HARDWARE IMPLEMENTATION OF THE OBSTACLE DETECTION SYSTEM

A. DM Calculation with Census Algorithm

There are sevral algorithms for DM calculation, in our application; the Census algorithm is selected as described above. Fig. 4 describes the overall system architecture based on Census algorithm for a 3x3 pixels window size and a disparity range of 20 pixels. CT module is the bit string construction module, and HD is the Hamming distance calculation module. Stereo matching is divided into two stages, the Census transform stage and the disparity calculation stage. For the Census transform stage, the left and right images are transformed into images with Census vector pixel values, instead of gray-level intensity pixel values. For a 3x3window size, the Census transform generate an 8 bits long transformed bit vector per each pixel. Fig. 5 shows how to access simultaneously to the 9 pixels of a 3x3 window using a shift register, the length of this register is equal to 2N+3, N represents the numbers of columns of the image. The disparity calculation stage evaluates the similarity between the Census vectors generated by the left and right Census transform. At this stage, Hamming distances are evaluated using a template vector string for a pixel in the left image and the corresponding vector strings for pixels in the right image. After the parallel comparisons, the order of the two pairs with the shortest Hamming distances determines the resulting disparity.

B. Obstacle Detection

VDI construction and HT for obstacle detection and distance calculation are carried out after DM calculation using the hardware architecture represented in Fig. 3. The proposed architecture uses a set of parallel accumulators and comparators to construct and binarize VDI. Another set of accumulators and comparators is used to perform HT.

A carefully selected Hough threshold is used to detect vertical lines in VDI. The coordinate of the detected vertical line is the corresponding disparity. As shown in equation (1), the distance between the obstacle and the camera is calculated for each disparity and saved in a Luck Up Table (LUT). The LUT facilitates the distance calculation after identifying the obstacle corresponding disparity. A 7-segments displaying system is used to display distances.

IV. EXPERIMENTAL RESULTS

The obstacle detection system is designed using VHDL language for coding, and Xilinx Integrated Software Environment ISE 9.2i CAD software for synthesis, technology mapping, placement, and routing. The system is implemented on Celoxica RC200 development board based on Xilinx Virtex-II Xc2v10004FG456C FPGA and interfaces two 330 line CCD cameras. The stereoscopic bench used is shown in Fig. 6. An example of the complete system showing detected obstacle at 5.5 m distance is shown in Fig. 7. The implemented system is also capable to detect

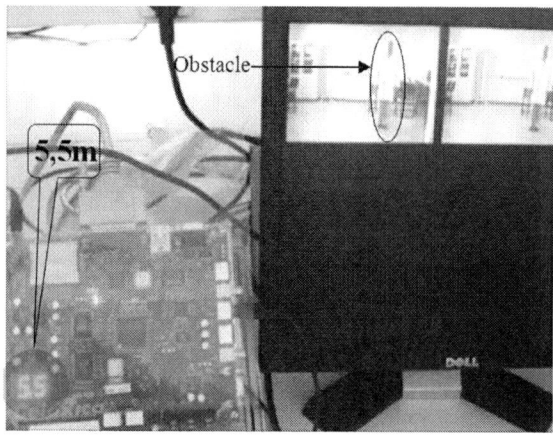

Figure 7. Example of the obstacle detection system showing detected obstacle at 5.5 m.

multiple obstacles, but it display only the distance of the nearest obstacle.

Table I summarizes the device utilization reports from the Xilinx synthesis tool. The window size of the Census algorithm used to calculate disparity is 7x7 pixels, with a maximum disparity range of 20 pixels. Due to blocks RAM available on the used FPGA, the maximum image size we can treat is limited to 280x180 pixels. The maximum achieved frequency of the system is 28,2 MHz, the system is capable to treat 560 frames per second and the minimum time required to detect an obstacle is 1,85 ms. For a 640x480 image, our design was synthesized for the Xilinx XC4VLX200-Ff1513-11 FPGA. The maximum achieved frequency is 51,7 MHz which correspond to a rate of 180 f/s and the processing time to detect an obstacle is 5,5 ms (i.e. between the obstacle apparition and its detection by the system, a vehicle at speed of 100 Km/h will travel only 0,15 m).

Table II shows the frame rate performance comparison result. As shown, the new system gives significant frame rate improvements over the previous obstacle detection systems.

V. CONCLUSION

In this paper, we have built an FPGA-based real time obstacle detection system using Census stereovision algorithm, V-disparity image construction and Hough transform. The proposed system was implemented within a single FPGA including all the processing stages, DM calculation, VDI construction, HT and distance calculation of the detected obstacle. The real-time system performances are tested and validated in indoor environment. The system is capable to treat up to 180 images of 640x480 pixels per second. The processing time to detect an obstacle is 1,85 ms for 280x180 pixels image and 5,5 ms for 640x480 pixels image. The system is also able to detect generic and multiple obstacles.

In this work we proved the detection system feasibility, future works will be focused more on the detection quality (e.g. distance accuracy and resolution, volume of detected obstacle, false alarm rate, etc.). The proposed system can be used for different applications such as intelligent mobile robots, automotive, and robots navigation.

TABLE I. LOGIC UTILIZATION FOR REAL TIME DETECTION SYSTEM IMPLEMENTATION (7X7 PIXELS WINDOW, 20 PIXELS DISPARITY RANGE AND 280X180 PIXELS IMAGE).

Resources	Used	Available	Utilization
CLB Slices	4756	5120	92%
Slices FFs	2439	10240	23%
LUTs	7183	10240	70%
Blocs RAM	39	40	97%
Maximum Frequency	28,229MHz		

TABLE II. FRAME RATE COMPARISON OF REPORTED OBSTACLE DETECTION SYSTEMS

Obstacle detection system	Image size (pixels)	Rate (f/s)
FPGA-DSP Virtex 4-ADSP-TS201S [3]	640x480	22
MP-SoC (32 MIPS R3000) [4]	256x512	40
Software program (P4, 2.4 GHz) [5]	320x240	4.4
Software program (P4, 2.8 GHz) [6]	640x480	5
FPGA (Virtex II) [This paper]	320x240	560
	640x480	180

REFERENCES

[1] C. Pantilie, S. Bota, I. Haller, S. Nedevschi, "Real-time obstacle detection using dense stereo vision and dense optical flow," in Proc. IEEE 6th International Conference on Intelligent Computer Communication and Processing, Cluj-Napoca, Romania, Aug. 2010, pp. 191-196.

[2] J. Seunghun, C. Junguk, P. Xuan Dai, L. Kyoung Mu, P. Sung-Kee, K. Munsang, J. Jae Wook, "FPGA Design and Implementation of a Real-Time Stereo Vision System," IEEE Trans. on Circuits and Systems for Video Technology, vol. 20, no. 1, Jan. 2010, pp.15-26.

[3] N. Ventroux, S. Renaud, P. Frédéric, V. Pierre-Emmanuel, G. Stéphane, "Stereovision-based 3D obstacle detection for automotive safety driving assistance," in Proc. of IEEE 12th International Conference on Intelligent Transportation Systems, St. Louis, MO, Oct. 2009, pp. 1-6.

[4] A. Greiner, F. Petrot, M. Carrier, M. Benabdenbi, R. Chotin-Avot, R. Labayrade, "Mapping an obstacles detection, stereo vision-based software application on a multi-processor system-on-chip," IEEE Intelligent Vehicles Symposium, Tokyo, Japan, 2006, pp. 370-376.

[5] M. Bai, Y. Zhuang, W. Wang, "Stereovision based obstacle detection approach for mobile robot navigation," International Conference on Intelligent Control and Information Processing, Dalian, China, Aug. 2010, pp. 328-333.

[6] Y. Cong, J. Peng, J. Sun, L. Zhu, Y. Tang, "V-disparity Based UGV Obstacle Detection in Rough Outdoor Terrain," Acta Automatica Sinica, vol. 36, no.5, 2010, pp. 667-673.

[7] K. Ambrosch, M. Humenberger, W. Kubinger, A. Steininger, "SAD-Based Stereo Matching Using FPGAs," Embedded Computer Vision, Part II, Springer, London, UK, 2009, pp. 121-138.

[8] R. Zabih, J. Woodfill, "Non-parametric local transforms for computing visual correspondence," in Proc. of the Third European Conference on Computer Vision, Secaucus, NJ, USA, 1994, pp. 151-158.

[9] J. Woodfill, B. Von Herzen, "Real-time stereo vision on the PARTS reconfigurable computer," IEEE 5th Annual Symposium on FPGAs for Custom Computing Machines, Napa Valley, CA, USA, Apr. 1997, pp. 201-210.

[10] R. Labayrade and D. Aubert, "Robust and Fast Stereovision Based Road Obstacles Detection for Driving Safety Assistance," Workshop on Machine Vision Applications, Nara, Japan, Dec. 2002, pp. 624-627.

[11] P.V.C. Hough, "Machine analysis of bubble chamber pictures," in Proc. of International Conference on High Energy Accelerators and Instrumentation, 1959, pp. 554-556.

978-1-4673-5289-5/12 $31.00 © 2012 IEEE

Design of a C-Element Based Clock Domain Crossing Interface

Zaid Al-bayati, O. Ait Mohamed
ECE Department,
Concordia University,
Montréal, QC, Canada
{z_albaya, ait}@ece.concordia.ca

S. Rafay Hasan
ECE Department,
Tennessee Tech. University,
Cookeville, TN
shasan@tntech.edu

Y. Savaria
EE Department,
Ecole Polytechnique de Montréal,
Montréal, QC, Canada
yvon.savaria@polymtl.ca

Abstract—Circuit failures due to metastability and single event transients are increasing in deep sub-micron technology. Technology scaling is also causing degradation in reliability of bi-stable circuits. Synchronization circuits that are robust to metastability are important especially with the increased use of multi-clock domain designs in SoCs. In this paper, we address the design and verification of clock domain crossing interfaces (CDCs) that connect mutually asynchronous clock domains. We propose the use of C-element based synchronizers in clock domain crossing interfaces. In order to enhance the Mean Time Between Failures (MTBF) of the interface, a new C-element design with an improved metastability recovery time is proposed. The new design is implemented in 90nm CMOS technology and simulated using SPICE.

I. INTRODUCTION

Today's Systems-on-Chip (SoC) usually consist of several IP blocks integrated on a single chip. These blocks typically operate at different clock frequencies. A critical issue when designing a SoC is to safely transfer data and control signals between different clock domains. This phenomenon is commonly known as Clock Domain Crossing (CDC). A major concern with the CDC interface design is the inevitable possible timing violation constraint, which may cause metastability in the receiver clock domain. If metastability is not handled properly, it can cause problems ranging from transient errors to system failure.

There are several CDC interfacing techniques that have been developed over the years. Authors in [1] provide a survey of some successful and unsuccessful design attempts. The most common and simple CDC technique is to use a req/ack handshake protocol with two-flop synchronizers [2]. Synchronization is performed on single-bit control signals instead of data signals that might consist of several bits. An example circuit illustrating this technique is shown in Fig. 1. In this CDC, the incoming request signal is sampled by two flip-flops clocked by the receiver, with the assumption that metastability is resolved within one clock cycle. However, this is not always the case especially in modern deep-submicron technologies (DSM), where many different clock domains transfer data at high frequencies. Metastability has been extensively studied in literature. It is observed that synchronizer's reliability, measured in Mean Time Between Failures (MTBF), decreases in conditions such as low temperature, and high operating frequency [3, 4].

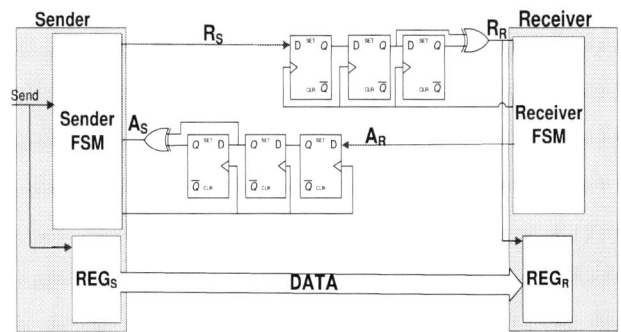

Fig. 1. Two-flop synchronizer based handshake circuit

In this work, we propose to use Muller C-elements, which are storage elements commonly used in asynchronous circuits, to synchronize incoming signals. In order to increase the reliability of the design, a new robust C-element that has a fast convergence time design is proposed. SPICE simulations have shown that at the limit of the simulator, the new C-element has a maximum convergence time of about 101 ps even at worst case timing differences. Simulations showed that the C-element's delay remains constant while the typical master-slave flip-flop's delay continues to increase exponentially beyond the C-element's maximum delay when the data and clock signals get closer. Detailed simulations are performed to test the overall design's reliability under worst case timing relations and extreme temperatures (-50°C to 50°C). The proposed design exhibits robust behavior under all simulated scenarios.

The rest of this paper is organized as follows: Section II discusses the CDC interface design. The CDC protocol is discussed first followed by the overall circuit design and finally the C-element design. Section III presents the obtained SPICE simulation results. Section IV presents our conclusions.

II. PROPOSED C-ELEMENT BASED CDC INTERFACE DESIGN

In this section, we discuss the proposed CDC interface design. We start from a high level of abstraction by giving an overview of the protocol that the circuit uses. Then, the circuit design of the synchronizer is elaborated. Finally, the proposed C-element design is discussed.

978-1-4673-5289-5/12 $31.00 © 2012 IEEE   187

A. The CDC Handshake Protocol

CDC using handshake protocols can be divided into two schemes: full handshake (4-way handshake) and partial handshake (2-way handshake). We will focus on the partial handshake protocol in this paper since it provides higher performance [5]. The circuit shown in Fig. 1 uses a partial handshake protocol. The protocol is illustrated as follows:

- The sender places a data item on the data bus and makes a request by changing the level of the sender request signal (R_S signal in Fig. 1).

- The change in the level of R_S signal gets synchronized into the receiver using the two-flop synchronizer. This change generates a single cycle pulse at the output of the XOR gate (R_R signal in Fig. 1).

- The generated pulse latches the data into the receiver domain (REG_R in Fig. 1) and at the same time, it notifies the receiver FSM that a data transfer is ongoing.

- The receiver FSM sends an acknowledgement to the sender by changing the level of A_R. A_R is synchronized by the sender's two-flop synchronizer.

- This generates a single cycle pulse at the output of the XOR gate (A_S signal in Fig. 1) which informs the sender FSM that the transfer was successfully completed.

The whole sequence of events takes a maximum of two sender clock cycles and two receiver clock cycles in addition to the time taken by the two FSMs. Metastability can occur in the circuit if the request and clock signals of the first flip-flop change at the same time. The first flip flop becomes metastable and this metastability might propagate to the second flip-flop and through the XOR gate to the receiver.

In order to achieve higher reliability, we propose to replace the first flip-flop with a custom-designed robust circuit. The proposed circuit resolves to a stable logic value faster due to its low time constant, Hence the circuit's MTBF increases according to the MTBF equation [6]:

$$MTBF = \frac{e^{tr/\tau}}{f_D \, f_{CLK} \, T_0}$$

Where:

τ: The circuit's time constant;

tr: The resolution (reconvergence) time;

f_D: The data frequency;

f_{CLK}: The clock frequency;

T_0: A parameter related to the width of the metastability window.

The reduction in τ can be achieved more effectively if the circuit design is optimized to handle only one type of transitions (falling or rising). With careful protocol development, one can restrict the 'useful' transition only to the rising (or falling) transition. In order to do this restriction without losing the performance benefit of a 2-way handshake protocol, we propose to use the design configuration illustrated in Fig. 2.

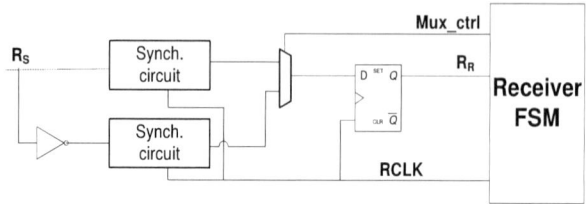

Fig. 2. The circuit configuration (signal names are for a receiver synchronizer)

The "Synch. circuit" block is our proposed reliable synchronizer design and is discussed in detail in the next section. This block is optimized to handle rising transitions only. We employ two parallel instances of this block such that only one instance is used at a time.

At the start of operation, Rs is at logic '0' and the block instance that has R_S at its input is the one that is selected by the output mux. This instance will undergo a 0-to-1 transition at the next request and the "Synch. circuit" is optimized to handle this transition. Once the receiver knows that a request is made, it toggles the mux select signal. In this way, the next request (which is signaled by a 1-to-0 transition on R_S and a 0-to-1 transition on R_S' signal in Fig. 2) is handled by the "Synch. circuit" instance with R_S' at its input. The receiver's FSM toggles the mux control after each transfer request.

The "Synch. circuit" block also deals with the generation of the one cycle pulse internally thereby eliminating the need for the third flip-flop and the XOR gate shown in Fig. 1. The circuit in Fig. 2 replaces the traditional edge detect synchronizer shown in Fig. 1 at both ends. In the next section, we will discuss the "Synch. circuit" block design.

B. Proposed Circuit Design

The synchronizer circuit design is shown in Fig. 3. For brevity, this section explains the circuit in terms of the receiver synchronizer. As explained in the previous section, we only have to consider synchronizing the 0-to-1 transition as the circuit with the 1-to-0 transition is masked out by the mux in Fig. 2.

The core of the circuit is composed of two Muller-C elements and an XNOR gate. These three gates make sure that the asynchronous input (R_S in Fig. 3) is retimed to be synchronous to the receiver clock. The two Muller C-

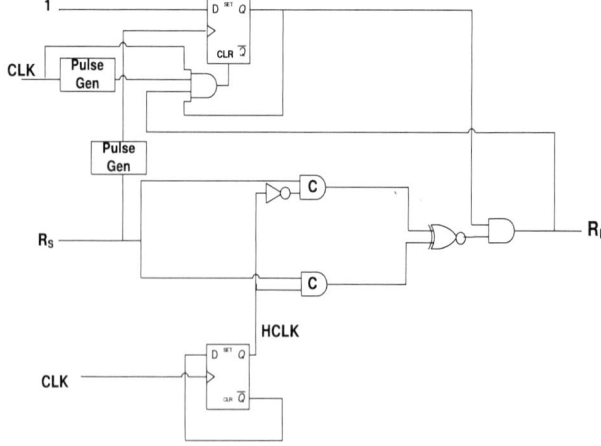

Fig. 3. The proposed design (signal names are for a receiver synchronizer)

978-1-4673-5289-5/12 $31.00 © 2012 IEEE

elements have the asynchronous R_S signal as one input. The other inputs to the two C-elements are a clock signal running at half the frequency and the same phase as the receiver clock (we denote this slow clock as HCLK) and its complement (HCLK'). The outputs of the C-elements are connected to the XNOR gate.

When the R_S signal rises signaling a data transfer request, only one C-element immediately propagates logic '1' to its output. The other C-element stays low (since it has logic '0' at the HCLK or HCLK' input). Subsequently, the XNOR gate output stays at logic '0'. When the clock toggles, the other C-element will turn high driving the output of the XNOR gate high. This event occurs with a small deterministic delay with respect to the clock edge and is therefore deterministic in the receiver clock domain. The clock inputs of the C-elements toggle values every clock cycle. This means that the circuit output is only allowed to change once per receiver clock cycle at the start of the cycle.

The flip-flop and other companion gates at the top of figure have two functions. First, they make sure that the output of the circuit remains low when there is no pending request. The second function is that the flip-flop and gates are responsible for bringing down the output R_R signal after it is latched into the receiving domain, When the R_R signal goes high, the 4-input AND gate brings it down immediately after the next positive clock edge through the DFF reset input. This ensures that the circuit generates a one-cycle pulse to comply with the handshake protocol.

If R_S and HCLK change level at the same time, one of the C-elements may produce a metastable output. This can even propagate to the output of the following XNOR gate. To reduce the probability of metastability reaching the next stage, the C-element is custom designed and optimized at the circuit level. The C-element design is discussed next.

C. C-element Design

In our design, the C-element is the circuit part where metastability may occur, therefore, it has to be designed carefully. A good synchronizing C-element should have a low time constant (τ) so that the circuit's MTBF increases as discussed in Section II-A. We have used the C element implementation presented by Martin in [7] as a basis for the new C-element design. SPICE simulations have shown that it has the lowest τ among static CMOS C-element implementations. However, as the performance of the C-element determines the reliability of the interface, it was optimized for additional reliability. This was achieved by increasing the current in the C-element when it is operating in metastability. The additional current decreases the convergence time of the C-element and hence increases the design's MTBF. The C-element design is shown in Fig. 4.

The additional circuitry added to Martin's C-element is single pull-up PMOS transistor at the top of the circuit and its enabling circuit. The extra current in the C-element is provided through this path. This idea is similar to earlier works [3, 8, 9] in which the same concept is applied to latches used in two-flop synchronizers. Here C-elements are targeted. The PMOS is turned on using the enable signal (en

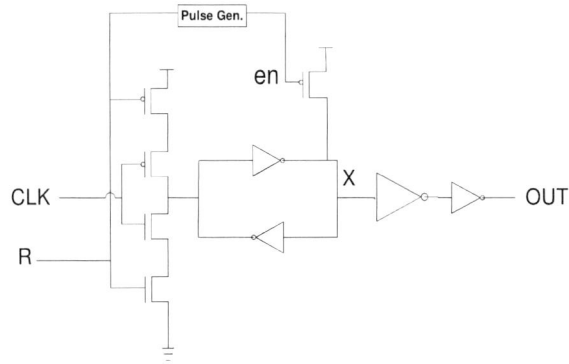

Fig. 4. The C-element design

in Fig. 4) which is generated with a deterministic delay with respect to the input regardless whether metastability has occurred or not. This is in contrast to [8], where the pull-up transistors are enabled only after metastability is detected and the metastable signal is filtered and used to enable the pull-up path. In the proposed design, a non-metastable signal (the input) is used to enable the PMOS for higher reliability. This also allows the PMOS to turn on much earlier hence reducing the convergence period. In [3], the pulse is applied with a delay after the clock to two pull-up transistors. Our simulations have shown that for our case, it provides faster convergence to a wider range of scenarios if the en signal is related to the input. Moreover, only one pull-up transistor was used instead of two. If no metastability occurs, the pull-up transistor does not affect the circuit's operation.

To further increase reliability, two back-to-back inverters are added at the output of the C- element. The first inverter has a high threshold. This high threshold prevents intermediate voltage levels associated with metastability from propagating to the output. The first inverter switches only when the voltage at node X gets close to VDD masking metastable voltage levels.

III. SIMULATION RESULTS

The proposed design was implemented with TSMC 90nm CMOS technology. To test the robustness of the proposed approach, this implementation was comprehensively simulated using HSPICE. Simulations under worst case timing differences and temperature variations from -50 to 50 degree Celsius were performed and confirmed the correct operation and the robustness of the design. Fig. 5 shows a simulation snapshot of the overall operation of the design.

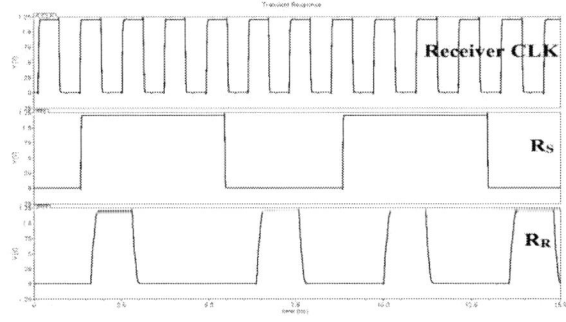

Fig. 5. SPICE simulation

978-1-4673-5289-5/12 $31.00 © 2012 IEEE

A major factor contributing to the robustness of the design is the additional current provided by the pull-up transistor in the C-elements. The effect of this current is illustrated in Fig. 6 which shows the voltage at node X (of the C-element design in Fig. 4) when the inputs of the C-element have the worst case timing relationship at fs resolution. While the C-element is entering metastability, the PMOS is turned on and provides the necessary pull-up to force the C-element to quickly converge to a stable value. The PMOS effectively sets an upper bound of 101 ps on the delay of the C-element in our simulations which were performed at 0.1fs resolution.

The comparison with the two-flop synchronizer is illustrated in Fig. 7. The figure shows the resolution times of the C-element (blue) and the master-slave flip-flop (red) of similar sizes. These times were obtained with parametric analysis in HSPICE. The resolution time is plotted versus the log of the timing displacement between the input and the metastability point. The metastability point is defined as the last point in time in which data is sampled by the bi-stable element within the same clock cycle. The figure shows that initially the flip-flop resolves faster, however, its resolution time increases exponentially as the input gets close to the metastability point (moving left to right in the figure). The proposed C-element has a higher resolution time initially but, as the input gets nearer to the metastability point, the C-element resolution time becomes almost constant around 100 ps. This means that metastability is likely to resolve before the next clock edge. At the same time, the high threshold inverter makes sure that during this resolution time, the output is kept stable and no metastability is propagated. The average time constant over the plotted region is τ=18.69 ps for the flip-flop and τ=10.33 ps for the proposed C-element.

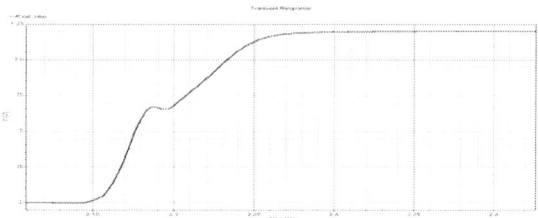

Fig. 6. Worst case response of the C-element

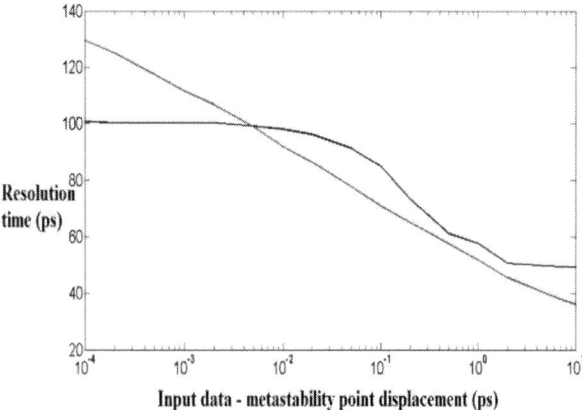

Fig. 7. Resolution time of the C-element (blue) and the master slave flip-flop (red)

IV. CONCLUSION

In this paper, we presented a new reliable CDC interface design. The interface uses two parallel C-elements as a sampling mechanism. The C-element is custom designed for high reliability. This is achieved through reducing the metastability convergence period, and by utilizing high threshold inverters to mask the output while the intermediate signals are metastable. Simulations have shown that the new C-element resolves within 101 ps even at worst case timing differences and that it has an average time constant that is 45% better than a similar-sized master-slave flip-flop.

REFERENCES

[1] R. Ginosar, "Fourteen ways to fool your synchronizer," in Proc. 9th IEEE Int. Symp. Asynchronous Circuits and Systems (ASYNC'03), 2003, pp.89–97.

[2] Cummings C., Clock Domain Crossing (cdc) Design and Verification Techniques Using System Verilog, SNUG-2008, Boston, MA, 2008.

[3] M. Kayam, R. Ginosar, and C.E. Dike, "Symmetric Boost Synchronizer for Robust Low Voltage, Low Temperature Operation," EE Tech. Rep., Technion, 2007.

[4] D. J. Kinniment, Synchronization and Arbitration in Digital Systems, John Wiley & Sons, Ltd, 2007.

[5] M. Stein, "Crossing the Abyss: asynchronous Signals in a synchronous World", EDN, pp. 59–69, July 2003.

[6] C. L. Portmann et al., "Metastability in CMOS library elements in reduced supply and technology scaled applications," in IEEE J. Solid-State Circuits, vol. 30, pp. 39–46, Jan. 1995.

[7] A. J. Martin, "Formal progress transformations for VLSI circuit synthesis," in Formal Development of Programs and Proofs, E. W. Dijkstra, Ed. Reading, MA: Addison-Wesley, 1989, pp. 59–80.

[8] J.Zhou, D.J.Kinniment, G. Russell, and A. Yakovlev, "A Robust Synchronizer Circuit", in Proc. IEEE Comp. Soc. Ann. Symp. on VLSI (ISVLSI'06), pp. 442-443, 2006.

[9] R. L. Cline. "Method and circuit for improving metastable resolving time in low-power multi-state devices" US patent 5,789,945, February 27, 1996.

Design and Implementaiton of a Fall Detection System using Compressive Sensing and Shimmer Technology

H. Rabah[1], A. Amira[2,3] and A. Ahmad[4]

[1]University of Lorraine, France
[2]College of Engineering, Department of Electrical Engineering,
Qatar University Doha, Qatar
[3]Nanotechnology and Integrated, Bio-Engineering Centre,
University of Ulster, United Kingdom
Department of Computer Engineering, Faculty of Electrical and Electronic Engineering
Universiti Tun Hussein Onn Malaysia, Johor, Malaysia
Contact: (hassan.rabah@univ-lorraine.fr,abbes.amira@qu.ed.qa, afandia@uthm.edu.my)

Abstract— **Falls are a major problem in older adults worldwide with an estimated 30% of elderly adults over 65 years of age falling each year. Timely detection of medical emergencies can reduce treatment latency and significantly improve healthcare services. This paper presents the design and implementation of a fall detection system using compressive sensing. The proposed system is built around Shimmer technology and deploys advanced data compression technique using the orthogonal matching pursuit (OMP) algorithm. The detection approach is based on the manipulation of three axial accelerometer data acquired from the Shimmer wireless medical device and reconstructed using the OMP algorithm. The main system building blocks have been simulated and implemented on the Virtex-5 and Zynq 7 field programmable gate array (FPGA) using Vivado high level synthesis tool for system evaluation and IP core generation, area, power and computation time estimation. Results obtained have shown promising results for the detection of falls with different scenarios.**

Terms—**Fall detection, Compressive sensing, FPGA, Orthogonal matching pursuit, Vivado high level synthesis.**

I. INTRODUCTION

Falls are a major problem in older adults worldwide with an estimated 30% of elderly adults over 65 years of age falling each year. Timely detection of medical emergencies can reduce treatment latency and significantly improve healthcare services. For example, if a patient experiencing ventricular fibrillation is treated within three to five minutes, his chance of survival is as high as 48% to 74%. In contrast, if early treatment is not available, his chance of survival is only about five percent [1].

Unfortunately, in uncontrolled environments outside a hospital, the time to detect an emergency could be unbounded. For example, a person who experiences a stroke or a heart attack at home could go undiscovered until well after he/she has died. To reduce the time for intervention and detection in such environments, we aim to develop an automatic real-time fall detection system using Shimmer wireless platform for data acquisition, compressive sensing

techniques for data compression, and field programmable gate arrays (FPGAs) to accelerate the main system's building blocks.

There are three main categories of fall detectors based on the technological approach used. The vision-based approach, the environmental approach, and in the wearable approach where one or more wireless wearable movement sensors such as accelerometers and gyroscopes are worn by the patient, whose values are transmitted via wireless technology to remote logging and processing platform. It is the later approach we have adopted in our work because it offers advantages such as low installation cost (indoor and outdoor), small size and offers the possibility to acquire also physiological data (electrocardiography (ECG), blood pressure and electroencephalography (EEG)).

In the wearable sensing systems composed of a set of wireless sensing units, the most challenging issue is the achievement of efficient power management. This issue is generally addressed by compression of sensed data to reduce the number of transmission and take into account the energy consumption in the acquisition chain composed of accelerometer processing element and transmission part. The novel concept of compressed sensing (CS) allows a signal to be acquired and accurately reconstructed with significantly fewer samples than required by Nyquist-rate sampling [2]. Compressive sampling relies on the sparsity of a signal, or its representation in the transform domain. This sparsity can be found in many bio-signals such as ECG, accelerometer data [3]. One of the difficulties in CS is the reconstruction of the compressed signals which is a computationally intensive process. OMP technique can be used for the reconstruction of three dimensional (3D) axial accelerometer data which will be acquired from the Shimmer platform for our proposed fall detection system.

In this paper, we focus on the implementation and acceleration of automatic real-time fall detection system based on compressive sensing reconstruction, orientation and acceleration estimation of three axial accelerometer data, and fall detection decision. The paper is organized as follows: in Section II we give a detailed description of fall detection system. The proposed architecture is described in Section III. FPGA implementation results and performances are presented in Section IV. Concluding remarks and future work are outlined in Section V.

II. FALL DETECTION ENVIRONMENT

The proposed fall detection environment is illustrated in Figure 1 which consists of the Shimmer [4] wearable wireless platform communicating via Bluetooth with the Xilinx ML505 FPGA prototyping board. The preprocessed sparse 3D accelerometer data are acquired at the FPGA and reconstructed using the OMP block. The resulted signals on x, y and z are manipulated afterwards by the detection block to detect different falls.

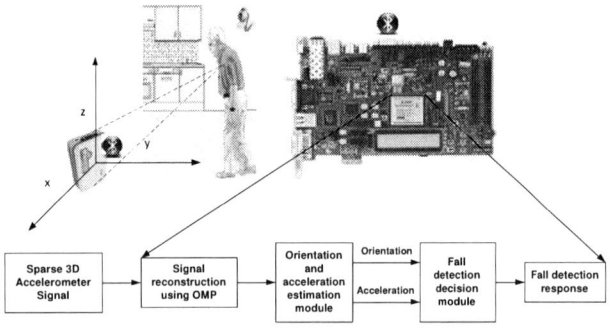

Fig. 1. Proposed fall detection environment using OMP compressive sensing technique and Shimmer platform

A. Acquisition Device (Shimmer Platform)

The three-axis-accelerometer signals are continuously analyzed algorithmically to determine whether the patient's body is falling or not. Body attached accelerometers have been used to detect human movement, including falls. The placement of a wearable 3D-accelerometer at the waist has been thought to be optimal, when compared to wrist and knee. Waist attached accelerometers are located near the body's center of gravity, providing reliable information on subject body movements. Existing fall detection solutions may only analyze the acceleration magnitude to detect falls as described in [5], [6].

B. Compressive Sensing Reconstruction Block

The compressive sampling technique is used in shimmer device for data acquisition and compression. The measured signal is defined by:

$$y = \Phi * x \qquad (1)$$

where Φ is the measurement matrix, y measured signal and x the original signal. The CS reconstruction block implements a reconstruction algorithm consisting on finding an accurate estimate of x from its inputs y and Φ. Orthogonal matching pursuit (OMP) algorithm is utilized in the proposed system. OMP algorithm has been demonstrated to be an efficient and

reliable reconstruction algorithm [7]. Compared to other techniques, OMP requires the least computation time and it is simple to implement. However, this computation time remains relatively high due to the involved complex operations such as dot product, vector matrix multiplication and matrix inversion. In the proposed reconstruction system, OMP is implemented as a hardware core.

C. Fall Detection and Decision Blocks

At every 100 ms during minimum time interval of about 1.5 second, the falls detection decision module memorizes the estimated angle θ of the patient's body orientation (upright, fall threshold, non-upright or reclining) and registers the patient's body acceleration if it crosses the impact threshold or not. Angle $\theta(0)$ is the first angle to cross the fall threshold and is defined to occur at time 0. Angles $\{\theta(-10), \theta(-9)$ to $\theta(-1)\}$ and angles $\{\theta(+1), \theta(+2)$ to $\theta(+4)\}$ correspond to angles estimated for sample points relative to 0. Once angle $\theta(0)$ is greater than fall threshold the look-up-table verfies that a change in the angle θ is at least the difference between an upright threshold and non-upright threshold occurs within maximum time interval of one second, and that the acceleration magnitude crosses the impact threshold. If these conditions are not verfied, the fall confirmation criteria are not met and the fall is not detected.

III. PROPOSED SYSTEM ARCHITECTURE

Figure 2 illustrates two variants of the proposed system architecture for compressed sensing reconstruction and fall detection and decision. The system is mainly composed of an embedded processor and hardware accelerators. The processor is used to manage and control the different operations and also execute software tasks such as decision making. The hardware cores are dedicated to accelerate CS reconstruction based on orthogonal matching pursuit and fall detection process.

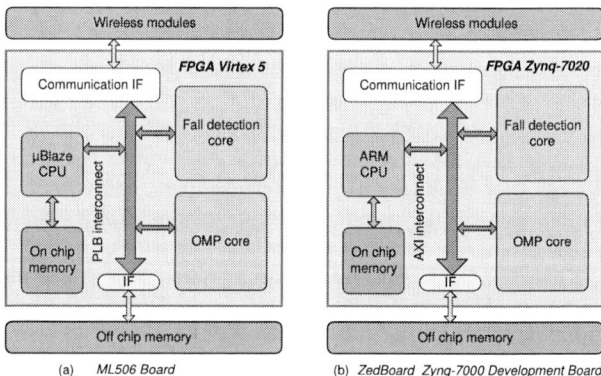

(a) ML506 Board   (b) ZedBoard Zynq-7000 Development Board

Fig. 2. Proposed architectures

A. OMP Algorithm and Architecture

In compressive sensing, the sampling is performed by multiplying the original signal $x \in \mathbb{R}^N$ by a rectangular matrix $\Phi \in \mathbb{R}^{K*N}$, with its columns denoted $\varphi_1,...,\varphi_N$ $(K < N)$. The obtained measurement vector $y \in \mathbb{R}^K$, is smaller than the signal x. The OMP algorithm requires two inputs, the sampling matrix Φ and the measurement vector y, and provides an approximation $\hat{x}$ of the original signal x, which is

978-1-4673-5289-5/12 $31.00 © 2012 IEEE

m sparse. The algorithm used in this paper is described in details by Tropp in [8]. During iteration, the algorithm chooses the column of Φ that is most strongly correlated with the residual of measurement *y*, and then it removes the contribution of this column to compute a new residual. It also computes a new estimate of the original signal, after *m* iterations, the algorithm will calculate the finale estimate of original signal. The procedure for OMP reconstruction is shown in Algorithm 1 [8].

Algorithm 1: OMP for Signal Recovery.

1- Initialize the residual $r_0 = y$, the index set $\Lambda = \{\emptyset\}$ and iteration index $l = 1$
2- Find the index λ_i that solves the optimization problem
$$\lambda_i = arg \ \max_{j=1...N} |\langle r_{i-1}, \varphi_j \rangle|$$
3- Augment the index set $\Lambda_i = \Lambda_{i-1} \ \cup \ \{\lambda_i\}$ and the matrix of chosen columns (or atoms) $\widetilde{\Phi}_i = [\widetilde{\Phi}_{i-1} \ \varphi_{\lambda_i}]$.
4- Solve a least square problem to obtain a new estimate $\hat{x}$ of signal *x:*
$$\hat{x}_i = arg \ \min_x \| y - \widetilde{\Phi}_i \ x \|$$
5- Compute the new residual:
$$r_i = y - \widetilde{\Phi}_i \hat{x}_i$$
6- Increment index *i* and return to Step 2 if $i < m$.
7- Retrieve the final estimate $\hat{x}$.

B. Fall Detection Architecture Using 3axial Accelerometer

Optimized hardware implementation has been carried out for the proposed fall detection technique presented in Algorithm 2.

Algorithm 2: Fall Detection based on Acceleration and Orientation Estimation.

1- Apply Butterworth filter on Shimmer reconstructed data.
   $Y n = A * (X n + X n-1) + B * Y n-1, A=0.0155$ and $B=0.9687$
2- Caluclate the Acceleration (Euclidian norm) of the three axes *x, y* and *z*
3- Estimate the patient's orientation relative to the reference vector
$$cos^2 (\theta cur) = (xref \ x0 + yref \ y0 + zref \ z0)^2 \ / \ (x^2_0 + y^2_0 + z^2_0)$$
4- Calculate the variables α and β
   $\alpha = (xref \ x0 + yref \ y0 + zref \ z0)^2$
   $\beta = (x^2_0 + y^2_0 + z^2_0)$
5- Set a threshold on angle and acceleration to detect the fall

---

IV. IMPLEMENTATION RESULTS AND SIMULATIONS

A. Implementation Strategy and Target Architectures

In this study, we targeted the ML506 board equipped with the Virtex5 FPGA from Xilinx (XC5VSx50T-1FF1136, and development board equipped with Zynq-7000 FPGA from Xilinx (XC7Z020-CLG484). The reconstruction algorithm is prototyped using Vivado high level synthesis tool [9]. Vivado HLS takes as its input a C, C++ or SystemC description of functionality at a high level of abstraction. It then generates a device-specific Verilog or VHDL register-transfer-level (RTL) description of a hardware implementation targeting an FPGA. RTL description is co-simulated using Modelsim. Finally an IP core is generated (Xilinx Pcore) and integrated using Xilinx platform studio (XPS) tool.

B. High Level System Validation

The validation of reconstruction is performed by describing the OMP algorithm and a testbench in C++. The testbench reads a data file obtained from the shimmed and performs a sparsisation using wavelet decomposition and measurement. The OMP algorithm is defined as a function called with measurement data as argument and returns reconstructed sparse wavelets coefficients. Figure 3 shows the simulation results obtained for the reconstruction of a fall front signal. The signal is reconstructed with a PSNR of 137 dB.

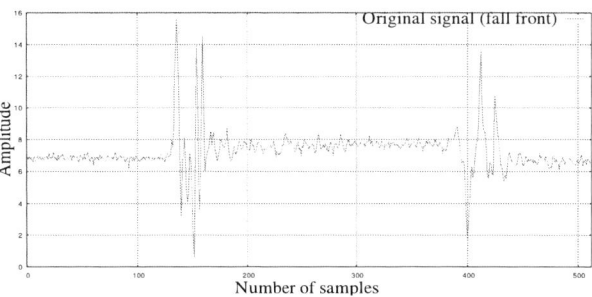

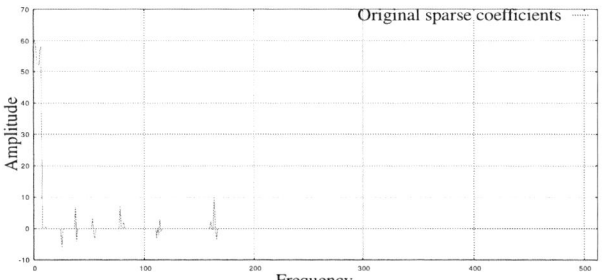

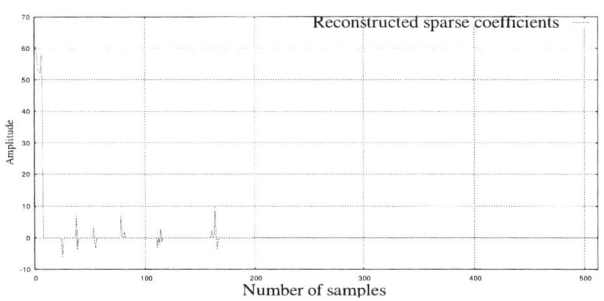

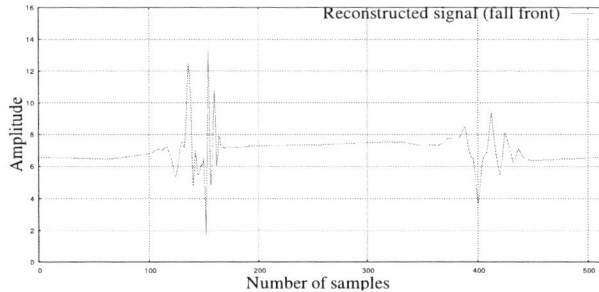

Fig.3. Fall front (component *X*), original signal, wavelet coefficients, reconstructed wavelet coefficients and reconstructed signal

C. IP Core Generation and Implementation

Vivado HLS tool is used to generate the RTL description of the reconstruction algorithm. For this first study, we have opted of the generation of an IP core using floating point representation with a single format (32 bits complying with *IEEE-754 Standard*). The only constraint imposed in this study is a clock period of 10ns. Obviously, others constraints can be applied for further and efficient architectural exploration, which is not the scope of the paper. We have targeted two FPGA families, Virtex 5 and Zynq 7. Table I gives the resource usage and comparison for these two targets. The components represent sub-block created by sub-functions. The expression can be any arithmetic operators. Table II gives the power estimate for a working frequency if 128 MHz. The obtained RTL description was validated using ModelSim and C co-simulation. After that, we have exported the design as Xilinx Pcore to be integrated in XPS flow with interfaces for PLB and AXI interconnects. Figure 4 gives the routing results for the Virtex 5 and Zynq 7 targets.

TABLE I. RESOURCE USAGE COMPARISON (VIVADO HLS RESULTS)

	BRAM		DSP48E		FF		LUT	
	VIRTEX5	*ZYNQ7*	*VIRTEX5*	*ZYNQ7*	*VIRTEX5*	*ZYNQ7*	*VIRTEX5*	*ZYNQ7*
Component	5	5	13	27	4,978	3,182	6,399	5,594
Expression	N/A	N/A	N/A	N/A	N/A	N/A	266	321
Memory	14	14	N/A	N/A	N/A	N/A	N/A	N/A
Multiplexer	N/A	N/A	N/A	N/A	N/A	N/A	662	690
register	N/A	N/A	N/A	N/A	757	494	N/A	N/A
Total	19	19	13	27	5,731	3,776	7,327	6,605

TABLE II. POWER ESTIMATION (VIVADO HLS RESULTS)

	POWER (MW)	
	VIRTEX 5	*ZYNQ7*
Component	3,180	2,220
Expression	26	32
Memory	1	1
Multiplexer	66	69
register	75	59
Total	3,348	2,381

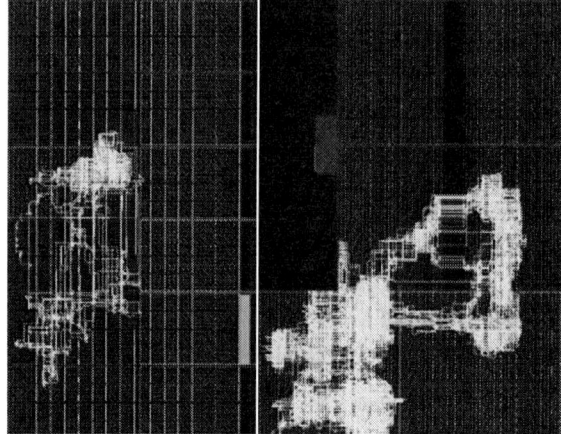

Virtex 5 Layout   Zynq7 Layout

Fig.4.Routed OMP core targeting Virtex 5 and Zynq 7

V. CONCLUSION

This paper presented an embedded system for fall detection using compressive sensing reconstruction. Different building blocks of the system have been described and a system architecture based on embedded processor and hardware accelerators is proposed targeting FPGA technology. System level design methodology with Vivado HLS flow is used for system validation and IP core generation and simulation.

A hardware implementation of the OMP algorithm for signal reconstruction in compressive sensing is presented. The architecture is prototyped and verified using high level design tool. The OMP IP core is generated using floating point representation allowing a good tradeoff between precision, flexibility and footprint targeting Virtex 5 and Zynq FPGA.

VI. REFERENCES

[1] D.Litvak Y.Zigel and I.Gannot "A Mmethod for automatic fall detection of elderly people using floor vibrations and soundproof of concept on human mimicking doll falls" *IEEE Transactions on Biomedical Engineering,* 56:2858–2867, 2009.

[2] E. J. Candes and M. B. Wakin, "An introduction to compressive sampling," *Signal Processing Magazine, IEEE,* vol. 25, no 12, pp. 21-30, Mar. 2008.

[3] A. M. Dixon, E. G. Allstot, A. Y. Chen, D. Gangopadhyay and D. J. Allstot, "Compressed sensing reconstruction: Comparative study with applications to ECG bio-signals," *in Circuits and Systems (ISCAS), IEEE International Symposium on,* May 2011, pp. 805 –808. 2011.

[4] www.shimmerresearch.com (Online)

[5] J. Reina-Tosina M. Prado and L. Roa. "Distributed intelligent architecture for falling detection and physical activity analysis in the elderly" *In Proceedings of the 2nd Joint EMBS/BMES Conference,* pages 1910–1911, 2002.

[6] M. Stuber W. Keck U. Lindemann, A. Hock and C. Becker. "Evaluation of a fall detector based on accelerometers: A pilot study" *Medical and Biological Engineering and Computing,* 43:548–551, 2005.

[7] P. Dymarski, N. Moreau and G. Richard, "Greedy sparse decompositions: A comparative study," *Journal on Advances in Signal Processing,* pp. 1-16, August 2011

[8] J. A. Tropp, Anna and C. Gilbert, "Signal recovery from random measurements via Orthogonal Matching Pursuit," *IEEE Trans. Inform. Theory,* vol. 53, pp. 4655-4666, 2007.

[9] http://www.xilinx.com/products/design-tools/vivado/integration/esl-design/hls/index.htm (Online)

978-1-4673-5289-5/12 $31.00 © 2012 IEEE

Virtual Shared Memory Architecture for Inter-Task Communication in Partial Reconfigurable Systems

Chuan Hong, Khaled Benkrid, Ali Ebrahim, Xabier Iturbe
System Level Integration Group, School of Engineering
University of Edinburgh
Edinburgh, Scotland, UK
{C.Hong, K.Benkrid, A.Ebrahim, X. Iturbe}@ed.ac.uk

Abstract—This paper presents a virtual shared memory architecture for inter-task communication in partial reconfigurable systems. The hardware tasks communicate with each other using the same content shared by physically separated Block RAMs (BRAMs). The coherence of the content is ensured by the Internal Configuration Access Port (ICAP), rather than conventional on-chip logic. The benefit of this approach resides in the flexibility of partial task reconfiguration that results from the ICAP-based synchronization mechanism, allowing hardware tasks to behave like software tasks, as they can be swapped in/out of the chip arbitrarily without any area boundary constraints. Moreover, a fast synchronization method which uses compressed bitstream is presented in this paper. The result shows significant improvements in synchronization speed at a low area overhead.

Keywords-FPGA; Partial Reconfiguration; Reconfigurable System; Inter-Task Communication; Fast Configuration

I. INTRODUCTION

Field Programmable Gate Arrays (FPGAs) have developed rapidly in the latest thirty years, as they offered custom hardware performance with reprogrammability [1]. With process technology scale-down, modern FPGAs have achieved significant levels of hardware complexity, reconfiguration speeds, efficiency and convenience. The dynamic partial reconfiguration (DPR) feature in particular gained considerable interest, mainly by the academic community, in the late 90's as they offered the possibility of hardware tasks being swapped in and out of FPGA on demand and on the fly [2]. This effectively brings three main advantages. 1) silicon resources can be exploited more efficiently; 2) fault tolerance can be achieved by reallocating hardware tasks away from damaged resources; and 3) power is saved by removing non-running or idle tasks off chip. Encouraged by these, various approaches have been proposed to exploit DPR advantages. Xilinx Corp. [3], the leading FPGA provider, contributed a series of innovations in this area, from hardware support, e.g. Internal Configuration Access Port (ICAP) [4], via design flow solutions, to software APIs (e.g. PR-PlanAhead, MicroBlaze ICAP driver) [5]. With such support, academic researchers focused more on system level design methodologies [6] and design tools [7], with practical applications as demonstrative case studies. However, DPR has not been widely adopted mainly because improvements in performance fail to compensate for the design complexity, tool immaturity, and sometimes lack of proper support. To counter this, techniques usually range from enhancing existing solutions e.g. with more functionality and reliability, or develop brand new techniques with ambitious new targets. Most of the state of the art is inclined towards the former [6][7][8]. Our approach however consists in the development of a software-like Reliable Reconfigurable Real-time Operating Systems (R3TOS) on FPGAs for we believe that the widespread of this technology will not materialize until a software-like programming model is developed for it [9].

One particular problem with many previous DPR techniques is its reliance on static predefined regions for reconfigurable hardware tasks, whereby static boundaries are straddled with static communication interfaces, e.g. Bus Macros or Proxy Logic [10]. Therefore, the resources of the reconfigurable regions are wasted since the static area of any region is made to fit the biggest possible task [11]. In R3TOS, however, hardware tasks are not bound to static reconfigurable regions but can be placed freely in any suitable area of the chip. For this, the ICAP port is harnessed for task intercommunication and synchronization [12]. In this paper, we present a novel virtually shared memory mechanism for inter-task communication and synchronization, which is inspired by the symmetric multiprocessing (SMP) paradigm in in multiprocessor programming. In it, physically separated on-chip Block Memories (BRAMs), which belong to hardware tasks, are synchronized by the ICAP Port as shared memory, thus reflecting the same content at all times. To minimize ICAP communication overhead, we also present a fast configuration method using compressed bitstream, together with a custom designed ICAP controller, which achieves two orders of speed-up improvement, compared with Xilinx HWICAP IP core [5].

The remainder of this paper is organized as follows. First, section II presents an overview of R3TOS Then, the detailed shared memory inter-task communication architecture and its implementation are presented in section III. After that, section IV introduces the fast configuration using compressed bitstream and our custom designed ICAP controller. Finally, section V gives implementation result, before conclusions are drawn in section VI.

II. BACKGROUND

In R3TOS, FPGA chip area is divided into two partitions: 1) a static region, where the R3TOS kernel resides, and 2) a

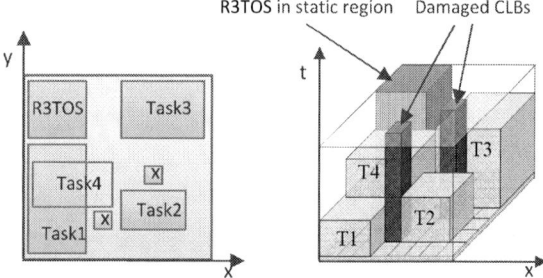

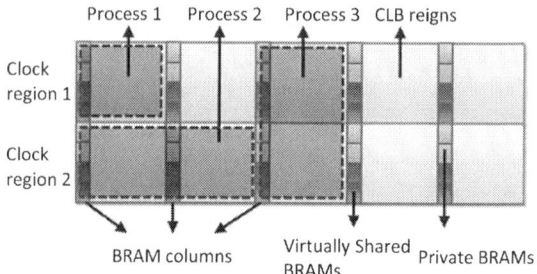

Figure 1. Task placement in R3TOS

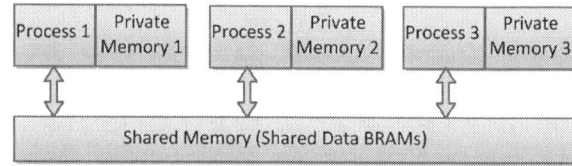

Figure 2. (a) BRAM Columns organisation on Xilinx FPGAs, (b) Proposed shared memory program architecture

reconfigurable region, where hardware (and software) tasks are flexibly allocated. The reconfigurable region is a large contiguous and uninterrupted region, rather than fragmented regions with static communication interfaces, such as Bus Macro or Proxy Logic. Fig.1 illustrates chip resource utilization in both 1-D and 2-D views. The R3TOS is located at the corner of the chip, leaving other resources to be used for reconfigurable tasks. Damaged resources (marked with a cross) are avoided by reallocating tasks around it.

The static region is composed of three main kernels: 1) a task scheduler, used for scheduling tasks in real time e.g. using the Earliest Deadline First algorithm (EDF) [13]; 2) a task allocator, which is responsible for keeping chip fragmentation minimum while allocating the upcoming tasks using an allocation algorithm e.g. EAC-BF algorithm [13]; and 3) ICAP controller, to access the chip configuration memory with maximum speed (detailed in Section. IV). All of these three components are implemented on a Xilinx soft processor PicoBlaze [14] programed in assembler. Additionally, the R3TOS has an API implemented on a MicroBlaze processor, which gives the FPGA a software look-and-feel since processes created by users can run either in pure software (on the MicroBlaze) or hardware depending on the resource efficiency. In the work presented in this paper, shared memory is used for inter-task communication (see section III) coordinated by R3TOS, essentially using a shared memory controller (see section IV).

III. VIRTUAL SHARED MEMORY ARCHITECTURE

Modern FPGAs are not homogeneous as they contain various resources e.g. Block RAMs (BRAMs), DSP Blocks or embedded processors, which are added to conventional CLB logic for the sake of the higher performance. BRAMs in particular occupy a column for each few CLB columns, providing distributed data storage capacity across the whole chip. In our architecture, task data and synchronization is based on BRAMs, namely each BRAM column of a clock region represents one single reconfigurable cell, and each task has to span at least one BRAM column of a clock region (see Fig.2a).

In most Xilinx FPGAs, each clock region's BRAM column consists of four BRAMs. In our context, the upper two are used as private data BRAMs and the lower two are used as shared data BRAMs. All shared data BRAMs in all columns keep the same content, which is used for inter-task communication and synchronization. Fig.2b shows the resulting shared memory-

based abstraction, which is inspired by a similar abstraction in the multicore programming domain. To keep all shared BRAMs coherent without incurring any logic and routing overheads, we use the internal ICAP port to synchronize data across all BRAMs through the configuration layer.

Based on this shared memory architecture, traditional SMP programming directives including barriers, critical sections, and lock variables are available in our environment. Barriers allow tasks to be synchronized at any stage of a program execution, whereas in critical sections, only one task is enabled. Using locks, one task can prevent all other tasks from accessing a particular shared memory column. These mechanisms are enabled by assigning one bit (or several bits) in the configuration memory as a flag (or flags) for every task (see Fig.3). All flags are supervised by a shared memory controller which acts as a master controller and used ICAP for this purpose. They are stored in Look-Up Tables (LUTs) residing in the first CLB column of a particular task. LUT content is accessible from both the task (functional layer) and the ICAP (configuration layer). Thereby a task can raise a request by modifying the flag-LUT content then wait for the response from the ICAP, which periodically polls the flags of each task.

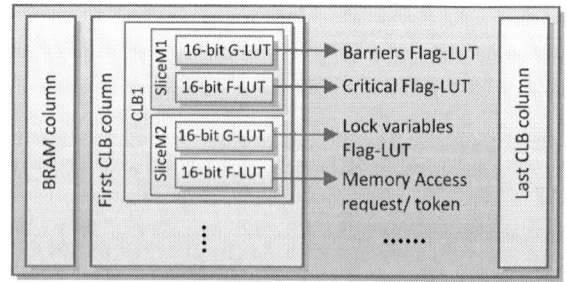

Figure 3. Directives flags stored in LUTs

978-1-4673-5289-5/12 $31.00 © 2012 IEEE   196

To maintain the coherence of the shared memory, a token is used to ensure only one shared memory write is allowed at any particular time. In order to write to shared memory, the task has to raise the request bit in its LUT, and wait for the token to be granted by the shared memory controller. Only the task holding the token has the right to write into shared memory. The request bit is polled by the shared memory controller periodically, as with other flags (see Fig.3). The polling frequency depends on the number of running tasks. To reduce the communication overhead, frequent writes are not advisable and each task is should write large data sizes of data after token is authorized, if possible.

IV. COMPRESSED BITSTREAM CONFIGURATION

A. Block RAM Configuration

To access shared memory BRAM content, the ICAP is used to configure or readback the data (in the form of partial bitstream). Indeed, the bitstream stored in the configuration memory determines both the functionalities of the CLBs and the contents of BRAMs. In general, a bitstream is composed of frames. For instance, a CLB column consists of 21 frames, whereas a BRAM column of 4 BRAMs has 64 frames, with each frame consisting of 41 32-bit words. To write or readback a frame (or frames), a series of commands need to be sent to a set of configuration registers through the ICAP, including Frame Address Register (FAR), Command Register (CMD), Frame Data Register (FDR) among others [4]. The ICAP port controller uses a finite state machine (FSM) to maximize the configuration/readback speed, with the command sequences stored in an internal memory, which is flexibly editable by a PicoBlaze processor in our implementation (see Fig.4).

The mapping between the frame word and the BRAM content is illustrated in Fig.5. Before accessing the BRAM frames, the clock of the BRAM has to be disabled to ensure it is writing/ reading safely. Four "save bits" are used as write protection bits for each BRAM in this frame. To write into a BRAM, the save bit has to be cleared. Therefore, the upper two BRAMs of a BRAM column are masked out/protected by setting the corresponding two save bits. Note that the bitstream in the bottom half of the FPGA has an inversed version compared with the top half. Therefore when copying BRAM contents from the top half to the bottom, not only the order of 41 words has to be reversed, but also the 32-bit word itself.

The shared memory controller uses the aforementioned token-based mechanism to administer the BRAM writing permissions. Only tasks that are authorized with the token can write to the shared memory. The "finish" flag is set after write completion to indicate to the shared memory controller that synchronization is needed to update all other shared memory BRAMs. The 64 frames of the modified BRAM column are then readback, masked by the save bit, and broadcasted to all other shared BRAMs.

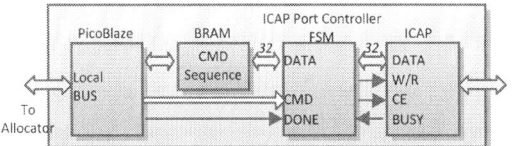

Figure 4.   Shared memory controller datapath

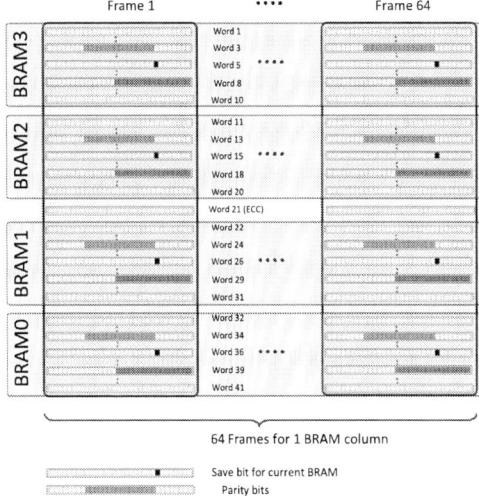

Figure 5.   Bitstream mapping with BRAM content

B. Compressed Configuration

The conventional bitstream is composed of a number of setup commands, followed by a pair of Frame Address Register (FAR) and Frame Data Register (FDR) and a start command. After the starting command, frames are written sequentially 1 word per clock cycle. Therefore, configuration time is proportional to the configuration area, or the number of frames. However if multiple frames have identical data, they can be configured by only changing the FAR using the same FDR (Fig. 6). Hence, not only the bitstream size is highly reduced, but also the configuration speed is considerably improved, with orders of magnitude speed-up (depending on the number of the identical frames).

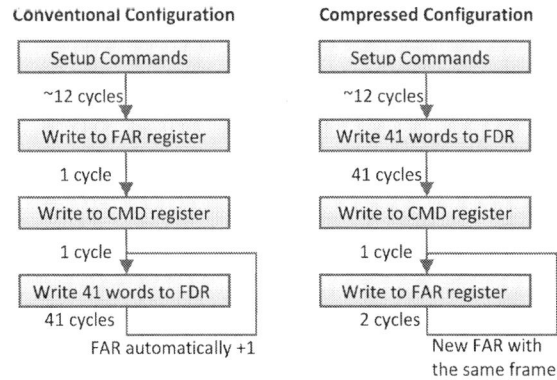

Figure 6.   Conventional vs. Compressed bitstream configuration

978-1-4673-5289-5/12 $31.00 © 2012 IEEE   197

By using compressed bitstream, the speed of broadcasting shared memory data is significantly increased, with up to ×41 improvement. Although the compressed configuration flow has been integrated to Xilinx synthesis tools for the full configuration, it has not been applied to partial reconfiguration. The compressed configuration is not only suitable for the synchronizing the BRAM content, but also useful in replicating multiple identical tasks in different locations, as well as blanking the task region after a task is removed from the chip.

V. IMPLEMENTATION RESULTS AND ANALYSIS

We have captured our proposed shared memory architecture (including the share memory controller implemented on a PicoBlaze soft processor and ICAP port controller with compressed bitstream functionality) in Verilog HDL and PicoBlaze assembly. Table I gives the resource consumption breakdown on a Xilinx XC4VFX60 FPGA showing a very low overhead. Table II gives timing performance results on the same chip. Here, the command sequence included 12 command words (32-bit) for setup, 41 words/frame for configuration data, and 2 command words for post-transmission. The XC4VFX60 FPGA supports the latest ICAP (32-bit width). Compared with the Xilinx HWICAP IP core, which typically spends more than 300 µs to configure/readback one single frame, when using a MicroBlaze processor, our ICAP port controller achieves a significant improvement in speed as it is implemented on a low footprint PicoBlaze processor. Moreover, clocked at 100MHz, our ICAP port controller exploits the maximum throughput of the ICAP port (i.e. 400 MB/s) and can even achieve better bandwidth thanks to the compressed bitstream technique. Indeed, with compressed bitstream, the configuration speed can approximate 41x400MB/s (=1.64GB/s). Note that the time it takes to configure/ readback 1 BRAM column (64×41 words) is equal to 26.81 µs, and the time it takes to configure/readback 1 CLB frame (41 words) is equal to 1.14 µs. With this, the time it takes to broadcast 10 BRAM columns, for instance, is 39.18 µs. Broadcasting to all BRAMs on the XC4VFX60 FPGA (58 columns) takes 92.94 µs.

TABLE I. SHARED MEMORY CONTROLLER RESOURCE BREAK-DOWN (ON A XC4VFX60 FPGA)

Component	Slices	BRAMs
ICAP Port Controller	357	1
Shared memory controller (PicoBlaze)	96	1
Total	453	2
Percentage of Virtex4 FX60	1.7%	0.8%

TABLE II. TIMING PERFORMANCE (100 MHZ CLOCK FREQUENCY)

Process	Cycles	Time
Readback from 1 BRAM Column	2681	26.81 µs
Configure to 1 BRAM Column	2681	26.81 µs
Broadcast to 10 BRAM Column	3918	39.18 µs
Broadcast to all BRAM Columns (58)	9294	92.94 µs
Read flags	114	1.14 µs
Release token	114	1.14 µs

VI. CONCLUSION

In this paper, we presented an efficient virtual shared memory architecture for inter-task communication on FPGAs. The physically distributed BRAMs share the same content at all times, synchronized by a shared memory controller which uses the internal ICAP port for this purpose. Therefore, no logic and routing resources are consumed reducing static logic and increasing dynamically configurable logic area for computing tasks. To reduce the communication overhead, compressed bitstream configuration was harnessed and implemented on a custom high performance and low ICAP controller. This architecture is part of a bigger project (called R3TOS) which aims to develop a reliable reconfigurable real-time operating system on FPGAs.

REFERENCES

[1] I.Kuon, R.Tessier, and J.Rose, "Fpga architecture: Survey and challenges," Foundations and Trends in Electronic Design Automation, vol. 2, pp. 135--253, 2008

[2] G. J. Brebner, "A Virtual Hardware Operating System for the Xilinx XC6200," Proc. of the International Workshop on Field-Programmable Logic, Smart Applications, New Paradigms and Compilers, 1996.

[3] Xilinx Inc., "Xilinx Partial Reconfiguration User Guide," UG702, October 2010.

[4] Xilinx Inc., "Xilinx Virtex-4 FPGA Configuration Guide," UG071, June 9, 2009.

[5] Xilinx Inc., "LogiCORE IP XPS HWICAP," DS586 July 23, 2010

[6] B.Blodget, S.McMillan, and P.Lysaght, "A lightweight approach for embedded reconfiguration of FPGAs," in Proc. Design, Automation and Test in Europe Conference and Exhibition, pp. 399--400, 2003

[7] E.L.Horta, J.W.Lockwood, D.E.Taylor, and D.Parlour, "Dynamic hardware plugins in an FPGA with partial run-time reconfiguration," in Proc. 39th annual Design Automation Conference, pp. 343--348, 2002

[8] A. Donato, F. Ferrandi, M. Redaelli, M. D. Santambrogio, and D. Sciuto, "Caronte: a complete methodology for the implementation of partially dynamically self-reconfiguring systems on FPGA platforms," in 13th Annual IEEE Symposium on Field-Programmable Custom Computing Machines (FCCM), pp. 321—322, 2005

[9] X.Iturbe, K.Benkrid, A.T.Erdogan, T.Arslan, M.Azkarate, I.Martinez, and A.Perez, "R3TOS: A reliable reconfigurable real-time operating system," in Proc. NASA/ESA Conference on Adaptive Hardware and Systems, pp. 99--104, 2010.

[10] P. Sedcole, B. Blodget, J. Anderson, P. Lysaght, T. Becker, "Modular partial reconfiguration in Virtex FPGAs," in Proc. International Conference on Field-Programmable Logic and Applications, pp. 211—216, 2005

[11] X. Iturbe, K. Benkrid, T. Arslan, I. Martinez, M. Azkarate and M. D. Santambrogio, A Roadmap for Autonomous Fault-Tolerant Systems, Proc. of the International Conference on Design and Architectures for Signal and Image Processing, pp. 311-321, 2010.

[12] X. Iturbe, K. Benkrid, T. Arslan, R. Torrego, and I. Martinez, " Methods and Mechanisms for Hardware Multitasking: Executing and Synchronizing Fully Relocatable Hardware Tasks in Xilinx FPGAs ". International Conference on Field-Programmable Logic and Applications (FPL'11), Crete, 2011.

[13] C.Hong, K.Benkrid, X.Iturbe, A.Ebrahim, and T.Arslan, "Efficient On-Chip Task Scheduler and Allocator for Reconfigurable Operating Systems," in IEEE Embedded Systems Letters, vol. 3, issue. 3, pp. 85--88, 2011.

[14] Xilinx Inc., "PicoBlaze 8-bit Embedded Microcontroller User Guide," UG129, 2011.

Memristors-based NMOS Logic Circuits

Z. Abid

Department of Electrical and Electronics
Engineering Technology
HCT-ADMC
Abu Dhabi, UAE
zeabid@ieee.org

Dirar Homouz, Baker Mohammad
KUSTAR, Abu Dhabi, UAE

Wei Wang
IOT Center, Chinese Academy of Science,
WuXi, Jiangsu Province, China

Abstract— NMOS Logic family based on memristors and NMOS transistors is proposed. The used memristor is a special metal-insulator-metal junction with unipolar switching characteristics. The suggested design can have more than 4X density improvement, with similar delay and power performance compared to CMOS logic family. The proposed NMOS Logic family can be implemented and fabricated using nMOS process technology only, leading to a significant cost reduction since the pMOS devices are not required.

Index Terms— NMOS, memristor, resistive junction, unipolar switching, CMOS.

I. INTRODUCTION

Digital circuits based on memristors rely on either dense gate Arrays of CMOS transistors and memristors (Crossbars / FPGA-like structure) or discrete components [1,2]. The used memristor devices can be unipolar or bipolar switches. These switches are usually two terminals devices leading to a structured nanoelecronic circuits [1] but can be designed with four terminals, such as memristive Si nanowire devices [3], opening broader window for various types of Logic circuits and high-density hybrid-integration with CMOS technology

The NMOS transistor logic family, augmented with two-terminal memristors, can provide a reduced foot print , and consequently higher density, compared to CMOS and pseudo-NMOS logic family. However, the speed and power dissipation of this suggested logic family need to be simulated and optimized. The N-type transistors, with high electrons mobility, integrated with a memristor fabricated in a Via, can lead to a high-density logic circuits with high-speed operation.

In this paper, we introduce a new junction-nMOS structure to implement the new NMOS logic. This device integrates the metal-oxide-metal (MIM) junction with nMOS transistors to establish the CMOS-like operation. In particular, the proposed design replaces the p-type transistor of the pseudo nMOS gate with an MIM unipolar anti-junction memristor (Fig. 2). In the pseudo NMOS logic, the p-type transistor is always in ON state and the standby power is large. Whereas, in the proposed junction-nMOS device, the nMOS devices and the MIM structure work in a complementary manner, exhibiting a CMOS-like ON/OFF switching, leading to a CMOS-like operation with minimum standby power and small footprint.

The proposed design requires a simple device structure and its fabrication is based on the standard nMOS process with the additional deposition step for the MIM junction during the via filling. This fabrication has already been used for resistive

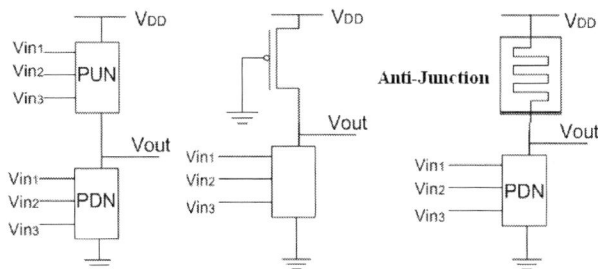

Fig. 1: a) CMOS devices with p-type pull-up (PUN) and n-type pull down (PDN) networks. b) Pseudo-nMOS device, c) Proposed logic family with nMOS transistors (for PDN) and unipolar memristors. V_{in1-3} are the input signals

memory applications [4-6], which is available for the proposed logic devices.

To compare the performance of the proposed device with CMOS logic family, we developed a compact model of the memristor MIM junction for circuit simulations

II. UNIPOLAR MEMRISTOR JUNCTION

The MIM memristance junction is a two-terminal nanoscale device consisting of two metal electrodes separated by an oxide layer [5-16]. It exhibits a reproducible hysteretic switching of bistable resistance states that can be used for information storage [4-5] (Fig. 2). By using CMOS-compatible oxides such as HfO_2 and ZrO_2, the junctions can be integrated as a Via on top of the source or drain terminal of a MOS transistor [4-6].

The integration of CMOS transistors and MIM memristor junctions can lead to efficient memory and FPGA circuits. Recently, the MIM junction has been considered as a typical memristor device, the fourth element in addition to resistor, capacitor, and inductor [7,8], that can be used to design various electronic systems. As shown in Fig. 2, the MIM junction can be bipolar (V_{set} and V_{reset} have different polarities) and unipolar (V_{set} and V_{reset} have the same polarity), depending on the electrodes, oxide, and doping materials. The unipolar memristor junction requires $V_{set} > V_{reset}$, V_{set} and V_{reset} have short and long pulses respectively. Recently, a unipolar device demonstrated an inverse switching behavior $V_{set} < V_{reset}$, compared with the standard unipolar junction [10], denoted here as a unipolar anti-junction memristor.

978-1-4673-5289-5/12 $31.00 © 2012 IEEE

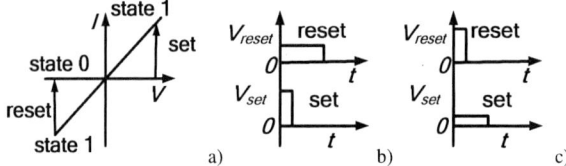

Fig. 2. a) The hysteretic characteristics of the bipolar junction with positive V_{set} and negative V_{reset}. b) Unipolar junction with positive V_{set} and V_{reset} ($V_{set} > V_{reset}$). c) Unipolar anti-junction. ($V_{reset} > V_{set}$). Note that states 1/0 are the low/high resistance states.

Based on the memristor or filament theories [7-10], we developed a compact IV model of the unipolar anti-junction memristor:

$$v(t) = \left(R_{ON} w + R_{OFF}(1-w) \right) i(t)$$

$$w = \begin{cases} 1 & t \geq T_{set} & V_{reset} \geq V \geq V_{set} \\ 0 & t \geq T_{reset} & V \geq V_{reset} \\ w & otherwise \end{cases}$$

where R_{ON} and R_{OFF} are the low and high resistances (corresponding to states 1 and 0 respectively), T_{set} and T_{reset} are the SET and RESET switching times, and V_{set} and V_{reset} are the switching voltages. Note that the switching factor w is normalized to 0 and 1, depending on the time and level of the applied voltage to the device. The unipolar junction requires $V_{set} > V_{reset}$ and $T_{set} < T_{reset}$, while the anti-junction requires $V_{set} < V_{reset}$ and $T_{set} > T_{reset}$.

III. UNIPOLAR MEMRISTOR NMOS LOGIC SIMULATION

The proposed logic circuits use one anti-junction as PUN, and nMOS devices as PDN, to establish CMOS-like operations. we consider an anti-junction-nMOS inverter (one nMOS device in the pull down network) as the case study. The inverter has a fan-out of 4 and a 10µm metal-1 interconnect as a load. The switching sequence of the anti-junction and nMOS device is shown in Table I. The anti-junction and the nMOS device work in a complementary manner.

Table I.: Operation of the proposed inverter; Logic 1 or 0 input/output correspond to approximately 0 and 1.2V voltage levels

Input	nMOS	Output	Anti-Junction
0	OFF	1	ON
$0 \rightarrow 1$	OFF $\rightarrow$ ON	$1 \rightarrow V_{T1} \rightarrow 0$ (falling)	ON $\rightarrow$ OFF (reset, leakage)
1	ON	0	OFF
$1 \rightarrow 0$	ON $\rightarrow$ OFF	$0 \rightarrow V_{T2} \rightarrow 1$ (rising)	OFF $\rightarrow$ ON (set, no leakage)

A. Fall Time:

When the input switches from logic 0 to 1 causing the output to switch from logic 1 to 0, the fall (switching) time of the output is: $T_{nMOS} + T_{reset} + T_{RC}$, where T_{nMOS} is the nMOS intrinsic time in fs ($\ll T_{RC}$) and T_{reset} is the anti-junction reset time ($< T_{RC}$). Thus, the fall time depends mainly on T_{RC} which is the RC delay due to the discharge of the load capacitor

through $R_{sd,ON}$ expected to be in the ps range, similar to a CMOS inverter. Note that this switching mechanism, giving rise to T_{RC}, is slightly different from a CMOS inverter. The reason is that anti-junction resistance resets (its resistance changes from R_{ON} to R_{OFF}) during this switching.

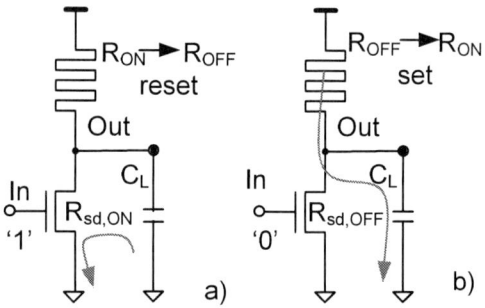

Fig. 3. The switching of the memristor-based NMOS inverter. a) The fall time (1 to 0 switching, the anti-junction memristor resets). b) The rise time (1 to 0 switching, the anti-junction memristor sets).

The switching of the memristor-based NMOS inverter from logic 1 to 0 and from logic 0 to 1 are shown in Fig. 3a and 3b respectively, where the anti-junction switches to either OFF-state (R_{ON} to R_{OFF}) or to ON-state (R_{OFF} to R_{ON}), A small portion of T_{RC} is the leakage period T_1: the time to switch the output voltage V_{out} from V_{DD} to $V_{DD} - V_{reset}$ during which both nMOS and anti-junction are ON. After time T_1, the anti-junction is OFF and the output voltage will continue its fall (to close to 0V):

$$V_{out,low} = R_{sd,ON} V_{DD} / \left(R_{sd,ON} + R_{OFF} \right)$$

with a low RC time constant and $R_{sd,ON} \ll R_{OFF}$.

In order to reduce T_1 and its associated leakage power, V_{reset} is increased. However, V_{reset} value affects the R_{ON} of the anti-junction through: $V_{DD} - V_{reset} = R_{sd,ON} V_{DD} / \left(R_{sd,ON} + R_{ON} \right)$. Thus, V_{reset} and R_{ON} values need to be optimized together. Furthermore, after the anti-junction resets (switch off) a higher R_{OFF} allows a logic "0" voltage level to be closer to 0V.

B. Rise Time:

When the output switches from logic 0 to 1, the rising time is: $T_{nMOS} + T_{set} + T'_{RC}$, and is mainly dominated by T'_{RC} since T_{nMOS} and T_{set} are much smaller than T'_{RC}. T'_{RC} is the time needed for a load C to charge initially through a resistor R_{off} then R_{on}. While T_{RC} considers the ON state resistances of both nMOS and anti-junction, T'_{RC} considers the OFF state resistances of both nMOS and anti-junction (C is the same) during the output voltage V_{out} changing from '0' to $V_{DD} - V_{set}$. Thus, it is expected that $T'_{RC} > T_{RC}$. During T'_{RC}, the circuit does not have high leakage current since initially both nMOS and anti-junction are OFF, then only NMOS stays OFF.

978-1-4673-5289-5/12 $31.00 © 2012 IEEE   200

The first portion of the rising time T_2 has a large time constant (Fig. 4a) since the load capacitance is charging through R_{OFF}, and the output voltage slowly increases to $V_{DD} - V_{set}$. Once, the anti-junction is set, the second portion T_3 has a much smaller time constant since the load capacitance is charging through R_{ON} (Fig. 4a) until the output voltage reaches $V_{out,high} = R_{sd,OFF} V_{DD} / R_{sd,OFF} + R_{ON}$. This value is close to V_{DD} (1.2V) since $R_{ON} \ll R_{sd,OFF}$

In order to reduce the time T_2, we need to increase V_{set} and reduce R_{OFF}. However, a high R_{OFF} is required to get a strong '0' output. Fig. 4a shows the switching of the memrsitor-based nMOS inverter while Fig. 4b shows the switching of a standard CMOS inverter for comparison purpose.

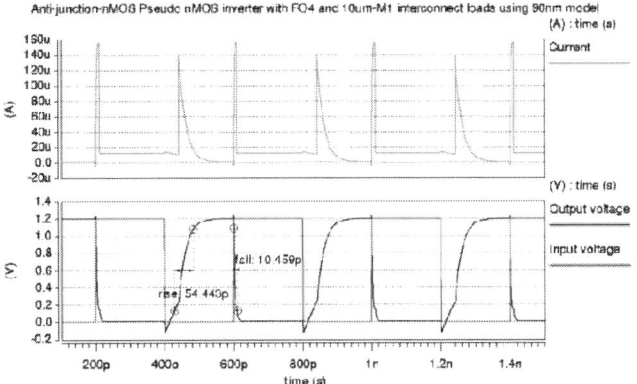

Fig. 4a: Simulation results (current I, input and output Voltages) of a memristor-NMOS inverter.

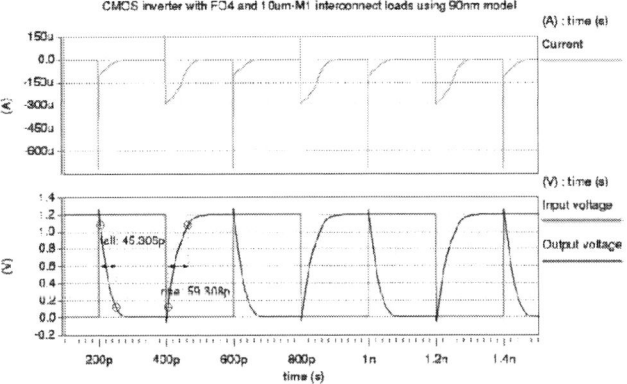

Fig 4b) Simulation results (current I, input and output Voltages) of the corresponding CMOS inverter.

Given a 90nm nMOS transistor (180nm width), V_{DD} =1.2V, its ON-resistance $R_{sd,ON}$ = 750Ω, assuming $V_{reset} - V_{set} = 0.1$V and $R_{OFF} = 100$kΩ, for V_{reset} = 0.8, 0.9, 1.0, 1.1V, then R_{ON} = 1.51, 2.26, 3.78, and 8.31kΩ, respectively. The minimum fall time is at $V_{reset} = 1.06$V, V_{set} =0.96V and $R_{ON} = 6.8$kΩ extracted from circuit simulations (as shown in Fig. 5). Consequently, $V_{out,low}$ = 9.01mV. Based on these parameters and T_1, the leakage power can be estimated.

By using the Verilog language to model the anti-junction memristor and HSPICE to simulate CMOS and NMOS inverters, the switching curves of the proposed device in the form of current I and voltage V are shown in Fig. 4. The nMOS transistor model is based on 90nm technology [19] with fan-out of 4, a 10μm metal-1 interconnect loads and V_{DD} = 1.2V. The anti-junction memristor's parameters are as follows: R_{OFF} is 100kΩ, R_{ON} is 6.8kΩ, V_{set} =0.96V and V_{reset} =1.06V. As shown in Fig. 4, the rise time is 54.4ps and the fall time is 10.5ps. The peak switching current is 155μA which corresponding to a switching energy of 2.92fJ. For comparison, the CMOS inverter has 59.3ps rise time and 45.3ps fall time with a peak current of 300μA which corresponds to a switching energy of 16.5fJ.

In order to optimize the performance of the proposed inverter, circuit simulations were conducted based on various values of Vset and Vreset. The rise time and the noise margins were extracted (Fig. 5 and 6). Fig. 6 also shows the plot of Vout versus Vin, for Vrest=1.06V and Vset=0.96V. Since the memristor has two different states (Ron and Roff), the noise margins for the rising and the falling output voltage are different (Fig. 6). The worst noise margin of the two sets were adopted as 0.43V and 0.3V for the high and low outputs respectively. Simulation results suggests that Vset and Vrest values have no effect on the noise margins, fall time, peak leakage current, and switch energy, while showing large effect on the fall time and consequently the delay.

The noise margin and other performance properties of the proposed device are summarized in Fig. 5 and 6. The design parameters of the proposed design can be adjusted to obtain similar noise margin, rise time, fall time, power dissipation and energy consumption as the CMOS device.

The values of T_{reset} and T_{set} are critical to determine the delay and power performance of the proposed inverter. If the T_{reset} and T_{set} are in the range of ps, similar to the CMOS transistor intrinsic switching time, the proposed device can have similar delay and power as the CMOS device. However, if T_{reset} and T_{set} are similar to T_{RC}, the delay and power of the proposed device will significantly increase. The results show superior performance than pseudo-nMOS but worse than the CMOS. The recent experimental results demonstrated that the T_{reset} and T_{set} values can be less than 5ns [13] for a filament size of 30nm by 50nm. By setting/resetting a small segment of the filament or using a thinner oxide layer, the switching time may be significantly improved, reaching the pS range.

C. Comparison:

Based on the above logic inverter case study, we compare the area, delay and power performance of the proposed device with CMOS and pseudo-nMOS devices [17-19] for general logic circuits with nMOS pull down network (see Table II). The results show that the proposed design reduces the area by 4X, while providing similar delay and standby power performance compared with CMOS. It also achieves more than 10X standby power reduction, while requiring similar area and delay compared with pseudo nMOS gate.

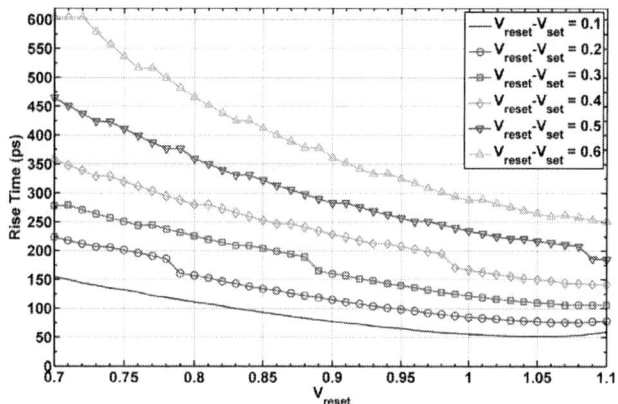

Fig. 5. Rise time of the memristor-NMOS inverter with various Vreset and Vset values obtained from HSPICE simulations.

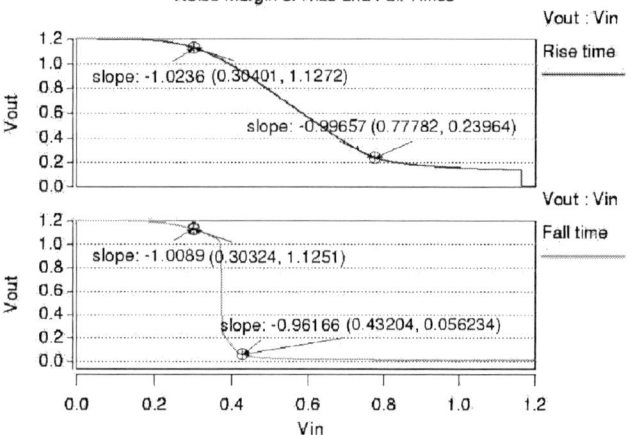

Fig. 6. Noise margin of rise and fall times. For rise time, the noise margin at high and low outputs are 0.43V and 0.3V, respectively. For fall time, the noise margin at high and low outputs are 0.77V and 0.3V, respectively.

Table II: Comparison of CMOS, pseudo nMOS and proposed design

	CMOS	Pseudo-NMOS logic	Proposed NMOS-Memristor Logic
Area	$4\,A_{PDN}$	$A_{PDN} + A_p$	A_{PDN}
Delay	$2(T_{nmos}+T_{RC})$ $\approx 2T_{RC}$	$<2T_{RC}$	$2T_{nMOS}+T_{set}+T_{reset}$ $+2T_{RC} \approx 2T_{RC}$
Standby Power	$P_{short\text{-}circuit}$	$P_{leakage}>10P_{short\text{-}circuit}$	$\sim P_{short\text{-}circuit}$

Note: In CMOS device, P type transistors require 3X area of the n-type transistor to achieve the same switching speed.

IV. CONCLUSION

In this paper, we introduced a novel memristor-based NMOS logic. It consists of nMOS transistors with one memristor operating as CMOS logic with a small standby power that can. be reduced by further design optimization. The proposed approach may revive the performance advantages of the nMOS logic that has been abandoned in the last century, while significantly reducing the power penalty. Furthermore,

the proposed device can simplify the fabrication by removing the p-type layouts and material processes, leading to a significant cost reduction. The memristor can be fabricated during the via preparation without additional lithography and mask cost. These advantages will enable the proposed structure to establish efficient nMOS digital circuits. Future work will include the use of three and four terminal memristor devices to improve the performance of the proposed logic family.

REFERENCES

[1] Wei Fei, Hao Yu, Wei Zhang, and Kiat Seng Yeo, " Leveraging Memristive Systems in the Construction of Digital Logic Circuits," Proceedings of the IEEE, vol. 100, No-6, pp. 2033-2049, 2012

[2] M. M. Ziegler and M. R. Stan, B "Design and analysis of crossbar circuits for molecular", nanoelectronics, Proc. IEEE Conf. Nanotechnol. , pp. 323–327, 2002.

[3] Davide Sacchetto, Giovanni De Micheli and Yusuf Leblebici, "Multiterminal Memristive Nanowire Devices for Logic and Memory Applications: A Review," Proceedings of the IEEE, vol. 100, No-6, pp. 2008-2020, 2012

[4] Gilberto Medeiros-Ribeiro, Janice H. Nickel, J. Joshua Yan, " Progress in CMOS-memristor integration," IEEE Conference, pp. 1-4, 2011

[5] I. G. Baek, M. S. Lee, S. Seo, M. J. Leo, D. H. Seo, D. S. Suh, J. C. Park, S. O. Park, H. S. Kim, I. K. Yoo, U. I. Chung and J. T. Moon, "Highly scalable nonvolatile resistive memory using simple binary oxide driven by asymmetric unipolar voltage pulses," IEDM Tech. Dig., pp. 587-590, 2004.

[6] An Chen, S. Haddad, Wu Yi-Ching, Fang Tzu-Ning, Lan Zhida, S. Avanzino, S. Pangrle, M. Buynoski, M.Rathor, Cai Wei, N. Tripsas, C. Bill, M. VanBuskirk, M. Taguchi, "Non-Volatile Resistive Switching for Advanced Memory Applications," IEDM Tech. Dig., pp. 746-749, 2005.

[7] D.B. Strukov, G.S. Snider, D.R. Stewart and R.S. Williams, "The missing memristor found," Nature (London), vol. 453, pp. 80-83, 2008.

[8] L.O. Chua, "Memristor-the missing circuit element," IEEE Trans. on Circuit Theory, vol. 18, pp. 507-519, 1971.

[9] R. Waser and M. Aono, "Nanoionics-based resistive switching memories," Nature Materials. vol. 6, pp. 833-840, 2007.

[10] Bin Yu, Sanghyun Ju, Xuhui Sun, Garrick Ng, Thuc D. Nguyen, M Meyyappan, David B. Janes, "Indium selenide nanowire phase-change Memory," Applied Phys. Letters, vol. 91, p. 133119 , 2007.

[11] K. Tsunodaa, Y. Fukuzumi, J. R. Jameson, Z. Wang, P. B. Griffin, and Y. Nishi, "Bipolar resistive switching in polycrystalline TiO2 films," Applied Physics Letters, vol. 90, pp. 113501/1-3, 2007

[12] Weihua Guan, Ming Liu, Shibing Long, Qi Liu, and Wei Wang, "On the resistive switching mechanisms of Cu/ZrO2: Cu/Pt," Appl. Phys. Lett., vol. 93, pp. 223506/1-3, 2008.

[13] Y.C. Yang, F. Pan, Q. Liu, M. Liu, F. Zeng, "Fully room-temperature- fabricated nonvolatile resistive memory for ultrafast and high-density memory application," Nano lett., vol. 9, pp. 1636-1643, 2009.

[14] Q. Liu, W. Guan, S. Long, M. Liu, S. Zhang, Q, Wang, J. Chen, "Resistance switching of Au-implanted-ZrO2 film for nonvolatile memory application ," J. Appl. Phy., vol. 104, p. 114514, 2008.

[15] Q. Liu, W. Guan, S. Long, R. Jia, M. Liu, J. Chen, "Resistive switching memory effect of ZrO2 films with Zr+ implanted," Appl. Phy. Lett., vol. 92, pp. 012117/1-3, 2008.

[16] W. Guan, S. Long, Q. Liu, M. Liu, W. Wang, "Nonpolar Nonvolatile Resistive Switching in Cu Doped ZrO2," IEEE Electron Device Lett., vol. 29, pp. 434-437, 2008.

[17] Neil Weste and David Harris, "CMOS VLSI Design," 3rd edition, Adisson Wesley, 2004.

[18] Semiconductor Industry Association (SIA), International Technology Roadmap for Semiconductors (ITRS), 2011 edition [online] available at: http://www.itrs.net/

[19] Predictive Technology Model, http://www.eas.asu.edu/~ptm

FPGA Implementation of a Feature Extraction Technique based on Fourier Transform

Mohammed Bahoura
Department of Engineering,
Université du Québec à Rimouski,
300, allée des Ursulines, Rimouski, Qc, Canada.

Hassan Ezzaidi
Department of Applied Sciences,
Université du Québec à Chicoutimi,
555, boul. de l'Université, Chicoutimi, Qc, Canada.

Abstract—In this paper, a feature extraction method based on short-time Fourier transform (STFT) was implemented on FPGA for real-time pattern recognition. The proposed technique was implemented using Xilinx System Generator (XSG) in MATLAB/SIMULINK environment. The response signals obtained during feature extraction process of blue whale call are represented. Also, the classification performance based on the fixed-point XSG implementation is compared to that based on the floating-point MATLAB one using isolated blue whale calls.

I. INTRODUCTION

The feature extraction is an important step in pattern recognition systems. It transforms originally high-dimensional patterns into lower dimensional vectors by capturing the essential of their characteristics [1]. Various feature extraction techniques have been proposed in the literature for different signal applications. In speech and speaker recognition, they are essentially based on Fourier transform, cepstral analysis, autoregressive modeling, and wavelet transform. These feature extraction techniques were also used in the recognition of musical instruments, biomedical signals, marine mammal vocalizations, etc.

Automatic recognition systems were firstly proposed and evaluated using software platform such as MATLAB. However, their hardware implementation remains a great challenge that requires a tradeoff between complexity, computation speed and efficiency of these systems. Most of the hardware-based architectures are proposed for speech and speaker recognition using digital signal processor (DSP) and field programmable gate array (FPGA) [2]–[4].

In this paper, we propose an FPGA-implementation of a feature extraction technique for real-time passive acoustic monitoring (PAM) system that can be used to identify and localize underwater mammals.

II. FEATURE EXTRACTION

Fig. 1 represents the spectrograms of three typical blue whale calls followed by the used feature extraction technique. The vocalization signal is split into successive frames of N samples, $s(m, n)$, where m is the frame index and n is the sample time index within the given frame. Then, a short-time Fourier transform is applied to each frame to extract D-dimensional feature vector $\mathbf{x}_m$. Feature extraction can be seen as a mapping $f : \mathbb{R}^N \to \mathbb{R}^D$, where $D \ll N$.

A. Short-time Fourier transform

The short-time Fourier transform (STFT) of a given frame $s(m, n)$ is a Fourier transform performed in successive frames:

$$S(m, k) = \sum_n s(m, n)e^{-j2\pi nk/N} \qquad (1)$$

where $s(m, n) = s(n)w(n - mL)$, and $w(n)$ is a windowing function of N samples. This function is located at mL, where L is the shift-time step in samples. N represents also the number of discrete frequencies that is usually chosen to be a power-of-2 for using the fast Fourier transform (FFT). The overlap ratio between successive frames is $(N - L)/N$. The power spectrum density (PSD) is then computed [5], [6]

$$P_s(m, k) = \frac{1}{N}|S(m, k)|^2 \qquad (2)$$

At the sampling frequency f_S, each windowed segment (frame) is represented by N-points PSD covering the frequency range $[-\frac{f_S}{2}, \frac{f_S}{2}[$. As power spectrum is symmetric, it can be described by only $N/2$ discrete frequencies. Each frequency index k represents a discrete frequencies $f_k = kf_S/N$, where $0 \leq k < N/2$.

B. Feature vector extraction

As shown in Fig. 1, this method consists in extracting features from two subbands, (15.137-20.508 Hz) and (38.574-84.961 Hz) corresponding to the AB and D calls frequency ranges, respectively. The first six components of the feature vector $\mathbf{x}_m$ are obtained by averaging PSD points between $P_s[m, 31]$ and $P_s[m, 42]$ by bins of 2 points. The last six features are similarly obtained for the PSD interval from $P_s[m, 79]$ to $P_s[m, 174]$ but with bins of 16 points. For a given frame m, the feature vector components are defined by

$$x_{m,n} = \begin{cases} \dfrac{1}{2} \displaystyle\sum_{k=29+2n}^{30+2n} P_s(m, k) & 1 \leq n \leq 6 \\[3mm] \dfrac{1}{16} \displaystyle\sum_{k=63+16(n-6)}^{78+16(n-6)} P_s(m, k) & 7 \leq n \leq 12 \end{cases} \qquad (3)$$

Hence, a 12-dimensional feature vector is constructed, $\mathbf{x}_m = [x_{m,1}, x_{m,2}, \ldots, x_{m,12}]^T$.

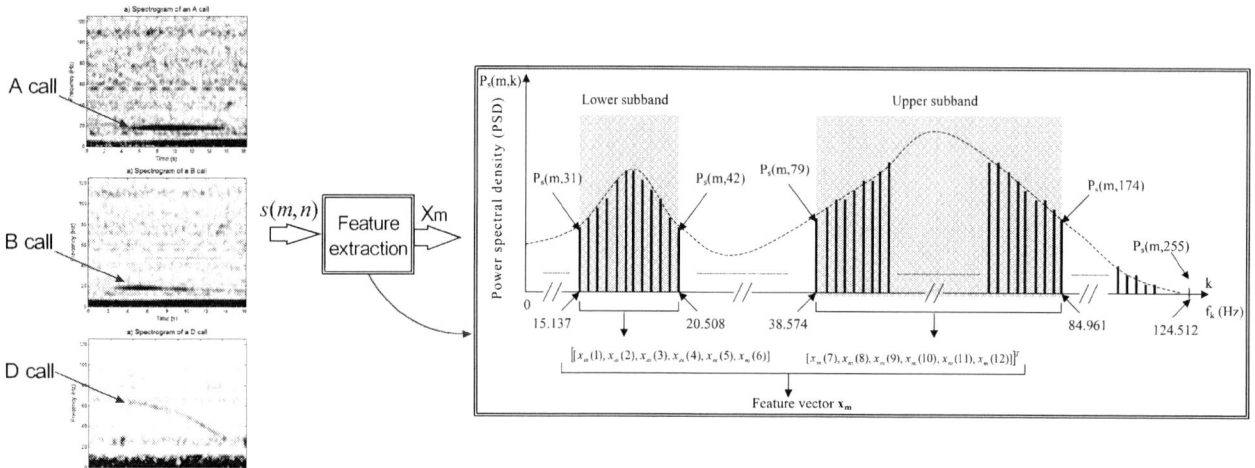

Fig. 1.   Feature extraction process using short-time Fourier transform (STFT). Spectrograms of A, B, and D calls are given in the left followed by the computation of the feature vector from spectrum, which is described in the right [5]. For each signal frame $s(m,n)$, the feature vector $\mathbf{x}_m$ is constructed by extracting six components from lower subband (15.1375-20.508 Hz) and six components from upper subband (38.574-84.961 Hz). The discrete frequencies are obtained with a sampling frequency of 250 Hz and frame a length of 512 samples.

III. CLASSIFICATION

A multi-layer perceptron (MLP) neural network implemented on MATLAB is used to compare performances of the fixed-point architecture to the floating-point one. It is characterized by 12 inputs, 25 hidden neurons and 3 output neurons. We use an *hyperbolic tangent* function for hidden layer and a *logistic sigmoid* function for output layer [5]. The used database contains 100 isolated calls per class, where 90 calls are used for training and 10 calls for testing.

IV. FPGA IMPLEMENTATION

The described feature extraction technique was implemented on FPGA using Xilinx System Generator (XSG) and the Virtex-6 FPGA ML605 Evaluation Kit. Fig. 2 presents the proposed architecture that is mainly based on the algorithm described in section II.

A. Signal windowing

The windowed signal $s(m,n)$ was obtained by multiplying $s(n)$ by Hamming window $w(n)$ of $N = 512$ samples stored in a ROM block driven by a cyclic modulo-N counter (Fig. 2).

B. Sort-time Fourier transform

The STFT of a given input frame, $s(m,n)$, is computed using the Xilinx FFT block.

$$S(m,k) = \sum_{n=0}^{N-1} s(m,n) e^{-jnk2\pi/N} \qquad (4)$$

where N is the transform size, k is the frequency bin index ($0 \leq k \leq N - 1$), and $j = \sqrt{-1}$. Pipelined streaming input/output option has been chosen to achieve continuous computation of the short-time Fourier transform. The Xilinx FFT block provides two outputs corresponding to real, $S_r(m,k)$, and imaginary, $S_i(m,k)$, parts of $S(m,k)$.

C. Power Spectrum

The power spectrum block computes power spectrum using the real and imaginary parts provided by the Xilinx FFT block.

$$P_s(m,k) = \left(S_r^2(m,k) + S_i^2(m,k)\right)(1/N) \qquad (5)$$

where N is the frame length of 512 samples.

D. Feature extraction

For each frame $s(m,n)$, the feature extraction block in Fig. 2 computes its 12 feature components according to Eq. (3). It can be noted that division by 2 and 16 and replaced by multiplication by $1/2$ and $1/16$, respectively. Each feature component block uses a register enabled by the frequency index (k) to save it. A second register enabled by the "done" signal of the FFT block is used to synchronize the feature component values at the processing end of each frame.

E. One value per frame

An optional subsystem in Fig. 2 based on the standard Simulink blocks is used to take one feature vector per frame rather than repeated N identical vectors.

Table I gives the resource requirement and the maximum operating frequency as reported by Xilinx ISE tool.

TABLE I

RESOURCE UTILIZATION AND MAXIMUM OPERATING FREQUENCY OF THE VIRTEX-6 XC6VLX240T CHIP USING CLB LOGIC STRUCTURE.

Resource utilization	
Flip Flops (301,440)	13,2385 (4.4%)
LUTs (150,720)	12,014 (8.0%)
Bonded IOBs (600)	306 (51.0%)
RAMB18E1s (832)	62 (0.2%)
DSP48E1s (768)	9 (1.2%)
Maximum Operating Frequency	66.061 MHz

978-1-4673-5289-5/12 $31.00 © 2012 IEEE

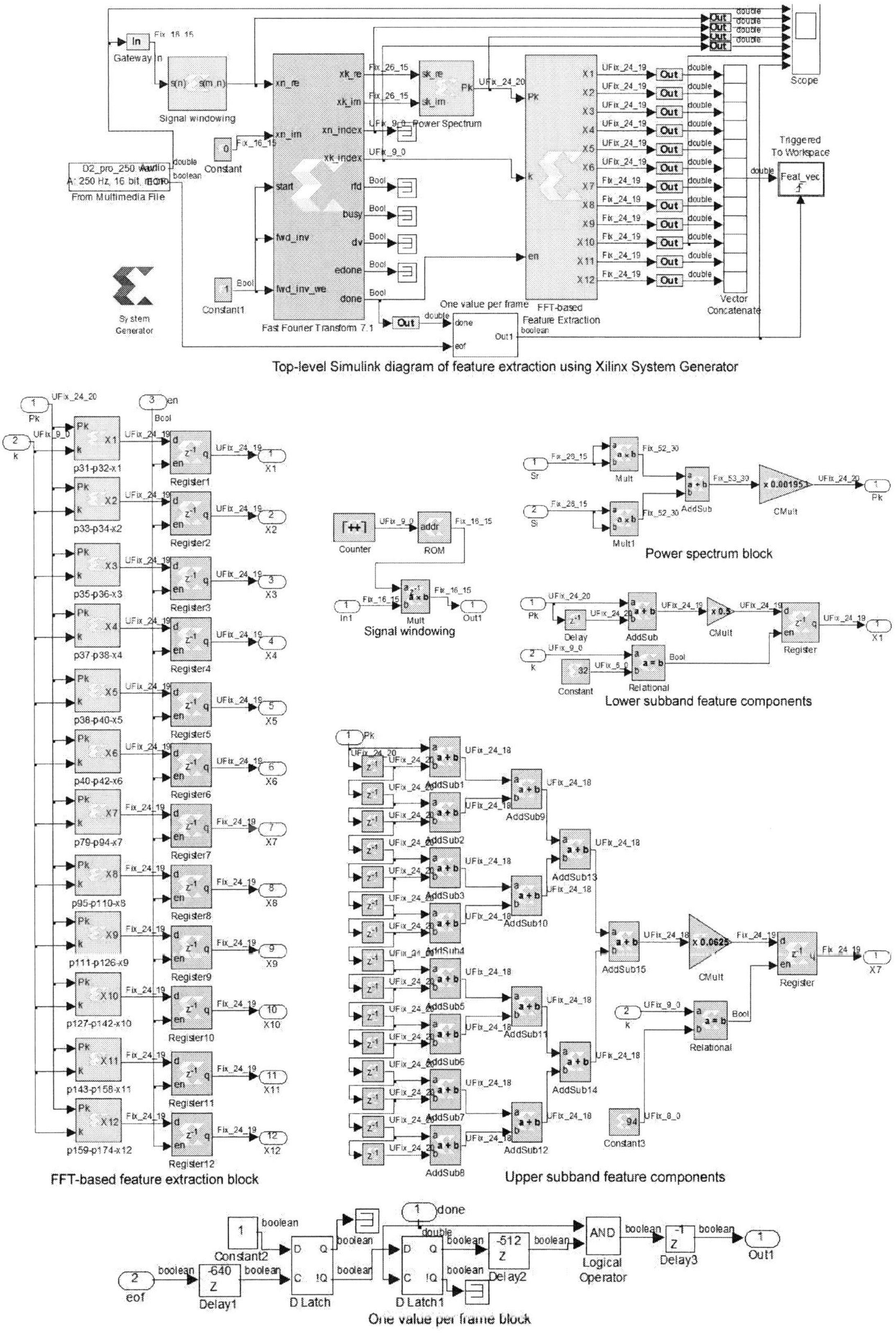

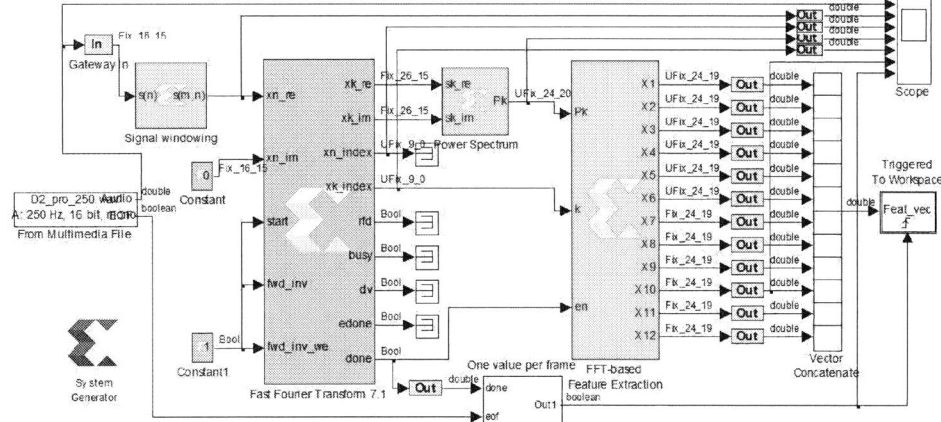

Fig. 2. Feature extraction technique implemented on FPGA using Xilinx System Generator. The top-level Simulink diagram is given on the top followed by details of different subsystems. The green blocks are designed using the XSG blocks (blue). The white blocks are the standard Simulink blocks.

978-1-4673-5289-5/12 $31.00 © 2012 IEEE

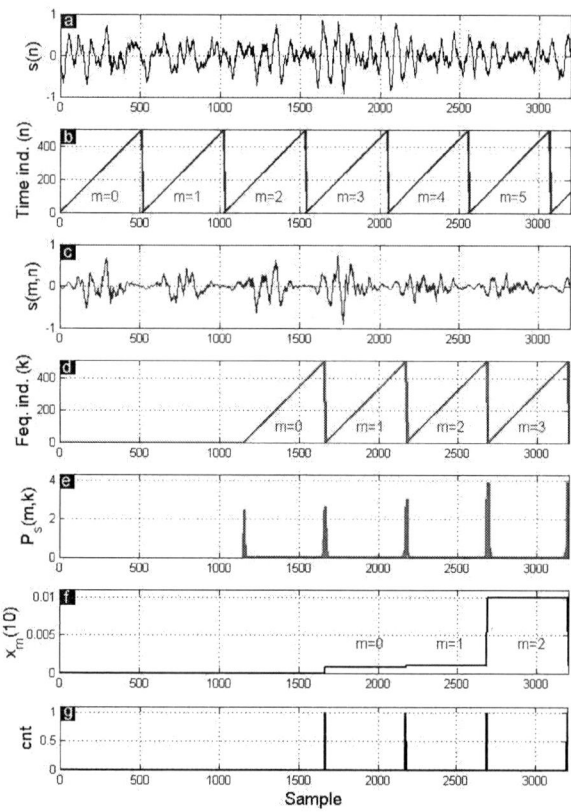

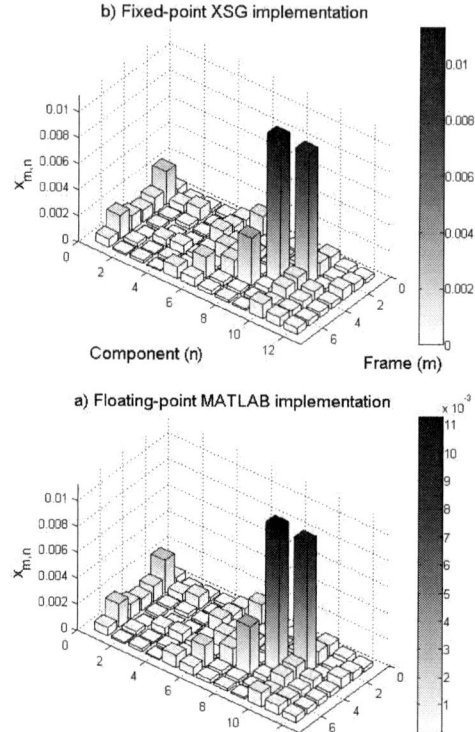

Fig. 3. Response signals obtained during feature extraction process of a D call. a) input signal $s(n)$, b) Time index (n) that allows to delimit successive frames $s(m, n)$, c) windowed signal $s(m, n)$, d) Frequency index (k) that allows to delimit successive power spectra $P_s(m, k)$, e) Power spectra $P_s(m, k)$, f) 10th component of the feature vector, and g) control signal that send one feature vector per frame to workspace.

Fig. 4. Example of feature vectors obtained from a D call using fixed-point XSG implementation (a), and floating-point MATLAB implementation (b).

V. RESULTS

Fig. 3 presents response signals at different levels of this architecture. It can be seen the FFT computing latency of 1150 samples that represents the delay between the first sample of the input frame $s(m, n)$ and the first sample of its corresponding spectrum $P_s(m, k)$. As pointed previous in subsection IV-D, the feature components are available just from the end of their corresponding power spectrum $P_s(m, k)$. Fig. 4 shows that fixed-point XSG implementation gives the same performance as floating-point MATLAB implementation. Table II gives the confusion matrix of the true classes versus assigned classed for fixed-point and floating point implementations of the STFT-based feature extraction method. These results present the averaged values of 50 repeated tests.

TABLE II
CONFUSION MATRIX OF MATLAB AND XSG BASED IMPLEMENTATIONS.

True class	Assigned class (XSG)			Assigned class (Matlab)		
	A	B	C	A	B	C
A	6.72	3.02	0.26	6.72	3.02	0.26
B	0.24	8.20	1.56	0.24	8.20	1.56
D	0.00	0.00	10.00	0.00	0.00	10.00

VI. CONCLUSION

In this paper, feature extraction technique based on Fourier transform has been implemented on FPGA using Xilinx System Generator. The fixed-point XSG implementation provides similar performances than the floating-point MATLAB one. As future work, a classifier as a neural network will be added to this architecture to implement complete pattern recognition system on FPGA.

REFERENCES

[1] M. Bahoura, "Pattern recognition methods applied to respiratory sounds classification into normal and wheeze classes," *Computers in Biology and Medicine*, vol. 39, no. 9, pp. 824–843, 2009.
[2] J.-C. Wang, J.-F. Wang, and Y.-S. Weng, "Chip design of MFCC extraction for speech recognition," *Integration, the VLSI Journal*, vol. 32, pp. 111–131, 2002.
[3] M. Staworko and M. Rawski, "FPGA implementation of feature extraction algorithm for speaker verification," in *17th International Conference "Mixed Design of Integrated Circuits and Systems", MIXDES 2010*, 2010, pp. 557–561.
[4] R. Ramos-Lara, M. López-García, E. Cantó-Navarro, and L. Puente-Rodriguez, "Real-time speaker verification system implemented on reconfigurable hardware," *Journal of Signal Processing Systems*, pp. 1–15. [Online]. Available: http://dx.doi.org/10.1007/s11265-012-0683-5
[5] M. Bahoura and Y. Simard, "Blue whale calls classification using short-time fourier and wavelet packet transforms and artificial neural network," *Digital Signal Processing*, vol. 20, no. 4, pp. 1256–1263, 2010.
[6] ——, "Serial combination of multiple classifiers for automatic blue whale calls recognition," *Expert Systems with Applications*, vol. 39, no. 11, pp. 9986–9993, 2012.

Modified Circular Hough Transform using FPGA

F.Ferhat-taleb Alim, K.Messaoudi, S.Seddiki and O.Kerdjidj
Centre de développement des technologies avancées
CDTA
Baba Hassen, Algiers
kmessaoudi@cdta.dz

Abstract—**The main objective of our paper is to implement on FPGA (Field Programmable Gate Array) an architecture of a method used to detect circles from edge images based on Hough transform algorithm. At first, an algorithm has been developed and tested using Matlab software. The modified algorithm that we have followed to realize our architecture simplifies trigonometric calculations to just simple additions by using Cordic algorithm. Then architecture has been elaborated and implemented under Xilinx ISE 9.2i Foundation tool and tested on real-time. We will discuss the performance and parallelism of the algorithm and we will show its efficiency.**

Keywords-HoughTtransform; iris detection; cordic algorithm; FPGA; V2MB1000.

I. INTRODUCTION

One of the major challenges in computer vision is determining the location, orientation of a particular object in an image. Usually objects may come in different sizes and shapes not pre-defined in an arbitrary object detection program. A solution to this problem is to provide an algorithm that can be used to find any shape within an image, and then classify the objects according to parameters needed to describe the shapes. A commonly used technique to achieve this; is the Hough Transform (HT).

The Hough transform [1] is one of the most common methods used for detecting parametric curves in a binary image. A parametric curve here means a curve or surface which can be described by an equation with number of free parameter, e.g. a line, a circle, an ellipse, etc. The general principle of Hough transform is to establish a projection between the image space and the parameter space representing the desired curve.

The Hough transform (HT) has a broad application today, e.g. it is used in robotics, biometry, and image processing. Although, the HT is a robust technique for finding curves in image, the hardware complexity increases with the number of parameter. So, the computational complexity and requirement of excessive memory are the main obstacles for the hardware implementation of HT. Recently, the superiority of the Hough transform is challenged by the detection of complex curves like circles [2, 3], ellipse, or arbiter curves and the optimization of the treatment delay and memory space on the implementation of the Hough transform on programmable circuit.

Therefore, different architectures and algorithms [4, 5, 6,7 and 9] of HT have been proposed. Most of these architectures

have met the problem of implicit evaluation of trigonometric functions that makes the implementation difficult. To overcome this problem, we propose in this paper architecture based on cordic algorithm. The architecture of modified circular Hough transform (CHT) of has been first simulated in Matlab and obtained good results. Then, it has been implemented on FPGA circuit.

In this paper, we present how we detect circle on image with Hough transform. Section III explains the modified CHT using the basic idea of Cordic algorithm. Results of software implementation of modified CHT are given in section IV. The Architecture which has been developed for this algorithm is describes in section V. Section VI discusses the implementation results and section VII presents some conclusions.

II. CIRCULAR HOUGH TRANSFORM

The CHT has been recognized as robust techniques for curve detection. This method can detect object even polluted by noise. The CHT was sketched by Duda et al. [8]. The CHT is one of the modified versions of the HT. The CHT aims to find circular patterns within an image. The CHT is used to transform a set of feature points in the image space into a set of accumulated votes in a parameter space.

Then, for each feature point, votes are accumulated in an accumulator array for all parameter combinations. The array elements that contain the highest number of votes indicate the presence of the shape. A circle pattern is described by (1).

$$r^2 = (x-a)^2 + (y-b)^2 \qquad (1)$$

Where a and b are the coordinates of the center and r is the radius of the circle.

In contrast to the linear HT [8], the Circular Hough Transform (CHT) relies on 3 parameters, which requires a larger computation time and memory space, increasing the complexity of extracting information from our image. To simplify calculation we can fix the value of the radius r (r become known) then the circle will depend only on parameters of the center (a, b).

For a given radius, each point of a circle in an image will itself be a circle of radius r centered on the given point in Hough space. The intersection of these circles is the center of the initial circle, and this is shown in Fig. 1.

978-1-4673-5289-5/12 $31.00 © 2012 IEEE

If we vary the radius r (r is unknown), each edge point in image space will have a volume of cone in the Hough space, and this is shown in Fig. 2.

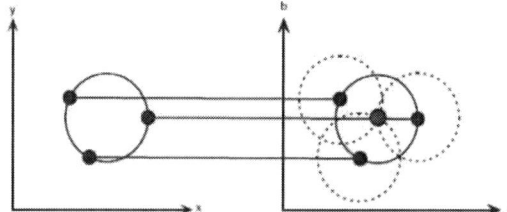

Figure 1. Detection of a circle with a fixed radius using HT

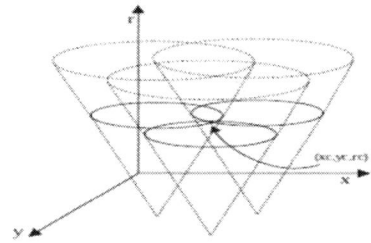

Figure 2. Detection of a circle with a fixed radius using HT

The intersection of all these cones gives us the coordinates of the center and the radius of the circle researched.

A. Circular Hough transform algorithm

Let Ic be the binary image, the circular Hough transform algorithm is:

- Initialize accumulator H to zero.

- For each edge point (x,y) in the image I

- For each radius r.
   - Draw a circle by taking this point as a center.
   - Increment the value of all points on the accumulator where the circle passes through.

- Search peak value in the accumulator.

III. MODIFIED CIRCULAR HOUGH TRANSFORM

The modified circular Hough transform is based on Cordic algorithm. This one is used to calculate trigonometric functions.

A. Cordic algorithm

In 1957 Jack E. Volder [10] described the Coordinate Rotation Digital Computer or CORDIC for calculation of trigonometric functions, multiplication, division and conversion between binary and mixed radix number system. The CORDIC- algorithm provides an iterative method of performing vector rotations by arbitrary angles using only shift and adds. To explain CORDIC principal, we choose two points (x,y) and (x',y') belonging to a circle, and this is shown in Fig. 3.

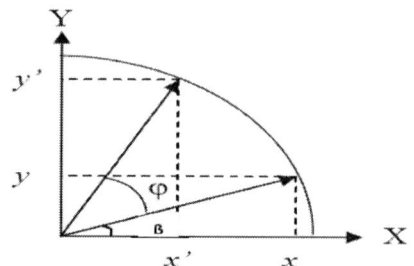

Figure 3. Rotation of vector by the angle φ

$$\begin{cases} x = R \sin \beta \\ y = R \cos \beta \end{cases} \quad (2)$$

and

$$\begin{cases} x' = R \sin (\beta+\varphi) \\ y' = R \cos (\beta+\varphi) \end{cases} \quad (3)$$

Combining equations (2) and (3) we get:

$$\begin{cases} x' = x \cos\varphi - y \sin\varphi = (x - y \tan \varphi) \cos\varphi \\ y' = y \cos\varphi + x \sin\varphi = (y + x \tan \varphi) \cos\varphi \end{cases} \quad (4)$$

We have $\varphi_0 = \operatorname{atan}2^{-j} \approx 2^{-j}$, $j > 3$. If we choose $j = 4$ so φ << and $\cos\varphi \approx 1$ we get:

$$\begin{cases} x' = x - y \times 2^{-4} \\ y' = y + x \times 2^{-4} \end{cases}$$

To generalize for all points:

$$\begin{cases} X_{i+1} = X_i - Y_i \times 2^{-4} \\ Y_{i+1} = Y_i + X_i \times 2^{-4} \end{cases} \quad (5)$$

Where **i** is the iteration index. So, to sweep one quarter of the circle, the number of micro rotation = $(\pi/2) / 2^{-4} \approx 25$ iterations.

B. Modified algorithm

In order to explain the modified algorithm we have to remind the equation (1) of circle.

$$r^2 = (x-a)^2 + (y-b)^2$$

The coordinates of the radius r (x_r, y_r) are given by equations:

$$x_r = x \cos(\theta) - y \sin (\theta)$$

$$y_r = y \cos (\theta) + x \sin(\theta)$$

While θ is the angle of the vector from the origin to the radius.

As we have seen in circular Hough Transform, having a radius unknown, we apply the algorithm of HT on a limited radius domain. For each radius we have an accumulator table in which we have drawn a circle for each edge point.

In fact it is the same circle which is drawn for each edge point since it has the same radius; this is why in order to simplify calculations, we have proposed to take this circle

978-1-4673-5289-5/12 $31.00 © 2012 IEEE   208

in a virtual space where the center of the circle is coincident with the origin of this space (0.0).

Then we just made shifts for each edge point in order to fill the accumulator tables, instead of sweeping the entire circle. It was sufficient to sweep the quarter of the circle, and then get the rest from it. We have repeated those steps for different radius varying from 4 to 16.

C. Modified Circular Hough transform algorithm

The modified circular hough transform algorithm is:

- Initialize the accumulator to zero.

- For each radius R.

- For each micro-rotation (xGi , yGi).

- For each edge counter point (xi , yi).

$$A_{1i} = x_i + x_{Gi}, B_{1i} = y_i + y_{Gi}$$
$$A_{2i} = x_i - y_{Gi}, B_{2i} = y_i + x_{Gi}$$
$$A_{3i} = x_i - x_{Gi}, B_{3i} = y_i - y_{Gi}$$
$$A_{4i} = x_i + y_{Gi}, B_{4i} = y_i - x_{Gi}$$

- Increment the accumulators that have the address (A1i, B1i), (A2i, B2i), (A3i, B3i), (A4i, B4i) in the RAM.

- Look for peaks in all accumulators.

- RESET the RAM

IV. SIMULATION RESULT USING MATLAB

Since Hough Transform is applied only on a binary image, we have to convert our color image to binary edge image before applying the Hough transform.

Introducing our image it passes through different applications that we have programmed using Matlab. Steps of circular Hough with Matlab are shown below in Fig. 4:

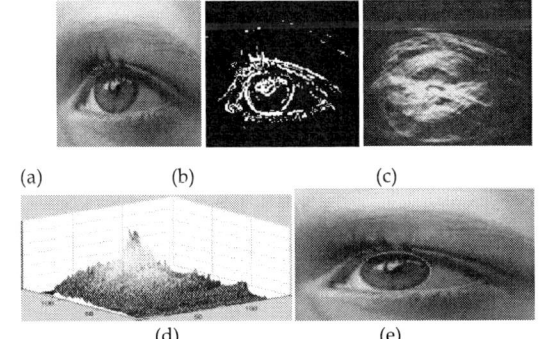

Figure 4. a) initial Image, (b) Image after Canny-Deriche filter., (c) Hough Plan of 2D image, (d) 3D Hough plan , (e) Detection of circle on the iris.

V. HARDWARE IMPLEMENTATION

In this part, we will describe on VHDL language, our architecture of modified Circular Hough Transform. This architecture is composed of different blocs. The global architecture is shown in Fig. 5:

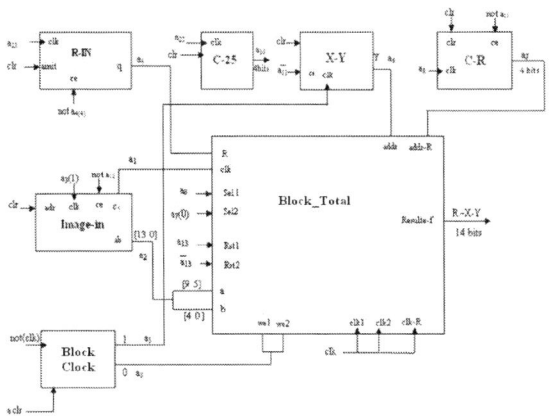

Figure 5.   Global architecture

A. Total bloc

This bloc's architecture is shown in Fig. 6. It is composed of:

- Generator: Generate addresses of the virtual circle where its center is coincident with the space origin.

- Shift: Used for the changes of reference in order to give us the right address that will be accumulated

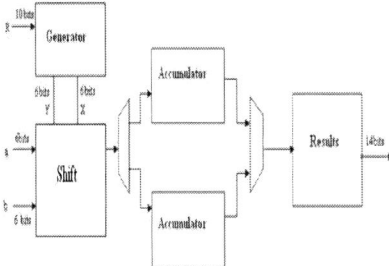

Figure 6.   Total bloc architecture

- Accumulator: Two accumulators that work in parallel, the first is used for finding results and the second for filling up.

- Multiplexer: Used to switch between the two accumulators.

- D–MUX: Used to switch between the two accumulators, with inverse select line as the Multiplexer.

- Results: find the final result (R: radius, X, Y: the coordinate of the circle center).The word "data" is plural, not singular.

B. Simulation of Global architecture

The result of the functional simulation of our global architecture that gives the final result is represented on Fig. 7.

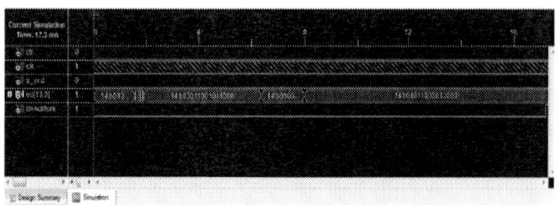

Figure 7. Simulation result of the global architecture.

The table I show the resources used for the implementation of our architecture on XC2V1000_4fg 456 circuit.

TABLE I.   IMPLEMENTATION RESULT

Number of Slices	578 out of 5120	11%
Number of External IOBs	19 out of 324	5%
Number of GCLKs	3 out of 16	18 %
Number of BRAMs	8 out of 40	20%
Delay execution time	10, 028 ns	

C. Real time execution

This is the final step of implementation, so we have generated a file [.bit], which will be sent through the JTAG, and this is used to program the FPGA circuit and be able to test and verify the credibility and the performance of our architecture on the board V2MB1000 in real time.

In order to display the result (R: radius, X, Y: the coordinate of the circle center), we use 7 segments display, we use also dip switch to select which value will be displayed (Fig. 8).

- If dip switch '11100000' : display R

- If dip switch '10100000' : display X

- If dip switch '11000000': display Y

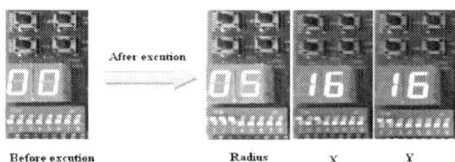

Figure 8.   Real time results

VI.   CONCLUSION

The purpose of the work presented in this paper is the implementation on FPGA of the Hough Transform method used to detect circles from images. The algorithm performed is not applied directly on real image but on binary edge image. In order to obtain a binary edge image, the image passed through several steps of pretreatment that have been performed using Matlab, then we have developed and tested the algorithm of our architecture using Matlab and obtained accuracy results.

The architecture that we have proposed has been described in VHDL language and through synthesis which allows a time reduction design. The simulation results show the good functioning of our architecture. For the synthesis, the results indicate that the proposed architecture occupies 11% of XC2V1000-4fg456C circuit of the Virtex-II with a frequency of 66.914 MHz.

As a perspective, it could be interesting that this work can be used in a biometric system via an implementation on a map-based FPGA. For the realization of such a system it should be taken into consideration the stages of acquisition and preprocessing of the iris.

REFERENCES

[1] P.V.C. Hough, "Method and means for recognizing complex patterns", 1962.

[2] M.Nosrati , R.Karimi, H.Nosrati, A. Nosrati," A method for detection and extraction of circular shapes from noisy images using median filter and CHT", Journal of American Science, pp: 84-88, 2011

[3] H. Ruppertshofen , C.Lorenz Sarah Schmidt , P.Beyerlein, Z.Salah, R.Georg, H.SchrammL "Discriminative Generalized Hough transform for localization of joints in the lower extremities" , journal of computer science, vol.26, pp: 97-105, 2010

[4] T.Moravčík," An Approach to Iris and Pupil Detection in Eye Image", XII International PhD Workshop OWD, pp : 239-242, 23–26 Oct. 2010

[5] T.Duan, D.Duc, T.Hong Du, "Combining Hough Transform and Contour Algorithm for detecting Vehicles License-Plates", Proceedings International Symposium on Intelligent Multimedia, Video and Speech Processing , pp:747-750, Oct. 20-22, 2004

[6] O.Djekoune, K. Achour, "Incremental Hough Transform: An Improvement Algorithm for Digital Devices Implementation", Journal real-time imaging, vol. 10, N° 6, pp: 351-363, Dec.2004.

[7] S. Tagzout, K. Achour, O. Djekoune, "Hough transform for FPGA implementation", IEEE Workshop on Signal Processing Systems, pp. 384-393, oct.2000.

[8] R.O. Duda , P.E. Hart, "Use of the Hough transform to detect lines and curves in pictures", ACM Communications magazine, vol. 15, pp:11–15, 1972

[9] K. Maharatna et Swapna Banerjee «A VLSI array architecture for Hough transform», Pattern Recognition, vol.34, pp: 1503-1512. 2001

[10] J.E. Volder. "The CORDIC Trigonometric Computing Technique". IRE Transactions on.Electronic Computers, vol. EC-8, N° 3, pp: 330-334, 1959

Montgomery Modular exponentiation on FPGA

Anane Nadjia[1], Anane Mohamed [2] and Issad Mohamed[1]

[1] CDTA (Centre de Développement des Technologies Avancées), BP 17, Baba Hassen)

[2] ESI (Ecole nationale Supérieure d'Informatique), BP 68M Oued Smar.

Alger, Algérie

e-mail: anane@cdta.dz

Abstract—Modular exponentiation is the main operation of RSA-based public-key cryptosystems. It is implemented by repeated modular multiplications which are time consuming for large operands. Accelerating the RSA requires reducing the number of modular multiplications, thus reducing the time to perform one modular multiplication.

In this paper, we present an architecture designed to implement a fast modular exponentiation using the right to left binary method (R-L), which allows the parallel execution of modular operations "squares and multiplications". The fast modular multiplication used is based on Montgomery algorithm.

This architecture has been implemented on an FPGA circuit of Xilinx, the XC4VLX25-12ff668 of Virtex-4. The implementation results showed that the architecture computing 1024 bits modular exponentiation presents good performance in terms of speed and occupied area with the possibility of changing the key size by reprogramming the FPGA according to the security level and the performance to attain.

Keywords-component: *cryptosystem, modular exponentiation, Montgomery modular multiplication, RSA, FPGA.*

I. INTRODUCTION

The rapid and continuous development in communications through open networks such as Internet has created a growing need to encrypt sensitive or confidential data before their transfer.

The RSA (Rivest, Shamir, Adleman) is a secure and efficient public-key cryptosystem [1]. However, it is very intensive computing since it operates on large operands: 1024 bits or greater to provide sufficient security.

Software implementation of RSA is too slow for real-time applications [2]. However hardware implementation is required to achieve performances of such applications [3].

The key operation of RSA is the modular exponentiation which is used to encrypt a message M by using a public key e or recovers the original message from the encrypted message by using the private key d such as:

$$C= (M^e \bmod N), \qquad M= (C^d \bmod N),$$

where N is the modulus of n bits.

Accelerating the RSA cryptosystem requires reducing the modular multiplications number hence the execution time of the modular multiplication.

The goal of this paper is to present an FPGA implementation of a fast modular exponentiation, by exploiting the parallelism offered by the R-L binary modular exponentiation, the high speed of Montgomery modular multiplication and the FPGA performances to provide more security with the possibility of changing the key size by reconfiguration of the FPGA depending on the desired security and performance.

The remainder of this paper is organized as follows. Section 2 presents the binary modular exponentiation. In section 3, the Montgomery modular multiplication is presented. Section 4 describes the architecture developed for computing the Montgomery modular exponentiation. Section 5 summarizes the implementation results of the architecture on an FPGA of Virtex-4. Finally, a conclusion and some perspectives are given.

II. BINARY MODULAR EXPONENTIATION

Modular exponentiation, realized by a series of modular multiplications, is very costly in computation time for large operands.

The simplest and easy method to compute 1024 bits modular exponentiation is the binary method [4], known as the "Square and multiply". It is based on scanning the bits of the binary exponent, then a squaring is performed at each step and depending on the scanned bit value, a subsequent multiplication is performed.

To increase the computation performances of this operation:

1. The R-L binary exponentiation is used, which scans the exponent bits from the least significant (LSB) and allows the squaring and multiplication to run in parallel [5].

2. The Montgomery algorithm is used, which replaces the division by the modulus by a series of addition and shift operations in order to reduce the execution time of the modular multiplication [6].

III. MONTGOMERY MODULAR MULTIPLICATION

Montgomery modular multiplication used to calculate the R-L binary exponentiation is fast and suitable for hardware implementation. It replaces the division by the modulus with simple right shifts, easy to implement but requiring some pre and post calculations.

978-1-4673-5289-5/12 $31.00 © 2012 IEEE

The application of Montgomery algorithm on two numbers A and B of n bits gives the following result:
$$C = MM (A, B) = A \times B \times r^{-1} \bmod N.$$

Where $r = 2^n$, and n is an integer with $2^{n-1} < N < 2^n$, such as gcd $(r, N) = 1$.

Since $r = 2^n$, it suffices to the modulus N to be an odd integer, where this condition is satisfied for the RSA.

The Montgomery algorithm used is without final subtraction represented by algorithm 1, which allows reduction of the execution time in plus of the occupied area.

Algorithm 1: Montgomery without final subtraction.

Inputs: $A = a_{n+1}a_n...a_0$, $B = b_{n+1}b_n...b_0$, $N = n_{n-1}n_{n-2}...n_0$,
With A, B < 2N, $a_{n+1} = b_{n+1} = 0$.

Output: $S_{n+2} = A \times B \times 2^{-n-2} \bmod N$

Begin
    $S_0 = 0$
    For i from 0 to n+1 do
        1. $q_i = (S_i + a_i \times B) \bmod 2$
        2. $S_{i+1} = (S_i + a_i \times B + q_i \times N)/2$
    End For.
End.

The subtraction has been omitted and replaced by two iterations which let the operand sizes equal to (n+1) instead of (n-1) [7].

The architecture of the Montgomery modular multiplication without final subtraction is represented on figure 1.

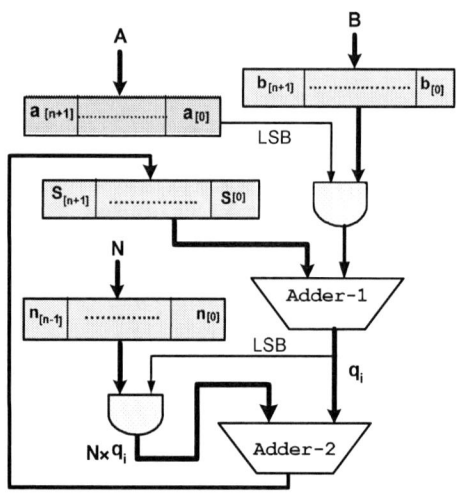

Figure 1. Montgomery modular multiplication architecture

IV. MONTGOMERY MODULAR EXPONENTIATION ARCHITECTURE

To adapt the Montgomery multiplication algorithm to the R-L binary modular exponentiation, we must make a mapping in order to enter in the Montgomery domain, where a series of modular multiplications are performed, then a re-mapping is required to recover the true result in the last iteration of the modular exponentiation out of the Montgomery domain.

Algorithm 2 shows the R-L binary modular exponentiation based on Montgomery modular multiplication.

Algorithm 2 : Montgomery based R-L binary exponentiation

Inputs: e, M, N, r^2; e = $(e_{n-1} e_{n-2}... e_3 e_2 e_1 e_0)_2$
Output: C = $M^e \bmod N$;

C = Montg $(1, r^2, N)$, S = Montg (M, r^2, N);
Begin
    For i = 0 to n - 1 do
        Begin
            a. If $(e_i = 1)$ then C = Montg (C, S, N); [multiply]
            b. S = Montg (S, S, N); [square]
        End;
    C = Montg $(C, 1, N)$.
End.

The proposed architecture for fast computing the binary modular exponentiation based on Montgomery multiplication is presented on figure 2.

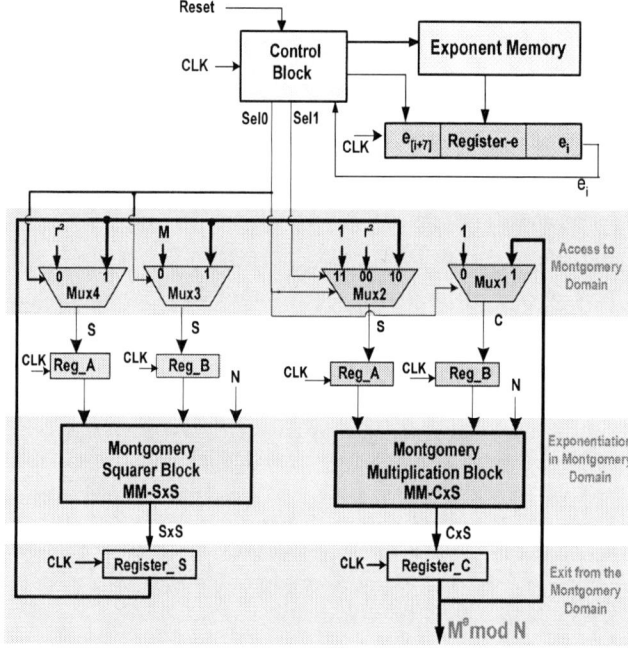

Figure 2. Architecture of the Montgomery R-L binary exponentiation

It consists of :

1. A control block to manage squaring and multiplication operations. It has as inputs the Reset and the clock CLK.

2. A memory to store the exponent e connected to a shift register for shifting the exponent bits.

3. Two Montgomery modular multiplication blocks MM-SS for squares and MM-CS for multiplications with two registers S and C at their outputs to store respectively the intermediate results of squares and multiplications.

978-1-4673-5289-5/12 $31.00 © 2012 IEEE

4. Two registers A and B at the input of each multiplication block to store the inputs data (M, r^2) or the intermediate results.

5. Two multiplexers at the input of each multiplication block to select the inputs data (M, r^2,1) or the intermediate results.

Entering in the Montgomery domain needs to perform two Montgomery modular multiplications in parallel:

$$MM\ (M, r^2 \ (mod\ N))\quad and\quad MM\ (1, r^2(mod\ N)),$$

which gives as outputs : $M \times r\ (\ mod\ N)$ and $1 \times r\ (mod\ N)$.

In the Montgomery domain, multiplications and squares are performed in parallel by two Montgomery multipliers and successively until the exhaustion of all the exponent bits.

The multiplexers select the inputs ($M \times r\ mod\ N$) and ($1 \times r\ mod\ N$) or the outputs of the multipliers according to the logic control to give the result ($M^e \times r\ mod\ N$).

Exiting from the Montgomery domain means performing a last Montgomery multiplication in order to eliminate the r factor from the result to finally obtain ($M^e\ mod\ N$).

The two Montgomery multiplication blocks used are optimized in terms of execution time and area, where the blocks Select Ram (SRAM) of 18 Kbits available in the FPGA were used to store the operands.

Operands are introduced word per word which has an advantage to reduce the size of the adders and internal registers (1024 bits) to the size of the selected word. That has reduced the hardware and routing resources required to implement these registers.

The first eight bits of the exponent e are introduced into an eight bits serial shift register. Once the calculus is completed, eight successive bits are sent from the memory to take place in the shift register to continue the computing until the last bit of the exponent e.

Our architecture performs Montgomery modular exponentiation based on three phases:

- **Phase 1:** Entry in the Montgomery domain.
- **Phase 2:** Computation of Montgomery modular exponentiation by modular squares and multiplications.
- **Phase 3 :** Exit from the Montgomery domain.

At each multiplication block, a pair of multiplexers is used to select the operands in the inputs of the two multiplication blocks according to the running phase of the exponentiation algorithm.

These multiplexers are controlled by the selection signals Sel0 and Sel1 as shown in Table 1.

TABLE I. SELECTION OF THE THREE EXPONENTIATION PHASES

Sel1	Sel0	Phase	Block MM-CS	Block MM-SS
0	0	1	Montg ($1,r^2$,N)	Montg (M,r^2,N)
0	1	2	Montg (C,S,N)	Montg (S,S,N)
1	1	3	Montg (C,1,N)	

V. IMPLEMENTATION RESULTS

The hardware architecture to perform the binary modular exponentiation in parallel was designed by using Xilinx ISE 9.2i and implemented on a Virtex-4 FPGA circuit of Xilinx, the XC4VLX25 12ff668 [8].

For best use of the resources offered by FPGAs, the exponent e of 1024 bits has been stored in Select RAM blocks of 18 K-bits, available on the FPGA to minimize the logic and routing resources necessary for the implementation of our architecture. This last was simulated using the ModelSim simulator PE 6.0 and the results were compared to those provided by Maple 12 tool [9], where two procedures were developed: one for key generation and encrypting the message and the other was reserved for the decomposition of the exponent e of 1024 bits in 8 bit words for storing them in the memory in order to reduce the size of the shift register.

We encrypt a message M represented by "mess" in figure 3 which is equal to:

mess:=1075784381388936074359059546347304415077448622514571005685651967999358456333915068546583350371526993264591044491526132504302815998972099372258786251387240758993659410618314143546063746131832042854039261569700178938371760904881035693538331926792176566772051194365074736926509718

The public key or the exponent e, the modulus N, the private key d and the constant r^2:=2**(2*1026) mod N were calculated by Maple 12 and are equals to:

e =11239,

N=82206061353103504512232870195738121018797639464769714322397441571279092086187528073413435182164916487067159376750226007872188485249576843937689870391450921580177437207188993844911556404643407098465351347461782589831203880068348141865079384143237698051906909239383264774070326646085964075162324033422113014621.

d=56678954482178045775006006864202749334875247638802430490635979601018034040027329392373050024610369059372134354518863006940260572310612239849911807604177685913254290312713759862258997723334297878481104918467499480456837973144471337467918896301987183090281548462637700021189745040224614369079380331226059975211.

r^2=74631413297914341936125959271525042613966027591627380666142813463861711146760703324097198633705718074356015032071638856867287396958844317332829774215338073167965039164622488648217077569665628757533940102521243535511731067452045345282229612750258947440907084322952132333209726638744220168639658107174997596445

These values have been loaded into the designed architecture which performs the modular exponentiation provided by our hardware to obtain the encrypted message C by: C= (mess)^e mod (N).

At the same time this message was encrypted by Maple with computing the modular exponentiation by the following function:

$$C = (mess \ \& \ ^\wedge e) \ mod \ N,$$

which gives the following value for the encrypted message :

C-Maple :=357559624665207117011828988852509663883
69145220924402971549147999524402195534176512791451
19290240981912531376872562018597015257036345462767
49415077291562410219648019872239970454495025692816
54961936154923397845084483423324618187445485127276
67601183958333758017776916504011987871589127479914
2692038917469484859

We can find this value in the fifth line and the original message in the third line of the timing diagram of figure 3.

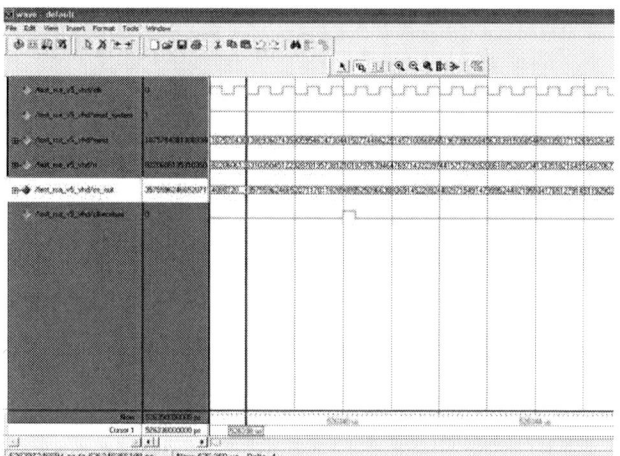

Figure 3. Results of encrypting the message mess

The computation results of the modular exponentiation provided by MAPLE for encrypting the message "mess" were equivalent to those provided by our architecture for computing the R-L binary modular exponentiation of the same message. That justifies the good working of our architecture.

The proposed architecture allows the computation of modular exponentiation for keys up to 1024 bits. The computation time for 1024-bits exponentiation is 85.63 ms.

$T_{total} = T \times 1026 \times 1026 = 85.63$ ms, where the first 1026 represents the number of Montgomery multiplications and the second 1026 represents the iterations number of the Montgomery multiplications.

The period $T = 2 \times Tc = 81.35$ ns, where Tc is the time of critical path, Tc = 40, 675 ns.

The implementation results are summarized in table 2, where the numbers of slices, memory blocks and I/O blocks are given with the area occupancy percentage of our architecture on the FPGA.

TABLE II. OCCUPANCY RATE ON THE FPGA

Resources	Occupancy rate	%
Slices	10170 of 10752	94%
IOBS	3076 to 4486	68%
18Kbits BRAMS	2 of 72	3%

VI. CONCLUSION

This paper has described the architecture for computing the R-L binary modular exponentiation and its implementation on FPGA.

The binary modular exponentiation is computed by a series of squaring and multiplications.

Montgomery modular multiplication has been chosen to realize the two operations, which has greatly reduced the computation time of the modular exponentiation.

The architecture allows the parallel execution of squaring and multiplications.

The proposed architecture can support key sizes up to 1024 bits using the same circuits and making some changes in the control block with increasing the memory which is still available on the FPGA as only 3% were used.

The resources offered by FPGAs such as memory blocks and Carry Save Adders (CSA) have been used advantageously to speed-up the execution time of the modular exponentiation.

However this time can further be reduced by using the m-ary method which is a generalization of the binary method but instead of breaking the exponent into single bits, it breaks it in m-bits windows and then performs as many multiplications as there are non-zero windows that reduce the total number of operations needed by the modular exponentiation [10]. This method requires a certain amount of pre-computations ((m-1) powers of the message) which can be stored on the blocks Select Ram available on FPGAs.

REFERENCES

[1] Muhammad I. Ibrahimy, Mamun B.I. Reaz, Khandaker Asaduzzaman and Sazzad Hussain, "Hardware Prototyping of an Efficient Encryption Engine", International Journal of Electrical, Computer and Systems Engineering 4:4 2010.

[2] N. Nedjah L.M Mourelle, "Parallel computation of modular exponentiation for fast cryptography", Int. Journal of. High Performance Systems Architecture, Vol. 1, No. 1, 2007, Copyright © 2007 Inderscience Enterprises Ltd .

[3] Ersin Oksuzoglu and Erkay Savas. "Parametric, Secure and Compact Implementation of RSA on FPGA", International Conference on Recofigurable Computing and FPGAs, Dec.2008, pp.391-396.

[4] Ming-Der Shieh, Jun-Hong Chen, Hao-Hsuan Wu, and Wen-Ching Lin, "A New Modular Exponentiation Architecture for Efficient Design of RSA Cryptosystem", IEEE Trans. On VLS I Systems, Vol 16 N° 9 Sept. 2008.

[5] Colin D. Walter, "Right-to-Left or Left-to-Right Exponentiation?" First International Workshop on Constructive Side-Channel Analysis and Secure Design COSADE 2010, pp. 40-46.

[6] Nathaniel Pinckney and David Money Harris, "Parallelized Radix-4 Scalable Montgomery Multipliers", Journal Integrated Circuits and Systems 2008; v.3 / n.1:39-45.

[7] Xia Hong, Hu Wenhao, Yan Jiangyu, "Design and Implementation of High-Performance Modular Exponentiation Arithmetic Unit", First International Conference on Information Science and Engineering (ICISE), 2009.

[8] "Virtex-4 FPGA User Guide", Xilinx .Dec. 2008.

[9] M.B.Monagan, K.O.Geddes, K.M. Heal, G. Labahn, S. M. Vorkoetter, J.McCarron, P.DeMarco, "Maple Introductory Programming Guide", Maplesoft, a division of Waterloo Maple Inc. 1996-2008.

[10] N. Nedjah L.M Mourelle, "High-performance SoC-based implementation of modular exponentiation using evolutionary addition chains for efficient cryptography", Applied soft computing , Elsevier journal, 11 (2011) 4302–4311.

GA-based Modeling of a Non-ideal Second Order low-pass ΣΔ Modulator

[1,2]A. Dendouga, [2]S. Barra, [2]S. Kouda, [2]M. Yekhlef, and [2]N. Bouguechal

[1]Microelectronic and Nanotechnology Division,
Centre de Développement des Technologies Avancées (CDTA),
20 Août 1956, Baba Hassan, BP17, Algiers 16303, Algeria
[2]Advanced Electronics Laboratory
Département d'Électronique, Université de Batna
05 avenue Chahid Boukhlouf Mohamed, 05000 Batna, Algérie

Abstract—**An approach is presented for the high-level simulation and synthesis of discrete-time modulators based on a simulation-based optimization strategy. The high-level synthesis approach determines both the optimum modulator topology and the required building block specifications, such that the system specifications mainly signal to noise ratio (SNR). A genetic-based differential evolution algorithm is used in combination with a fast dedicated behavioral simulator to realistically analyze and optimize the modulator performance. The approach has been implemented in a MATLAB Optimization tool. Simulation results are shown for both the analysis and optimization capabilities, illustrating the effectiveness of the approach. The selected range of optimized modulator topologies as a function of the modulator specifications for a wide range of values indicate the capabilities of and the performance range covered by the tool.**

Keywords-oversampling modulators;sigma delta modulator; genetic algorithm; switched capacitor.

I. INTRODUCTION

Sigma-delta modulators are widely used in A/D and D/A conversion [2]. With the advent of deep-submicron CMOS processes having diminishing voltage supply, the dynamic range in the analog amplitude domain is diminishing with each process node. On the other hand, the switching speed of the transistors is increasing [12]. ΣΔ modulators were traditionally used for a multitude of applications such as instrumentation, telecommunications, and audio application [3].

Modeling of analog-to-digital converter (ADC) components, as well as of digital measuring systems based on ADCs, allows the device behavior to be predicted with a few of preventive experiments. A model turns out to be useful for investigating the ADC metrological behavior in several operating conditions during the main phases of development: design, evaluation, and improvement. In ADC design, the pre-eminent intrinsic error source is the quantization, and theoretical fundamental studies have been devoted to this topic by specialized research groups.

Genetic algorithms (GAs) have been successfully applied to a wide range of optimization problems including design, scheduling, routing, and signal processing. In Sigma delta (ΣΔ) modulator design, GA can be effectively used to optimize the scaling coefficients in order to achieve the desired signal-to-noise ratio [14].In this work we have proposed a method to

accomplish simultaneously the optimal topology selection and high-level modulator design based on a simulation-based optimization approach using a genetic algorithm.

In Section II we will define in the first section the deferent parameters (noises and non-idealities) that can affect on the performances of the Sigma-Delta modulator. And the detailed analytical models of these non-idealities are presented, followed by SIMULINK models of them. In the next Section, the models of an ideal modulator and a non-ideal one are presented. In Section III we will illustrate the effectiveness of the optimization methodology of modulator. Meaningful results are presented showing

the effect of performance-degrading non-idealities. Finally, in Section VII, conclusions are drawn.

II. BEHAVIOR MODEL OF INTEGRATOR

Analog circuit implementations of an integrator deviate from its ideal behavior due to several non-ideal effects. One of the major causes of performance degradation in SC ΣΔ modulators, indeed, is due to incomplete transfer of charge in the SC integrators.

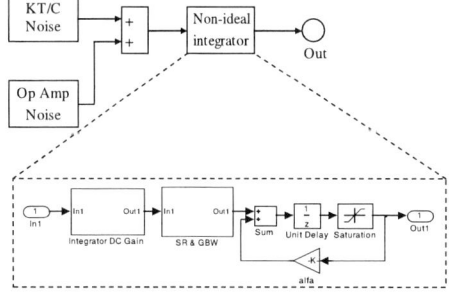

Figure 1. Noisy real integrator model.

The non-ideal effect of integrator is a consequence of following main non-idealities:

- Clock jitter
- Amplifier finite DC gain
- Amplifier band-width
- Amplifier slow-rate
- Slew-limiting

- Amplifier DC gain non-linearity
- Switch ON-resistance non-linearity
- KT/C noise

A. Imperfections of the integrator

- Finite DC gain

The dc gain of an ideal integrator described by Eqn. (1) is infinite.

$$H(z) = \frac{z^{-1}}{1 - z^{-1}} \tag{1}$$

In practice, however, the gain is limited by circuit imperfections. The consequence of this integrator "leakage" is that only a fraction α of the previous output of the integrator is added to each new input sample (parameter 'alfa' in Figure. 1). The limited dc gain of the integrator increases the in-band noise. The transfer function of the integrator with leakage becomes

$$H(z) = g \frac{z^{-1}}{1 - \alpha z^{-1}} \tag{2}$$

where g and α are the integrator's gain and leakage, respectively [4]. The dc gain of the integrator H_0, therefore, becomes

$$H_0 = H(1) = g \frac{1}{1 - \alpha} \tag{3}$$

- Slew rate and unity-gain bandwidth

The effects of the bandwidth (BW) and slew rate (SR) are related to each other, and may be interpreted as a nonlinear gain [6][4]. In fact, finite BW and SR in SC circuits lead to a non-ideal transient response within each clock cycle, thus producing an incomplete or inaccurate charge transfer to the output at the end of the integration period. In the settling behavior of SC circuits, the BW determines the linear settling behavioral, and the SR governs the nonlinear settling behavioral. Eqns. (4) and (5) recovers the settling behavior [13].

$$t \le t_0 \quad V_{out} = SR.t \tag{4}$$

$$t > t_0 \quad V_{out} = V_{out}(t_0) + (G - SR.t_0) \times \left(1 - e^{-\frac{t - t_0}{\tau}}\right) \tag{5}$$

Where t_0 and G are the transition point between the slewing and linear settling period and the ideal gain of SC circuit respectively. Imposing the conditions for the continuity of the derivatives of (4)-(5) in t_0. We suppose the slopes of the (4) and (5) at t_0 are the same. As a result, we obtain

$$t_0 = \frac{G}{SR} - \tau \tag{6}$$

- Non linearity of amplifiers

Theoretically its transfer function is:

$$V_s = A V_e \tag{7}$$

A is the amplification factor. However, its transfer function is approximated given by (8)[7].

$$A \cong A_0 \left(1 + \alpha_1 |V_0| + \alpha_2 |V_0| + \alpha_3 |V_0| + \cdots\right)^{\frac{1}{2}} \tag{8}$$

($\alpha 1$, $\alpha 2$, $\alpha 2$ ...) are the amplification factors parasites. Thus, for a pure sinusoidal signal of frequency f in the input of the amplifier, we find the output of the amplifier not only the output signal of departure, amplified, same frequency of V_e but

also other parasitic signals with higher frequency and proportional to the frequency f_r, in this case we say that there is harmonic distortion, because this spectrum of frequencies 2f,3f,ets....[13].

B. Noise on the integrator

The inherent thermal noise in integrated circuits is caused by the random fluctuation of carriers due to thermal energy [4]. Flicker noise, or 1/f noise, is dominant at low frequencies because the power spectral density is nearly proportional to the inverse of the frequency [10]. Generally, each device (transistors, resistors, etc.) contributes to the overall noise (input referred noise), with uncorrelated manner with respect to other devices, constituting the same system.

Hence, the $\Sigma\Delta$ modulators exhibit thermal noise from the op-amp and the finite switch resistance of the SC integrator during the sampling as well as the integration phases. Thermal noise has a white spectrum and a wide band, limited only by the time constants of the switched capacitors or the bandwidths of the amplifiers [11].

Obviously, the thermal-noise estimation of op-amp dependents on the op-amp architectures. For simplicity sake we assume (based cadence simulation) that the noise sources only include the noises produced by input transistors in the differential pair.

C. Switch noise

Switches are one of the major elements in SC circuits. The ideal role of them is to have zero or infinite resistance when they are ON or OFF. However, as switches in CMOS technology are realized by using nMOS and pMOS transistor, they manifest some non-idealities such as nonlinear on-resistance, clock- feed-through, and charge injection [10].

- Switch thermal noise

A simplified representation of the input sampling network to the switched-capacitor $\Sigma\Delta$ modulator is shown in Figure. 2 (a) during the sampling phase. The modeling of the MOS switch as a resistor produces the circuit shown in Figure. 2 (b), where Ron is the equivalent on-resistance of the switch and $V_n(f)$ is a noise voltage equal to the thermal noise voltage of a resistor with the power spectral density specified in Eqn. (9).

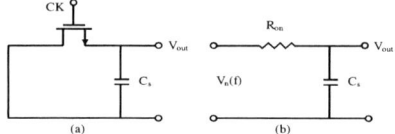

Figure 2. (a) Simplified input sampling network, and (b) the circuit model for calculating the switch noise.

$$\frac{\overline{V_n^2(f)}}{\Delta f} = 4kT R_{on} \tag{9}$$

The RC circuit in Figure. 2 (b) creates a low-pass filter, with magnitude-squared value of the frequency response given in Eqn. (10), that filters the resistor thermal noise, $V_n(f)$. The total noise power filtering is provided in Eqn. (11).

$$|H(j\omega)|^2 = \frac{1}{1 + (\omega R_{on} C_S)^2} \tag{10}$$

978-1-4673-5289-5/12 $31.00 © 2012 IEEE   216

$$\overline{V_{out}^2} = \frac{kT}{C_s} \qquad (11)$$

- Charge Injection

When a MOS switch is on, it operates in the triode region and its drain-to-source voltage, V_{DS}, is approximately zero. During the time when the transistor is on, it holds mobile charges in its channel. Once the transistor is turned off, these mobile charges must flow out from the channel region and into the drain and the source junctions. [8-10].

The charge injection is given by Eqn. (12):

$$\Delta V_{out} = \frac{\Delta Q_{ch}}{C_s} = \frac{-kWLC_{ox}(V_{DD} - V_{th} - V_{in})}{C_s} \qquad (12)$$

Notice that ΔV_{out} is linearly related to V_{in} and V_{th}. However, V_{th} is nonlinearly related to Vin [8-11]. Therefore, charge injection introduces nonlinear signal-dependent error into the S/H circuit.

- Clock Feedthrough

Clock feed-through is due to the gate-to-source overlap capacitance of the MOS switch. For the S/H circuit of Figure. 3, the voltage change at V_{out} due to the clock feed-through is given by Eqn. (13) [8]:

$$\Delta V_{out} = \frac{-C_p(V_{DD} - V_{SS})}{C_p + C_L} \qquad (13)$$

Where Cp is the parasitic capacitance.

- Nonlinear ON-resistance

Is a signal-dependent variation of the on-resistance of the switch introduces harmonic distortion into the circuit. The finite turn-on resistance is existed between the source and drain terminations of MOS, which is controlled by the gate voltage and usually nonlinear. Based on transistor large signal operation equations in linear region, on-state resistance for NMOS, PMOS and CMOS switch can be derived as given by Eqn. (14).

$$I_d = \frac{1}{2} \mu_n C_{ox} \frac{W}{L} \left[2(V_{gs} - V_{th})V_{ds} - V_{ds}^2 \right] \qquad (14)$$

The resistance of switch can be expressed

$$R_{on} = \left(\frac{\partial I_d}{\partial V_{ds}} \right)^{-1} = \frac{1}{\mu_n C_{ox} \frac{W}{L}(V_{gs} - V_{th})} \qquad (15)$$

D. Clock jitter

Clock jitter, the intrinsic uncertainty in the transition time of the clock, increases the in-band noise of the modulator [5]. The error introduced when a sinusoidal signal with amplitude A and frequency fin is sampled at an instant which is in error by an amount δ is given by

$$x(t + \delta) - x(t) \approx 2\pi f_{in}\delta A\cos(2\pi f_{in}t) = \delta \frac{d}{dt}x(t) \qquad (16)$$

III. SECOND ORDER LOW-PASS SIGMA DELTA MODULATOR MODEL

To demonstrate the design procedure we present a second order low-pass discreet time sigma delta modulator The main specifications the modulator are listed in Table 1. The SIMULINK model, shown in Figure 4, is a 2nd order non-ideal sigma delta modulator with distributed feedback architecture.

TABLE I.   PARAMETERS OF THE MODULATOR SHOWN IN FIGURE. 4

Parameter	Value
Oversampling Ratio (OSR)	256
Clock Frequency (MHz)	11.28
Input Sinusoidal Frequency (kHz)	2.06
Samples Number	65536

In this example the four parameters shown in Table 2 are assumed as design parameters which the GA will work on. The table also shows the range over which the parameters are varied. The lower and upper boundary need to be specified by the user and should be chosen by circuit implementation considerations.

TABLE II.   DESIGN PARAMETERS FOR THE GENETIC ALGORITHM

GA design parameter	Parameter range
Integrator gain a1	0.1-2
Integrator gain a2	0.1-2
Integrator gain a3	0.1-2
Integrator gain a4	0.1-2

For the $\Sigma\Delta M$ the oversampling ratio (OSR) needs to be specified. The OSR is related to in signal bandwidth, BW and the sampling frequency f_s by OSR = fs/(2 × BW). Here, we choose OSR = 256. Furthermore, the criteria (i.e., the goal values) for the

GA to optimize need to be defined; in this example these is the SNR (to be maximized). Finally, the GA needs to have values for the number of individuals in each generation, and the number of generations; we choose here 100 and 50, respectively.

The choice of population size and number of generations is a trade-off between simulation time, and the degree of design space exploration and individual diversity that will be achieved during the GA evolution process.

IV. GENETIC ALGORITHM

The GA is then run using 100 individuals, which are, within the specified range, randomly chosen parameter sets; for each individual a simulation is carried out and the SNR is calculated. This calculation is performed by a function 'calcSNR' available through the Delta Sigma Toolbox for Matlab [1].

I. SIMULATION RESULTS

Simulations were performed   for a real noisy integrator blocks. Real integrator block takes into account the main circuit non-idealities like op-amp finite dc gain, slew rate, gain-bandwidth product, and amplifier saturation voltage, there are also noises which affect the integrators like thermal noise, jitter noise, clock feed-through, charge injection, and flicker noise.

The second-order modulator shown in Figure 3 with the parameters listed in Table 1 and a selected combination (Table 3) was performed and simulated. Figure 4 shows the PSD of the modulator output and Figure 5 shows the SNR as function of the input level.

978-1-4673-5289-5/12 $31.00 © 2012 IEEE

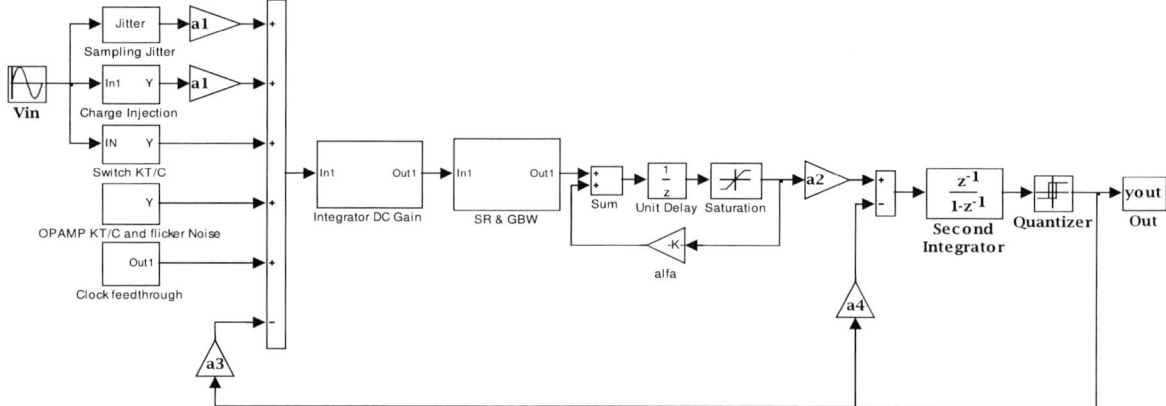

Figure 3. non-ideal model of a second order sigma delta modulator.

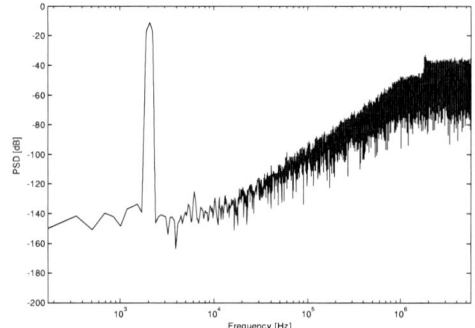

Figure 4. Output PSD of the SIMULINK model of the modulator.

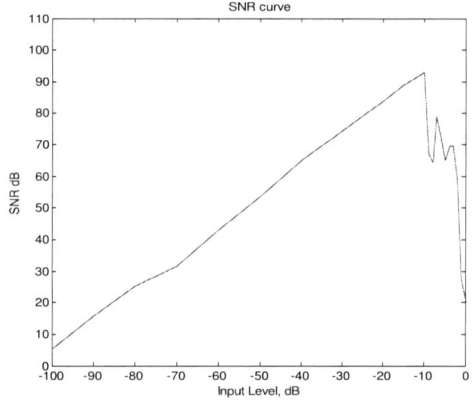

Figure 5. SNR of the low-pass second-order sigma delta modulator as a function of the input signal amplitude

TABLE III.   CHOSEN OPTIMAL PARAMETER

a1	a2	a3	a4
0.2784	0.5012	0.2238	0.2745
0.4309	0.5237	0.3079	0.3886
0.3038	0.3784	0.2321	0.2165

II.   CONCLUSION

In this paper, a more complete and precise behavioral model of a second order low-pass SC sigma delta modulator is presented and discussed including most sources of noise. The usefulness of the methodology has been illustrated through the modeling of a second order low-pass $\Sigma\Delta M$. The described methodology delivers near optimum system level design parameters.

REFERENCES

[1] Category: Control Systems, File: SD Toolbox [Online]. Available:http://www.mathworks.com/matlabcentral/fileexchange

[2] Schreier R., Temes G. C, "Understanding Delta-Sigma Data Converters," New York : IEEE Press, 2005.

[3] S. R. Northworthy, R. Schreier, and G. C. Temes, "Delta-Sigma Data Converters," Piscataway, NJ: IEEE Press, 1997.

[4] H. Zare-Hoseini and I. Kale, "On the effects of finite and nonlinear DC-gain on the switched-capacitor delta-sigma modulators," in Proc. IEEE Int. Symp. Circuits Systems, May 2005, pp. 2547–2550.

[5] B. E. Boser and B. A. Wooley, "The design of sigma-delta modulation analog-to-digital converters," IEEE J. Solid-State Circuits, vol. 22, no. 12, pp. 1298–1308, Dec. 1988.

[6] F. Medeiro, B. Perez-Verdu, A. Rodriguez-Vazquez, J.L. Huertas, "Modeling Opamp-Induced Harmonic Distortion for Switched-Capacitor SD Modulator Design," in IEEE Int. Symp. Circuits Systems, vol. 5, May-Jun. 1994, pp. 445–448.

[7] L. Dai, R. Harjani, "CMOS Switched-Op-Amp-Based Sample-and-Hold Circuit" IEEE Journal of Solid-State Circuits, vol. 35, no. 1, pp. 109-113, January 2000.

[8] Wu Xiaofeng, Liu Hongxia, Su Li, Hao Yue, Li Di and Hu Shigang "A bootstrapped switch employing a new clock feed-through compensation technique" Journal of Semiconductors, vol. 30, no. 12, pp. 125007-1 - 125007-10, Dec 2009.

[9] J. Shieh, M. Patil, and A. J. Sheu, "Measurement and analysis of charge injection in MOS analog switches," IEEE J. Solid-State Circuits, vol. SSC-22, no. 4, pp. 277–281, Apr. 1987.

[10] J. Shieh, M. Patil, and A. J. Sheu, "Measurement and analysis of charge injection in MOS analog switches," IEEE J. Solid-State Circuits, vol. SSC-22, no. 4, pp. 277–281, Apr. 1987.

[11] D.A. Johns, K. Martin, "Analog Integrated Circuit Design, " John Wiley & Sons, Inc., Toronto, 1997

[12] Y. Wang, K. Muhammad, and K. Roy, "Design of sigma-delta modulators with arbitrary transfer functions," Ieee Transactions on Signal Processing, vol. 55, pp. 677-683, Feb 2007.

[13] A. Dendouga, N. Bouguechal, S. Kouda, S. Barra, and B. Lakehal, "Contribution to the modeling of a non-ideal Sigma-Delta modulator," Journal of Computational Electronics, pp. 1-9, 2012.

[14] R. Wilcock and M. Kraft, "Genetic Algorithm for the Design of Electro-Mechanical Sigma Delta Modulator MEMS Sensors," Sensors, vol. 11, pp. 9217-9232, Oct 2011.

Embedded Implementation of an IP-PBX /VoIP Gateway

Faroudja Abid*, Nouma Izeboudjen, Mohamed Bakiri, Sabrina Titri, Fatiha Louiz and Dalila Lazib
Microelctronic and Nanotechnologie Laboratory
Advanced technology development center
Algiers, Algeria
fabid@cdta.dz, abidfaroudja@yahoo.fr

Abstract— This paper deals with an embedded IP-PBX (IP-Private Branch exchange) /VoIP (Voice over IP) gateway, around the OpenRISC processor and an open source Asterisk PBX running on an embedded Linux kernel. In this work we leverage the Opencores reuse strategy, which provides a lower cost solution to realize a portable SoC (System on Chip) for VoIP application. In this approach both hardware and software components are Opensources. The achieved part of the system has been implemented on a Virtex5 FPGA circuit. The hardware part uses 50% of BRAM memories, 27 % of slice registers and 48% of slice LUTs, the power consumption is 4.233 W. Regarding the software part, a Linux kernel has been ported to the Virtex5 FPGA, and the network functionality of the system is successfully tested.

Keywords—Asterisk, Embedded system, FPGA, SoC, OpenCores, OpenRISC, Opensource, VoIP application.

I. INTRODUCTION

Embedded systems are defined as special purpose computer-systems designed to perform one or more dedicated functions, where all essentials parts are integrated into a single chip. In this work we expose the idea of designing and implementing an embedded IP-PBX/VoIP gateway based on a single SoC, around the OpenRISC processor and an Opensource Asterisk PBX which allows connection to the public switched telecommunication network (PSTN) and telephone services through packet networks. The VoIP technology uses the IP network for voice services as an alternative to the PSTN. The voice transmission through packet networks offers lower infrastructure and lower calls cost compared to the PSTN. To develop our own platform for VoIP application we leverage the Opencores design methodology which is based on publishing all necessary information about the hardware. The purpose is to realize a portable system while reducing power consumption, device size and product cost by using embedded Linux kernel and embedded Asterisk. The system integrates the hardware components into FPGA and embeds the software of the application into the external memory. To design and implement the system, we have specified the software and the

hardware parts of the project [1] and [2]. The proposed SoC [3] and [4] includes a 32bit Reduced Instruction Set Computer (RISC) processor [5] and a minimum set of elements needed to provide VoIP functionality. These cores are connected through the wishbone bus interface. Regarding the software part, we aim to build an embedded system around Asterisk [6], an Opensource which implements a mini PBX in Linux environment. The achieved part of the system has been implemented on a Virtex5 FPGA circuit. The hardware part uses 50% of BRAM memories, 27 % of slice register and 48% of slice LUTs. Concerning the software part, a Linux kernel has been ported to the Virtex5 FPGA and a network functionality of the system has been successfully tested.

In section II, a background and related work are given. Section III deals with the proposed system overview; herein we focus on a hardware part. The system prototyping and synthesis results are presented in this section. Section IV gives the software test results, since the network functionality of the system constitutes the main step in a VoIP application; we expose the test result of the achieved part for embedded network application. Finally, section V concludes this paper.

II. BACKGROUND AND RELATED WORK

Asterisk is an Opensource which implements a mini PBX in Linux environment. It is a phone switch allowing connection to the PSTN and telephone services through packet networks created by Mark Spencer of Digium. Asterisk supports SIP, IAX, H323 signaling and the RTP protocols. The idea behind embedding Asterisk is to give a lower and portable solution since most of implementations are done on a computer. A few works emerged from industrial and academic researches to design such system. Among these implementations, the IP04 from Rowetel [7] a four IP-PBX based Blackfin processor which runs Asterisk and uClinux. Asterisk was ported to Blackfin processor by David Rowe. He gives an open hardware implementation which is used by Dan Amarandei and al for Wideband VoIP Middleware [8]. Another implementation based Xilinx PowerPC processor is given by Stelios Koroneos from Digital OPSIS [9]. In our work we

978-1-4673-5289-5/12 $31.00 © 2012 IEEE

have adopted an Opencores approach to build an OpenRISC processor based implementation. This choice is justified by the following points:

- OpenRISC is an open source IP-core freely available as a Verilog model. The processor is intended for embedded, portable and network applications.
- Availability of the Register Transfer Level (RTL) descriptions for all the IPs components included in the proposed system based OpenRISC and tools at free cost.
- The whole system can be mapped into FPGA or ASIC.

III. SYSTEM OVERVIEW

A. System architecture

Figure 1 illustrates the architecture of an IP-PBX implementation done on an x86 PC plus a PCI card (a getaway which allows connection to the PSTN) for analog ports.

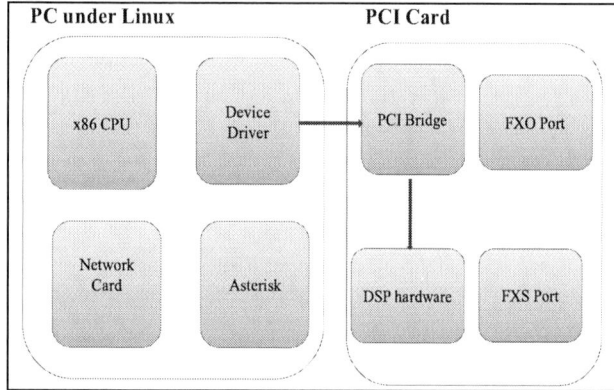

Fig. 1. IP-PBX Hardware configuration

Figure 2 presents the architecture of an embedded implementation based OpenRISC soft processor and Asterisk. This implementation allows functionality of an IP-PBX on embedded system designed for this purpose; the system hold the software and the hardware part of the system in single chip, hence no PC and no PCI card are required.

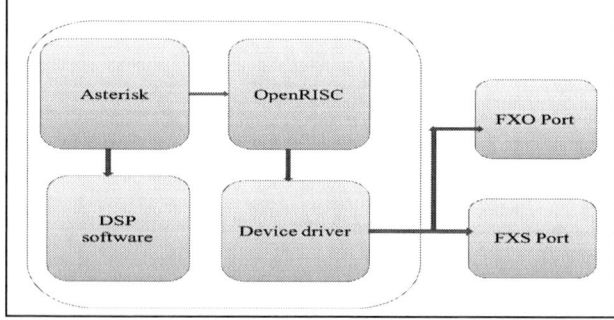

Fig. 2. Embedded implementation of an IP-PBX

Figure 3 shows the board block diagram around the FPGA Virtex5 circuit. It includes some peripherals needed for the test of the system implemented on the FPGA for VoIP

application. For the hardware part, an HDL file describing all the cores of the library is created. The proposed architecture is based on the OpenRISC processor and a minimum set of elements. These elements are the debug unit for debugging purpose, a memory controller that controls an external memory that carries Linux and Asterisk, a Universal Asynchronous Receiver Transmitter (UART), the Internet connection is established by the Opencores MAC/Ethernet. A SoC Verilog description is created for the integration of all the cores. Considering the complexity of the system, we proceed to test communication protocol of each subsystem composed of set of IPs connected to the bus wishbone with the OpenRISC processor. Figure 4 shows the architecture of the achieved subsystem, for network functionality test [10].

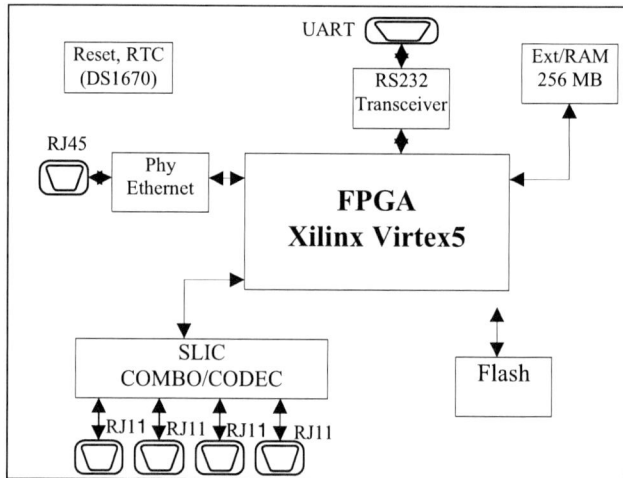

Fig. 3. Board block diagram based FPGA

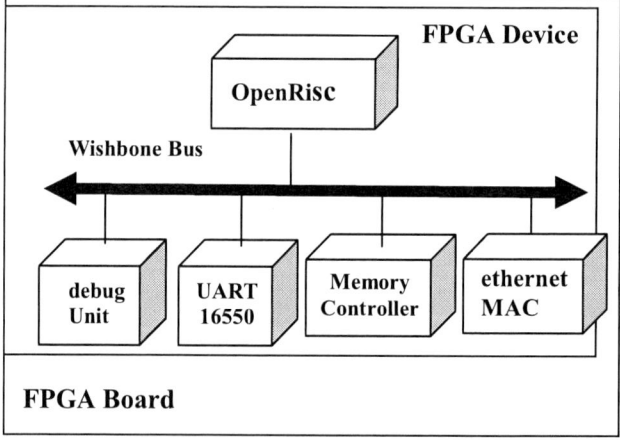

Fig. 4. SoC subsystem architecture

B. OpenRISC processor

In this section we present the OpenRISC processor since it is the basic IP in the designed SoC.

The OpenRISC 1200 is a synthesizable 32-bit RISC processor developed and managed by a team of developers at Opencores as a Verilog model. It is freely available from the Opencores

978-1-4673-5289-5/12 $31.00 © 2012 IEEE   220

website, licensed under the GNU LGPL license. An overview of the OpenRISC processor architecture can be seen in figure 5.

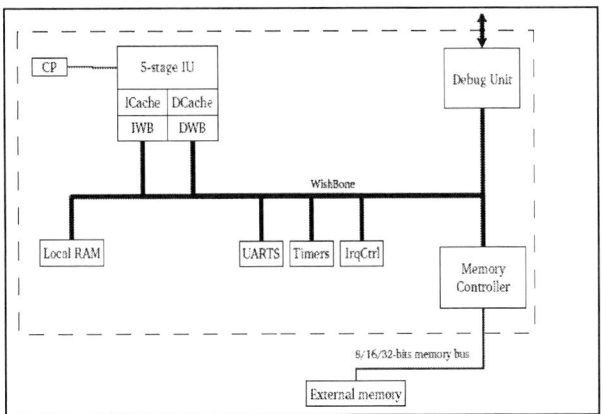

Fig. 5. OpenRISC 1200 processor architecture

C. System Prototyping

The achieved subsystem has been prototyped using the Xilinx development platform ML501-Virtex5, The operating frequency is 66 MHz. The development tool used for the design and implementation is ISE 12.1 tool by Xilinx [11]. Table 1 shows the synthesis results.

The achieved part of the SoC uses 50% of BRAM memories, 27 % of slice registers and 48% of slice LUTs. It is noticeable that the BRAM memories are the most used resources, the use of other resources remain low. The power consumption is 4.233 W.

TABLE I.   SYNTHESIS RESULTS. TARGET DEVICE XC5VLX50-1FF676

Slice Logic resources	Used Slice Logic	Available Slice Logic	% Occupied resources
Number of Slice registers	7864	28800	27%
Number of Slice LUTs	13872	28800	48%
Number of bonded IOBs	192	440	43%
Number of BRAMs	24	48	50%

TABLE II.   POWER CONSUMPTION

Total estimated power consumption	POWER (W)
Logic	0.009
IO (W)	3..346
BRAM (W)	0.053
Quiscent	3.389
Dynamic (W)	0.394
Total(W)	4.233

IV. SOFTWARE TESTS RESULTS

A. Software development

The Software development of the embedded system includes two parts:

1. Configuration and compilation of embedded Linux operating system. In this part the kernel is adjusted to the target system and compiled.
2. Programming the VoIP application.

Embedded Linux Configuration

First we have ported the Linux kernel 3.0.rc1 to the Ml501 FPGA platform by defining a new *dts* file (device tree file *ml501.dts*)) [12]. Mainly we have configured linux to use the Opencores MAC/Etehrnet as a network controller and include the TCP/IP stack that provides the IP connection. At first we have run Linux kernel on Or1ksim simulator. Then we have booted it on the virtex5 FPGA. The Internet protocols needed for network functionality (ICMP (Internet Control Message Protocol) for ping, TCP (Transmission Control Protocol) and UDP (User Datagram Protocol)) are available under Linux.

CPU Software

The CPU software is based on the embedded Linux operating system. Figure 6 presents the general software architecture. The achieved subsystem runs the Linux kernel 3.0 with the HTTPD web server and other application. The TCP/IP stack that provides the IP conection is also included. The communication between the operating system and the system hardware is implemented by the device drivers through standardized Linux interfaces. The software includes device drivers for the Ethernet and UART.

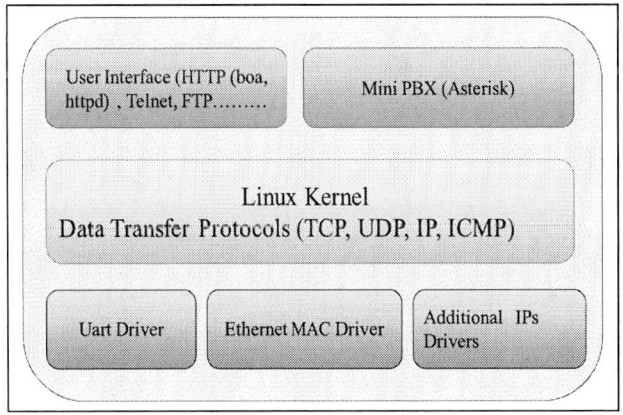

Fig. 6. CPU software architecture

The network application test (Web server)

For the network application an HTTPD web server application running on the embedded Linux is chosen as test of the subsystem. The purpose is to test the network functionality of the proposed SoC and the web server is just used as a network application example. Figure 7 shows the test environment, which consists of an Ethernet connection of the evaluation ML501 board that represents the web server to the remote web-client. The serial link is used as user interface in order to communicate with the ML501, where the OpenRISC-based system running the embedded Linux kernel and the web server is embedded. The network configuration of the embedded server is set according to CDTA/LAN (IP address 10.1.70.75, netmask 255.255.0.0).

The test of the HTTPD web server has been successfully done on the ML501 FPGA platform. The remote web-client with web browser displays the static web page hosted in the embedded web server, which enable http file transfer as shown in figure 7.

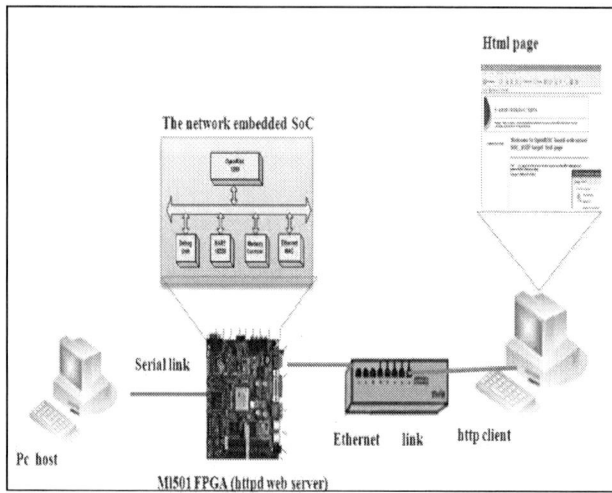

Fig. 7.   Test environment

V.   CONCLUSION

In this work an embedded IP-PBX/VoIP gateway system around OpenRISC processor and a PBX Asterisk is exposed. The results given are related to the achieved part of the SoC, developed using the Opencores IP reuse strategy for hardware and software components. This part concerns the network functionality test which constitutes the main step in a VoIP application. The hardware part uses 50% of BRAM memories the most used resources, while the use of the other resources remains low. Hence the Virtex5 FPGA can include the remaining part of the SoC. Concerning the software part, the network application running on embedded Linux operating system is first embedded and tested on the ML501 board. Further, as perspective we aim to embed the Asterisk PBX and finalize the implementation of the whole system around OpenRISC on the FPGA for a portable system.

REFERENCES

[1] Abid Faroudja, Nouma. Izeboudjen, Sabrina. Titri, Leila.Salhi, Fatiha.Louiz, Dalila.Lazib "Hardware /Software Development of System on Chip Platform for VoIP Application ", ICM, International Conference on Microelectronics, December 19-22, 2009, Marakech, Morocco, pp62-65.

[2] N. Izeboudjen, K. Kaci, M. Bengherabi, S. Titri, F. Louiz, L. Sahli, D. Lazib, *N. Idirene, "An Open Hardware Architecture based System on

Chip for Voice over Internet", JFAAA05, 18-21 January 2005, Dijon, France.S.

[3] Titri, N. izeboudjen, F. Louiz, M. Bakiri, F. Abid, D. lazib, L. rekab, "An Opencores /Opensource based Embedded System-on-Chip platform for Voice over Internet", book: VOIP Technologies, pp:145-172, INTECH, ISBN: 978-953-307-549-5,2011.

[4] F.Abid, F.Louiz, N.Izeboudjen, D.Lazib, S.Titri, M.Bakiri "Design and Implementation of an Embedded VoIP System on Chip ", CIAM'2011, International Conference on Automation and Mechatronics, November 22-24, 2011.Oran (Algeria), pp 561-565.

[5] Mjan Lampret "OpenRISC 1200 IP Core Specification", Rev. 0.7 Sep 6, 2001.

[6] Jim Van Meggelen, Leif Madsan and Jared Smith, "Asterisk the future of telephony, second edition ", Oreilly, August 2007. ISBN-13: 978-0-596-51048-0

[7] Free Telephony Project, Open Telephony Hardware, IP04 Four Port IP-PBX retrieved February 14, 2010 from

[8] Dan Amarandei and Mostafa El-Said, "Implementation of Wideband VoIP Middleware Using Embedded Systems", IEEE2010

[9] http://www.digital-opsis.com

[10] F. ABID, N. IZEBOUDJEN, L. SAHLI, D. LAZIB, S. TITRI, F. LOUIZ, M. BAKIRI, "Towards an Open Embedded System on Chip for Network Applications" WSEAS 2010, Latest Trends on Circuits, Systems and Signals. The 4th international conference on Circuits, Systems and Signals, CSS'10. Greece, July 22-25, 2010, pp 163-168. ISSN: 1792-4324, ISBN: 978-960-474-208-0.

[11] XILINX ISE 12.1 user manual.

[12] Mohammed BAKIRI, Sabrina TITRI, Nouma IZEBOUDJEN, Faroudja ABID, Fatiha LOUIZ and Dalila LAZIB "Embedded System with Linux Kernel Based on OpenRISC 1200-V3",SETIT2012,March 21-24, 2012, Sousse, Tunisia.

A Fast Hardware/Software Co-Verification Method using a real hardware acceleration

Mossaad Ben Ayed[1], Faouzi Bouchhima[2], Mohamed Abid[3]

National Engineering School of Sfax

University of Sfax, Tunisia

[1] mossaad_benayed@yahoo.fr, [2] f_bouchhima@yahoo.fr, [3] mohamed.abid@enis.rnu.tn

Abstract— Due to the number and the nature of components integrated in them, Systems-On–a-Chip (SoC) have become increasingly complex. To solve the problem of cost, flexibility and the time-to-market, systems designed with mixed hardware software systems has increased and the verification method has become a key position of the design process. This paper describes a new hardware/software co-verification methodology for SoC, based on the integration of a SystemC simulator and an FPGA accelerator. Between the SystemC simulator [1] [2] and the FPGA board, a shared communication was established to accelerate the simulation via flexible interfaces. The key issue is the synchronization between the two parts.

Keywords-component; Co-Verification; SystemC ; Transaction Level Modeling; Synchronization.

I. INTRODUCTION

Embedded systems are mostly heterogeneous devices. Their design is based on hardware and software components. These parts cannot be developed independently, since their interaction is a key point of the system behavior. Each part needs to be aware of the characteristics of other parts, in order to provide optimized components. The best strategy adopted is co-design, since it allows us to develop HW/SW component concurrently [3].

Co-simulation is a key methodology in co-design that allows verification of the hardware, the software, and their interaction. The essential aim of the co-simulation is to validate and to cover the performance as well as the functionality. The main problem appears when the system complexity grows and the validation becomes more and more time consuming. To overcome this challenge, and speed up HW/SW co-simulation, Transaction Level Modelling (TLM) is adopted. This paper is organized as follows. Section 2 summarizes existing work on HW/SW co-simulation. Section 3 explains the proposed solution to accelerate the co-simulation. Section 4 and 5 present a detailed description of the communication model and the synchronization model between SystemC and the FPGA platform. Experimental results are discussed in section 6. Finally, concluding remarks are given in section 7.

II. RELATED WORK

Several simulation frameworks have been proposed in the literature. They can be classified into two main categories: homogenous and heterogeneous.

Homogenous frameworks use a single simulator for the simulation of both hardware (HW) and software (SW) components. The main advantage of this category is the simplification of the design modeling and the good simulation performance. However, homogenous frameworks, usually based on extended existing languages suffer from lack of libraries and synthesis tools and they are suitable only in a very initial phase of the design, prior to HW/SW partionning.

Inversely, heterogeneous frameworks, which are based on integrating existing simulators (using co-simulation), warrant a more accurate tuning between HW/SW components and benefit from the existing libraries and tools. The major problem in this category is the communication and synchronization models between the different simulators.

Several frameworks [4] [5] [6] are mainly focused on Multilanguage system description, that is, a HDL for hardware and a programming language for software. All these heterogeneous co-simulations are based on solving the problems of controlling and synchronizing several simulation engines. These frameworks are adopted because of the best simulation performance and the easiest integration but it was the only possible choice when VHDL or Verilog simulation was the highest possible level of abstraction for simulating hardware.

The advantage of design with SystemC [7] [8] [9] is the use of the bus at different abstraction layer to obtain more efficient co-simulation. HW and SW are described by using C. This approach simplifies the implementation of the initial model as well as the HW / SW partitioning. In fact, HW components are simulated by using the SystemC simulation kernel, while SW programs run on an Instruction Set Simulator (ISS). Thus, more accurate performance estimation could be obtained.

The frameworks based on the last approach use two essential steps. The first is the Inter Process Communication (IPC). It is used to make the communication between the ISS and the SystemC simulator. The second is the Bus Wrapper. It ensures synchronization between SystemC simulation and the ISS.

These frameworks still suffer from some performance bottlenecks, caused by the use of the ISS. However, ISS gives the best simulation accuracy. To accelerate simulation, in spite of the accuracy, the native SW simulation is adopted using SystemC and time annotations. Some works try to improve the performance estimation accuracy in native simulation by modeling and simulating the OS behavior essentially the interruptions and preemption mechanism [10].

Other works [15] is based on using multi-ISS to accelerate the simulation. But this framework suffers from a complex synchronization scheme that increases the overhead.

Our solution replaces the ISS by real processor. The main advantages are:

- Our framework gives a high speed SW validation without any lost of accuracy since SW will be executed by the target microprocessor.

- The hardware part will be described and simulated using SystemC.

So, this framework represents a very useful platform for software engineers to validate their code before the hardware components become available.

III. CONVENTIONAL APPROACHES

As known, if all modeled blocks are implemented in hardware emulation, the system cost, as well as the running and debugging cost, will become expensive. Therefore, a combined method using an emulator and a simulator is the most adopted to model SOCs.

To increase the verification speed while maintaining clock accuracy, an FPGA type ALTERA DE2-70 is used. Thus the verification framework uses SystemC simulator to simulate the hardware components and the NIOS II processor to execute the software applications. The main idea is based on the replacement of the ISS by a real processor, which complicates the synchronization task between the HW and SW. Two issues are essential: the communication and the synchronization models. The next section describes the communication model.

IV. COMMUNICATION MODEL

This section gives a brief introduction to the communication model. A USB link is used in the communication between PC and FPGA because this kind of communication has better speed than PCI which it adopted in emulation [11]. This communication is based on packets which are constructed by the communication interface between simulator and emulator.

Two forms of exchanged packets are used to perform the synchronization scheme between the simulator and the emulator, figure 1.

Interruption packet is the first form. It consists of two parts: a header and a body. The last one contains the routine number and the interruption time stamp. The header of this form presents the type of synchronization and the routine number indicates the routine task to be executed. The time stamp represents a synchronization point and it is used to execute the interrupt routine at the appropriate instant.

Data packet is the second form. It comprises a header and the data. The header in this case contains the synchronization type, the size of data to send and the time stamp to synchronize when it is necessary.

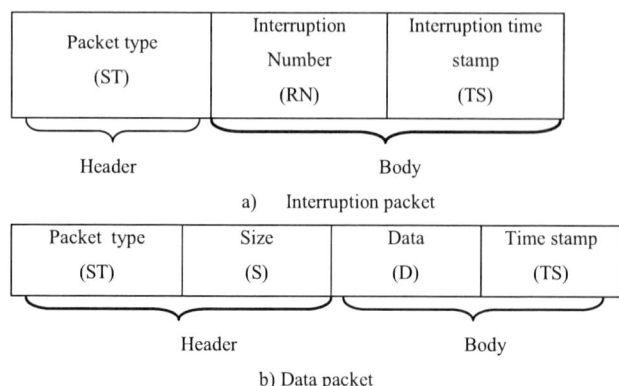

a)   Interruption packet

b) Data packet

Figure 1: Synchronization forms

Note that any packet received by the NIOS II side generates an USB interruption that can be exploit in the implementation phase to interrupt the NIOS processor each time a packet is received.

V. SYNCHRONIZATION MODEL

A key issue of the proposed verification framework is the time synchronization between the SystemC simulator and the NIOS II processor emulated on the FPGA board. The verification method is based on the following synchronization schemes which respect the interaction style that can be involved between HW and SW components. Note that, in the same design, HW and SW components may use different synchronization schemes

➢ **Scheme 1: The SW task receives data periodically from the hardware task.**

This scheme is based on FIFO memory between SW task and HW task. The main idea consists on fixed synchronization time between simulator and emulator (see figure 2). Because of the difference of speed, the HW imposes a synchronization Time (T_{sync}). This T_{sync} must be more than HW or SW tasks time.

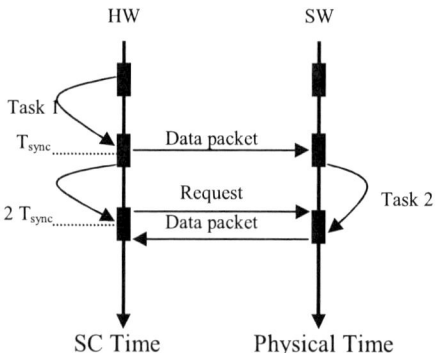

Figure 2: Synchronization model: scheme 1

➢ **Scheme 2: The SW task waits the end of the hardware task.**

When a hardware component is simulated by SystemC, the SW task uses a waiting loop for data (see figure 3). Once the hardware task (task1) is finished, the simulator sends data to the SW task and a switch context from SystemC to board is taken. At this time, the SW task receives data and resumes the execution. Here, the execution time of task1 is modeled by the SystemC wait() function. The amount of time used by the wait function is sent to the SW part to inform it about the duration of the waiting loop (see figure 3). Note that the SystemC and the emulator need to usually exchange information about the time.

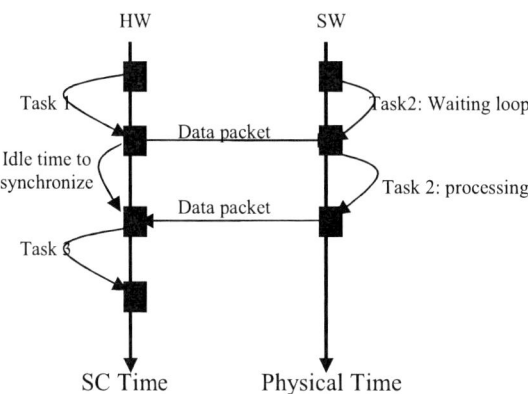

Figure 3 : Synchronization model: scheme 2

> **Scheme 3: The SW task receives an interruption to indicate the end of the hardware task**

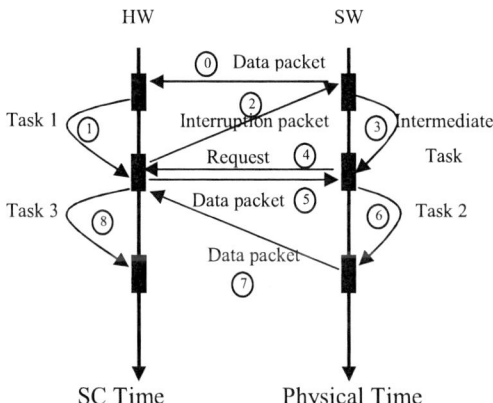

Figure 4 : Synchronization model: scheme 3

This scheme is illustrated by the figure 4. In this case, the software does not use a waiting loop but the end of the task is indicated by interruption, so the software can execute the task instead of waiting. The Simulation scheduler, running on NIOS II, sends data to the simulation interfaces (arrow 0), which activates the hardware task1. At the end of task1 process, and before sending data to SW task, the *wait_for_interrupt(sc_time)* function is called (see figure 5), so the simulator advances its time (arrow 1) and sends an interrupt packet to inform emulator for the next time stamp (arrow 2). At this time, the simulation scheduler activates a NIOS II timer with a period that coincides with the received interruption time stamp and begins the execution of an

intermediate task (an eventual user background task). When the timer is reached, it interrupts the background task. Thus the simulation scheduler activates the task 2 (the number of the interruption is received with the interrupt packet). The last one may request data, thus the task 1 resumes execution and sends data packet (arrow 4), which activates task 2. Figure 6 shows the template of the code.

```
void wait_for_interrupt(sc_time t)
{
wait(t);
send_interruption_packet(…..) ;
}
```

Figure 5 : Wait_for_interrupt code

Where t is an estimation of the task1 duration

```
            /* Task1 code */
Instructions
….
….
Wait_for_interrupt (t);
Switch_context();/* switch context to SC*/
```

Figure 6 : Template of synchronization code

> **Scheme 4: The SW task may receives a random interruption resulting from externally data reception**

This scheme is illustrated by the figure 7. The SystemC begins the execution of the task 1 and, when finished, sends a data packet to the SW task. The task 2 starts and the SystemC executes the *Hardware_Input_Interface* : a process that models the input interfaces of the hardware subsystem (its execution do not advances the SystemC local time). The process may generate a random interrupt packet which informs of the reception of a new data. The sent packet via USB generates an USB interruption which will interrupt the task 2. Thus, the USB interruption plays the same role as the hardware interruption. Once the interruption is occurred, we need only to know, thanks to the received interrupt packet, the interruption routine to execute (here is task 3)

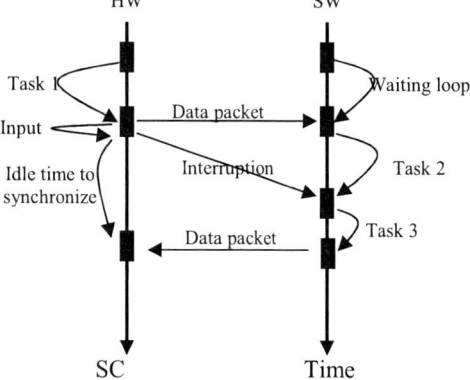

Figure 7 : Synchronization model: scheme 4

To ensure communication and to save synchronization context, an array of shared registers is used.

VI. EXPERIMENTAL RESULTS

In this section, we propose a fingerprint recognition system [14] to validate our co-verification framework. At first, a communication model is implemented. The model is divided into receiving/ transmitting drivers. Figure 10 shows the different components of the model.

- SW driver is made by Windows Driver Kit (WDK). This driver contains two main functions Read and Write.

- Channel: the USB communication offers a transaction speed of 480 Mbits/s; unlike PCI which offers only 133 Mbits/s [12] [13].

- HW driver is based on Philips ISP1362 controller.

Then, Based on the native execution of the fingerprint recognition on a 2 GB RAM, 1.66 GHz Intel Core 2 Duo processor with Windows XP operating system, we notice that the time execution of the minutia extraction is the minimum. We divided our system on hardware components and software applications based on time execution (see figure 8).

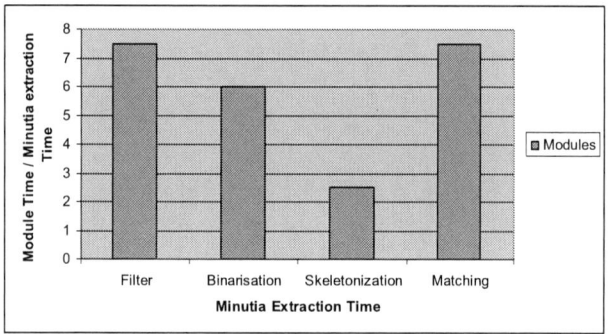

Figure 8: Time execution of each modules as a function of minutia extraction time

The co-verification time of each module is given in table 1.

TABLE I : SIMULATION TIME

	Module	Time (s)
HW Components	Read of fingerprint	0.03
	Filter	
	Binarization	
	Matching	
Interface	Interface	0.5
SW Applications	Skeletonization	0.01
	Minutia extraction	
All Modules		0.54

Result shows that our simulation/emulation environment reduces the time simulation.

VII. CONCLUSION

A methodology to perform early design stage validation of hardware/software systems is proposed in the paper. The co-verification framework is based on synchronization between SystemC simulator and FPGA board emulator. The main idea is to accelerate the simulation by replacing the ISS with NIOS II as a real processor. Experiments with a real example proved the effectiveness of the proposed framework. We need others type of examples to validate all synchronization aspects. As a future work, the adding of OS support by the software part will be considered.

VIII. REFERENCES

[1] OSCI; "Functional Specification for SystemC 2.0", available at www.systemc.org

[2] T. Grotker, S Liao, G. Martin, S Swan; "System Design with SystemC"; Kluer Academic Publishers, ISBN 1-4020-7072-1

[3] D. Micheli, D.Ernst, R. Wolf, W. Eds. Readings in Hardware/Software Co-design, Morgan Kaufmann, 2001

[4] Liem C., Nacabal F., Valderrama C., Paulin P., And Jerraya A.. System-on-chip cosimulation and compilation. IEEE Design and Test of Comput. 14, 2, 16–25.1997.

[5] Valderrama C., Nacabal F., Paulin P., And Jerraya A. Automatic VHDL-C interface generation for distributed cosimulation: Application to large design examples. Design Autom. Embed. Syst. 3, 2/3, 199–217.1998.

[6] Coste, P., Hessel, F., Marrec, P. L., Sugar, Z., Romdhani, M., Suescun, R., Zergainoh, N., AND Jerraya, A. Multilanguage design of heterogeneous systems. In Proceedings of IEEE International Workshop on Hardware-Software Codesign. 54–58.1999.

[7] Liu, J., Lajolo, M., AND, A. Software timing analysis using HW/SW cosimulation and instruction set simulator. In Proceedings of the IEEE International Workshop on Hardware/Software Co-design. 65–69.1998.

[8] Fummi, F., Martini, S., Perbellini, G., AND Poncino, M. Native ISS-SystemC integration for the cosimulation of multi-processors SoC. In Proceedings of the IEEE Conference on Design Automation and Test in Europe. 564–569. 2004.

[9] Moussa, I., Grellier, T., AND Nguyen, G. Exploring SW performance using SoC transactionlevel modelling. In Proceedings of the IEEE Conference on Design Automation and Test in Europe. 120–125. 2003.

[10] Bouchhima, A. Yoo, S. Jarraya A.., "Fast and accurate timed execution of high level embedded software using HW/SW interface simulation model", Design Automation Conference: ASP-DAC, pp. 469 – 474, 2004.

[11] Soha Hassoun, Senior Member, IEEE, Murali Kudlugi, Duaine Pryor, and Charles Selvidge "A Transaction-Based Unified Architecture for Simulation and Emulation" IEEE transactions on very large scale integration (VLSI) systems, vol. 13, no 2, 2005.

[12] Jan Exalson, "USB COMPLETE Everything You Need to Develop Custom USB Peripherals" book, third edition, 2005

[13] ISP1362 Embedded Programming Guide Version 9 June 2002

[14] Mossaad Ben Ayed, Faouzi Bouchhima and Mohamed Abid, "Automated Fingerprint Recognition Using the DECOC Classifier", International Journal of Computer Information Systems and Industrial Management Applications. ISSN 2150-7988 Volume 4 pp. 546-553. 2012.

[15] S. Cordibella, F. Fummi, G. Perbellini, D. Quaglia, "A HW/SW Co-Simulation Framework for the Verification of Multi-CPU systems", IEEE transactions, 2008

Design and Implementation of an FPGA-based Fuzzy MPPT Controller for PV Standalone Systems

A. Messai, A. Mellit and A. Guessoum

Laboratoire *LATSI*
USD Blida
Blida, Algeria
messai_adnane@yahoo.fr

Abstract— **In this paper, the design of an FPGA based digital fuzzy MPPT controller with the typical DC-DC boost converter is described. We have optimally designed the DC-DC converter intended for PV applications, developed the MPPT digital control function and implemented it on an FPGA circuit. Simulation results show that, the developed system was able to seek the instantaneous maximum power being produced by the PV in all environmental conditions.**

Keywords- PV systems, DC-DC boost converter, digital fuzzy controller, Field Programmable Gate Array (FPGA), Maximum Power Point Tracking (MPPT).

I. INTRODUCTION

To maximize a photovoltaic (PV) system's output power, continuously tracking the maximum power point (MPP) of the system is necessary. As it is known, the position of this point (i.e. MPP) depends on the irradiance conditions, the panel's temperature, and the load connected to the PV system. To really seek this point, we generally develop maximum power point tracking (MPPT) algorithms which provide the theoretical means to achieve the MPP of solar panels; these algorithms can be realized in many different forms of hardware and software. The objective of this work is to make a contribution to develop a working solution to the MPP problem [1].

In this research, Field Programmable Gate Array (FPGA) has been used as to implement the digital fuzzy MPPT controller for the PV DC-DC converter. It combines the high-speed operation which is feature of the hardware, and adaptability which is feature of the software offering a good performance, and flexibility to the entire system. Taking into consideration the availability constraint, the Memec Design VMB-1000 board model was chosen.

Using current and voltage sensors, the MPPT designed fuzzy controller, implemented inside the FPGA circuit, will be able to adjust the duty cycle of the DC-DC boost converter to draw the maximum possible power out of the PV array at a given insolation and temperature conditions.

To optimize the energy yield from the PV panel, a high efficiency step-up DC/DC converter was also designed.

The main architecture of the considered standalone PV system is given by the figure bellow:

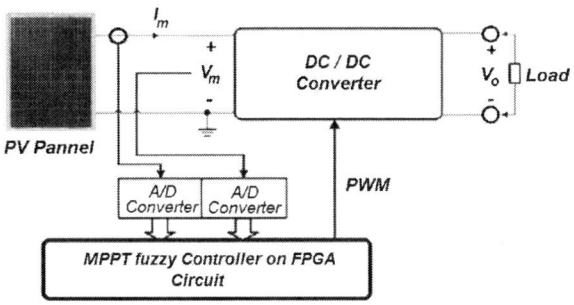

Figure 1. Generic stand alone PV system

It consists of PV panel for energy extraction from the sun, a control MPPT system and a power DC/DC converter for the load interface. Through this work, BP-solar photovoltaic module (BP SX150) was chosen because of its low cost and high power output. At STC the module produces 150 W at 34.5 V and 4.35 A [1].

In the sections that follow, the paper first deals with the DC/DC boost converter. After we will see the design, the simulation and the implementation on a FPGA of the being developed MPPT controller.

II. DC-DC CONVERTER DESIGN

The considered electronic switch-mode DC/DC converters operate by storing the input energy temporarily and then releasing that energy to the output at a different voltage and current maintaining the integrity of the system specifications. In the context of PV applications, this type of converters present an electrical load to the solar panel that varies as the output voltage of the converter varies. This load variation in turn causes a change in the operating point (current and voltage characteristics) of the panel. Thus by intelligently controlling the operation of the DC-DC converter, the power output of the panel can be controlled and made to output the maximum possible. In our case, a fuzzy logic control algorithm is suggested for the MPPT task. We use the combination of 25 fuzzy rules and pulse-width-modulation control (PWM) signal to seek the instant maximum power point. To do this, the controller continuously senses the current and the voltage at the output of the PV panel, calculates the correspondent electrical power and then automatically generates the control (PWM) signal to be applied to the switching device (MOSFET transistor) of the used DC/DC converter.

Up to now, various DC/DC converters have been proposed for PV applications, however, they generally suffer from high power losses due to its hard-switching operation. To address this problem, we have designed a high efficiency step-up DC/DC converter [2].

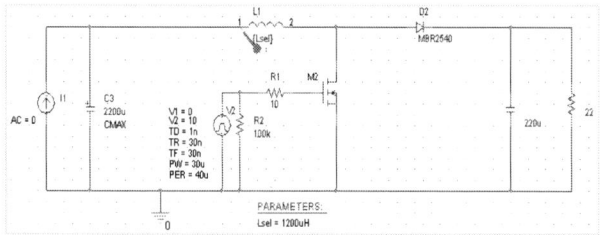

Figure 2. Electrical schematic of the used DC/DC converter

To achieve designing task, we have started with the rule design hypothesis stipulating that the lost power in the interface must not exceed 10% of the whole produced power in the PV system [3]. Assuming that the lost power in this stage is divided on its constituent electronic components as depicted in the figure below:

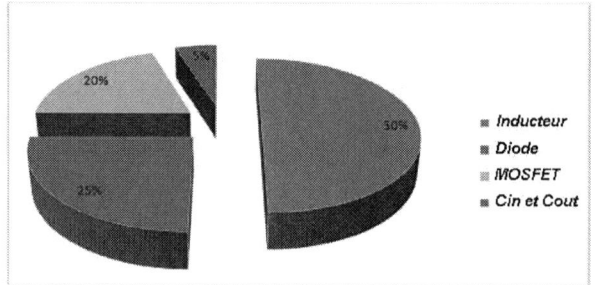

Figure 3. Ratios of dissipated Power in the DC/DC converter.

Through this figure, we can note that the important loss of energy in the inductor (50%) is associated with high losses in the (parasitic) unavoidable series resistance. Whereas, the losses in the input and output capacitors are minimal (5%), mainly when they are chosen with low equivalent series resistances (*ESR*). The remaining energy losses concern the semiconductors elements (Diode and MOSFET).

Taking into consideration the input voltage range of the converter is which is ranging from 9 to 36 Vdc, a switching frequency of 25 KHz, a duty cycle ratio varying in the interval [0, 0.4] And other practical design rules, we have use the mathematical equations [4]:

$$Vout = Vin/(1-D)$$
$$Iout = \eta(1-D)Iin$$
(1)

In which: *D* is the duty cycle

And η is the power efficiency.

Which leads to the optimal values related to the various electronic component of being designed DC/DC converter. In the final step we obtain:

L(self value) = 1.2 mH ;

Cin = 5.74 µF

Cout : 22.8 µF

Diode : Shottky

Power MOSFET : IRF840

III. FUZZY MPPT CONTROLLER DESIGN

To control the above designed DC/DC converter, we have also proceeded to the design of an MPPT fuzzy controller. This controller will be able to extract the most power from the sun, given the limitations of today's silicon-based photovoltaic cells, and implement said power into the connected load. To implement this task, a two-inputs one-output digital Fuzzy Logic Controller (FLC) was carried out [5].

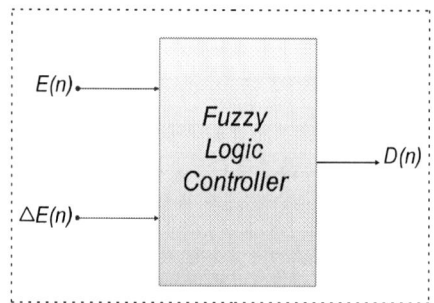

Figure 4. two-inputs one-output digital Fuzzy Logic Controller.

The power-voltage characteristic of a PV module is characterized by a MPP which is unique and climate conditions dependent (irradiation and cell temperature). As an example, in Fig. 5 the MPP corresponding to an irradiance of 1000 W/m2 and a cell temperature of 25 °C is shown.

The MPPT using the FLC approach is designed in a manner that the control task try continuously to move the operation point of the solar array as close as possible to the maximum power point (MPP).

$$E(n) = \frac{P(n) - P(n-1)}{V(n) - V(n-1)}$$
$$\Delta E(n) = E(n) - E(n-1)$$
(2)

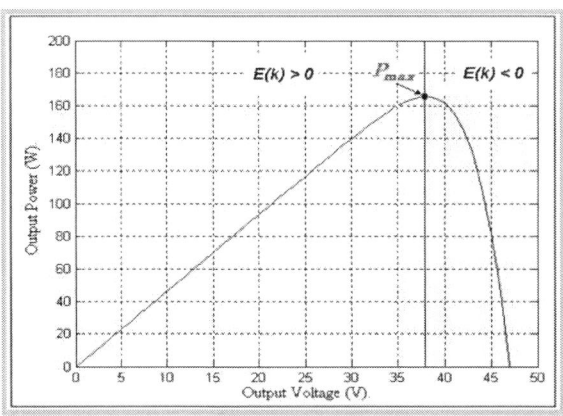

Figure 5. Power-voltage characteristic at 25 °C and 1000 W/m2 [6].

Where E and DE are the error and the change in error respectively, n is the sampling time, while p(n) is the instant power delivered by the PV generator; V is the instant voltage. The inputs are chosen so that the instant value of *E(n)* shown the load operation power point's direction. It is possible to know if the operating point stays in a zone where the derivative of the P–V characteristic is positive or negative, while *DE(n)* shows in which direction the load operation power point moves.

Figure bellow shows the membership functions of the input variables *(E, DE)* and the output variable *(D)* adopted for the design and the implementation of our FLC.

Five linguistic variables (NB, NS, ZE, PS, PB) are adopted for each of the three input/output variables. Where NB stands for Negative Big and NS: Negative Small, ZE: Zero, PS: Positive Small, PB: Positive Big.

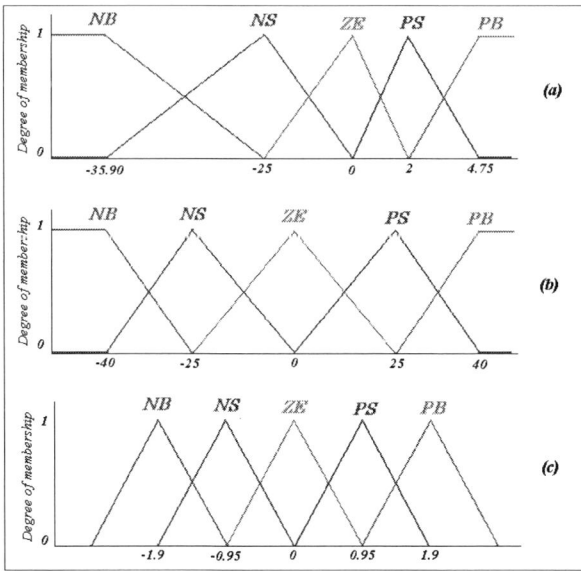

Figure 6. Membership functions for inputs and output of Fuzzy Logic Controller. (a) Error, (b) error change, (c) duty cycle ratio.

TABLE I.   FUZZY RULES

		Error Variation				
		NB	NS	ZE	PS	PB
Error	NB	PB	PB	PS	PB	PB
	NS	PB	PS	PS	PS	PB
	ZE	NS	NS	ZE	PS	PS
	PS	NB	NS	NS	NS	NB
	PB	NB	NB	NS	NB	NB

In order to facilitate the hardware implementation of the considered MPPT controller, some constraints to reduce the hardware complexity of the FLC have been respected [13].

(1) For each input variable, the overlapping degree of its membership function is at most two,
(2) Membership function in the output variable (D) has a symmetric triangular shape.

IV.   IMPLEMENTATION ON AN FPGA CIRCUIT

The designed MPPT controller was configured in a ready-made FPGA main board. The selected implementation platform was the Virtex-II (XC2v1000–4fg456) Xilinx FPGA family included in the Virtex- II™ V2MB1000 Development Board [7]. This chip was largely sufficient to implement all the constituents of the MPPT controller addressed in this paper. This was possible since the chip contains 5120 slices and 10,240 logic cells as well as forty 18 x 18 multipliers, and etc.

Figure 7. The used implementation plateforme : V2MB1000 developpement board.

The implementation of the Fuzzy Logic Controller is straightforward by coding each component of the fuzzy inference system in VHDL hardware description language integrated with the Xilinx foundation ISE 7.1i tool [8], which supports the ModelSim Xilinx Edition-III (MXE-III) v6.0a, used here for the simulation purposes. The design of the FLC is highly flexible as the membership functions and rule base can be easily changed by simply changing some parameters in the codes and in the design constraints on specific synthesis tool [7].

978-1-4673-5289-5/12 $31.00 © 2012 IEEE

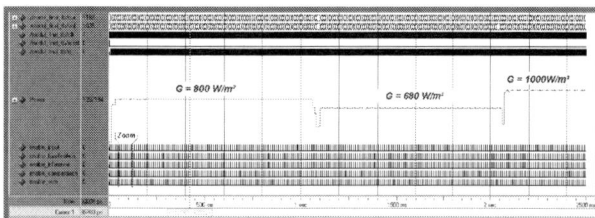

Figure 8. The evolution of the output power of the solar array (W) versus time (ms) for different insolation (800, 680 and 1000 W/m², 25 °C).

The simulations were carried out so that to each value of the duty cycle ratio *(D)* generated by the fuzzy controller corresponds a single point of coordinates *(I, V)* on the memorized current–voltage considered characteristic. The digitized values of currents and voltages are in their turn injected to the outputs of the controller to generate the next duty cycle ratio value and vice versa. The timing simulation waveforms obtained by ModelSim Xilinx are presented in Fig. 8.

V. SIMULATION RESULTS

In order to show the effectiveness of the developed MPP tracker, the whole PV system, with the above designed constituents was encoded into MATLAB Simulink software as shown in the figure bellow. The various parts of the PV system have been modeled by separate blocks. The main parts are:

- photovoltaic panel "BP SX150"
- the GA designed fuzzy-based MPPT,
- the PWM generator, and
- the DC/DC boost converter connected to a resistive load.

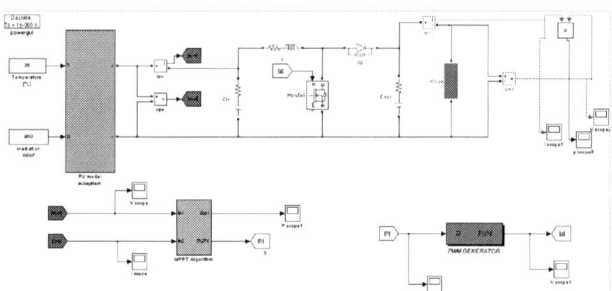

Figure 9. Matlab-Simulink simulation of the fuzzy MPPT with resistive load.

We have run the simulations under the two situations. In the first, we have removed the MPP controller and connected directly the panel to the load. In the second case, we have connected our controller to the remaining of the system. Figure 10 shows that the system behaves exactly as expected and that there is an important enhancement power transfer

from the source to the load in the presence of the developed MPPT controller.

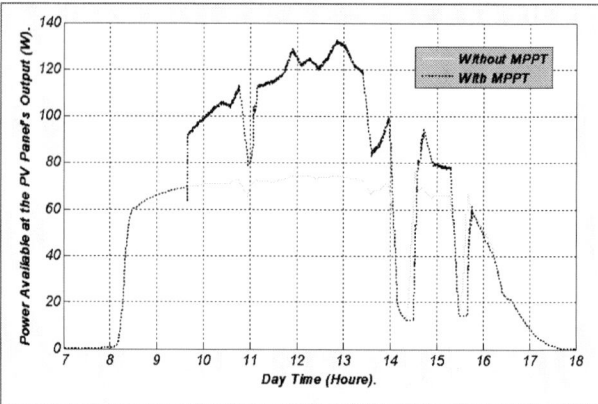

Figure 10. MPP tracking with and without the controller

VI. CONCLUSION

This paper has presented a prototype of the digital MPPT fuzzy controller based on FPGA for DC/DC converter. It consists on a stepwise adaptive search, leads to fast convergence and performs well with respect to change in atmospheric conditions. In addition to the software simulations, an FPGA chip was also used to implement the hardware prototype of the designed controller. This work is the first step before a final field installation to experimentally validate the effectiveness of the proposed MPP tracker in a stand-alone photovoltaic system.

REFERENCES

[1] D. P. Hohm *et al., "Comparative Study of Maximum Power Point Tracking Algorithm Using an Experimental,Programmable, Maximum Power Point Tracking Test Bed*", in proc 28th IEEE Photovoltaic SpecialistsConference(PVPC-28), pp1699-1702, Anchorage, Alaska, September, 2000

[2] J. M. Enrique, E. Duràn, M. Sidrach-de-cardona, J. M. Andùjar, "*Theoretical assessment of the maximum power point tracking efficiency of photovoltaic facilities with different converter topologies*", Solar Energy, 2007, pp: 31-38.

[3] George Chryssis, "High Frequency Switching Power Supplies : Theory and design", McGraw-Hill edition, 1989.

[4] Muhammad H. Rashid, "*Power Electronics Handbook*", Academic Press (2001)

[5] Homaifar, A., McCormick, E., 1995. Simultaneous design of membership functions and rule sets for fuzzy controllers using genetic algorithms. IEEE Transactions on Fuzzy Systems 3 (2), 129–139.

[6] Messai, A. *et al.," Maximum power point tracking using a GA optimized fuzzy logic controller and its FPGA implementation"*, Solar Energy 85 (2011) 265–277.

[7] Virtex-II™ V2MB1000 Development Board User's Guide, Version 1.4, May 2002.

[8] Xilinx ISE 7.1i Software Manuals: http://www.xilinx.com/support/sw_manuals/xilinx7/.

A Comparative Performance Evaluation Study of The Basic Binary Tree and Aloha Based Anti-collision Protocols For Passive RFID System

Ahmed RENNANE, Hadjer SAADI, Rachida TOUHAMI
LINS, Faculty of Electronics and Informatics
USTHB,
Bab Ezzouar, Algéria
Rennaneahmed;hadjer_saadi; rtouhami@yahoo.fr

Mustapha C.E. YAGOUB
ERML Dept. of Electronics and Informatics
University of Ottawa
Ottawa Ontario, Canada
myagoub@uottawa.ca

Abstract— **RFID (Radio Frequency Identification) technology is one of the most important branches of the automatic identification technology which achieves the identification of targets by contactless communications. One of the major challenges for RFID systems is to avoid or solve collisions due to interference that might occur between tags which decrease the efficiency of RFID system. This paper focused on a detailed comparative performance evaluation study and an adoption of certain algorithms of anti-collision for passive RFID system, for the two families of solving collision: deterministic (according to the: QTA, CTTA, BS-QTA, BS-CTTA) and stochastic (ALOHA, SLOTTED ALOHA, np-CSMA). The simulation results for the two families that we get with protocols evaluation are based on their own proper evaluation parameters. Thus, for evaluating the simulation of the stochastic family, we have proposed a new performance metric which judges and correctly evaluates this anti-collision protocol; which is: the number of identified tags versus to the time. This metric provides a direct measure of the time taken to read a group of tags.**

Index Terms— **RFID, Anti-collision algorithm, Protocols, Binary Tree, ALOHA, Probabilistic methods, deterministic methods.**

I. INTRODUCTION

RFID is a method that exhibits a variety of attracting advantages such as a longer reading distance, working capability in harsh and complicated conditions, no ambiguity in object identification, etc [1]. RFID is a technique used to identify objects by using electromagnetic waves. An object can be tagged with an electronic code responding label. An electronic tag consists of an antenna and an integrated circuit. Upon receiving any valid interrogating signal from any interrogating source, such as a reader, the tag will respond according to its designed protocol. In an RFID system according to readers and tags, there are two types of collisions: reader collisions and tag collisions. Tag collision is the event that the reader cannot identify the data of tag when more than one tag occupies the same communication channel simultaneously [2]. Reader collisions may happen due to frequency and tag interferences. When tags backscatter their

IDs to a reader, signals from adjacent readers cause interference. If more than one reader interrogates the same tag, tag interference occurs since the tag cannot correctly listen to the commands from the readers [2]. Most of the standards for the UHF RFID systems propose Aloha-based anti-collision algorithms that are probabilistic and binary tree search anti-collision algorithms that are deterministic.

This paper is focusing on a detailed comparative study of certain algorithms of anti-collision for passive RFID system. Several parameters are developed to evaluate the simulation of these algorithms, based on their own evaluation parameters such as the offered load, the system throughput, capture effect and the time needed to complete the identification process, etc.

II. BASIC TREE PROTOCOLS

The tag identification process is deterministic, and is based on iteratively querying a subset of tags which match a given property until all tags are identified [2-4].

A. Query Tree Algorithm

The QT protocol assumes that there are two tags with ID 01 and 10. The reader first transmits a null string, which matches the prefix of both tags. As a result, both tags reply to the reader simultaneously and cause a collision. The reader then pushes the prefix "0" and "1" onto its stack. The next read cycle begins with the reader popping and transmitting the prefix "0". This causes tags with ID matching the prefix "0" to reply, that is, tag 01. After that, reader transmits prefix "1". As only tags with prefix "1" can reply, the reader only receives a reply from tag 10. QT may require multiple iterations, that is, large tree depth, in order to identify tags [2-3].

B. Collision Tracking Tree Algorithm

In the collision tracking tree algorithm (CTTA), tags send their IDs from $(k + 1)$th bit to the end bit if the prefix is the same as tags' first k bits. Then the reader transmits a signal to stop sending IDs from tags if there is a collision. CTTA constructs the next prefix with the bits received before collision and reduces the waste of time caused by collisions occurred at the received bits [2].

C. BI-SLOTTED TREE BASED RFID TAG ANTI-COLLISION PROTOCOLS

They can reduce the identification time follows [2]:

a) REQUEST: A reader sends n−1 length inquiring bits (prefix) to tags.

b) GROUPING: Tags in the field of the reader respond their tag IDs to the reader if the inquiring bits are the same as the first n − 1 bits of tag IDs.

• When the tags respond their IDs to the reader, they choose one of two time slots depending on whether nth bit is '0' (first slot) or '1' (second slot).

• BSCTTA: tags send their IDs from (n + 1)th bit until receiving an ACK signal, which is sent from the reader when a collision occur.

c) DECISION: Depending on whether collision have occurred or not, the reader decides proper procedure.

d) Repeat identification process until all tags are identified.

III. ALOHA-BASED ANTI-COLLISION ALGORITHMS

In this category, we find many versions:

A. Pure ALOHA

The pure ALOHA (PA) algorithm is a simple anti-collision method based on TDMA technique. In PA based RFID systems, a tag respond with its unique ID randomly after being power supplied by a reader. It then waits for the reader to reply with, i) a positive acknowledgment (ACK), indicating that its ID has been received correctly, or ii) a negative acknowledgment (NACK), meaning that a collision has occurred. If two or more tags transmit, a complete or partial collision occurs, each tag then resolve by backing off randomly before retransmitting their ID. So the first disadvantage of this algorithm is the high probability of collision [1],[4-8].

Therefore, the throughput S of pure ALOHA for the offered traffic G is shown as follow [1]-[4-8]:

$$S = G * e^{-2*G} \tag{1}$$

Where the maximum throughput is 0.184 when G=0.5.

B. Slotted ALOHA

In Slotted ALOHA algorithm, the time is divided into several slots, and the tags must transmit their IDs in synchronous time slots. If there is a collision, tags retransmit after a random delay. The collision occurs at slots boundary only, hence there are no partial collisions. Thus, compared to pure ALOHA, this method will decrease the probability of collision. Therefore, the throughput S of slotted ALOHA for the offered traffic G is given as [1], [4-8]:

$$S = G * e^{-G} \tag{2}$$

Where the maximum throughput is 0.368 when G=1.

C. Framed Slotted ALOHA (FSA)

In FSA, the reader offers information to the tags about the frame size and the random number which is used to select a time slot in the frame. One time slot is a time interval in which tags transmit their serial number. Each tag selects a slot

number using the random number and responds to the slot number in the frame. Reader will identify tags with multiple frames, so it can solve the problem in S-ALOHA algorithm [1].

Since the frame size of FSA algorithm is fixed, its implementation is simple, but it has a weakness that drops efficiency of tag identification.

D. Dynamic Framed Slotted ALOHA (DFSA)

DFSA algorithm changes the frame size for efficient tag identification. To determine the frame size, it uses certain information such as the number of slots used to identify the tags and the number of the slots collided and so on. So DFSA algorithm can solve partially the problem of FSA. DFSA algorithm has several versions depending on the methods changing the frame size [1]-[4-8].

E. Non-persistent CSMA

In CSMA protocol, it is possible to judge whether other tags are transmitting their packets, because each tag does not transmit any carrier wave except for its packet transmission. If a carrier wave is sensed on the communication channel, the condition is called "busy"; otherwise, it is called "idle". The CSMA is a protocol that decides whether packet transmission should start as the result of a carrier sense. When the result is "busy" the next action to avoid collision is needed.

In a real communication system, when a tag transmits its packets, other tags detect the transmission or collision from the propagation delay time that depends on distance between tags.

The throughput S of non-persistent CSMA for the offered traffic G is shown as follows [5-6]:

$$S = \frac{G*e^{-d*t}}{G(1+2*d)+e^{-d*t}} \tag{3}$$

d is the normalized propagation delay. The throughput S is obtained under an ideal communication environment.

IV. INCREASING EFFICIENCY WITH CAPTURE EFFECT

The tag's responses in the reverse link can be identified from the reader even if they collide (occupy the same slot). This can happen if the strength of one signal is higher than the rest of the signals in the same slot. This is known as the capture effect. For this paper, it is possible to take advantage of the capture effect and increase the performances of the stochastic anti-collision algorithms by choosing the appropriate threshold (T) in the reader, which acts as a filter for the weak signals. As a result, the reader identifies the tag even if collision has occurred. The value of T which is called capture ratio indicates the amount by which a data packet must be stronger than others to be detected by the receiver without error [1], [5-6]. For example, the throughput S in the Slotted Aloha algorithm with the capture effect is given by:

$$S = G * e^{-\left(\frac{T*G}{1+T}\right)} \tag{4}$$

V. EVALUATION PARAMETERS OF RFID SYSTEMS

The most fundamental elements used to evaluate the stochastic protocols are offered traffic G, throughput S, and average transmission delay. In this simulation, we propose a new performance metric by which to judge these anti-collision protocols: the number of tags identified versus to the time. This metric provides a direct measure of the time taken to read a group of tags.

The *offered load G* corresponds with the number of tags transmitting simultaneously at a certain point at time *t0*. The average *offered load G* is the average over an observation period *T* and is extremely simple to calculate from the transmission duration θn of a data packet [1]

$$G = \sum_{1}^{n} \frac{\theta_n}{T} * r_n \qquad (5)$$

Where $n = 1, 2, 3 \ldots$ is the number of tags in the system and $r_n = 0, 1, 2 \ldots$ is the number of data packets transmitted by tag n during the observation period.

Throughput : the total amount of the packet successfully transmitted to the reader in a time interval is called throughput, and the normalized throughput by the transmission data rate is shown as S. If the transmission data rate and the quantity of information in a packet are defined R (b/s) and T(bit) and n packets are successfully transmitted in a time unit, S is [6]:

$$S = \frac{T * n}{R} \qquad (6)$$

*The number of identified tags :*This metric provides a direct measure of the time taken to read a group of tags. In the deterministic anti-collision algorithm family the most fundamental parameters used to evaluate the basic tree protocols are: a) The number of bits required to identify all tags with a length of tag identification of 12 bits instead of 96-bits fixed in EPC Class 1 Gen2 because of the limitation memory capacity of our available computers [2]; b) The number of iterations required for the identification of all tags with12-bit tag ID length. One iteration is determined by one request and one of the possible responding states: collision, no collision, and no response.

VI. SIMULATION RESULTS

A. ALOHA-BASED ANTI-COLLISION ALGORITHMS

The results of simulation are shown in the figures which follow: in a reader's interrogation zone which is equal to 5m radius, the number of tags increases from 2 to 1500. The channel data rate is 40 Kbps as specified by the EPCglobal standard. Tag IDs are k – 96 bits long, which is the most commonly used ID length [1]. The ID values are uniformly distributed and the packet generation follows a Poisson distribution. In order to identify the tag, the reader's sensitivity requires 6 dB of difference between the signal from the tag of interest and the channel noise (interference from other tags). When the capture effect is not considered, the throughput is close to the theoretical value even if the number of tags is 1500. Moreover, when the capture effect is considered, the throughput is larger than that of the first case. This is because a transmitted packet sometimes survives owing to the difference of received-power between transmitted tags. In addition, when the capture effect is not considered, the rapidity of tag identification process is reduced. In the case of pure ALOHA, collisions often occur, thus, the capture effect is the reason to increase the throughput. In the S-Aloha protocol, the influence of the capture effect is remarkable, as shown in Fig 2. So, we

see that the S-aloha protocol gives results more efficient compared to Aloha for the same environment of simulation.

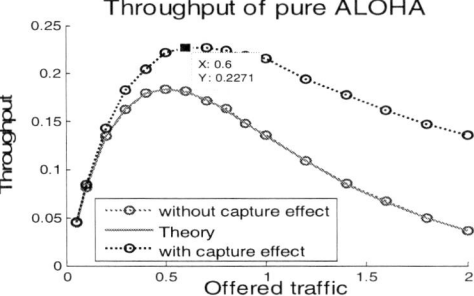

Fig. 1: Offered traffic and throughput of pure ALOHA

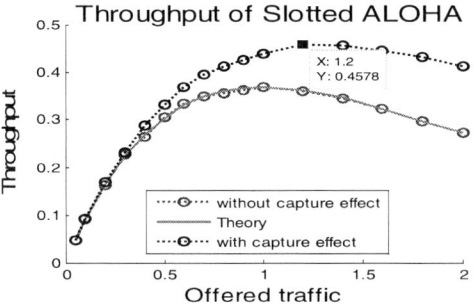

Fig. 2: Offered traffic and throughput of S-ALOHA

The simulation result of throughput of np-CSMA is shown in Fig.3. In the simulation, the normalized propagation delay was 0.01 or 0.1. The maximum throughput of np-CSMA was dependent on d. If d is small, the maximum throughput of np-CSMA is higher than that of pure ALOHA. However, if d is large, the performance is close to pure ALOHA. This is because other tags transmit their packets during a large propagation delay time even if a tag transmits its packet after sensing the carrier. As a result, collision occurs. Thus, the capture effect influences in the same way as in the other protocols. The evaluation in term of time of these protocols, as shown in Fig. 4, the np-CSMA protocol gives results more efficient compared to S-aloha and Aloha protocol which has weaker performances for the same environment of simulation. Thus the application of capture effect in improves the performances by reducing the probability of collision.

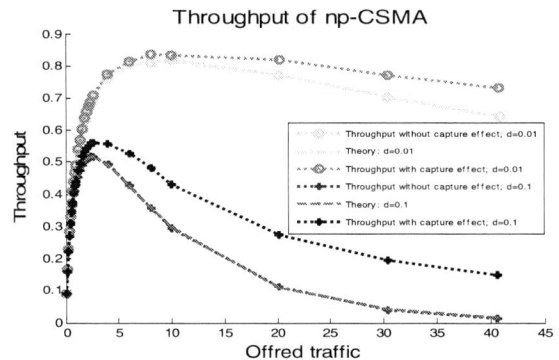

Fig. 3: Offered traffic and throughput of np-CSMA

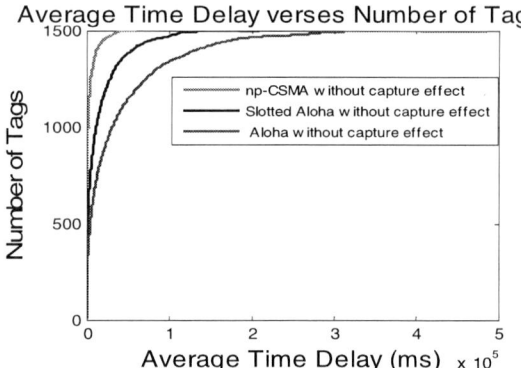

Fig.4: Average delay time versus number of tag in stochastic anticollision protocols

B. TREE PROTOCOLS

This simulation is focused on the performance comparison between query tree algorithm (QTA) and bi-slotted query tree algorithm (BSQTA) in one side. On the other side, the collision tracking tree algorithm (CTTA) and bi-slotted collision tracking tree algorithm (BSCTTA).

The simulation condition is as follows. In the field of the reader, the number of tags increases from 2 to 1024. At the radio level, we use the collision model for multiple access channel, which means that if two or more tags that are in the reader range transmit simultaneously, then the reader does not receive any one of the transmitted packets.

In order to take comparison between these algorithms studied, an example is given as follows.

Let's assume that there are five tags, which the tag IDs are '000000101001', '100000100011', '100010111110', '011110000100', '011110000100', and '101011100001' in the field of an RFID reader. Then we observe the total required bits for identifying all five tags to compare QTA with BSQTA and to compare CTTA with BSCTTA. Here, the tree searching order is from LSB ('0') to MSB ('1'). Since we use LIFO, we don't search from the root level every time, but search from the closest node at the tree. Simulation results of these algorithms are summarized in the table 1.

TABLE I : SIMULATION RESULTS OF QTA, CTTA, BS-QTA, BS-CTTA FOR AN EXAMPLE OF 5 TAGS

Protocols	Parameters		
	Number of iterations	Number of bit prefix	Response bit number
QTA	12	34	92
BS-QTA	6	11	92
CTTA	8	22	48
BS-CTTA	4	7	48

We start with the comparison between BSQTA and QTA. According to Table 1, for identifying all five tags, QTA requires 34 bits for the prefixes and 92 bits for the responses, so 126 bits in total are needed for tag identification. On the other hand, BSQTA requires 11 bits for the prefixes and 92

bits for the responses, so 103 bits in total are needed for tag identification. Thus, BSQTA reduces the prefix overhead more than QTA.

For all five tags identification, CTTA Protocol requires 70 bits in total for tag identification. On the contrary, BSCTTA requires 55 bits in total for tag identification. Hence, BSCTTA reduces the prefix overhead more than CTTA. Consequently, RFID systems when either BSQTA or BSCTTA is applied achieve faster tag identification than RFID systems when either QTA or CTTA is applied.

VII. CONCLUSIONS

In this paper, we addressed the two families of collision resolution protocols (deterministic and stochastic) used for RFID systems. We defined a new metric for evaluating the stochastic anti-collision protocols, the time needed to complete the identification process, which provides a direct measure of identification speed process to read a group of tags. Based on this metric, we evaluated the performance of existing stochastic anti-collision protocols to make a comparison between them.

According to the simulation results of tree based RFID tag anti-collision protocols: QTA, BSQTA, CTTA and BSCTTA, RFID systems when either BSQTA or BSCTTA is applied achieve faster tag identification than RFID systems when either QTA or CTTA is applied. In deterministic protocols, we can predict the maximum duration of the identification process, and as the collision error can be detected during the phase of anti collision, the speed can be optimized.

The stochastic protocols performances are strongly affected by the size of the frame used. Because the total number of tags to be identified is not known, therefore the design of frames is a big problem for stochastic protocols.

REFERENCES

[1] Finkenzeller. RFID handbook, 2nd Ed., J.Wiley & Sons, 2003.

[2] Miodrag Bolic, David Simplot-Ryl, Ivan Stojmenovic: RFID SYSTEMS RESEARCH TRENDS AND CHALLENGES, John Wiley & Sons 2010.

[3] EPC radio-frequency identification protocols class-1 generation 2 UHF RFID protocol for communications at 860 MHz–960 MHz version 1.0.9, Jan. 2005.

[4] H. Saadi, R. Touhami, M. CE. Yagoub, "Performance Analysis of Basic Tree Protocols and ALOHA-Based Anti-Collision Algorithms Used in RFID Systems", 11th IEEE ISSPIT, 978-1-4673-0753-6/11 Bilbao, Spain, December 2011.

[5] H. Saadi, R. Touhami, M. CE. Yagoub, "Simulation of the Anti-Collision Process of RFID Systems Based on Multiple Access Protocol Modeling", IEEE Proc, 11th IEEE ISSPIT, 978-1-4673-0753-6/11Bilbao, Spain, December 2011.

[6] Harada, H., Prasad, R., Simulation and Software Radio for Mobile Communication, Norwood, MA, Artech House, 2000.

[7] Thomas F. La Porta, Gaia Maselli, Chiara Petrioli, "Anticollision Protocols for Single-Reader RFID Systems: Temporal Analysis and Optimization", IEEE transaction on Mobile Computing, Vol. 10, No. 2, February 2011.

[8] Weilian Su, Nikolaos V. Alchazidis, Tri T. Ha, "Multiple RFID Tags Access Algorithm", Vol. 9, No. 2, February 2010.

Power Conservation for Wireless Mesh Network

Sarra Mamechaoui
STIC Laboratory
Abou Bekr Belkaid University
Tlemcen, Algeria
sarra.mamechaoui@mail.univ-tlemcen.dz

Fedoua Didi
Dept of Computer engineering
Abou Bekr Belkaid University
Tlemcen, Algeria
f_didi@mail.univ-tlemcen.dz

Guy Pujolle
Pierre et Marie Curie University
Paris 6, FRANCE
Guy.Pujolle@lip6.fr

Abstract—**Nowadays, Wireless mesh networks are being widely used as a cost-efficient means for coverage extension and backhaul relaying between IEEE 802.11 access points (APs). Recently, there has been a proliferation of application development that involves wireless access networks which is responsible for the raising in consumption and increasing demand for energy which results in the increase in carbon dioxide levels in the environment. This paper focuses mainly on different existing energy conservation methods in different networks. Several approaches are presented and discussions on the details related to energy management WMN are also presented. The classification of the layers of TCP / IP model of the largest existing approaches spent on energy conservation is treated.**

Keywords-WMN; 802.11s; Energy consumption; TCP/IP layers, Routing Protocol; Sleep.

I. INTRODUCTION

Since several wireless networks evolve into the next generation to provide better services, a key technology, wireless mesh networks (WMNs), has emerged recently [1]. Characteristics of WMN such as rapid deployment and self configuration make WMN suitable for transient on-demand network deployment scenarios such as disaster recovery, hard-to-wire buildings, conventional networks and friendly terrains [2]. The architecture of WMNs can be classified into three types [2]: Infrastructure/Backbone WMNs, in this architecture, mesh routers form an infrastructure for clients. Client WMNs, Client meshing provides peer-to-peer networks among client devices. Hybrid WMNs, this architecture is the combination of infrastructure and client meshing. Mesh clients can access the network through mesh routers as well as directly meshing with other mesh clients. While the infrastructure provides connectivity to other networks such as the Internet, Wi-Fi, WiMAX, cellular, and sensor networks, the routing capabilities of clients provide improved connectivity and coverage inside WMNs.

Energy consumption in network exchanges is rising as higher capacity network equipment becomes more power-hungry and requires greater amounts of cooling. Combined with rising energy costs, this increasing demand for energy in wireless field related with the increase in carbon dioxide levels.

The goal of power management for WMNs varies; power management aims to control connectivity, interference, spectrum spatial-reuse, and topology. If a single channel is used in each network node, the interference among the nodes directly impacts the spectrum spatial-reuse factor. Reducing transmission power level decreases the interference and increases the spectrum spatial-reuse efficiency. However, more hidden nodes may cause performance degradation in MAC protocols. Thus, power management schemes are closely coupled with MAC protocols. Moreover, since connectivity affects performance of a routing protocol, power management is also crucial for the network layer [1].

In the related literature, there are several works that jointly perform routing, scheduling and channel assignment in order to improve the performance of the network.This paper addresses the incorporation of energy conservation at three layers lower of the protocol stack for wireless mesh networks.

The remainder of this paper is organized as follows. Section II introduces the protocols that save energy in the physical layer are considered in this paper. Section III describes work dealing with energy efficient protocols in the MAC layer of wireless mesh networks, and power conserving protocols in the network layer are addressed in section IV. Finally, section V summarizes and concludes the paper.

II. PHYSICAL LAYER APPROACHES

Physical layer techniques advance rapidly as communication theories [1]. Several contributions have been tailored towards studying the physical layer problems in WMN. Thus, research problems are pointed out in this Section.

A. Sleeping and Rate-Adaptation

Nedevschi et al. [3] developed two approaches to save energy. The first puts network interfaces to sleep during short idle periods. To make this effective they introduce small amounts of buffering, much as 802.11 APs (access points) do for sleeping clients; this collects packets into small bursts and thereby creates gaps long enough to profitably sleep. Potential concerns are that buffering will add too much delay across the network and that bursts will exacerbate loss. Their algorithms arrange for routers and switches to sleep in a manner that ensures the buffering delay penalty is paid only once (not per link) and that routers clear bursts so as to not amplify loss noticeably. The result is a novel scheme that differs from 802.11 schemes in that all network elements are able to sleep when not utilized yet added delay is bounded. The second approach adapts the rate of individual links based on the utilization and queuing delay of the link.

B. Connectivity strategies

Mudali et al. proposed [4]; various connectivity strategies based on the Critical Number of Neighbors approach are evaluated via simulation to determine the relationship between transceiver power savings and the network lifetime. The evaluation indicates that the selected connectivity strategies are able to produce cumulative transceiver power savings. The extent of the power savings produced by individual backbone nodes is largely dependent upon the location of the node relative to the

978-1-4673-5289-5/12 $31.00 © 2012 IEEE

(imaginary) center of the backbone network. The evaluation also suggests that cumulative transceiver power savings do not automatically translate into corresponding extensions of network lifetime.

C. Greening of the Internet

Gupta and Singh [5] thought that sleeping is an appropriate way to maximize internet's energy conservation. However, in order to implement algorithms for sleeping, firstly a redesigning of the network equipment's hardware was done to allow software-enabled sleeping, secondly the routing protocols was modified so as to adapt energy consumption and allow its loading via aggregation and sleeping, thirdly, for more options of route selection allowing aggregation and sleeping, the Internet topology was amended as well, and, finally studying the impact of sleeping on protocols such as TCP with an eye on changing the protocol so as to adapt to the presence of sleeping nodes.

Summary

In this section, several contributions have been tailored towards studying the way of achieving energy efficiency in WMNs. Sleeping approach appears to be the appropriate way and very effective for saving energy with some enhancement in the hardware and routing protocols.

III. MAC LAYER APPROACHES

The MAC (Media Access Control) layer is a sub layer of the data link layer, which is responsible for providing reliability to upper layers for the point-to-point connections established by the physical layer [1]. This section presents the details of power saving in MAC sub layer and a discussion of recent work that has started to address them.

A. PSM (Power Saving Mode) IEEE 802.11 Protocol

IEEE 802.11 [6] specifies a simple Power Save Mode (PSM). Nodes are assumed to be time-synchronized and awake at the beginning of each beacon interval. After waking up, each node stays on for a period of time known as the ATIM window. During the ATIM window, since all nodes are guaranteed to be on, packets are advertised that have arrived since the previous beacon interval (or could not be sent in the previous beacon interval). These advertisements take the form of ATIM packets. More formally, when a node has a packet to advertise, it sends an ATIM packet to the intended destination during the ATIM window, following the rules of IEEE 802.11's CSMA/CA mechanism. In response to receiving an ATIM packet, the destination will respond with an ATIM-ACK packet (unless the ATIM specified a broadcast or multicast destination address). When this ATIM handshake has occurred, both nodes will remain on after the ATIM window and attempt to send their advertised data packets before the next beacon interval, subject to CSMA/CA rules. If a node remains on after the ATIM window, it must keep its radio on until the next beacon interval. If a node does not receive an ATIM or ATIM-ACK (assuming unicast advertisements) it will enter sleep mode at the end of the ATIM window until the next beacon interval. This process is illustrated in Figure 1. The dotted arrows indicate a "causes" relationship. Node A sends a data packet to B while C, not receiving any ATIM packets, returns to sleep for the rest of the beacon interval.

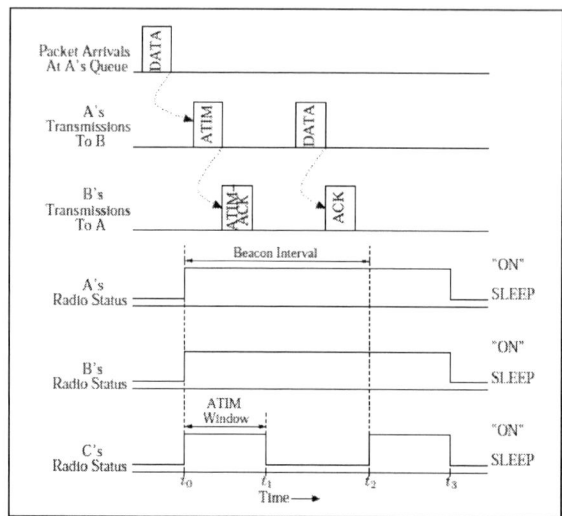

Figure 1.   IEEE 802.11 power save mechanism [6].

While 802.11 PSM is a simple protocol to implement, there are some disadvantages like : The IEEE 802.11 standard defines power saving for clients devices, by allowing the nodes to switch between the Awake and sleep states. But the standard does not provide mechanism for placing APs into power saving mode and requires that the access point be continuously powered. Also, nodes must remain on for the entire beacon interval if they have only one packet to send and/or receive. This can be wasteful in terms of energy.

B. SOFA (A Sleep-Optimal Fair-Attention scheduler)

In SOFA [7] authors proposed a downlink traffic scheduler on the AP of a WLAN, called SOFA, which help its PSM clients to save energy by allowing them to sleep more, hence to increase battery lives. If a client has buffered packets at the AP in a beacon period, and that client decides to receive it, it has to remain awake from the beginning of the beacon period till the last packet scheduled for it in the beacon period is delivered. Therefore a large portion of energy wastage (for the client) comes from the AP transmitting other clients' a packets before it finishes transmitting the client's last packet to it. SOFA manages to reduce such energy wastage and maximizes the total sleep time of all clients. SOFA favors clients with smaller attention requests by allowing them to spend less energy to get one unit of attention, while still helping other clients with larger attention requests to sleep more compared with other popular scheduling policies like round-robin and FCFS.

C. PRCW(Physical Rate and Contention Window based admission control)

In [8], the paper presents a class based admission control algorithm for 802.11e based wireless local area networks. The strengths of this admission control is dynamicity and flexibility of the algorithm, which adapts to the situation of the BSS, like global load, number of best effort AC, and position of QSTA by report of QAP, thing that have never been taken together, but each solutions have used a point of sight separately. Thus it achieves higher throughput than other admission control for 802.11

e. The idea which consists of changing parameters [AIFSN, CWmin and CWmax] of best effort flows, for decreasing collisions, is used, so they think that it is an efficient way to protect QoS flows from best effort flows, and to allow reducing the number of collisions and power consumption. So they increase AIFSN [best effort flow], CWmin and CWmax only at 70% of load of network, to prevent starvations of best effort flows, and so they increase rate utilization of channel. They also use the current rate transmission of QSTAs, according to their positions, instead of the minimum rate transmission used by standard 802.11e, for calculate the load of network and derived the TXOPi necessary for all the stations, with i=1 to number of active stations. The 802.11e standard starves the low priority traffic in case of high load, and leads to higher collision rates, and did not make a good estimate of weight of queues, so there is an unbalance enters the flows with high priorities

D. LEACH(Low-Energy Adaptive clustering Hierarchy)

LEACH is scheduled MAC protocol with clustered topology [9]. The nodes organize themselves into local clusters with one node acting as the local cluster head. The cluster heads are chosen randomly according to a specific algorithm based election a probability function which takes into account various criteria such as energy available nodes. To distribute energy consumption evenly, LEACH propose to rotate cluster heads randomly. Each node determines a cluster to associate with by choosing the cluster head that requires minimum communication energy. Once all the nodes are organized into clusters each cluster head creates a schedule for the nodes in the cluster. This allows that each cluster member node can switch to sleep mode at all times except during its transmit time; thus, minimizing its energy consumption.

Summary

Research in power save protocols at the MAC layer has taken several directions. First, there is a scheduling approach whereby packet transmission times are carefully chosen to avoid collisions. This approach lends itself well to power save since nodes can sleep when they are not scheduled to communicate. Second, there are protocols which attempt to allow all but a small subset of nodes to sleep. The members of this subset are periodically rotated and chosen to preserve multi-hop paths in the network.

Different mechanisms of MAC sub layer were introduced, allowing substantial energy savings in this section. We have noted that a large majority of work in Mac layer, was based on the proposal of the 802.11 standard, in this case the protocol PSM (Power saving mode) and extended it to improve the functioning.

The protocols of power saving for MAC sub layer presented in this section are summarized in this Table I.

TABLE I.   SUMMARY OF MAC PROTOCOLS

Protocol	Network	Topology	Contribution
LAECH	802.15.4	Clustered	Low energy clustering
PRCW	802.11e	Flat	Reduce collision
PSM	802.11	Flat	Active/sleep period
SOFA	802.11	Flat	a round-robin scheduling and sleep/wakeup

The next Section we will discuss approaches that conserves energy in the network layer.

IV.   NETWORK LAYER APPROACHES

The network layer is responsible for network self-configuration and data routing. For configuring network topology, the network layer select an appropriate mode for a node and determines the most suitable neighbors with which to associate and form communication links. The network topology is updated after link failures or at regular intervals for assuring network connectivity and optimizing network lifetime by balancing energy consumption among other nodes in the network.

A. Connected Active Subset

The intuition behind a connected active subset protocol, such as SPAN [10] or GAF[11], is that when there are many nodes close together in a multi-hop wireless network, only a subset of these nodes need to be active in order to maintain network connectivity. These protocols strive to keep only a small subset of nodes awake in the network to provide network connectivity, and then place the rest of the nodes in a sleep state for the vast majority of the time. Often, the members of the active subset are rotated in order to distribute the energy consumption more evenly between different network nodes and to accommodate network topology changes due to mobility. The main advantage of the connected active subset strategy is that there is little impact on communication. Packets primarily travel through nodes that are always on, and thus experience low delay. One main disadvantage of the active subset strategy is that it is inherently dependent on node density for energy savings [12]. The basic premise is that there are enough nodes that only a small number of them are needed at any one time. In low density networks, almost no power can be saved using this strategy because almost every node must stay active. Another main disadvantage of this strategy is the overhead required to maintain an effective subset.

B. Clustering in Wireless Mesh Networks

Clustering makes the network fast, more efficient and reliable. In this approach Yogesh Birla et al [13] use a stable election protocol for clustering and also select a cluster head in the cluster it is reduced the indusial over head of mesh nodes and also save the battery power by selecting cluster head. Only the cluster heads are responsible for establishing connection to other cluster heads, only cluster heads are communicate each other, then they save the battery power of all other nodes was present in the cluster.

C. Energy and Throughput-aware Routing (ETR).

In this approach [14] authors proposed a novel routing algorithm for wireless mesh networks. The proposed algorithm has been specially devised for mesh networks owned by operators as: first it relies on carefully planned mesh networks that, by doing proper channel assignment, do not suffer from interference between different link groups, second it is based on a centralized algorithm that is executed at a central location and is responsible for the entire routing in the network, and finally it has been carefully designed to satisfy operator requirements in terms of service guarantees and energy consumption. One of the key objectives of the proposed routing algorithm is that it provides users with throughput guarantees. This is performed by taking into account that the number of flows admitted at a given link group does not exceed the available resources as given by the linearized capacity region. Another key objective of the proposed algorithm is that it minimizes the overall energy consumed by the mesh network. Based on existing models of the node energy consumption, those show that the energy consumed by an active node is approximately constant independent of its transmission behavior, their routing solution aims at switching off as many nodes as possible.

D. Green Framework

In [15] Boutaba et al. proposed a new green framework that provides the WMN administrator with a parameterized objective function to choose to yield the desired tradeoff between the achieved network throughput and energy consumption. Specifically, they first proposed an Optimal Green Routing and Link Scheduling, called O GRLS, which aims at finding the optimal tradeoff. In this approach, they formulate the problem as an integer linear program (ILP) and proposed a simple yet efficient algorithm based on Ant Colony, called Ant Colony Green Routing and Link Scheduling (AC-GRLS) to solve the formulated ILP problem. Through extensive simulations, they showed that theirs proposed framework can achieve significant gains in terms of energy consumption as well as achieved throughput.

Summary

In the network layer, we think that the tendency of clustering is a good approach because it has already proven itself in the MANET. The selection of the cluster head in each cluster on the basis of energy level in wireless mesh network can reduce the rate of energy consumption by scheduling activities in the cluster.

The discussed approaches in network layer are summarized and classified in Table II. As routing protocols may use hybrid techniques to achieve their intended target application, some of the protocols are classified to several categories.

TABLE II.   SUMMARY OF ROUTING PROTOCOLS

Routing Protocol	Network	Topology	Classification
Connected Active Subset	Ad hoc	Flat	Node centric routing
Clustering	Wireless mesh network	Clustered	Node centric
Green Framework	Wireless mesh network	Flat	Green routing and link Scheduling
Energy and Throughput-aware Routing	Wireless mesh network	Flat	Cost field based

V.   CONCLUSION

Various existing energy conservation methods proposed by different studies but the energy is still a challenging issue for wireless mesh network. Based on the different studies that have been conducted by various researchers. In this paper, we have summarized some research results, which have been presented in the literature on energy saving methods in wireless mesh network. Although many

of these energy saving techniques look promising, there are still many challenges that need to be solved in wireless mesh networks. Therefore, further research is necessary for handling these kinds of situations.

REFERENCES

[1] Ian F. Akyildiz and Xudong Wang. "Wireless Mesh Networks". John Wiley & Sons, Ltd, 2009.

[2] Akyildiz IF,Wang X and Wang W. "Wireless mesh networks: a survey". Computer Networks Journal (Elsevier), 445–487, 2005.

[3] S. Nedevschi, L. Popa, G. Iannaccone, Reducing Network Energy Consumption via Sleeping and Rate-Adaptation, Proceedings of the 5th USENIX Symposium on Networked Systems Design and Implementation (2008), pp. 323-336.

[4] P. Mudali, T.C. Nyandeni, N. Ntlatlapay, and M.O. Adigun, Design and Implementation of a Topology Control Scheme for Wireless Mesh Networks, IEEE AFRICON 2009.

[5] M. Gupta and S. Singh, Greening of the Internet, SIGCOMM'03, August 25–29, 2003, Karlsruhe, Germany. Copyright 2003 ACM

[6] IEEE 802.11, Wireless LAN Medium Access Control (MAC) and Physical Layer (PHY) Specifications, 1999.

[7] Z. Zeng, Y. Gao, and P. R. Kumar, SOFA: A Sleep-Optimal Fair-Attention scheduler for the Power-Saving Mode of WLANs, 31st International Conference on Distributed Computing Systems, 2011 IEEE.

[1] F. Didi, H. Labiod, G. Pujolle, M. Feham, "Physical Rate and Contention Window based admission control (PRCW) for 802.11 WLANs", IEEE Symposium on Computers and Communications, Riccione, Italy, August, 2010.

[8] W.R. Heinzelman, A. Chandrakasan, and H. Balakrishnan. Energy-Efficient Communication Protocol for Wireless Micro Sensor Networks. In IEEE Proceedings of the Hawaii international Conference on System Sciences (HICSS '00), 2002.

[9] B.Chen, K. Jamieson, H. Balakrishnan and R. Morris, Span: An Energy-Efficient Coordination Algorithm for Topology Maintenance in Ad Hoc Wireless Networks, Kluwer Academic Publishers. Manufactured in The Netherlands, 2002.

[10] Y. Xu, J. Heidemann and D. Estrin, Geography-informed energy conservation for ad hoc routing, Proceedings of the Seventh Annual ACM/IEEE International Conference on Mobile Computing and Networking (MobiCom), 2001.

[11] Douglas M. Blough and Paolo Santi, "Investigating upper bounds on network lifetime extension for cell-based energy conservation techniques in stationary ad hoc networks," in Proceedings of the eighth annual international conference on Mobile computing and networking. 2002, pp. 183–192, ACM Press.

[2] Y. Birla, Esh Narayan and Kr. Saraswat, Clustering in Wireless Mesh Networks, VSRD-IJCSIT, Vol. 2 (5), 2012, 413-418.

[3] A. Oliva, A. Banchs and P. Serrano, Throughput and energy-aware routing for 802.11 based mesh networks, Comput. Commun. (2012), http://dx.doi.org/10.1016/j.comcom.2012.04.004

[12] A. Amokrane, R. Langar, R. Boutaba and G. Pujolle, A Green Framework for Energy Efficient Management in TDMA-based Wireless Mesh Networks, 8th International Conference on Network and Service Management (CNSM 2012).

978-1-4673-5289-5/12 $31.00 © 2012 IEEE

A Novel Cluster-based Fault-tolerant Scheme for Wireless Sensor Networks

Mohamed LEHSAINI
STIC Laboratory
University of Tlemcen
m_lehsaini@mail.univ-tlemcen.dz

Chifaa TABET HELLEL
STIC Laboratory
University of Tlemcen
tabet_chifaa@yahoo.fr

Abstract— **In wireless sensor networks, failures occur due to energy depletion, environmental hazards, hardware failure, communication link errors, etc. These failures could prevent them to accomplish their tasks. Moreover, in critical applications, these failures could lead to disaster. Thus, it is necessary to establish an efficient fault-tolerant scheme to guarantee data delivery to the base station when some nodes fail. In this paper, we proposed a Cluster-based Fault-tolerant Scheme (CFS) which could tolerate links failures to guarantee routing reliability. Finally, we conducted several simulations to illustrate CFS performance and compared obtained results to GRAB protocol.**

Keywords: Cluster-based, Failure recovery, Fault-tolerance, Node failure, WSN.

I. INTRODUCTION

Wireless sensor networks (WSN) are self-organized networks that typically consist of a large number of low-cost and low-powered sensor devices, called sensor nodes, which can be deployed over a geographical area for monitoring physical phenomena like temperature, humidity, vibrations, seismic events, and so on [1]. Now, WSN are permeating a variety of application domains such as avionics, environmental monitoring, structural sensing, telemedicine, space exploration, and command and control.

WSN should have a long lifetime to accomplish the application requirements. However, In addition to resource constraints in WSN, the failure of sensor nodes is almost unavoidable due to energy depletion since they have been usually deployed in hostile environments and their batteries cannot be recharged or replaced, hardware failure, communication link errors, and so on [2,3,4]. Therefore, fault tolerance has become as important as other performance metrics such as energy efficiency, latency and accuracy.

In general, the consequence of these failures is that a node becomes unreachable, violates certain conditions that are essential for providing a service or returns false readings which could cause a disaster especially in critical applications. Furthermore, the above fault scenarios are worsened by the multihop communication nature of WSN. It often takes several hops to deliver data from a sensor node to the remote base station; therefore, failure of a single node or link may lead to missing reports from the entire region of WSN.

Therefore, since sensors are prone to failure, fault tolerance should be seriously considered in many sensor network applications. Recently, several studies have dealt with fault tolerance in WSN, particularly in the routing process.

Moreover, these works focus on the detection and recovery of failures in WSN.

In this paper, we propose a Cluster-based Fault-tolerant Scheme (CFS) to tolerate faults in WSN while dissipating less extra energy and time. In CFS, each cluster has a main cluster-head and its vice. Sensor nodes with stronger capabilities are elected as cluster-heads which could aggregate sensor data before forwarding them to a remote base station, thereby saving energy. Furthermore, in clusters, main cluster-heads and their vice cooperate with each other to reduce extra costs by sending only one copy of sensed data to the sink.

CFS could tolerate links failures and therefore guarantee routing reliability. Finally, we conducted several simulations to demonstrate CFS performance and we compared obtained results with those of GRAB [5].

The rest of this paper is organized as follows: in Section 2, we present a survey of fault-tolerant approaches; Section 3 provides model parameters; in Section 4, we propose our fault-tolerant scheme; and Section 5 presents a performance analysis of the proposed scheme. Finally, we conclude our paper and discuss future research work in Section 6.

II. RELATED WORK

Many recent studies dealt with fault tolerance in WSN especially in terms of routing data to the base station. In this section, we summarize and compare existing fault tolerant techniques that enable to guarantee a reliable routing in WSN.

In [5], the authors have proposed a meshed multipath routing called GRAB that allows creating a forwarding mesh from the source to the sink based on the cost of delivering data at each node. Therefore, nodes farther away from the sink have the highest cost of delivering data. Sensor readings propagate along the path of least cost towards the base station. In this technique, the resulting mesh is based on a credit system in which the amount of credit assigned by the source node to the packet enable to determine the width of mesh. GRAB ensures reliable delivery of data to the base station but it consumes more energy which makes it undesirable for WSN deployed in hostile areas.

Node-disjoint multipath [5] generates a number of alternate paths that do not share any nodes with the primary path or other alternate paths except the source and the destination nodes. This scheme ensures that failures in some nodes on the primary path do not affect alternate paths therefore data delivery to the base station would be

978-1-4673-5289-5/12 $31.00 © 2012 IEEE

guaranteed. Creating multiple disjoint routes to the base station requires that the global network topology is known. Therefore, this technique consumes more energy because there is a redundancy of data sent to the base station.

In [6], the authors have proposed a braided multipath technique that consists to use braided or partially disjoint paths. For each node on the primary path, an alternate path not including that node is determined and these alternate paths are not much more expensive than the primary path in terms of latency and overhead. This technique guarantees recovery when a few nodes on the primary path fail. However, when most of the nodes on the primary path fail, new path discovery is required, which generates significant additional overhead.

The scheme presented in [7], is a dynamic fault-tolerant routing protocol that aims to prolong network lifetime. It enables to maintain network connectivity even if a node depletes its energy, thereby providing reliable data delivery to the sink. In this protocol, when a node exhausts its battery it would find a suitable alternative path to the sink by establishing new links with its neighbors.

This scheme ensures reliable data delivery to the base station. However, in this scheme, the time to find new parents when a parent node depletes its energy affects the time of data delivery to the sink.

In [8], the authors present a fault-tolerant multi-level routing protocol with sleep scheduling for WSN called FMS. FMS allows to maintain network connectivity even if a node exhausts its energy and guarantee reliable data delivery in event-driven applications. It is applied in dense WSN where sensors are randomly deployed and in which it is assumed that each node has a unique identifier, and the communication between neighboring nodes is bidirectional. Furthermore, in FMS, a random sleep scheduling is used to save energy.

Although FMS enables to guarantee reliable date delivery to the sink, it presents some limitations as the above protocol and it requires a synchronization of the sensors during sleep-scheduling phase.

RERP [9] is an adaptive fault-tolerant routing protocol. This protocol assumes that a node has at least two neighbors in the forward direction towards the sink. The ability of a node to tolerate faults depends upon the number of active neighbors. So, if a node has N neighbors it can tolerate N-1 faults by selecting one node as a main path and the other nodes (N-1) as backup paths for data delivery. Moreover, RERP ensures multiple paths towards the base station from each node of the network. However, it consumes a lot of energy when it broadcasts error reporting messages and data packets are lost when the route to the destination is not available from a node.

DMRF [10] is a dynamical jumping real-time fault-tolerant routing protocol which operates in two data transmission modes: hop-by-hop mode and jumping transmission mode. In DMRF, each node utilizes the remaining transmission time to relay data packets to the sink and the state of the forwarding candidate node set to choose the next hop. When node failure, network congestion or an empty region occurs, the transmission mode will switch to jumping mode, which might reduce the transmission delay. However, in DMRF, the jumping transmission mode doesn't guarantee that the destination node is available and it consumes more energy in case of a long jump.

III. MODEL PARAMETERS

Before heading into the technical details of our contribution, we first give some definitions and notations that will be used in our paper later.

A wireless sensor network is represented by an undirected graph $G=(V,E)$, where V represents the set of sensor nodes and $E \subseteq V^2$ is the set of edges that gives the available communications: an edge $e=(u,v)$ belongs to E if and only if the node u is physically able to transmit messages to v and vice-versa. At each sensor node $u \in V$ is assigned a unique value to be used as an identifier, so that the identifier of u is denoted by $Node_{Id}(u)$.

The 2-density of a node u represents the ratio between the number of links in its 2-neighborhood (links between u and its neighbors and links between two 2-neighbors of u) and the 2-degree of u $\delta_2(u)$, denoted by $\rho_2(u)$. It is expressed by the following formula:

$$\rho_2(u) = \frac{|(v,w) \in E| v, w \in N_2[u]|}{\delta_2(u)} \quad (1)$$

where $N_2[u]$ is the 2-neighborhood set of u (the nodes that are located not more than two hops from the node u) and $\delta_2(u)$ is the 2-degree of u (the size of $N_2(u)$).

The clusters formed should not exceed a certain threshold $Tresh_{Upper}$ to avoid exhausting quickly the energy of cluster-heads. This threshold depends on network topology and its value is calculated as follows:

- Let u the node that has the highest 2-degree in the network,

$$\delta_2(u) = Max\{\delta_2(u_i); \ u_i \in V\} \quad (2)$$

- Let Avg the average value of 2-degree of all nodes in the network,

$$Avg = \frac{\sum_{i=1}^{n} \delta_2(u_i)}{n} \quad (3)$$

Where n is the size of the network.

- $Thresh_{Upper}$ is the threshold size of clusters:

$$Thresh_{Upper} = \frac{(\delta_2(u) + Avg)}{2} \quad (4)$$

In this paper, we assume that the sensors are randomly distributed in a two-dimensional Euclidean plane P. Each sensor has an omni-directional antenna which allows it to reach with a single transmission all nodes within its vicinity and we consider that sensors have 2-hop positional information. We also assume that each sensor has a generic weight and it is able to calculate it. This weight represents the fitness of each node to be a cluster-head.

IV. CLUSTER-BASED FAULT-TOLERANT SCHEME FOR WSN

In this section, we present our proposed scheme which generates balanced clusters. It is performed in three consecutive phases:

A. Cluster formation

Clusters formation process generates 2-hop clusters (2-clusters) wherein each cluster has a primary cluster-head (CH_p) and its vice (CH_v) which are elected in 2-neighborhorhood based on the weights of sensors. The weight of each sensor is a combination of 2-density $\left(\rho_2(u)\right)$ and residual energy $\left(E(u)\right)$ as presented in Eq. (5). The weight parameter is periodically calculated by each node in order to illustrate the suitability of a node for playing cluster-head's role. We involve 2-density factor in the purpose to generate clusters whose members are linked with cluster-heads and remaining energy parameter to select the nodes with more energy in their 2-neighborhood.

$$Weight\ (u) = \alpha * \rho_2(u) + \beta * E(u) \ \wedge \ \alpha + \beta = 1 \quad (5)$$

The values of α and β are chosen depending on the application. For example if we want to favor the node that has more energy as cluster-head we would attribute great value to β.

Since cluster-heads are responsible to accomplish several tasks such as coordination among the cluster members, transmission gathered data to the remote base station, and management of their own cluster; we propose to set up periodically cluster-head election process. Therefore, cluster-heads do not rapidly exhaust their battery power.

At the beginning of each round, each sensor calculates its weight and generates a 'Hello' message including three extra fields addition to other regular contents: Weight, CH_p and CH_v, where CH_p and CH_v are set to zero. Then, it broadcasts it in its 2-neighborhood as well as it eavesdrops its neighbor's 'Hello' message. After these exchanges, the sensor with the greatest weight in its 2-neighborhood, is elected as primary cluster-head and the node with the second largest weight, is chosen as its vice. Each sensor node updates its state vector by assigning respectively to CH_p and CH_v the identifiers of the corresponding primary and secondary cluster-heads. Then, primary cluster-head broadcasts an advertisement message (ADV_CH) including its state vector to its 2-neighboors to request them to join it. Each sensor receiving the message and which is not yet a member of a cluster as well as its weight is respectively less than the weight of CH_p and its vice, transmits REQ_JOIN message to CH_p to join it. Corresponding cluster-head checks if its own cluster size does not reach $Thresh_{Upper}$ it would transmit ACCEPT_CH message to this sensor.

B. Routing paths

Cluster-heads construct a CH-to-CH routing paths to use them for data transmission and each cluster-head creates a time schedule, in which time slots are allocated for intra-cluster communication, data aggregation.

When a node detects a relevant event, it sends it to its corresponding cluster-heads (CH_p and CH_v). If the primary cluster-head does not send an acknowledgment to the sending node in a threshold time, its vice considers it that is down and it sends the ACK message. Moreover, to forward information to the base station, each cluster-head sends it to the corresponding primary cluster-head and its vice, and if primary cluster-head does not respond in time its vice does it.

C. Packet delivery

Sensor nodes in a cluster send their gathered data to their respective cluster-heads. Accordingly, the cluster-heads are responsible for coordination among the cluster members, aggregation of their data and transmission of the aggregated data to the sink, directly or via multi-hop transmission mode.

During this phase, sensors begin to send collected readings to their respective cluster-heads. The radio of each member of a cluster could be turned off until the sensor's allocated transmission time. After receiving of the data, cluster-heads aggregate them before sending them to the base station. Each cluster-head communicates using different CDMA codes in order to reduce interference from nodes belonging to other clusters.

V. EVALUATION AND SIMULATION RESULTS

In this section, we conduct several simulations to evaluate CFS performance and compare them with those of GRAB. For that, we utilize ns2 [11] to implement CFS and we select sensor parameters similar to Berkeley motes [12]. We use a field size of $150 \times 150m^2$ wherein 1200 nodes are uniformly distributed. The maximum transmission range of a node is 10 meters. The energy consumptions for transmitting, receiving and idling are respectively 60 mW, 12 mW and 12 mW. A random source node generates a report every 10 seconds and in each run 100 reports are generated.

We evaluate the impact of node failure rate on success ratio and energy consumption and we compare the obtained results with those of GRAB [5].

To evaluate CFS performance, we measure the success ratio, which is the ratio of the number of report packets successfully received at the sink and the total number generated by all sensors. This metric illustrates the robustness of CFS to forward data when some sensors fail. Furthermore, we also measure total energy consumption to illustrate the efficiency of CFS in terms of the cost of excessive overhead. The obtained results are averaged over 10 different runs.

To illustrate the CFS performance, we evaluate the success ratio of data delivery to the base station according to the number of failed nodes. For that, we vary the node failure rate from 5% to 50%.

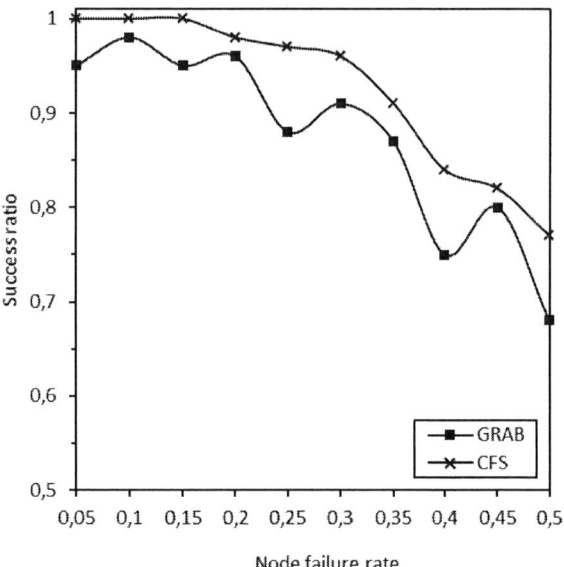

Figure 1: Success ratio for different node failure rate

Figure 1 shows the success ratio according to node failure rate. The success ratio is above 90% for node failure rates of up to 35%. As the node failure rate increases, the success ratio tends to decrease; CFS still maintains very high degrees of robustness compared to GRAB. The success ratio remains above 80% when 45% nodes fail, and is around 80% in the extreme case when half of the nodes fail. This shows that CFS is robust even with severe node failures compared to GRAB. The high success ratio also demonstrates that CFS is highly tolerate to inaccurate cost fields because the probability of having a cluster-head and its vice have failed and that they are involved in data forwarding, is small.

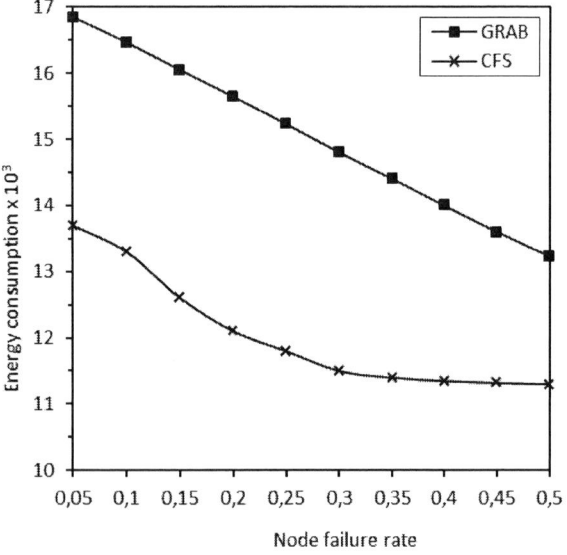

Figure 2: Energy consumption for different node failure rate

The energy consumptions are shown in Figure 2. When node failure increases, the energy decreases linearly. This is because the idle energy dominates the total energy consumption. A higher node failure rates means more node failures, thus proportionally less energy consumption. Furthermore, CFS consumes less energy than GRAB because CFS implies the cluster-heads and in the critical case their vices to forward data to the base station.

VI. CONCLUSION

In this paper, we have proposed an efficient scheme called CFS to deal with fault tolerance in WSN. The proposed scheme allows ensuring the reliability data delivery to the base station while minimizing energy consumption.

In CFS, we utilized a clustered architecture in which there are 2-clusters with primary cluster-heads and their vices to guarantee data delivery to the base station.

Simulation results showed that CFS provides good results compared to GRAB.

REFERENCES

[1] Akyildiz I.F., Su W., Sankarasubramaniam Y., Cayirci E.,"Wireless sensor networks: a survey". Computer Networks (Elsevier). Vol.38, No.4, pp 393-422, March 2002.

[2] A. Woo, T. Tong, and D. Culler,"Taming the underlying challenges of reliable multihop routing in sensor networks". Proceedings of the 1st International Conference on Embedded Networked Sensor Systems, pp.14-27, November 2003.

[3] L. Paradis and Q. Han, "A Survey of Fault Management inWireless Sensor Networks", Journal of Network and Systems Managemen, Vol.15, No.2, pp.171-190, 2007.

[4] H.M. Ammari and S.K; Das, "Fault tolerance measures for large-scale wireless sensor networks", Journal ACM Transactions on Autonomous and Adaptive Systems (TAAS), Vol. 4, No. 1, pp.1-28, 2009.

[5] Ye, G. Zhong, S. Lu, and L. Zhang. "Gradient broadcast: A robust data delivery protocol for large scale sensor networks". SPRINGER Wireless Networks, 11(2), pp.285-298, 2005.

[6] D. Ganesan, R. Govindan, S. Shenker, and D. Estrin. "Highly-resilient, energy-efficient multipath routing in wireless sensor networks". ACM Mobile Computing and Communications, Review, 1(2),pp. 251-254, October 2002.

[7] A. Ajay, N.Tarasia, S. Dash, S.Ray, A.R.Swain "A Dynamic Fault-tolerant Routing Protocol for Prolonging the Lifetime of Wireless Sensor Networks", International Journal of Computer Science and Information Technologies, 2(2), pp.727-734, 2011.

[8] Ajay, N.Tarasia, S. Dash, S.Ray, A.R.Swain "Fault Tolerant Multilevel Routing Protocol with Sleep Scheduling (FMS) for Wireless sensor Networks", European Journal of Scientific Research ISSN 1450-216X Vol.55 No.1, pp.97-108, 2011.

[9] K. KULOTHUNGAN, J. Angel Arul Jothi, A. Kannan, "An Adaptive Fault Tolerant Routing Protocol with Error Reporting Scheme for Wireless Sensor Networks", European Journal of Scientific Research, Vol.60 No.1, pp. 19-32, 2011.

[10] Guowei Wu , Chi Lin , Feng Xia , Lin Yao , He Zhang and Bing Liu, "Dynamical Jumping Real-Time Fault-Tolerant Routing Protocol for Wireless Sensor Networks", National Natural Science Foundation of China, 2010.

[11] http://www.isi.edu/nsnam/ns/

[12] J. Hill, R. Szewczyk, A.Woo, S. Hollar, D. Culler and K. Pister. "System architecture directions for networked sensors", International Conference on Architectural Support for Programming Languages and Operating Systems, 2000.

Successive Interference Cancellation Receiver (SIC) in DS-OCDMA System

Kada BITEUR
Department of Electronics. Faculty of Engineering
Sciences, University Djillali Liabes
Sidi Bel Abbes 22000, Algeria
Email: biteur99@gmail.com

Malika KANDOUCI
Department of Electronics. Faculty of Engineering
Sciences, University Djillali Liabes
Sidi Bel Abbes 22000, Algeria
Email : maikand04@gmail.com

Abstract— The CDMA (Code Division Multiple Access) is to spread or distribute the signal over a very large bandwidth up to the'' invisible'' to other users sharing the same band. The study presented in this work is an exploratory study on the application of direct sequence CDMA in optical transmission systems (OCDMA Optical Code Division Multiple Access or Multiple Access Code Division in Optics), where data to be transmitted is spread time, in our study we used sequences called quasi-orthogonal unipolar OOC, we considered the multiple access interference (MAI) due to multiple access technique is the only performance limitation. The multi-user detection, and in particular the structure SIC (Successive Interference Cancellation Receiver) has been studied in this work, the calculation of the probability of error shows an improvement over the conventional receiver (CR), for validate the theoretical studies, simulations are performed in MATLAB and presented in this work.

Keywords— OCDMA, Multiple Access Interference (MAI), Multi-User Detection, Receiver SIC.

I. INTRODUCTION

The OCDMA system studied is direct sequence (DS-OCDMA), non-coherent and synchronous [1], the bit time data to be transmitted is divided into a number of intervals called "bins chips" by direct multiplication of data with the code (Fig 1.): the direct-sequence CDMA (DS-CDMA). The number of "intervals chips" is the length of the code sequence OCDMA. The number of pulses or "chips" amplitude unit in the code sequence, corresponds to the weight of the code. The distribution of these pulses in the "chips intervals' is related to the family code used.

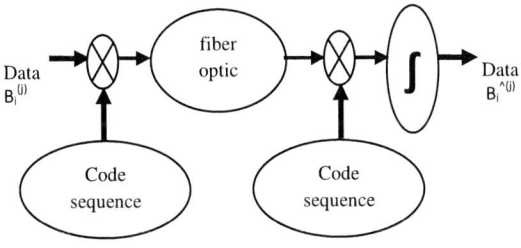

Fig 1. Diagram of a DS-OCDMA system

II. TIME CODES UNIPOLAR

With unipolar codes, we can not have a strict orthogonality. These codes must satisfy the properties of "quasi-orthogonality" following [2]:

➤ The condition of self-correlation codes:

$$Z_{C_k,C_k}(1) = \sum_{j=0}^{L-1} C_j^{(k)} C_{j+1}^{(k)} = \begin{cases} = W & \text{for } L = 0 \\ \leq \lambda_a & \text{also} \end{cases} \quad (1)$$

with: W is the weight of the code sequence $\{c_j^{(k)}\}$

➤ Intercorrelation of the condition codes:

$$Z_{C_k,C_p}(1) = \sum_{j=0}^{L-1} C_j^{(k)} C_{j+1}^{(p)} \leq \lambda_c \quad \forall 1 \quad (2)$$

with :
- L is the length of the code sequence OOC
- λ_a is the value maximum auto-correlation of the codes.
- λ_c is the maximum value of cross-correlation codes.

At least the constants λ_a and λ_c can be equal to 1.

Many pseudo-orthogonal codes and their application to unipolar Optical CDMA have been studied since 1988 [3]. These codes are the codes Optical Orthogonal (OOC) [4].

A. Optical Orthogonal Codes (OOC)

OOC codes were developed in 1989 by JASalhi [5] These codes are characterized by four parameters (L, W, λ_a, λ_c) :

- L is the length of the sequence
- W is the weight of the code, which represents the number of chips to "1"
- λ_a and λ_c are respectively the constraints of self and cross-correlation.

For values of self and cross-correlation $\lambda_a = \lambda_c = \lambda$, the number of users(N) is limited by the band called Johnson, given by the relation [6]:

$$N(L,W,\lambda_c,\lambda_a) \leq \left\lfloor \frac{1}{W} \left\lfloor \frac{L-1}{W-1} \left\lfloor \frac{L-2}{W-2} ----- \left\lfloor \frac{L-h}{W-h} --------- \right\rfloor \right\rfloor \right\rfloor \right\rfloor \quad (3)$$

The symbol $\lfloor X \rfloor$ represents the integer value of a lower value X. In the case where λ_a and λ_c minimum ($\lambda_a = \lambda_c = 1$), JASalehi [5] showed that the maximum number of code sequences is:

$$N(L,W,1,1) \leq \left\lfloor \frac{L-1}{W(W-1)} \right\rfloor \quad (4)$$

For our study, we will use the codes as OOC $\lambda_a = \lambda_c = 1$.
Several methods of generating OOC codes can be implemented. In this study, we used the method "BIBD" (Balanced Incomplete Block Design) [7].
Consider the code (97,4,1,1) , length L=97, weight W=4, such as $\lambda_a = \lambda_c = 1$.
According to (4), N = 8 (number of users).

Were generated in MATLAB as determined by the BIBD, unipolar sequences consisting of 97 chips which are 4 to 1 (Table 1).

Table 1. Chips in a position

Code	p_1	p_2	p_3	p_4
#1	0	3	35	61
#2	0	4	43	50
#3	0	17	75	6
#4	0	64	9	24
#5	0	62	36	96
#6	0	54	47	93
#7	0	22	91	81
#8	0	88	73	33

III. SCCESSIVE INTERFERENCE CANCELLATION RECEIVER (SIC)

It is considered that the desired user is the user # 1, and we also consider that the system has N active users, and there is no near-far effect [8] (all users have the same energy);

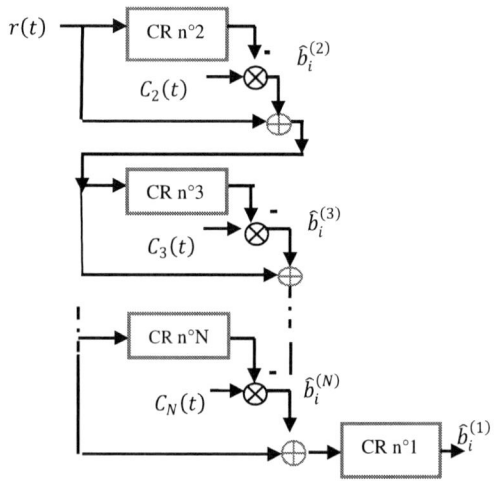

Fig 2. Diagram of the receiver SIC

The SIC involves several steps (Fig 2.):

The first step is to detect the data sent by a user unwanted set, the received signal to subtract the estimated contribution of that user. For this, the data transmitted by the user is estimated by a CR with a decision threshold S2. Then, the reconstructed signal is estimated by multiplying by the user code unwanted, and subtracted from the received signal. We obtain a new signal:

$$r_1(t) = r(t) - \hat{b}_i^{(2)} C_2(t) \qquad (5)$$

The second step is to repeat the previous step to a second user unwanted by considering the signal obtained in the previous step as the received signal. We obtain a new signal:

$$r_2(t) = r(t) - \hat{b}_i^{(2)} C_2(t) - \hat{b}_i^{(3)} C_3(t) \qquad (6)$$

➤ This process can be repeated either on all unwanted users (in this case, the SIC has N -1 floor) or on some unwanted users.

➤ At the end of the procedure, the signal $r_{(N-1)}(t)$ is applied to the input of the receiver of the user desired conventional 1:
➤

$$r_{N-1}(t) = r(t) - \hat{b}_i^{(2)} C_2(t) - \hat{b}_i^{(3)} C_3(t) - \cdots - \hat{b}_i^{(N)} C_N(t) \qquad (7)$$
$$= b_i^{(1)} C_1(t) + \sum_{j=2}^{N} \left(b_i^{(j)} - \hat{b}_i^{(j)} \right) . C_j(t)$$

In Figure 4, the first user is detected undesired user # 2, then the second user is # 3 ... etc ... User choice to detect unwanted does not affect performance. However, if users do not have the same power, it should detect unwanted users starting with one that has the greatest power, and then continue with decreasing powers.

A. probability of error receiver interference cancellation in series (SIC):
1) Case of a series cancellation:

$$r(t) = \sum_{k=1}^{N} b_i^{(k)} C_k(t) \qquad (8)$$

$$r_1(t) = r(t) - \hat{b}_i^{(k)} C_N$$
$$= b_i^{(1)} C_1(t) + \sum_{k=1}^{N} b_i^{(k)} C_k(t) \left(b_i^{(1)} - \hat{b}_i^{(k)} \right) C_N(t) \qquad (9)$$

The decision variable of the desired user can be written

$$Z_i^{(1)} = W b_i^{(1)} + I_1 + A_1 \qquad (10)$$

With I_1 : the interference term,

$$I_1 = \sum_{k=2}^{N-1} \int_0^{T_b} b_i^{(k)} C_k(t) C_1(t) dt \qquad (11)$$

And A_1 The term cancellation of the first floor

$$A_1 = \left(b_i^{(N)} - \hat{b}_i^{(N)} \right) \int_0^{T_b} C_N(t) C_1(t) dt \qquad (12)$$

The cancellation term A1 can take only two values 0 and -1, because the conventional receiver detected the user makes a mistake when $b_i^{(N)} = 0$
The probability of error of user n°1 is

$$P_{e0} = \frac{1}{2} P_{e10} + \frac{1}{2} P_{e11} \qquad (13)$$

with $P_{e10} = prob\left(\hat{b}_i^{(1)} = 1/b_i^{(1)} = 0\right)$ (14)

and $P_{e11} = prob\left(\hat{b}_i^{(1)} = 0/b_i^{(1)} = 1\right)$

P_{e11b} the probability of error if the user desired sent data « 1 » ; and P_{e10} if to send data « 0 ».

- Underline: User # 1 sends $\boldsymbol{b_i^{(1)} = 1}$

$$P_{e11} = prob\left(Z_i^{(1)} < S_1/b_i^{(1)} = 1\right)$$ (15)

$$= prob\left(W + I_1 + A_1 < S_1/b_i^{(1)} = 1\right)$$

So with the approximation we can write

$$P_{e11} = prob\left(A_1 = -1/b_i^{(1)} = 1\right).prob\left(I_1 < S_1 - W + 1/b_i^{(1)} = 1\right)$$
$$+ prob\left(A_1 = 0/b_i^{(1)} = 1\right).prob\left(I_1 < S_1 - W + 1/b_i^{(1)} = 1\right)$$ (16)

$$Prob\left(I_1 < S_1 - W + 1/b_i^{(1)} = 1\right) = \sum_{i=0}^{S_1-W} C_{N-2}^i \left(\frac{W^2}{2L}\right)^i \left(1 - \frac{W^2}{2L}\right)^{N-2-i}$$ (17)

We define the function

$$f(a,b,k) = \sum_{i=a}^{b} C_k^i \left(\frac{W^2}{2L}\right)^i \left(1 - \frac{W^2}{2L}\right)^{k-i}$$ (18)

f(a,b,k) = 0 si a > b

Then $prob\left(I_1 < S_1 - W + 1/b_i^{(1)} = 1\right) = f(0, S_1 - W, N - 2)$

Similarly, we obtain:

$prob\left(I_1 < S_1 - W/b_i^{(1)} = 1\right) = f(0, S_1 - W - 1, N - 2)$

On the other hand:

$$prob\left(A_1 = -1/b_i^{(1)} = 1\right) = prob\left(b_i^{(N)} - \hat{b}_i^{(k)} = -1/b_i^{(1)} = 1\right)$$
$$.prob\left(\int_0^{T_b} C_N(t)C_1(t)\,dt = 1\right)$$
$$= \frac{W^2}{L}.\frac{1}{2}prob\left(\hat{b}_i^{(N)} = 1/b_i^{(N)} = 0/b_i^{(N)} = 1\right)$$
$$= \frac{1}{2}.\frac{W^2}{L}f(S_N - 1, N - 2, N - 2)$$ (19)

$$prob\left(A_1 = 0/b_i^{(1)} = 1\right) = 1 - prob\left(A_1 = -1/b_i^{(1)} = 1\right)$$
$$= 1 - \frac{1}{2}.\frac{W^2}{L}f(S_N - 1, N - 2, N - 2)$$

$$\boldsymbol{P_{e11}} = \frac{1}{2}.\frac{W^2}{L}f(S_N - 1, N - 2, N - 2).f(0, S_1 - W, N - 2)$$
$$+ \left(1 - \frac{1}{2}.\frac{W^2}{L}f(S_N - 1, N - 2, N - 2)\right).f(0, S_1 - W - 1, N - 2)$$
$$= \frac{1}{2}.\frac{W^2}{L}f(S_N - 1, N - 2, N - 2).f(0, S_1 - W, N - 2)$$ (20)

- Similarly we calculate the probability of error in case the user # 1 sends $\boldsymbol{b_i^{(1)} = 0}$

Similarly, we obtain :

$$\boldsymbol{P_{e10}} = \frac{1}{2}.\frac{W^2}{L}f(S_N, N - 2, N - 2).f(S_1 + 1, N - 2, N - 2)$$
$$+ \left(1 - \frac{1}{2}.\frac{W^2}{L}f(S_N, N - 2, N - 2)\right).f(S_1, N - 2, N - 2)$$ (21)

Fig 3. shows the evolution of the probability of error based on the detection threshold of S1 for the desired user receiver with a SIC cancellation stage compared to the CR receiver for code OOC (97,4,1,1), with a number of users N = 8.

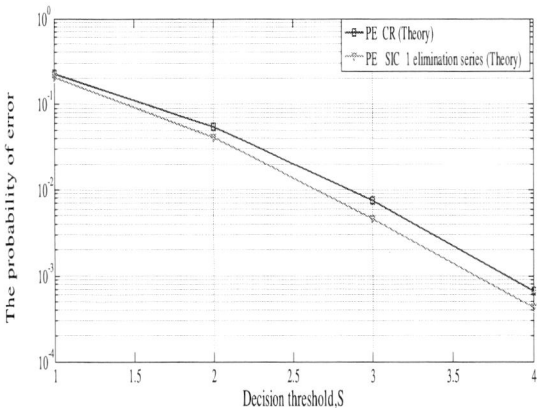

Fig 3. the probability of error for an elimination series compared to the probability of error for the RC code OOC (97,4,1,1), N = 8.

According to Fig 3. , we see that for a cancellation stage for the receiver SIC, there is a better chance of error from the CR receiver

2) The case of two serial cancellations :
It has two floors of cancellations A_1 and A_2

- A_1 : the term to cancel the first floor

$$A_1 = \left(b_i^{(N)} - \hat{b}_i^{(N)}\right) \int_0^{T_b} C_N(t)C_1(t)\,dt$$ (22)

- A_2 : the term cancellation of the second floor

$$A_2 = \left(b_i^{(N-1)} - \hat{b}_i^{(N-1)}\right) \int_0^{T_b} C_{N-1}(t)C_1(t)\,dt$$ (23)

For the second cancellation stage was 2×3 possible combinations for the cancellation terms of A1 and A2, so the calculation of the probability of error still requires approximations relative to the first cancellation stage

Fig 4. shows the evolution of the probability of error based on the threshold of detection of the desired user S1 for receiver SIC with single and two stage cancellation compared to the CR receiver.

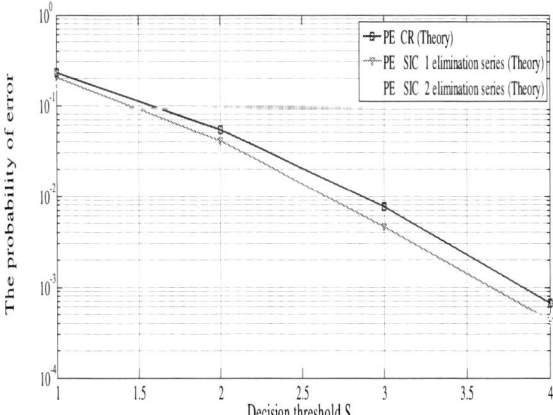

Fig 4. the probability of error for one and two eliminations in series compared to the probability of error for the RC code OOC (97,4,1,1), N = 8.

From Fig 4. we can observe that for two floors of cancellations there is an improvement for the error probability compared to a single elimination series.

B. validation of the theoretical expression of the probability of error SIC:

To validate the theoretical expression PeSIC, we simulated in MATLAB with a link OCDMA receiver SIC.

It was considered an OOC code (97,4,1,1), with N = 8 active users.

Fig 5. shows the error probability theory and simulated based on the detection threshold of S for the desired user and two floors of cancellations

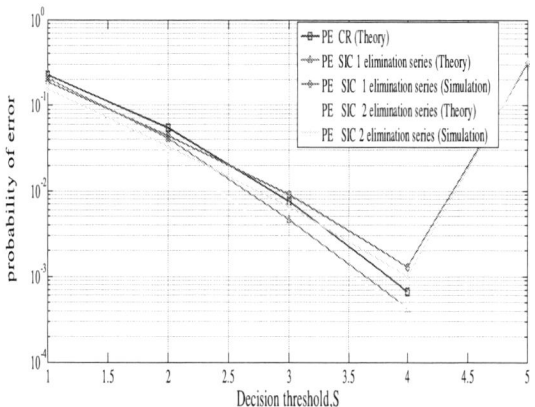

Fig 5. The error probability of SIC two cancellations and a series based on the detection threshold of the OOC code (97,4,1,1), N = 8.

We can observe according to Fig 5., a gap between the simulation curve and the theoretical curve. This difference is due to the approximation made to simplify obtaining the theoretical expression.

Therefore, our theory underestimates the number of errors as possible, because the cancellation errors create errors in the subsequent stages, for it was considered against the previous stages to detect because unwanted users were properly canceled.

VI. CONCLUSION

In this work, we were interested in the technical division multiple access direct sequence code for optical transmission systems (DS-OCDMA) for the family code called: Optical Orthogonal Codes (OOC).

We then studied the multiuser receiver with interference cancellation (SIC) for one and two cancellations in series in the case where AMI was the only performance limitation. This study shows that the receiver SIC provides improved performance compared to the Conventional receiver (CR). However, this structure is very complex to analyze theoretically accurately.

A numerical simulation of a DS-OCDMA chain has been developed for the receiver SIC in the case of an OOC code (97,4,1,1) for 8 active users. The obtained results allow to

confirm the theoretical estimate of the number of possible error.

REFERENCES

[1] P.R. Prucnal « Optical Code Division Multiple Access : Fundamentals and Applications». CRC; Har/Cdr edition, ISBN : 084933683X (2006).

[2] J. A.Salehi « Code division multiple-access techniques in optical fiber networks - part I: Fundamental principles and part II: Systems performance analysis», IEEE Transactions on Communications, vol. 37, pp. 824-842, (Août 1989).

[3] M. Azizoglu, J.A. Salehi, and Y. Li. «Optical CDMA via temporal codes. IEEE Transactions on Communications», 40(7) :1162–1170, (July 1992).

[4] K. Jamshidi and M. Abtahi. «Performance analysis of various optical CDMA systems using OOC's with correlation bounded by two», Proc. of International Symposium on Telecommunications (IST), pages 115–120, (Sep. 2005).

[5] J.A. Salehi and C.A. Brackett «Code division multiple-access techniques in optical fiber networks-part I : Fundamental principles», IEEE Transactions on Communications, 8(37) :824–833, (Aug. 1989).

[6] SM Johnson « A new upper bound for error-correcting codes », IRE Transactions on Information Theory, vol IT-8, pp 203-207, (April 1962).

[7] H.Chung, P.Kumar, «Optical orthogonal codes - new bounds and an optimal construction», IEEE Transactions on Information theory, vol. 36, pp. 866-873, (July 1990).

[8] Goursaud-brugeaud C, «Réception multi- utilisateurs par annulation parallèle d'interférences dans les systèmes CDMA optiques », Thèse Université de Limoges, (juillet 2006).

Coverage and Connectivity Protocol for Wireless Sensor Networks

Abdelkader Khalil

Faculty of Sciences
University Abderrahmane Mira of
Béjaïa 06000, Algeria
khalilabdelkader@gmail.com

Rachid Beghdad

Faculty of Sciences
University Abderrahmane Mira of
Béjaïa 06000, Algeria
rachid.beghdad@gmail.com

Abstract—**A fundamental challenge in Wireless Sensor Networks (WSNs) is the *coverage problem*. To solve this problem, several coverage solutions exist in literature based on different methods (protocol, geometrical and analytical solutions). In this paper, we suggest a solution to solve the problem of coverage and connectivity, called Connected Cover Set based on IDentity of node (CCSID). The idea was to adapt a concept from graph theory of connected dominating sets having minimum cardinality MCDS, to build cover sets. The solution CCSID divides the set of deployed nodes into subsets. In each subset, a minimum number of active nodes are selected to ensure coverage and connectivity. Simulation results show that CCSID achieves a high coverage ratio and uses fewer active nodes.**

Keywords- Wireless Sensor Networks; coverage; Connectivity; Minimum Connected Dominating Set; Cover Set.

I. INTRODUCTION

With the recent advances in micro-electronics technologies and wireless communications, a new type of networks has emerged: Wireless Sensor Networks (WSNs). This type of networks includes a large number of devices called *sensors* deployed over a geographical area to be monitored. A sensor is able to sense, process and transmit data over a wireless communication channel. The applications of WSNs include battlefield surveillance, healthcare, environmental and home monitoring, industrial diagnosis and so on [1].

A fundamental issue in WSNs is the *coverage problem* [2, 3] that mainly consists in ensuring continuous and effective observation of an area of interest (AI) while taking into account some constraints, in particular the number of active sensors and their limited energy. The coverage can be considered as a measure of the monitoring quality produced by a sensor network [4].

WSNs are usually dense and redundant (more than 20 nodes/m² [5]). So, the coverage of AI can be done, but it is not optimized if all nodes contribute for observing this AI. So, one may exploit the nodes redundancy to periodically select a subset of active nodes among all deployed nodes. This will maximize coverage of AI while extending the WSN lifetime Determining cover sets is usually done using analytical approach for solving the coverage problem in WSN. It consists first in decomposing of all deployed sensors into subsets, then finding for each subset, the necessary and sufficient number of active nodes that ensure the coverage and connectivity in WSN.

In the literature many solutions to the coverage and connectivity problem were presented [6, 7, 8, 9, 10, 11, 12, 13, 14, 15, 16, and 17] but they still suffer from some weaknesses, this is the reason why we focused on the solution of such a problem.

In this paper, we suggest a solution for determining cover sets in WSN. More precisely, a connected cover set based on identity of node (CCSID) that uses a graph theory concept: Minimum Connected Dominating Sets (MCDS), in order to build coverage sets, is proposed. The main contributions of our approach are: (1) high coverage ratio, (2) small number of active nodes, (3) connectivity guaranteed.

After describing our connectivity and coverage approach (Section II), Section III describes our simulation results. Finally Section IV concludes this paper.

II. CCSID SOLUTION

A. CCSID Overview

Connected Cover Set based on IDentity of nodes (CCSID) is proposed as a basis for a solution to the problem of cover set construction. It is based on analytical approach and uses the concept of minimum connected dominating set from grapgh theory. The WSN is modeled by a graph $G (S,E)$ (see Section III.A) CCSID is operated by two phases; The duration of the first phase determines a minimum dominating set MDS as a cover set for assuring the coverage in WSN. And in the second phase, it makes this MDS connected for obtaining a MCDS, for assuring more area coverage and the connectivity in WSN.

B. CCSID Description
* Phase 1: Construction of MDS.

The determination of a minimum dominating set is an NP-hard problem [18]. Therefore, we propose a heuristic based solution to solve it. We obtain the MDS by determining the Maximal Independent Set (MIS) having the minimum cardinality. So this phase contains two steps:

- *Step1: Determination of MISs*

In this step, we determine all possible MISs, such that the first Maximal Independent Set (MIS₁) is determined as that:

1- we choose node u having the less identity node; this node u becomes the first member of MIS_1 (line 5).

2-All neighbours of u (neighbours dependent a SR) are excluded from the next selection (line 6), and then we update the set S.

3- we choose the second member of MIS_1 by selecting the less identity node in updating set S (line 7).

The other members of $MIS1$ are determined by the same process which continues until the set S will be empty (line 4). Subsequently, the above process is repeated to obtain members of others MISs {$MIS2, MIS3...$} corresponding the 2nd, 3rd ... less identity node in S (line 3).

- Step 2: Selection of MDS.

In this step, we select the MIS having the minimum cardinality among others $MISs$ as minimum dominating set (MDS) (line 11).

The Algorithm's pseudo code of this phase is:

Algorithm of heuristic MDS

```
Begin

1- i=1; V=S;  /* S: a set of deployed nodes, V: parameter saving  S */

/* Step1: (In: S; Out: MISs). */

2-   While (i ≤ │S│)

3-     j= i ; S= V;   /* j: variable which determines index of MIS */

4-       While S ≠ ∅

5-         s_i ∈ MIS_j ;  /* node i is member of  MIS j */

6-         S = S − N(s_i) ;  /*to remove neighbours dependent a SR of node i/

7-         s_i =  Min_ID(S);  /*to select the  node having less identity from S /

8-       End While;

9-     i ++

10- End  While;

/* Step 2: (In: MISs; Out: MDS) */

11-  MDS = Min_card (MIS_j);  /* to select the set having the min. cardinality /

End.
```

* Phase 2: Construction of MCDS.

In this phase, $CCSID$ assures the connectivity of MDS, it makes disconnected elements of MDS connected by adding the minimum number of node. So this phase contains two steps:

- Step1: Determination of disconnected node.

In first time, the $MCDS$ contains only the elements of MDS (line 1), then we test every node s_i in $MCDS$ for we know that it is connected or no (line 7). This test is the following:

- If there are other node s_j in $MCDS$, such that the link between s_i and s_j is less than CR. And

- If this link is void, i.e. the parameter $a_{ij}=0$.

Then we say that the node s_i is connected and we make the link between s_i and s_j visited ($a_{ij}=1$) (line 8). Else we say that the node s_i is disconnected.

- Step 2: Addition a new node in MCDS.

When we determine the disconnected node s_i in the pervious step, we add a new passive node s_k from the neighbours of node s_i in $MCDS$, such that s_k is close at base station (line 13); and we make the link between s_i and s_k is visited (line 14).

The pervious steps are repeated for every element of $MCDS$ until all active nodes in $MCDS$ will be connected (lines 4 and 5).

The Algorithm's pseudo code of this phase is:

Algorithm of heuristic MCDS

```
Begin

/* Step 1 */

1-  MCDS = MDS; /*the set MDS is included in a set MCDS */

2-  i =1;

3-  While i ≤ │MCDS│

4-     j = 1; s_i = MCDS(i); s_i-connect=false;

5-       While (j ≤ │MCDS│) ∧ (no s_i-connect)

6-         s_j = MCDS(j);

7-         If (i≠j) ∧ (dist(s_i,s_j) ≤ Rc) ∧ (a_ij==0) then /* to test the node s_i
is  connected or not */

8-           s_i-connect= true; a_ij=1;  /* s_i is connected by  s_j and the link
                                          (i, j)is  visited  */

9-         End If ;

10-        j ++;

11-      End While;

/* Step 2 */

12-  If (no  s_i-connect) then /*to add a new node in MCDS */

13-      MCDS=MDS ∪ {s_k/ s_k∈N(s_i) ∧ Min_dist(dist(s_k,s_SB)) };

14-      a_ik =1; /* we make the link between s_i and visited  s_k */

15-  End If;

16-  i ++;

17- End While;

END.
```

III. SIMULATION

We simulate our solution by using MATLAB 7 language to evaluate $CCSID$ and compare its performance with two solutions: the first is a centralized approach proposed by Pazand and Datta [17] and the other is a distributed solution proposed by Khanouche [9]. In [17], the base station, and after the neighbors discovery step, selects periodically the set of *active* nodes to cover the AI. These selected nodes are those having maximum degree. In [9], each node generates a random timeout. At the expiration of the timeout, the "*active*" node transmits a domination message to its neighbors. Each node receiving such a message will become "*passive*". The active nodes cover AI during a period. This process will be repeated until there is no node in working. Nevertheless, these two approaches do not guarantee the connectivity aspect.

A. Simulation Parameters

Experimental results were obtained from randomly generated networks in which nodes are deployed over a $50\times50(m^2)$ square sensing field. Simulations were carried over densities varying from 100 to 450 nodes. For each network deployment, we assume that the deployed nodes can fully cover the sensing field with the given sensing range. The coverage ratio is evaluated by dividing the deployment area to $1x1$ (m^2) cells. A cell is considered covered if its centre is covered. The results presented hereafter are the average of 100 iterations for each simulated scenario. The performance metrics include: *percentage of active nodes* and *coverage ratio*.

B. Performance Comparison

In this subsection, we compare the performances of *CCSID* with two solutions: the first is centralized algorithm proposed by Pazand and Datta [17]. This algorithm will be noted *CDSC* (*Centralized Dominating Set for Coverage*). And the other is distributed algorithm proposed by khanouche [9]. This algorithm will be noted *DCovPDS* (*Distibuted Coverage preserving based on Domanting Set*). We will consider here two cases depending the *CR/SR* ratio. For that we fixe the sensing range at 10m, while the communication range is equal to *SR* for the first case, and $2\times SR$ for the second (*CR>SR*).

1) Percentage of Active Nodes: It is important to have as few active nodes as possible. Fig. 1 shows the average percentage of active nodes while varying densities in case *CR=SR*.

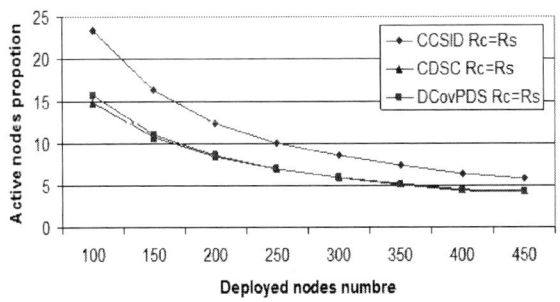

Figure 1. Average percentage of active nodes selected by CCSID, DCovPDS and CDSC in case CR=SR.

We can observe that, for *CR=SR* and 100 deployed nodes, *CDSC*, *DCovPDS* and *CCSID* have 14.89%, 15.74% and 23.43% of active nodes respectively. Then as the density increases *CCSID* generates less active nodes but it is more than *CDSC* and *DCovPDS* (5.76% *Vs* 4.30% and 4.27% respectively for WSN consist 450 deployed nodes). This difference of active nodes propotion is exploited by *CCSID* for assuring more connectivity in WSN, such that only 23 to 26

nodes are active in the first case for different sensors populations.

The Fig. 2 shows the average percentage of active nodes while varying densities in case *CR>SR*.

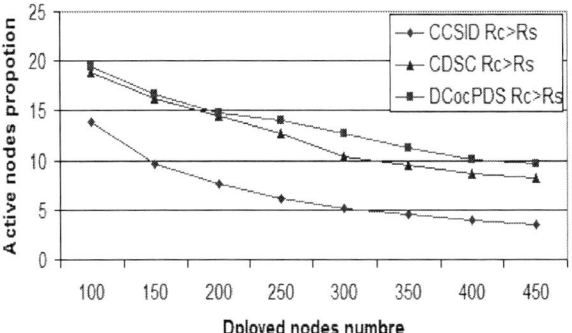

Figure 2. Average percentage of active nodes selected by CCSID, DCovPDS and CDSC in case CR>SR.

When *CR>SR*, the active nodes proportion is increased significantly by *CCSID*, it varies from 14 to 16 nodes. This is due to the fact that the links of communication between nodes grow proportionally to the number of deployed nodes, then the probability to find a disconnected node in MDS is decreased. So we add the little new nodes in the second phase of *CCSID*. Contrary, *CDSC* and *DCovPDS* product of active nodes more than *CCSID* in this case (19 to 36 nodes, 20 to 43 nodes respectively).

2) Coverage Ratio: It is defined as the percentage of interest area covered by active nodes. This percentage is calculated by using the method described in Section V.A.

The Fig. 3 represents the average coverage ratio of *CCSID*, *DCovPDS* and *CDSC* while varying densities in case *CR=SR*.

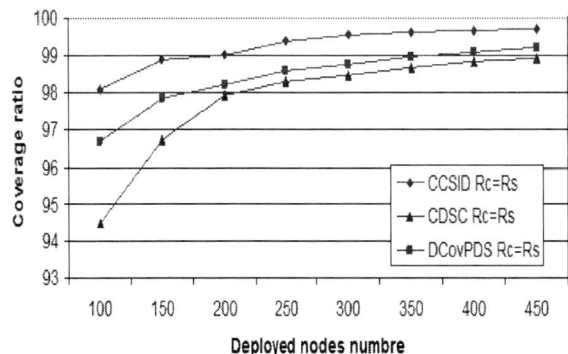

Figure 3.Average coverage ratio against density achieved by CCSID, DCovPDS and CDSC in case CR=SR.

When *CR=SR*, *CCSID* provides a better coverage ratio with 98.11% for the lowest density, this ratio increases gradually until it exceeds 99.70% for the highest density. For 100 deployed nodes, it is shown in Fig. 3 that *CCSID* provides an improvement of coverage ratio equal to 3.6% and 1.44% compared to *CDSC* and *DCovPDS* recpectively. And this improvement continues for a highest density (0.76% and 0.47% compared to *CDSC* and *DCovPDS* recpectively for 450 deployed nodes).

The Fig. 4 represents the average coverage ratio of *CCSID*, *DCovPDS* and *CDSC* while varying densities in case *CR>SR*.

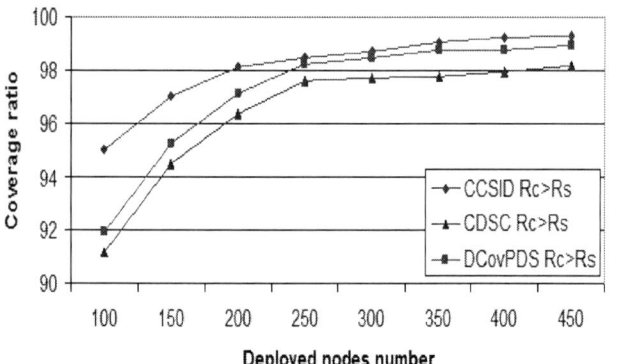

Figure 4. *Average coverage ratio against density achieved by CCSID, DCovPDS and CDSC in case CR>SR.*

In the case of *CR>SR*, *CCSID* provides too a better coverage ratio than *CDSC* and *DCovPDS*. It is 95.03%, 91.17% and 91.91% for *CCSID*, *CDSC* and *DCovPDS* respectively for 100 deployed node, so there is an improvement of coverage ratio equal to 3.86% and 3.12% compared to *CDSC* and *DCovPDS* recpectively. The coverage ratio for *CCSID*, *CDSC* and *DCovPDS* grow proportionally to the number of deployed nodes. It is 99.32%, 98.21% and 98.96% for *CCSID*, *CDSC* and *DCovPDS* respectively for 450 deployed nodes. So there is an improvement by our solution.

VI. CONCLUSION

In this paper we have presented a centralized solution *CCSID* to maintain coverage and to assure the connectivity in Wireless Sensor Networks. That is based on a graph theory formulation using a Minimum Connected Dominating Sets concept, in order to construct cover sets. Simulation has been done to validate the effectiveness of the suggested solution. The results show that, *CCSID* outperforms the centralized solution *CDSC* [17] and the distributed solution DCovPDS [9] in terms of coverage ratio which is the most important metric. It also competes perfectly in terms of selected active nodes.

As future work, we plan to study the coverage and connectivity in case of mobile nodes, damaged nodes, and with the presence of obstacles.

REFERENCES

[1] I. F. Akyildiz, W. Su, Y. Sankarasubramaniam, and E. Cayirci, "Wireless sensor networks: a survey," Computer Networks Journal, vol. 38, no. 4, pp. 393–422, March 2002.

[2] C. F. Huang and Y.C. Tseng, "A survey of solutions to the coverage problems in wireless sensor networks," Journal of Internet Technology, vol. 6, no. 1, pp. 1–8, 2005.

[3] M. Cardei and J. Wu, "Energy-efficient coverage problems in wireless ad hoc sensor networks," Computer Communications Journal, vol. 29, no. 4, pp. 413–420, February 2006.

[4] S. Meguerdichian, F. Koushanfar, M. Potkonjak, and M. B. Srivastava, "Coverage problems in wireless ad-hoc sensor networks," 20th Annual Joint Conference of the IEEE Computer and Communications Societies, vol. 3, pp. 1380–1387, 2001.

[5] V. Rajavavivarme, Y. Yang, and T. Yang, "An overview of wireless sensor network and applications," Proceedings of the 35th Southeastern Symposium on System Theory, pp. 432–436, March 2003.

[6] A. Gallais, J. Carle, D. Simplot-Ryl, and I. Stojmenovic, "Localized sensor area coverage with low communication overhead," IEEE transactions on mobile computing, vol. 7, no. 5, pp. 661–672, 2008.

[7] D. Tian and N. Georganas, "A coverage-preserving node scheduling scheme for large wireless sensor networks," Proceedings of the 1st ACM international Workshop on Wireless Sensor Networks and Applications, pp. 32–41, 2002.

[8] F. Ye, H. Zhang, S. Lu, L. Zhang, and J. Hou, "A randomized energy-conservation protocol for resilient sensor networks," Wireless Networks, vol. 12, no. 5, pp. 637–652, Octobre 2006.

[9] M.E.Khanouche, treatment of coverage problem in WSN, magister thesis, University of Bejaia, Algeria, 2010.

[10] N.A.A. Aziz, K.A. Aziz, W.Z.W. Ismail. Coverage Strategies for Wireless Sensor Networks. World Academy of Science, Engineering and Technology 50, 2009.

[11] M. Berg, O. Cheong, M.V. Kreveld, M. Overmars. Computational geometry: algorithms and applications, 3rd edition, Springer Press, 2008.

[12] M. A. M. Vieira, L. F. M. Vieira, L. B. R. Ruiz, A. A. F. Lureiro, A. O.Fernandes, and J. M. S. Nogueira, Scheduling Nodes in Wireless Sensor Networks: A Voronoi Approach. In 28th Annual IEEE International Conference on Local Computer Networks. IEEE, 2003.

[13] S. Slijepcevic and M. Potkonjak. Power Efficient Organization of Wireless Sensor Networks. *IEEE International Conference on Communications*, 2: 472-476, 2001.

[14] M. Cardei, D. MacCallum, X. Cheng, et al. Wireless Sensor Networks with Energy Efficient Organization, *Journal of Interconnection Networks*, Vol 3, No3-4, pp 213-229, December 2002.

[15] S. Yang, F. Dai, M. Cardei and J. Wu. On Multiple Point Coverage in Wireless Sensor Networks. In *Proceedings of the 2nd IEEE Intern Conference on Mobile Ad-hoc and Sensor Systems*, 2005.

[16] Y. Ye. o(n3l) potential reduction algorithm for linear programming. Mathematical Programming,vol.50, no.2, pp.239-258, 1991..

[17] B. Pazand and A. Datta, "Minimum Dominating Sets for solving the coverage problem in wireless sensor networks," Proceedings of the International Symposium on Ubiquitous Computing Systems, pp. 454–466, 2006.

[18] J. Carle, A. Gallais and D. Simplot-Ryl, Preserving area coverage in wireless sensor networks by using surface coverage relay dominating sets. In Proceedings of 10th IEEE Symposium on Computers and Communications (ISCC'2005), pages 347–352, 2005.

[19] http://en.wikipedia.org/wiki/Maximal_independent_set.

Differential Dual Amplitude-Width PPM Coding for FSO Communication Systems

Mehdi ROUISSAT*, Ahmed R. BORSALI, Mohammad E. CHIKH-BLED
Laboratory of Telecommunications of Tlemcen (LTT)
Dept Electronic, Faculty of thechnology
Abou Bekr Belkaid University, PB 119
Tlemcen, Algeria
*mehdi.rouissat@mail.univ-tlemcen.dz

Abstract — **In this paper, a hybrid modulation scheme called DDAWPPM (Differential Dual Amplitude-Width Pulse Position Modulation) suitable with wireless optical systems has been proposed, on the basic of DPPM (Differential Pulse Position Modulation), PAM (Pulse Amplitude Modulation) and PWM (Pulse Width Modulation). The combinations (DPPM-PAM) and (DPPM-PWM) have been proposed in previous works, but the proposed DDAWPPM and beside the symbol synchronization ability, it shows more improvement in term of data rate and spectral efficiency, on the other side it shows a degradation in term of power efficiency.**

We present theoretical expressions of spectral efficiency, power requirements, and the normalized data rate improvement, and we present comparison results to DPPM and the hybrids PAM-DPPM and PWM-DPPM.

Keywords - DDAWPPM; DPPM; FSO; hybrid modulation; performance analysis.

I. INTRODUCTION

In Free Space Optical Systems (FSO), the channel is usually treated as an Intensity Modulation and Direct Detection (IM/DD)[1]. For IM/DD, the modulation OOK (On Off Keying) is the simple and widely used modulation scheme in commercial FSO systems, because of its easy in implementation, simple transceiver design, bandwidth efficiency and cost effectiveness. On the other hand, higher average power efficiency can be achieved by employing pulse time modulation schemes, in which a range of time dependent features of a pulse may be used to convey the information. Digital (Discrete) Pulse Time Modulation (PTM) techniques fall into two categories, namely isochronous and anisochronous. Isochronous modulation schemes encode data by varying the position or width of a pulse, but the overall symbol structure remains constant. In contrast, anisochronous schemes have no fixed symbol structure [2].

The PPM (Pulse Position Modulation) is a popular isochronous PTM format for free-space optical (FSO) communications [3], PPM based systems is more power efficient than OOK, but it has several drawbacks such as the complexity in implementation due to higher level of accuracy required in slot and symbol synchronization and the rapid decline of bandwidth efficiency with increasing power efficiency. For these reasons, anisochronous schemes (DPIM [4], DPPM [5]) have been proposed as alternative modulation formats to PPM.

DPPM (Differential Pulse Position Modulation) has solved the problem of symbols synchronization. However, it shows a modest data rate and spectral efficiency. In order to improve those parameters, several alternative modulations have been proposed. The hybrid modulation DPPM-PAM (DAPPM) [6] and DPPM-PWM (DWPPM) [7] have been proposed in other previous works, but they are still not strong candidates when the data rate and the spectral efficiency are of a great importance. For these reasons, the combination DPPM-PAM-PWM (DDAWPPM) will be presented in this work.

In this paper, symbol structure of the proposed modulation and comparison results about average power requirement, bandwidth efficiency and data, to DPPM, DAPPM and DWPPM rate will be presented.

II. DIFFERENTIAL PULSE-POSITION MODULATION (DPPM)

DPPM is an anisochronous modulation technique; it is simple modification of the existing PPM that is also used in wireless optical communications systems. The symbols in DPPM are obtained from the corresponding PPM symbol by deleting all of the "off" slots following the "on" slots, as shows Tab.1.

One of the advantages of DPPM over PPM is the symbol synchronization ability, as every symbol finish with a pulse. In order to avoid the case where the time between adjacent pulses is zero, an additional guard slot may be added in each symbol immediately after the pulse. Thus, a symbol which encodes b bits of data is represented by a pulse in one slot, after k slots of zero power, where $1 \leq k \leq M$, ($M = 2^b$). The minimum and the maximum symbol lengths are 2Ts and (M+ 1)Ts respectively, so the mean symbol length is (M + 3)Ts/2

The averages power requirements by the DPPM and PPM schemes normalized to OOK modulation (On off Keying) are given respectively by:

$$\frac{P_{DPPM}}{P_{OOK}} = \frac{2M}{M+3} \sqrt{\frac{2}{M \log_2 M}} \qquad (1)$$

$$\frac{P_{PPM}}{P_{OOK}} = \sqrt{\frac{2}{M \log_2 M}} \qquad (2)$$

The bandwidth required to support communication at a bit rate based on the average symbol duration relative to OOK, is given in [7].

$$B_{DPPM} = \frac{(M+3)R_b}{2 \log_2 M} \qquad (3)$$

The band utilization efficiency is given in [7]:

$$\eta_{DPPM} = \frac{2 \log_2 M}{(M+3)} \qquad (4)$$

III. DIFFERENTIAL AMPLITUDE PULSE-POSITION MODULATION (DAPPM)

DAPPM is also an asynchronous modulation technique; it is a combination of PAM and DPPM. Therefore, the symbol length and the amplitude of the pulse vary to represent the information being transmitted.

In this paper we allow the pulses to take only two levels, thus, the pulse can take one of two levels (A_1, A_2) to convey one of the L symbols, where the number of possible symbols L is given by:

$$L_{DAPPM} = 2M \qquad (5)$$

The relationship between the two levels is a design parameter. In this paper we take:

$$A_2 = 2.A_1 \qquad (6)$$

The same as in DPPM modulation scheme, an additional guard slot may be added in each symbol immediately after the pulse in order to avoid the case where the time between adjacent pulses is zero. Thus, the mean symbol length is the same as in DPPM.

The average power requirement by the DAPPM scheme normalized to OOK is given in [10]:

$$\frac{P_{DAPPM}}{P_{ook}} = \frac{2(M+1)}{3} \frac{1}{2^M} \sqrt{\frac{2}{M \log_2 M}} \qquad (7)$$

The minimum and the maximum power requirements by DAPPM modulation scheme are $1A_1$ and $2A_1$ respectively, in each symbol length. Thus, the average power requirement based on the average symbol length is given by the following formula:

$$p_{DAPPM} = \frac{3/2}{(M+3)/2} = \frac{3}{M+3} \qquad (8)$$

The relationship between the two averages powers: DAPPM and PPM is given by:

$$\frac{P_{DAPPM}}{P_{PPM}} = \frac{\frac{3}{(M+3)}}{1/M} = \frac{3M}{M+3} \qquad (9)$$

In order to find the average normalized power requirement for DAPPM, we multiply (9) by (2):

$$\frac{P_{PPM}}{P_{OOK}} \times \frac{P_{DAPPM}}{P_{PPM}} = \frac{P_{DAPPM}}{P_{OOK}} \qquad (10)$$

Consequently, the average normalized power requirement for the DAPPM is given by:

$$\frac{P_{DAPPM}}{P_{OOK}} = \frac{3M}{M+3} \sqrt{\frac{2}{M \log_2 M}} \qquad (11)$$

Based also on average symbol length, the band utilization efficiency of DAPPM is given by:

$$\eta_{DAPPM} = \frac{2 \log_2 (2M)}{(M+3)} \qquad (12)$$

IV. DIFFERENTIAL WIDTH PULSE-POSITION MODULATION (DWPPM)

DWPPM is a hybrid asynchronous modulation technique; it is a combination of PWM and DPPM, where the symbol length and the pulse width vary to represent the information being transmitted. It's proposed mainly to improve the bandwidth efficiency and the data rate over DPPM, but it didn't show a remarkable improvement.

In this paper we allow the pulses to take two widths. Thus, the pulse can take one of two widths (w_1, w_2) to convey one of the L symbols, where the number of possible symbols is the same as in DAPPM, and it's given by (5).

The relationship between the two widths is a design parameter. In this paper we take:

$$W_2 = 2.W_1 \qquad (13)$$

The minimum and the maximum symbol lengths are $2Ts$ and $(M+2)Ts$ respectively, so the mean symbol length is $(M+4)Ts/2$.

To find the average power requirements, we use the same method as we used for DAPPM. The minimum and the maximum power requirements per symbol by DWPPM modulation are $1A_1$ and $2A_1$ respectively. Thus, the average power requirement based on the average symbol length is given by:

$$p_{DWPPM} = \frac{3/2}{(M+4)/2} = \frac{3}{M+4} \qquad (14)$$

The relationship between the two averages powers: DWPPM and PPM is given by:

$$\frac{P_{DWPPM}}{P_{PPM}} = \frac{\frac{3}{(M+4)}}{1/M} = \frac{3M}{M+4} \qquad (15)$$

By using the same steps as in DAPPM, the average normalized power requirement for the DWPPM scheme is given by:

$$\frac{P_{DWPPM}}{P_{OOK}} = \frac{3M}{M+4}\sqrt{\frac{2}{M\log_2 M}} \qquad (16)$$

Also based on average symbol length, the band utilization efficiency of DWPPM is given by:

$$\eta_{DWPPM} = \frac{2\log_2(2M)}{(M+4)} \qquad (17)$$

V. DIFFERENTIAL DUAL AMPLITUDE WIDTH PULSE-POSITION MODULATION (DAWPPM)

The DDAWPPM is a modified anisochronous hybrid form of the existing DPPM. The data in DDAWPIM are presented by the combination Amplitude-Width and the time interval between the previous and the present pulse of two adjacent symbols. It's proposed to improve the data rate and the bandwidth efficiency over DAPPM and DWPPM.

In order to avoid symbols in which the time between adjacent pulses is zero, an additional guard slot will be added to each symbol immediately following the pulse (Fig.1). The minimum and the maximum symbol lengths are 2Ts and (M+ 2)Ts respectively, so the mean symbol length is (M + 4)Ts/2.

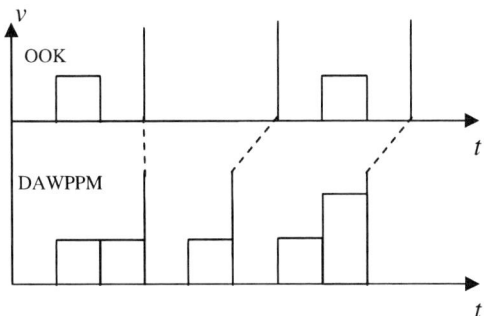

Figure 1. Encoding Example of a serial data bit to DAWPPM

By the combination Amplitude-Width, the number of possible symbols in DAWPPM modulation scheme is given by:

$$L_{DAWPPM} = 4M \qquad (18)$$

Examples of mapping between source bits and transmitted slots for PPM, DPPM, DAPPM and DDAWPPM are shown in Table.1.

TABLE I. MAPPING BETWEEN SOURCE BITS (OOK) AND TRANSMITTED SLOTS FOR PPM, DPPM, DAPPM AND DDAWPPM

bits	PPM	DPPM	DAPPM	DDAWPPM
000	10000000	01	0a	0a
001	01000000	001	00a	00a
010	00100000	0001	000a	0aa
011	00010000	00001	0000a	0A
100	00001000	000001	0A	00A
101	00000100	0000001	00A	0AA
110	00000010	00000001	000A	0aA
111	00000001	000000001	0000A	0Aa

Where:

a: is a pulse with A_1 as amplitude
A: is a pulse with A_2 as amplitude
aa: is a pulse with A_1 as amplitude, and W_2 as width.
AA: is a pulse with A_2 as amplitude, and W_2 as width.

A. Data Rate:

The data rate that can be achieved by a modulation scheme is very important, but it is not enough to judge the performance of such modulation. The data rate that can be achieved with DDAWPPM is:

$$D_{DDAWPPM} = \frac{\log_2(4.M)}{T_{mean}} = \frac{2.\log_2(4.M)}{(M+4)}(\text{bit}/s) \qquad (19)$$

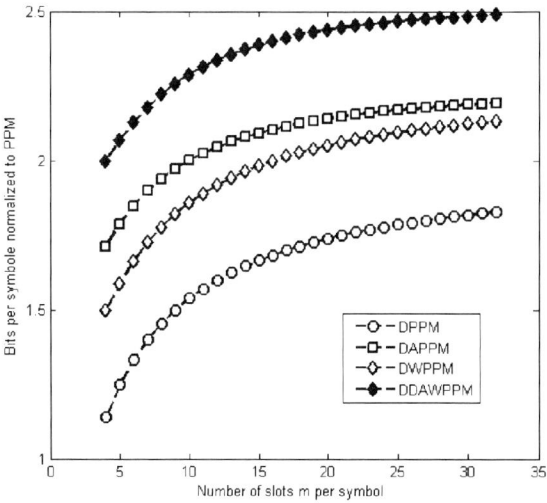

Figure 2. Normalized data rate for DPPM, DAPPM, DWPPM and DDAWPPM

To show the improvement in data rate, we define the parameter R, which presents the ratio in term of data rate of any modulation scheme to that achieved by PPM.

$$R = \frac{D_M}{D_{PPM}} \qquad (20)$$

D_M: D_{DPPM} or $D_{DDAWPPM}$

$$D_{DDAWPPM} = \frac{2M.\log_2(4.M)}{(M+4)\log_2 M} \qquad (21)$$

The Fig.2 shows the normalized data rate of DPPM, DAPPM, DWPPM and DDAWPPM for different values of M. The Figure shows that the propose scheme (DDAWPPM) present the higher data rate among the four modulation techniques, for all the values of *M*. And DAPPM presents higher data rate compared to DPPM and DWPPM. Even though DWPPM has the same number of symbol as DAPPM by using two width instead of two levels, it shows lower data rate, this is because of it has longer symbols, which has not improve the data rate of DPPM as in DAPPM.

B. Power Requirements and Bandwidth Efficiency

We use the same method as we used for DPPM and DAPPM to find the average power requirements. The minimum and the maximum power requirement by DDAWPPM modulation are $1A_1$ and $4A_1$ respectively. Thus, the average power requirement based on the average symbol length is given by:

$$p_{DDAWPPM} = \frac{5/2}{(M+4)/2} = \frac{5}{M+4} \quad (22)$$

DAPPM and DDAWPPM require the same power for a given M, but with different time length.

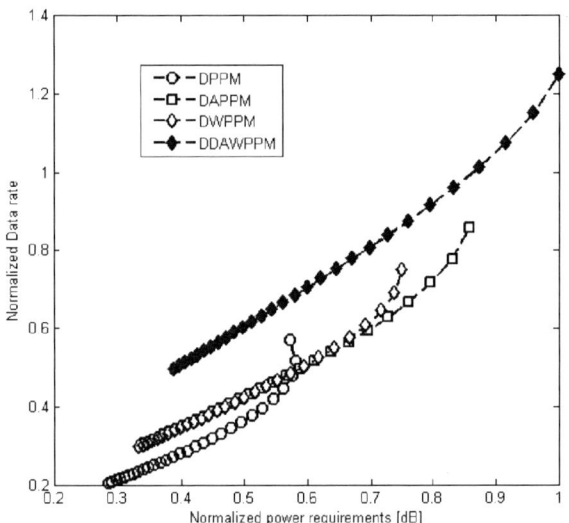

Figure 3. Normalized Power Requirements based on Bandwidth efficiency for DPPM, DAPPM, DWPPM and DDAWPPM

By the same method, the normalized average power requirement of DDAWPPM is given by:

$$\frac{P_{DDAWPPM}}{P_{OOK}} = \frac{5M}{M+4} \sqrt{\frac{2}{M \log_2 M}} \quad (23)$$

Also based on average symbol length, the band utilization efficiency of DDAWPPM is given by:

$$\eta_{DDAWPPM} = \frac{2 \log_2 (4.M)}{(M+4)} \quad (23)$$

The Fig.4 shows the normalized power requirement based on the spectral efficiency for DPPM, DAPPM, DWPPM and DDAWPPM for different values of *M*, from 4 to 32. The figure shows that DDAWPPM modulation presents the best spectral efficiency among the four modulation schemes, at the same time it presents the lowest power efficiency for all the values of *M*.

These results, and the previous results show that the modulation DDAWPPM is a good alternative for FSO systems that require high spectral efficiency and data rate, beside the

advantage of the symbol synchronization, and the only drawback of this modulation technique is the degradation in the power efficiency.

VI. CONCLUSION

In this paper, a new modified modulation scheme called DDAWPPM "Differential Amplitude-Width Pulse Position Modulation" has been presented on the basis of Differential Pulse Position Modulation), PAM (Pulse Amplitude Modulation) and PWM (Pulse Width Modulation), this modulation scheme is proposed essentially to improve the data rate and the bandwidth efficiency over DPPM.

The combination PAM-DPPM and PWM-DPPM were introduced as hybrid modulation schemes in a previous works, but they steel show moderate data rate and spectral efficiency. Our proposed modulation scheme has improved the data rate and the spectral efficiency over PAM-DPPM and PWM-DPPM, but it shows degradation in term of power efficiency.

The proposed concept may be well suited to use in Free Space Optical Communications Systems that require high data rate with simple transceiver, by virtue of its increased data rate, high spectral efficiency and the symbol synchronization ability.

REFERENCES

[1] J. R. BARRY. Wireless infrared communications", London Kluwer, 1994.

[2] B. WILSON, Z. GHASSEMLOOY, Pulse time modulation techniques for optical communications: a review," IEE proceedings, 1993. vol. 140, no. 6, p. 346–357.

[3] S. J. DOLINAR, J. HAMKINS, B. E. MOISION, and V. A. VILNROTTER. Optical modulation and coding in Deep Space Optical Communications. Edition. Wiley-Interscience. 2006.

[4] Z. GHASSEMLOOY and A.R. HAYES. Digital Pulse Interval Modulation for Optical Communication. IEEE Commun. Magazine.98. Vol. 36, p. 95-99.

[5] DA-SHAN SHIU and J.M. KAHN. Differential Pulse Position Modulation for Power-Efficient Optical Communication. IEEE Trans. on Commun.99. p. 1201-1210.

[6] U. SETHAKASET and T.A. GULLIVER. Differential Amplitude Pulse Position Modulation for Indoor Wireless Optical Channels. IEEE Communications Society. 2004

[7] HUANG AI-PING, FAN YANG-YU, LIU YUAN-KUI, JIA. MENG, BAIBO, TAN QING-GUI. A Differential Pulse Position Width Modulation for Optical Wireless Communication. IEEE conference on Industrial Electronics and Application. ICIEA. 2009.

[8] M. ROUISSAT, A.R BORSALI, M. CHICK-Bled . AH-MPPM a new modulation scheme for Free Space Optical communication systems. IEEE conference on High Capacity Optical Networks and Enabling Technologies (HONET), 2011.p 106 – 110.

[9] H. PARK and J.R. BARRY. Modulation Analysis for Wireless Infrared Communications," presented at IEEE International Conference on Communications, ICC 95, Seattle, 1995, pp. 1182-1186.

[10] NAZMY AZZAM, MOUSTAFA H. ALY and A.K. ABOUISEOUD. Bandwidth and Power Efficiency of Various PPM Schemes for Indoor Wireless Optical Communications. 26th National Radio Science Conference (NRSC2009) Egypt.

Influence of Noisy Channel on Acoustic Echo Cancellation in Mobile Communication

Mahfoud HAMIDIA

Speech Communication and Signal Processing Laboratory
Faculty of Electronics and Computer Science, USTHB
P.O. Box 32, Bab Ezzouar, Algiers, Algeria
mhamidia@usthb.dz

Abderrahmane AMROUCHE

Speech Communication and Signal Processing Laboratory
Faculty of Electronics and Computer Science, USTHB
P.O. Box 32, Bab Ezzouar, Algiers, Algeria
namrouche@usthb.dz

Abstract – **In this paper, we investigate the influence of noisy channel on the performance of acoustic echo cancellation system. In the mobile communications, acoustic echo is mainly caused by the coupling between the loudspeaker and the microphone of the mobile device. So, an Acoustic Echo Canceller (AEC) should be used locally inside this device. This paper evaluates the performances of AEC, generally based on adaptive filtering, where the transmitted speech is encoded and decoded by AMR-WB (Adaptive Multi-Rate Wide Band) speech codec. The encoded speech is transmitted over a transmission channel modeled by AWGN (Additive White Gaussian Noise) and Rayleigh fading channel. The simulation results show the strong degradation of the AEC performances, due to the noisy channel.**

Keywords – **Acoustic echo cancellation, Adaptive filtering, AMR-WB, AWGN channel, Rayleigh channel, Misalignment.**

I. INTRODUCTION

In the last decade, mobile communication systems such as GSM (Global System for Mobile communications) and UMTS (Universal Mobile Telecommunications System) have seen an increasing interest. So, the mobile phone became one of the most used devices for human to human communication. Unfortunately, the quality of the transmitted speech is degraded by the acoustic echoes. These echoes occur due to direct loudspeaker to microphone coupling, and from reflections within the enclosed environment [1]. Hence, acoustic echo cancellation can enhance the speech quality of speakerphones as well as cell phones considerably by reducing the power of the far-end signal that is retransmitted to the far-end speaker [2]. The principal structure of AEC is an adaptive filter, which generally uses a FIR (Finite Impulse Response) filter. The GSM communication system uses a speech coder to reduce the bit-rate needed to represent the speech signal while keeping an acceptable quality of the decoded speech [3].

AMR-WB (Adaptive Multi Rate Wide Band) speech codec is standardized in the second generation (2G) and the third generation (3G) of cellular system. The coding speech is transmitted over a channel transmission. However, the quality of speech at destination output depends on the channel conditions. Bad conditions in the channel transmission will produce many errors in the speech output, and hence, the speech quality falls [4].

This paper is organized as follows: After this brief introduction in Section I, acoustic echo cancellation is discussed in the Section II. Speech transmission over GSM is presented in the Section III, then; the simulation results are given and commented in the Section IV. Finally, conclusions are drawn in the Section V.

II. ACOUSTIC ECHO CANCELLATION

The acoustic echo cancellation is a typical application of system identification. The acoustic coupling between the loudspeaker and the microphone at the near-end will cause the far-end speech to be transmitted back to the far-end and from the acoustic echo. This will severely deteriorate the quality of speech communication. The goal of an echo canceller is first to identify the echo path between the loudspeaker and the microphone by the adaptive filter. Then, the adaptive filter can produce a replica of the echo signal, which will be used to remove the echo, before the signal is delivered to the far-end [5]-[6], as is shown in Fig 1. For simplifying, we considered the single-talk scenario, where the near-end speaker is in the state of silence.

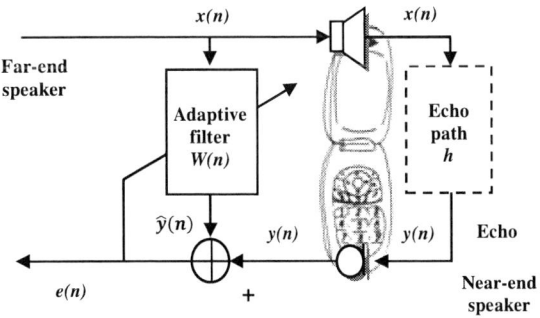

Fig 1. Acoustic echo canceller system.

The acoustic echo signal is modeled by the equation:

$$y(n) = X^T(n).h \qquad (1)$$

where

$$X(n) = \lfloor x(n) \; x(n-1), \ldots \; x(n-L+1)\rfloor^T \qquad (2)$$

$$h = [h_0 \quad h_1 \quad h_2 \dots \quad h_{L-1}]^T \qquad (3)$$

$X(n)$ is the length-L history of the received signal, or far-end input signal. h is the true echo path impulse response, it is assumed to be finite impulse response (FIR) filter of length L, where superscript $(.)^T$ denotes vector transposition. The residual echo signal which represents the estimation error is:

$$e(n) = y(n) - \hat{y}(n) \qquad (4)$$

At each sample time n, the adaptive filter output $\hat{y}(n)$ or the echo replica is:

$$\hat{y}(n) = X^T(n) . W(n) \qquad (5)$$

where

$$W(n) = [w_0(n) \quad w_1(n), \dots \quad w_{L-1}(n)]^T \qquad (6)$$

is the vector of adaptive filter coefficients at time n.

The NLMS (Normalized Least Mean Squares) adaptive filtering algorithm is an extension of the standard LMS algorithm [7]. This algorithm is widely used for adapting the adaptive filter coefficients, due to its simplicity and low computational complexity and it is defined by:

$$W(n + 1) = W(n) + \frac{\mu}{\varepsilon + \|X(n)\|} X(n) \, e(n) \qquad (7)$$

Where $\|.\|$ is the Euclidian norm, $W(n + 1)$ is the next tap weight value and $W(n)$ is the present tap weight value of the adaptive filter. μ, $0 < \mu < 2$, is the step size used in the weight vector updating, and $\varepsilon > 0$ is a regularization factor which avoids division by zero.

III. SPEECH TRANSMISSION ON GSM

A. AMR-WB Speech coding

The AMR-WB speech codec [8] was selected in December 2000 by the Third Generation Partnership Project (3GPP) for GSM and third generation mobile communication WCDMA system for providing wideband speech services [9]. It also was selected as recommendation G 722.2 by the ITU-T. AMR-WB deals with the voice sample with 16 KHz sampling frequency, 16 bit accuracy, and signal bandwidth ranged from 50 Hz to 7 KHz. Compared to the traditional narrow-band speech codec with a bandwidth range from 300 to 3400 Hz, 8 KHz sampling, and 8 bit accuracy, the output signal of AMR-WB codec has a higher quality [4].

The AMR-WB speech codec consists of nine speech coding modes with bit-rates of 23.85, 23.05, 19.85, 18.25, 15.85, 14.25, 12.65, 8.85 and 6.6 Kbit/s. AMR-WB includes also a background noise mode, which is designed to be used with the dependant background noise generator at low bit-rate in the discontinuous transmission using VAD (Voice Activity Detector) module of the GSM. AMR-WB has a high granularity of bit-rates, between 6.6 and 23.85 Kbit/s, this

makes it possible to maximize speech quality by adapting the codec bit-rate to increase robustness against transmission errors.

Fig 2 shows a general structure of AMR-WB speech codec. This codec is encoded 20 ms per frame. The original speech is divided into two frequency bands of 50 Hz to 6400 Hz and 6400 Hz to 7000 Hz and is encoded by next procedure to simplify the complexity. The low frequency band is encoded using the Algebraic Code Excited Linear Prediction (ACELP) algorithm. The linear predictive analysis is carried out 20 ms per frame and uses the LPC coefficient of 16 orders. The high frequency band is reconstructed using the parameters of the low frequency band and the random excitation signal. The gain of high frequency band is adjusted by the speech information of the low frequency [10]. The spectrum of the high frequency band is reconstructed using the LP filter which is generated from the low frequency band.

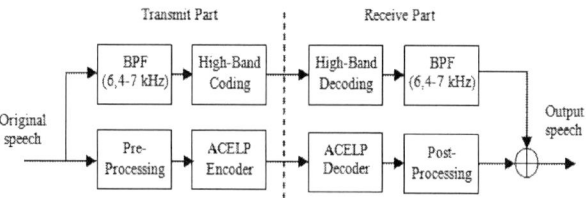

Fig 2. General structure of AMR-WB speech codec.

B. Transmission channel

In mobile communication systems, the area covered by the system is divided into smaller areas, with base stations in their centers. Communication with mobile stations is performed through base stations.

1) AWGN channel

One of the most important channel models considered in channel coding is the Additive White Gaussian Noise (AWGN) channel. The values n_i added by the AWGN channel can be considered as a Gaussian distributed random variable with zero mean, and variance σ^2. The probability density function of added noise $P(n_i)$ can be written as:

$$P(n_i) = \frac{1}{\sqrt{2\pi\sigma^2}} e^{-\frac{n_i^2}{2\sigma^2}} \qquad (8)$$

If c_i is the code symbol, the noisy code symbol s_i is modeled by the following equation:

$$s_i = c_i + n_i \qquad (9)$$

2) Rayleigh fading channel

Rayleigh fading is a statistical model for the effect of a propagation environment on a radio signal such as that used by wireless devices. It assumes that the power of a signal that has passed through such a transmission medium (also called a communications channels) will vary randomly, or fade, according to a Rayleigh distribution [11]. Rayleigh fading is most applicable when there is no line of sight between the transmitter and receiver. The code symbol passed through a Rayleigh channel which is given by:

$$s_i = a_i\, c_i + n_i \qquad (10)$$

were a_i is the amplitude of code symbol.
The density of probability $P(a_i)$ can be written as:

$$P(a_i) = \begin{cases} \dfrac{a_i}{\sigma^2} e^{-\frac{a_i^2}{2\sigma^2}} & if \quad a_i > 0 \\[2mm] 0 & otherwise \end{cases} \qquad (11)$$

IV. SIMULATION RESULTS

In this section, we present the obtained results with the simulation model of transmitted speech over GSM, as shown in Fig 3. Thus, the signal of far-end speaker $x(n)$ is encoded in MSC (Mobile Switching Center) and is transmitted over a transmission channel. In this paper, the transmission channel is modeled by the AWGN channel and the Rayleigh fading channel using MATLAB. The MS (Mobile Station) receives and decodes the transcoded speech $\tilde{x}(n)$. We simulated the coding speech using AMR-WB codec of the standard ITU-T [12]. The GSM AMR-WB uses three modes of bit-rates (6.60, 8.85 and 12.65 Kbit/s).

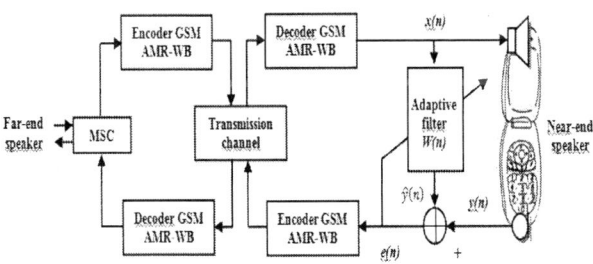

Fig 3. Simulation model of transcoded speech in GSM standard.

For the simulation of AEC system, we used the acoustic echo path plotted in Fig 4. It is real impulse response of cockpit car signal consists of 1024 points, sampled at 16 KHz. We have used two speech signals for far-end speakers, which are drawn from NOIZEUS database [13] (a male speaker (sp01) and a female speaker (sp11)). These signals are sampled at 8 KHz.

The AEC system based on adaptive filter uses NLMS algorithm, with, μ=0.7, ε = 2.2204 × 10⁻¹⁶. The total number of iterations is $N = 18000$. The length of this filter is $L = 1024$ (equal to the length of the impulse response of the echo path).

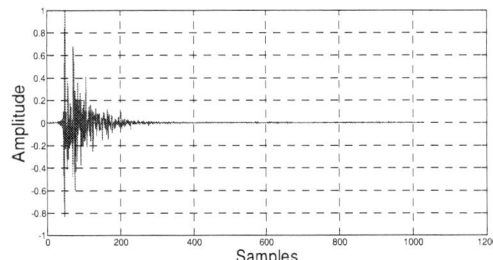

Fig 4. Car cockpit impulse response.

To evaluate the performance degradation of each AEC system, two error measures were used: Echo Return Loss Enhancement (ERLE) and the misalignment (mismatch system). These criteria are given in the following equations:

$$ERLE(dB) = 10 \log_{10} \left(\frac{E\{|y(n)|^2\}}{E\{|e(n)|^2\}} \right) \qquad (12)$$

$$misalignment\ (dB) = 10 \log_{10} \left(\frac{\|W(n) - h\|^2}{\|h\|^2} \right) \qquad (13)$$

where $\|W(n) - h\|$ is the Euclidian distance between the adaptive coefficients vector and the true echo path vector h.

We have used the two speech signal for evaluated the effect of the noisy channel. The speech of the far-end speaker is encoded and decoded using AMR-WB speech codec with 6.60 Kbits/s. We calculate the average of ERLE in the steady state phase, after the convergence of NLMS algorithm. The evaluation results are shown in the Table 1.

TABLE 1. EVALUATION OF ERLE FOR PERFECT AND NOISY CHANNEL

Perfect channel SNR Values of noisy channel		AWGN channel		Rayleigh channel	
		Average ERLE(dB)		Average ERLE(dB)	
		Male	Female	Male	Female
Perfect channel		17.31	19.82	17.57	19.93
Noisy channel	10 dB	16.57	17.29	11.64	13.48
	5 dB	15.28	13.02	11.45	11.74
	0 dB	9.17	11.34	10.92	10.99

Table 1 indicates the degradation of AEC system performances (in terms of maximizing the ERLE) by the noisy channel. This degradation is different for male and female speaker, AWGN channel and Rayleigh channel. These noisy channels cause ERLE to decrease due to the added noise. The comparative evaluation of ERLE between transcoded speech through perfect channel and noisy channel demonstrates the degradation of AEC system performance (decrease in the ERLE), as shown in Fig 5, which is evaluated for each 64 iterations. We have used GSM AMR-WB with bit-rate 6.60 Kbit/s for evaluating the effect noisy channel on the behavior of NLMS algorithm.

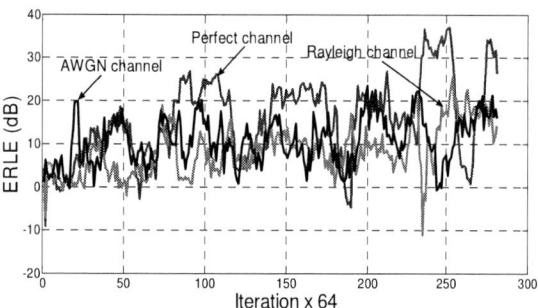

Fig 5. Curves of ERLE for perfect channel and noisy channel, SNR=0 dB, male speaker.

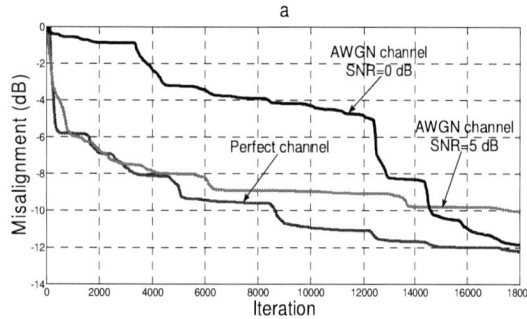

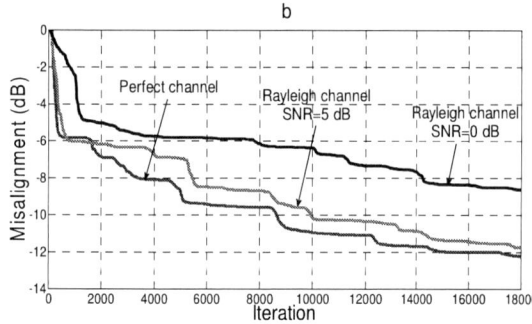

Fig 6. Misalignment evaluation of NLMS algorithm, a) AWGN channel, b) Rayleigh channel.

Fig 6 shows the misalignment evaluation of adaptive filter coefficients for perfect and noisy channel transmission, with two SNR values (5 dB and 0 dB). These results demonstrate the degradation of mismatch system, while the adaptive algorithm has a good performance, on term of error minimizing between adaptive coefficients and the real coefficients of echo path in the perfect channel. This error increases inversely proportional to the SNR value of the noisy channel (AWGN and Rayleigh channel). Thus, these noisy channels can seriously affect the acoustic echo estimation.

V. CONCLUSION

This paper investigates the evaluation of AEC performances and NLMS algorithm behavior for transcoded speech over a noisy channel transmission. The transcoded speech is encoded and decoded using the speech codec GSM AMR-WB and passed through the noisy channel modeled by AWGN channel and Rayleigh fading channel. Comparative evaluation of ERLE between transcoded speech through perfect and noisy channels shows that the degradation of AEC system performance occurs. The obtained results also demonstrate the degradation of mismatch system.

REFERENCES

[1] T. G. Burton, R. A. Goubran, "A generalized proportionate sub-band adaptive second-order Volterra filter for acoustic echo cancellation in changing 4nvironments," IEEE Trans on audio, speech and language processing Vol. 19, No. 08, pp. 2364-2373, 2011.

[2] C. Breining, T. Schertler, "Delay-free low-cost step-gain estimation for adaptive filters in acoustic echo cancellation," Elsevier Science B.V, Signal processing 80, pp. 1721-1731, 2000.

[3] M. Debyeche, A. Krobba, A. Amrouche, "Effect of GSM speech coding on the performance of speaker recognition system," IEEE proc, ISSPA, pp. 137-140, 2010.

[4] E. Pryadi, K. Gandi, H. Y. Kanalebe, "Speech compression using CELP speech coding technique in GSM AMR ," IEEE proc, WOCN, 2008.

[5] K. Chen , P.Y. Xu, J. Lu, B.L. Xu, "An improved post-filter of acoustic echo canceller based on sub-band implementation," Elsevier, Applied Acoustics 70, pp. 886-893, 2009.

[6] J. Arenas-García, A. R. Figueiras-Vidal, "Specch compression using CELP speech coding technique in GSM AMR ," IEEE Trans on audio, speech and language processing Vol. 17, No. 06, pp. 1087-1098, 2009.

[7] S. Haykin, "Adaptive Filter Theory," Third edition, Prentice Hall Inc, New York, 1996.

[8] 3GPP Technical specification TS 26.190. AMR Wideband Speech Codec : Transcoding functions. Available : http://www.3gpp.org

[9] S. K. Wang, Z. H. Yang, Z. Y. Wu, "Transcoding scheme between AMR-WB and VMR-WB," IEEE proc, National natural science foundation of china No. 00072094, 2009.

[10] D. J. Lee, G. H. Jeong, I. S. Lee, "Unbalanced multiple description coding by adding enhancement layer for AMR-WB speech codec in VoIP," IEEE proc, ICACT, pp. 1387-1390, 2008.

[11] O. P. Shaik, K. A. Babu, "OFDM Signal in Fading Transmission Channel of Communication System," International Journal of Advanced Engineering Sciences and Technologies, Vol. 10, No. 2, pp. 185-192, 2011.

[12] ITU-T G.722.2, "Wideband coding of speech at around 16 kbit/s using Adaptive Multi-Rate Wideband (AMR-WB)," Annex C, Fixed-point C-code, 2008.
Available : http://www.itu.int/rec/T-REC-G.722.2/en.

[13] Y. Hu, P. C. Loizou, "Subjective comparison and evaluation of speech enhancement algorithms," Elsevier B.V, speech communication 49, pp. 588-601, 2007.
Available : http://www.utdallas.edu/~loizou/speech/noizeus/.

Fault injection for verifying testability of fault tolerant structures at the Verilog level

G. Ait Abdelmalek
Department of Electronics
Mouloud Mammeri University
Tizi-ouzou, Algeria
e-mail: ghania_79@yahoo.fr

R. Ziani
Department of Electronics
Mouloud Mammeri University
Tizi-ouzou, Algeria
e-mail: ziani_r@yahoo.fr

M. laghrouche
Department of Electronics
Mouloud Mammeri University
Tizi-ouzou, Algeria
e-mail: larouche_67@yahoo.fr

Abstract— **An integrated circuit is said efficient if it is able to perform its intended function with a level of quality and reliability of the highest. To ensure the quality of operation, it is necessary to test and verify these circuits in the early steps of design. This article aims to present a specific test procedure, in order to alleviate the problem of testing the fault-tolerant of structures by minimizing the cost associated with it.**

Keywords- fault, test, testability, ATPG, fault tolerance.

I. INTRODUCTION

The general principle of the test of an integrated circuit is to highlight a possible malfunction due to a physical failure. Given the large number of possible physical failures in a circuit, it is impossible to develop fast and efficient tools that can directly manipulate these failures. The fault models have been developed to represent the defects which may occur in a circuit. The most testing techniques rely on these fault models. To reduce the effect of failures, the solutions proposed by technologists since the year 2000 were initially to replace the aluminum tracks by copper tracks which allow speeding up interconnections between transistors, reducing losses and increasing resistance to electromigration. Along with proposed solutions, traditional methods of fault tolerance "redundancy", and hardening of components are used.

Since, the fault tolerance inherent in the modular design provides a new testing strategy by integrating it in complex circuits. However, the test of these structures requires methods and tools for predictive evaluation of the tolerance fault level induced by the manufacturing process or those relative to their use. We were interested by tools able of performing faults injection in the cycle of design. These tools should allow for example, when one has only the high level model of a circuit, to obtain an upstream assessment of the reliability of this circuit.

Drawing on fault-tolerant TMR structures, we can improve and get a wide reliability with the designed circuits. The design of the test plan begins from the beginning of conception, as soon as the required functions" specifications "are known. This paper will be organized as follows:

Section 2 recalls the principles for assessing fault tolerance. In section 3, the set up an ATPG specific test procedure of benchmarks circuits (ISCAS85 and ITC99) is represented, allowing to bind the design of several electronic circuits, their simulation once realised, and to finish with the test in order to guarantee their robustness and therefore the reliability under multiple simultaneous faults injected at the Verilog level by analyzing experimental results of the calculation of T (fault tolerance probability). Finally, Section 4 gives some final results.

II. TEST OF FAULT TOLERANT IN DIGITAL STRUCTURES

A. General principles

The objective of fault-tolerant in digital structures is to place barriers trying to prevent the presence of a fault which can lead to system failure propagation [1]. These structures have been used and occupy a large place in sensitive areas such as avionics, aerospace, automotive, electronics or military. This type of structure is now used primarily to detect or to correct some faults occurring during operation of the circuit which requires no manufacturing defect. To avoid the weaker yield , the idea is to use these circuits to tolerate potential faults of manufacture.

B. Example of fault tolerant structures

Fault-tolerance of structures most commonly used are called self-testable circuits based on the use of error correcting codes. This tolerance is based on insurance redundancy in the representation of information or in adding bits to control word or by a representation of information words into a new form containing redundancy [2]. Another technique to achieve fault-tolerant structures is based on hardware redundancy [3]. The structure is composed of N identical modules. The outputs of these modules are then compared and the final output of the system is determined by a majority vote system. The simplest design and most widely used is the TMR (Triple Modular Redundancy). As shown in figure 1, it consists of M three identical modules and a voter.

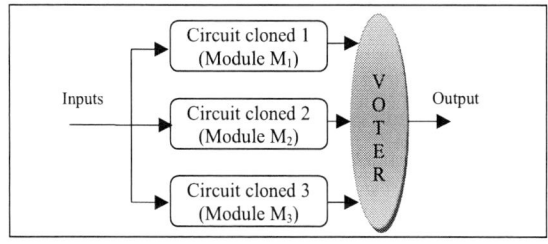

Figure 1: TMR structure

III. Specificity of the Test

To highlight the features of a test structure fault tolerant, we take the example of a TMR structure. In this type of structure, if a stuck at fault is present in a module, the wrong output of this module is masked by the other two non-faulty modules. It is therefore impossible to test errors of simple stuck at fault in a TMR structure since they are not observable output of the voter. The test manufacturing output should not inform us about the possible presence of one or more defects. We only want to know whether with or without defect, the circuit operates. In other words, if there were manufacturing defects, they are tolerated by the structure.

A. How to test these structures?

Generally, when we want to test a circuit, ATPG (Automatic Test Pattern Generator) generates test vectors which can show the presence of a fault. These test vectors are effective in detecting the presence of a manufacturing defect but not for the purpose of detecting multiple faults in a single structure. However, the redundancy brought in fault-tolerant circuits makes them tolerant to certain defects. For example, to make a structure TMR fault, it takes at least two bonding faults present in two different modules and the two faults propagate the error to at least one common outlet. In this case, one or more outputs of the voter majority will be wrong.

The ATPG gives us the FC (Fault Coverage) rate and T (Tolerance probability) or resistance as a pair of faults is tolerated by the structure, defined by:

$$T = \frac{Number\ of\ pairs\ of\ tolerated\ faults}{Total\ number\ of\ faults}$$

B. A new fault model

The ATPG generates test vectors that detect faults of simple collage being more effective, ATPG based on a new fault model is necessary. To continue the example of a TMR test, a fault model for detecting multiple faults such as manufacturing faults bilateral pairs (also called simply faults bilateral) must be introduced:

o   Let Sk, all stuck at faults of a module k (k = 1, 2, 3) of a TMR system.

o   Let R be the set of all pairs $<f_i, f_j>$ faults such as $f_i \in P_k$ and $f_j \in P_k$.

o $<f_i, f_j>$ is a pair of bilateral fault if f_i and f_j are in two differents modules.

C. Observation

All the bilateral faults are not observable at exit:
Let ψ_i, all outputs erroneous because the fault f_i
Let ψ_j, all outputs erroneous because the fault f_j
-   If $\Omega_i \cap \Omega_j = \phi$, then the faults f_i and f_j do not modify the primary output, the bilateral fault is undetectable.
-   If $\Omega_i \cap \Omega_j \neq \phi$, then faults f_i and f_j spread to at least one primary output common: bilateral fault is detectable.

The figure 2 shows three examples of the three modules of a TMR. The third module is considered without defect.

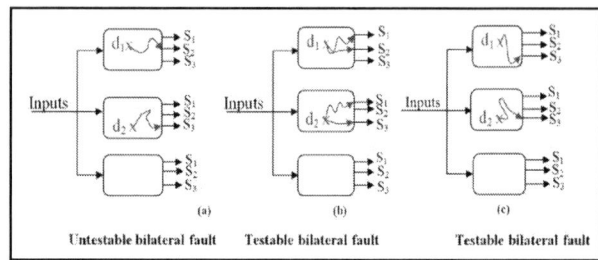

Figure 2: Observability of the pairs of bilateral faults

In the case a, the two faults are not propagated towards any common exit, therefore the bilateral fault is not observable at exit of the voting one. For the cases b and c, the faults are propagated towards at least an exit the common (S_1 and S_2) thus pair of bilateral faults is observable.

D. Characteristic

If each module in TMR structure contains n simple stuck at faults possible, while the total number of faults bilateral between two modules is equal to $2n^2$. If we consider the entire TMR structure with three modules, then the number of faults is equal to bilateral $6n^2$. This number can be reduced by taking into account the fact that the modules are identical and therefore $<f_i, f_j> = <f_j, f_i>$. Finally, the full number of bilateral faults is $3n^2$. The number of bilateral faults is thus quadratic compared to the number of simple faults. It is thus imperative to have a powerful ATPG detecting the bilateral faults effectively to test a structure TMR with a reasonable number of vectors of test.

IV. Experiments

To calculate the T probability of circuit, we used initially, the simulator "ModelSim PE Student Edition 6.6 b" of the company Mentor Graphics. We took the example of the "counter/decounter" presented by Figure 3 and described in VHDL description language using software ISE 9.2.

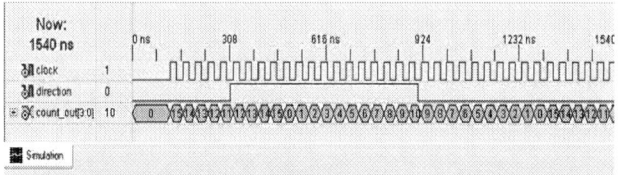
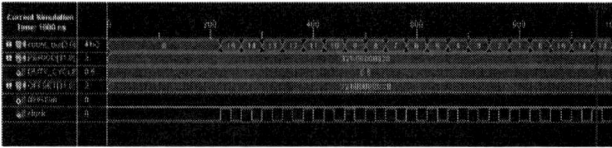

Figure 3. Circuit simulation without faults

To see the behavior of the circuit in the presence of faults, we injected the (s@0, s@1) stuck at fault by forcing the external points of the counter/decounter.

As shown on Figure 4, no information on the number of faults untestables is obtained; therefore it is impossible to calculate the probability T. This type of software also does not make it possible to inject faults on the level of the internal lines of the circuit. To calculate T probability we then used in the second time the software TetraMax Synopsys [3-6]. This works with a model in the form of primitives "AND, OR..."It is a very powerful tool for fault simulation and generation of test vector.

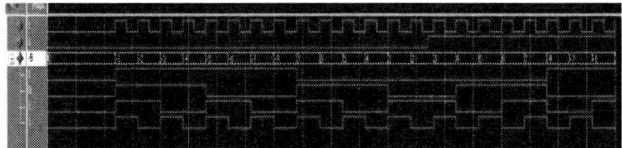

Figure 4. Circuit simulation with fault

To understand the functioning of the ATPG, we present its implementation on a simple example based on the c17 circuit of ISCAS85series.

A. The flood of using Tetramax

The generation of test vectors contains eight phases [5, 6].

- Phase 1: Starting Tetramax

```
Type stsyn 2005: the configuration script
Type tmax32: starting the Tetramax software in console mode
```

- Phase 2: reading the description of circuits

```
Read Netlist <file_name> [Format<Edif Verilog Vhdl>]
[-Sensitive  -INSensitive] [-Delete] [-Library][-Tolerant_circuits][-
Noabort][Verbose]
      Exemple: > read netlist. / ISCAS85/ c17 .v
```

The Figure 5 gives us the original c17 verilog format.

```
1 module c17 (E1, E2, E3, E4, E5, S1, S2);
2 input E1, E2, E3, E4, E5;
3 output S1, S2;
4 nand P1 (.Z (N1), .A (E1), .B (E3));
5 nand P2 (.Z (N2), .A (E3), .B (E4));
6 nand P3 (.Z (N3), .A (E2), .B (N2));
7 nand P4 (.Z (N4), .A (N2), .B (E5));
8 nand P5 (.Z (S1), .A (N1), .B (N3));
9 nand P6 (.Z (S2), .A (N3), .B (N4));
10     end module
```

Figure 5. Verilog format of the original c17 circuit (without DFR)

- Phase 3: Description of TMR structure

To manufacture TMR structures, we use verilog descriptions [4, 5] of the circuits. For example, the Figure 6 described the verilog format of c17 TMR structure. The verilog format of voter is described by Figure 7. A voter has three inputs and one output. It consists of three AND gates and an OR gate with three inputs: $S_1 = E_1.E_2 + E_1.E_3 + E_2.E_3$.

```
1 module TMRc17 (E1, E2, E3, E4, E5, S1, S2);
2 input E1, E2, E3, E4, E5;
3 output S1, S2;
4 c17 T1 (E1, E2, E3, E4, E5, S1M1, S2M1);
5 c17 T2 (E1, E2, E3, E4, E5, S1M2, S2M2);
6 c17 T3 (E1, E2, E3, E4, E5, S1M3, S2M3);
7voter V0 (S1M1, S1M2, S1M3, S1);
8 voter V1 (S2M1, S2M2, S2M3, S2);
9 end module
```

Figure 6. Verilog format of the c17 circuit modified in TMR

```
1 module voter (E1, E2, E3, S1);
2 input E1, E2, E3;
3 output S1;
4 and gate_1 (L1, E1, E2);
5 and gate_2 (L2, E1, E3);
6 and gate_3 (L3, E2, E3);
7 or gate_4 (S1, L1, L2, L3);
8 end module
```

Figure 7. Verilog format of voter

- Phase 4: Creation of the list of collage fault of a module is shown in Figure 8.

1	sa1	NC	P_5/Z	19	sa0	NC	P_3/Z
2	sa0	--	P_5/A	20	sa1	NC	P_3/B
3	sa0	--	P_5/B	21	sa1	NC	P_3/A
4	sa0	--	P_1/Z	22	sa1	NC	P_3/Z
5	sa1	NC	P_5/B	23	sa0	--	P_3/A
6	sa0	NC	P_5/Z	24	sa0	--	P_3/B
7	sa1	NC	P_1/B	25	sa1	NC	P_6/Z
8	sa1	NC	P_1/A	26	sa0	--	P_6/A
9	sa1	NC	P_1/Z	27	sa0	--	P_6/B
10	sa0	--	P_1/A	28	sa0	--	P_4/Z
11	sa0	--	P_1/B	29	sa1	NC	P_6/A
12	sa1	--	P_5/A	30	sa0	NC	P_6/Z
13	sa0	NC	P_2/Z	31	sa1	N	P_4/B
14	sa1	NC	P_2/B	32	sa1	NC	P_4/A
15	sa1	NC	P_2/A	33	sa1	NC	P_4/Z
16	sa1	NC	P_2/Z	34	sa0	--	P_4/B
17	sa0	--	P_2/A	35	sa0	--	P_4/A
18	sa0	--	P_2/B	36	sa1	--	P_6/B

Figure 8. List of simple stuck at fault of the circuit c17

- Phase 5: Injecting a fault pivot

As shown in figure 9 the fault pivot is a simple stuck at fault injected by modification of the netlist (gate array and connections).

```
1    /*stuck-at 0 of the output of P2 */
2    module c17modifie_13 (E1, E2, E3, E4, E5, S1, S2);
3    input E1, E2, E3, E4, E5
4    output S1, S2;
5    nand P1 (.Z (N1), .A (E1), .B (E3));
6    nand P2 (.Z (Nremp), .A (E3), .B (E4));
7    nand P3 (.Z (N3), .A (E2), .B (N2));
8    nand P4 (.Z (N4), .A (N2), .B (E5));
9    nand P5 (.Z (S1), .A (N1), .B (N3));
10   nand P6 (.Z (S2), .A (N3), .B (N4));
11   assign N2=0;
12   end module
```

Figure 9. Injection of S@0 on the P_2 gate exit

- Phase 6: Commands Tetramax script

The Tetramax script is created to automate the testing of the entire TMR structure. A part of c17 circuit script is shown in

Figure 10. This is the first ATPG to start treating all pairs of faults {f_1, f_2} with f_1 = S@1 of P_5/Z. Also, the Figure 11 shows the verilog format of TMR structure.

```
1    read netlist /auto/iscas85/circuits/c17/c17modifie_1.v
2    read netlist /auto/iscas85/circuits/c17/c17.v
3    read netlist /auto/iscas85/circuits/c17/TMRc17.v
4    read netlist /auto/iscas85/circuits/lib_comb.v
5    run build_model TMRc17
6    run drc
7    add faults T2/P5/Z -stuck 1
8    add faults T2/P5/B -stuck 1
9    add faults T2/P5/Z -stuck 0
10   add faults T2/P1/B -stuck 1
11   add faults T2/P1/A -stuck 1
12   add faults T2/P1/Z -stuck 1
13   add faults T2/P2/Z -stuck 0
14   add faults T2/P2/B -stuck 1
15   add faults T2/P2/A -stuck 1
16   add faults T2/P2/Z -stuck 1
17   add faults T2/P3/Z -stuck 0
18   add faults T2/P3/B -stuck 1
19   add faults T2/P3/A -stuck 1
20   add faults T2/P3/Z -stuck 1
21   add faults T2/P6/A -stuck 1
22   add faults T2/P4/A -stuck 1
23   set atpg -abort limit 100
24   set atpg -decision NOR
25   set atpg -norandom _fill
26   run atpg -ndetects 1
27   write patterns
/auto/iscas85/circuits/patterns/c17/pat1.txt –  Internal -format
verilog_ single _file -parallel 0
28   drc -force
```

Figure 10. Tetramax script describing an ATPG

```
1    module TMR c17 (E1, E2, E3, E4, E5, S1, S2);
2    input E1, E2, E3, E4, E5;
3    output S1, S2;
4    c17modifie_1 T1 (E1, E2, E3, E4, E5, S1M1, S2M1);
5    c17 T2 (E1, E2, E3, E4, E5, S1M2, S2M2);
6    c17 T3 (E1, E2, E3, E4, E5, S1M3, S2M3);
7    voteur V0 (S1M1, S1M2, S1M3, S1);
8    voteur V1 (S2M1, S2M2, S2M3, S2);
9    end module
```

Figure 11. Verilog description of TMR structure

- Phase 7: Launch of ATPG

The stage of launching of the totality of the ATPG is the longest stage. It can vary from ten seconds for the circuit C 17 at several days for larger circuits, such as C 7552.

- Phase 8: Analysis and synthesis of results

The Figure 12 shows the results of c17 circuit ATPG, the T probability can thus be known in following: T= number of pairs of faults tolerated / total number of faults = 9/16=56.25%. We repeated the same test procedure for the other benchmark circuits (ISCAS85 and ITC99).

```
// Module tested: TMRc17
//    Uncollapsed Stuck Fault Summary Report
// fault class          code  #faults
// Detected             DT       7
// possibly detected    PT        0
// Undetectable         UD       9
// ATPG untestable      AU        0
// Not detected         ND        0
// total faults         16
// test coverage        100.00%
//          Pattern Summary Report
// #internal patterns            4
//    #basic scan patterns       4
```

Figure 12. Result of an ATPG

V. CONCLUSION

The testing technique presented in this article demonstrates the important role that can have the fault injection early in the design process in future and current CMOS technology to the analysis and evaluation of predictive reliability. In this way, such an analysis is essential to determine the test strategy and tolerance of the most suitable (standard mechanisms to implement and to protect sites) to avoid the risk of discovering that after making the level of reliability of the circuit is not in conformity with the specifications and allow to reduce the cost and TTM (time to market).

The redundancy introduced by the fault tolerance of a structure makes the test different from a conventional structure where all redundancy is normally absent and for which a structural test can be applied effectively. On the other hand, for structures that have a significant level of hardware redundancy as 5 MR, 7MR or more, the number of no testable faults it is important. This requires a new ATPG working with new fault models to test these structures efficiently. Today, with the goal of design for reliability and fault tolerance, the testing technique must have a different role. It must accept the presence of one or more defects in the structure and tolerates them to participate in the final goal of increasing performance products.

REFERENCES

[1] L. Anghel : Les limites technolgiques du silicium et tolérance aux fautes, Rapport de thèse, 2000.

[2] M. Hafezparast, "tolerant Fault hardware designs and to their reliability analysis", thesis of doctorate, Brunel University of west London, 1990

[3] http://www.vtvt.ece.vt.edu/vlsidesign/cadtools.php

[4] Layout and parasitic information for ISCAS circuits, dropzone.tamu.~edu/.../iscas. html-Etats-Unis, 1980.

[5] http://www.cad.polito.it/download/tools/itc99.html

[6] A. Khouas, "Test de systèmes électroniques" fascicule de cours, École polytechnique Montréal, Automne 2007.

Design and Simulation Methodology for Switch-Cap Circuits Used in Data Converter Applications

Sanad Kawar, Mahmood Mohammed, Khaldoon Abugharbieh
Electrical Engineering Department
Princess Sumaya University for Technology
Amman, Jordan

Abstract— **This paper provides a full methodological approach to designing and verifying differential sample and hold switched-capacitor circuits used in A/D converters. It provides a step-by-step process for translating system requirements such as signal-to-noise ratio (SNR) and sampling frequency into A/D requirements and subsequently into op-amp specifications. It also includes the design process of a switched-capacitor common mode feedback circuit (CMFB) to control the common mode output voltage. Furthermore, this paper provides practical methods for verifying the stability of the system. A design and simulation example for a differential sample and hold switched-capacitor circuit operating in a system requiring 5 MHz sampling frequency and a 6-bit A/D converter is provided. Mentor Graphics CAD tools were used in the design and simulations process using 180nm CMOS device models. This paper can be used as a resource for design engineers in the industry as well as universities teaching graduate level advanced electronics and data converter courses.**

Keywords-component; sample and hold; A/D ; op-amp; switched-capacitor ; common mode feedback ; stability

I. INTRODUCTION

Sample and hold circuits are considered the core components of A/D converters [1]. Their main role is tracking an analog input and storing its value at sampling frequency instances for a specific time length until the A/D converter can complete the conversion of the input value [2]. Due to their critical role in proper and stable data conversion, care must be taken in the design of each of their components. The op-amp used in the circuit must be stable and capable of operating within system requirements. The common mode feedback circuit and the sample and hold circuit itself are switched-capacitor circuits. Due to their non-linear nature, they require special attention in their design and stability verification methods.

The paper is organized as follows. Section II discusses a general sample and hold circuit. Section III contains a design flowchart and explains how to translate system requirements into A/D specifications. Section IV explains the design of an open-loop amplifier capable of satisfying system requirements as well as DC and AC design parameters. Section V explains the common mode feedback design, whereas section VI shows the sample and hold system. Section VII explains in detail how to check the system's stability. To illustrate the design process, an example sample and hold circuit is designed to operate in a 6-bit pipeline A/D converter. The A/D converter

will be used in a system requiring 37dB SNR and 5MHz sampling frequency. It has a maximum differential input of 1V peak to peak. Each section contains the corresponding design example. Simulation results are shown in section VIII.

This paper provides a detailed resource for design engineers and universities teaching graduate level advanced electronics courses. It guides engineers and students starting with system requirements in choosing and designing all components of a sample and hold circuit. It also provides two clear methods for checking the stability of the system in transient simulations.

II. SAMPLE AND HOLD CIRCUIT

For any A/D system to work properly, a sample and hold circuit is likely used as the first stage. Fig. 1 below shows a block diagram of a switched-capacitor sample and hold circuit, including common mode feedback circuitry. During the sample stage (ϕ1), capacitor C_{IN} will hold a charge equal to the product of its value and the input voltage. During the hold stage (ϕ2), the charge stored on C_{IN} will be transferred to C_F due to charge conservation. This causes the output to track the input with a closed-loop gain as shown in (1). The bias block provides bias currents and voltages for the op-amp, whereas the CMFB controls the common mode output voltage.

$$V_{o,differential} = \frac{C_{IN}}{C_F} V_{in,differential} \qquad (1)$$

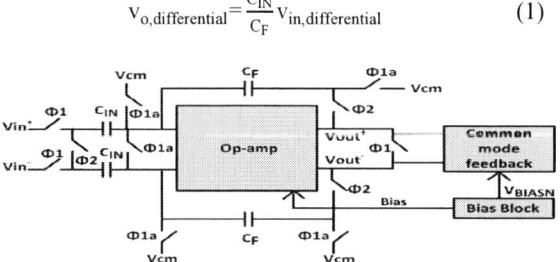

Figure 1: System block diagram

III. DESIGN PROCEDURE

Design engineers are usually given system specifications in terms of SNR, sampling frequency and input voltage range. It is vital to know how to translate these requirements into a specific A/D topology and resolution. Subsequently op-amp specifications, like open-loop gain and unity-gain frequency are determined.

Fig. 2 below is a full design flowchart, showing a step-by-step guide of the design process. The first step is determining

978-1-4673-5289-5/12 $31.00 © 2012 IEEE

the number of bits, which is determined by the minimum SNR requirement of the system.

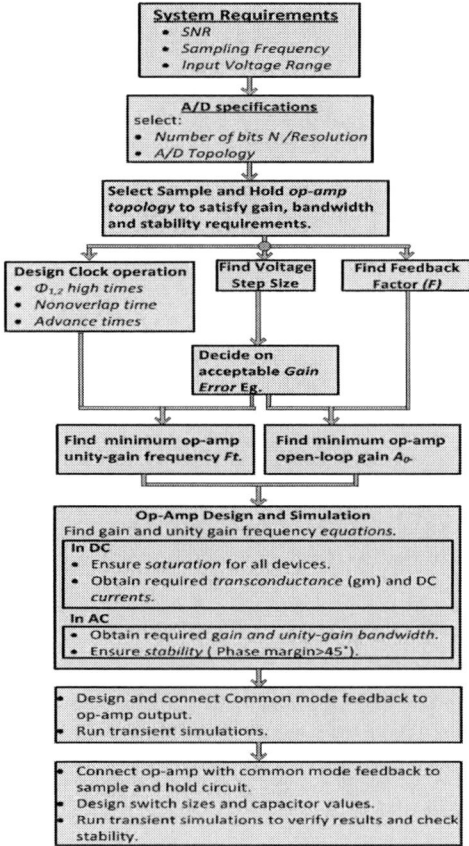

Figure 2: Design procedure flowchart

Equation (2) relates the system SNR requirement directly to the number of bits N [2]. Following that, an A/D topology is selected based on the required resolution and sampling frequency as shown in Fig. 3 [3]. For the example in this paper, a 6-bit pipeline A/D converter is used, because of its ability to satisfy system requirements.

$$SNR(dB) = 6.02N + 1.76dB \qquad (2)$$

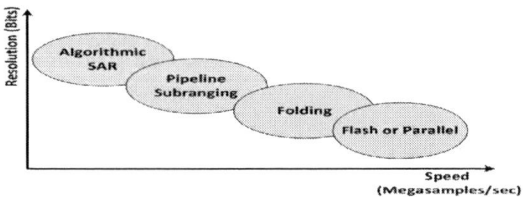

Figure 3: A/D topology selection

IV.   OP-AMP DESIGN

A. Topologies

The first step in op-amp design is to select a topology that is capable of satisfying gain and bandwidth requirements. Commonly used topologies include the two-stage amplifier

and the folded cascode amplifier. For the design example, the folded cascode shown in Fig. 4 was used. This is due to its ability to provide high gain while easily maintaining stability without the use of any additional compensation circuitry. It is also necessary at this point to select a current source to provide bias currents and voltages for both the op-amp and the CMFB. The wide-swing cascode current mirror in [4] was used in the example. It provides the bias voltages v_{BIASP}, bp2 and bn1 to the op-amp and therefore determines the currents passing through its branches. It also provides bias voltage v_{BIASN} to the CMFB circuit to be discussed in Section V.

After selecting topologies and deriving gain and bandwidth equations for the selected circuits, a key step in the design is to translate the system and A/D requirements into minimum open-loop op-amp specifications. This must start with proper clocks operation.

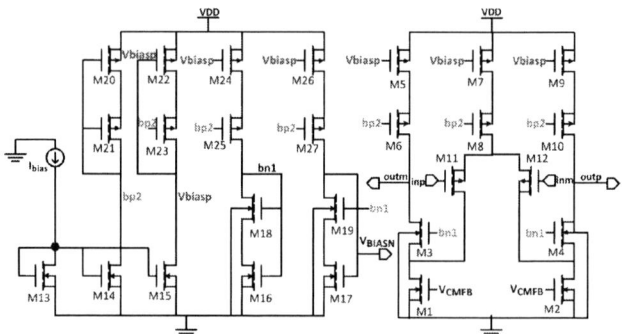

Figure 4: Folded Cascode op-amp with biasing

B. Clocks Operation

There are three clocks operating the circuit: $\phi1$, $\phi1_A$ and $\phi2$. $\phi1$'s high time is the sampling (tracking) time, whereas $\phi2$'s high time is the hold time. $\phi1_A$ is an advanced $\phi1$ clock and is used to minimize non-ideal charge sharing issues. The circuit operates in an entirely different way between these two stages. Therefore, it is the designer's job to make sure $\phi1$ and $\phi2$ never overlap. $\phi1$ high time is calculated according to (3) below. Suitable non-overlap and advanced times are selected.

$$t_{\phi1\text{-high}} = \frac{t_{period}}{2} - t_{non\text{-}overlap} - t_{advanced\ time} - t_{rise\ and\ fall} \qquad (3)$$

C. Open-loop amplifier specifications

After determining the sampling time, the least significant bit (LSB), the gain error (Eg) and the feedback factor (F) must be calculated. The LSB is calculated according to (4), where N is the number of bits and $V_{pp\text{-differential}}$ is the maximum differential input voltage. Following that, Eg must be defined as a fraction (1/x) of the LSB. The value selected for x depends on speed and accuracy requirements of the system. The next step is using (5) to calculate τ_{max} as shown in (6). Then, the minimum unity-gain bandwidth (ft_{min}) can be found using (7). It must be noted that the time (t) in (5) and (6) is the settling time of the sample stage only, and is considered to be 80% of $t_{\phi1\text{-high}}$ found in (3) (i.e. excluding 20% for slew time). A_0 is the open-loop op-amp gain.

Finally, using the right side of (5), and substituting F from (8), the minimum op-amp open-loop gain A_o can be calculated. For the design example, a closed-loop gain of 2 is desired, and x in (5) is selected to be equal to 2. $t_{\phi1\text{-high}}= 96.8nS$, LSB= 15.625mV and F=0.3125. $A_{o,min}$ and ft_{min} were then calculated to be 46.2dB and 19.94 MHz respectively.

$$LSB = \frac{V_{PP\text{-dfferential}}}{2^N} \qquad (4)$$

$$e^{-\frac{0.8 t_{\phi1-high}}{\tau}} \leq Eg = \frac{1}{A_0 F} = \frac{1}{x} LSB \qquad (5)$$

$$\tau_{max} = \frac{-0.8\, t_{\phi1-high}}{\ln\left(\frac{1}{x}lsb\right)} \qquad (6)$$

$$ft_{min} = \frac{1}{2\pi\tau_{max}} \text{x Closed-loop gain} \qquad (7)$$

$$F = \frac{Cf}{Cf + Cs + Cparasitic} \qquad (8)$$

D. Design and Simulations

Now that the op-amp specifications are set, the design process must start at DC. The FETS are sized to ensure all devices are operating in saturation, and obtain the required transconductances and DC currents. At first, devices M5-M10 in Fig. 4 are sized such that their ratios to the corresponding FETS in the current mirror can set the current passing through each branch. The rest of the FETS are sized afterwards. Following that, AC simulations are performed, and the design is tweaked to obtain the required gain and bandwidth. It is important to verify that the open-loop system is stable by verifying that the phase margin is above 45° at the unity-gain frequency. In order to obtain a valid AC simulation, the CMFB must not be connected at this stage of the design process for the reasons explained in section VII. To properly bias the op-amp without including the CMFB, the gates of M1 and M2 in Fig. 4 should be connected directly to the bias voltage V_{BIASN}. Example AC simulations are shown in section VIII.

V. COMMON MODE FEEDBACK DESIGN

The op-amp and the sample and hold design will affect the differential signal, but will not set common mode voltages. Therefore, it is necessary to add additional common mode feedback circuitry to control the common mode output voltage at a required value, usually halfway between the power supplies [2]. An example of a switched-capacitor CMFB circuit is shown in Fig. 5. The switched capacitor Cs is switched between fixed bias voltages V_{CMREF} and V_{BIASN} during $\phi1$ and being in parallel with the non-switched capacitor Cc during $\phi2$. This allows Cs to determine the DC voltage across Cc by altering the control voltage V_{CMFB} to keep the common mode output voltage constant [2]. The bias voltage V_{CMREF} is a low impedance reference voltage that sets the value of the common mode output voltage. V_{BIASN} is a fixed bias voltage supplied by the bias block. The switched capacitor along with the switches on its sides will operate as an equivalent resistance determined by (9), where fs is the sampling frequency. As a result, the equivalent circuit shown in Fig. 6 will effectively act as an RC circuit. The common mode feedback may possibly control the dominant pole of the system. Therefore, care must be taken when selecting capacitor values. The 3-dB bandwidth of the

common mode feedback is given in (10), where f_S is the sampling frequency.

$$R_{equivalent} = \frac{1}{fsCs} \qquad (9)$$

$$f_{3\text{-dB}} = \frac{fs\ Cs}{2\pi Cc} \qquad (10)$$

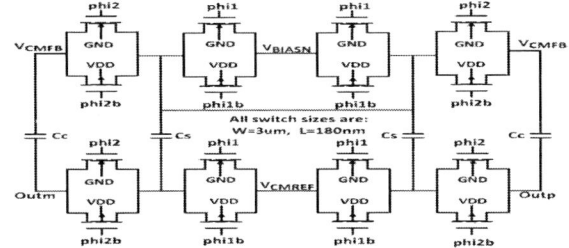

Figure 5: Common mode feedback circuit

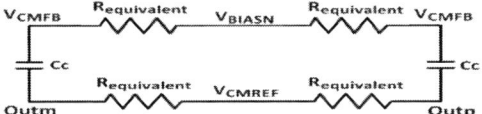

Figure 6: Common mode feedback equivalent circuit

VI. FINAL SAMPLE AND HOLD CIRCUIT

The current mirror, op-amp and common mode feedback should now be inserted into the sample and hold configuration. Fig. 7 below is the final sample and hold schematic used in the design example. The Capacitors' values must be selected to satisfy (1).

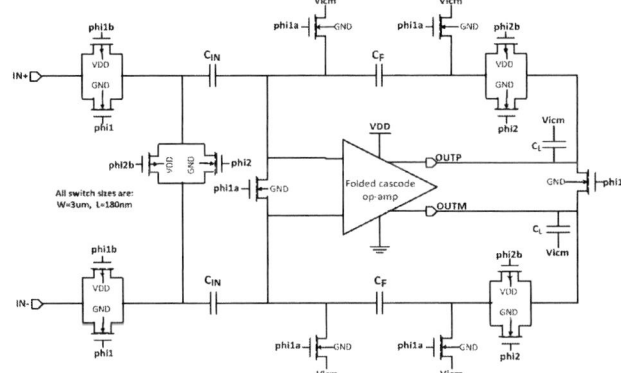

Figure 7: Sample and Hold schematic

VII. SYSTEM STABILTY

Stability describes the ability of a system to withstand variations in the conditions under which the circuit is operating [5]. Stability is typically checked using AC simulations [6]. However, such methods cannot provide correct stability information about sample and hold circuits because AC simulations do not account for the switching operation in the switched-capacitor circuits. Moreover, the AC simulation stability check used in [6] will face an issue with the capacitors in the feedback loop, since these capacitors will block the DC signal. To verify the stability of the switched-capacitor sample and hold circuit, two methods are presented

978-1-4673-5289-5/12 $31.00 © 2012 IEEE   265

using transient simulations. The first method is applying a step voltage to the power supply (VDD). The second method is applying a pulse step current at the input of the op-amp. In both methods, the system is considered stable if the output responds smoothly to the changes in the operating conditions, i.e., without overshooting and ringing. In order to differentiate between the system's response to variations in the input and its response to the applied changes, the inputs are replaced with DC voltage sources. Section VIII includes simulations for the first method verifying the stability of the design example.

VIII. SIMULATIONS

A. Open-Loop Simulations

Fig. 8 shows the differential input connected directly to the op-amp as should be done for open-loop simulations. Fig. 9 shows the AC simulations of the open-loop op-amp's output magnitude and phase. The gain and unity-gain frequency are 53.59dB and 68.24MHz respectively. The phase margin is 48.21° indicating the stability of the open-loop op-amp.

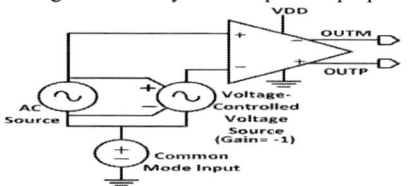

Figure 8: Open-loop Simulation Schematic

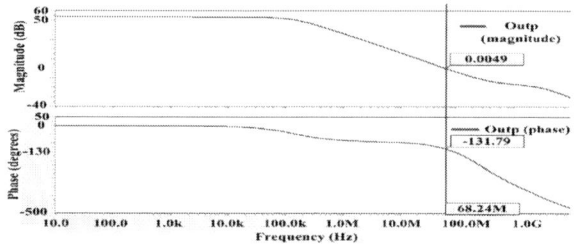

Figure 9: AC simulation of open-loop op-amp output magnitude and phase

B. Closed-loop Simulations

Fig. 10 shows transient simulations for the sample and hold differential and single ended outputs. The circuit is operating at 5MHz and provides a closed-loop gain of 2. Fig. 11 graphically represents both stability methods on the sample and hold schematic. An arrow on the output waveform indicates where the stability verification methods should be applied and observed. Fig. 12 shows the effect of the VDD step on the output. Stable and less stable responses are shown.

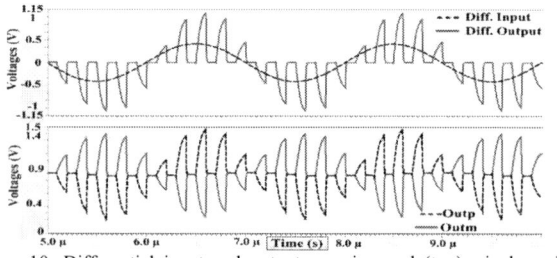

Figure 10: Differential input and output superimposed (top), single ended outputs superimposed (bottom)

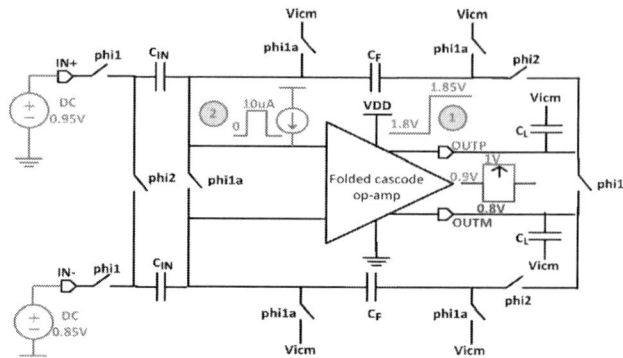

Figure 11: Stability verification methods

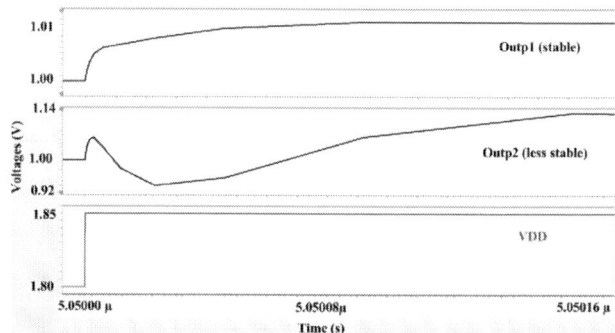

Figure 12: VDD step (bottom), stable (top) and less stable responses (middle)

IX. CONCLUSIONS

In this paper, the design process of a sample and hold circuit was fully presented. A/D topology selection and translation of system requirements into op-amp specification were discussed. In addition, the design of the op-amp, the common mode feedback and the sample and hold circuit were detailed. Furthermore, two methods for checking the stability of the closed-loop system using transient simulations have been implemented. Finally, a design example including simulations has been provided. This paper provides a guide for design engineers and graduate students on how to approach non-linear switched-capacitor circuit design.

REFERENCES

[1] S. Haddadian, and R. Hedayati, 'A Unity Gain Fully-Differential 10bit and 40MSps Sample-And-Hold Amplifier in 0.18μm CMOS' *Proc. World Academy of Science, Engineering and Technology 47*, pp. 467–472, 2008.

[2] David A. Johns, and Ken Martin, 'Analog Integrated Circuit Design'. United State: John Wiley & sons, 1997., pp. 287, 291, 342, and 451.

[3] P. Prasad Rao, K. Lal Kishore 'Optimizing the Number of Bits/Stage in 10-Bit, 50Ms/Sec Pipelined A/D Converter Considering Area, Speed, Power and Linearity' *Proc. World Academy of Science, Engineering and Technology 62*, pp. 491–497, Feb. 2012.

[4] T. Fukumoto, H. Okada, and K. Nakamura, 'Optimizing Bias-circuit Design of Cascode Operational Amplifier for Wide Dynamic Range Operations' *Proc. ISLPED '01*, pp.305-309, California-USA, Aug. 2001.

[5] Arieh L. Shenkman, 'Transient Analysis of Electric Power Circuits Handbook'. Netherlands: Springer, 2005., pp.517-518.

[6] Adel S. Sedra, and Kenneth Smith, 'Microelectronic Circuits'. 5th ed., New York: Oxford University Press, 2004., pp.855-859.

Particle Swarm Optimization for the Identification of Worst Case Test Vectors of Total-Dose Induced Leakage Current Failures in ASICs

M. M. Abdel-Aziz, H. A. Abdel-Aziz, A. A. Abou-Auf , and A. G. Wassal

Abstract—We develop a methodology for identifying worst-case test vectors necessary to detect leakage current failures in standard-cell based ASIC devices exposed to a total ionizing dose. We developed a novel search methodology based on the Particle Swarm Optimization technique.

Index Terms— Particle Swarm Optimization, PSO, CMOS, total dose, leakage current, worst-case, test vectors.

I. INTRODUCTION

The total-dose testing standard, MIL-STD-883, method 1019, emphasizes the use of worst-case test vectors (WCTVs). However, they are typically not used in total-dose testing of ASIC devices because they are known to be almost impossible to identify for most VLSI devices. We have already developed a methodology to identify WCTV for leakage-current failure induced in standard-cell ASICs [1]-[5]. This methodology was based on gate-level or cell-level fault models which enabled us to represent the device under test (DUT) at those levels instead of the regular method of transistor-level presentation. This methodology follows the typical design flow of ASIC devices using standard cell library. It also does not depend on the hardware descriptive language (HDL) such as VHDL used for design entry, ASIC design tools, target cell library, or the fabrication process. However, the methodology was limited to ASIC devices with small number of gates and/or input pins because exhaustive search methods for WCTV are used and the corresponding search time for WCTV is exponentially proportional to twice the number of the primary inputs of the DUT. In this paper, we apply the Particle Swarm Optimization (PSO) search technique to generate the WCTV for ASIC devices exposure to total ionizing dose. The execution time of the applied search technique is reasonable even with ASIC devices with number of gates and/or large number of input vectors. We prefer to use PSO in this paper than genetic algorithm (GA) because PSO is easy to implement and there are few parameters to adjust. PSO has been successfully applied in many areas: function optimization, artificial neural network training, fuzzy system control, and other areas where GA can be applied. Also, GA

has many problems like stuck at local optimum.

II. TOTAL IONIZING DOSE EFFECT ON MOS TRANSISTOR

Total ionizing dose (TID) affects MOS transistors and causes several types of damage in the circuit behavior due to TID induced charges. These charges change the physical parameters of MOS transistor such as the transistor threshold voltage V_{th}. The TID effect is strictly depending on the transistor biasing during its exposure to TID, where the transistor counters a significant degradation, such as threshold voltage shift, if it is biased high "1" during the TID radiation.

III. TID LEAKAGE CURRENT FAILURE ANALYSIS

The leakage current for a MOS transistor corresponds to the drain current, when the device is in the cutoff or subthreshold region of operation as expressed in the following equation.

$$I_{ds} = I_o W/L[1 - \exp(-V_{ds}/v_t)]\exp((V_{gs} - V_{th} - V'_{off})/nv_t) \quad (1)$$

where V_{th} is the threshold voltage, I_o and V'_{off} are constants, and v_t is the thermal voltage and n is the subthreshold swing parameter.

We note that I_{ds} increases exponentially with a decrease in V_{th} and also has linear dependency of the leakage current on channel width W (channel length L is usually set to the minimum feature size). Contemporary deep submicron CMOS devices exhibit negative shifts in V_{th} due to oxide-trap charge built up due to total ionizing dose radiation found in extraterrestrial space environment [2]. The total-dose-induced leakage current can be several orders of magnitudes higher than the pre-irradiation leakage current as the example shown in Fig. 1 for the CMOS inverter.

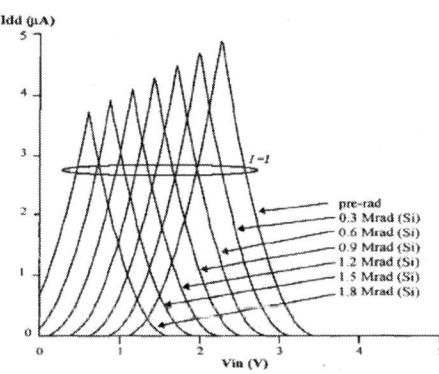

Fig. 1 Negative shifts in Vth in CMOS Inverter

Mostafa M. Abdel-Aziz, mostafa_m@aucegypt.edu, (O): +20 (2) 2615-2727
Hamzah A. Abdel-Aziz, hamzah@aucegypt.edu, (O): +20 (2) 2615-4364
Ahmed A. Abou-Auf, aabouauf@aucegypt.edu, (O): +20 (2) 2615-3048
Electronics Engineering Dept., American University in Cairo, AUC Avenue, P.O. Box 74, New Cairo, 11835, Egypt, , Fax: +20 (2) 2795- 4956

Amr G. Wassal, a.wassal@ieee.org, (O): +20 (10) 871-5559
Computer Engineering Department, Cairo University, Cairo, Egypt

978-1-4673-5289-5/12 $31.00 © 2012 IEEE

IV. FAILURE MODELING

Fault modeling is an abstraction of a given failure at a certain level of circuit representation that is performed to simplify the process of generating test vectors [6].

In earlier work [3][4][5], we introduced gate-level fault model for leakage current failure to be used in the generation of WCTVs. We modeled the leakage current failure for standard ASIC CMOS cells (such as INV, NOR, and NAND) used in combinational circuits. Assuming a CMOS logic cell of m inputs then $\mathbf{I} = [I_1 I_2 ... I_m]$ and $\mathbf{P} = [P_1 P_2 ... P_m]$, where $I_j \in \{0,1\}$ and $P_j \in \{0,1\}$ are irradiation and postirradiation vectors; respectively. For example, the fault model for inverter and 2-input NOR gate (nor2) expresses a normalized measure of the leakage current induced by total dose, I_L, at the logic gate level as

$$I_L(\mathbf{I},\mathbf{P})_{inv} = \left(I \; \overline{P} \; W_n\right) \quad (2)$$

$$I_L(\mathbf{I},\mathbf{P})_{nor2} = \left(I_1 \overline{P_1} \; \overline{P_2} \times W_{n1} + I_2 \overline{P_2} \overline{P_1} \times W_{n2}\right) \quad (3)$$

where W_n is the channel width of the NMOS transistors. In the recent effort [7], we also discussed field-oxide leakage current which is caused by the inversion of the substrate under the field oxide and how it can be modeled as parasitic transistors connected parallel to the main transistor and its parameters are extracted from the geometry of the cells. Conceptually, the above equations can be modified to include the effect of field-oxide leakage current failure as

$$I_L(\mathbf{I},\mathbf{P})_{inv} = \left(I \; \overline{P} \times \sum_i C(i) W_n\right) = C \times \left(I \; \overline{P} \times W_n\right) \quad (4)$$

$$I_L(\mathbf{I},\mathbf{P})_{nor2} = \left(I_1 \overline{P_1} \; \overline{P_2} \times \sum_i C(i) W_{n1} + I_2 \overline{P_2} \overline{P_1} \times \sum_i C(i) W_{n2}\right)$$
$$= C \times \left(I_1 \overline{P_1} \; \overline{P_2} \times W_{n1} + I_2 \overline{P_2} \overline{P_1} \times W_{n2}\right) \quad (5)$$

where $C = \sum_i C(i)$ is a factor that varies exponentially with total dose. And $C(i)$ is a function of the threshold voltage $V_t(i)$ and other transistor parameters in the subthreshold region for the ith parasitic transistor.

We can summarize the fault model condition of the leakage current failure as:

1) At least one input of the cell has irradiation input $I = 1$ and postirradiation input $P = 0$. The condition $I = 1$ ensures worst-case bias during irradiation and consequently maximum V_{tn} shift of the corresponding NMOS. The condition $P = 0$ attempts to turn off the corresponding NMOS. If those transistors are looked upon as switches, total dose degradation will cause them to leak and thus modeled as stuck-on transistors.

2) Those NMOS-PMOS pairs that have no stuck-on NMOS transistors are set so that a connection path between V_{DD} and ground is established. This means that series transistors should be ON, and parallel transistors should be OFF. This condition manifests the leakage current in stuck-on pairs due to total dose.

3) In order to manifest the leakage current failure in the CMOS gate, the postirradiation input vector $\mathbf{P}$ should be set to attempt to force the output to logic 1.

Those fault models are then validated against SPICE simulation using transistor parameters from the target process and transistor parametric degradation from total-dose experiments [9]. We then develop a package of VHDL/Verilog functions that implements those models. The leakage-current fault model for the cells is validated using transistor-level SPICE simulation. We use the transistor model extracted from TSMC 0.35µ process and the total-dose effect on this process reported earlier [9].

V. PSO FOR GENERATING THE WORST-CASE TEST VECTORS

A. Introduction to Particle Swarm Optimization

Particle Swarm Optimization (PSO) is a population based stochastic optimization technique inspired by social behavior of bird flocking or fish schooling [11]. PSO shares many similarities with evolutionary computation techniques such as Genetic Algorithms (GA), where the system is initialized with a population of random solutions and searches for optimal solution by updating its generations. However, unlike GA, PSO has no evolution operators such as crossover and mutation in GA. In PSO, the potential solutions, called particles, fly through the problem space by following the current optimum particles. PSO optimizes problems by having a population of candidate solutions (particles). Each particle has position/velocity which represents magnitude/direction of the solution. A possible solution to the numeric optimization problem under investigation is represented by the position of a particle. Additionally, the current velocity $v(t)$ represents the direction toward a new solution. A particle also has a measurement of the quality of its current position (called fitness), the particle's best known position (that is, a previous position with the best known quality), and the quality of the best known position. In past several years, PSO has been successfully applied in many research and application areas. It is demonstrated that PSO gets better results in a faster, cheaper way compared with other methods. Another reason that PSO is attractive is that it has few parameters to adjust. One version, with slight variations, works well in a wide variety of applications. Particle swarm optimization has been used for approaches that can be used across a wide range of applications, as well as for specific applications focused on a specific requirement.[12]

PSO optimizes a problem by iteratively trying to improve particles with regard to a given measure of quality. These particles move around in the search-space according to simple mathematical formulae over the particle's position and velocity. Each particle's movement is influenced by its local best known position and is also guided toward the best known positions in the search-space, which are updated as better positions are found by other particles. This is expected to move the swarm toward the best solutions. We can briefly describe the steps for PSO as follows:

1. Create a 'population' of agents (called particles).
2. Evaluate each particle's position according to the objective function.
3. If a particle current position is better than its previous best position, update it.
4. Determine the best particle (according to the particle's previous best positions).
5. Update particles' velocity according to:

$$v(t+1) = \big(w * v(t)\big) + \Big(c_1 r_1 * \big(p(t) - x(t)\big)\Big) + \Big(c_2 r_2 * \big(g(t) - x(t)\big)\Big) \qquad (6)$$

6. Move particle to their new positions according to:

$$x(t+1) = x(t) + v(t+1) \qquad (7)$$

7. Go to step2 until stopping criteria are satisfied.
8. Generating the Worst-Case Test Vectors

B. Identifying Worst-Case Test Vectors

The key element in our problem is the particle with its failure effect on the DUT. The position of particles can be simply described as an integer number representing the input combination of '0's and '1's. The level of failure (normalized leakage current value) for each particle represents how fit is this particle to the target case "worst failure".

1) Generating the initial swarm

Each population (swarm) has a number of particles. Each particle has a position and a velocity which initially generated by random numbers. The input test vector is combination of '0's and '1's of the primary input of chip which is represented by an equivalent integer called "the particle." After we use PSO algorithm to identify the WCTV using integer particles, we convert this integer to a combination of '0's and '1's.

2) *Evaluate each particle's position according to the* objective function

We start by calculating the fitness value IL–the amount of the normalized leakage current using the developed fault model for a given input vector or the particle's position.

3) If a particle current position is better than its previous best position, update it

After we calculate the particle position, we compare the current position of particle with the previous position and store the greater in P$_{localPosition}$

4) *Determine the best particle (according to the particle's* previous best positions)

We compare the position of all particles with any new particle position. And if the new particle position is the largest so we store it in P$_{globalposition}$

5) Update the velocity of the particles

Each particle has velocity which directs the particle to the solution (search goal). The particle's velocity is initiated by random number. And it is updated according to Equation (6). The velocity calculation depends on three terms. The first term is $w * v(t)$, where w is the inertia weight and is simply a

constant value; $v(t)$ is the current velocity at time t. The second term is $c_1 r_1 * \big(p(t) - x(t)\big)$, where the c_1 factor is a constant called the cognitive (or personal or local) weight and r_1 is a random variable in the range $[0,1)$, which is greater than or equal to 0 and strictly less than 1, the $p(t)$ vector value is the particle's best position found so far, and the input vector $x(t)$ is the particle's current position. The third term in the velocity update equation is $c_2 r_2 * \big(g(t) - x(t)\big)$ where the c_2 factor is a constant called the social—or global—weight, the r_2 factor is a random variable in the range $[0,1)$, and the $g(t)$ vector value is the best known position found by any particle in the swarm so far. Once the new velocity, $v(t+1)$, has been determined, it will be used to compute the new particle position $x(t+1)$.

We set the value of w (the inertia weight) to 0.729 which recommended by previous research effort investigating the effects of various PSO parameter values on a set of benchmark minimization problems[13]. We set the values for both c_1 and c_2, the cognitive and social weights, to 1.49445. Again, this value was recommended by previous research study. If you set the value of c_1 to be larger than the value of c_2, you place more weight on a particle's best known position than on the swarm's global best known position, and vice versa. The random variables r_1 and r_2 add a random component to the PSO algorithm and help prevent the algorithm from getting stuck at a non-optimal local minimum or maximum solution.

6) Move particle to their new positions

We calculate the new particle position using Equation (7) where the new position equals the old position plus the new velocity and calculate the quality (leakage current) of the new position.

7) Go to Step 2 until stopping criteria are satisfied

We go to Step 2 of the PSO algorithm and keep iterate until stopping criteria is reaching. The stopping criteria is either reaching estimated worst-case leakage current or completing the number of iteration we initially set.

8) Generating the Worst-Case Test Vectors

Finally, we convert the WCTV we get from applying PSO to a combination of '0's and '1's which are the primary inputs of chip.

C. Algorithm Validation

In order to validate the using of PSO technique, we compared its results with the exhaustive search technique in a reasonable circuit such as 8×8 multiplier chip. The results obtained from PSO are closer to the results from the exhaustive search. PSO is an evolutionary computation technique so it doesn't grantee the optimal WCTV. However, the difference between the near optimal vector and the absolute optimal is not significant. We verified this concept by applying experimentally WCTV and near WCTV on a test chip exposed to total ionizing radiation. The experimental is applied to 8×8 multiplier chip fabricated through MOSIS MEP fund [10]. The results of this experiment are shown in Fig. 2 where we can find the worst-case and near worst have

978-1-4673-5289-5/12 $31.00 © 2012 IEEE

the same effect. We implement this algorithm using Microsoft Visual Studio C# and we use Mentor Graphic Questa for the exhaustive search.

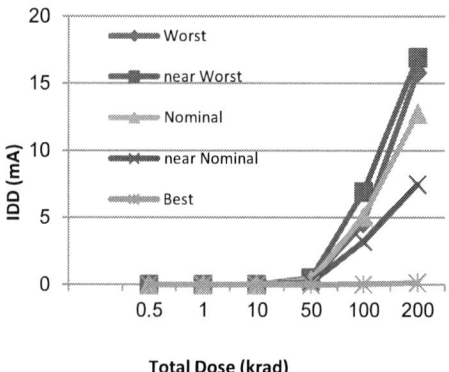

Fig. 2 The leakage current effect for different test vector type at different total dose levels

VI. ALGORITHM RESULTS

The objective of PSO techniques is to identify the particle (solution) that can result in either the global maximum of IL (worst-case) or near maximum (near worst-case). We apply the PSO on the test chip we used before in the validation experiments. This chip has a 8×8 multiplier which results in 16 primary inputs and an IP vector size of 32 bits. The test vector (the particle) is represented by integer number. We arbitrarily chose the initial number of particles in swarm to 10000 particles and the algorithm run 1000 iteration. At the last iteration, the best particle with the highest particle quality (IL_{max}) value was representing the near worst-case test vector. After the 1000 iteration, the best position (solution) was 4062343133 which has the highest normalized leakage current (particle quality) equal to 3787.5.

The execution of this search technique was completed in few minutes using regular PC workstation and resulted in a near worst-case test vector. The search time is much smaller than exhaustive search time which took two days to complete the searching for the worst case test vectors.

The exhaustive search is strictly depending on the complexity of test chip, in which, its complexity is $(2^{4N} * M)$ where N is the number of input ports of the DUT and M is the number of cells in the DUT. On the other hands, PSO technique complexity is independent of the number of DUT ports. However, the exhaustive search finds the exact WCTV for the DUT because it scans all vectors which may not be reached in the PSO approach.

VII. CONCLUSION

In this paper, we provide alternative way to get WCTV of leakage current using nondeterministic search algorithm. We have applied PSO technique to improve the search time of WCTV for leakage current failures induced by total dose in ASIC devices for its simplicity. The main advantage of this algorithm is that it requires only the gate-level netlist of the devices under test, which can be provided from a conventional ASIC synthesis tool such as Mentor Graphics Leonardo

Spectrum. Another advantage of the algorithm is that it scales easily and can be applied to very large circuits which are usually impossible to find the WCTV using exhaustive search method. The main limitation of this technique is that it may produce a near worst-case vector and does not guarantee the optimal worst-case vector. Nevertheless, the leakage effect of near worst-case can be considerable as the optimal WCTV.

VIII. REFERENCES

[1] A. A. Abou-Auf, "Failure Analysis and Worst-Case Generation of Test Vectors for CMOS Circuits Exposed to Ionizing Nuclear Radiation," Ph.D. Thesis, Electrical Engineering Department, University of Maryland at College Park, 1993.

[2] A. A. Abou-Auf, D. F. Barbe, and H. A. Eisen, "A Methodology for the Identification of Worst-Case Test Vectors for Logical Faults Induced in CMOS Circuits by Total Dose," IEEE Trans. Nucl. Sci., NS-41, No. 6, pp. 2585-2592, 1994.

[3] A. A. Abou-Auf, "Gate-Level Modeling for Leakage Current Failure Induced by Total Dose for the Generation of Worst-Case Test Vectors," IEEE Trans. Nucl. Sci., NS-43, No. 6, pp. 3189 - 3196, December 1996.

[4] A. A. Abou-Auf, "Total-Dose Worst-Case Test Vectors for Leakage Current Failure Induced in Sequential Circuits of Cell-Based ASICs," IEEE Trans. Nucl. Sci., NS-56, No. 4, pp. 2189 - 2197, August 2009.

[5] A. A. Abou-Auf "Total-Dose Worst-Case Test Vectors for Leakage Current Failure Induced in Combinational Circuits of Cell-Based ASICs" Proc. of NATIONAL RADIO SCIENCE CONFERENCE, NRSC'2009. Future University, 5 th. Compound, New Cairo, Egypt, March 17 – 19, 2009. URSI-26 Egypt 2009.

[6] K. M. Butler, Ph.D. Dissertation, "Techniques for assessing fault model and test quality in automatic test pattern generation for Integrated Circuits," University of Texas at Austin, 1990.

[7] A. A. Abou-Auf, H. A. Abdel-Aziz, Mostafa M. Abdel-Aziz, T. A. Abdul-Rahman "Significance of Worst-Case Test Vectors for Leakage Current Failures Induced by Total Dose in ASICs " in Proc. of The Nuclear and Space Radiation Effects Conference (NSREC 2010), Denver - USA, July 2010.

[8] H. J. Barnaby, "Total-Ionizing-Dose Effects in Modern CMOS Technologies," IEEE Trans. Nucl. Sci., Vol. NS-53, No. 6, pp. 3103- 3121, December 2006.

[9] R.C. Lacoe, J.V. Osborn, D.C. Mayer, S. Brown, and J. Gambles, "Total-Dose Tolerance of the Commercial Taiwan Semiconductor Manufacturing Company (TSMC) 0.35-pm CMOS Process," IEEE Radiation Effects Data Workshop, pp. 72 – 76, 2001.

[10] https://www.mosis.com/

[11] Kennedy, J.; Eberhart, R. (1995). "Particle Swarm Optimization". Proceedings of IEEE International Conference on Neural Networks. IV. pp. 1942–1948. DOI:10.1109/ICNN.1995.488968.

[12] http://msdn.microsoft.com/en-us/magazine/hh335067.aspx

[13] Eberhart, R.C.; Shi, Y. (2000). "Comparing inertia weights and constriction factors in particle swarm optimization". Proceedings of the Congress on Evolutionary Computation. 1. pp. 84–88.

Study and modeling of defects in integrated circuits for their reliability analysis

G. Ait Abdelmalek
Department of Electronics
Mouloud Mammeri University
Tizi-ouzou, Algeria
e-mail: ghania_79@yahoo.fr

R. Ziani
Department of Electronics
Mouloud Mammeri University
Tizi-ouzou, Algeria
e-mail: ziani_r@yahoo.fr

M. Laghrouche
Department of Electronics
Mouloud Mammeri University
Tizi-ouzou, Algeria
e-mail: larouche_67@yahoo.fr

Abstract— **Nanometric technologie is the way for the realization of numerical structures of considerable size gathering several hundreds of million transistors on one chip. This miniaturization makes manufacturing processes more complex and less reliable. By consequence, the output of manufacture is likely to drop considerably. To improve the output, the first thing is the optimization of the manufactoring processes. On the other hand, when the latter reach their limits, it will be necessary to find a solution on the level of the design. The faults tolerance techniques can be the solution to these problems. However, the higher expectation of reliability can only be met by more thorough and comprehensive testing of this structure. This paper analyzes the ability of these structures to tolerate manufacturing defects and the conditions to improve the yield and therefore their reliability.**

Keywords - test, yield, reliability, fault tolerance

I. INTRODUCTION

Since 2010, the resolution of integrated circuits increased to 22 nm [1]. The miniaturization of manufacturing processes and the increasing of the number of interconnect levels have helped integrate real systems on a chip. Current systems integrate more and more heterogeneous functional blocks (digital, analog, memory, RF, MEMS, etc ...). However, if the early integration of circuits had intended only to integrate more functions in the least possible surface, today's research performance becomes a strategic argument. Indeed, with the advent of submicron technologies heavily, some physical phenomena that were previously negligible become dominant and these phenomena can induce failures and have a direct impact on the reliability and manufacturing yield. This implies that it becomes very difficult to manufacture a circuit without any defects. Therefore, testing of integrated circuits has become an essential task in the semiconductor industry.

The motivation of our work is to propose robust structural solutions included in the design flow of integrated circuits that improve and ensure reliability even in the defects presence. The proposed methodology consists of two approaches: technical testing of circuits in order to detect their failures by using the ATPG (Automatic Test Pattern Generator) and fault tolerance technique by using the TMR (Triple Modular Redundancy) structure. This methodology is applied to combinational circuits of ISCAS85 and ITC99 benchmarks modified into TMR structures. The tests show that, although the reliability is improved, the TMR structures are not fault-tolerant enough to offset their cost in silicon area. We propose to optimize these structures by increasing their redundancy

with the method of partitioning. The results show that partitioning greatly improves the fault tolerance of the majority of tested circuits and therefore their reliability and the manufacturing yield. Also, this paper show that the manufacturing yield, and consequently the reliability depend on the cost in silicon area and consumption.

II. TEST OF INTEGRATED CIRCUITS

The advent of integrated circuit technology has introduced electronics in many aspect of present-day life. As the use of electronic components increases, the expectation of lower cost, better accuracy, and higher reliability increases. Lower cost and better accuracy is achieved by putting more transistors per unit of silicon, using design automation, increasing device operation speed, and reducing its power consumption. However, these design steps cannot guarantee reliability. In fact, as the circuit density increases, the probability of a manufacturing defect increases. The expectation of reliability can only be met by the comprehensive of testing. Classical fault models (stuck-at, stuck-open, stuckon ...) have been proved to be efficient for the analysis of many of these faults and the majority of the techniques of test are based on these models of faults [2, 3]. However, these fault models cover only partially the spectrum of the real failures. In today's integrated circuits, the functional test tends to being replaced by the structural test [2-4].

The effectiveness of the test depends on the Yield (Y), the Fault Coverage (FC) and the Defect Level (DL) corresponding respectively to the ratio of the number of circuits that pass the test on the total number of circuits, the ratio of the number of faults detected on the total number of errors and the report of the number of circuits on the number of faulty circuits that pass the test. Studies in [4] show that:

$$DL = [1 - Y^{(1-FC)}] \times 100\% \qquad (1)$$

III. DESIGN FOR RELIABILITY AND MANUFACTURING YIELD

A. General principles

The tolerant numerical structures with the faults were mainly conceived to tolerate faults appearing during the circuits operation. Their principle is to use resources of redundancies to detect and correct the faults. There are several faults tolerance structures [5, 6] which are classified according to the resources of redundancy which they use: material, information, time, software or hybrid (composition of various resources). These structures have been designed to tolerate transient or temporary faults but they can also tolerate manufacturing defects and thus increase the yield. The figure 1 shows the TMR (Triple Modular Redundancy) structure, which is the simplest design and most widely used structure.

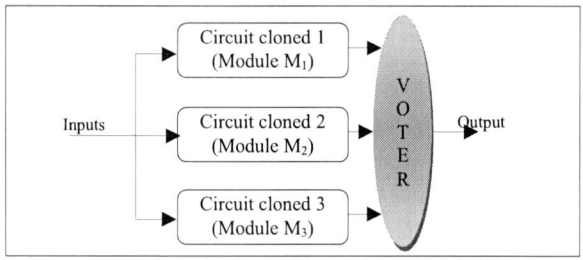

Figure 1: TMR structure

The TMR consists of M three identical modules and a Voter. The outputs of these three identical modules are voted by a majority voter to give a single output (Figure 2).

The reliability of TMR structure is given by the equation (2) [6]:

$$R_{TMR}(t) = R_{Voter} \times \sum_{i=2}^{3}(C_3^i \times R(t)^i [1 - R(t)]^{3-i}) \qquad (2)$$

Where, R_{voter} and R are respectively the reliability of the voter and the reliability of one module.

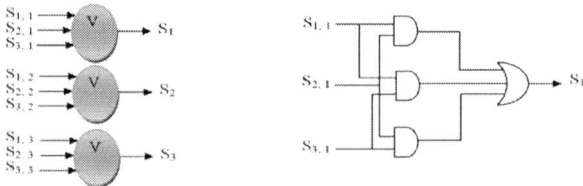

Figure 2: Voter bit by bit

However, the technical realization of such structures is very expensive. Therefore the motivations of designers are not making so many designs for reliability, but to determine the manufacturing process that aims to develop reliable circuits with a production cost of the lowest possible.

B. Calcul of manufacturing yield of TMR structure

Considering that A_c is the original surface of the circuit without redundancy and A_v is the surface of the voter, if we neglect the size of the interconnections, the TMR structure surface A_{TMR} is given by the equation (3).

$$A_{TMR} = 3A_C + A_V \qquad (3)$$

The area cost A_O (Area Overhead) of the implementation of a circuit structure transformed into TMR is given by the equation (4) [7, 8].

$$A_{O=} \frac{3A_C + A_V}{A_C} = 3 + \frac{A_V}{A_C} \qquad (4)$$

By calling Y_{TMR} the manufacturing yield of the TMR structure and Y_C the manufacturing yield of the structure without redundancy, the condition so that TMR structure increases the yield and therefore the reliability is as follows:

$$Y_{TMR} > A_O \times Y_C \qquad (5)$$

Like $Y_{TMR} \leq 1$, this condition becomes:

$$Y_C \leq \frac{1}{A_O} \qquad (6)$$

IV. APPLICATION

In our application we have used combinational circuits of benchmarks ISCAS85 and ITC99 [9, 10]. Next, we transform their architecture into TMR structure. The figure 3 represents the principal phases for the test of these circuits simulated under multiple simultaneous faults of stuck fault.

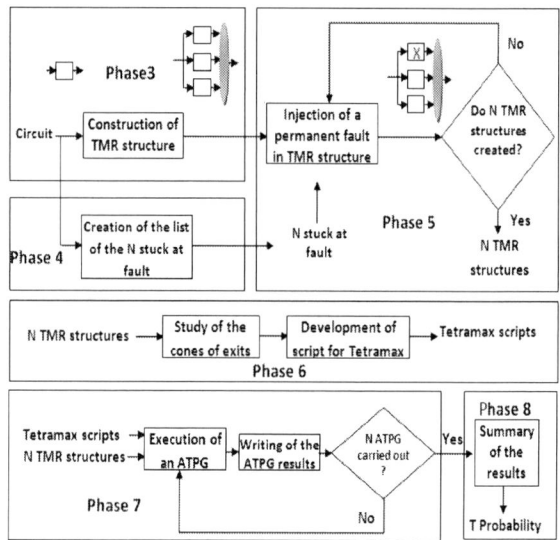

Figure 3: Summary of the principal phases of the ATPG

For an effective test of TMR, a new fault models should be introduced into the ATPG, allowing detection of several manufacturing faults. To see the behavior of circuits in presence of multiple simultaneous faults, we injected a pair of stuck faults (s@0, s@1). The ATPG gives us the rate of fault coverage (FC) and the probability T that a pair of faults either tolerated by the structure. T is defined by the equation (7).

$$T = \frac{\text{Number of pairs of faults tolerated}}{\text{Total number of pairs of faults}} \qquad (7)$$

To calculate the probability T, we used the simulation software Tetramax [9]. Table 1 summarizes the characteristics of the tested circuits.

978-1-4673-5289-5/12 $31.00 © 2012 IEEE

TABLE 1 : CHARACTERISTICS OF THE TESTED CIRCUITS

	simulated circuit	In/Out	Gates	Number of Stuck-at faults	Number of pairs of Faults	Pairs of faults reduced by ATPG	A_O (%)	T (%)	R (%)
Benchmarks ISCAS85	c432	36/7	160	392	689725	75.10^3	3.10	40,00	35,2
	c499	41/32	202	486	1062153	95.10^3	3.39	52,73	54,09
	c1908	33/25	880	1826	15001503	$1.3.10^6$	3.08	56,45	59,62
	c2670	233/140	1193	2852	36598290	$1.87.10^6$	3.33	75,96	85,44
	c3540	50/22	1669	3438	53184141	$4.91.10^6$	3.04	54,09	56,12
	c5315	178/123	2307	4970	111146595	$3.4.10^6$	3.16	93,20	98,67
	c6288	32/32	2416	6250	175771875	$18.2.10^6$	3.03	38,03	32,38
	c7552	207/108	3512	7438	248946141	$7.40.10^6$	3.09	84,93	93,87
Benchmarks ITC99	b02	1/1	25	64	18336	$1.5.10^3$	3.47	86,37	94,93
	b03	4/4	150	382	656085	290.10^3	3.57	87,93	95,98
	b04	11/8	480	1477	9814665	535.10^3	3.32	84,30	93,37
	b05	1/36	608	2553	29326311	$1.17.10^6$	3.16	88,66	96,43
	b06	2/6	66	155	107880	$7.73.10^3$	3.68	87,50	95,70
	b07	1/8	382	1120	5643120	399.10^3	3.38	81,90	91,35
	b09	1/1	131	417	781875	$57.3.10^3$	3.45	83,07	92,37
	b10	11/6	172	468	984906	$63.5.10^3$	3.31	89,40	96,86
	b11	7/6	366	1308	7696926	703.10^3	3.19	74,50	83,80
	b12	5/6	1000	2777	34698615	857.10^3	3.30	95,46	99,40
	b13	10/10	309	835	3136260	$59.6.10^3$	3.49	96,96	99,72

A. Impact of cost in silicon area on T tolerance probability

Figure 4 show that the transformation of circuits in TMR structures does not increase their manufacturing yield, and thus their reliability. The cost in silicon area due to their achievement is very high and tolerance is not enough to offset. The realization of TMR structures will lead to increased performance of a circuit if the circuit is above the curve ($T>T_{min}$) which is not the case in Figure 4.

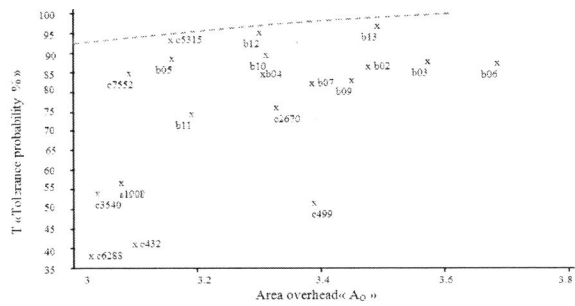

Figure 4: Faults tolerance of circuits depending on the cost (A_O)

B. Optimization of TMR tolerance probability

Therefore, for the TMR structures can increase reliability through increased manufacturing yield, other solutions must be found. There are two possible ways: either reduce the cost in silicon area, by using another type of structure, or use one of the design methods of DFT (Design for testability) like the partitioning [11] shown in figure 5. The partitioning reinforces the redundancy of TMR structures it is this second solution which we chose to implement.

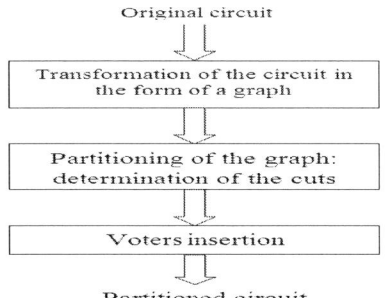

Figure 5: Principle of partitioning

The partitioning has two positive effects on faults tolerance. First, it reduces the combinatorial depth of circuits, which reduces the number of paths of errors propagation. Second, it makes each partition independently. Thus, a manufacturing defect in a part of the partition has no impact on another part of the partitioning. TMR structure is modified by partitioning each channel into two or three partitions. A voter is placed between each partition. Figure 6 shows the three TMR structures: Simple TMR (a), Double TMR (b) and Triple TMR (c).

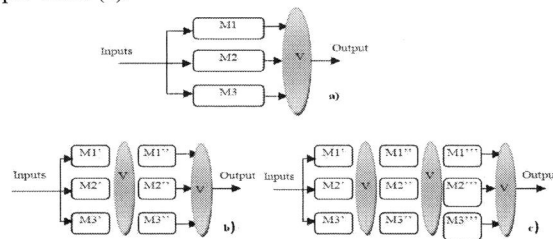

Figure 6: Partitioning of TMR structure

Figure 7 illustrates the partitioning of the c17 circuit. As this circuit has two exits, the number of voting of simple structure TMR is two (V_1, V_2) The partitioning carried out using the sh-METIS tool produced two of the same partitions cuts (3 logical gates in each partition) for two cuts. Two voters additional (V_3, V_4) is thus added. A total of four voters are thus necessary for structure TMR Double.

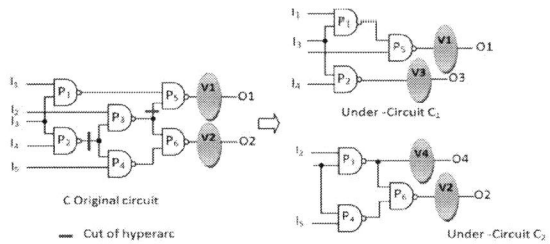

Figure 7: Partitioned c17 circuit

From figure 8, it is clear that with the help of partitioning techniques, the tolerance of TMR structures can be improved and the majority of the tested circuits have their characteristics ranging between the curves corresponding to the two cases: not robust voter and robust voter. However, if the design effort is made to make the strongest possible voters, the real curve T_{min} corresponding to these efforts would be between the two extreme cases presented in Figure 8.

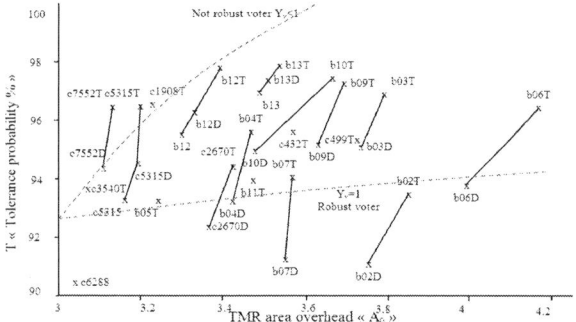

Figure8: Reliability and yield improved through the partitioning

Table 2 summarizes the characteristics of the partitioned TMR structures. We find that the area overhead of partitioning of TMR structure is lower for larger circuits. Indeed, less than 2% for the largest ISCAS85 circuits and less than 5% for the largest ITC99 circuits. This low area overhead is a positive point for the T_{min} probability which can thus satisfy the condition $T > T_{min}$ required. Indeed several structures, in particular TMR triple satisfied the conditions for increasing the reliability and manufacturing yield.

TABLE 2: RESULTS OF PARTITIONING

Simulated circuit	Simple TMR	Double TMR		Triple TMR						
							Robusts voters		Not robusts voters	
	Voters number	Voters #	A_O (%)	voters #	A_O (%)	T (%)	T_{min} (%)	R (%)	T_{min} (%)	R (%)
c432	7	29	10.55	38	14.86	95,53	93,59	99,41	99,61	88,48
c499	32	49	6.14	59	9.75	95,11	93,79	99,30	NA	88,38
c1908	25	53	3.03	68	4.65	96,50	93,05	99,64	96,29	88,68
c2670	140	160	1.40	179	2.72	93,90	93,37	98,93	98,35	88,04
c3540	22	56	1.90	95	4.09	93,79	92,91	98,89	95,36	88,01
c5315	123	149	1.05	161	1.53	96,38	92,99	99,61	95,90	88,65
c6288	32	49	0.58	64	1.10	84,14	92,73	93,25	93,93	82,99
c7552	108	134	0.69	158	1.32	96,22	92,86	99,58	94,85	88,62
b02	5	8	8.15	9	10.87	93,28	93,95	98,70	NA	87,84
b03	34	43	4.25	47	6.14	96,75	93,88	99,68	NA	88,72
b04	74	99	3.23	109	4.52	95,53	92,44	99,41	98,79	88,48
b05	60	73	1.08	91	2.57	93,30	93,06	98,71	96,41	87,85
b06	15	22	8.64	26	13.58	96,26	94,30	99,59	NA	88,63
b07	57	80	4.58	84	5.38	94,11	92,59	99,00	99,66	88,11
b09	29	40	4.96	44	6.77	97,13	93,75	99,75	NA	88,78
b10	23	36	5.28	50	10.97	97,34	94,73	99,79	NA	88,81
b11	37	76	6.14	94	8.97	93,91	93,40	98,93	98,77	88,04
b12	127	139	0.85	167	2.84	97,79	93,37	99,85	98,05	88,87
b13	63	66	0.67	70	1.56	97,83	93,54	99,86	99,40	88,88

V. CONCLUSION

This paper analyzes the behaviour of integrated circuits designed for test (DFT) and reliability in presence of manufacturing defects and their ability to tolerate manufacturing defects. The modified circuits into TMR structures provide a good compromise between manufacturing yield, reliability and area overhead. It is well known, that the impact of the realization of fault-tolerant structure of the manufacturing yield and reliability can be positive when two conditions are respected (i) the manufacturing yield must be less than $1/A_O$, and (ii) the T probability must be greater than a value T_{min} which depends on technological parameters of manufacture. But also, the optimization of fault tolerance through a judicious choice of the partitioning, and key locations within the circuit would be more appropriate to vote. However, the redundancy consumes more energy and increases the propagation delay through the circuits. We must therefore analyze the interesting compromise including assessing system-level redundancy to what percentage it is possible to go not to lose the expected benefits for new technologies.

REFERENCES

[1] International Technology Roadmap for Semiconductors (ITRS), Edition 2007

[2] M. Cimino, "Design of circuits radio frequencies under constraints of extended reliability", thesis of doctorate, University of Bordeaux I, 2007

[3] A.Machouat, "Development and application of a method of analysis of functional failures and contribution to the improvement of the use of the static and dynamic optical techniques", thesis of doctorate, of the static and dynamic optical techniques", thesis of doctorate, University of Bordeaux I, 2008.

[4] A. Bounceur, " Plateforme CAO pour le test de circuits mixtes", thèse de doctorat, Institut National Polytechnique de Grenoble, 2007

[5] J. Han and P. Jonker,"Toward hardware redundant, fault tolerant logic for Nanoeletronics", IEEE Design & Test off computer, Vol.22, No.4, 2005

[6] M. Hafezparast,"tolerant Fault hardware designs and to their reliability analysis", thesis of doctorate, Brunel university of west London, 1990.

[7] C.H. Stapper "Yield model for fault clusters within integrated circuits", IBM Newspaper off Research and Development, vol. 28, N°5, the USA 1984.

[8] D.P. Siewiorek, R.S.Swarz" Applicable Systems Computer, Design and Evaluation " ED. DIGITAL Close 1992

[9] Web site,www.vtvt.ece.vt.edu/vlsidesign/cadtools.php

[10] A. Khouas, "Test de systèmes électroniques" fascicule de cours, École polytechnique Montréal, Automne 2007.

[11] C. Edmond Bichot, "Development of a metheuristic news for the airspace division", thesis of doctorate, Institut National Polytechnique of Toulouse, 2007.

A New Computer Based Auto-recloser Framework

Hamid Bentarzi and Mahfoud Chafai

Signals and systems Laboratory
Department of Control and Power Engineering
University of *M'hamed* Bougara
Boumerdes, Algeria

Abderrahmane Ouadi and Boubakour Harhati

Department of Electronic

University of *M'hamed* Bougara
Boumerdes, Algeria

Abstract - It is so important to clear all kind of electrical faults that take place in the electrical network, otherwise, this may lead to blackout. For this reason, many protective devices have been developed among them an auto-recloser. Auto-recloser is a circuit breaker associated with a control circuit that may automatically re-close the breaker after opening it due to a fault detection. Since more than 80% of the faults are self clearing in their nature hence, power outage time can be reduced if the reclosing function is automated and carefully planned. In this paper, a new frame work is proposed using PC associated with Acquisition card.

Index Terms –Fault types, self clearing fault, auto-recloser, Data Acquisition Card.

I. INTRODUCTION

Power system protection includes all functions that can clear types of abnormal disturbances leading to power system failure. Different methods of detection and clearing time of these abnormalities are used to restore normal operation. System protection has evolved, over the years, from relatively primitive devices with limited capability to complex systems that involve complexhardware(numerical relays). There are different devices that are important in power system protection. Such devices, the relays those are available in many types, serving a host of different purposes and having different design characteristics [1,2].

The function of protective relaying is to promptly remove from service any element of the power system that starts to operate in an abnormal manner. In general, relays operate after some detectable damage has already occurred. Their purpose is to limit further damage to other equipment, to minimize danger to people, to reduce stress on power grid and, above all, to remove the faulted equipment from the power system as quickly as possible. So the integrity and stability of the remaining system is maintained.

Auto-recloser relay, which forms the basis of our work, is the most common element of protection used to deal with excessive currents in power system. It is designed and implemented based on IDMT function provided with three shots.

TABLE I TYPES OF FAULT

	Transient	Semi-permanent	permanent
Causes	Lightning, Conductor swing[a],	a tree branch falling on the line	a broken conductor
feature	disappear after a first short dead time	requires more than one shot	removed manually

II. AUTO-RECLOSER PRINCIPLE OPERATION

When a fault occurs, the auto-recloser detects overcurrent and breaks the circuit, instantaneously or after an intentional time delay. After a first preset interval of time (usually 1/3 ec) the recloser recloses called first shot. If the fault persists, the recloser again trips, and waits a second preset interval (usually 15 sec). It recloses again, only to trip if the fault is still there (second shot). After a third open interval (usually 45 sec), the recloser closes one last time. If the fault persists, the recloser opens and locks out (third shot), requiring manual resetting. If any recloser is successful, the entire cycle is reset automatically (See Fig. 1) [1].

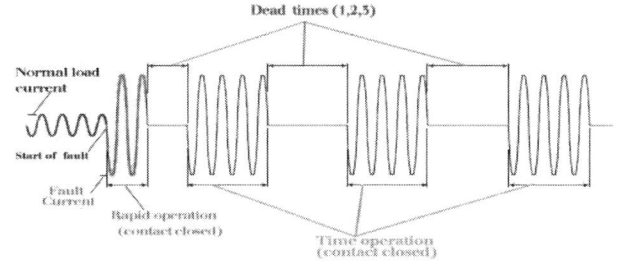

Fig.1 Typical sequence for recloser operation[6].

Any Auto-recloser consists of three principle parameters which are:

1) Dead time:is the time during the fault detection and the power line opening or the amount of time necessary for opening the circuit breaker(circuit breaker coil is energized).

2) Reclaim time: is defined in IEEE Std. C37.100-1992 as the time required, after one or more counting operations, for the counting mechanism to return to the starting position. In an Auto-reclosing relay, the reset time is the time following a successful closing operation, measured from the instant the auto-reclose relay closing contacts make, which must elapse before the auto-reclose relay will initiate a new reclosing sequence in the event of a further fault incident

3) Number of shots: is the predetermined number of attempt for the auto-reclosing to clear the faults being sensed, there is no specific rule to select the number of shots but so often Auto-reclosers have three (3) shots.

Today almost all Auto-recloser are designed based on IDMT function [3]. IDMT that is one type among the inverse time over current relays fits very well this application because with thethe more severe fault, the faster it should be cleared to avoid damage to the apparatus. The reclaim time of an Auto-Recloser represents in fact the time delay of the IDMT which is based on the following equation:

$$t_{delay} = \frac{0.14}{(\frac{I}{I_S})^{0.02} - 1} * \text{TMS} \quad (1)$$

where : t_{delay} : relay operation time (time delay),
TMS :   time multiplier setting,
I :   current detected by relay,
I_S :current set-point.

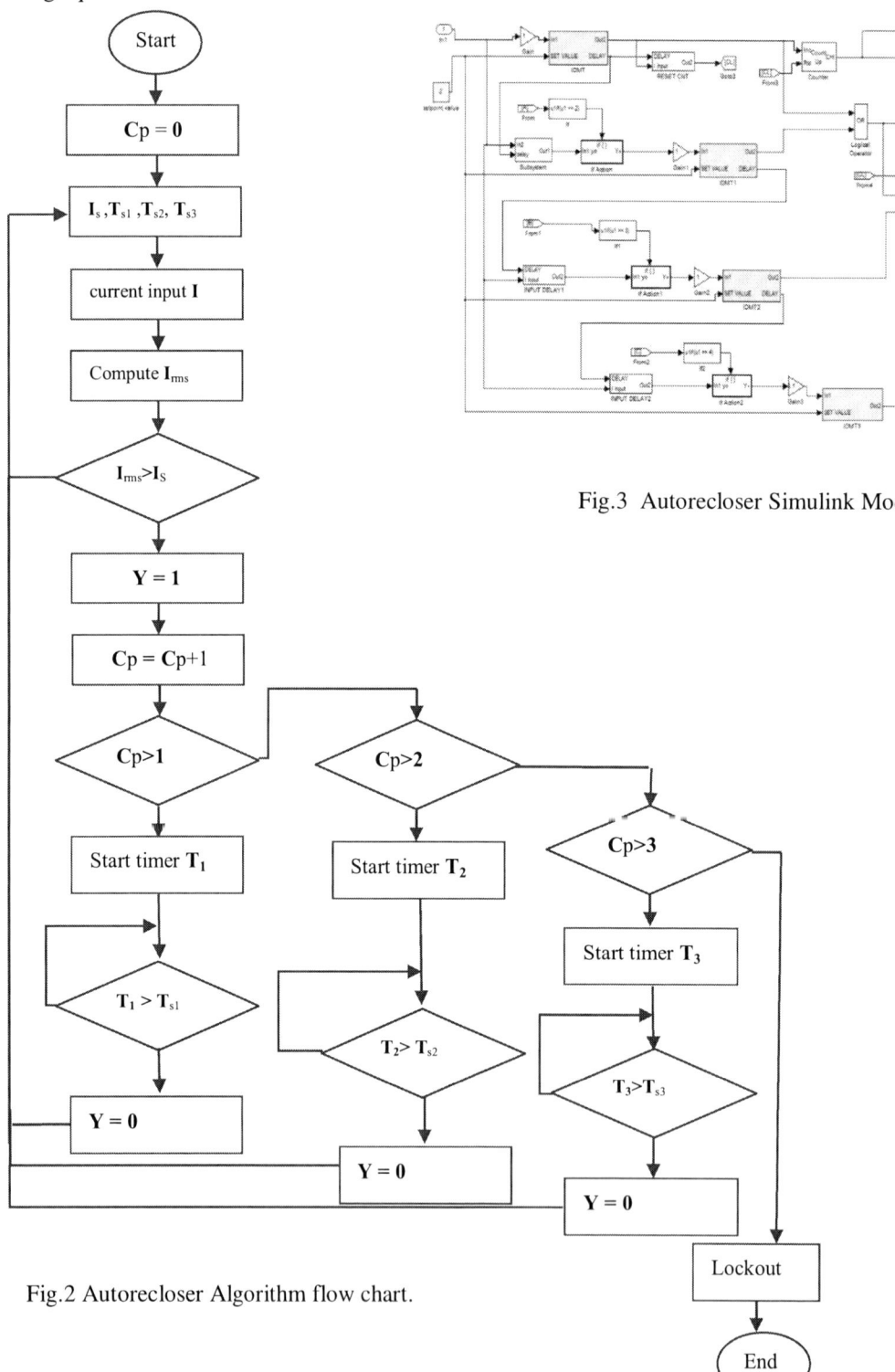

Fig.3 Autorecloser Simulink Model.

Fig.2 Autorecloser Algorithm flow chart.

978-1-4673-5289-5/12 $31.00 © 2012 IEEE   276

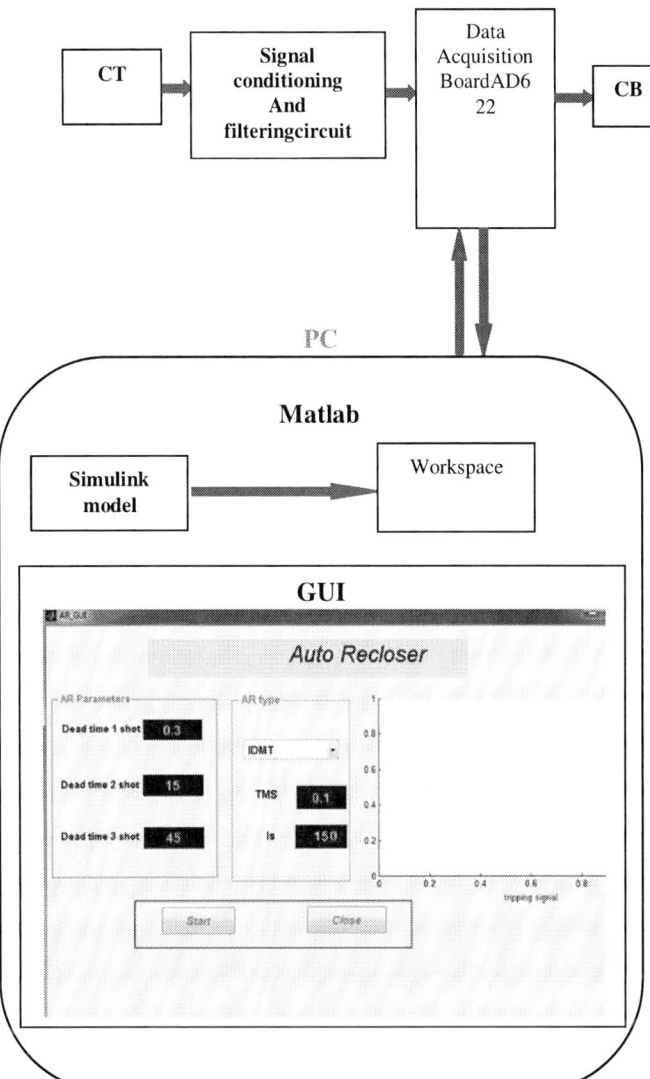

Fig.4 General block diagram of Auto-recloser.

III. SOFTWARE STRUCTURE

Autorecloser algorithm, which has been implemented using Simulink/ Matlab, its flow chart is shown in Fig.2.Figure 3 shows Autorecloser Simulink Model. Besides, Graphical User Interface (GUI) has been developed using the same tool as software development, the auto-recloser user can select and set his wanted parameters, and make a test by running simulation and displaying the tripping signal (see Fig.4).

IV. HARDWARE ARCHITECTURE

In protection field, current transformers are used to sense the current and provide the measured quantity as voltage signal to be the input of relay. Circuit breaker is used as actuator.

The autorecloser hardware whose block diagram shown in Fig.4 consists of:
- Signal transformation: current transformer (CT) transform currents of the power system to low safe values.
- data acquisition boards: the measured values of the power system parameters fed from CT in analog forms are passed through an anti-aliasing filter amplifier (low pass filter). Sample and hold circuits and analog multiplexed are used to sample the three different signals supplied by instrument transformers at the same time. The sampled signals are converted into digital form.
- PC: these digital signals are fed from data acquisition boardto the PC where they will be processed.

The developed auto-recloser has been implemented in PC associated with acquisition card AD 622 [4, 5]

V.TEST RESULTS AND DISCUSSION

We tested the Auto-recloser Simulink model for different values of the input current amplitude A=150, 227, 250, 450 and 520, setting TMS=0.1 sec, Is=150A, Dead time(1)=0.3sec, Dead time(2)=15sec, Dead time(3)=45sec.
The table II represents the simulation and theoretical results of the reclaim times (Operation time of the IDMT).

TableII Reclaim time of the Auto-recloser

Amplitude input current (A)	Calculated from IEC standard	Simulation result (sec)
150	-	-
227	4.019	4.019
250	2.874	2.875
450	0.838	0.840
520	0.675	0.677

The output tripping signals for Auto-recloser model are as shown in Fig.5.Figure shows no tripping signal for Auto-recloser with amplitude 150A
we did not get any tripping signal since the ratio PSM almost equal 1.As the input current goes high, the reclaim time of auto-recloser decreases, and this satisfies the previous equation.

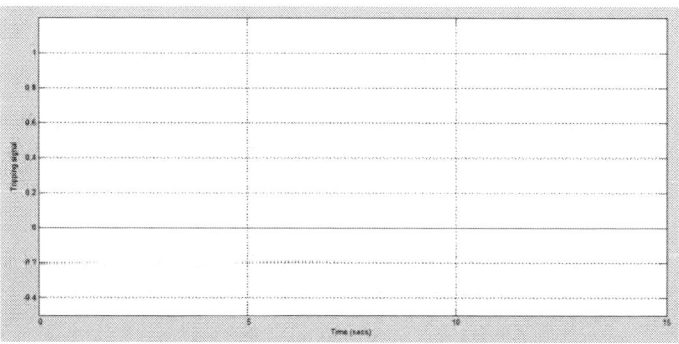

Fig.5 Tripping signal for Auto-recloser with amplitude 150 A.

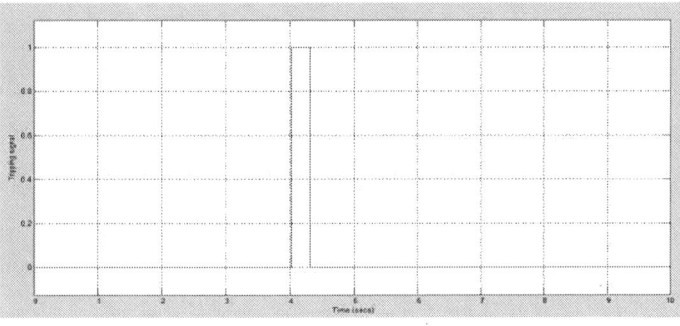

Fig 6 tripping signal for Auto-recloser with amplitude 227A.

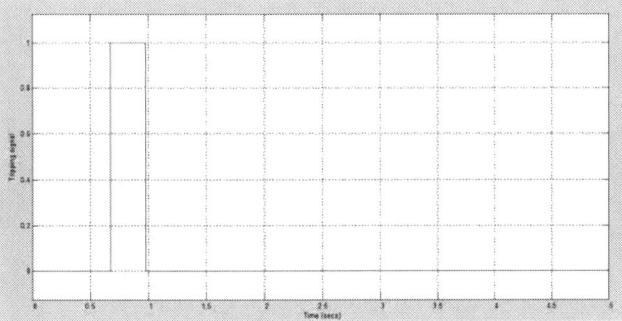

Fig.7 tripping signal for Auto-recloser with amplitude 520A.

A. Test for transient fault

Tripping signal for Auto-recloser with amplitude 227A is shown in Fig.6.

Tripping signal for Auto-recloser with amplitude 520A is shown in Fig.7.

For transient faults in Fig.6 and Fig.7 the auto-recloser trips after a reclaim time of 4.019sec and 0.675sec respectively then waits for dead time1(0.3sec) after that the fault is cleared.

B. Test for Semi-permanent fault:

Figure 8 shows tripping signal for Auto-recloser with amplitude 250 A. For semi-permanent faultsin Fig.8 and Fig.9the faults are cleared after second shot.

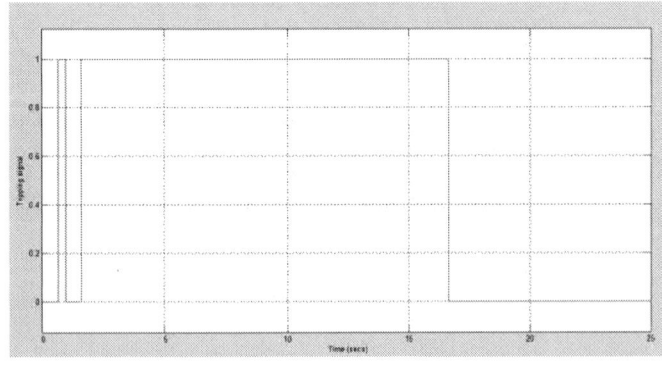

Fig.9 tripping signal for Auto-recloser with amplitude 250 A.

C. Test for permanent fault

In the last figure (Fig.10), since the fault persists, the Auto recloser exhausts its maximum number of shots (3) without

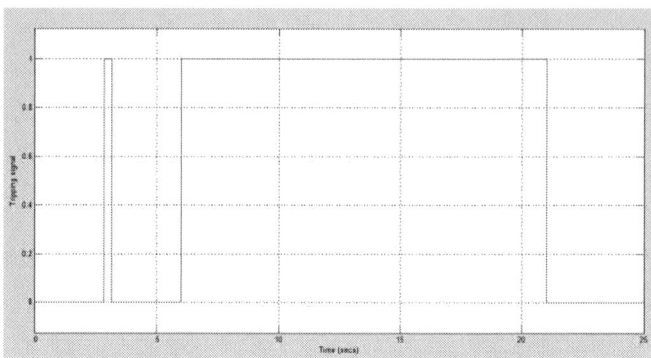

Fig.8 tripping signal for Auto-recloser with amplitude 520 A.

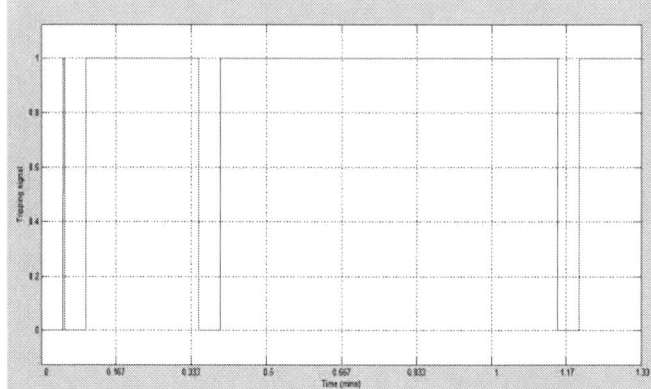

Fig.10 tripping signal for Auto-recloser with amplitude 250 A.

clearing the faults then Auto- recloser locks out and requires a manual resetting . All tests results agree the previous works results[6].

VI. CONCLUSION

A PC-based Auto-recloser prototype has been developed in this work, where we have implemented the algorithm using the Simulink/Matlab and interfaced with the real world via a data acquisition card AD622, which is used to allow acquiring real time signal, processing it, and send it back out.

After testing, it can be noticed that the obtained results satisfy the principle of operation of auto-recloser and its characteristics using new frame work.

REFERENCES

[1] H. Bentarzi, Some Aspects of Protective Relaying in Modern Electric Power System, unpublished.
[2] P.M. Anderson, Power System Protection, IEEE Press, NY, 1999.
[3] P. Rush, Network Protection and Automation Guide, 1sted, Library of Cayfosa, Barcelona, 2002.
[4] IEEE Power System Relaying Committee, "Understanding Microprocessor-Based Technology Applied Relaying", WG I-01 Report, IEEE Organization, January 2009.
[5] AD622 Data Acquisition card user's manual.
[6] A.G. Phadke, T. Hlibka, M.,Ibrahim, " A digital computer system for EHV Substation: Analysis and Field Tests", IEEE Trans. Power Apparatus and System, Vol. PAS-95, pp. 291-301, 1976.

A Non-Convex Classifier Support for Abstraction-Refinement Framework

Samir Ouchani
Computer Security Laboratory
Concordia University
Montreal, Canada
Email: s_oucha@ece.concordia.ca

Otmane Ait'Mohamed
Hardware Verification Group
Concordia University
Montreal, Canada
Email: ait@ece.concordia.ca

Mourad Debbabi
Computer Security Laboratory
Concordia University
Montreal, Canada
Email: debbabi@ece.concordia.ca

Abstract—The main challenge of the counterexample guided abstraction/refinement model checking is the separation of real and spurious counterexamples. This goal is achieved by the classification. In this paper, we reduce the complexity of classification by targeting the problem of feature selection for a considered data set. To do so, we develop a Support Vector Machine (SVM) extended by a Smoothly Clipped Absolute Deviation (SCAD) penalty, to improve the classification scalability by selecting the most important features. The obtained model leads to solve a non-convex optimization problem. The latter is solved by a successive linear programming algorithm with finite convergence. Preliminary computational experiments on different benchmarks demonstrate that our methods accomplish the desired goal of selecting the most important features with a minimum error.

I. INTRODUCTION

Classification gains popularity by its application on different areas that can help mainly in decision-making. In verification, it is used to separate visible and invisible variables, to pass from abstraction to refinement, in the Counter Example Guided Abstraction and Refinement (CEGAR) procedure [1]. Generally, Integer Linear Programming (ILP) has been proposed to solve this classification problem.

In this paper, we use support vector machine as a learning machine to classify a data set. Datasets are used to understand, analyze, monitor data and select the important features in many areas such as form and character recognition, genomic sequences, and security by detecting and diagnosing Spams. Support vector machine (SVMs) [4] have demonstrated a good performance in classifying high dimensional and low sample data size. However, the standard SVM suffers from redundant variable [2] since its decision rule depends on all the variables. In addition, the feature selection aims at picking out some original features to facilitate data collection, reduce storage space and classification time, and also to improve the prediction accuracy by avoiding the "curse of dimensionality" [3]. Furthermore, feature selection often leads to a compact classifier with better accuracy and interpretability [2]. In our work, we target the efficiency of SVM by showing the important feature for classification.

Background. The Support Vector Machines (SVM) are a class of learning algorithm defined initially for discrimination, which means the prediction of a binary classification of qualitative variables. Then, they are generalized to cover multi-

clustering. The basic idea is to find the optimal hyperplane where observable data supports the maximum margin within the hyperplane.

The optimal hyperplane can be found by two solution paradigms. The first considers the hyperplane as a solution of an optimization problem (Fig. 1). The second one looks for the non linear discriminant surface obtained by a kernel function (Fig. 2). The kernel function models a new approximation of the classification problem and of the data structure.

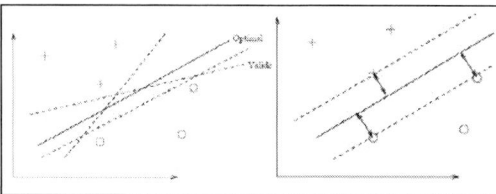

Fig. 1. the Maximum Margin Hyperplane.

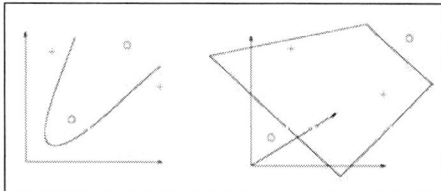

Fig. 2. Kernel Transformation function.

In a feature space H, finding a hyperplane separator of an orthogonal vector w is defined by the following optimal problem:

$$
\begin{aligned}
&\min_{w} \quad (1/2)\|w\|^2 \\
&\text{s.t.:} \\
&\qquad \forall i > 0 : y_i <w, x_i> +b \geq 1.
\end{aligned}
\tag{1}
$$

The case where data are not separable by a plan, requires relaxation of constraints by introducing penalty terms (ξ_i). The main SVM optimization problem, tuned by the parameter δ,

---

978-1-4673-5289-5/12 $31.00 © 2012 IEEE　　　279

will be written as follows:

$$\min_{w} \quad (1/2)\|w\|^2 + \delta \sum_{i=1} (\xi_i) \tag{2}$$

s.t.:

$$\forall i > 0 : y_i < w, x_i > +b \geq 1 - \xi_i.$$

In our work, we are interested in the last case, where data are difficult to be separated by a plan. We formulate SVM as a regularization problem with a non-convex penalty. In this case, the optimization problem merges two parts: the data fit is represented by the hinge loss function, and the regularization is defined as the Smoothly Clipped Absolute Deviation (SCAD-SVM). The SCAD-SVM leads to both objective feature selection and data classification. To solve this non-convex problem we apply the Differential Convex function Algorithm (DCA) [3]. Its performance on real microarray data set from different sources is shown in the numerical result part of the paper.

Related work. [4] applied the basic SVM to deal with multi-targets segmentation on caudatum, putamen and pallidum data sets. In [5], the SVM is used with qualitative analysis approaches to be applied on agricultural product such as tea data set. In [6], the SVM is initialized by particle swarm optimizer (PSO) to classify Chinese text. The datasets from the UCI repository are classified by SVM in Combination Vector Projection Method (CVPM) to cover the unclassified region [7]. To detect pedestrians and to classify them according to their moving direction and relative speed, the SVM classifier is applied with a kernel [8]. Most of the related work avoid the non convex SVM which is classified as NP-hard problem [3].

Paper organization. The remainder of this paper is organized as follows. The proposed approach is detailed in Section II. Section III describes the experimental results. Finally, Section IV concludes this paper and provides hints on the possible future works.

II. APPROACH

To construct an SVM, we consider the training set $\{(x_i, y_i)\}_{i=1}^{n}$ where $x_i \in \mathbb{R}^d$ is the input vector, and $y_i \in \{+1, -1\}$ indicates the class label of a bi-classification. The classification problem is to learn a discrimination function $f : \mathbb{R}^d \to \{+1, -1\}$ that permits to assign to any new observed element its related class label. For a given data set X, x_i represents the expression levels of d elements of the ith sample tissue and y_i it's normal or not; often we have $d >> n$.

A. The SCAD SVM

The SCAD penalty is a multiple function composed of a constant, linear and quadratic function . It is expressed as follows:

$$p_\lambda(|w|) = \begin{cases} \lambda|w| & \text{if} \quad |w| \leq \lambda, \\ -\frac{|w|^2 - 2\alpha\lambda|w| + \lambda^2}{2(\alpha-1)} & \text{if} \quad \lambda < |w| \leq \alpha\lambda, \\ \frac{(\alpha+1)\lambda^2}{2} & \text{if} \quad |w| > \alpha\lambda. \end{cases} \tag{3}$$

where $\alpha > 2$ and $\lambda > 0$ are two tuning parameters. The function 3 is a quadratic spline function with two knots at λ and $\alpha\lambda$ except that being singular at the origin, as in Figure 3 where $\alpha = 3$ and $\lambda = 0.4$. The function $p_\lambda(|w|)$ has a continuous first-order derivative.

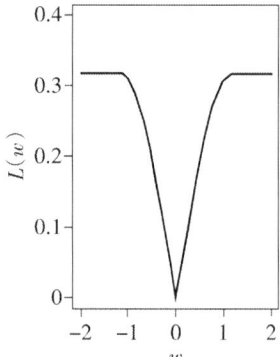

Fig. 3. the SCAD penalty function.

To achieve the feature selection and clustering in time. We add, to the original SVM, the SCAD penalty. It is called Scad-SVM and it is expressed as follows:

$$\min_{b,w} \quad (1/n) \sum_{i=1}^{n} [1 - y_i(b + w.\phi(x_i))]_+ + \sum_{j=1}^{q} p_\lambda(|w_j|), \tag{4}$$

The objective function in 4 consists of the hinge loss part and the SCAD penalty on w. The parameter λ balances the trade-off between data fitting and model parsimony. If λ is too small, the procedure tends to over-fit the training data and gives a classifier with a little sparsity. If λ is too large, the produced classifier can be very sparse but have a poor discriminating power.

B. DCA for SCAD SVM

To apply DC algorithm into the SCAD SVM described by the objective function in (2) as a non convex problem to be minimized. We must show that it is a DC programming, and this means a difference of two convex functions. To find its two convex function components, we begin by eliminating the absolute value in (1). To do so, we define a new function f_λ equal to the SCAD penalty function, where it can be rewritten by:

$$f_\lambda(t) = \begin{cases} \lambda t & \text{if} \quad 0 \leq t \leq \lambda, \\ -\frac{t^2 - 2\alpha\lambda t + \lambda^2}{2(\alpha-1)} & \text{if} \quad \lambda < t \leq \alpha\lambda, \\ \frac{(\alpha+1)\lambda^2}{2} & \text{if} \quad t > \alpha\lambda, \\ f_\lambda(-t) & \text{if} \quad t < 0. \end{cases} \tag{5}$$

It is clear that f_λ is concave in $]-\infty, 0[$ and $]0, +\infty[$. So, the main objective problem (2) can be written as:

$$\min \quad (1/n) \sum_{i=1}^{n} \xi_i + \sum_{j=1}^{q} f_\lambda(w_j)$$

s.t.: $\qquad\qquad\qquad\qquad\qquad\qquad\qquad\qquad$ (6)

$$y_i(b + w.\phi(x_i)) \geq 1 - \xi_i \quad (i = 1, \ldots, n)$$
$$\xi_i \geq 0 \quad (i = 1, \ldots, n).$$

Secondly, we find the convex component of $f_\lambda(t)$ to form a DC function, where: $f_\lambda(t) = g(t) - h(t)$.
Also, we have the following:

$$\begin{aligned}
\varphi(t) &= -\frac{t^2 - 2\alpha\lambda t + \lambda^2}{2(\alpha-1)} \\
&= -\frac{1}{2(\alpha-1)}[(t-\lambda)^2 + 2\lambda(1-\alpha)t] \\
&= \lambda t - \frac{1}{2(\alpha-1)}(t-\lambda)^2.
\end{aligned} \qquad (7)$$

Also, we need to define the function g_t as a polyhedral convex:

$$g(t) = \begin{cases} \lambda t & \text{if } t \geq 0, \\ -\lambda t & \text{if } t < 0. \end{cases} \qquad (8)$$

and, h by:

$$h(t) = \begin{cases} 0 & \text{if } 0 \leq t \leq \lambda, \\ \frac{1}{2(\alpha-1)}(t-\lambda)^2 & \text{if } \lambda < t \leq \lambda\alpha, \\ \lambda t - \frac{(\alpha+1)\lambda^2}{2} & \text{if } t > \lambda\alpha, \\ h(-t) & \text{if } t < 0. \end{cases} \qquad (9)$$

By noting :

$$G(\xi, w) = (1/n) \sum_{i=1}^{n} \xi_i + \sum_{j=1}^{q} g(w_j), \qquad (10)$$

and

$$H(\xi, b, w) = H(w) = \sum_{j=1}^{q} h(w_j). \qquad (11)$$

Finally, problem (4) can be written as a DC polyhedral problem as follows:

$$\min \quad G(\xi, w) - H(w)$$

s.t.: $\qquad\qquad\qquad\qquad\qquad\qquad\qquad\qquad$ (12)

$$y_i(b + w.\phi(x_i)) \geq 1 - \xi_i \quad (i = 1, \ldots, n)$$
$$\xi_i \geq 0 \quad (i = 1, \ldots, n).$$

To apply DCA on problem 12, we must define the sub gradient v of $H(w)$ where $v \in \partial H(w)$. Then:

$$v_j = \begin{cases} 0 & \text{if } -\lambda \leq w_j \leq \lambda \\ (\alpha-1)^{-1}(w_j - \lambda) & \text{if } \lambda < w_j \leq \alpha\lambda \\ (\alpha-1)^{-1}(w_j + \lambda) & \text{if } -\alpha\lambda < w_j \leq -\lambda \\ \lambda & \text{if } w_j > \alpha\lambda \\ -\lambda & \text{if } w_j < -\alpha\lambda \end{cases} \qquad (13)$$

C. Algorithm

The proposed algorithm is a four steps based procedure. It is described as follows.

- Initialization. Let $k = 0$ and Choose $u^0 = (\xi^0, b^0, w^0) \in \mathbb{R}^n \times \mathbb{R} \times \mathbb{R}^q$,
- $\forall k => 0$:
  - Compute $v^k \in \partial H(w^k)$ via (13).
  - Compute u^{k+1} by solving the convex optimization problem

    $$\min \quad \{G(\xi, b, w)\} - \langle v^k, w \rangle$$

    s.t.: $\qquad\qquad\qquad\qquad\qquad$ (14)

    $$y_i(b + w.\phi(x_i)) \geq 1 - \xi_i;$$
    $$\xi_i \geq 0 \quad (i = 1, \ldots, n).$$

  - If $(\|\xi^k + 1 - \xi^k\| + |b^k + 1 - b^k| + \|w^k + 1 - w^k\| \leq \epsilon$, then, Stop;
    Otherwise, set $k = k + 1$; goto Step 1.

The problem (14) is equivalent to :

$$\min \quad (1/n) \sum_{i=1}^{n} \xi_i + \sum_{j=1}^{q} t_j$$

s.t.: $\qquad\qquad\qquad\qquad\qquad\qquad\qquad$ (15)

$$y_i(b + w.\phi(x_i)) \geq 1 - \xi_i \quad (i = 1, \ldots, n)$$
$$\xi_i \geq 0 \quad (i = 1, \ldots, n)$$
$$g(w_j) \leq t_j \quad (j = 1, \ldots, q)$$

Since $g(w_j) = max(\lambda w_j, -\lambda w_j)$, we obtain the following linear problem which is equivalent to (14)

$$\min \quad (1/n) \sum_{i=1}^{n} \xi_i + \sum_{j=1}^{q} t_j - \langle v^k, w \rangle$$

s.t.: $\qquad\qquad\qquad\qquad\qquad\qquad\qquad$ (16)

$$y_i(b + w.\phi(x_i)) \geq 1 - \xi_i \quad (i = 1, \ldots, n)$$
$$\xi_i \geq 0 \quad (i = 1, \ldots, n)$$
$$\lambda w_j \leq t_j \quad (j = 1, \ldots, q)$$
$$-\lambda w_j \leq t_j \quad (j = 1, \ldots, q)$$

III. EXPERIMENTAL RESULTS

In this section, we compare our approach Scad-SVM with L1-SVM [1], on different real dataset of bi-classification type.

A. Data sets

Different public datasets are used for *training* then *testing* phases. They are considered as benchmarks by data mining community. Table I describes four datasets which are Data1 [1], Data2 [2], Data3 [3] and Data4 [4]. The feature column contains the number of feature for each one of datasets. Train A indicates the number of elements that belong to the class A in training sets, and similarly for the Train B. Test A and Test B mention the size of elements that belong to class A and to class B in testing datasets, respectively.

[1] http://www.rii.com/publications/2002/vantveer.html
[2] http://www.broad.mit.edu
[3] http://www-genome.wi.mit.edu/mpr
[4] http://www.nipsfsc.ecs.soton.ac.uk

TABLE I
DATASETS STATISTICS
2012 24th International Conference on Microelectronics (ICM)

Data sets	Feature	Tain A	Train B	Test A	Test B
data 1	24481	44	34	12	7
data 2	7129	27	11	20	14
data 3	12600	52	50	8	13
data 4	10000	44	56	310	390

B. Parameters

For our experiments, we consider the following parameters to run our algorithm for the proposed datasets.

- *The function* ϕ : For the simplicity, we take the fixed point of ϕ which means $\phi(x) = x$,
- *The parameters* α *and* λ: After different tuning test, we found that 3.4 and 0.4 are the best values for α and λ, respectively.
- *The initial points* : Arbitrary, we take the w and b between -5, and +5.

C. Numerical results

We implemented both of the approaches in V.S C++ *v6.0* environment and performed the experiments on a machine running Windows XP Professional edition. Figure 4 shows the complexity of the selected datasets. Figure 5 presents the training error for Scad-SVM versus L1-SVM, and the testing error is illustrated in Figure 6. From Figures 5 and 6, we observe that SCAD-SVM selects the potential features and converges rapidly with minimum error of both phases training and testing.

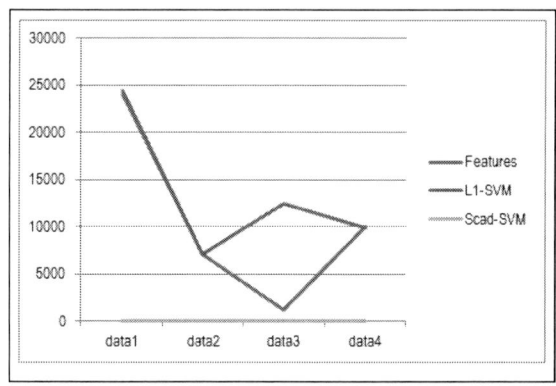

Fig. 4.   Feature Selection Size.

IV. CONCLUSION

In this paper, we proposed an extendible non convex support vector machine to support abstraction refinement procedure for spurious counter examples. To ensure the scalability of the proposed classifier, we applied it successfully on real benchmarks. Also, it is compared with the well known L1-SVM. As future work, we intend to extend our technique to support kernels, proposing new penalties and apply it to divers area such as security datasets.

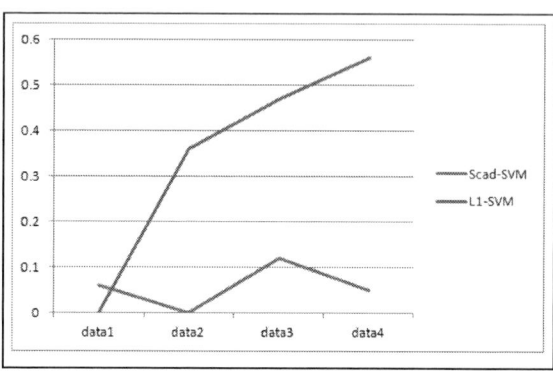

Fig. 5.   DataSets Training Error.

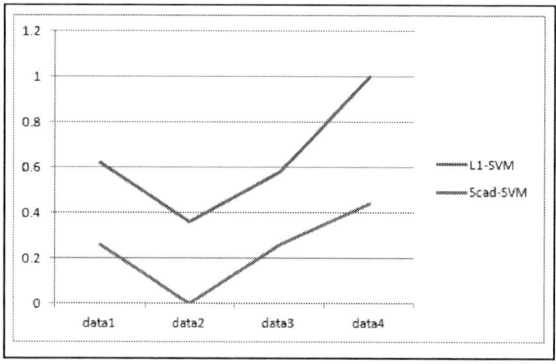

Fig. 6.   DataSets Testing Error.

REFERENCES

[1] E. Clarke, A. Gupta, and O. Strichman, "Sat based abstraction-refinement using ilp and machine learning techniques," in *IN PROCEEDINGS OF CAV*. Springer-Verlag, 2002, pp. 265–279.
[2] C. J. C. Burges, "A tutorial on support vector machines for pattern recognition," *Data Min. Knowl. Discov.*, vol. 2, pp. 121–167, June 1998.
[3] R. Horst, P. Pardalos, and N. Thoai, *Introduction to global optimization*, ser. Nonconvex optimization and its applications. Kluwer Academic Publishers, 1995. [Online]. Available: http://books.google.ca/books?id=w6bRM8W-oTgC
[4] S. Yanhui, D. Enqing, L. Zhenzhi, L. Chenglin, C. Bo, and L. Zhenguo, "Research on the segmentation of tiny multi-target in brain tissues based on support vector machines," in *Complex Medical Engineering (CME), 2011 IEEE/ICME International Conference on*, may 2011, pp. 478–482.
[5] Y. Chen, R. Wang, and H. Chen, "A non-standard-substance pesticide residue qualitative analysis method based on svm," in *Cloud Computing and Intelligence Systems (CCIS), 2011 IEEE International Conference on*, sept. 2011, pp. 89–93.
[6] Y. Zhang, M. Jiang, and D. Yuan, "Chinese text mining based on distributed smo," in *Communication Software and Networks (ICCSN), 2011 IEEE 3rd International Conference on*, may 2011, pp. 175–177.
[7] R. Li, A. Li, T. Wang, and L. Li, "Vector projection method for unclassifiable region of support vector machine," *Expert Systems with Applications*, vol. 38, no. 1, pp. 856–861, 2011.
[8] A. P. Grassi, V. Frolov, and F. P. Len, "Information fusion to detect and classify pedestrians using invariant features," *Information Fusion*, vol. 12, no. 4, pp. 284–292, 2011.

Numerical modeling of MOS transistor using implicit finite different-time domain method

Samir. Labiod, Saida. Latreche, Billel. Smali, M. R. Beghoul, C. Gontrand

Laboratory of Hyperfrequency and Semiconductor (LHS), Electronic department faculty of sciences engineering, Mentouri, Constantine University, 25000, Algeria

E-mail : samir.labiod@gmail.com   latreche.saida@gmail.com

Abstract- **In this work, we present a numerical modeling for a MOS transistor device. This motivated the present comprehensive study of its operations by accurate 2-D numerical simulations. All simulations codes are implemented using MATLAB code simulator. The numerical model is based on a finite-difference approximation of drift-diffusion model (DDM), which contains the Poisson equation and the carrier transport equations. The proposed algorithm provides the time and space distribution of the unknown functions electrostatic potential, carriers' concentration, current density for the MOS transistor. Finally, the obtained results are presented and show a good agreement with numerical simulations using finite element software (ISE-TCAD software).**

Index Terms—Drift-Diffusion Model; Finite difference method; Gummel's method

TERMINOLOGY

q : Electronic charge.

n_i : Intrinsic carrier concentration.

V_G: Gate voltage

V_D: Drain voltage

U_T : Thermal voltage.

μ_n, μ_p: Electron and hole mobility.

D_n, D_p : Electron and hole diffusion coefficient.

J_n : Electron current density.

J_p : Hole current density.

τ : Dielectric relaxation Time.

τ_n, τ_p : Carrier lifetimes of electron and hole.

L_D: Debye Length.

I. INTRODUCTION

In this work, the 2D numerical simulation of MOS transistor using a semiconductor model is adopted, while the drift-diffusion equations are loosely coupled through successive updates of the variables V, n and p. Finite differences time domain is used to solve this problem.

The discretization uses a first and second order finite difference scheme with upwinding based on the characteristic variables [1]. The Gummel and Newton methods are most commonly used to solve the nonlinear steady-state problem arising from the three semiconductor device equations [2], where these equations can be solved in a decoupled manner [3].

Poisson's equation and carrier concentrations equations are solved at all grid point to update the electrostatic potential, electrons and holes concentration, this procedure must be repeated until convergence, where each cach matrix system will be solved using Newton's method.

Euler forward scheme are used to discretize the drift-diffusion model in time. This later leads to have an implicit system which can be solved at each time step using Gummel's method.

Results of steady state regimes are presented and compared with other results obtained using ISE-TCAD software (Sentaurus). However, interesting physical phenomena arise from the manner in which charge fluctuations. The variation of the drain current with drain and gate voltage are also shown and discussed.

The variation of electrons charges in channel with time are represented which confirm several physic phenomenal.

II. SEMICONDUCTOR EQUATIONS

A. Active device model

In this study, the standard drift-diffusion equations coupled with Poisson's equation are used [4]. This model formulates the problem using three dependent variables V (the electrostatic potential), n (electron concentration), and p (hole concentration). Even in this model's simplest form, strong nonlinear dependencies are also present.

The three basic semiconductor equations appear below:

$$\nabla^2 V = -\frac{q}{\varepsilon}(p(x) - n(x) + Dop(x)) \tag{1}$$

Dop accounts for the net ionized impurity concentration

The charge conservation equations formulated for electrons and holes, are respectively:

$$\frac{\partial n}{\partial t} - \frac{1}{q}\vec{\nabla}\vec{J}_n = -q.r_{SRH} \tag{2}$$

$$\frac{\partial p}{\partial t} + \frac{1}{q}\vec{\nabla}\vec{J}_p = -q.r_{SRH} \tag{3}$$

Where r_{SRH} represents the Shockley-Read-Hall recombination ratio, which is a general recombination process using traps in the forbidden band gap of the semiconductor.

$$r_{SRH} = \frac{n.p - n_i^2}{\tau_p.(n + n_i) + \tau_n.(p + n_i)} \tag{4}$$

Current densities are expressed here through the "drift-diffusion" approximation:

$$\vec{J}_n = -q.n.\mu_n.\vec{\nabla}V + q.D_n.\vec{\nabla}n \tag{5}$$

$$\vec{J}_p = -q.p.\mu_p.\vec{\nabla}V - q.D_p\vec{\nabla}p \tag{6}$$

In thermal energy, carrier's concentrations n and p can be assumed by the following expressions:

$$n = n_i . e^{\frac{V-V_n}{U_T}} \qquad (7)$$

$$p = n_i . e^{\frac{V_p-V}{U_T}} \qquad (8)$$

III. NUMERICAL APPROACH

B. Discretization scheme

The discretization uses a first and second order of (1) in 2D-finite difference mesh, leads to have equations:

$$V'(i+1,j)+V'(i-1,j)+V'(i,j+1)+V'(i,j-1)-2.V'(i,j) = \qquad (9)$$
$$\frac{q.h}{\varepsilon}\Big[n'(i,j)-p'(i,j)-Dop(i,j)\Big]$$

Euler implicit method [5, 6] seeks to approximate the derivatives in (2) and (3) with regard to the discrete solutions points defined by spatial and temporal cells. Then the carrier continuity equations may be discretized in implicit form as follows:

$$a_1(i,j).n^{t+1}(i-1,j)+a_2(i,j).n^{t+1}(i+1,j)+a_3(i,j).n^{t+1}(i,j-1)+$$
$$a_4(i,j).n^{t+1}(i,j+1)-[a_5(i,j)+1].n^{t+1}(i,j) =$$
$$\frac{\tau.U_T.h^2.\mu_0}{\Delta t.L_D^2.\mu_n}\Big[R(i,j)-n'(i,j)\Big] \qquad (10)$$

$$a_6(i,j).p^{t+1}(i-1,j)+a_7(i,j).p^{t+1}(i+1,j)+a_8(i,j).p^{t+1}(i,j-1)+$$
$$a_9(i,j).p^{t+1}(i,j+1)-[a_{10}(i,j)+1].p^{t+1}(i,j) =$$
$$\frac{\tau.U_T.h^2.\mu_0}{\Delta t.L_D^2.\mu_p}\Big[R(i,j)-p'(i,j)\Big] \qquad (11)$$

Where μ_0 present the normalized value of the electron and hole mobility.

The expression of the coefficients a_{1-10} can be found in [7].

h and Δt are limited respectively by the Debye length and the dielectric relaxation time.

In order to derive the iteration procedure at each time step, we first relate Gummel's method to Newton's method which is well known to converge quadratically.

C. Initial solution

In this case to get the initial solution we must solve Poisson's equation but with the difference that it replaces (4) and (5).

The whole numerical procedure can be summarized as follows:

Start with $V^{k=1}=0$
Begin iterations
Step 1: Calculate n^k and p^k using using (4) and (5)
Step 2: Solve Poisson's equation
End iterations
Save V, n and p.

D. Final solution

At each time step, equations (9), (10) and (11) are solved using Gummel's iterations [8]. Poisson equation is solved at all grid points, followed by electron continuity equation, and then by hole continuity equation, each matrix system is solved using successive over-relaxation (SOR) method [9].

The whole numerical procedure to calculate the final solution can eventually be summarized as follows:

Step 1: Initialization
For k from 0 to final time
Step 2: compute electrostatic using (9)
Step 3: compute electrons density using (10)
Step 4: compute holes density using (11)
End time iterations.

In practice, the iterative process is stopped at the minimum value of k such that abs $(max(\delta n))<\varepsilon$, where ε is a fixed tolerance.

However, since the exact solution is obviously not available, it is necessary to introduce suitable stopping criteria to monitor the convergence of the iteration.

IV. THE CONSIDERED STRUCTURE

The 2-D structure used in the simulation of the active device is shown in Fig.1. It is an N-channel MOS with a substrate doping of $N_A = 2.10^{16}$ $At.cm^{-3}$, source and drain junction doped $N_D = 2.10^{20}$ $At.cm^{-3}$, an oxide thickness of $T_{ox} = 5nm$. The device width is 1μm. The separation between drain and source is $L_{eff} = 0.5\mu m$, with gate length $L_G = 0.55\mu m$, and a source-drain junction depth $X_j = 0.05\mu m$.

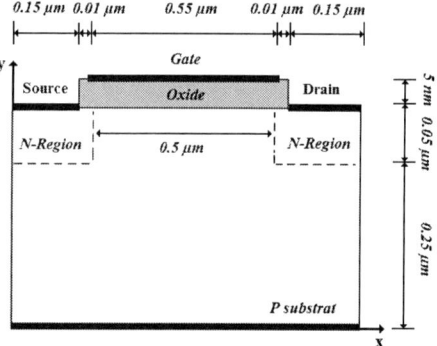

Fig. 1. MOS structure: The Case of study

V. SIMULATION RESULTS

The device is biased with a direct potential of $V_G = 0.8$ V and $V_D = 1V$. The DC parameter distributions are obtained by solving the drift-diffusion model where time derivatives for the three unknown are directly set to zero.

Fig.2 shows the potential distribution inside the device. We notice the potential barriers source-substrate and drain-substrate.

The first is due to the diffusion potential between different types of regions, the second at the three ohmic contacts. Fig.3

shows the computed electrons distribution in the entire device in logarithmic scale. It can be seen that there is an important quantity of electrons which are attracted at oxide-semiconductor interface. The figure shows also the onset of the pinch-off effect which indicates the lack of channel region near the drain junction. Contour plot of current density lines with *0.65* steps in logarithmic scale is presented in Fig.4. Notice the correct flow of electrons which injected from source into the drain contacts. The current density matrix has been obtained by using equations (5) and (6).

The static current-potential characteristics for the n-channel MOS transistor are finally presented in Fig.5. For a given bias, drain current is calculated by integrating current density over drain contact. A good agreement between the proposed model and finite element simulator (ISE-TCAD) is observed.

Fig.6 shows the gate C-V characteristics of nMOS, we notice that the gate capacitance increase to the oxide capacitance for the direct and reverse gate voltage. The capacitance for the three contacts are also calculated and presented for the direct and reverse gate voltage.

The dynamical behavior of the MOS transistor is defined by calculating the matrix capacitance C_{ij} which describes the dependence of the charge at the terminal i with respect to the voltage applied at the terminal j with all other voltages held constant [10].

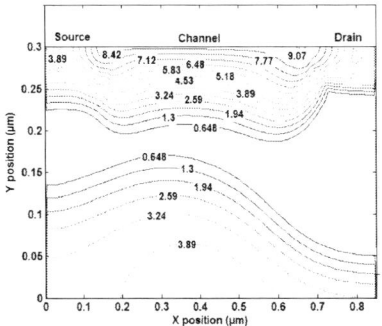

Fig.4. Contour plot of current density in logarithmic scale

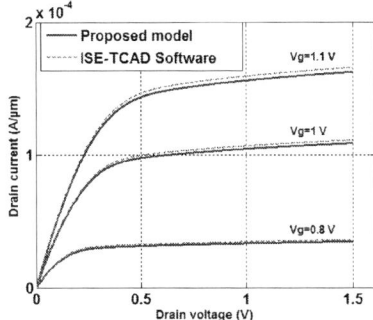

Fig.5. Calculated drain current (output characteristic)

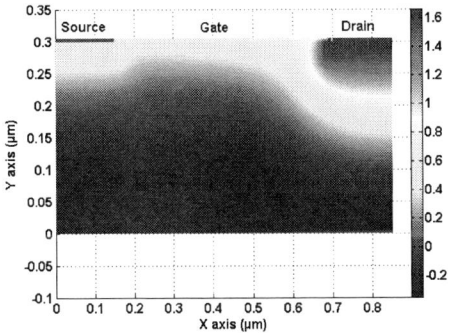

Fig.2. Electrostatic potential for the entire device

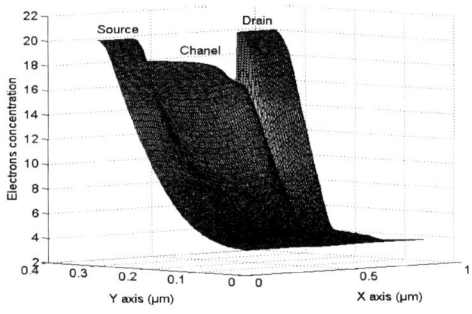

Fig.3. Electron concentration for the entire device

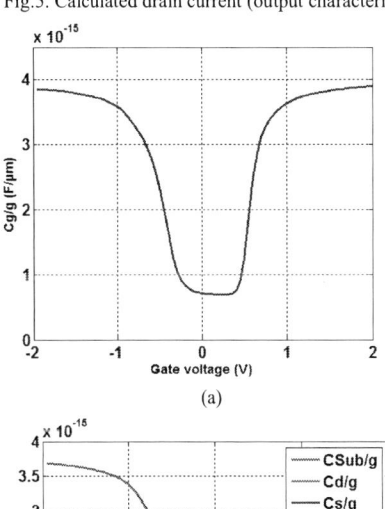

(a)

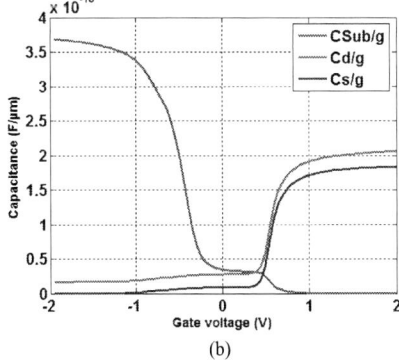

(b)

Fig.6. (a) Gate capacitance of the MOS transistor; (b) C-V curves for the three terminal MOS structure

Transient simulations were simulated at $V_D=1V$ with Vg switched from *0 V* to *0.8* V with a non linear variation of duration *0.01 ps* (see Fig. 7).

Drain, source and substrate current variations versus time are shown in fig. 8. When the channel is clearly formed, the stability of the drain and source current appears above *10 ps* whereas substrate current above 1ns. We can show clearly that the transient solution converge towards results obtained in the steady state solution, by this we can confirm the validity of the proposed method. Fig. 9 shows the variation of electron concentration in the channel-oxide interface versus time, we note that electrons are injected from substrate into the channel region. These charges do not recombine immediately which results the instability of currents for the three contacts drain, source and substrate.

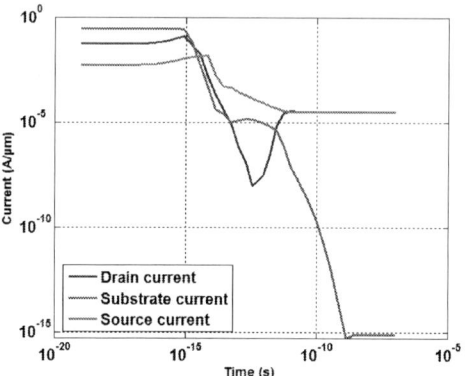

Fig.9. Calculated transient currents

VI. CONCLUSION

In this paper, we have developed a two-dimensional time-dependent simulator that solves the three semiconductor equations using backward Euler scheme for a MOS transistor. The discretization in first and second order finite difference scheme is used on nonuniform grids in two space dimensions, whereas Euler forward method to accomplish time domain, the implicit system obtained is solved by Gummel's method at each time step.

The modeling results are qualitatively in agreement with those obtained using commercial software (ISE-TCAD), also the transient solution converges towards the steady state solution which confirms the validity of the proposed algorithm.

REFERENCES

[1] L. Ballestra, S. Micheletti, and R.Sacco, On a Viscous-hydrodynamic model for semiconductors: numerical simulation and stability analysis. Comput Visual Sci, vol. 4, pp. 79–86, May 2001.

[2] M. Patil,U. Ravaioli, and T. Kerhoven, Numerical evaluation of iterative schemes for Drift-Diffution simulation. Gordon. Breach Science Publisher imprint, India, vol. 8, pp. 337-341, 1998.

[3] D. L. Schafetter and H. K. Gummel, Large-signal analysis of a silicon Read diode oscillator, IEEE Transaction on Electron Devices (1969) 64-77.

[4] M, Kwok k. Ng, Physics of semiconductor devices, Wiley, 2007.

[5] J. C. Butcher, Numerical Methods for Ordinary Differential Equations, 2rd Edition, John Wiley & Sons Ltd, England, 2008.

[6] D. Vasileska and S. M. Goodnick, Computational Electronics, 1st Edition, Morgan & Claypool, USA, 2006.

[7] R. Mirzavand, A. Abdipour, and G. Moradi, Full-Wave semiconductor devices simulation using ADI-FDTD method, Progress in Electromagnetic Research M (2010) 191-202.

[8] A. Quarteroni, R. Sacco, and F. Saleri, Numerical Mathematics, Springer-Verlag press, New York, 2000.

[9] A. Quarteroni, R. Sacco, and F. Saleri, Numerical Methods, Springer-Verlag Press, Milano, 2007.

[10] N. Arora, "MOSFET Modeling for VLSI Circuit Simulation: Theory and Practice", World Scientific, ISBN-13 978-981-256-862-5, 1993.

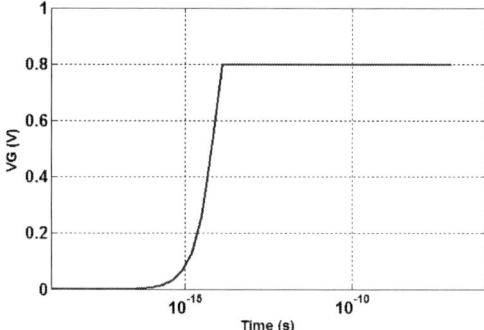

Fig.7. Applied gate bias versus time

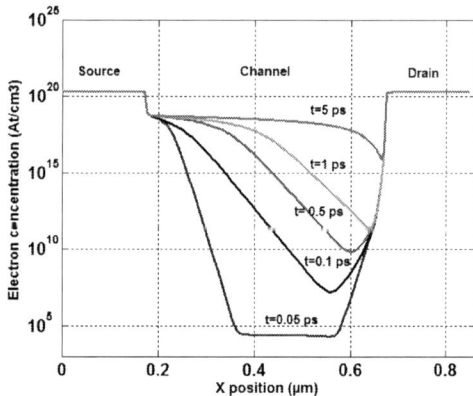

Fig. 8. Electron charge quantity variation in the channel for different time values

Self- heating effects in SOI MOSFET transistor and Numerical Simulation Using Silvaco Software

FZ .RAHOU
Faculty of Technology
Tlemcen University
Tlemcen, Algeria
f_rahou@yahoo.fr

A.GUEN.BOUAZZA /M.RAHOU
Faculty of Technology
Tlemcen University
Tlemcen, Algeria
am_rahou@yahoo.fr

Abstract—**In this paper we briefly present SOI MOSFET transistor and problems generated at high-temperature and self-heating effects, then we present simulation results we obtained using SILVACO TCAD tools relating to SOI n-MOSFET structures we have consider. We will also exhibit some simulation results we obtained relating to the influence of temperature variation on our structure, that having a direct impact on their drain current.**

Index Terms—**SOI technology, SOI MOSFET, Self-heating effects, Silvaco Software.**

I. INTRODUCTION

Integrated circuits that are used in military, automobile, nuclear and well-logging industry require high temperature operation (above 150 °C). The excellent physical and electronic properties of silicon make it an important semiconductor material for high-temperature applications [1].

Several technologies have been explored as a possible choice for high-temperature operation. These technologies include CMOS [1], SOI [1], and GaAs [1].

The use of bulk CMOS device at high temperatures is limited by the presence of latch-up and high leakage current through the well junction.

Nowadays, SOI technology is considered to take the CMOS processing to its ultimate scalability. In order to highlight the qualities and also the defects of SOI technology. SOI devices have no latch-up and low leakage current due to the absence of the well .It is for these advantages that SOI technology is largely used at high temperatures applications.

We propose in this work to present, simulation results we obtained using SILVACO software for an SOI n-channel MOSFET with static biased.

II. SOI MOSFET TRANSISTOR

SOI (Silicon-On-Insulator) is initially invented for application in many special environments, such as radiation-hardened or high-voltage integrated circuits. It is only in recent years that SOI has emerged as a serious contender for low-power and high-performance applications [2]. SOI MOSFET is different from the traditional bulk MOSFET. For bulk MOSFET, the silicon channel region is built on the substrate directly. For SOI MOSFET, a buried oxide layer is formed on the bulk silicon substrate. On the top of the buried oxide layer there is a silicon thin film, where active MOS devices and circuits are located. The cross section of a basic n-type MOSFET on SOI is shown in Fig. 1 (a) and a photograph of a fabricated SOI MOSFET example is shown in Fig. 1 (b).

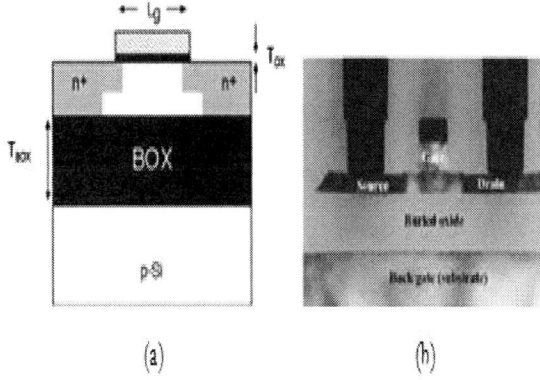

Fig. 1. (a) Structure of SOI MOSFET, (b) Photograph of real SOI MOSFET [3-4].

SOI MOSFET has many advantages over bulk MOSFET in device and circuit level. Because of the buried oxide (BOX) layer, the parasitic capacitances of SOI MOSFET devices are smaller than those of bulk MOSFET. Thus, the delays of digital CMOS circuit due to the junction capacitances can be reduced by using SOI MOSFET, which therefore increase the speed of the digital CMOS circuit. In another aspect, the power delay product of SOI CMOS circuit is much smaller as compared to the bulk counterpart, again owing to the smaller parasitic capacitances in SOI MOSFET as well as reduced leakage currents through BOX. So we can say that, SOI MOSFET technology has high speed and low-power

978-1-4673-5289-5/12 $31.00 © 2012 IEEE   287

properties. In device aspect, SOI MOSFET has no latch-up due to the buried oxide isolation and device isolation is much simpler for the SOI MOSFET as compared to the bulk MOSFET, which make SOI CMOS technology a higher device density and an easier device isolation structure.

Although these advantages of SOI technology are well known, the successful introduction of SOI technology for large-scale applications faces some key challenges across the entire spectra of material, process, manufacturing, devices, and designs. The SOI manufacturing processes are just becoming mature enough for mass production of low-cost, low-defect-density substrates. Another major concern is the control of silicon film thickness to accurately control the threshold of fully depleted devices.

SOI MOSFET can be further divided as partially depleted (PD) Fig. 2 (a) and fully depleted (FD) SOI device Fig. 2(b). Also, elements that have a thin SOI layer (normally <50 nm) and have all body areas under the channel depleted, are called fully depleted type SOI. Conversely, elements that have a thick SOI layer (normally >100 nm) and have some areas at the bottom of the body area that are not depleted, are called partial partially depleted SOI [5].

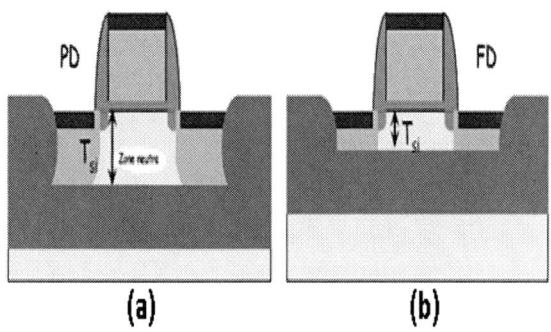

Fig. 2. (a) Structure of partially depleted SOI MOSFET, (b) Structure of fully depleted SOI MOSFET [6].

III. DEVICE SIMULATION

Numerical simulations of the SOI n MOSFET were performed by using the SILVACO TCAD tools. The different parameters of our structure are assumed as follows:

TABLE I. PARAMETERS OF SOI n MOSFET TRANSISTOR.

Symbol	Designation	Value
L_D, L_S, L_G	Drain length, Source length and Gate length	1[um]
L	Channel length	0.7[um]
T_{OX}	Gate oxide thickness	0.017[um]
Tsi	Silicon film thickness	0.2[um]
T_{BOX}	Buried oxide thickness	0.4[um]
	substrate thickness	1.2[um]
	Depth junction	0.52[um]
N_A	Substrate concentration	1×10^{17}[cm-3]
N_D	Drain and Source concentration	1×10^{20} [cm-3]

The below structure is obtained using ATLAS device simulation using. The thickness of the silicon film is 0.2 um. This ensures that the channel is partially depleted.

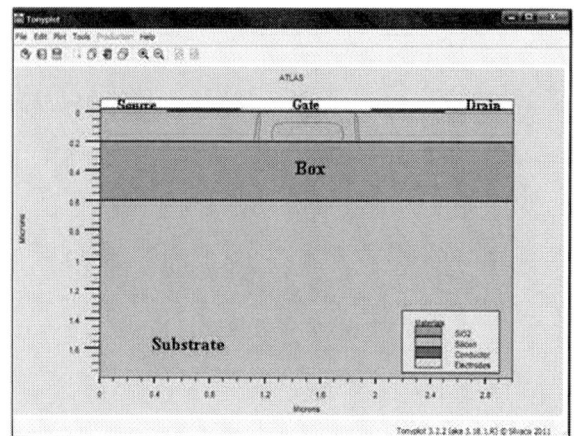

Fig. 3. Device structure of the n-MOSFET with effective channel length 0.7μm, channel doping is 1E17cm-3, drain and source doping concentration is 1E20cm-3, gate oxide thickness is 0.017um, Silicon film thickness 0.2μm.

IV. SELF-HEATING EFFECTS

During the operation of the MOSFET transistor, the electric power generates a quantity of heat per Joule effect. More the power will be raised; the temperature of the channel will increase.

However, the physical parameters such as mobility, the threshold voltage or the saturation speed are temperature dependent. These three parameters are related to the temperature by the following empirical relations (1), (2) and (3) [7]:

$$\mu_{eff} = \mu_{eff,Tamb}(T/T_{amb})^{-k1} \tag{1}$$

$$V_{th} = V_{theff,Tamb} - k_2(T - T_{amb}) \tag{2}$$

$$V_{sat} = V_{sateff,Tamb} - A_T[(T/T_{amb})/T_{amb}] \tag{3}$$

Where: $k_1 \in [1,5 ; 1,7]$, $k_2 \in [0,5 ; 4]$, $A_T = 3,3 \cdot 10^4$.

μ_{eff} ,Tamb , V_{theff} ,Tamb et V_{sateff} ,Tamb are respectively the effective mobility, the effective threshold voltage and the effective saturation speed at the ambient temperature, T_{amb}.

The reduction of the effective mobility is the restrictive factor dominating. When the temperature or the dissipated power increases, mobility decrease involving the decrease of the drain current I_{DS}. Thus, the dissipated power will be lower, which will involve the reduction in the temperature. A phenomenon self maintained is then set up connecting the temperature of the channel and I_{DS}. It is **Self-heating effects.** Consequently, I_{DS} - V_{DS} characteristics presents in saturation a decrease similar to the behavior of a negative resistance [7].

Generated heat is evacuated by the whole of the device according to the type of material and its thermal conductivity. This last quantity varies linearly according to the temperature of the crystal. According to the values given to TABLE II for pure materials [7], the thermal conductivity of silicon is 100 times larger than that of SiO_2. This means that heat will be evacuated more easily by silicon than by the dioxide of silicon. This last act like a thermal insulator compared to silicon.

TABLE II. SOME TYPICAL VALUES OF THE THERMAL CONDUCTIVITY FOR THE PURE CRYSTAL AT 300 K [11]

Typical values of K at 300 K ($W.m^{-1}.K^{-1}$)	
Silicon	148
SiO_2	1.4
Aluminum	237
Copper	401

Generated heat is thus evacuated with difficulty in the case of a MOSFET comprising a buried oxide compared to its counterpart on massive silicon. This insulation involves the increase in the temperature in the channel. Consequently, the effects of **Self-heating effects** are more significant in the SOI MOSFET transistor.

Fig. 4 present the temperature distribution within a PDSOI n-MOSFET and the thin film's temperature appears to be higher than the external temperature by about 140 K.

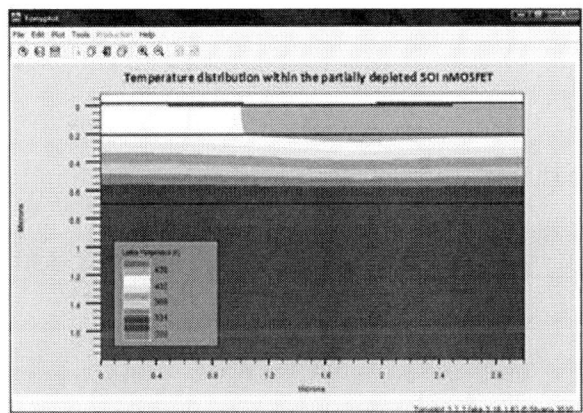

Fig. 4. Temperature distribution within a partially depleted SOI n MOSFET. The external temperature is 300 K.

V. SIMULATION RESULTS AND DISCUSSION

Fig. 5 and Fig. 6 illustrate respectively I_{DS}-V_{DS} characteristics for FD SOI n MOSFET transistor and PDSOI n MOSFET for gate bias of 3V.
The current drain I_{DS} decreases with increasing temperature. At high temperature, the channel mobility decreases. This reduction in mobility leads to a reduction of the drain current.

For PDSOI device, the Kink effect decreases for high temperatures (Fig. 5). Kink effect is one of the principal effects of the floating substrate, started by the accumulation of carriers produced by impact ionization in the silicon film. This effect is translated in PDSOI transistors by the increase of drain current and an electric noise in its saturation region.

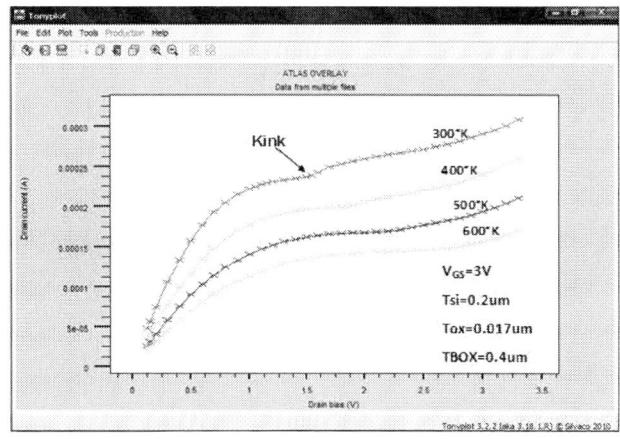

Fig. 5. Simulated I_{DS}–V_{DS} plot for a partially depleted SOI n-MOSFET at operating temperatures of 300, 400,500 and 600 K.

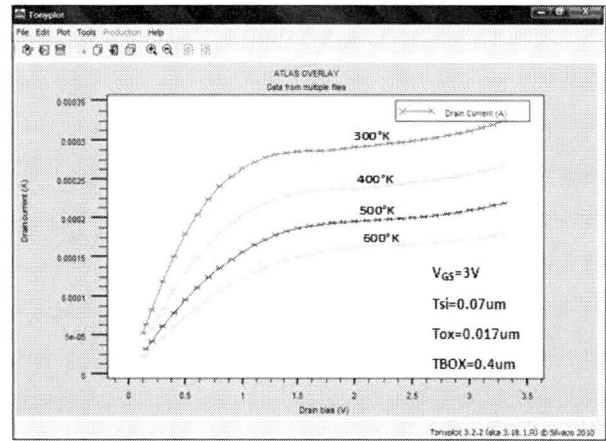

Fig. 6. Simulated I_{DS}–V_{DS} plot for a fully depleted SOI n-MOSFET at operating temperatures of 300, 400,500 and 600 K.

The dependence of the drain current I_{DS} with the temperature is influenced by the threshold voltage and the channel mobility $I_D(T) \approx \mu(T)[V_{GS}-V_{th}(T)]$:. The influences of the reduction of these two parameters in current are opposed.

The $[V_{GS}\_V_{th}(T)]$ term caused the drain current to increase with increasing temperature because the threshold voltage decreased with temperature. On the other hand, the $\mu(T)$ term caused the drain current to decrease with increasing temperature because at high temperatures, lattice scattering dominates and caused a reduction in the channel mobility. At high gate bias, the $\mu(T)$ term dominates while at low gate bias, the $[VGS\_Vth(T)]$ term dominates[1]. (Fig. 7 and Fig. 8).

Gate bias V_{GS} for which current I_{DS} does not vary according to the temperature, in a beach of temperature given is called V_{GS}(**ZTC**)(Zero Temperature Coefficient).

In addition, we can see the advantage of SOI technology, which exhibit a ZTC points over a wide range of temperature up to 600 K. Shoucair was able to identify the ZTC bias point of bulk CMOS transistors in both the linear and the saturation regions up to only 200 °C [1].

Beyond that, the drain current is offset by a large amount with respect to the ZTC drain current. This is because at high temperature beyond 200 °C, the leakage current is comparable to the drain current which limits the device operation. The extended temperature range of SOI devices is due to the suppression of leakage current by the buried oxide between the active thin film and the substrate [1].

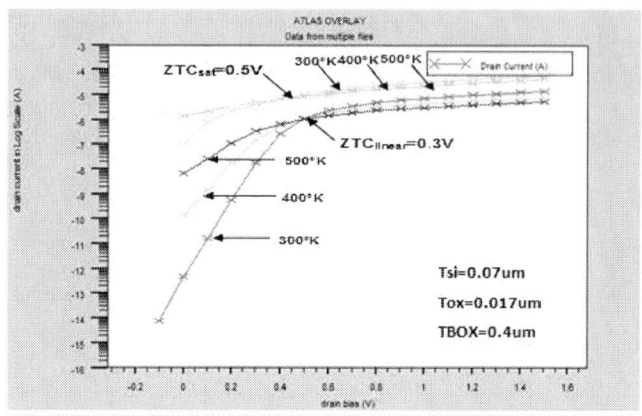

Fig. 9. Simulated I_{DS}–V_{GS} plot for a fully depleted SOI n-MOSFET: in the linear region **ZTC- linear** : $V_{DS}=0.1$ V and the saturation region **ZTC- sat** : $V_{DS}=2$ V.

VI. CONCLUSION

From the simulations results of the SOI n MOSFET structure worked via Atlas-SILVACO tool, we could note the following remarks:

The temperature is one of the essential parameters to be taken into account. Indeed the temperature makes it possible to modify the components performances and consequently circuits.

The use of CMOS devices on bulk substrate at high temperatures is limited by the presence of the latch-up and high leakage currents.

CMOS devices on SOI substrate functions at high temperatures. With this technology the latch-up is eliminated and the leakage current are unimportant, it is for these advantages that SOI technology is largely used at high temperatures applications.

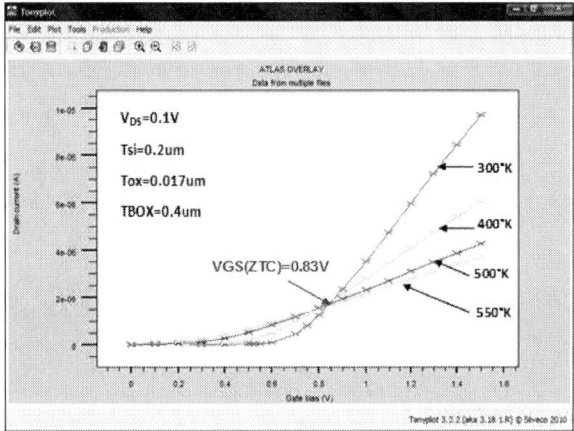

Fig. 7. Simulated I_{DS}–V_{GS} plot for a partially depleted SOI n-MOSFET at operating temperatures of 300, 400,500 and 600 K.

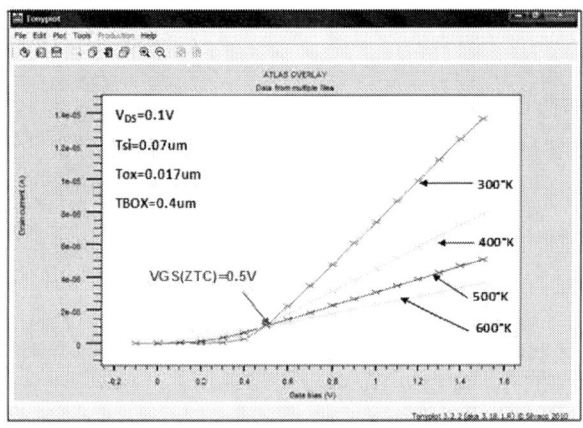

Fig. 8. Simulated I_{DS}–V_{GS} plot for a fully depleted SOI n-MOSFET at operating temperatures of 300, 400,500 and 600 K.

ZTC bias point exists in both linear and saturation regions of I_{DS} -V_{GS} characteristics according to the value of the V_{DS} biasing applied, which is well illustrated on the graph of Fig. 9.

In the short channel devices , the potential barrier to form the conduction channel depends by both the transversal field (controlled by the gate-to-source bias)and the longitudinal field (controlled by the drain-to-source bias), at high drain bias, we see a lower ZTC bias point, the potential barrier decreases leading to drain-induced-barrier- lowering (DIBL) that causes the threshold voltage to drop. Hence we see a lower ZTC bias point at high drain bias for a short channel SOI MOSFET.

REFERENCES

[1] A.K. Goel_, T.H. Tan, High-temperature and self-heating effects in fully depleted SOI MOSFETs Microelectronics Journal 37 (2006) 963–975.

[2] J. P. Colinge, Silicon-on-Insulator Technology: Materials to VLSI. Boston, MA: Kluwer, 1991.

[3] Wei Ma thesis, Linearity Analysis of Single and Double-Gate Silicon-On-Insulator Metal-Oxide-Semiconductor-Field-Effect-Transistor, Ohio University, August 2004.

[4] Alexandre SILIGARIS « Modélisation grand signal de MOSFET en hyperfréquences : application à l'étude des non linéarités des filières SOI » Thèse de doctorat 2004-UNIVERSITE DES SCIENCE ET TECHNOLOGIES DE LILLE.

[5] http://docinsa.insa-lyon.fr/these/pont.php?id=daviot

[6] G.K. Celler, S. Cristoloveanu, Journal of Applied Physics, Vol. 93, no. 9, p. 4955,2003.

[7] F.S. Shoucair, W. Hwang, Electrical characteristics of large scale integration (LSI) MOSFETs at very high temperatures part II: experiment, Microelectron. Reliab. 24(3) (1984) 497–510.

Multi-Objective Genetic Algorithm Optimization of CMOS Operational Amplifiers

[1]Samir Barra, [1,2]Abdelghani Dendouga, [3]Souhil Kouda, and [1]Nour-Eddine Bouguechal

[1]Advanced Electronics Laboratory -LEA
Department of Electronics
University of Batna
Chahid Boukhlouf M. El-Hadi Avenue
Batna, Algeria
barrasamir@hotmail.com

[2]Microelectronics and Nanotechnology Division
Center for Development of Advanced
Technologies - (CDTA)
August 20 1956 City, BP 17, Baba Hassen
Algiers, Algeria

[3]Department of Electronics
University of M'sila
B.P 166 Ichbelia
M'sila, Algeria

Abstract—this work studies the problem of CMOS operational amplifiers (op-amps) optimization. A front Pareto based-MOGA (Multi-Objective Genetic Algorithm) methodology is proposed to optimize the operational amplifier. The proposed approach is used to find the optimal dimensional transistor parameters in order to obtain operational amplifier performances for analog and mixed CMOS-based circuit applications. To evaluate the proposed approach, an example in both time and frequency domains for a two-stage CMOS Operational Transconductance Amplifier (OTA) is presented in 0.18µm process. The simulation results confirm the efficiency of MOGA in determining the device sizes in an analog circuit.

Keywords— Multi-objective Genetic Algorithm (MOGA), Optimization, CMOS Analog Circuit, Operational Amplifier.

I. INTRODUCTION

Nowadays, progresses made in the VLSI technology are leading to the full integration of mixed analog/digital circuits such as Pipelined ADC [1] [2] and Sigma-Delta modulator [3]. Even though the analog part presents a small portion of the entire circuit, its design is a very complicated task that generally relies on the experience of the skilled designer. Despite its importance, analog design automation still lags behind that of digital circuits [4]. Besides, analog circuit design is a hard and tedious work due to the large number of parameters, constraints and performances that the designer has to handle.

Genetic Algorithms (GAs) has been well known for their optimization abilities. They are particularly suited for this application since the operational amplifier (op-amp) design problem can be easily converted into a multi-objective search task [5]. Much research work has been devoted to the field of sizing transistors and synthesis of the analog circuit by GA [6-12], and multiple-objective optimization methods using genetic algorithms for multiple-objective problems are applied to the problem of op-amp design optimization [13-23], which is a problem of practical interest.

Our contribution in this work is the investigation of the hybrid evolutionary-based design system for automated sizing of analog integrated circuits (ICs). A front Pareto based-MOGA approach is proposed to design analog ICs. On the basis of the combination of Spectre and MATLAB, the system links circuit performances, evaluated through electrical simulation, to the optimization system in the MATLAB environment, once a circuit topology is selected.

In this paper, we present the applicability of multi-objective genetic algorithm optimization (MOGAs) approach based on Weighted Sum Strategy to optimize performance parameters of operational amplifier. The front Pareto based-MOGA approach is employed to automatically detect this front in a process that efficiently finds optimal parameterizations and their corresponding values in an aggregate fitness space. The problem of sizing an operational amplifier is addressed. The Pareto front is introduced as a useful analysis concept in order to explore the design space of such analog circuit.

II. OPTIMIZATION METHODOLOGY

Optimal design of analogue circuits consists of finding a variable set x ={x_1, x_2, ..., x_n} that optimizes a performance functions, such as gain, offset, signal to noise ratio, maximum operating frequency etc., while meeting imposed specifications and/or inherent constraints, for example, saturation conditions of transistors, technology limits, impedance matching, etc. Vector x may encompass biases, lengths and widths of MOS transistors, component values etc. [12]. Many multi-objective optimization methods have been developed over the past years [13-14]. These methods can generally be classified under the two main categories; weighted or aggregated approaches [15-23] and the Pareto-based approaches [11] [13].

A. Performance specifications

To simplify the discussion (as well as to save space) we concentrate on one op-amp topology, the two-stage op-amp of figure 1. The main electrical parameters of the circuit are low-frequency voltage gain (A_v), gain-bandwidth product (GBW), slew-rate (SR), dissipated power (P_{diss}), phase margin (PM) and area (A), among others. The design variables are the transistor sizes (width and length), the value of the passive components (capacitors and resistors), and the value of bias currents and bias voltages. For this particular two-stage op-amp, there are fourteen design variables.

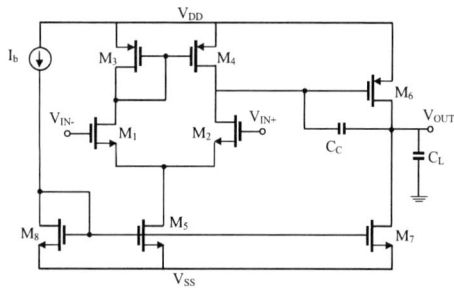

Fig. 1. Two stage op-amp

1) Unity-gain bandwidth

The unity gain bandwidth of two stage OTA is given by the expression [19]:

$$GBW = \frac{g_{m1}}{C_C} \quad (1)$$

2) Phase margin

The phase margin of op-amp depends on the sum of phase shifts, at the unity-gain frequency, contributed by the nondominant poles and zeros.

$$PM = \mp 180 - \tan^{-1}\left(\frac{GBW}{p1}\right) - \tan^{-1}\left(\frac{GBW}{p2}\right)\tan^{-1}\left(\frac{GBW}{z}\right) (2)$$

where GBW is the unity-gain bandwidth, p_1, p_2 and z are poles and zero of the operational amplifier open-loop transfer function, respectively.

3) Open-loop DC gain

For the two-stage op-amp, the open-loop voltage gain is given by [9]:

$$A_V = \frac{g_{m1}}{g_{ds2} + g_{ds4}} \cdot \frac{g_{m6}}{g_{ds7} + g_{ds6}} \quad (3)$$

4) Slew Rate

For the two-stage op-amp the conditions to ensure a minimum slew rate SR are [19]:

$$SR = \frac{I_5}{C_C} \quad (4)$$

5) Power consumption

For the two-stage op-amp the power consumption has the form [16]:

$$P = (V_{DD} - V_{SS})(I_5 + I_7) \quad (5)$$

where V_{DD} and V_{SS} are the power supply of op-amp.

6) Area

$$Area = \sum_{i=1}^{k} W_i . L_i \quad (6)$$

where W_i and L_i are the width and length of the transistors M_i, respectively.

7) Current equations

In the two-stage op-amp, if we define the bias currents I_5 and I_7 through transistors M_5 and M_7, respectively, we have [13]:

$$I_5 = \frac{W_5 L_8}{L_5 W_8} I_b \quad (7)$$

$$I_7 = \frac{W_7 L_8}{L_7 W_8} I_b \quad (8)$$

B. Multi-Objective Genetic Algorithm Optimization

In this work the optimizing program is written by MATLAB language, and we used MATLAB Optimization toolbox to implement GA Multi-objective optimization. At the beginning we generate the individual randomly n times (n represents the population size). The individual is made up of binary code string encoding a particular sized op-amp fitness of every individual can be got, and then the GA can be used to choose the better individuals as the parents of the next generation. After crossing and mutating, the new generation is produced. Performing the above works iteratively the goal will be achieved in the end.

1) Representation

In the program, every individual is presented by a binary code string. From Fig. 1 we can see that there are 8 transistors and a miler capacitor to be adjusted. As a total there are 14 parameters to be adjusted and each gene of the chromosome stands for one parameter. Thus the parameter vector is compressed to [11]: [W_1, L_1, W_3, L_3, W_5, L_5, W_6, L_6, W_7, L_7, W_8, L_8, C_c, I_{bias}].

The GA is used to manipulate the transistor sizes, biasing current, I_{bias} and the compensation capacitance, C_c [21]. A summary of the objectives and desired specifications can be found in Table 1. The bounds used for the transistor width, length, I_{bias}, C_c are given in Table I.

2) Fitness function

In this stage the parameter space is explored and the design improved with respect to the objective functions. The optimization, which is called multi-objective optimization, is based on an evolutionary algorithm known as weight-based genetic algorithm [15-23]. A weighted approach has been used to optimize op-amps. It uses adaptive weights along the optimization process to determine the overall fitness of an individual [16].

In order to reduce the calculation, we can mix every sub-objective into one general function, so the problem can be changed into one-objective optimization. The overall fitness can be achieved from the equation (9) [15]:

$$F = \sum_{i=1}^{n} \omega_i . f_i \quad (9)$$

In the above equation (9), ω_i is the weight coefficient of every sub-objective, f_i is the overall fitness of every performance considered and i is the number of the performance considered.

III. RESULTS AND DISCUSSIONS

In this simulation six performances are considered. They are the DC gain, bandwidth of unity-gain, phase margin, power area, and Slew Rate. The optimization process optimizes the individual to improve its fitness score. This process will continue until the total number of generations is reached.

Fig. 2 shows the variation in normalized overall objective function as a function of generation number where the minimum objective function can be reached for 100 iterations. As it can be seen the algorithm is converged to the optimized point, after 85 generations. Thus all the given specifications are satisfied and total fitness function is equal to 0.37.

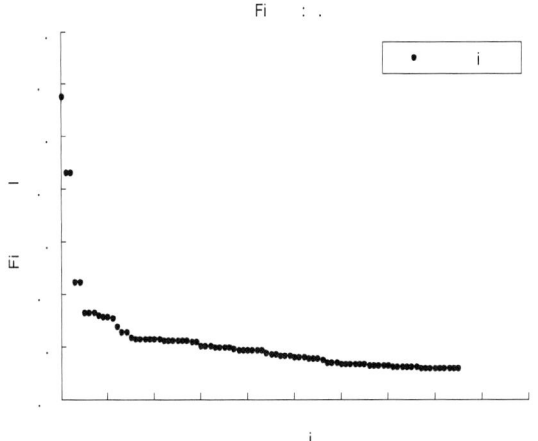

Fig. 2. Illustrates the Pareto optimal solutions of a two-objective design problem among the feasible optimal design space after 100 populations.

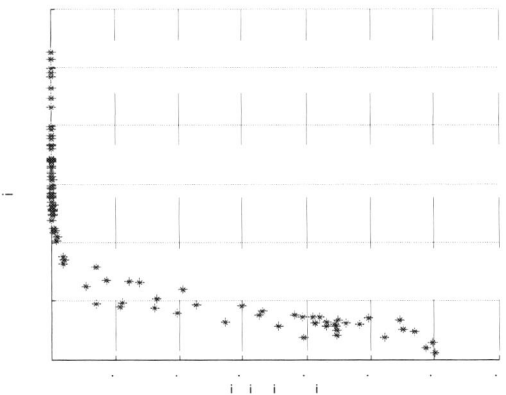

Fig. 3. Pareto optimal solutions for two-objectives of power consumption and DC gain after 208 iteration.

Fig. 3 indicates a direct relationship between DC gain and Power Consumption. It can be concluded that op-amp consumes too much power; if the DC gain is too small. One of the most interesting interpretations that one can have from obtained Pareto solutions is to understand the relationship between different objectives when the design parameters vary

in the design space. Hence, based on the design requirements, the designer can select each of the Pareto-optimal results.

Also by using variables obtained from GA, the OTA circuit is simulated by using Cadence Virtuoso Spectre in 0.18μm process and simulation results are shown in Table II and Fig. 4.

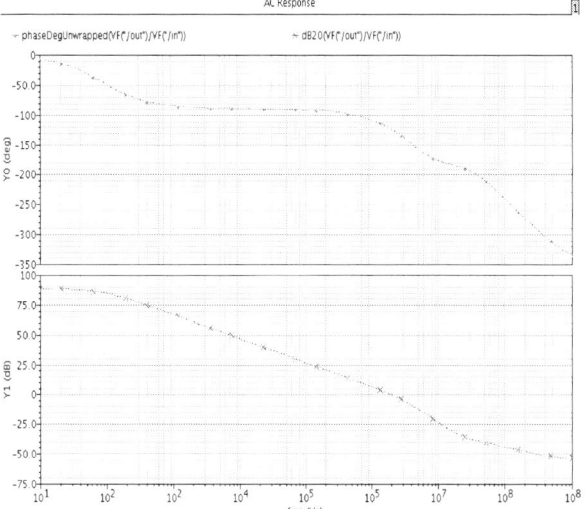

Fig. 4. Frequency response of the Two-stage Op-amp.

TABLE I. PERFORMANCE COMPARISON.

Variable	Value
$W_1=W_2$ (μm)	1,34
$L_1=L_2$ (μm)	1,5
$W_3=W_4$ (μm)	8,5
$L_3=L_4$ (μm)	1,11
W_5 (μm)	1,4
L_5 (μm)	0,44
W_6 (μm)	48
L_6 (μm)	0,92
W_7 (μm)	6
L_7 (μm)	0,8
W_8 (μm)	23,5
L_8 (μm)	0,68
C_c (pF)	1,5
I_{bias} (μA)	50

Table II shows the performance of the design obtained by MATLAB Optimization Tools, The objective was to maximize the unity gain bandwidth and minimize power consumption subject to the other given constraints. The simulation results confirm the efficiency of GA in determining the device sizes in an analog circuit. Also a comparison has been made between the results of the presented algorithm in this work and references [13], [19] in Table 2. According to the simulation results, the performance of the op-amp optimized by the proposed method is much better than those in the reference (particularly low power, low area). In comparison with the design presented in [13] and [19], the achieved Pareto design has lead to better aspects (96dB gain, 0,047mW, 89μm², and 70° PM).

TABLE II. COMPARISON WITH PREVIOUS WORKS.

Performance measure	[13]	[19]	This work MATLAB	This work Spectre
DC gain (dB)	103	78,38	96	87
Bandwidth of unity gain (MHz)	10,6	12.2	2,5	1,11
Phase Margin (°)	57,1	62	70	64
Slew rate (V/μs)	5,9	5.5	2,25	2,19
Area (μm2)	/	236,25	89	89
Power (mW)	0,993	0,060	0.047	0,051
Technology (μm)	0,25	0,18	0,18	0.18
Simulation run -time	10 hours	/	10 minutes	2 minutes

The OTA design optimization stage took 10 minutes on a 2.6GHz Intel processor, which compares well with a previously reported optimization time of 4 hours for the same circuit [15]. Our approach based on hybrid tools (optimization MATLAB tools and Cadence Spectre) for implantation of the MOGA optimization and the Spice simulations, and to complete all simulation about 10 minutes are needed. We observe the superiority of our method against other work [13] [15] in terms of simulation run-time and in terms of circuits' performances (Power, Area, DC gain and PM).

IV. CONCLUSION

In this paper Genetic Algorithm has been used for analog integrated circuit design. The Genetic Algorithm and equation based optimization were combined to produce an accurate tool in order to determine the device sizes in an analog circuit. A MOGAs-based approach is proposed to optimize the performances of two stage OTAs.

The results prove the effectiveness of the approach in the applications of Analog and Mixed-signal design where the design space is too complicated to be done with the classical methods within a short time. It can be concluded that the proposed MOGA-based approach is efficient and gives the promising results for circuits design and optimization problems.

REFERENCES

[1] Samir Barra, Souhil Kouda, Abdelghani Dendouga, and Nour-Eddine Bouguechal, "Simulink Behavioral Modeling of a 10-bit Pipelined ADC", *International Journal of Automation and Computing*", Vol 10 No 2, Apr. 2013.

[2] Samir Barra, Abdelghani Dendouga, Souhil Kouda, and Nour-Eddine Bouguechal, "Contribution to The Analysis and Modeling of The Non-Ideal Effects of Pipelined ADCs using MATLAB", *Journal of Circuits Systems and Computers*, Vol. 22, No. 2, Feb. 2013.

[3] Abdelghani Dendouga, Nour-Eddine Bouguechal ,Souhil Kouda , Samir Barra, and Brahim Lakehal, "Contribution to the modeling of a non-ideal Sigma-Delta modulator", Journal of Computational Electronics, Vol. 11, Issue 4, Dec. 2012, pp 321-329.

[4] Rutenbar, R.A., Gielen, G., Roychowdhury, J, "Hierarchical modeling, optimization, and synthesis for system-level analog and RF designs", *Proc. of the IEEE*, Vol. 95, Issue 3, 2007, pp. 640–669.

[5] Li Zhiyuan, Ye Yizheng, and Ma Jianguo, "Design Procedure for Optimizing CMOS Low Noise Operational Amplifiers", *Journal of Semiconductors*, Vol. 30, No. 4, April 2009.

[6] Jan M´lchal, And Josef Dobe`, "An Application of Multiobjective Optimization in Electronic Circuit Design", *In the Proceedings of the 5th*

International Conference on Systems Theory and Scientific Computation , WSEAS/IASME, Malta, Sep. 15-17, 2005, pp.213-219.

[7] Abdullah Konaka, David W. Coitb, and Alice E. Smith,"Multi-objective optimization using genetic algorithms: A tutorial", *Reliability Engineering and System safety Journal*, Vol.91, 2006, pp. 992-1007.

[8] Gabriel Oltean, Sorin Hintea, and Emilia Sipos, "A Genetic Algorithm-based Multi-objective Optimization for Analog Circuit Design", *In the Proceedings of the 13rd International Conference on Knowledge-Based and Intelligent Information & Engineering Systems*, KES, 28-30 September 2009, Santiago, Chile, pp. 506–514.

[9] Güner Alpaydın, Sina Balkır, and Günhan Dündar, "An Evolutionary Approach to Automatic Synthesis of High-Performance Analog Integrated Circuits", *IEEE Transactions on Evolutionary Computation*, Vol. 7, No. 3, June 2003, pp. 240-252.

[10] Bo Liu, Yan Wang, Zhiping Yu, Leibo Liu, Miao Li , Zheng Wang, Jing Lu, Francisco V. Fernández," Analog circuit optimization system based on hybrid evolutionary algorithms", *the VLSI journal*, Vol. 42 ,2009, pp. 137– 148.

[11] Elisenda Roca, Manuel Velasco-Jime´nez, Rafael Castro-López, Francisco V. Fernández, "Context-dependent transformation of Pareto-optimal performance fronts of operational amplifiers", *Analog Integrated Circuits and Signal Processing journal*, Vol. 73, Issue 1, October 2012, pp 65-76,

[12] Mario Köppen, Gerald Schaefer, and Ajith Abraham, "*Intelligent Computational Optimization in Engineering*", Edition Springer 2011, pp.300-331.

[13] Maria Del Mar Hershenson, Stephen P. Boyd, Thomas H. Lee, "GPCAD:A Tool For CMOS Op-Amp Synthesis", *In the Proceedings of IEEE/ACM International Conference on Computer-Aided Design, ICCAD 98* , Nov. 8-12,1998, San Jose, California, USA, pp. 296 – 303.

[14] R. S. Zebulum, M. A. Pacheco, and M. Vellasco, " Synthesis of CMOS operational amplifiers through Genetic Algorithms", *In Proceedings of the Brazilian Symposium on Integrated Circuits, SBCCI'98*, Sep.30-Dec.03,1998, Rio de Janeiro, Brazil, pp.125–128.

[15] Milton Jonathan, Marco Aurélio, Cavalcanti Pacheco , Ricardo Salem Zebulum, and Marley B.R. Vellasco, "Multiobjective Optimization Techniques: A Study Of The Energy Minimization Method And Its Application To The Synthesis of OTA Amplifiers", *In the Proceedings of the Second NASA/DoD Workshop on Evolvable Hardware* , Palo Alto, California, USA, July 13-15, 2000 , pp. 133–140.

[16] Jianhai Yu, Zhigang Mao," Automated Design Method for Parameters Optimization of CMOS Analog Circuits Based on Adaptive Genetic Algorithm", *In the Proceedings of the 7th International Conference on ASIC ASICON*, 22-25 Oct,2007, Guilin, China, pp. 1217 - 1220.

[17] Jianhai Yu, Zhigang Mao,"A Design Method In CMOS Operational Amplifier Optimization Based On Adaptive Genetic Algorithm", *Wseas Transactions On Circuits and Systems*, Issue 7, Vol. 8, July 2009, pp.548-558.

[18] Sawal Ali, Reuben Wilcock, Peter Wilson, Andrew Brown, "Yield Model Characterization for Analog Integrated Circuit Using Pareto-Optimal Surface", *In the Proceedings of the 15th IEEE International Conference on Electronics, Circuits and Systems, ICECS*, Aug.31-Sept. 3 ,2008,Malte , Malta, pp.1163– 166.

[19] Jafari, A, Zekri M, Sadri S, Mallahzadeh A.R, "Design of Analog Integrated Circuits by Using Genetic Algorithm", *In Proceedings of the IEEE 2nd International Conference on Computer Engineering and Applications, ICCEA* , Mar.19-21 2010, Bali Island, Indonesia, Vol. 1, pp. 578 - 581.

[20] M. Taherzadeh-Sani, R. Lotfi, H. Zare-Hoseini And O. Shoaei, "Design Optimization Of Analog Integrated Circuits Using Simulation-Based Genetic Algorithm", *in Proceedings of Signal Circuit and Systems Conference(SCS)*, Romania, 2003, pp. 73-76.

[21] C. Goh And Y. Li, Multi-Objective Synthesis Of CMOS Operational Amplifiers Using A Hybrid Genetic Algorithm", *in the Proceedings of the 4th Asia-Pacific Conference on Simulated Evolution And Learning (SEAL 2002)*, Singapore, Nov. 2002, pp. 214-219.

[22] B. K. Mishra and Sandhya Save, "CAD Design of Operational Amplifiers with Noise Power Balance for SoC Application", *International Journal of Computer Theory and Engineering*, Vol. 2, No. 4, August, 2010.pp. 608-612.

[23] P. Prem Kumar, K. Duraiswamy, and A. Jose Anand, "An Optimized Device Sizing of Analog Circuits using Genetic Algorithm", European *Journal of Scientific Research*, Vol.69, No.3,2012, pp. 441-448.

Simulation of Ion Implantation for CMOS 1μm Using SILVACO Tools

Mohamed BOUBAAYA, Fayçal HADJ LARBI, Slimane OUSSALAH

Microelectronics and Nanotechnology Division
Centre for Development of Advanced Technologies, CDTA
Baba Hassen, Algiers, Algeria
mboubaaya@cdta.dz

Abstract— **One micron gate-length LDD-CMOS (Lightly Doped Drain - Complementary Metal Oxide Semiconductor) technology uses N and P-MOSFETs, realized on the same substrate, in a way to benefit simultaneously from a combination of their characteristics. Some regions of these transistors like the Wells and the Source/Drain are created by a reliable technique called** *ion implantation*. **The aim of this paper is to simulate the ion implantation steps included in the LDD-CMOS 1 micron process which has been acquired by CDTA (Centre for Development of Advanced Technologies) from ISiT (Fraunhofer-Institut für Siliziumtechnologie). The process simulation framework ATHENA of the TCAD (Technology Computer Aided Design) SILVACO's software was employed to simulate our process and extract some process technology parameters, such as P+ and N+ doping concentrations as well as the ion implantation energies and doses, and the corresponding junction depths.**

Keywords-CMOS; ion implantation; TCAD simulation; ATHENA

I. INTRODUCTION

CMOS is by far the most wide-spread technology in the integrated circuits fabrication industry. TCAD simulation is occurring as a crucial step before device manufacturing. Ion implantation is the main technology for creation of the p-n junction and a specific thermal '*drive-in*' is required for the redistribution of the doping elements [1].

The process technology acquired by CDTA from ISiT institute is a double layer metal twin-well technology on p^+-type (Boron) 12 μm epi-layer (10 Ωcm) deposited on p-doped (10 mΩcm) Silicon <100>-oriented substrate. The use of a lightly doped epitaxial layer on a heavily doped substrate allows achieving high latch-up equivalent circuit immunity. For process simulation, we have used the ATHENA tool of the commercially available TCAD software package of SILVACO. There are four different analytical implant models that are considered for implantation according to the temperature, impurity, time and particles energy [2]. Gaussian implant model uses Gaussian distribution, Pearson implant model calculates the asymmetrical ion implantation profile and more better Dual Pearson model extends toward profiles heavily affected by channeling [3-4]. ATHENA process simulator of TCAD SILVACO software uses analytical (Gaussian, Pearson,

Dual Pearson and SIMS-Verified Dual Pearson distributions) and statistical techniques (Monte Carlo based on the Binary Collision Approximation) to model ion implantation. By default, the analytic models are used and are based on range concepts from "Ranges Concepts and Heavy Ion Ranges" [1].

Process simulation output includes two and three dimensional structure profiles, and the extraction of structural parameters. Structural information includes doping profiles and impurity concentrations.

II. PROCESS SIMULATION

A. Pearson Implant Model

Generally, the Gaussian distribution is inadequate because real profiles are asymmetrical in most cases. The simplest and most widely approved method for calculation of asymmetrical ion-implantation profiles is the Pearson distribution [5]. The Pearson function refers to a family of distribution curves that result as a consequence of solving the following differential equation:

$$\frac{df(x)}{dx} = \frac{(x-a)f(x)}{b_0 + b_1 x + b_2 x^2} \tag{1}$$

in which $f(x)$ is the frequency function. The constants a, b_0, b_1 and b_2 are related to the moments of $f(x)$ by:

$$a = -\frac{\Delta R_p \gamma (\beta + 3)}{A} \tag{2}$$

$$b_0 = -\frac{\Delta R_p^2 (4\beta - 3\gamma^2)}{A} \tag{3}$$

$$b_1 = a \tag{4}$$

$$b_2 = -\frac{2\beta - \gamma^2 - 6}{A} \tag{5}$$

where $A = 10\beta - 12\gamma^2 - 18$, with γ and β are the skewness and kurtosis, respectively.

B. Simualtion Techniques

Process simulators provide quantitative and qualitative information regarding the relationship between final device characteristics and the process fabrication parameters. SSUPREM4 of ATHENA is a file based simulation program, using text files to simulate process steps, and doping profile

978-1-4673-5289-5/12 $31.00 © 2012 IEEE

information. Physical processes are simulated numerically using the finite difference and finite element methods.

The structure is generated on a grid of non-overlapping triangular elements (fig.1.a) creating mesh nodes which the total number of must not exceed 20.000 [6]. Up to forty regions representing different materials can be represented including single crystal silicon, poly-silicon, silicon dioxide, silicon nitride, aluminum and photo resist.

The structure is then initialized, specifying all relevant information including initial concentration of impurities, initial resistance and crystalline orientation. The implantation stages including wells annealing process are introduced in Fig. 1 (b-e).

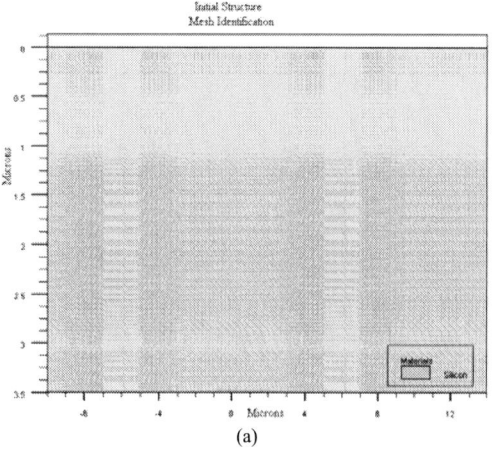

(a)

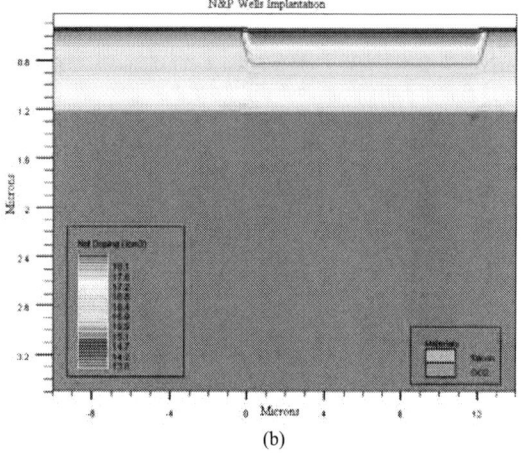

(b)

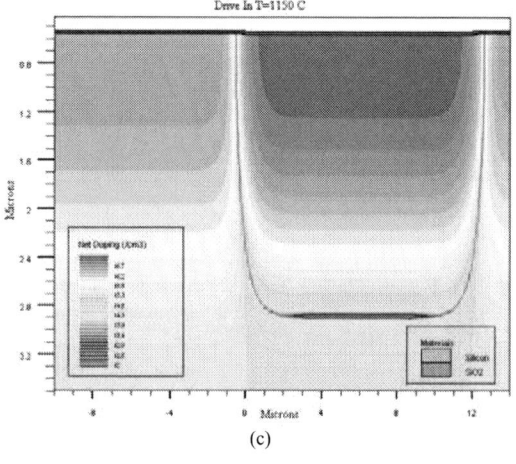

(c)

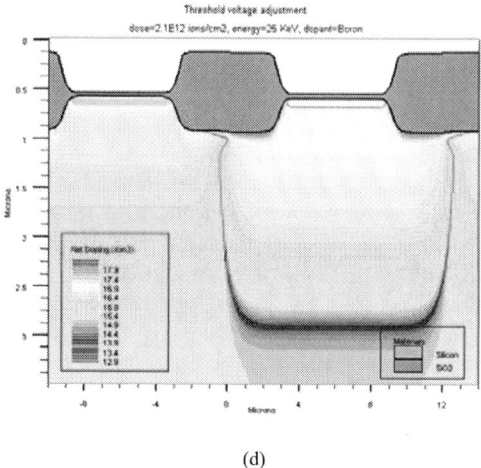

(d)

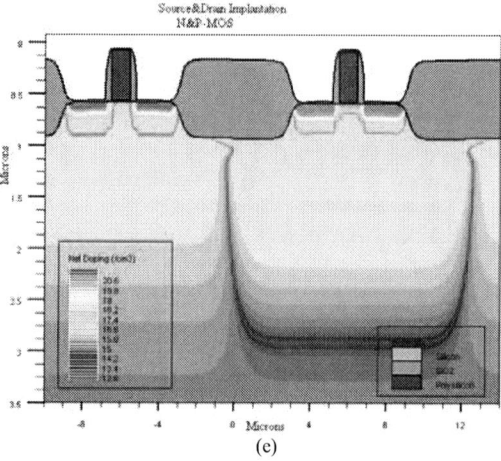

(e)

Figure 1.   Different output stages in N&P-MOS simulation including (a) mesh definition, (b) N and P well formation, (c) annealing, (d) threshold voltage adjustment, and (e) source and drain imlpants for n- and p-MOS transistors.

III. STRUCTURAL CHARACTERISTICS

A. Dopants profile (NMOS transistor)

Figure 2 illustrates the concentration for three impurities in the n-MOS source drain regions. We find a peak implant concentration for Arsenic, a concentration for Phosphorus as a result of the Lightly Doped Drain ion implantation step, and a concentration for Boron as result of P-Well implantation, after Arsenic source/drain implantation. However, the concentration of Boron was decreased and we have a junction depth of about 0.30 µm.

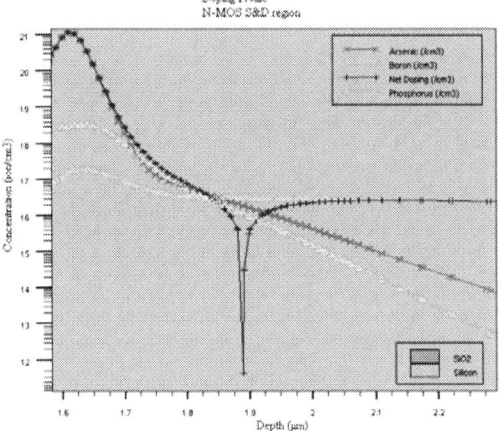

Figure 2.   Doping profile after the n-MOS source and drain ion implantation.

B. Dopants profile (PMOS transistor)

Figure 3 shows the concentration of Boron and Phosphorus in the p-MOS transistor. The graph depicts the significant increase in concentration of Boron as a result of source/drain implant. The junction depth in the source and drain regions is approximately 0.30 µm, and it is about 2.31 µm in the N-well.

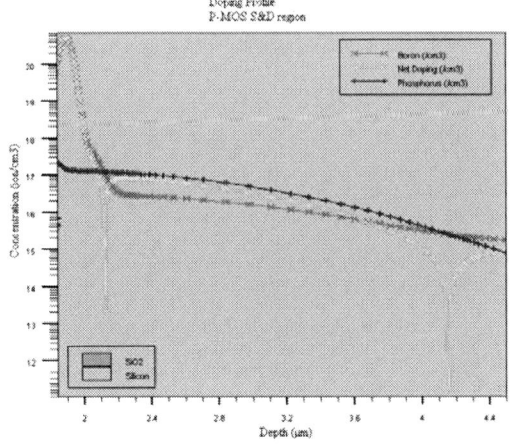

Figure 3.   Doping profile in the p-MOS transistor.

C. Sheet resistance

The implantation of the different CMOS regions contributes to the sheet resistance variation of the doped layers. Table I represents the values of the sheet resistance for N-Well and source/drain of each transistor (n- and p-MOS transistor).

TABLE I.   EXPERIMANETAL AND SIMULATED SHEET RESISTANCE

Region	Sheet resistance ($\Omega/\square$)	
	Simulation results	Experimental results (ISiT)
N-Well	$1.024\ 10^3$	$1\ 10^3 - 1.4\ 10^3$
S/D n-MOS	20.35	42 - 68
S/D p-MOS	42.75	55 - 95

D. Threshold voltage

The threshold voltage (Vth) of a device is the voltage at which the MOS transistor channel begins to conduct the current. This voltage was adjusted by ion implantation. After all the LDD-CMOS 1 µm ion implantation steps, Vth is extracted for n- and p-MOS transistors from the drain-source current versus the gate voltage characteristic using the ATLAS device simulation module of SILVACO. Table II represents a comparison between simulated and experimental values of Vth for n- and p-MOS transistors.

TABLE II.   THRESHOLD VOLTAGE VALUES

Vth (V)	Simulated	Measured
NMOS	-1.09	-1.12
PMOS	0.63	0.64

IV. CONCLUSION

This paper demonstrates the application of the process simulation software, ATHENA, to the CMOS 1 µm baseline and particularly to the ion implantation. Pearson model approach is used to the ion implantation simulation because it gives good shape of doping profile and time consumption. The threshold voltage values extracted from the simulation for n- and p-MOS transistors are close to the experimental ones.

REFERENCES

[1] J. Lindhard, M. Scharff, and H.E. Schiott. "Range Concepts and Heavy Ion Ranges", Kgl. Dan. Vid. Selsk. Mat.-fys. Medd., v. 33, 1963.

[2] A. F. Tasch, "An Improved Approach to Accurately Model Shallow B and BF2 Implants in Silicon", J. Electrochem. Soc., v. 136, p. 810, 1989.

[3] A.F. Burenkov, F.F.Komarov, and M.M.Temkin. "Analytical Calculation of Ion Implantation through Mask Windows" (in Russian), Microelektronika, v. 16, p. 15, 1987.

[4] D.G. Ashworth, M.D.J. Bowyer, and R. Oven, "Representation of Ion Implantation Distributions in Two and Three Dimensions", J. Phys. D, v. 24, p. 1120, 1991.

[5] D.G. Ashworth, R. Oven and B. Mundin, "Representation of Ion Implantation Profiles be Pearson Frequency Distribution Curves",J.Phys. D, v. 23, p. 870, 1990.

[6] Athena User's Manual, March 4, 2011.

Modeling CMOS PIN Photodiode using COMSOL

Mohamad HAMADY, Ehsan KAMRANI, and Mohamad SAWAN

Polystim Neurotechnologies Laboratory
Electrical Engineering Department, Polytechnique Montréal
Québec, Canada
mohamad.hamady@polymtl.ca

Abstract— Modeling semiconductor devices has become mandatory in most challenging research activity. Finding a powerful tool that models these devices represents a goal of these users. In this work, Silicon PIN photodiode is designed using complementary metal-oxide semiconductor (CMOS) Technology. COMSOL Multiphysics is the selected challenging tool for simulation and characterization of this design. The adjustment of the proposed model as well as the different outputs like Electric field and Electric potential distribution, I-V characteristics and other parameters are presented. The output current has shown an allure of a current that consists of three regions: Trap-Assisted Tunneling, Band-to-Band Tunneling and avalanche. These current regions in addition to the high value (10^5 V/cm) of the obtained electric field are typical for Silicon Avalanche Photodiodes (SiAPD). The possibility to use this tool for SiAPD analysis and simulation is therefore discussed.

Keywords: COMSOL, Semiconductors, PIN photodiode.

I. INTRODUCTION

Several tools are used to simulate and/or design semiconductor devices. TCAD is the most traditional tool that is currently used and it was adopted in different publications like in [1]. However, the process to make a full design by this tool is complicated and makes it difficult to use. In addition, this tool only takes into considerations the electric characteristics of the design and not all the physical aspects that exist in the model. This might produce results that did not exactly match with the experimental ones as stated in [2]. Another tool like Coventorware provides also some semiconductor simulations. Unfortunately, these simulations cannot be used to make a suitable design. Although, it permits to make a deposition process but there is no doping process (which is very essential in semiconductor design) can be provided by this tool. Hence, the simulations that are done by this software are likely to be static and never can be kinetic. Synopsis, Silvaco and Ansys are other available tools used in semiconductor simulations but they also suffer from non-adaptability with semiconductor process and functioning like previous cited tools.

COMSOL Multiphysics is a tool that, following to its title, takes into consideration all the physical aspects that exists in the design. The semiconductor model [3], [4] and [5] represent the only available model that treats the semiconductor devices using COMSOL. From this model we have designed a CMOS PIN photodiode after paying high attention in defining the geometry, physics and boundary conditions. The interest of using CMOS technology is its low cost and high integration capacity with the circuit. In addition, modeling a PIN photodiode represents a good background before going further in more complex design like modeling a Silicon Avalanche photodiode (SiAPD) with Premature Edge Breakdown Prevention (PEBP) techniques as in [2].

Silicon PIN photodiodes are used in different kinds of applications. In general, the bulk region of these photodiodes consists of intrinsic silicon. However, in practice this region is slightly doped n-type n⁻ silicon (usually called PvN), or slightly doped p-type p⁻ silicon (usually called PπN). A complete design of PvN in 2D and in 3D is presented in section II (the model also functions for a PπN photodiode). The model adjustment and the obtained results are explained in section III.

II. MODEL DESIGN

As all models designed by COMSOL, we have provided the PIN photodiode model the required physical modules, boundary conditions and the proposed geometry. The model is then meshed in order to obtain the required results. Maxwell's equations, Boltzmann transport theory and Neumann boundary conditions are employed to resolve the different equations that exist in semiconductor and to provide boundary conditions between the different junctions. The basic equations that exist in model are expressed in equation (1):

$$
\begin{aligned}
-\nabla.(\varepsilon \nabla \psi) &= q(p - n + N) \\
-\nabla.J_n &= -qR_{SRH} \\
-\nabla.J_p &= qR_{SRH}
\end{aligned}
\tag{1}
$$

where ψ equals to electrostatic potential, q indicates the elementary charge, n and p are the electron and hole concentrations respectively, N represents the fixed charge associated with ionized donors.

The electron and hole current densities, **J**n and **J**p (2), are expressed in terms of ψ, n and p:

$$
\begin{aligned}
J_n &= -qn\mu_n\nabla\psi + qD_n\nabla_n \\
J_p &= -qn\mu_p\nabla\psi - qD_p\nabla_p
\end{aligned}
\tag{2}
$$

where μ_n and μ_p are the carrier mobilities, D_n and D_p are the carrier diffusivities.

Shockley-Read-Hall recombination is represented by the term R_{SRH} (3), which is a general recombination process using traps in the forbidden band gap of the semiconductor.

$$R_{SRH} = \frac{np - n_i^2}{\tau_p(n + n_1) + \tau_n(p + p_1)} \quad (3)$$

where n_i is the intrinsic carrier concentration, τ_n and τ_p are the carrier lifetimes, n_1 and p_1 are equal to n_i.

The boundary conditions are expressed in equation (4)

$$\psi = V_a + \frac{\kappa T}{q}\ln\left(\frac{\frac{N}{2} + \sqrt{\left(\frac{N}{2}\right)^2 + n_i^2}}{n_i}\right), \quad n = \frac{N}{2} + \sqrt{\left(\frac{N}{2}\right)^2 + n_i^2} \quad (4)$$

$$p = -\frac{N}{2} + \sqrt{\left(\frac{N}{2}\right)^2 + n_i^2}, \qquad n_i^2 = np$$

where Va is the applied voltage which is going to be varied with a small step ΔV.

At each value of Va, the current densities J_n and J_p are calculated using Lagrange multipliers that already exists in COMSOL. A more detailed description about the physical equations and the boundary conditions in addition to complete Silicon properties are explained and shown in [3].

The two electrodes in [3] are placed opposite to each others. We have designed the two electrodes N^+ and P^+ to be at the top of the substrate to be adapted with CMOS technology. The p-type doping profile is considered to have a negative sign while the n-type doping profile is considered to have a positive sign. The doping concentration N is approximated with an exponential functions as in (5) in addition to the slightly doped type.

$$f(a,b,k) = e^{\left(-\left(\frac{a}{ch}\right)^2\right)} * \left(\left(|b| < \frac{k}{2}\right) + \left(|b| \geq \frac{k}{2}\right) * e^{\left(-\left(\frac{\left(|b|-\frac{k}{2}\right)}{ch}\right)^2\right)}\right) \quad (5)$$

Hence N can be expressed as:

$$N = N_v + NDn\,\max * f(y, x - d_{cathode}, e_{cathode})$$
$$+ NDn\,\max * f(y, x + d_{cathode}, e_{cathode}) - NAp\,\max * f(y, x, e_{anode}) \quad (6)$$

where N_v represents the slightly doped n⁻ silicon (replaced by $-N_\pi$ in case of slightly doped p⁻ silicon), NDnmax and NApmax represents N^+ (maximum n-type) and P^+ (maximum p-type) respectively, $e_{cathode}$ and e_{anode} represents the cathode and anode dimension respectively, $d_{cathode}$ corresponds to the distance from the midpoint of e_{anode} to the midpoint of $e_{cathode}$, ch is the doping fall-off constant and it is expressed by (7) as:

$$ch = \frac{ju}{\sqrt{\log\left(\frac{NAp\,\max}{N_v}\right)}} \quad (7)$$

where ju represents the junction depth.

The geometry of the 2D model is built in the xy plane. The PIN photodiode approximates 30 μm along x-axis and 3.4μm along y-axis. The anode approximates 10 μm at the middle of the substrate along x-axis and the cathode approximates 3 μm and designed to be some distance far from the anode extremity. The applied voltage Va is placed on the anode, and the cathode is supposed to be grounded. High attention is paid while variation of Va as the model might not converge. A small initial value (typically zero) and a small step ΔV (typically of order 10^{-2}) might avoid the non convergence of the model.

Once the geometry is defined, we mesh the design as shown in Fig. 1. We have done higher refinement to the meshes that are located near the cathodes and near the anode with respect to other meshes that exist in the system as they contain most of the physical and boundary conditions.

Most designers are satisfied in 2D simulations as they take the third dimension following to the technology limit (0.35 μm or 0.18 μm, etc..). COMSOL provides the possibility to make a 3D simulation. In our model, once the 2D design in the xy plane is done, we extrude the design along the z-axis. Then we add the physical and boundary conditions to the new design. Fig. 2 shows the overview of the model in 3D of PνN before and after meshing.

For the 3D design, we have reproduced exactly the same physical equations and boundary conditions in 2D as we haven't assumed any variations along the z-axis.

III. RESULTS AND DISCUSSIONS

The model consists of different input parameters (geometry, doping concentration, applied voltage, etc..) that influence the obtained results. In addition, the model can provide different output parameters (Electric field, Electric potential distribution, etc..). For simplicity, we will study the variations of PIN structure on the output electric field. Fig. 3 depicts the PIN and the associated variables to be varied.

Figure 1: Model mesh of PIN.

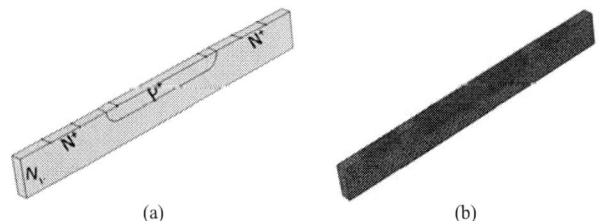

(a)   (b)

Figure 2: Overview of the PIN model in 3D (a) before meshing (b) after meshing.

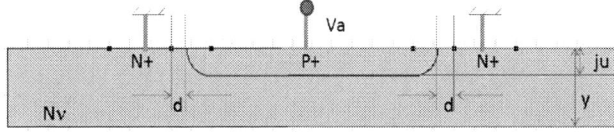

Figure 3: The PIN model where ju is the junction depth, d is the gap between two electrodes, y the thickness of the substrate under study, Va is the applied voltage, N^+ and P^+ are maximum n-type and p-type doping concentration, N_v represents the slightly doped n⁻ silicon

A. Model Adjustments

The model adjustment was studied without taking into account the state-of-the-art of any fabrication technology in order to have a complete overview whatever the technology adopted is. We have supposed $N_v = 10^{15}$ cm^{-3} and $N^+ = P^+ = 10^{17}$ cm^{-3} and we have applied a variable reversed bias voltage along Va. We have studied the influence of the gap between two electrodes (d), the junction depth (ju), and the PIN thickness along y as shown in Fig. 3 on the electric field.

- Influence of the gap between two electrodes (d)

The distance between the electrodes is governed by the technology that will be applied (0.35µm, 0.18µm, etc..). However, when technology permits, it is important to know what the best position of the electrodes is. The electric field was calculated for different distance (d) between P^+ and N^+ and the results are shown in Fig. 4. The electric field increases as the distance (d) decreases. In addition, Fig. 4-a shows an electric field near to an avalanche electric field.

- Influence of the junction depth (ju)

Following to Silicon absorption length, the light with longer wavelength can penetrate to the deeper junction which allows the incident light with longer wavelength to excite electron-hole pairs at the deep regions. In some applications of image sensor, it is important to have a photo receiver that detects all visual light. Following to [6], the blue, green and red light photodiodes must have a junction depth of 0.1 nm, 1.1 nm and 2.1 nm respectively. The electric field for these three different junction depths is shown in Fig. 5.

- Influence of different thickness (y)

Like precedent statement, it is also important to know what the best thickness of substrate is when technology permits. The electric field was calculated for different thickness (y) and the results are shown in Fig. 6. The electric field increases as the thickness (y) increases.

We have studied the influence of variation of the PIN structure on the electric field. The same variations were noticed while using a PπN photodiode. The high values of the electric field were seen at the junction peripheral edges and this will produce premature breakdown of the device [7]. The resolution

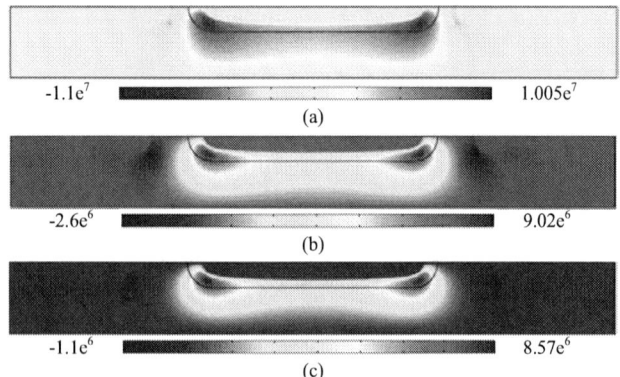

Figure 4: The electric field for (a) d=1 nm (b) d=2 nm (c) d=3 nm and the corresponding electric field bar in V/m.

of this problem represents our future work.

Other tools are more powerful than COMSOL in semiconductor simulations as shown in Table 1. Hence, it is preferable to use this tool in combination with others tools in order to have a complete device design.

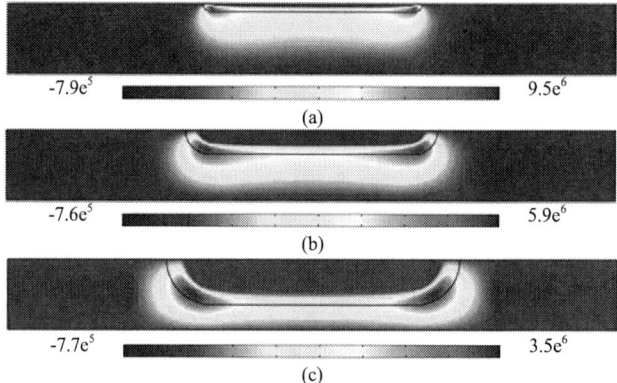

Figure 5: The electric field for (a) ju=0.1 nm (b) ju=1.1 nm (c) ju=2.1 nm and the corresponding electric field bar in V/m.

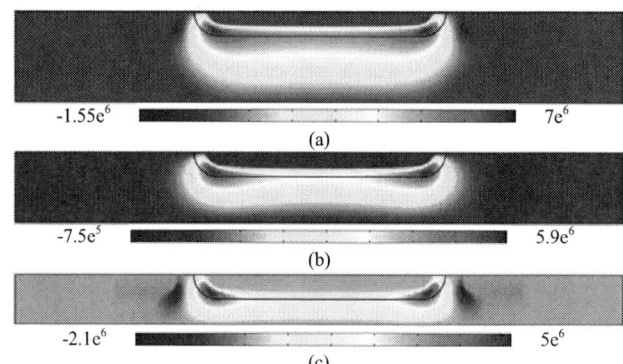

Figure 6: The electric field for (a) y=4.2 nm (b) y=3.2 nm (c) y=2.2 nm and the corresponding electric field bar in V/m.

TABLE I. COMPARISON BETWEEN DIFFERENT TOOLS IN SEMICONDUTOR SIMULATIONS

Tool / Parameter	COMSOL	TCAD	MATLAB	ANSYS
Facility to produce the model	Excellent	GOOD	POOR	GOOD
Calculate all parameters together	GOOD	POOR	Excellent	GOOD
Speed	Excellent	GOOD	POOR	GOOD
Complexity Level	Low	Very High	Low	Moderate
Cost	High	Moderate	Low	Moderate
Reliability	GOOD	Moderate	Moderate	Moderate
Matching with CMOS Technology	POOR	Excellent	POOR	POOR
Optics adaptability	Excellent	POOR	GOOD	Moderate
Optoelectronics adaptability	Moderate	Moderate	GOOD	POOR
Hardware Link	GOOD	POOR	GOOD	GOOD

B. Simulated Outputs

In this section we present some of the output results that can be obtained by the model. We have studied the PIN photodiode of Fig. 5-b (Green PIN).

- • Doping concentrations

We have introduced a doping profile as in equation (6). The model has reproduced the doping profile n and the doping profile p of equation (1) exactly as they are introduced as shown in Fig. 3. The two maximum n-type (N^+) are at both extremities and are shown in Fig.7-a. The maximum p-type (P^+) at the middle of the substrate is shown in Fig. 7-b

- • Electric Potential

We have calculated the electric potential distribution of the PIN in 3D. The distribution almost disappears around 0.7V (silicon threshold voltage) as shown Fig. 8-b, while it is very high in forward bias around P^+ as shown in Fig. 8-a and also very high in reversed bias around N^+ as shown in Fig. 8-c .

- • The I-V characteristics

The reverse bias voltage is only considered while studying the I-V characteristic of the photodiode. Fig. 9 shows a breakdown voltage of about 22V. This potential is relatively high to be used in CMOS technology. However, it shows an allure of a current that consists of the following zones: (from 0V to -15V) Trap-Assisted Tunneling, (from -15V to -22V) Band-to-Band Tunneling, and (after -22V) the avalanche zone.

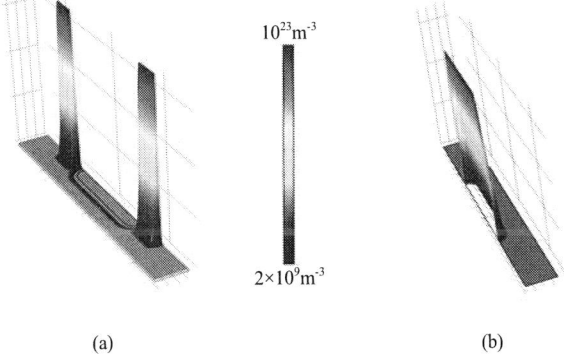

(a)   (b)

Figure 7: The doping concentration of (a) n-type (b) p-type of the PIN.

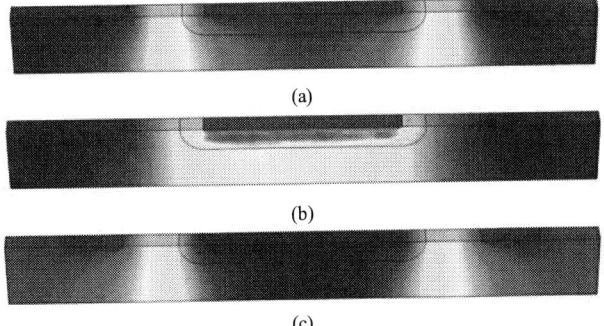

Figure 8: The variation of the electric potential distribution (a) in forward bias (b) around the threshold voltage (c) in reversed bias.

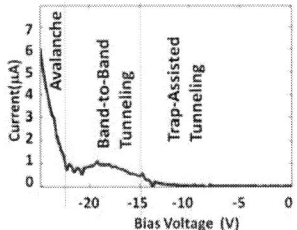

Figure 9: The I-V characteristics of the PIN in reversed bias.

These zones are very interesting as our future work is to design a SiAPD photodiode. In addition, the value of the electric field is in order of 10^5V/cm as shown in Fig. 4-a. This high value represents the beginning of the avalanche zone of Silicon photodiodes. The introduction of n/p-well or deep n/p-well will reduce the premature breakdown and also the breakdown voltage till around 10V as in [2].

IV. CONCLUSION

We have modeled and characterized a PνN photodiode in CMOS technology using COMSOL. As long as the physical and boundary conditions are well defined, this tool can provide an easy way to simulate a very complex subject like semiconductors design. In addition, it allows creating or modifying the model in relatively quick time compared to other tools. Modeling a Silicon avalanche photodiode for Near InfraRed Spectroscopy (NIRS) will be the perspective work. Therefore, a good declaration of the Premature Edge Breakdown Prevention (PEBP) techniques must be employed in order to obtain a low breakdown voltage and a very high electric field situated just under the active area and not only at the junction peripheral edges.

ACKNOWLEDGMENT

We gratefully acknowledge the financial support from the Canadian Institutes of Health Research (CIHR) and CMC Microsystems.

REFERENCES

[1] A. Sultana, E. Kamrani, M. Sawan, "CMOS Silicon Avalanche Photodiodes for NIR Light Detection: A Survey", Journal of Analog Integrated Circuits and Signal Processing, 17 April 2011

[2] E. Kamrani, F. Lesage, M. Sawan, "Premature Edge Breakdown Prevention Techniques in CMOS APD Fabrication" , The 10th IEEE International NewCAS'12 conference , Montreal, Canada, pp. 345-348 June 2012.

[3] COMSOL Model Librairy Version 3.5a, "Semiconductor Diode", 2008.

[4] R. Pryor, "Modeling PIN Photodiodes" , 2010 COMSOL conference, Boston, USA, 2010.

[5] R. Millett, J. Wheeldon, T. Hall, H. Schriemer, "Towards Modelling Semiconductor Heterojunctions", 2006 COMSOL Conference, Boston, USA, 2006.

[6] O. Chen, W. Liu, "Color-Selective CMOS photodiodes Based on Junction Structures and Process Recipes", Advances in Photodiodes, Intech Press, March 2011.

[7] M.A. Karami, M. Gersbach, H.J. Yoon, E. Charbon, "A new single-photon avalanche diode in 90nm standard CMOS technology", Optics Expr. 18, 2010.

A Novel Method based On Capacitance-Voltage for Negative Bias Temperatures Instability Studies: Concept and Results

Abdelmadjid BENABDELMOUMENE, Boualem
DJEZZAR, Hakim TAHI, and Amel CHENOUF
Microelectronics and Nanotechnology Division
Centre de Développement des Technologies Avancées
20 Août 1956, Baba Hassen, Algiers 16303
Email: abenabdelmoumen@cdta.dz.

Leonard TROMBETTA
Dept. of ECE, N308 Engineering Building 1
Houston, TX 77204-4005, USA,
Email: Ltrombetta@uh.edu

Mohamed KECHOUANE,
USTHB–Physics Faculty,
Bab Ezzouar, 16111 Algiers, Algeria

Abstract— **In this paper, a novel method in MOS capacitors is suggested for negative bias temperature instability (NBTI) measurement. This method is based on C-V technique and allows to independently extracting the interface (ΔN_{it}) and oxide traps (ΔN_{ot}). The method permits a broad investigation in the reliability study by exploiting capacitance. It is based on a simple theoretical concept and consists to measure the evolution of capacitance in two points; the first at the flat-band voltage (V_{fb}) and the second at V_{fb} - 100mv. The relations of voltage shifts components (V_{fb}, V_{mg}, and V_{it}) are developed by considering a linear CV characteristic variation between V_{fb} voltage and mid-gap voltage (V_{mg}). The experimental results have shown that the proposed approach allows reducing the recovery amount compared to full C-V characteristics. The trapped charge (ΔN_{eq_vfb}) calculated from ΔV_{fb}, ΔN_{ot} and ΔN_{it} present a linear stress-time-dependence which is in good agreement with those found in literature. In addition, the results have shown a similar kinetics of interface state generation as well as oxide trapped charges.**

Keywords- Capacitance-Voltage, NBTI, oxide-trap, interface-trap.

I. INTRODUCTION

Negative bias temperature instability (NBTI) in MOS transistors has become one of the major device-reliability problems. Several techniques have been suggested to understand the degradation mechanisms, among them we quote; JEDEC standard technique [1], New Fast-switching NBTI characterization [2], on-the-fly techniques (interface trap OTFIT using charge pumping [3, 4], one-point [5], three-point [5] OTF-V_{th} using IV).

Moreover, all the carried out studies showed that the NBTI degradation is induced by two components; interface-trap (ΔN_{it}) and oxide-trap (ΔN_{ot}). These latter contribute to threshold (or flat band) voltage degradation via ΔV_{it} and ΔV_{ot}

respectively. Therefore, it is very important to accurately estimate and separate these two components in order to understand the degradation mechanisms, but their extraction presents a real challenge, because of relaxation phenomenon.

In the present study, we suggest a new method enabling to simultaneously extract both components responsible of the NBTI degradation. The method is based on a simple theoretical concept and appropriated to the capacitor devices. It has major advantages, not only permits the extraction of the two components, hence expanding the NBTI investigation in MOS capacitor, but also it improves the accurate determination of the both components by reducing the recovery amount. The capacitor devices were not previously used in such kind of degradation due to relaxation effect.

This paper is outlined as follows. In section II, we present the theoretical approach and concept of the proposed method and briefly describe the extraction of the two component contributions to flat-band (V_{fb}). The approach for extracting ΔV_{it} and ΔV_{ot} will be also shown. Afterwards we give, in section III, the experimental details, such as samples, experimental setup, operating conditions, and stress measure stress (SMS) protocol. The result analysis and discussion are given in Section IV. Finally, conclusions are summarized in section V.

II. EXTRACTION AND CONCEPT OF THE PROPOSED METHOD

A. Separation of the two flat-band shift components

The voltage shift in a MOS capacitor can be expressed as the sum of two voltage shifts; one is caused by the increase in the oxide charge (Q_{ot}), while the second is caused by the increase in the interface charge (Q_{it}) and can be written as [6]:

$$\Delta V_{fb} = \Delta V_{it} + \Delta V_{ot} \qquad (1)$$

This work was supported by the High School Educational and Scientific Research Ministry of Algeria under the National Funding of Research (FNR) contract number 22/CDS/DMN/ CDTA/2011.

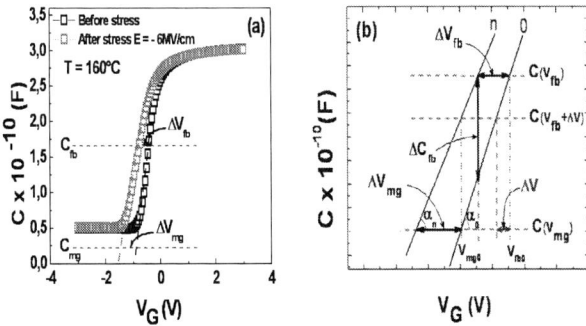

Figure 1. -a- Capacitance-Voltage curve shift and stretch-out under negative bias temperature stress. -b- Illustrative approach (linear variation approximate between V_{fb} and V_{mg}) for extracting of the two flat band shift contributions, interface-trap ΔN_{it} and oxide trap ΔN_{ot}.

The oxide and interface trapped charge density are calculated based on the assumption that the interface traps are neutral charges when the Fermi level is at mid-gap [7, 8]. In this case, the mid-gap voltage shift (ΔV_{mg}) is entirely due to the oxide trapped charges, which causes a parallel shift of the C-V characteristics, while the interface traps (ΔV_{it}) are responsible for the "stretch-out" of a C-V curve by changing its slope. The interface trap ΔN_{it} (lied from mid-gap to flat-band) and oxide trap ΔN_{ot} densities are determined from the mid-gap charge separation method of MOS capacitors [9] using equations given in tab I. It is worth noting that the shift caused by interface states at flat band mode has an opposite sign compared to the shift at inversion mode.

B. Theoretical approach of the proposed method to extract ΔV_{it} and ΔV_{ot}

Figure 1-a shows the capacitance-voltage curve obtained on n-substrate MOS capacitors, before and after negative bias temperature stress (E = - 6MV/cm and T = 160°C) at 1MHz frequency. The C-V measurement after stress should indicate a flat-band voltage shift and a curve slop change. These effects are a consequence of trapped charges present in the interface and oxide.

As mentioned above, the proposed method consists of assuming a linear characteristic variation between V_{fb} and V_{mg} voltages, as schematically illustrated in Figure 1-b. Taking two points on linear characteristic curve, one at V_{fb} and the second near V_{fb} ($V_{fb}+\Delta V$, where ΔV = - 100 mV). This allows the determination of ΔV_{it}, ΔV_{fb}, and ΔV_{mg} as follows:

$$\Delta V_{it} = \left(\frac{tg(\alpha_0)}{tg(\alpha_n)} - 1 \right) . \left(V_{fb0} - V_{mg0} \right) \quad (2)$$

$$\Delta V_{fb} = \frac{C_{fb0} - C_{fbn}}{tg(\alpha_n)} \quad (3)$$

$$\Delta V_{mg} = \Delta V_{fb} - \Delta V_{it} \quad (4)$$

$$tg(\alpha_0) = \frac{C_{fb0} - C'_{fb0}}{\Delta V} , \quad tg(\alpha_n) = \frac{C_{fbn} - C'_{fbn}}{\Delta V} \quad (5)$$

where V_{fb0} and V_{mg0} are flat-band and mid-gap voltages before stress, respectively. C_{fb0} and C_{fbn} are capacitances associated with V_{fb0} before and after stress, respectively. C'_{fb0} and C'_{fbn} are capacitances associated with $V_{fb0}+\Delta V$ before and after stress, respectively. From these relationships, we can easily extract ΔN_{it} and ΔN_{ot} through the relations given in Table.1.

III. SAMPLES AND EXPERIMENTS DETAILS

A. Samples

The samples used to validate this method are n-substrate MOS capacitors, fabricated by Houston Wafer Fabrication (HWF) using conventional furnace oxidation. The oxide was grown in dry O_2 at 1100°C to a thickness of 30 nm. Aluminum gates were deposited by thermal evaporation using shadow mask with a gate area of 2.7×10^{-3} cm^2. The gate capacitance, C_{ox} is about 3.17×10^{-7} F.cm^{-2}.

B. Experimental setup and conditions

The SMS protocol has been performed using fully automated bench. The bench includes a *HP 4284 A precision LCR meter*. Using LabVIEW software, we have been able to control the SMS protocol by sequentially applying a constant voltage DC stress and capacitance measurement at different frequencies (1Hz to 1MHz). The probe station is *Signatone wafer probe-station*, which is equipped with a hot plate allowing measurements at different temperatures. The interval of stress time and the whole stress time used in SMS protocol were fixed at 100s and 3600s, respectively. During the sensing

TABLE I. EXPRESSIONS FOR ΔN_{IT} AND ΔN_{OT} CHARGE DENSITIES AT FLAT-BAND AND INVERSION FOR TWO SUBSTRATE TYPES.

		N-substrate	P-substrate
At flat-band	ΔN_{it}	$C_{ox} \dfrac{\Delta V_{it}}{q}$	$-C_{ox} \dfrac{\Delta V_{it}}{q}$
	ΔN_{ot}	$-C_{ox} \dfrac{\Delta V_{ot}}{q}$	$-C_{ox} \dfrac{\Delta V_{ot}}{q}$
At inversion	ΔN_{it}	$-C_{ox} \dfrac{\Delta V_{it}}{q}$	$C_{ox} \dfrac{\Delta V_{it}}{q}$
	ΔN_{ot}	$-C_{ox} \dfrac{\Delta V_{ot}}{q}$	$-C_{ox} \dfrac{\Delta V_{ot}}{q}$

Where C_{ox} (F/cm^2) is the gate oxide capacitance per unit area, q (C) is the electron charge, ΔV_{it} (V) is the interface trap voltage shift, ΔV_{ot} (V) is the mid-gap (oxide trap) voltage shift, and ΔV_{fb} (V) is the flat-band voltage shift.

procedure, two capacitance measurements were performed; the fist one at V_{fb}, and the second at V_{fb} - 100mV, using 1 MHz frequency. Different capacitors were used for each couple of temperature /electric field. NBTI degradation was observed at elevated temperature in the range of 120°C to 160°C, and at different fields -3MV/cm, -4MV/cm to -6MV/cm.

IV. RESULT ANALYSIS AND DISCUSSION

Figure 2.a and 2.b show the as-measured $|\Delta V_{fb}|$ as a function of stress time at different electric field at 120°C. The voltage shifts are extracted using equation (2) developed in our approach. In the same figure $|\Delta V_{fb}|$ obtained from the shift of the full C-V characteristic at flat band voltage are also presented (see the cross-symbols). We note that both voltages show a negative shift, contrarily to the NBTI effect in inversion where the shift is positive. This effect is explained by the difference in the scanned band energy. The increasing shift with time is an indication of augmentation of trapped charge at the interface and in the oxide.

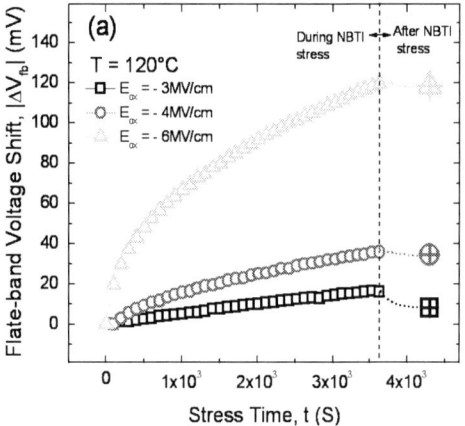

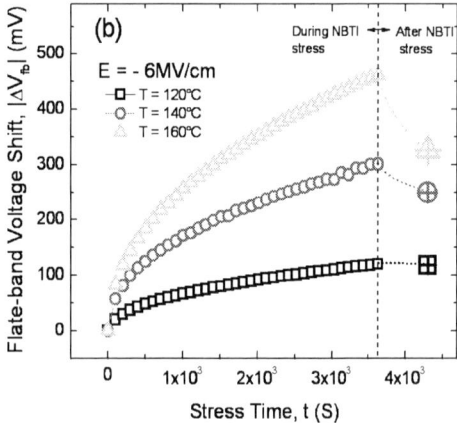

Figure 2. . Time evolution of flat band voltage shift, ΔVfb as-measured, (non Cross-symbols), (a) for different electric field stress at 120°C and (b) for different temperatures at -6MV/cm. Cross-symbol illustrates ΔVfb extracted from full C-V characteristics after stress.

The comparison of the flat-band shift extracted by our approach with that obtained from full C-V characteristics presents some discrepancies. This latter is induced by recovery phenomenon. The recovery clearly appears for stress electric field of -6 MV/cm and temperature of 140°C and increases with increasing temperature as shown in Figure 2.b. Therefore, these results indicate that using our approach allows reducing the recovery amount (by reducing measurement time duration).

Some works attributed recovery amount to the fast-detrapping of positive oxide charge generated during NBTI [10]. *Reaction-Diffusion* (R-D) model [11, 12], based on the dissociation of hydrogen at the Si/SiO$_2$ interface and diffusion into SiO$_2$, explains the recovery by the interface dangling bonds re-passivation.

However, this model can explain the results at low temperature and low field obtained by the assumption that the amount of created hydrogen is not well enough to re-passivate the interface dangling bands, or by the fact that the created hydrogenated species are positively charged. In this case, the drift is more consistent than neutral hydrogenated species. This proposed analysis may be not entirely accurate because there may be an amount of recovered charges, but the SMS procedure (setup) can not detect this relaxed amount.

ΔN_{eq_Vfb}, ΔN_{it} and ΔN_{ot}, which are calculated from voltage shifts ΔV_{fb}, ΔV_{it} and ΔV_{ot} and given by equations (3), (2) and (4), respectively, are illustrated in Figure 3. The time evolution of these components is presented for different stress temperatures and fields, in log-log scale. The ΔN_{eq_Vfb} extraction is given by:

$$\Delta N_{eq\_Vfb} = C_{ox} \frac{\Delta V_{fb}}{q} \qquad (6)$$

We found that the temperature and/or electric field stress increase leads to trapped charge density increase. Therefore the NBTI effect is well appeared on our measurements. We notice moreover that this evolution presents a linear time variation in log-log scale. The evolution of these components shows an evident power-law time-dependence. It is in good coherence with the author's power-law time-dependence ($\sim t^n$) suggestions.

We notice that the power-law exponent n of ΔN_{it} is independent of temperature and field, it is approximately 0.5. If we considering that there is no recovery during measurement delay, according to the RD model, the exponent n is often attributed to H$^+$ [13]. Noting that at low field (-4MV/cm) for different temperatures, n of ΔN_{it} and ΔN_{eq_Vfb} are approximately 0.7 and 0.6, respectively and decrease to approximately 0.5 for high field (-6MV/cm) at different temperatures. As mentioned above, the RD model [11, 12] is a well-known model that could explain the creation of interface traps. The positive trapped charge is attributed to the trapping of H$^+$ in the gate oxide [14] or to the trapping of holes at preexisting oxide defects, assumed to reach saturation quickly (~1 ms) after the stress was applied [15]. The H$^+$ interaction with Si-O plays also some role in the creation of positive

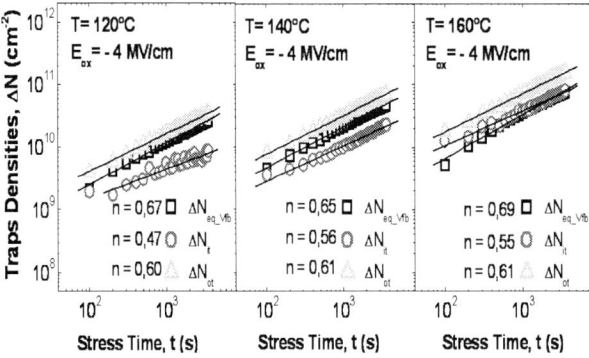

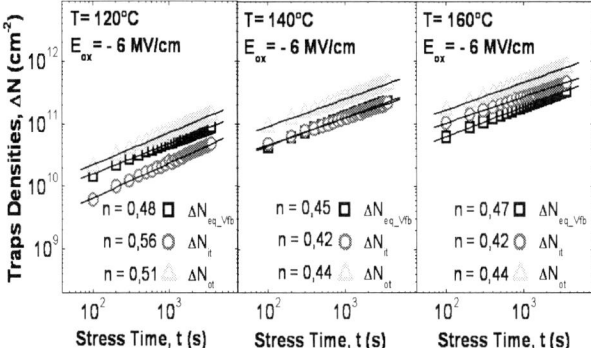

Figure 3. Time evolution of equivalent trapped charge ΔN_{eq_Vfb}, interface trap ΔN_{it} and oxide trap ΔN_{ot} calculated from ΔVfb, ΔVit and ΔVot, respectively for different temperature and different electrical field stress, in log-log scale.

trapped charge [16, 17]. In other words, the H^+ interaction with Si-O bond results in creation of fixed positive charge Si^+ and OH specie.

Through this analysis, the oxide trap creation at low field for different temperatures could be issued from combination of H^+ trapping in the gate oxide and hole trapping at preexisting oxide defects while at high field hole trapping at preexisting oxide defects could saturate and only H^+ trapping in the gate oxide predominates.

V. CONCLUSION

A novel NBTI measurement method was proposed to extract interface and oxide traps, employing only one technique (CV). The method permits improving the extraction accuracy of the two components by reducing the relaxation time during the measurement and expanding the investigation to reliability study case exploiting capacitor devices. The method is based on a simple theoretical concept using SMS protocol.

The result analysis has shown that ΔN_{eq_Vfb}, ΔN_{ot} and ΔN_{it} present a linear stress-time-dependence which is in good agreement with the one found in literature. In addition, the results have shown also a similar kinetics of interface state generation as well as oxide trapped charges. The independence power-law exponent $n=0.5$ of ΔN_{it} with temperature and fields

indicates the same interface traps mechanism creation contrary to the oxide traps. However, for a good interpretation and modeling, it will be interesting to reduce the stress-to-measure-delay time to avoid recovery effect.

REFERENCES

[1] JEDEC 14.2.2, A Procedure for Wafer Level DC Characterization of P-MOSFET Negative Bias Temperature Instabilities.

[2] D. Brisbin and P. Chaparala, "A New Fast-Switching NBTI Characterization Method That Determines Subthreshold Slope Degradation," IEEE Transactions on Nuclear Science, vol. 9, no. 2,. pp. ? Jun. 2009.

[3] W. J. Liu, Z.Y. Liu, D. Huang, C. C. Liao, L. F. Zhang, Z. H. Gan, W. Wong, C. Shen, and M. F. Li, "On-The-Fly Interface Trap Measurement and Its Impact on the Understanding of NBTI Mechanism for p-MOSFETs with SiON Gate Dielectric," IEEE International Electron Devices Meeting (IEDM'07), Washington, DC , USA, 10-12 Dec. 2007, pp. 813-816.

[4] D. Huang, W. J. Liu, Z. Liu, C. C. Liao, L. F. Zhang, Z. Gan, W. Wong, and M. F. Li, "A Modified Charge-Pumping Method for the Characterization of Interface-Trap Generation in MOSFETs," IEEE Transactions on Electron Devices, vol. 56, no. 2, pp. 267-274, Feb. 2009.

[5] V. Huard, M. Denais, C. Parthasarathy, "NBTI Degradation: From Physical Mechanisms to Modelling," Microelectronics Reliability, Elsevier, vol. 46, no. 1, pp. 1-23, Jan. 2006.

[6] D. Braunig and F. Wulf, "Radiation effects in electronic components," Elsevier Science, 1999 (Chapter 10), pp. .

[7] P. M. Lenahan and P. V. Dressendorfer, "Hole traps and trivalent silicon centers in metal/oxide/silicon devices," Journal of Applied Physics, vol. 55, no. 10, pp. 3495-3499, May 1984.

[8] T. P. Ma, G. Scoggan, and R. Leone, "Comparison of interface-state generation by 25-keV electron beam irradiation in p-type and n-type MOS capacitors," Applied Physics Letters, vol. 27, no. 2, pp. 61-63, Jul. 1975.

[9] J. A. Felix, D. M. Fleetwood, R. D. Schrimpf, J. G. Hong, G. Lucovsky, J. R. Schwank, and M. R. Shaneyfelt, "Total-dose radiation response of hafnium-silicate capacitors," IEEE Transactions on Nuclear Science, vol. 49, no. 6, pp. 3191-3196, Dec. 2002.

[10] Huard V and Denais M. Hole trapping effect on methodology for DC and AC negative bias temperature instability measurements in PMOS transistors, " in Proc. Int. Reliab. Phys. Symp., 2004, pp. 40–45.

[11] Ogawa S, Shiono N. Generalized diffusion–reaction model for the low-field charge build up instability at the Si–SiO2 interface," Phys Rev B;51(7):4218–30, 1995

[12] Alam M. A. "A Critical Examination of the Mechanics of Dynamic NBTI for PMOSFETs," IEDM Tech. Dig., pp. 346-349, 2003.

[13] S. Chakravarthi, A. T. Krishnan, V. Reddy, C. F. Machala, and S. Krishnan, "A comprehensive framework for predictive modeling of negative bias temperature instability," in Proc. Int. Reliab. Phys. Symp., 2004, pp. 273–282.

[14] S. Tsujikawa and J. Yugami, "Evidence for bulk trap generation during NBTI phenomenon in pMOSFETs with ultrathin SiON gate dielectrics," IEEE Trans. Electron Devices, vol. 53, no. 1, pp. 51–55, Jan. 2006.

[15] A. E. Islam, H. Kufluoglu, D. Varghese, S. Mahapatra, and M. A. Alam "Recent issues in negative-bias temperature instability: Initial degradation, field dependence of interface trap generation, hole trapping effects and relaxation," IEEE Trans. Electron Devices, vol. 54, no. 9, pp. 2143-2154, Sep. 2007.

[16] J. H. Stathis and D. J. DiMaria, "Reliability projection of ultra thin oxides at low voltage," in IEDM Tech. Dig., 1998, pp. 167-170.

[17] D. J. DiMaria, "Explanation for the polarity-dependence of breakdown in ultrathin silicon dioxide films," Appl. Phys. Lett., vol. 68, no.21 pp.3004-3006, May 1996.

Impact of the Inhomogeneous Structure of the Active Layer on the Transfer Characteristic of Polysilicon TFT's

TAYOUB Hadjira[*], BENSMAIN Asmaa, ZEBENTOUT Baya and BENAMARA Zineb

Electronic Departement
UDL university
Sidi Bel Abbès, Algeria
E-mail: hadjira.tayoub@gmail.com[*]

Abstract— Recently polycrystalline silicon (pc-Si) thin film transistors (TFT's) have emerged as the devices of choice for many applications. The TFT's made of a thin un-doped polycrystalline silicon film deposited on a glass substrate by the Low Pressure Chemical Vapor Deposition technique LPCVD have limits in the technological process to the temperature < 600°C. The benefit of pc-Si is to make devices with large grain size. Unfortunately, according to the conditions during deposition, the pc-Si layers can consist of a random superposition of grains of different sizes, where grains boundaries parallels and perpendiculars appear. In this paper, the transfer characteristics I_{DS}-V_{GS} are simulated by solving a set of two-dimensional (2D) drift-diffusion equations together with the usual density of states (DOS: exponential band tails and Gaussian distribution of dangling bonds) localized at the grains boundaries. The impact of thickness of the active layer on the distribution of the electrostatic potential, the effect of density of intergranular traps states and grain size on the TFT's transfer characteristics I_{DS}-V_{GS} have been also investigated.

Index Terms—Transistor TFT, 2D simulation, heterogeneous structure, grain size, transfer characteristic.

I. INTRODUCTION

Transistors made of thin layers (Thin Film Transistor: TFT) are used primarily for the realization of active matrix flat panel displays liquid crystal [1]. For flat screens, it is impossible to integrate transistors on a monocrystalline silicon wafer in the pixel array for technology reasons. On one hand, the maximum size of the wafer cannot exceed about 30 cm while it is now standard for flat screens to work on substrates of more than 2m [2]. On the other hand, the substrate of the transistor in monocrystalline silicon is typically conductor while that of TFT's is therefore glass or plastic insulation. In addition, the maximum temperature of the manufacturing process of monocrystalline silicon transistors is about 1000 ° C while that of the TFT's is a few hundred ° C, which is compatible with glass substrates with low cost or even some plastic substrates.

At present, the TFT's are mostly made of the hydrogenated amorphous silicon (a-Si:H) [3]. This material may indeed be deposited on large surfaces in low temperature (T = 200 °C). The choice of this temperature is dictated by the desire to use inexpensive glass substrates. The insufficient

electrical properties of this material limited the use of TFT (a-Si: H) in pixels. The control circuits of the matrix in this case are realized of monocrystalline silicon outside the glass substrate. The necessity for a connection between the matrix and the control circuits causes therefore an increase in weight, size, cost of the device and reduced reliability. The connection between the two entities mentioned above becomes very difficult. The electrical properties of polysilicon much higher than amorphous silicon, allow its use in transistors of pixels and in the control circuits also. The main objective of this paper is to use a program based on the numerical solution of two-dimensional equations (Poisson equation and the two equations of continuity of electrons and holes) to study the impact of: thickness of active layer on the distribution of the electrostatic potential and the effect of density of intergranular traps states and grain size on the TFT's transfer characteristics I_{DS}-V_{GS} for a drain bias equal to 1V.

II. MODELING DESCRIPTION

The structure of polycrystalline silicon is highly dependent on technology used to form the material. When the polycrystalline film is obtained after solid phase crystallization of amorphous silicon film, its structure is columnar. The crystallized material is assumed formed by parallel single-crystalline grains, L_G sized, separated by physically thick amorphous grain boundaries with a thickness of 1 nm.

A. Simulated structure

Figure 1 shows the geometrical model adopted in the two-dimensional simulation of the layer of poly-Si where the grain boundaries are perpendicular to the growth surface. Because of the complexity of the structure model and the limited number of points in programming language, only five perpendicular grain boundaries are considered here. The geometrical model assumes that Poly-Si layer is composed by crystalline grains separated by amorphous silicon transition zones commonly called grain boundaries which are perpendicular to the growth surface and which include the usual density of states (DOS) with exponential band-tails and Gaussian distributed deep levels.

978-1-4673-5289-5/12 $31.00 © 2012 IEEE

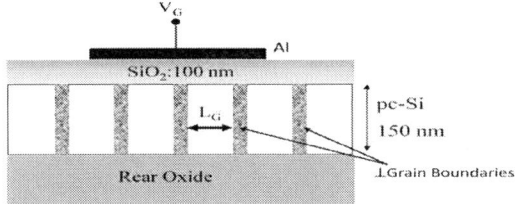

Fig. 1. Two dimensional geometrical of poly-Si layer.

B. Density of states at the grain boundaries

The usual density of states (DOS) model in the amorphous silicon is used to describe the distribution of the states inside the forbidden band-gap of the amorphous grain boundaries. It is consisted of two decreasing exponential distributions for the valence and the conduction band tail and two correlated Gaussian distributions for deep defects introduced by dangling bonds.

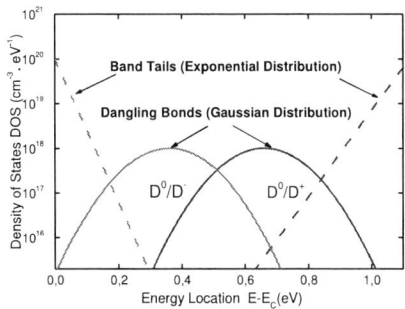

Fig. 2. Density of States (DOS) in the gap of the amorphous grain boundaries regions.

C. Electrical conduction

Simultaneous numerical resolution of the coupled Poisson "Eq. 1", electron "Eq. 2" and hole "Eq. 3" continuity equations are numerically solved in two dimensions:

$$\nabla\left(\varepsilon_S\nabla\Phi(x,y)\right) = - q(p(x,y) - n(x,y) + N_D - N_A + \sum N_T(x,y)) \tag{1}$$

$$\nabla J_n = qU\left(\Phi(x,y), \Phi_n(x,y), \Phi_p(x,y)\right) \tag{2}$$

$$\nabla J_p = -qU\left(\Phi(x,y), \Phi_n(x,y), \Phi_p(x,y)\right) \tag{3}$$

The electron and hole conduction currents, J_n and J_p, are then given by:

$$J_n(x,y) = -qn(x,y)\mu_n\nabla\left(\Phi_n(x,y)\right) \tag{4}$$

$$J_p(x,y) = -qp(x,y)\mu_p\nabla\left(\Phi_p(x,y)\right) \tag{5}$$

In these equations, the symbols used are:
ε_S: The permittivity of silicon material;
$n(p)$: The free electrons (holes) density;
Φ, Φ_n, Φ_p: The electrostatic potential, the electron and hole electrochemical Fermi potentials;
μ_n, μ_p: The band microscopic electron and hole mobilities;
$N_D(N_A)$: The ionized donor (acceptor) density;
$\sum N_T$: The sum of the different trap states present in the material.
In the both continuity equation, U represents the net active generation-recombination rate.

Due to the non-linearity of the three coupled partial equations of the system formed by the first three equations, the determination of unknown Φ, Φ_n, Φ_p values requires a numerical method. We have used the standard finite difference defined by Gummel's decoupling method [4]. Therefore, each equation is discretized on a variable mesh (strongly refined at interface and grain boundaries) then linearized to the first order. The numerical resolution is based on the method of relaxation by line and by column. The tridiagonal character of the matrix leads us to use the Gauss's elimination method.

The Hypotheses and parameters values of the density of states in the grain boundaries are the same that reported in reference [5].

III. RESULTS OF 2D-NUMERICAL SIMULATION

A. Simulation of Al/SiO₂/Poly-Si (N) structure at thermodynamic equilibrium and under gate bias

Figure 3 shows the 2D electrostatic potential distribution in the structure, we focus our study here just in the active layer of poly-Si for a thickness equal to 0.15 and 1 μm.

These figures demonstrate the appearance of a height of potential barrier at the grain boundaries with crystallites partially depleted due to the presence of a localized density of intergranular trap at grain boundaries. In this case many electrons will be trapped at grain boundaries leading to increase the distance between conduction band and Fermi level except for extreme areas. This potential barrier limits the passage of free carriers of a crystallite to another. Also the potential variations are identical from one grain to another grain except the ones near the free surfaces or metal contacts. These heights of potential barriers are more important in the poly-Si layer having less thickness of active layer. Certainly, the increase in thickness causes a decrease in resistivity, which implies a decrease in height of potential barrier at the grain boundaries [6].

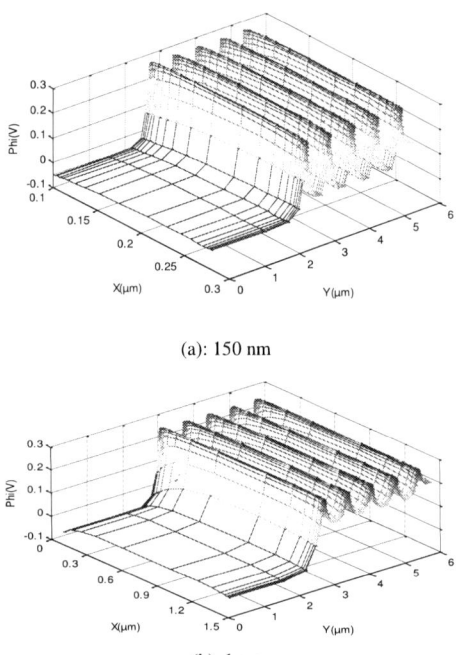

(a): 150 nm

(b): 1µm

Fig. 3. 2D-Distribution of electrostatic potential in poly-Si for two layer thickness.

The Application of a positive gate voltage equal to 5V causes an accumulation of electrons (majority carriers) at the interface Poly-Si/SiO$_2$ shown in the following figure.

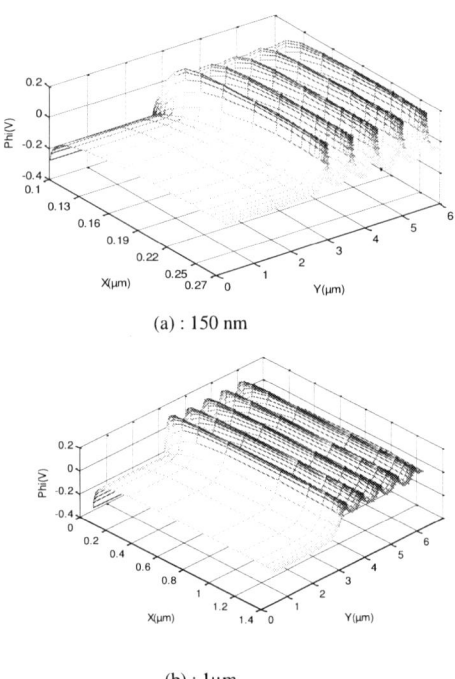

(a) : 150 nm

(b) : 1µm

Fig. 4. 2D-Distribution of electrostatic potential in poly-Si for two layer thickness for a bias voltage Vg positive.

The following figures show a desertion of electrons at the interface for a negative gate voltage equal to - 5V in the case of a thickness of 1 µm "Fig. 5.a" while an inversion of population is reached for a thickness of 150 nm "Fig. 5.b".

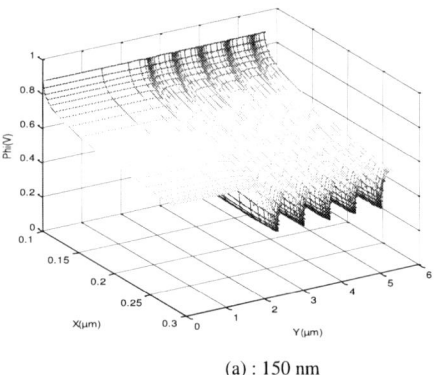

(a) : 150 nm

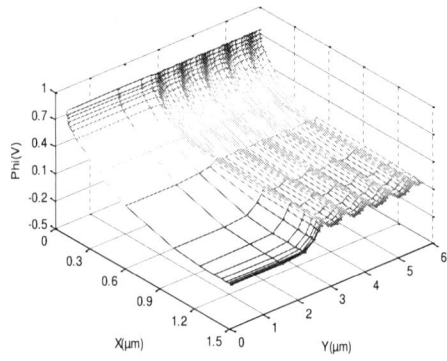

(b) : 1µm

Fig. 5. 2D-Distribution of electrostatic potential in poly-Si for two layer thickness for a bias.

B. Simulation of transfer characteristic IDS-VGS of Poly-Si TFT

An extension of two-dimensional geometric model was used to simulate the transfer characteristic I$_{DS}$ (V$_{GS}$) of Poly-Si TFT's. we use the structure already described above "Fig. 1" with the addition of source and drain contacts. All simulations are performed for a grain size L$_G$ of 300 nm, a thickness of 150 nm and drain bias V_{DS} equal to 1V.

- Effect of intergranular density of states

The presence of intergranular traps states in the band gap increases the value of the current I_{OFF} "Fig. 6" This change is explained by the fact that these traps states will behave in the blocking state as states permit, they will facilitate the transition of carriers from one band to another by creating a gateway intermediary for conduction, significantly increasing the current of the transistor in the blocking state. The carriers will no longer need to spend energy of 1.12eV to pass from a band to another but they can do that with small successive jumps requiring a lower energy for each jump. Thus the creation electron-hole pairs will be assisted by the allowed states in the band.

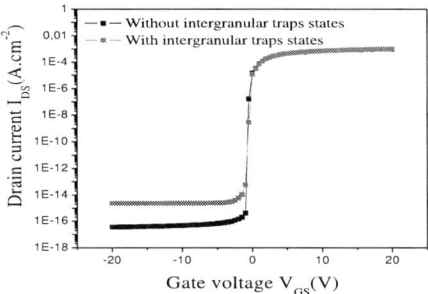

Fig. 6. Simulated transfer characteristics $I_{DS}(V_{GS})$ with and without density of intergranular traps states in poly-Si TFT's.

- Effect of grain size

Since the granular structure of the active layer has a direct influence on the transfer characteristic of the TFT Poly-Si, we are interested to represent more closely in "Fig. 7", the evolution of the drain current as a function of gate voltage for several grain sizes L_G ranging from 0.15 µm to 1µm.

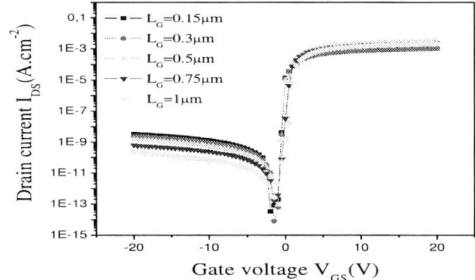

Fig. 7. Effect of grain size on the transfer characteristics I_{DS}-V_{GS} in poly-Si TFT's.

We note that in regime of accumulation the drain current I_{DS} is a little affected by the grain sizes. This is due to an important contribution of electron by the tension V_{GS} that fills those trapped in the grain boundaries by a lowering of the intergranular barrier. In regime of inversion, the current decreases with the increasing of grain sizes. It was shown that the process of growth of grain improves the physical and electrical properties of Poly-Si layers and consequently devices made from it [7].

CONCLUSIONS

In this paper, a numerical modeling was developed to simulate the effect of grain boundaries in the electrostatic potential and the characteristic electric of Poly-Si TFT's.
It has been found that:
- The presence of grain boundaries in the structure creates heights potential barriers limiting the passage of free carriers of a crystallite to another;
- The introduction of intergranular traps states causes a shift of the characteristic $I_{DS}(V_{GS})$. However, the reverse current tends to increase with V_{GS};
- The effect of density of intergranular traps states is important just for a negative gate voltage;
- Trap states behave in the blocking state as states permit that will facilitate the transition of carriers from one band to another by creating through a gateway for the conduction;
- The introduction of intergranular traps states increases just the current I_{OFF} .

REFERENCES

[1] S.D.Brotherton, J.R.Ayres, M.J.Edwards, C.A.Fisher, C.Glaister, J.P.Gowers, D.McCulloch, and M.Trainer, "Laser crystallised poly-Si TFTs for AMLCDs,' Thin Solid Films 337,188(1999).

[2] Van Diep BUI, "Conception et modélisation detransistors TFTs en silicium microcristallin pour des écrans AMOLED," Doctorate thesis, 2006.

[3] P.G.LeComber, W.E.Spear and A.Ghaith," Amorphous silicon field-effect device and possible application," Electron. Lett. Vol. 15 pp. 179-181 (1979).

[4] H. K. Gummel, "A self-consistent iterative scheme for one dimensional steady state transistors calculations," IEEE Transactions on electron devices, Vol. ED-11, pp. 455-465, October 1994.

[5] B. Zebentout, Z. Beanamara and T. Mohammed-Brahim, "Dependence of photovoltaic parameters on grain size and density of states in n+-i-p+ and p+-i-n+ polycrystalline silicon solar cells," Thin Solid Films 526, 84 (2007).

[6] A.AMROUCHE, "Modélisation des transistors en couches minces sur silicium polycristalline petits grains non dopé, en mode d'accumulation," Doctorate thesis, Rennes 1 University, N°549,1990.

[7] M. Imaizumi, T. Ito, M. Yamaguchi and K. Kaneto, "Effect of grain size and dislocation density on thin film polycrystalline silicon solar cells," 14 the European Photovoltaic Solar Energy Conference, 30 june-4 July 1997, Barcelona, Spain, pp.1385-1388.

ANFIS-based approach to study the subthreshold swing behavior for nanoscale DG MOSFETs including the interface trap effect

T. Bentrcia

Department of Physics
University of Batna
05000 Batna, Algeria
toufikmit@yahoo.com

F. Djeffal and E. Chebaaki

Department of Electronics
University of Batna
05000 Batna, Algeria
faycaldzdz@hotmail.com, elamhar@hotmail.com

Abstract— Since the DG MOSFET performance at nanoscale level is highly affected by geometrical parameters such as the channel length, reliable models which describe the ageing degradation of the device should be developed. The accurate description of this degradation is a need to estimate the device lifetime. In this work, a new model to predict the relative degradation of subthreshold swing factor is developed using the Adaptive Network Fuzzy Inference System (ANFIS) approach. The obtained results are discussed in order to get more information about the device degradation mechanism. The found results show good agreement with the numerical simulations (2D-ATLAS) making the proposed approach an efficient alternative approach to study the CMOS-based circuits degradation behavior.

Index Terms—DG MOSFET; short channel effects; hot-carrier-degradation effect; fuzzy modeling

I. INTRODUCTION

In the last few years, an accelerated race towards increasing the integration capacity of electronics components has been launched. The need for higher ratios of performance/dimensions is considered as a privileged goal that can not be ignored by the design engineers' community. The Metal Oxide Semiconductor Field Effect Transistor (MOSFET) paradigm has reached an advanced stage of development, where it is adopted now as the backbone of almost today's information processing and storage supports as indicated by the International Technology Roadmap of Semiconductors (ITRS) [1]. Based on this assertion, more attention should be focused on the performance analysis of new proposed structures. The multi-gate transistor called Double Gate (DG) MOSFET has been proposed as a promising candidate to remedy the packaging anomalies generated by Short Channel Effects (SCEs). However, the same situation reappears again when the device is scaled below the 100 nm technology node because SCEs are aggressively amplified and affect various device characteristics significantly [2].

Another reliability aspect that is closely related to the reduction of the device dimensions consists in the formation of interface traps under the hot carrier injection effect. In fact, with the tremendous advance of high speed and density of MOSFET devices, it is worthy to get extremely short channel lengths. Since the associated power supply voltages scale at a slower rate than channel lengths in the miniaturization process, the applied electric field increases consequently. The large electric fields created along the channel accelerate carriers from the source to the drain where a portion attends high energies and overcomes the $Si-SiO_2$ potential barrier [3]. So it may be injected into the gate oxide, leading to interface trap buildup and the trapping of carriers in the dielectric. These traps at or near the semiconductor/gate dielectric in interface can cause the degradation of device parameters such as transconductance or threshold voltage and may lead to dielectric breakdown [4, 5]. For this reason, the carrier trapping in gate oxides is considered as the principal cause of instability for short channel MOSFETs and becomes one of the most important concerns for next generation MOS integrated circuit development [6].

Despite that the real description of symmetrical DG MOSFET behavior with both ultra thin Silicon body and Short channel length requires the extension of the classical semiconductor equation to account for the quantum effects, the quantum mechanical modeling has received less interest since the obtain of a self-consistent solution from both the Schrodinger/Poisson's equation is very expensive in computing time [7]. Therefore, many alternative approaches used to study the immunity of performance criteria against the hot carrier degradation and subject to quantum effects are expressed using compact models deduced by imposing simplifying assumptions in order to reduce the complexity of the resolution approach [8, 9]. However, the obtained models, in this case, are questionable because of the simplifications introduced during the development of these models [10]. Therefore, the main goal of this work is to investigate the efficiency of predicting the relative degradation of DG MOSFET subthreshold swing including interface traps and quantum effects using an adaptive-network-based fuzzy inference system in order to study the nanoscale devices including the degradation effect. Two geometrical parameters are selected as input variables to the fuzzy system in order to predict the degradation in the swing factor. In this framework, we tried several types of membership functions for analysis in ANFIS training and selected the best one in terms of accuracy rate of prediction. The obtained results indicate that the Gaussian combination membership function rather than other MFs has a higher correct rate of prediction in testing.

978-1-4673-5289-5/12 $31.00 © 2012 IEEE

The remainder of this paper is organized as follows: Section 2 presents the architecture and the hybrid learning algorithm of an ANFIS. Section 3 presents the device under study in addition to the numerical simulation procedure. The fourth section is dedicated to the interpretation of the main obtained results. The last section contains some concluding remarks.

II. ANFIS ARCHITECTURE

The modeling of electronics devices can be interpreted as a kind of interpolative input-output mapping, where an equivalent schema is elaborated and an associated set of parameters are tuned up in order to reflect the correct behavior of the device. In practice, a training database is used to adjust these parameters so that both predicted and measured responses coincide.

The ANFIS structure introduced initially as a combination between fuzzy logic and artificial neural network methodologies is considered a prominent approach because of its ability to learn nonlinear relationships within complex systems especially in the presence of uncertainties in the training pattern data [11]. In this formalism, the input space is divided into many local subregions that means several of them can be activated simultaneously by a single input, and then a simple local model given by a linear function with adjustable coefficients is attached to each subregion. These subregions are obtained by a partition procedure of the input space based on the employment of fuzzy membership functions (MFs).

Since the principal parameters that have a major impact on short channel and quantum effects are geometrical parameters, the channel length L and thickness tsi are selected as the input variables of our ANFIS model as depicted in figure 1. In this representation, there are two adaptive layers, the first layer has modifiable parameters pertaining to the input MFs and are known as the premise parameters. The fourth layer includes modifiable parameters called consequent parameters pertaining to the first-order-polynomial model [12].

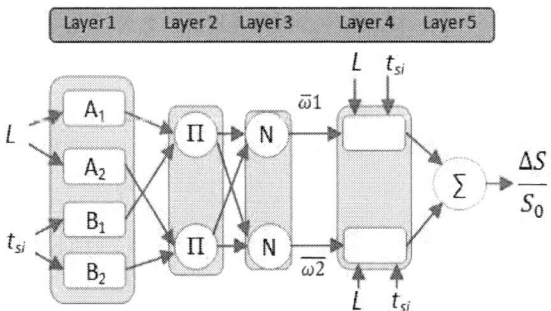

Figure 1. Illustrative representation of ANFIS with 2 inputs and 2 MFs.

So, ANFIS uses a feed forward network to search for fuzzy decision rules that perform well on a given task based on a given input–output data set, ANFIS creates a FIS where a typical rule set with n fuzzy IF-THEN rules can be expressed in our case as

$$IF \ (L \ is \ A_i) AND \ (t_{si} \ is \ B_j) THEN \ \left(\left(\frac{\Delta S}{S_0} \right)_l = p_l L + q_l t_{si} + r_l \right)$$

A_i and B_j are the linguistic terms of the precondition part with membership functions $\mu A_i(L)$ and $\mu B_j(tsi)$ respectively. The parameter $\left(\frac{\Delta S}{S_0} \right)_l$ denotes the output variables and p_l, q_l and r_l are the consequent parameters. The inferred ANFIS relative degradation of the swing factor is represented as a linear combination of the consequent parameters

$$\frac{\Delta S}{S_0} = \sum_{l=1}^{n} \left(\frac{\mu_{A_i}(L)\mu_{B_j}(t_{si})}{\sum_{l=1}^{n} \mu_{A_i}(L)\mu_{B_j}(t_{si})} \right) (p_l L + q_l t_{si} + r_l)$$

The adaptive parameters of an ANFIS model are of vital importance for the obtained performance. Despite that several alternatives exist for the optimization process in learning, the hybrid learning algorithm remains widely used in applications. This algorithm consists of two phases as follow [13]

- Computing error signals from the output layer backward to the input nodes.

- Finding a feasible set of consequent parameters using the least squares method.

In ANFIS, a new composite parameter α is formed by the union of the premise and consequent parameters. Its update formula is given by

$$\Delta \alpha = -\eta \frac{\partial E}{\partial \alpha}$$

where E is the overall error and η is a learning rate adjusted according to

$$\eta = \frac{\delta}{\sqrt{\sum_{\alpha} \left(\frac{\partial E}{\partial \alpha} \right)^2}}$$

where δ is the step-size. The consequent parameters of α are updated first using a least square algorithm then the premise parameters are tuned up by backpropagating the errors.

III. DEVICE DESIGN AND SIMULATION

In this work, we use the DG MOSFET structure to carry out simulation regarding its immunity against the hot carrier degradation effect. Various channel lengths and thicknesses are considered in order to evaluate the range of effects that may be more interesting for analysis. Subthreshold characteristics corresponding to a structure with a fixed channel length and thickness are simulated by using the 2D-Atlas simulator [14].

The simulations are performed based on a number of parameter values describing the geometrical and electrical configurations. The set of values corresponding to employed parameters is given in table I.

TABLE I.   DEVICE DESIGN AND SIMULATION PARAMETERS.

Parameter	Value

Parameter	Value
Oxide thicknesse	1.5 nm
Drain/Source doping	1×10^{20} cm^{-3}
Channel doping	1×10^{15} cm^{-3}
Workfunction	4.55 eV
Interface trap density	5×10^{12} cm^{-2}
Drain voltage	0.1 V
Gate voltage	0.7 V

The simulated structure have a uniform doping concentrations for both channel and source/drain regions. Simulations are carried out using two carrier types, the drift-diffusion model without impact ionization, doping concentration-dependant carrier mobility and electric field-dependant carrier model. SRH recombination/generation is also included in the simulation to account for leakage currents. The 3-D schematic cross-sectional view of the DG MOSFET device used in this work is presented in figure 2.

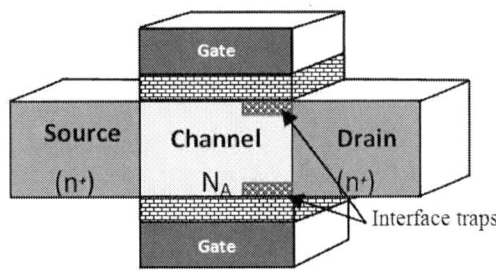

Figure 2. Cross sectional view of the symetrical DG MOSGET including interface traps.

IV. RESULTS AND DISCUSSION

The accuracy of the trained ANFIS depends on the accuracy and the effective representation of the data, which will be used for the training process. The training data sets used in this work are obtained by using the numerical simulator ATLAS. A total of the 91 set data are selected by meshing the channel length and the channel thickness ranges with a step of 5 nm and 0.5 nm, respectively, for the purpose of training tasks in the ANFIS. Membership functions that can have a strong influence on the behavior of our fuzzy system should be optimally determined. However, there is not any well-established method to determine membership functions in the ANFIS, so they are specified by the trial and error method. In this paper, MFs associated with channel length and thickness input parameters are selected heuristically and verified empirically. Therefore, the best fuzzy MF configuration which gives the best result is chosen for the calculation of the relative degradation of the swing factor.

In this work, the number of MFs for the inputs is fixed as 8 which gives a number of fuzzy IF-THEN rules for ANFIS equals to 64 (8×8). The results seem to suggest that using the Gaussian combination membership function gives superior performance compared to other types of membership functions, the best fuzzy MFs for both input parameters are illustrated in figures 3 and 4.

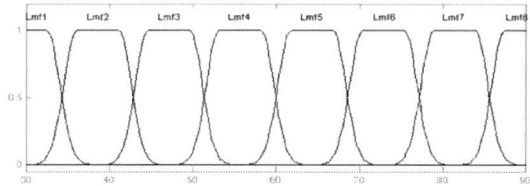

Figure 3. Gaussian combination membership functions of the channel length (L).

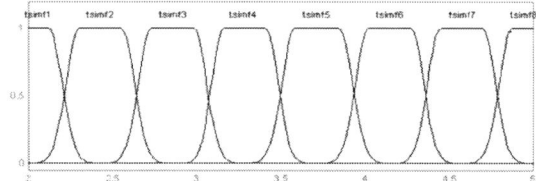

Figure 4. Gaussian combination membership functions of the channel thickness (t_{si}).

In order to well show the obtained fuzzy rule responses, figure 5 depicts the ANFIS controller rule surface for the variation of the relative degradation of swing factor as a function of both L and tsi. It can be noted that the highest degradation is located at a channel thickness equals to 3.5 nm and decreases considerably for extreme values of the interval [2: 5]. However, there are some local optima which make the analysis of the device immunity against the hot carrier injection a very hard task when subject to quantum effects and it is well-known in literature that such situation is not clearly interpreted. Besides this behavior, the degradation decreases rapidly for long channel values which is a logical consequence for the increase of the effective channel length. In addition, the degradation effect is more pronounced for the nanoscale devices. This observation can be explained by the impact of the increased applied electric field at the drain side because of the short channel effect.

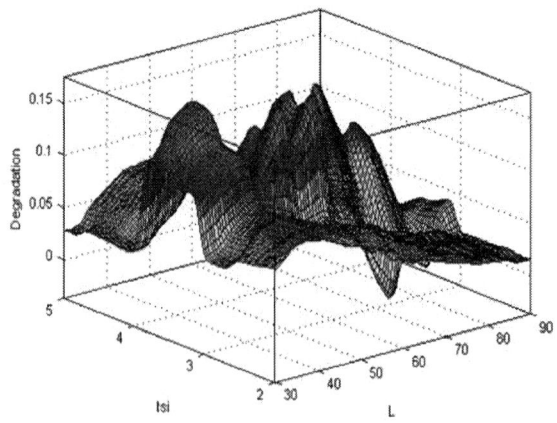

Figure 5. ANFIS controller rule surface.

978-1-4673-5289-5/12 $31.00 © 2012 IEEE

Figures 6 and 7 show the scatter diagrams of the predicted and numerical simulated values of the swing factor relative degradation for the training and testing data. From these figures, it is easy to note that the predicted values of the relative degradation follow the 45° line very closely for both categories where the slope of the regression curve is equal to 0.79 in the case of testing data, which means that a sufficient agreement is satisfied between the predicted and numerical results.

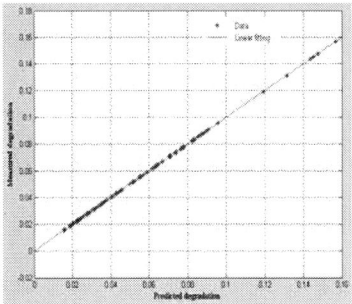

Figure 6.   Scatter diagram of measured and predicted relative degradation of swing factor for the training data.

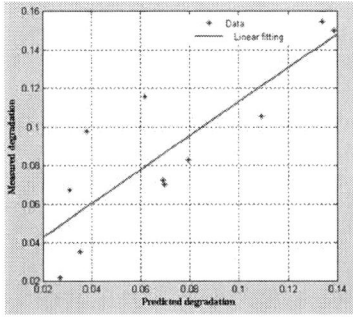

Figure 7.   Scatter diagram of measured and predicted relative degradation of swing factor for the testing data.

A summary of the main obtained results for both phases (training and testing phases) is provided in table II.

TABLE II.   PERFORMANCE CRITERIA FOR BOTH PHASES.

Criterion	Training	Testing
Average error	4.49×10^{-7}	2.76×10^{-2}
Correlation coefficient	1	0.83

V.   CONCLUDING REMARKS

In this paper, an ANFIS based approach has been proposed in order to predict the relative degradation of the subthreshold slope of the nanoscale DG MOSFET. Two geometrical parameters (channel length and thickness) have been considered as input parameters due to their high impact on short channel effects. The use of 2D-ATLAS simulator enabled

us to build the required knowledge database. A hybrid learning algorithm has been used to adjust various premise and consequent parameters. The behavior modeling process was completed in a relatively short time, with no need for user intervention during the computation. The type of membership function was effective to achieve further improvement of S model. The obtained performances have demonstrated that the developed fuzzy logic based approach is particularly suitable to be implemented in electronic device simulators to study the CMOS circuits at nanoscale level. Therefore, it is worthy to profit from the advantages of fuzzy logic based framework in building simulation tools that permit integrated circuit engineers to deal with uncertainties generated by quantum effects in new generations of nanoscale electronic devices. It is to note that the proposed approach can be extended to include other device parameters like: channel doping, insulator dielectric and thickness values.

REFERENCES

[1] International Technology Roadmap for Semiconductors (ITRS), Published online at http://public.itrs.net, 2009.

[2] R. Difrenza, P. Linares, G. Ghibaudo, "The impact of short channel and quantum effects on the MOS transistor mismatch," Solid-State Electronics 47 (2003) 1161–1165.

[3] T. Bentrcia, F. Djeffal, "Compact modeling of multi-gate MOSFET including hot-carrier effects," CMOS Technology: Electrical Engineering Developments Series, Vol. 1, (Nova Publishers, New York, 2011), chap. 4.

[4] F. Djeffal, T. Bentrcia, T. Bendib, "An analytical drain current model for undoped GSDG MOSFETs including interfacial hot-carrier effects," Physica Status Solidi C 8 (2011) 907-910.

[5] F. Djeffal, T. Bentrcia, M.A. Abdi, T. Bendib, "Drain current model for undoped gate stack double gate MOSFETs including the hot-carrier degradation effects," Microelectronics Reliability 51 (2011) 550-555.

[6] Z. Ghoggali, F. Djeffal, N. Lakhdar, "Analytical analysis of nanoscale double-gate MOSFETs including the hot-carrier degradation effects," International Journal of Electronics 97 (2010) 119–127.

[7] F. Prégaldiny, C. Lallement, D. Mathiot, "Accounting for quantum mechanical effects from accumulation to inversion, in a fully analytical surface-potential-based MOSFET model," Solid-State Electronics 48 (2004) 781–787.

[8] D. Munteanu, J-L. Autran, X. Loussier, S. Harrison, R. Cerutti, T. Skotnicki, "Quantum Short-channel Compact Modelling of Drain-Current in Double-Gate MOSFET," Solid-State Electronics 50 (2006) 680–686.

[9] T.P. Wen, A. K. Singh, "A comprehensive analytical study of an undoped symmetrical double-gate MOSFET after considering quantum confinement parameter," Microelectronics Journal 41 (2010) 162–170.

[10] T. Bentrcia, F. Djeffal, A. Benhaya, "Continuous analytic I-V model for GS DG MOSFETs including hot-carrier degradation effects," Journal of Semiconductors 33 (2012) 014001:1-014001:6.

[11] J. Jang, "ANFIS: adaptive-network-based fuzzy inference system," IEEE Transactions on Systems, Man and Cybernytics 23 (1993) 665–685.

[12] L-C. Ying, M-C. Pan, "Using adaptive network based fuzzy inference system to forecast regional electricity loads," Energy Conversion and Management 49 (2008) 205–211

[13] A. Azadeh, S.M. Asadzadeh, A.Ghanbari, "An adaptive network-based fuzzy inference system for short-term natural gas demand estimation: Uncertain and complex environments," Energy Policy38 (2010) 1529–1536.

[14] Atlas User's Manual: Device Simulation Software, 2008.

Influence of vertical scaling and temperature on impact-ionization effects in SiGe HBTs

Grazia Sasso, Vincenzo d'Alessandro, Maurizio Costagliola, Niccolò Rinaldi

Department of Biomedical, Electronics and Telecommunications Engineering,
University of Naples Federico II
via Claudio 21, 80125 Naples, Italy
Email: grazia.sasso@unina.it

Abstract—The temperature influence on impact-ionization mechanisms in advanced silicon-germanium heterojunction bipolar transistors is analyzed over the temperature range from 300 to 380 K for different technological nodes. Accurate results are obtained by performing simulations through a deterministic solver of the Boltzmann Transport Equation based on the spherical harmonics expansion of the distribution function. In particular, the impact of vertical scaling and lattice temperature on the open-base breakdown voltage BV_{CEO} and the $BV_{CEO} \times f_{T,PEAK}$ product is investigated.

Keywords—Boltzmann Transport Equation (BTE); hydrodynamic (HD) models; impact ionization; lattice temperature; Safe Operating Area (SOA); scaling; silicon germanium heterojunction bipolar transistors (SiGe HBTs)

I. INTRODUCTION

Silicon germanium hetero-junction bipolar transistors (SiGe HBTs) have acquired great importance in recent years due to their good noise and high frequency (HF) performance; furthermore, applications in the emerging HF markets steadily use SiGe HBTs for cost reasons. As demonstrated by recently published results [1], the aggressive technology development effort of the DOTFIVE project [2] has pushed the maximum cut-off frequency of these devices toward the terahertz range. The improvement of HF figures-of-merit (FoMs) in SiGe HBTs is accomplished by reducing device sizes and increasing doping levels [3]. On the other hand, downscaled devices are particularly vulnerable to reliability issues due to the higher electric fields and current densities, as well as to the rising thermal resistances, which have grown to thousands of K/W [4]. The physical limitations coming from device shrinking have therefore received attention in a number of works. The trade-off between breakdown voltage and high frequency performance has been discussed in [5], [6]. The temperature impact upon current gain and maximum cut-off frequency for different technological nodes has been analyzed in [7], where advanced simulation approaches were exploited to overcome the drawbacks of commercially available tools in describing aggressively scaled devices [8]. In particular, a deterministic solver of the Boltzmann Transport Equation (BTE) relying on the spherical harmonics expansion (SHE) of the distribution function [9], [10] was suitably adopted for the high-frequency nodes (i.e., the most scaled transistors), while the lowest-frequency node was simulated with an improved version of a commercial hydrodynamic (HD) simulator [11], [12]. Interestingly, it was found that the temperature sensitivity of the aforementioned device parameters is weakened by the vertical scaling.

This work is aimed to extend the analysis developed in [7] to impact-ionization mechanisms so as to offer a complete picture of the influence of vertical scaling and lattice temperature upon the behavior of modern SiGe HBTs. The SHE-based approach is adopted in order to guarantee more reliable results.

The paper is organized as follows. Section II is devoted to outline the main features of the technological nodes under test. In Section III, the simulation strategy is presented. Section IV discusses the results. Conclusions are drawn in Section V.

II. DEVICE STRUCTURES

The analysis is conducted by simulating three SiGe HBTs characterized by various scaling levels and denoted as #1, #2, and #3. The related doping profiles and germanium mole fractions are sketched in Fig. 1. The vertical scaling is emulated as follows: all the dimensions are shrunk by 4 and 8 so as to obtain HBTs #2 and #3 from the non-scaled HBT #1, respectively, while the doping concentrations are steadily increased, as summarized in Table I. The values of the peak cut-off frequency $f_{T,PEAK}$ were determined to be 66, 434, and 837 GHz [7]. This proves that the devices are associated to different technological nodes, which will be thereinafter referred to as "low-frequency node" (HBT #1), "intermediate node" (HBT #2), and "terahertz node" (HBT #3).

III. SIMULATION APPROACH

The simulation of semiconductor devices requires the description of distribution and transport of charged carriers, which are modeled by the Poisson equation and BTE. This problem can be accurately solved by resorting to an experimentally-verified deterministic 1-D software based on the SHE of the distribution function. In particular, the BTE is expanded with spherical harmonics on equi-energy surfaces; the set of the resulting balance equations is then solved with conventional numerical techniques [13]. This tool has been recently enriched to include the anisotropic structure of the conduction band, which ensures a reliable description of quasi-ballistic transport, as well as of the avalanche multiplication factor and local generation rate in extremely scaled devices, without any modifications of the scattering mechanisms and parameters [9], [10]. The impact-ionization model exploited for SiGe HBTs was taken from [14]. Differently from [7], the SHE-based solver was adopted for all technological nodes so as to avoid issues related to the calibration of the parameters associated to the impact-ionization model included in the commercial HD tool, thus guaranteeing reliable results regardless of the node.

The authors wish to acknowledge the support of the European Commission in the frame of the FP7 IST project DOTFIVE (IST-216110).

978-1-4673-5289-5/12 $31.00 © 2012 IEEE

TABLE I.   DETAILS OF THE DEVICES UNDER TEST.

	HBT #1	HBT #2	HBT #3
Maximum emitter doping [cm^{-3}]	5.00×10^{19}	3.20×10^{20}	5.00×10^{20}
Maximum base doping [cm^{-3}]	2.13×10^{19}	9.00×10^{19}	1.50×10^{20}
Epitaxial collector doping [cm^{-3}]	1.00×10^{18}	2.00×10^{18}	3.00×10^{18}
Maximum collector doping [cm^{-3}]	3.10×10^{19}	1.00×10^{20}	2.00×10^{20}
Emitter length [nm]	166	42	21
Base length [nm]	56	14	7
Collector length [nm]	418	104	52
Total device length [nm]	640	160	80

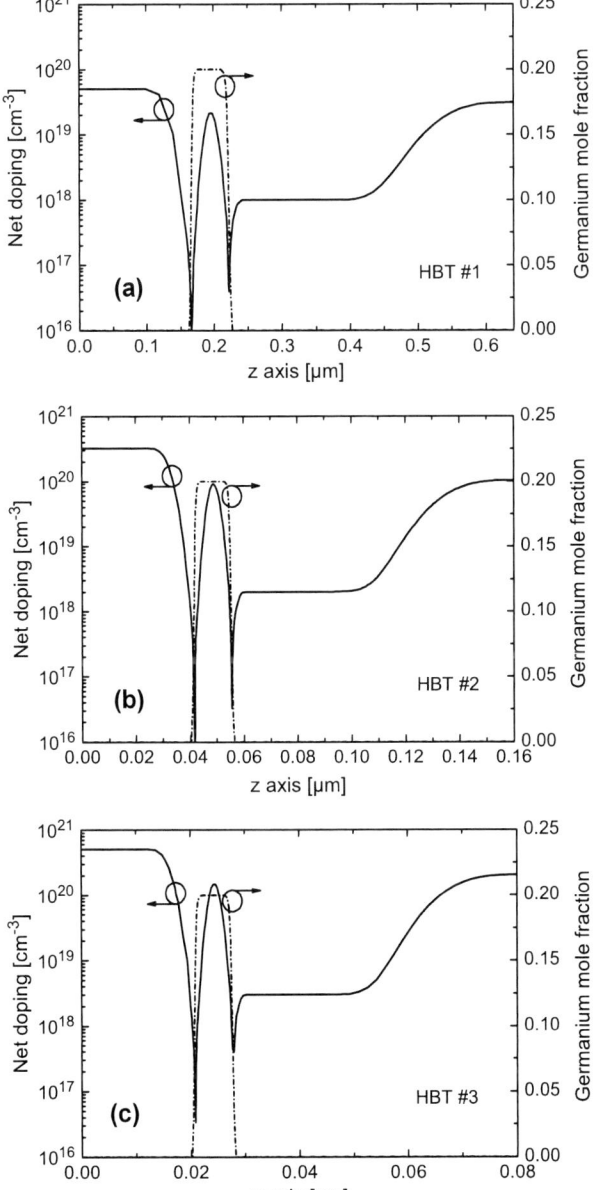

Figure 1.   Net doping and germanium profile of the "low-frequency-node" HBT #1 (a), the "intermediate-node" HBT #2 (b), and the "terahertz node" HBT #3 (c).

IV.   RESULTS

Due to the trade-off between HF FoMs and breakdown voltage in SiGe HBTs, the definition of SOA limits is a critical issue for designers and technologists. This has motivated an effort to explore the effect of vertical scaling and lattice temperature on avalanche multiplication. A parameter commonly used to define the operating limits of bipolar transistors is the open-base breakdown voltage BV_{CEO}. Although the maximum operating voltage in practical circuit configurations is actually higher than BV_{CEO} [6], [15]-[17], this voltage represents a critical boundary beyond which the base current reverses its sign due to avalanche multiplication in the collector-base space-charge region (SCR).

BV_{CEO} was extracted at various lattice temperatures by simulating the J_B–V_{CE} characteristics at V_{BE}=0.7 V, J_B being the base current per unit area. Fig. 2a shows the results obtained for the "intermediate-node" HBT #2. As can be seen, BV_{CEO} is about 1.65 V at T=300 K.

Numerical results were validated with experiments performed on state-of-the-art HF SiGe:C HBTs analyzed within the DOTFIVE project. On-wafer measurements were performed by means of a PM5 Karl Suss probe station equipped with RF probes and an HP4142B parameter analyzer to handle the electrical signals. The I_B–V_{CE} characteristics reported in Fig. 2b for various V_{BE} refer to a transistor with effective emitter area $W_E\times L_E$=0.13×0.88 μm^2 and featuring f_T/f_{max}=260/350 GHz [1], i.e., with frequency performance comparable with the "intermediate-node" HBT #2. It is found that simulations are almost aligned with experimental data: the measured BV_{CEO} was indeed found to be about 1.6 V at V_{BE}=0.7 V. Fig. 2b also reveals that BV_{CEO} increases with V_{BE}, which is an expected result, since (i) electrothermal effects play an increasingly relevant role; (ii) the higher the base current, the higher the avalanche multiplication factor needed to reverse its sign.

Fig. 3 summarizes the behavior of BV_{CEO} as a function of vertical scaling and temperature. The reduction with scaling depicted in Fig. 3a is ascribed to the electric field growth in the collector-base SCR, while the increase with temperature is due to the enhancement in phonon scattering, which reduces the population of high-energy electrons responsible for impact ionization [18]. The voltage BV_{CEO} normalized to its value at T=300 K is reported in Fig. 3b as a function of temperature in order to more easily compare the temperature coefficients corresponding to the technological nodes. It is shown that the temperature influence on breakdown voltage monotonically reduces with vertical scaling, which means that the electron dynamics are less influenced by an increase in oscillations of atoms in a high electric field region [18]. In order to highlight the accuracy of SHE results in comparison to more conventional approaches, the BV_{CEO}-T curves were also determined by 1-D simulations carried out through an improved version of a commercial HD tool [11], [12] by enabling the widely adopted Okuto-Crowell model to account for avalanche generation [19]. Fig. 3c illustrates that, although the simulator enjoys calibrated transport models and HD equations, (i) the BV_{CEO} values are overestimated, and (ii) an unphysical BV_{CEO} reduction with increasing temperature is unexpectedly predicted for the high technological nodes.

Lastly, with the aim to offer a complete overview on all the obtained results, the breakdown voltages and maximum cut-off frequencies $f_{T,PEAK}$ are reported in Fig. 4, where BV_{CEO} is plotted as a function of the $f_{T,PEAK}$ determined for each nodes, which reduces with temperature [7]. It is found that the product $BV_{CEO} \times f_{T,PEAK}$ – usually monitored to quantify the technology performance – is characterized by a positive temperature coefficient.

It must be remarked that the analysis has not been extended to the region with negative base current bounded by BV_{CEO} and the open-emitter breakdown voltage BV_{CBO} since the 1-D SHE-based solver cannot describe the possible occurrence of the avalanche-induced current hogging known as pinch-in, which is 3-D in nature [6], [17].

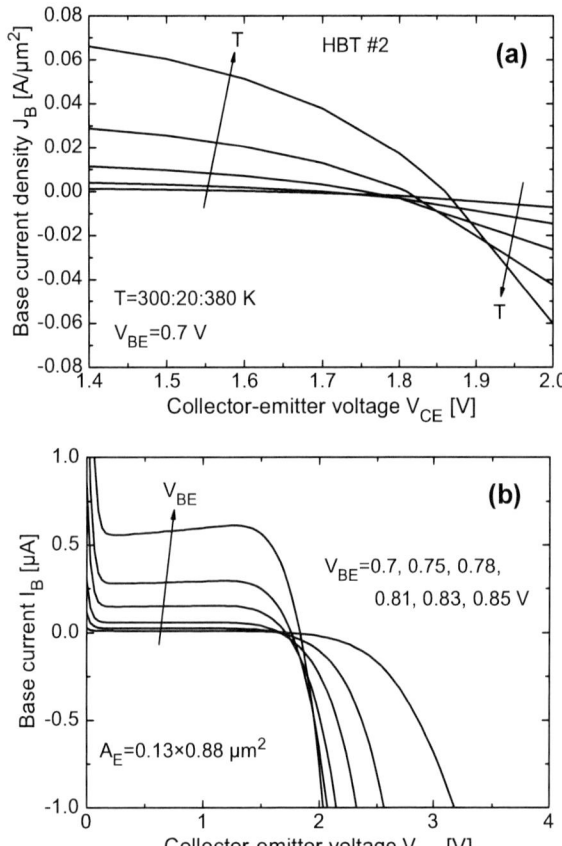

Figure 2. SHE-simulated J_B-V_{CE} characteristics @ V_{BE}=0.7 V of HBT #2 for different lattice temperatures (a) and experimental I_B-V_{CE} curves at various base-emitter voltages V_{BE} for an "intermediate-node" SiGe:C device (b).

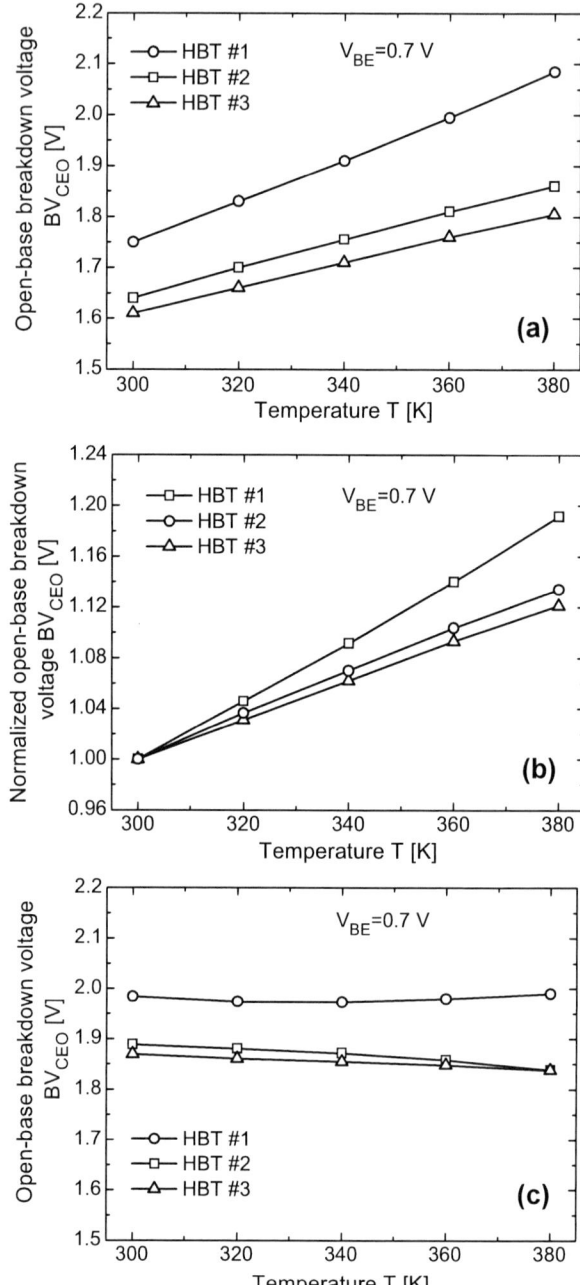

Figure 3. SHE-simulated open-base breakdown voltage BV_{CEO} (a) and BV_{CEO} normalized to the value @ T=300 K (b); HD-simulated BV_{CEO} (c) as a function of lattice temperature for devices #1, #2, and #3.

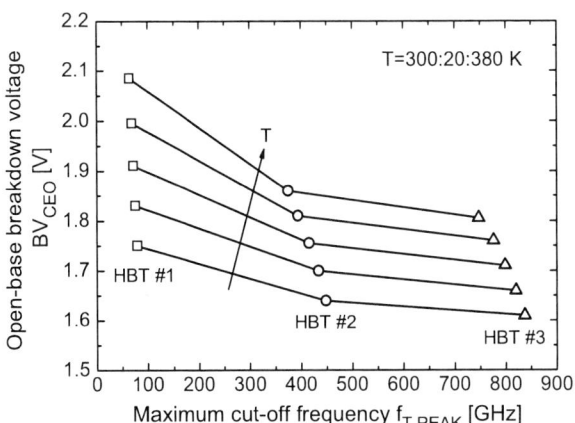

Figure 4.   Open-base breakdown voltage BV_{CEO} as a function of maximum cut-off frequency $f_{T,PEAK}$ by varying the lattice temperature.

V.   CONCLUSIONS

The influence of vertical scaling and temperature upon impact-ionization phenomena of SiGe HBTs over different technology nodes has been determined by invoking an advanced simulation strategy based on the spherical harmonics expansion of the distribution function, which has been recently found to be accurate for modern nanoscale transistors. This is useful to provide a clear insight into the physical limitations induced by downscaling, as well as for modeling purposes. BV_{CEO} results have been corroborated by experiments performed on state-of-the-art SiGe:C HBTs featuring f_T/f_{max}=260/350 GHz. It has been shown that the vertical shrinking contributes to weaken the breakdown voltage sensitivity to temperature. Moreover, the $BV_{CEO} \times f_{T,PEAK}$ parameter has been found to exhibit a positive temperature coefficient.

ACKNOWLEDGMENT

The authors are grateful to Prof. C. Jungemann for providing the SHE simulator.

REFERENCES

[1]   P. Chevalier, F. Pourchon, T. Lacave, G. Avenier, Y. Campidelli, L. Depoyan, G. Troillard, M. Buczko, D. Gloria, D. Céli, C. Gaquière, and A. Chantre, "A conventional double-polysilicon FSA-SEG Si/SiGe:C HBT reaching 400 GHz f_{MAX}," in Proc. IEEE Bipolar/BiCMOS Circuits and Technology Meeting, 2009, pp. 1-4.

[2]   [Online] Available: http://www.dotfive.eu/.

[3]   M. Schroeter, G. Wedel, B. Heinemann, C. Jungemann, J. Krause, P. Chevalier, and A. Chantre, "Physical and electrical performance limits of high-speed SiGeC HBTs – Part I: vertical scaling," IEEE Transactions on Electron Devices, vol. 58, no. 11, pp. 3687-3696, 2011.

[4]   V. d'Alessandro, I. Marano, S. Russo, D. Céli, A. Chantre, P. Chevalier, F. Pourchon, and N. Rinaldi, "Impact of layout and technology parameters on the thermal resistance of SiGe:C HBTs," in Proc. IEEE Bipolar/BiCMOS Circuits and Technology Meeting, 2010, pp. 137-140.

[5]   J.-S. Rieh, D. Greenberg, A. Stricker, and G. Freeman, "Scaling of SiGe heterojunction bipolar transistors," Proceedings of the IEEE, vol. 93, no. 9, pp. 1522-1538, 2005.

[6]   G. Sasso, M. Costagliola, and N. Rinaldi, "Avalanche multiplication and pinch-in models for simulating electrical instability effects in SiGe HBTs," Microelectronics Reliability, vol. 50, no. 9-11, pp. 1577-1580, 2010.

[7]   G. Sasso, V. d'Alessandro, M. Costagliola, S. Russo, and N. Rinaldi, "Impact of scaling on the DC/RF thermal behavior of SiGe HBTs for high-frequency applications," Materials Science and Engineering B, vol. 177, no. 15, pp. 1233-1238, 2012.

[8]   M. Al-Sa'di, V. d'Alessandro, S. Frégonèse, S.-M. Hong, C. Jungemann, C. Maneux, I. Marano, A. Pakfar, N. Rinaldi, G. Sasso, M. Schröter, A. Sibaja-Hernandez, C. Tavernier, and G. Wedel, "TCAD simulation and development within the European DOTFIVE project on 500GHz SiGe:C HBT's," in Proc. IEEE European Microwave Integrated Circuits Conference, 2010, pp. 39-32.

[9]   S.-M. Hong and C. Jungemann, "A fully coupled scheme for a Boltzmann-Poisson equation solver based on a spherical harmonics expansion," Journal of Computational Electronics, vol. 8, no. 3, pp. 225-241, 2009.

[10] S.-M. Hong, G. Matz, and C. Jungemann, "A deterministic Boltzmann equation solver based on a higher order spherical harmonics expansion with full-band effects," IEEE Transactions on Electron Devices, vol. 57, no. 10, pp. 2390-2397, 2010.

[11] G. Sasso, G. Matz, C. Jungemann, and N. Rinaldi, "Accurate mobility and energy relaxation time models for SiGe HBTs numerical simulation," in Proc. IEEE International Conference on Simulation of Semiconductor Processes & Devices, 2009, pp. 241-244.

[12] G. Sasso, G. Matz, C. Jungemann, and N. Rinaldi, "Analytical models of effective Dos, saturation velocity and high-field mobility for SiGe HBTs numerical simulation," in Proc. IEEE International Conference on Simulation of Semiconductor Processes & Devices, 2010, pp. 279-282.

[13] S.-M. Hong, A.-T. Pham, and C. Jungemann, Deterministic Solvers for the Boltzmann Transport Equation. Springer-Verlag, 2011.

[14] R. Thoma, H. J. Peifer, W. L. Engl, W. Quade, R. Brunetti, and C. Jacoboni "An improved impact-ionization model for high-energy electron transport in Si with Monte Carlo simulations," Journal of Applied Physics, vol. 69, no. 4, pp. 2300-2311, 1991.

[15] M. Rickelt, H.-M. Rein, and E. Rose, "Influence of impact-ionization-induced instabilities of the maximum usable output voltage of Si-bipolar transistors," IEEE Transactions on Electron Devices, vol. 48, no. 4, pp. 774-783, 2001.

[16] J. Kraft, D. Kraft, B. Löffler, H. Jauk, and E. Wachmann, "Usage of HBTs beyond BV_{CEO}," in Proc. IEEE Bipolar/BiCMOS Circuits and Technology Meeting, 2005, pp. 33-36.

[17] C. M. Grens, J. D. Cressler, and A. J. Joseph, "On common-base avalanche instabilities in SiGe HBTs," IEEE Transactions on Electron Devices, vol. 55, no. 6, pp. 1276-1285, 2008.

[18] M. Ershov, and V. Ryzhii, "Temperature dependence of the electron impact ionization coefficient in silicon," Semiconductor Science and Technology, vol. 10, no. 2, pp. 138-142, 1995.

[19] Y. Okuto, and C. R. Crowell, "Threshold energy effect on avalanche breakdown voltage in semiconductor junctions," Solid-State Electronics, vol. 13, no. 2, pp. 161-168, 1975.

A New Eye on NBTI-Induced Traps up to Device Lifetime Using on the Fly Oxide Trap Method

Boualem DJEZZAR, Hakim TAHI, Abdelmadjid BENABDELMOUMENE, and Amel CHENOUF
Microelectronics and Nanotechnology Division
Centre de Développement des Technologies Avancées (CDTA)
Algiers, Algeria
bdjezzar@cdta.dz

Abstract— In this work, we present a novel methodology to independently extract the interface-trap and border-trap generated by negative bias temperature instability (NBTI). It is mainly based on charge pumping technique (CP) at low and high frequencies. We emphasize on the easy-use of this approach and demonstrating its viability to characterize the NBTI. Using alternatively on the fly high and low frequencies CP measurements, it can separate the interface-traps and border-trap (switching oxide-trap) densities as well as their contributions to the threshold voltage shift in the same timeframe, without needing additional methods. The results show a power-law time-dependence of interface- and border-trap with an average exponent *n* of 0.16 and 0.07, respectively. In addition, it allows device lifetime projection at working conditions.

Index Terms— NBTI stress, Border-trap, Interface-trap, Lifetime, charge pumping.

I. INTRODUCTION

The lifetime of metal oxide semiconductor devices is strongly affected by NBTI degradation. The latter degrades the quality of Si/SiO_2 interface as well as silicon dioxide, SiO_2, inducing shift of threshold voltage (V_{th}). The current understanding links this shift to interface- and oxide-trap creation with temperature and electric field [1-3]. However, the relative contribution of interface and oxide traps to the V_{th} shift (ΔV_{th}) is still under debate and needs more experimental and theoretical investigations of both NBTI stress and recovery phases [4, 5]. The relaxation has been explained by hole trapping/detrapping in preexisting oxide-trap and/or oxide-trap creation near Si/SiO_2 interface, while interface-trap does not relax [6].

However, Ming Fu et al. [1] have proposed that both components recover and attributed them to a fast oxide-trap recovery and a slow interface-trap recovery. Nevertheless, the microscopic nature of those components and which component dominates the NBTI stress and relaxation effects is till now questionable. In order to find the relative contribution of interface- and oxide-trap to ΔV_{th} after NBTI stress and recovery, researchers combine on-the-fly interface-trap (OTFIT) [7] using CP technique to extract ΔV_{it} [induced by interface-trap increase (ΔN_{it})] and on-the-fly V_{th} (OTF-V_{th}) [8] to extract ΔV_{th}. Subsequently, ΔV_{ot} [induced by oxide-trap increase (ΔN_{ot})] is found by subtracting ΔV_{it} from the amount of ΔV_{th}. Other research group has used the cyclic dynamic of

NBTI which is supposed governed solely by hole trapping/detrapping mechanism with assumption that ΔN_{it} is permanent [9, 10].

Despite the advantages of OTFIT and OTF-V_{th}, their combination can not directly provide information on interface- and oxide-trap independently. The problem with this approach is not only using two methods, which causes more recovery, but also the difference in band energy scanned by OTFIT and OTF-V_{th}. The extraction of ΔV_{ot} is indirect and depends on ΔV_{it} and ΔV_{th}. The misestimating of trap generation induces misunderstanding of the fundamental mechanisms behind the NBTI degradation and subsequently restricts elaborating a reliable predictive model for device lifetime evaluation.

Therefore, it is worthy to develop a method which can separately and independently assess both components. For this purpose, we suggest a method, named On-the-fly oxide-trap (OTFOT). It allows measuring both interface-trap and switching oxide-trap [border-trap (ΔN_{bt})] individually. However and contrary to existing methods [7-9], the present method extracts ΔN_{it} and ΔN_{bt} using stress and CP in the same timeframe without stopping the stress. This method is derived from Oxide-trap charge-pumping (OTCP) method, developed for radiation effect [11, 12]. It can give further information on the behavior of NBTI-induced ΔN_{it} and ΔN_{bt}. In addition, it can also assess the NBTI-induced ΔV_{th} and the device lifetime related to NBTI reliability.

This paper is organized as follows. In section II, we present the experimental setup and condition details as well as the measure/stress/measure (MSM) protocol of OTFOT method. Result analysis and discussion are described in section III. Finally, conclusions are summarized in section IV.

II. EXPERIMENTAL PROCEDURE

A. Experimental setup

To validate the OTFOT method, we have characterized the NBTI effects on PMOS transistors. They were fabricated at ISiT (Institute for Silicon Technology) of Fraunhofer, Germany. The gate oxide thickness is 20 nm and C_{ox} is 2.12 x 10^{-7} F.cm^{-2}. MSM sequences have been performed using fully automated benches available at CDTA. The benches include a sensitive Agilent HP 4156C for current measurement and trigging Keithley 3940 to generate signal during. The Keithley

3940 multifunction synthesizer is used to apply trapezoidal signal of 1 MHz and control rise time, fall time and stress voltage. It offers more capabilities regarding the signal frequency than Agilent HP 4156C which is limited to 500 kHz. A hotplate is used, inside Karl Suss PA300 micromanipulator probe station, to vary the temperature of the chip. The test circuit chip (non-packaged) within the probe station were isolated from vibration and enclosed in a grounded faraday cage to avoid both RF and light effects.

B. Experimental conditions

Figure 1 shows the MSM protocol of OTFOT. During the stress interval, the voltage stress (V_{Gstr}) is applied onto the gate of the device through a DC voltage. After each stress time, a gate pulse train is applied, without modifying the experimental setup. The gate trapezoidal signal with an amplitude of $\Delta V_G = V_H - V_{GSstr}$ [where $V_L = V_{GSstr}$, a duty cycle of 50%, an equal rise and fall times, and two different frequencies (high, f_H and low, f_H)] are applied alternatively to measure the maximum CP currents, I_{CPH} and I_{CPL}. These currents result from recombination of electrons and holes at the interface-trap (for high frequency) and at both interface- and border-traps (for low frequency). At each stress time, one point I_{CP} is measured and the second one is obtained by interpolation (i.e. when I_{CPH} value is measure, the corresponding I_{CPL} value is interpolated and vise versa). The low frequency value has to be chosen as low as the CP signal is distinguishable from leakage current. Obviously, this limits the scanning depth into the oxide layer, but it is not an issue in NBTI degradation since the latter is meanly located near the interfacial region [13].

To avoid polluting CP measurements with parasitic leakage current and tunneling current, we have used devices with a large gate surface, W/L of 10/10 and 5/10, and thick gate oxide of 20 nm, since the basic mechanisms behind the NBTI are assumed to be the same in both thin and thick oxides, because defects induced by NBTI are located at the interface of Si/SiO_2 and in the oxide near the interface [13]. The stress interval is kept the same during all MSM cycles for different temperatures and electric fields. The whole stress time cycle is fixed at 3600s. During the NBTI stress, we used temperatures of 120, 140, and 160 °C and voltages of 7, 8, and 10 V, while keeping V_H at 1 V in all experiments.

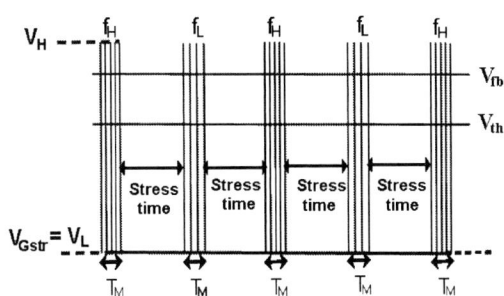

Fig.1. Measure/stress/measure (MSM) protocol used to extract the NBTI components without additional methods. It is based upon the on the fly CP, applied alternatively at low at high frequencies during measurements.

III. RESULTS AND DISCUSSION

Figure 2 shows the NBTI as-measured CP current normalized to the frequency and gate area at low (10 kHz) and high (1 MHz) frequencies. They are extracted using MSM protocol described in Fig. 1. The CP current increases in absolute value with stress voltage, indicating an increase of interface-trap at high frequency [Fig. 2 (a)], while at low frequency [Fig. 2 (a)] both interface- and border-trap are enhanced.

The separating approach of NBTI-induced ΔN_{it} and ΔN_{bt}, used in this work, is only based upon the as-measured I_{CP} of Fig. 2 and can be extracted using:

$$\Delta N_{it} = \frac{\Delta I_{CPH}}{q A_G f_H} \tag{1}$$

$$\Delta N_{bt} = \left| \frac{\Delta I_{CPL}}{q A_G f_L} - \frac{\Delta I_{CPH}}{q A_G f_H} \right| \tag{2}$$

where q (C) is the electron charge, A_G (cm^2) is the gate area, I_{CPH} (A) and I_{CPL} (A) are the maximum CP current at high, f_H (Hz) and low, f_L (Hz) frequencies, respectively. Hence, the amount of ΔN_{it} and ΔN_{bt} is accurately determined during stress under different electric fields and temperatures. Figure 3 illustrates their quantities as a function of stress time for 9 and 10 V and 120, 140, and 160°C in log-log scale. On the right axis, we present the voltage shifts induced by interface and border traps using relations: $\Delta V_{it} = q\Delta N_{it}/C_{ox}$ and $\Delta V_{bt} = q\Delta N_{ot}/C_{ox}$. The NBTI-induced ΔN_{it} and ΔN_{bt} time

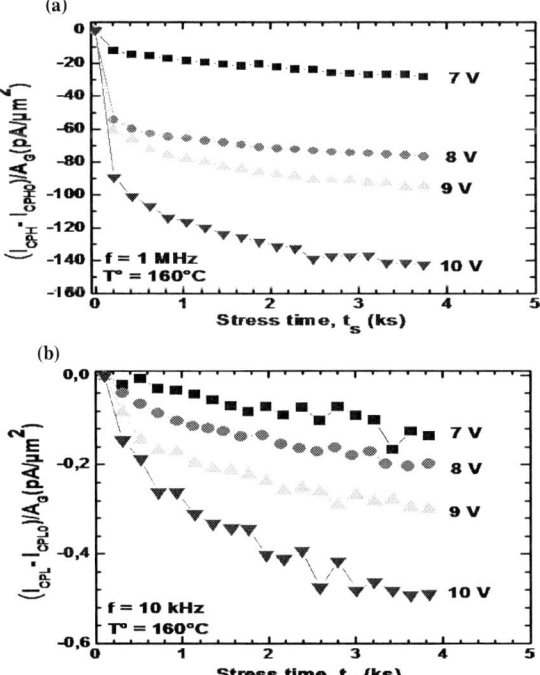

Fig.2. As-measured CP current (a) at high frequency and (b) at low frequency. Data are extracted at 160°C for different voltages.

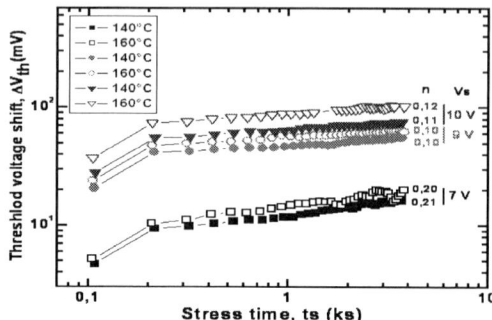

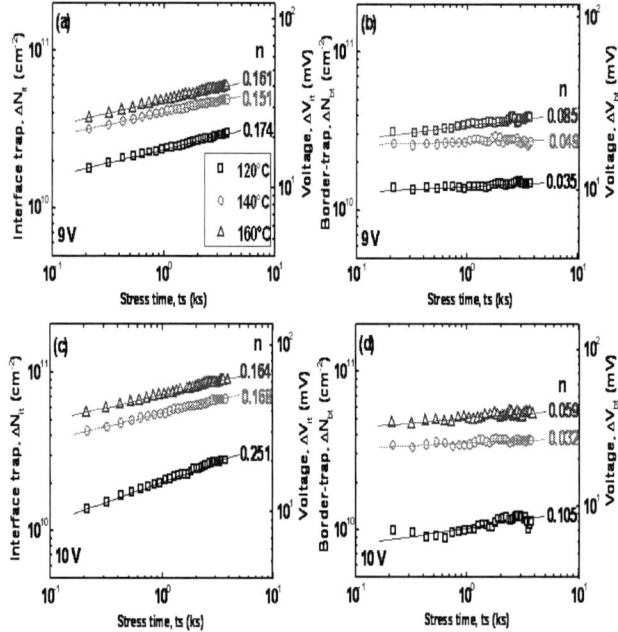

Fig.3. NBTI stress-induced interface- and border-trap ΔN_{it} and ΔN_{bt}, respectively (on the left axis) as well as their contribution to voltage shift ΔV_{it} and ΔV_{bt} (on the right axis).

Fig.4. NBTI-induced threshold voltage shift, ΔV_{th} for different voltages at 140°C and 160°C vs. stress time. ΔV_{th} curves exhibit power law with average exponent of 0.11.

TABLE I. COMPARISON OF OTFOT METHOD TO OTF METHODS

Method	ΔV_{it}	ΔV_{ot}	ΔV_{th}	Effect on the measure
OTF-V$_{th}$			X	- Initial pre-stress I_0 degrades I_d. - Under-estimation of ΔV_{th}.
OTFIT	X			- Initial pre-stress I_0 corresponds to relaxation phase. - Over-estimation of ΔV_{it}.
OTFOT	X	X	X	- Initial pre-stress I_0 corresponds to first point of stress phase. - At f_L a part of oxide traps is sensed.

characteristics show an evident power-law time-dependence t^n with exponent n of 0.16-0.17 and 0.03-0.08, respectively. These results are in perfect agreement with those previously published by different groups [1, 9, 10]. The first exponent is often attributed to H_2 diffusion mechanism in the R-D framework [14], while in conventional CP, n is about 0.3 [1]. Therefore, $n = 0.16$ demonstrates reduction of recovery during stress and subsequently the accuracy of the method used in this work. The second n is much less than that of ΔN_{it} indicating two different processes of NBTI degradation. It is attributed to hole trapping/detrapping process at the border-trap [15]. Dispersion in n values of trapping/detrapping mechanism at pre-existing border traps could be explained by the dispersive energetic distribution of border-trap and/or due to further generated hole trapping sites.

Not only our experimental method shows the same exponents as those found elsewhere [1, 9, 10, 14], but it is easy to use without any pre-assumption and/or additional method as well. In fact and contrarily to this work, Ming Fu et al. [1] combine two methods OTF-V$_{th}$ and OTFIT to extract ΔN_{bt} from ΔV_{th} and ΔN_{it} (see table I). The latter can in some way influence ΔN_{bt}. Other research group [9, 10] used the cyclic dynamic of NBTI which is supposed governed solely by hole_trapping/detrapping mechanism with assumption that ΔN_{it} is permanent. But, it has been shown that under certain conditions, the permanent component can relax as the recoverable component [1, 16]. Further, the cyclic dynamic method gives a large exponent (0.493). To obtain 0.16, the authors assume hole saturation within 1s. However, our analysis of ΔN_{it} data is only based on as-measured degradation without pre-assumption. They show a good agreement with H_2 paradigm and constitute an experimental proof for H_2 R-D interface-NBTI-based model. In addition, we have extracted

ΔV_{th} as $\Delta V_{it} + \Delta V_{bt}$. Figure 4 gives NBTI-induced ΔV_{th} for different electric fields at 140 and 160°C. All curves show the same behavior with stress time, electric field, and temperature. They exhibit a power law with n of 0.11 in log-log scale. Once again, the same ΔV_{th} exponent was depicted in previous experiment [10] using ultra-fast I-V method and which subsequently adds further proof supporting OTFOT method. In table I, we summarize the pros and cons of OTFOT method compared to OTFIT and OTF-V$_{th}$ methods.

After demonstrating the useful of OTFOT, we can use ΔV_{th} versus stress time for different voltages and temperatures (see Fig. 4) to estimate NBTI lifetime of studied devices. However, the projection lifetime must be performed at circuit normal operation conditions. To do so, we extrapolate ΔV_{th} data, which are obtained at accelerated temperatures, to ambient temperature. Figure 5 (a) illustrates ΔV_{th} at 25 °C (see grey circle symbol) for different stress times. These latter are extrapolated to operating supply voltage, 5 V and plotted for several stress times. Figure 5 (b) shows extrapolated data after 2600 s of stress. In addition, data for different temperatures are also illustrated.

Finally, the NBTI lifetime projection is given in Fig. 6. In this study, the device lifetime criterion due to NBTI is defined as ΔV_{th} reaches 10% of initial value (V_{th0}), which is equal to -1.2 V. We notice that the NBTI lifetime for these devices exceed 10 years for temperature 25°C, but decreases with temperature.

IV. CONCLUSION

We have proposed a new approach to independently extract and separate the NBTI-induced interface and border traps. This methodology allows extraction of these two kinds of traps

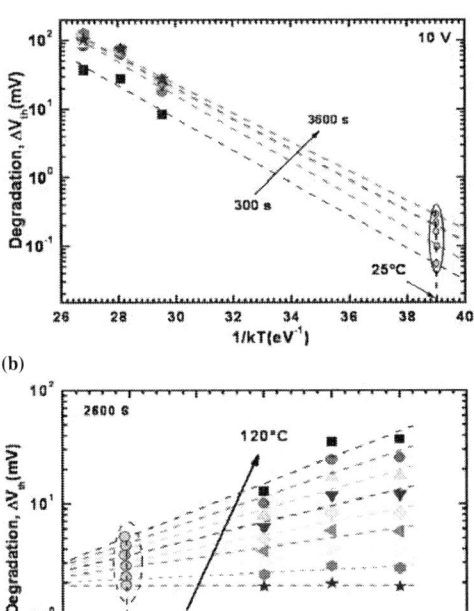

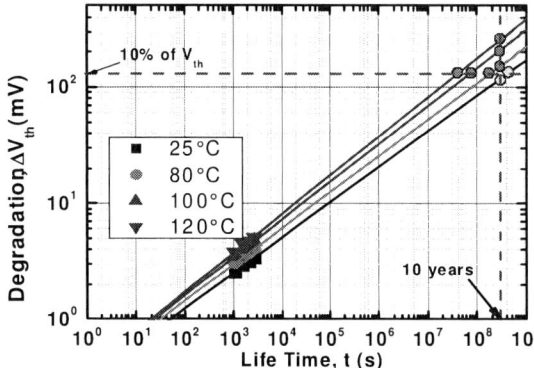

Fig.6. Threshold voltage shift vs. projection lifetime at different temperatures. Taking 10% of V_{th} as lifetime rule, our device lifetime exceed 10 year at 25°C and decreases with temperature.

(b)

Fig.5. Threshold voltage shift under NBTI stress (a) as a function of temperature at 10V for different stress times (b) as a function of volta e at 2600s for different temperature.

using only CP technique at low and high frequencies without combining other methods. In addition, ΔV_{th} is obtained from independently extracted ΔV_{it} and ΔV_{bt}. The experimental results show a good agreement with literature regarding power-law exponent. The device lifetime subjected to the NBTI effect is also estimated. Lastly, this method can contribute to further understanding of the NBTI degradation, especially through the threshold voltage shift components.

ACKNOWLEDGMENT

This work was supported by the High School Educational and Scientific Research Ministry of Algeria under the National Funding of Research (FNR) contract code 22/CDS/DMN/CDTA/2011.

We thank Mr. Boumediene Zatout for his assistance to perform measurements.

REFERENCES

[1] Ming Fu Li, D. Huang, C. Shen, T. Yang, W. J. Liu, and Z. Liu, "Understand NBTI mechanism by developing novel measurement techniques," IEEE Trans. on Dev. and Mater. Relia., vol. 8, no. 1, pp. 62-71 Mar. 2008.

[2] T. Grasser, B. Kaczer, W. Goes, Th. Aichinger , Ph. Hehenberger, and M. Nelhiebel, "Understanding negative bias temperature instability in the context of hole trapping," Micro. Eng., Elsevier, vol. 86, no. 7-9, pp. 1876-1882, July-Sep. 2009.

[3] H. W. Chen, C. H. Liu, "Impact of Hf content on positive bias temperature instability reliability of HfSiON gate dielectrics," Micro. Relia., Elsevier, vol. 50, pp. 614-617, 2010.

[4] T. J. J. Ho, D.S. Ang, A. A. Boo, Z. Q. Teo, and K. C. Leong, "Are interface state generation and positive oxide charge trapping under

negative-bias temperature stressing correlated or coupled?" IEEE Trans. on Elec. Dev., vol. 59, no. 4, pp.1013-1022, Apr. 2012.

[5] M. Duan, J. F. Zhang, Z. Ji, W. Zhang, B. Kaczer, S. De Gendt, and G. Groeseneken, "Defect loss: a new concept for reliability of MOSFETs" IEEE Trans. Elec Dev. Lett, vol. 33, no. 4, pp. 480-482, Apr. 2012.

[6] V. Huard, M. Denais, C. Parthasarathy, "NBTI degradation: from physical mechanisms to modelling," Micro. Relia., Elsevier, vol. 46, no. 1, pp. 1-23, Jan. 2006.

[7] W. J. Liu, Z.Y. Liu, D. Huang, C. C. Liao, L. F. Zhang, Z. H. Gan, W. Wong, C. Shen, and M. F. Li, "On-the-fly interface trap measurement and its impact on the understanding of NBTI mechanism for p-MOSFETs with SiON gate dielectric," IEEE Inter. Elec. Dev. Meeting (IEDM'07), Washington, DC , USA, 10-12 Dec., 2007, pp. 813-816.

[8] M. Denais, A. Bravaix, V. Huard, C. Parthasarathy, G. Ribes, F. Perrier, Y. R. Tauriac, and N. Revil, "On-the-fly characterization of NBTI in ultra-thin gate oxide PMOSFET's," in IEEE Inter. Elec. Dev. Meeting, San Francisco, CA, USA, 13-15 Dec., pp. 109-112, 2004.

[9] Z.Q. Teo, D.S. Ang, C.M. Ng, "Separation of hole trapping and interface-state generation by ultrafast measurement on dynamic negative-bias temperature instability" IEEE Trans. Elec. Dev., Lett., vol. 31, no. 7, pp. 656-658, Jul. 2010.

[10] D. S. Ang, Z.Q. Teo, T. J. J. Ho, and C. M. Ng, "Reassessing the mechanisms of negative-bias temperature instability by repetitive stress/relaxation experiments," IEEE Trans. on Dev. and Mater. Relia., vol. 11, no. 1, pp. 19–34, Mar. 2011.

[11] B. Djezzar, A. Smatti, and S. Oussalah, "Oxide-trap based on charge pumping (OTCP) extraction method for irradiated MOSFET devices: Part II (low frequencies)," IEEE Trans. on Nucl. Sci., vol. 51, no. 4, pp. 1732-1736, Aug. 2004.

[12] B. Djezzar, H. Tahi, and A. Mokrani, "Why is oxide-trap charge-pumping method appropriate for radiation-induced trap depiction in MOSFET?," IEEE Trans. on Dev. and Mater. Relia., vol. 9, no. 2, pp. 222–230, Jun. 2009.

[13] G. Pobegen, Th. Aichinger, M. Nelhiebel and T. Grasser, "Dependence of the negative bias temperature instability on the gate oxide thickness," in IEEE Inter. Relia. Phys. Sym. (IRPS'10), Garden Grove, CA, USA, May 2-6, 2010, pp. 1073-1077.

[14] A. E. Islam, H. Kufluoglu, D. Varghese, S. Mahapatra, and M. A. Alam, "Recent issues in negative bias temperature instability: initial degradation, field- dependence of interface trap generation, and hole trapping effects and relaxation", (Invited paper) IEEE Trans. Elec. Dev., vol. 549,no.1, pp. 2143- 2154, Sep. 2007.

[15] M. Denais, V. Huard, C. Parthasarathy, G. Ribes, F. Perrier, N. Revil, and A. Bravaix, "Interface trap generation and hole trapping under NBTI and PBTI in advanced CMOS technology with a 2-nm gate oxide," IEEE Trans. Dev. Mater. Relia., vol. 4, no. 4, pp. 715–722, Dec. 2004.

[16] T. Grasser, Th. Aichinger, H.Reisinger, J. Franco, P.-J.Wagner, M. Nelhiebel, C. Ortolland, and B. Kaczer, "On the 'permanent' component of NBTI," in IEEE Inter. Integ. Relia. Workshop (IIRW'08), S. Lake Tahoe, CA, USA, Oct. 12-16, 2008, pp. 1-7, 2010.

An Accurate Extraction Methodology for NBTI Induced Degradation Using Charge Pumping Based Methods

Hakim TAHI, Boualem DJEZZAR, Abdelmadjid BENABDELMOUMENE, and Amel CHENOUF
Microelectronics and Nanotechnology Division,
Centre de Développement des Technologies Avancées (CDTA), Algeria
email htahi@cdta.dz

Abstract— In this work, we present a new methodology to determinate and remove the geometric component in charge pumping (CP) based methods, such as on-the-fly interface trap (OTFIT) method. This methodology uses CP-current data of different gate length transistors (L_G) with fixed gate width (W_G) to obtain an empirical model for the remaining carriers in MOSFET channel after switch off. This allows investigating the geometric component (GC) as a function of device gate length during the negative bias temperature instability (NBTI) stress. We present the experimental evidence that GC increases during NBTI stress, which is most likely caused by Coulomb scattering of created traps. This implies that the precision of the estimation of NBTI stress induced-traps is affected as well as their creation dynamic. Consequently, the NBTI power-law time exponent, n is misestimated. We have also presented the experimental results after removing the GC. They show a same n for all transistors. This procedure is a valuable to correct CP-based method data for degraded devices.

Index Terms— NBTI stress, Geometric component, Lifetime, charge pumping.

I. INTRODUCTION

The negative bias temperature instability (NBTI) has been identified as one of the most serious reliability issues in modern CMOS technology. Therefore, many different measurement methods have been proposed in order to understand the NBTI physics based mechanisms and accurate estimation of the device life time. The most used methods are Charge Pumping (CP) based methods such as Conventional charge pumping (CCP) [1]-[2] and on-the-fly interface trap (OTFIT)[3]-[4]. An important source of errors for CP based methods is the carrier recombination in MOSFET substrate bulk, also known as the geometric component (GC) [1]-[2], [5]. Conventionally, this component is defined as an additional component to CP-current that arises when some electrons (holes) cannot reach the source or drain region during the transition from inversion to accumulation. These electrons (holes) recombine with incoming holes (electrons) from the substrate. The GC affects the accuracy of the CP based methods and therefore causes overestimation of interface density. In addition, this parasitic component can increase

before or during NBTI stress due to the coulomb scattering by the interface and oxide traps, generated by NBTI. Therefore, the virgin and degraded devices might not have the same transition time t_f (or t_r) to eliminate the GC [2]. On one side, the use of the same t_f (or t_r) for virgin and degraded devices leads to an overestimation of interface trap due to NBTI-induced trap effect on the GC. On the other side, using different t_f (or t_r) is not appropriate due to the variation of the CP scanned energy range in the silicon band gap.

In addition, Till now, there is no accurate method for the determination of GC from the CP-current. Moreover, the impact of NBTI stress on GC is still not well understood.

In this paper, we propose a new methodology to remove GC from CP-current data for NBTI stressed devices, without reducing t_f or using unconventional transistors [5]. This methodology is used to investigate the impact of NBTI on the GC.

This paper is organized as follows; Section II describes the experimental setup. In section III, an empirical model of GC is developed. The impact of the GC on CP-based methods for NBTI stresses is investigated in IV. Conclusions are drawn in the last section.

II. EXPERIMENTAL PROCEDURE

We characterized LDD-MOS transistors with different gate lengths (L_G) and fixed gate width (W_G), fabricated on the same chip at the Institute for Silicon Technology of Fraunhofer, Germany. The process is a dual-layer-metal 1-μm LDD-CMOS twin-well technology on P-type 12-μm epilayer on silicon (100) substrate with 16-nm-thick gate oxide layer grown in dry O_2. The gate capacitance per unit area, C_{ox} is about 2.12×10^{-7} F.cm^{-2}.

The NBTI is performed using on-the-fly interface tap method OTFIT [4], on both LDD-PMOS and LDD-NMOS transistors with different L_G and fixed W_G. The stress/measurement/stress (SMS) sequences have been performed at CDTA using fully automated bench. The bench includes a sensitive Agilent HP 4156C for current measurement and trigging Keithley 3940 to generate pulse trains during measurement by CP techniques. A hotplate is used, inside Karl Suss PA300 micro-manipulator probe station, to vary the temperature of the chip during the experiments. The test circuit chip (non-packaged) within the probe station are isolated from vibration and enclosed in a grounded faraday cage to avoid both RF and light effects.

The whole stress time cycle is fixed at 3600s. The stress temperature and voltage (field) are T=120°C and *Vs = -9 V* (5.6 V/cm.), respectively.

III. GEOMETRIC COMPONENT DETERMINATION METHODOLOGY

Fig.1 gives plots of the experimental OTFIT data, (see symbols) for different transistors PMOS [Fig.1.(a)] and NMOS [Fig.1.(b)] with fixed gate width and varied gate lengths, as a function of L_G for different stress times. All data are better fitted by the following expression:

$$
\begin{aligned}
I_{CP}\left(t\right)_{mea} &= I_0(t) + \alpha(t)EXP\left(\frac{L_G}{\lambda(t)}\right) \\
&= I_0(t) + \alpha(t)EXP\left(\frac{L_{EFF}+2\Delta L}{\lambda(t)}\right) \\
&= I_0(t) + \alpha(t)EXP\left(\frac{L_{EFF}}{\lambda(t)}\right)EXP\left(\frac{2\Delta L}{\lambda(t)}\right)
\end{aligned}
\tag{1}
$$

where L_{EFF} and ΔL are the effective channel and LDD sub-diffusion lengths, respectively. $I_0(t)$, $\alpha(t)$, and $\lambda(t)$ are the fitting parameters, independent of L_{EFF}, and are function of time, t. $\lambda(t)$ has a unit length (cm) and $\alpha(t)$ a unit current (A). The solid lines in Fig.1 present the fitting curves which are given by (1) for different stress times. $I_{CP}(t)|_{mea}$ in semi-log scale is shown in the inset of Fig.1. In addition, it is important to note here that the equation (1) fits well the CP data of both PMOS transistors of 1-μm CMOS process and NMOS with gate widths of 80 and 300μm, and gate length of 1, 2, 8, 80, 300μm of 2-μm CMOS process. These latter are not shown here, because of paper page limitation. This is why we have considered (1) as better fit for our data.

In previous work we have demonstrated that the charge pumping current in LDD-MOS transistors is the contribution of the pumping current from different regions constituting these transistors [6]-[7]. Therefore we can write CP current $I_{CP}(t)_{mod}$ during the NBTI stress as:

$$
\begin{aligned}
I_{CP}(t)_{mod} &= qfN_{it}^{C}(t)W_G L_{EFF} + 2qfN_{it}^{LDD}(t)W_G\Delta L \\
&+ 2qfN_{it}^{LOCOS}(t)\Delta W L_{EFF}
\end{aligned}
\tag{2}
$$

where q (C) is the electron charge, f (Hz) is the gate signal frequency. $N_{it}^{C}(t)$, $N_{it}^{C}(t)$, $N_{it}^{LDD}(t)$ and $N_{it}^{LOCOS}(t)$ are the interface-trap densities at stress time in the effective channel, LDD sub-diffusion under the gate oxide and LOCOS regions, respectively. W_G is the gate width, $2\Delta W$ is the CP-active width under the LOCOS region.

In addition, if we assume that, after transition from inversion to accumulation, the remaining carriers, $n_{L_G}(t)$ of the inversion layer recombine with substrate majority carriers, we can express GC during the NBTI stress $I_{Geo}(t)$ as:

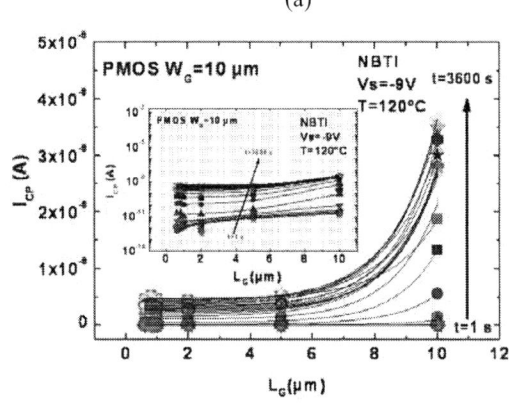

(a)

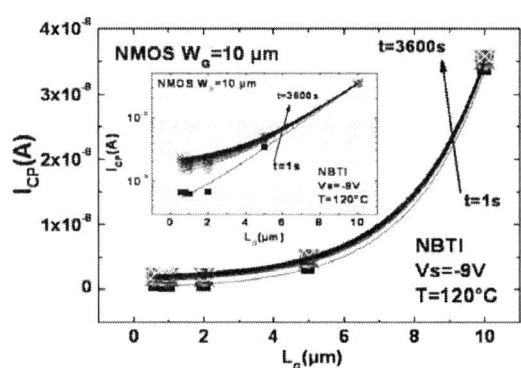

(b)

Fig.1. OTFIT data as a function of gate length for different stress times, (a) PMOS transistors, (b) NMOS transistors. The inset shows OTFIT data in semi-log scale.

$$
I_{Geo}(t) = qf n_{L_G}(t)\left[L_{EFF}+2\Delta L\right]\left[W_G+2\Delta W\right]
\tag{3}
$$

Hence, to obtain the GC we have to determine $n_{L_G}(t)$. First, we match (1) and (2)+(3):

$$
I_{CP}(t)|_{mea} = I_{CP}(t)|_{mod} + I_{Geo}(t)
\tag{4}
$$

then after, we differentiate both sides of (4) with respect to L_{EFF} to eliminate all length-independent terms. We obtain:

$$
\begin{aligned}
&qfN_{it}^{C}(t)W_G + 2qfN_{it}^{LOCOS}(t)\Delta W + qfn_{L_G}(t)\left[W_G+2\Delta W\right] \\
&+ qf\left[W_G+2\Delta W\right]L_{EFF}\frac{dn_{L_G}(V_L,V_H)}{dL_{EFF}} - \frac{\alpha(t)}{\lambda(t)}EXP\left(\frac{2\Delta L}{\lambda(t)}\right)EXP\left(\frac{L_{EFF}}{\lambda(t)}\right) = 0
\end{aligned}
\tag{5}
$$

In one side, for the negligible remaining carriers $n_{LG}(0) = 0$, (no GC), we have $\frac{dn_{L_G}(V_L,V_H)}{dL_{EFF}} = 0$. So, we can write (5) as:

$$
qfN_{it}^{C}(t)W_G + 2qfN_{it}^{LOCOS}(t)\Delta W = \frac{\alpha(t)EXP\left(\frac{2\Delta L}{\lambda(t)}\right)}{\lambda(t)}
\tag{6}
$$

$\alpha\, t\, EXP\!\left(\dfrac{2\Delta L}{\lambda(t)}\right)$ is the CP current of the transistor with effective gate length λ for which the GC is negligible.

In other side for, $L_{EFF} = 0$, (1) can expressed as :

$$I_{CP}\ t\ _{mea} = I_0\ t\ + \alpha\ t\ EXP\left(\frac{2\Delta L}{\lambda(t)}\right) \tag{7}$$

Since there is no CG for $L_{EFF}=0$ and $\alpha(t)EXP\!\left(\dfrac{2\Delta L}{\lambda(t)}\right)$ is found to be the CP current in the effective channel, therefore $I_0(t)$ corresponds to CP current in the LDD sub-diffusion under the gate oxide.

Substituting (6) in (5), we obtain the following first order differential equation:

$$\frac{EXP\left(\dfrac{2\Delta L}{\lambda(t)}\right)}{qf\ W_G + 2\Delta W}\ \frac{\alpha\ t}{\lambda\ t}\left(1 - EXP\left(\frac{L_{EFF}}{\lambda(t)}\right)\right) + n_{L_G}(t) + L_{EFF}(t)\frac{dn_{L_G}(t)}{dL_{EFF}} = 0 \tag{8}$$

Solving (8) yields the remaining carrier concentration $n_{L_G}(t)$, as a function of gate length, as:

$$n_{L_G}(t) = \frac{EXP\left(\dfrac{2\Delta L}{\lambda(t)}\right)}{qf\left[W_G + 2\Delta W\right]L_{EFF}}\ \frac{\alpha(t)}{\lambda(t)} \times \left(\lambda(t)\,EXP\left(\frac{L_{EFF}}{\lambda(t)}\right) - L_{EFF} - \lambda(t)\right) \tag{9}$$

Finally, substituting (9) in (3), the GC can be expressed as:

$$I_{Geo}(t) = \frac{\alpha(t)}{\lambda(t)}\,EXP\left(\frac{2\Delta L}{\lambda\ t}\right) \times \left(\lambda(t)EXP\left(\frac{L_{EFF}}{\lambda(t)}\right) - L_{EFF} - \lambda(t)\right) \tag{10}$$

IV. IMPACT OF GC ON CP-BASED METHOD FOR NBTI-INDUCED DEGRADATION

The evolution of GC during the NBTI stress for PMOS and NMOS transistors, extracted using (10), is given in Fig .2(a) and Fig 2(b), respectively. All extracted GC for different transistors increasing with time during NBTI stress. This due to the coulomb scattering by traps generated during this degradation. The impact of NBTI on GC is more pronounced in PMOS compared to NMOS transistors. This result is unsurprising, since the NBTI degradation is more important in PMOS than in NMOS transistors [8].

Fig.3 gives the CP current degradation, ΔI_{CP} with [Fig.3 (a, b)] and without [Fig.3 (c, d)] GC for PMOS and NMOS. The power-law time exponent (n) of ΔI_{CP} is strongly depending on the gate length during the NBTI stress, subsequently ΔN_{it} is also affected before GC removing. The GC impact is more significant as the gate length is long. After withdrawing the GC effect, all ΔI_{CP} curves exhibit the same exponent n of 0.12 and 0.28 for PMOS and NMOS transistors,

(a)

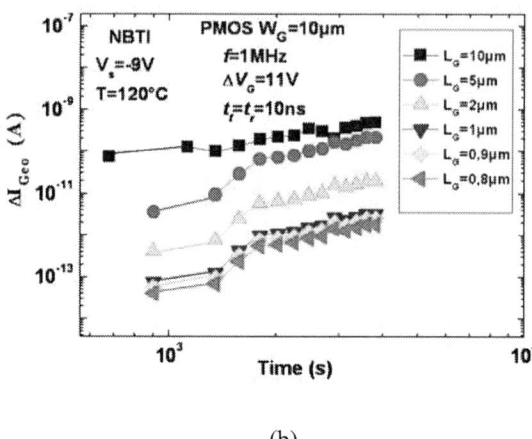

(b)

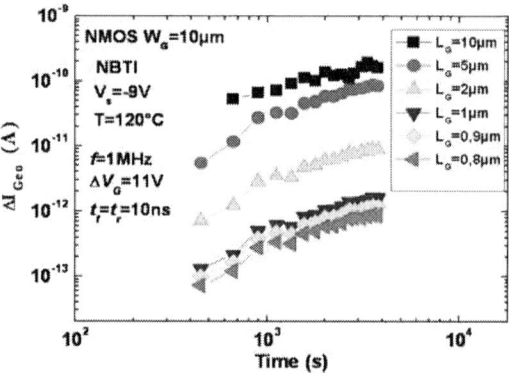

Fig.2 Geometric component evolution during NBTI stress, extracted using equation (10), (a) for PMOS transistors, (b) for NMOS transistors. The stress voltage is -9 V and the temperature is 120°C.

respectively. This implies that GC plays a key role in the assessment of the NBTI degradation as well as the determination of time exponent dependence of the degradation and consequently the life time of the devices.

In addition, After withdrawing GC effect, all ΔI_{CP} curves for both PMOS and NMOS transistors show a similar behaviour with stress time and have basically the same amount [Fig.3 (c, d)]. This means that the degradation sources could be the same for all devices. Indeed, according to (2) we can express the degradation as:

$$\Delta I_{CP}(t) = qf\Delta N_{it}^{C}(t)W_G L_{EFF} + 2qf\Delta N_{it}^{LDD}(t)W_G\Delta L \tag{11}$$
$$+ 2qf\Delta N_{it}^{LOCOS}(t)\Delta W L_{EFF}$$

Since the degradation of ΔI_{CP} is the same for all transistors, this degradation is only due to the second term of equation (11), which corresponds to LDD region, because this last is the same for all transistors.

(a)   (c)

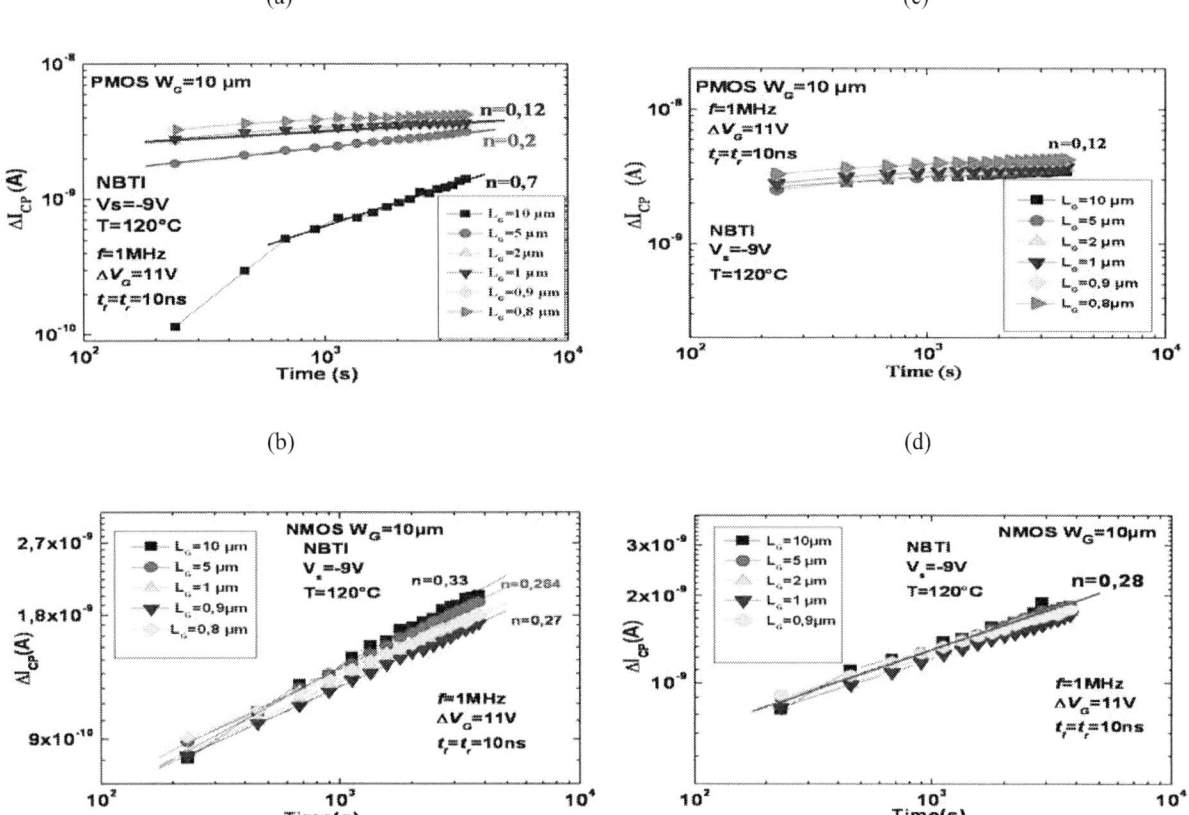

Fig .3 Degradation of CP current (ΔI_{CP}) before removing the geometrical component for (a) PMOS transistors, (b) NMOS transistors, (c) and (d) degradation of CP current (ΔI_{CP}) after removing the geometrical component for PMOS and NMOS transistor, respectively.

We believe that, up to one hour of stress at -9 V and 120°C, the degradation is mainly dominated by and located at the LDD regions, which are the same for all transistors, since these regions are more affect by mechanical constraints due to the process steps, such as etching, doping…, than the middle of the channel.

V. CONCLUSION

An experimental method for the determination and elimination of the GC in CP based methods is proposed. It uses an empirical model, obtained from the CP data using different gate length transistors. Furthermore, we have demonstrated that GC can affect NBTI degradation results in MOSFET, particularly the exponent n, which is a key parameter for any predictive time-like model. Also, we have shown how to avoid it and get the accurate data that can allow a good estimation of the degradation. On the other hand, by eliminating GC, we have demonstrated that the NBTI degradation may starts form laterals edges of the channel, which are more mechanically damaged during process fabrication, such as plasma etching, than propagates to the middle of the channel. This methodology is useful for CP-based method reliability evaluation of devices subjected to the NBTI stresses and for developing a predictive life time model.

REFERENCES

[1]   J. S. Brugler, P. G. A. Jespers, "Charge Pumping in MOS Devices," IEEE Trans. Elec. Dev. vol 16, no. 3, pp. 297-302, Mar 1969.

[2]   G. Groeseneken, H. E. Maes, N. Beltran, and R. F. De Keermaecker, "A Reliable Approach to Charge Pumping Measurement in MOS Transistors," IEEE Trans. Elec. Dev., vol 31, no.1, pp. 42-53, Jan.1984.

[3]   G. Van den Bosch, G. Groeseneken, "on Geometric component of charge-pumping current in MOSFET", IEEE Trans. Elec. Dev letter. vol 14, no.3, pp. 107-109, March.1993.

[4]   W. Liu, Z. Liu, D. Huang, C. Liao, L. Zhang, Z. Gan, W. Wong, C. Shen, and M.-F. Li, "On-The-Fly Interface Trap Measurement and its Impact on the Understanding of NBTI Mechanism for p-MOSFETs with SiON Gate Dielectric," in *Proceedings of IEEE International Electron Devices Meeting (IEDM)*, 2007.

[5]   Z. Liu, D. Huang, W. Liu, C. Liao, L. Zhang, , Z. Gan, W. Wong, and M.-F. Li, "Comprehensive Studies of BTI Degradation in SiON Gate Dielectric CMOS Transistors by New Measurement Techniques," in *Proceedings of IEEE International Reliability Physics Symposium (IRPS)*, 2008.

[6]   H. Tahi., B. Djezzar, and B. Nadji, "Radiation Effect Evaluation in Effective Short and Narrow Channel of LDD-Transistor with LOCOS-Isolation Using OTCP Method", IEEE Trans. Device Mater. Rel., Vol. 10 no 1, pp. 108-115, Mar. 2010.

[7]   B. Djezzar and H. Tahi, "Using Oxide-Trap Charge-Pumping Method in Radiation Reliability Analysis of Short Lightly Doped Drain Transistor," IEEE Tran. Dev. Mat. Rel. Vol.10, no1, pp. 18-25, Mar. 2010.

[8]   V. Huard, M. Denais, and C. Parthasarathy, "NBTI Degradation: From Physical Mechanisms to Modelling," Microelectron.Reliab., vol. 46, no. 1, pp. 1–23, 2006.

A study of volatile organic compounds diffusion in thin plasma polymerized TEOS thin film

A. Bougharouat and A. Bellel
Laboratoire des Etudes de Matériaux d'Electronique pour
Applications Médicales (LEMEAMED),
Faculté des Sciences de l'Ingénieur, Université Mentouri de
Constantine
Email: ali_boug@live.fr

S. Sahli
Laboratoire de Microsystèmes et Instrumentation (LMI),
Faculté des Science de l'Ingénieur,
Université Mentouri de Constantine

Abstract—A sensor based on quartz crystal microbalance (QCM) coated by TEOS film has been developed for the detection of volatile organic compounds (VOCs), such as methanol, ethanol, acetone and chloroform vapors. Plasma polymerized tetraethoxysilane (ppTEOS) thin films deposited on quartz crystal microbalance (QCM) electrode were used as sensitive layer. The measured isotherms on modified QCM electrode showed a good reproducibility and reversibility. The frequency shifts (Δf) of the QCM were found to be linearly correlated with the concentration of VOC vapor. The ppTEOS film is found to be highly sensitive and selective to methanol vapor than other vapors. The diffusion process of different alcohols vapor was studied and diffusion coefficients (D) were calculated. It is concluded that the diffusion of the vapors into the ppTEOS film follows Fickian kinetics. Fourier transform infrared spectroscopy (FTIR) analysis revealed the presence of CH_n and OH groups in the film structure, which suggests the formation of porous and low dense structure, resulting in the promotion of VOC vapor diffusion into the film.

Index Terms—quartz crystal microbalance, TEOS film, Volatile organic compounds, Sensor, Diffusions.

I. INTRODUCTION

Detection of toxic odorants and air-born volatile organic compounds (VOCs) becomes the world interest in the last few decades, owing to concerns about environmental protection, human health care, industrial processing, and quality control. The exposure to some VOCs for a long time may causes diseases such as eyesight disturbance, nasal mucous membrane, conjunctiva inflammation, nerve disease or serious irreversible effect [1, 2]. The development of chemical sensors with selectivity, rapidly, reliability and reproducibility becomes more interest. Among chemical sensor we find mass sensitive quartz crystal microbalance (QCM) [3], surface acoustic wave (SAW) [4] and ion sensitive field effect transistors (ISFET) [5]. There is a considerable interest in QCM for its low cost, simple, good enough for common use and can usually be operated at room temperature. The QCM has for long time been used in vacuum and gaseous environment as an ultra sensitive weighing device, especially for film thickness monitoring [6]. The QCM comprises a thin vibrating AT-cut quartz wafer sandwiched between two metal

excitation electrodes. When small amount of mass is absorbed at the surface of the quartz electrode, the resonance frequency of quartz is changed according to the well known Sauerbrey equation [7]. The QCM is an extremely sensitive sensor capable of measuring masse changes in the nanogram range. In the last few years, there has been an increasing amount of attention paid to the application of functionalized polymer-coated crystal microbalance (QCM) gas sensors. A number of coating materials have been successfully used in the detection with QCM of toxic gas [8], VOC [9] and DNA [10]. The selection of sensitive coating is a critical task in design and performance of chemical sensor. Polymers are widely used as chemically sensitive coating materials on QCM electrodes and are particularly suitable for detecting volatile organic compounds, because of the ability of the polymer to adsorb vapor reversibly [11]. There is an interest in using plasma polymerized films as sensitive layers in chemical sensors because they can be deposited on any substrate and feature excellent mechanical and thermal stability as well as insolubility in organic solvents, acids and alkalis [12]. Porous films can be very useful for QCM sensors development and has been a subject of research over the last decade. They can be obtained by spin coating, chemical or physical vapor deposition and Langmir-Blodgett film formation technique but only few of them have been obtained by plasma polymerization.

In this work, a simple coating method has been used for producing QCM chemical sensors. The method is based on plasma polymerization of TEOS (ppTEOS). The sensor was exposed towards a wide range of VOCs molecules (polar and non polar), such as ethanol, methanol, acetone and chloroform. The paper presents results on the effect of some discharge parameters on sorptive and diffusions properties of the coated QCM. We investigated the correlation between the interaction of the analyte with the sensitive layer and the chemical structures of the elaborated layers by means of Fourier transform infrared spectroscopy (FTIR) analysis.

978-1-4673-5289-5/12 $31.00 © 2012 IEEE

II. EXPERIMENTAL

A. Reagents and apparatus

Six megahertz, AT-cut QCM with polished gold electrodes (diameter 1.3cm) were purchased from International crystal Manufacturer (ICM). The frequency change of the sensor (adsorption measurements) was monitored by QCM measurement equipment. Tetraethylorthosilicate (TEOS) was purchased from Merck and used without further purification. The response of the sensor was tested using ethanol, methanol, acetone and chloroform as the analyte species.

B. Polymeric film production

Sensitive layers were deposited using plasma enhanced chemical vapor deposition (PECVD) at low frequency power from a pure vapor of TEOS. The system consists of a bell-jar chamber (310 mm diameter, 450 mm high), a pair of parallel symmetrical electrodes (120 mm diameter) separated by a distance of 2.5 cm, vacuum system (composed of Alcatel primary pump) and a monomer inlet system. The pressure in the reactor was monitored by a pressure measurement system (Pirani). Substrates were placed in the grounded lower electrode and the reactor chamber was pumped down to 1 Pa. A pressure of TEOS vapor (40 Pa) was injected in the reaction chamber without any carrier gas. The power during the polymerization was controlled by a 19 kHz generator. Chemical structure and composition of ppTEOS film were characterized by FTIR spectroscopy. All spectra were acquired in absorbance mode in 400-4000 cm^{-1} range using a Nicolet Avatar 360 FTIR spectrometer.

C. Gas sensing experiments

The liquid analyte of known volume and density was introduced in glass flask and heated to evaporate freely at about 50°C. The injected vapor was subsequently adsorbed onto the surface of the functionalized electrode of the quartz crystal which induced a frequency decrease. The recorded shift frequency (Δf) data were transferred to a computer via RS232 interface. The concentration of injected analyte in the cell was calculated in part per million (ppm) according to the following equation [13]:

$$c = \frac{22.4 \rho T V_s}{273 \, MV} x 10^3 \qquad (1)$$

where C is the concentration in ppm, ρ is the density of liquid sample in g/mL, T is the temperature of detection chamber in Kelvin, V_s is the volume of liquid sample in μL, M is the molecular weight of sample in g, and V is the chamber volume in liter.

III. RESULTS AND DISCUSSION

A. Isotherms properties

To check the reproducibility and the reversibility of ppTEOS-based sensor, the responses signal of QCM sensor coated with 690 nm thick film exposed to methanol, ethanol,

acetone and chloroform vapors at concentration of about 300 ppm have been recorded. The experiment was repeated for three times to insure a good reversibility of the sensor. After reaching steady state, the film was exposed to dry air until full desorption of the crystal was obtained and it was then exposed to previously used analyte concentration. From Fig. 1, it can be noted that the sensor attains the same Δf upon exposure to VOC vapor concentration at different times. In addition, the frequency of the crystal back shifted to its initial values indicates full desorption of analytes from the coated QCM electrode, which indicated that no chemical bond formed during the adsorption process. This behavior confirms that the sensing interaction between polymer coating and VOCs molecules is a physical absorption [14]. Furthermore, ppTEOS-based sensor is found to be reasonably selective and significantly sensitive to methanol vapor than other organic vapors. The response to methanol vapor is fast, large, reproducible and fully reversible. The reason for longer response time for higher molecular weight is because big molecule takes more time to diffuse within the film to reach equilibrium.

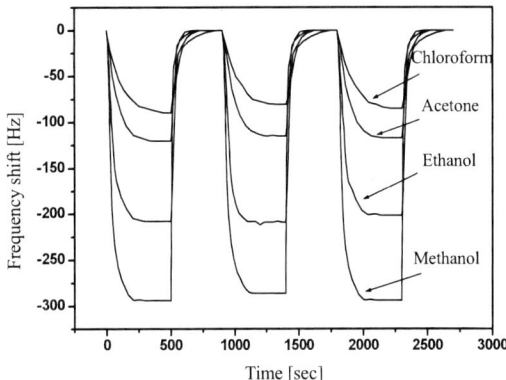

Fig. 1. Reproducibility and reversibility of coated QCM electrode when exposed to organic vapors

B. Organic vapor sensing properties of plasma-polymerization of pure TEOS films

The QCMs coated with 690 nm layers thickness, deposited from pure TEOS were exposed to four series of chemical gases: methanol, ethanol, acetone and chloroform. The maximum frequency changes (recorded after reaching the steady state) obtained for this series of vapors are plotted versus the corresponding analyte concentrations in Fig. 2. Generally, the frequency shift (Δf) of QCM gradually increases as the vapor concentration increases. It is clearly seen that the responses on QCM sensors present a linear relationship with varying analyte concentration. This is expected since the more analyte molecules provided in the test cell, more analyte would be adsorbed onto the sensitive layer. The linear correlation obtained with the slope equal to about 1.01, 0.46, 0.21 and 0.07 Hz per ppm for methanol, ethanol, acetone and chloroform, respectively. TEOS film is found to be reasonably selective and more sensitive to methanol vapor than other organic vapors. The responses are in following ascending order: chloroform<acetone<ethanol<methanol. The Δf of the QCM decreases with the increasing molecular weight of the analyte. For example at the concentration of about 300 mg/l, methanol induces a higher $\Delta f = 300$ Hz as compared to chloroform, for which $\Delta f = 23$ Hz, which is expected to induce a higher Δf due to its higher molecular weight. This result seems to indicate that a smaller molar mass alcohol has greater frequency response, which is not consistent with the findings in general reports for piezoelectric detectors [15,16]. However, these results are in good agreement with those reported by R. A. M. Cavalho [17] for plasma polymerized TEOS films. They have found that the adsorption capacity of the elaborated films was greater for more polar and/or smaller molecules due to the presence of microchannels with up to 200 Å of diameter as identified by atomic force microscopy. This behavior might be explained by the fact that smaller molecules easily diffuse into the polymeric film; consequently more molecules were adsorbed. In addition, the stronger polarity of the analyte molecules allows them to form stronger interaction with the sensing material. The results show that smaller molar mass molecules are more adsorbed than greater molecules indicating a strong affinity of ppTEOS film for methanol gas, but weak affinity for chloroform. The important observation is that ppTEOS-based sensors show different selectivity properties for each VOC analyte.

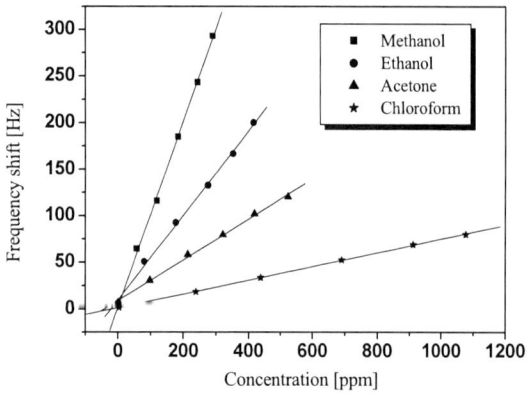

Fig. 2. Variation of shift frequency with analyte concentration

In order to confirm this assumption, FTIR analysis was carried out to investigate the chemical structure of the deposited layer. The resulting spectra recorded on ppTEOS film elaborated from pure TEOS at 8 W are displayed in Fig. 3. The spectra exhibit absorption peaks corresponding to Si-O-Si asymmetric stretching around 1049 cm^{-1} and Si-O-Si bending vibration around 795 cm^{-1}. The spectra show strong peaks of the methyl groups at 838 cm^{-1} due to Si-(CH$_3$)$_3$ vibration, which make the film more hydrophobic and by consequence the reversible adsorption/desorption will be easily achieved. The CH$_x$ peaks clearly observed at around 2962 cm^{-1} may induce more porous and low dense film structure as it has been reported in several works [20,21]. Increases in the microvoides volume usually promote gas sorption capacity. In addition, the deposited films contain the

surface hydroxyl (Si-OH) groups situated around 3660 cm^{-1}. Hydroxyl groups on ppTEOS film surface provide a main source for hydrophilic capability to interact with foreign molecules. It is believed that organic molecules interact via hydrogen bonding due to the presence of the –OH group.

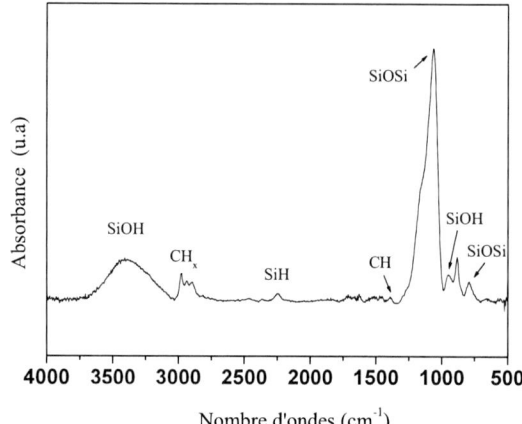

Fig. 3. FTIR spectra of the deposited layer

C. Vapours diffusion study in the ppTEOS film

The kinetics of VOC sorption is interpreted according to Fickian diffusion and adsorption processes. But if we consider that the two mode are completely independent phenomena (the part of VOC molecules which take part in surface adsorption equilibrium only take part in this equilibrium and will never diffuse), Fickian diffusion may be considered in this case significant than adsorption since in term of shift frequency the results indicated that larger molar masse VOC molecules induced low frequency response (greater molar mass of adsorbed molecule would produce a larger frequency shift than in case of smaller molecule). The experimental data were analysed using the following Fick's second equation which has been reviewed by Crank [18].

$$\frac{\Delta f_t}{\Delta f_s} = 4\sqrt{\frac{D}{\pi}}\frac{t^{1/2}}{L} \qquad (2)$$

Where Δf_t is the shift frequency due to vapour absorption into ppTEOS film at any time t, Δf_s is the frequency shift in the steady state (end of the absorption process), D is the diffusion coefficient and L is the film thickness. With the plot of $\Delta f_t/\Delta f_s$ as function of $t^{1/2}/L$, D can be calculated from the initial linear portion of the diffusion process (Fig. 4). The calculated values of D for the tested VOC of a concentration of about 300 ppm at the same film thickness of about 690 nm are given in Table 1. The calculated D values are within a reasonable order of magnitude similar to that reported for the diffusion processes of alcohol vapours in thin polyaniline salt film [19]. The D values are in the order: methanol> ethanol> acetone > chloroform. It seems that the physico-chemical properties of gas molecules such as polarity and chemical structure have an influence on the absorption behaviour into the sensing film. This may be explained on the basis of the differences in polar nature and molecular structure of different VOC. The lower D value of chloroform may be related to larger chemical structure. The differences in the calculated D values can be used as a basis for the selective determination of these VOC pollutants vapours.

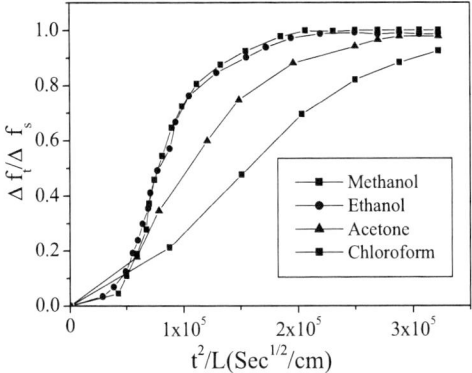

Fig. 4. Plot of $\Delta f_t/\Delta f_s$ as function of $t^{1/2}/L$ when ppTEOS coated electrode was exposed to 300 ppm of VOC.

TABLE I. *D* OF THE TESTED VOLATILE ORGANIC COMPOUNDS

volatile organic compounds	Diffusion coefficient (cm²/s)
Methanol	3.02×10^{-11}
Ethanol	2.15×10^{-11}
Acetone	7.91×10^{-12}
chloroform	2.81×10^{-12}

IV. CONCLUSION

A sensor based on quartz crystal microbalance coated with plasma polymerized TEOS thin film was elaborated for the detection of VOC molecules. TEOS films present interesting characteristics for the absorption of organic compounds, either polar or non polar. The responses in terms of frequency change are fast, reproducible and reversible. The frequency shifts versus analytes concentrations exhibited satisfactory linear relationship. The results showed that smaller molar weight molecules are more absorbed than greater molecules indicating a strong affinity of ppTEOS film for methanol gas. The diffusion coefficients (D) of different alcohols vapor were determined. It was found that the sensor follows Fickian kinetics.

ACKNOWLEDGMENT

This work was supported by the Algerian research agency ANDRU

REFERENCES

[1] G. Ziem and J. McTamney, "Profile of patients with chemical injury and sensitivity," *Environ Health Perspect*, Vol. 105, pp.417-436, 1997.

[2] R. Chem, S. Semple, F. Dick and A. Seaton, "Nasal, eye, and skin irritation in dockyard painters," *Occup. Environ. Med.*, Vol. 58, pp. 542-543, 2001.

[3] N. V. Quy, V. A. Minh, N. V. Luan, V. N. Hung and N. V. Hieu, "Gas sensing properties at room temperature of a quartz crystal microbalance

coated with ZnO nanorods," *Sens. Actuators, B*, Vol. 153, No. 1, pp. 188-193, 2011.

[4] C. Wen, C. Zhu, Y. Ju, H. Xu and Y. Qiu, () "A novel NO_2 gas sensor using dual track SAW device," *Sens. Actuators, A*, Vol. 159, No. 2, pp. 168-173, 2010.

[5] H. Tsai, C. F. Lin, Y. Z. Juang, I. L. Wang, Y. C. Lin, R. L. Wang and H. Y. Lin, "Multiple type biosensors fabricated using the CMOS BioMEMS platform," *Sens. Actuators, B*, Vol. 144, No. 2, pp.407-412, 2010.

[6] D. H. Yang, C. S. Park, J. H. Min, M. H. Oh and Y. S. Yoon, "Fullerene nanohybrid metal oxide ultrathin films," *Curr. Appl. Phys.*, Vol. 9, No. 2, pp.132–135, 2009.

[7] G. Sarerbrey, "Verwendung von Schwingquarzen zur Wägung dünner Schichten und zur Mikrowagüng,"*Z. Phys.*, Vol. 155, No. 2, pp.206–222, 1959.

[8] P. Sun, Y. Jiang, G. Xie, X. Du and J. Hu, "A room temperature supramolecular-based quartz crystal microbalance (QCM) methane gas sensor," *Sens. Actuators, B*, Vol. 141, pp. 104-108, 2009.

[9] J. Huang, Y.D. Jiang, X.S. Du and J. Bi, "A new siloxane polymer for chemical vapor sensor," *Sens. Actuators, B*, Vol. 146, pp. 388-394, 2010.

[10] S.R. Hong, H.D. Jeong and S. Hong, "QCM DNA biosensor for the diagnosis of a fish pathogenic virus VHSV," Talanta, Vol. 82, pp. 899-903, 2010.

[11] P. Si, J. Mortensen, A. Komolov, J. Denborg and P. J. Møller, "Polymer coated quartz crystal microbalance sensors for detection of volatile organic compounds in gas mixtures," *Anal. Chim. Acta.*, Vol. 597, No. 2, pp.223–230, 2007.

[12] K. Yamaoka, Y. Yohizako, H.Kato, D. Tsukiyama, Y. Terai and Y. Fujiwara, "Room-temperature plasma-enhanced chemical vapor deposition of SiOCH films using tetraethoxysilane," *Physica B*, Vol. 376-377, pp.399–402, 2006.

[13] Mohamad M. Ayad*, Gad El-Hefnawey, Nagy L. Torad , "A sensor of alcohol vapours based on thin polyaniline base film and quartz crystal microbalance," Journal of Hazardous Materials., Vol. 168, No.1, pp.85–88, 2009.

[14] Mohamad M. Ayad, Gad El-Hefnawey and Nagy L. Torad, "Quartz crystal microbalance sensor coated with polyaniline emeraldine base for determination of chlorinated aliphatic hydrocarbons," *Sens. Actuators, B*, Vol. 134, pp. 887-894, 2008.

[15] H. Lin and J. Shin, "Fullerene C60-cryptand coated surface acoustic wave quartz crystal sensor for organic vapors," *Sens. Actuators, B*, Vol. 92, No. 3, pp.243–254, 2003.

[16] L. Xu, X. Hu, Y. T. Lim, V. S. Subramanian, "Organic vapor adsorption behavior of poly (3-butoxythiophene) LB films on quartz crystal microbalance," *Thin solid films*, Vol. 417, No. 1-2, pp.90–94, 2002.

[17] R. A. M. Cavalho, R. R. Lima, A. P. Nascimento Filho, M.L.P.da Silva and N. R. Demarquette, "Plasma polymerized TEOS films for nanochannels formation and sensor development," *Sens. Actuators*, B, Vol. 108, No. 1-2, pp.955–963, 2005.

[18] Mohamad M. Ayad, Nagy L. Torad "Alcohol vapours sensor based on thin polyaniline salt film and quartz crystal microbalance," Talanta, Vol. 78, No. 4-5, pp.1280–1285, June 2009.

[19] J. Crank, The Mathematics of Diffusion, 2nd edn., Clarendon Press, Oxford, 1975, pp.414.

[20] I. S. Bae, S. J. Cho, W. S. Choi, B. Hong, H. D. Jeong, J. H. Boo, "Characterization on polymerized thin films for low-*k* insulator using PECVD," Prog. Org. Coating, Vol. 61, No. 2-4, pp. 245–248, 2008.

[21] Matsugushi, M., Uno, T., Aoki, T., Yoshida, M. "Chemically modified copolymer coatings for mass-sensitive toluene vapor sensors,"*Sens. Actuators, B*, Vol. 131, No. 2, pp.652–659, 2008.

Oxide Trap Annealing by H_2 cracking at E' Center under NBTI Stress

Cherifa TAHANOUT and Becharia NADJI
Department of Automation and Electrification of the industrial processes
Laboratory Microelectronics and Microsystems
University M'hamed Bougara of Boumerdes, Algeria
Email : ctahanout@yahoo.fr

Hakim TAHI, Boualem DJEZZAR, Abdelmadjid.
BENABDELMOUMENE and Amel chenouf
Microelectronics and Nanotechnology Division, Centre de
Développement des Technologies Avancées (CDTA),
Algeria
Email :htahi@cdta.dz

Abstract—**In this paper, we have modeled the results of negative bias temperature instability (NBTI) degradation obtained by using charge pumping and linear drain current techniques in the same time measurement setup. The proposed model is based on Hydrogen molecule (H_2) cracking at trapped hole center (E' center) to explain the decrease of oxide trapped charge at long NBTI stress time (t > 1000s). In fact, according to experimental data, which are obtained in the framework of the above cited setup, NBTI induces oxide trap generation at earlier stress time followed later by a decrease (annealing) at long stress time. This behavior looks like annealing in radiation effect on gate oxide in MOSFET, which is explained by H_2 cracking model. Based on this similarity, we extended this model to NBTI effect. The extended model presents a perfect correlation with the experimental data.**

Index Terms— NBTI, Positive oxide-trap, Hole trapping, H_2 cracking

I. INTRODUCTION

It is well known that the negative bias temperature instability (NBTI) arises from generation of interface trap (ΔN_{it}) and positive oxide trap charge (ΔN_{ot}). ΔN_{it} is resulting from dissociation of Si-H bonds at Si/SiO$_2$, this has been confirmed by electron-spin resonance (ESR) [1]-[3] showing the P_b center after NBTI stress. The formation of ΔN_{ot} is not yet understood. Early works have been reported that ΔN_{ot} is formed from interaction of released hydrogen species with the Si-O-Si bond resulting in the formation of Si$^+$ (positive charge) and Si-OH [4]. Tsujikaw and Yagami, have attributed ΔN_{ot} to the trap of proton (*H*+), in the gate oxide, resulting from the Si-H bond broking [5]. In the others studies, ΔN_{ot} has been attributed to the hole trapping at pre-existing gate oxide defect [6], this has been confirmed by the observation of *E'* center from ESR [7] indicates that the hole trapping at oxygen vacancies is mechanism for the formation of positive charge oxide trap.

Several authors have shown that hole trapping (ΔN_{ot}) at pre-existing E' center quickly reaches the saturation (1s) after stress is applied and remains constant with time.

In this work, we demonstrate by modeling the hole trapping (positive oxide trap) that this latter can decreases (annealing) using hydrogen molecule (H_2) cracking at E' center model, where H_2 splits into two H atoms. One H atom bonds with E' forming *E'-H* which is fixed in space and the second H atom diffuses into the oxide. Therefore, contrary to the recent work [8] which claims that the positive oxide charge trapping under NBTI is not correlated and not coupled to the Si–H bond dissociation that occurs at the interface, in our work we shown that hole trapping at pre-existing defect is coupled to the generation of interface trap by cracking of H_2 at E' center at long stress time.

II. EXPERIMENTS AND SAMPLES

We have characterized the NBTI effect on PMOS transistors with W_G/L_G =10/0.65 fabricated at ISiT (Institute for Silicon Technology) of Fraunhofer, Germany. The process is a conventional dual layer metal CMOS technology on P-type silicon <100> substrate with 16 nm thick gate oxide layer grown in dry O$_2$. The gate capacitance per unit area, C_{ox} is about 2.12×10^{-7} F.cm^{-2}.

The measurement/stress/ measurement (MSM) sequences have been performed using fully automated bench at Microelectronics and Nanotechnology Division of Centre de Développement des Technologies Avancées (CDTA). The bench includes a sensitive Agilent HP 4156C for current measurement and trigging Keithley 3940 to generate Pulse trains during measurement by Charge Pumping (CP) and linear drain current (I-V) techniques, see fig.1. A hotplate is used, inside Karl Suss PA300 micro-manipulator probe station, to vary the temperature of the chip during the experiments. The test circuit chip (non-packaged) within the probe station were isolated from vibration and enclosed in a grounded faraday cage to avoid both RF and light effects.

The stress interval is kept the same during all MSM cycle. The whole stress time cycle is fixed at 3600s (1 hour). The stress is periodically interrupted after 100s and pulse trains have been alternatively applied to measure CP current and linear drain current (I_{DS}). The measurement duration ($t_{CP}= t_{I-V}$) is about 6s.

This work was supported by the High School Educational and Scientific Research Ministry of Algeria under the National Funding of Research (FNR) contract number 22/CDS/DMN/ CDTA/2011.

978-1-4673-5289-5/12 $31.00 © 2012 IEEE

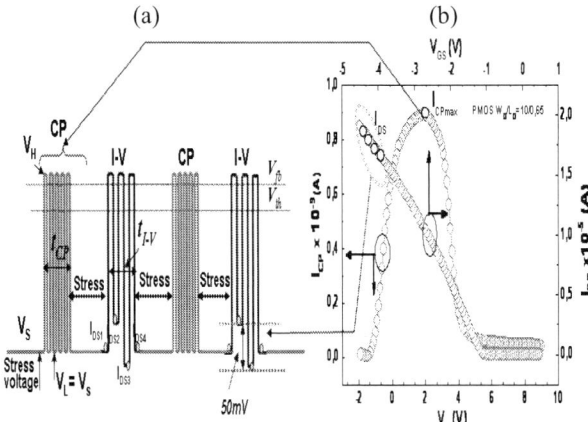

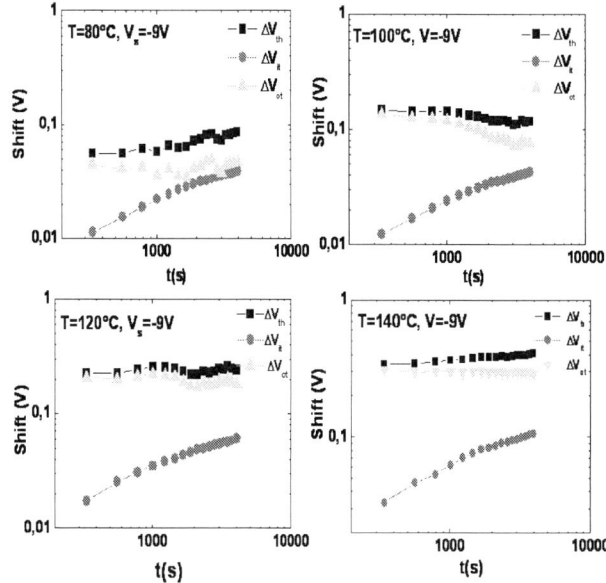

Fig.1 Measurement protocol of combined CP and linear drain current. a) Measurement /stress / Measurement waveform applied to the gate oxide, b) Illustration of CP (I_{CP}) and linear drain currents measurements (I_{DS1} to I_{DS4}) in PMOS transistor. During the stress interval, the voltage stress Vs=V$_L$ is applied onto the gate of the device, between two stress times (in this work 100 s) CP and linear drain current (I-V) techniques are alternatively applied in the same experimental run. During the CP measurement one point is measured, using high frequency signal, which correspond to CP current at (V_L=V$_S$, V_H). Through I-V measurement, four currents are measured I_{DS1}, I_{DS2}, I_{DS3} and I_{DS4} at V_L=V$_S$, V_L=V$_S$-25 mV, V_L=V$_S$+25 mV and V$_L$= V$_S$, respectively.

Fig.2. ΔV_{th}, and ΔV_{it} and ΔV_{ot} extracted by combining the CP and linear I_{DS} techniques at different temperatures at stress voltage of - 9V.

We varied the temperature, T° from room temperature (~25°C) to 140°C and stress voltage, $V_s = V_L$ was -9V, while keeping V_H at 2 V .

The contribution of ΔN_{it} to the threshold voltage shift (ΔV_{it}) is extracted using the CP current as follows:

$$\Delta V_{it} = \frac{\Delta I_{CP}}{f A_G C_{ox}} \qquad (1)$$

where A_G is the gate area (cm^2), C_{ox} (F/cm^2) is the capacitance per area. ΔI_{CP} is the difference between degraded (I_{CP}) and initial (I_{CP0}) CP current.

The contribution of the oxide trap to the threshold voltage shift (ΔV_{ot}) is extracted using:

$$\Delta V_{ot} = \Delta V_{th} - \gamma \Delta V_{it} \qquad (2)$$

where The threshold voltage shift (ΔV_{th}) is measured as like in on-the-fly two point method [9]. γ is the energy calibration factor to take account for differences between energies scanned by CP and I-V techniques in silicon band gap. It is given by:

$$\gamma = \frac{\Delta E_{CP}}{\Delta E_{I-V}} \qquad (3)$$

$$\gamma = - \frac{\ln \left(v_{th}^2 n_i^2 \sigma_e \sigma_p t_{em,e} t_{em,h} \right)}{\ln \frac{N_{A,D}}{n_i}}$$

where ΔE_{CP}(eV) and ΔE_{I-V}(eV) are the band energies scanned by CP and I-V techniques, respectively. v_{th} is the thermal velocity. n_i (cm^{-3}) is the intrinsic concentration. σ_e (cm^2) and σ_p

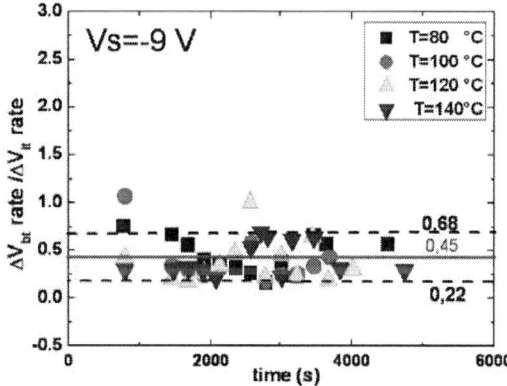

Fig.3. Estimation of the rater ratio of ΔV_{ot} and ΔV_{it}.

(cm^2) are the cross capture sections of electron and hole, respectively. They are determined with Groeseneken's method [10]. $N_{A,D}$ (cm^{-3}) is the acceptor (donor) concentration. $t_{em,e}$ (s) and $t_{em,h}$ (s) are the end of non-steady-state emission times for electron and hole, respectively.

III. RESULTS AND DISCUSSION

Fig.2 gives the extracted ΔV_{th}, ΔV_{it} and ΔV_{ot}. For stress time below 1000s, ΔV_{th} is quasi constant with time and approximately equal to ΔV_{ot}. For stress time greater than 1000s, ΔV_{ot} starts decreasing and diverging from ΔV_{th}. To investigate the origin of ΔV_{ot} decrease, we have calculated the ratio rate of ΔV_{ot} and ΔV_{it} ($d\Delta V_{ot}/dt$ devised to $d\Delta V_{it}/dt$). The results are given in fig.3. All calculated ratio rate are around

0.5 for different temperature at V_s=-9V. According to the reaction-diffusion (RD) model, this means that for two released hydrogen atoms from interface Si-H bond, one trapped hole is neutralized. We think that is due to the cracking of hydrogen molecule (H_2) at hole trapped site (E' center). Indeed, it has been reported that E' center could crack H_2, even at room temperature, turning them into hydrogen complex E' center (E'-H), through a reaction such as [11]:

$$E' + H_2 \rightarrow E' - H + H \qquad (4)$$

where E'-H is fixed in space, but H is mobile and can diffuse. We note here that H_2 cracking at E' center process is used to explain the annealing phenomenon in the post-irradiated devices [12]. In the post-irradiated devices the cracking step occurs before interface trap formation. However, in our NBTI stress device, the process is reversed i.e the interface trap generation takes place before cracking step. The released H from Si-H bond can react with another released H to form H_2 molecule. This latter diffuses and cracks at E' center.

IV. HOLE TRAPPING ANNEALING MODEL BY CRACKING OF H_2 AT E' CENTER

By addition the reaction (4) to reaction-diffusion (RD) NBTI framework developed by Islam *et al* [6], we have been able to couple between E' center and RD. The coupled equations are as follows:

$$\frac{d\Delta N_{it}}{dt} = k_f \left(N_0 - \Delta N_{it} \right) - k_r \Delta N_{it} \left(N_H^0 + N_{H1}^0 \right) \qquad (5)$$

$$\frac{dN_H}{dt} = D_H \frac{d^2 N_H}{dx^2} - k_H \left(N_H + N_{H1} \right)^2 + k_{H_2} N_{H_2} + k_E N_{H_2} N_{E'} \qquad (6)$$

$$\frac{dN_{H_2}}{dt} = D_{H_2} \frac{d^2 N_{H_2}}{dx^2} + \frac{1}{2} k_H \left(N_H + N_{H1} \right)^2 - \frac{1}{2} k_{H_2} N_{H_2} - k_E N_{H_2} N_{E'} \qquad (7)$$

$$\frac{dN_{H_1}}{dt} = k_E N_{H_2} N_{E'} \qquad (8)$$

$$\frac{d\Delta V_{ot}}{dt} = \frac{C_{ox}}{q} \left(\sigma_{HT} v_{th} p_h T_n \left(N_{E'} - \Delta N_{ot} \right) - \frac{dN_{H_1}}{dt} \right) \qquad (9)$$

Eq.(5) represents passivation/de-passivation effects of Si-H bond, where k_f, k_r, N_0, ΔN_{it}, N_H^0 and N_{H1}^0 are defined as Si-H bond-breaking rate, Si-H bond-annealing rate, initial bond density available before stress, interface defect density, hydrogen density at the Si/SiO$_2$ interface and hydrogen density generation by cracking of H_2 at E' center according to reaction eq.(4) at the Si/SiO$_2$ interface , respectively. Whereas eq.(6) and (7) describe diffusion (along *x* axis) of H and H_2 were k_H and k_{H2} present the generation and dissociation rates of H_2. k_E is the rate of H_2 cracking into atomic hydrogen. D_H and D_{H2} present, respectively, the diffusion coefficients of H and H_2. N_H and N_{H2} present the concentration of atomic and molecular hydrogen, respectively. N_{H1} is the concentration of

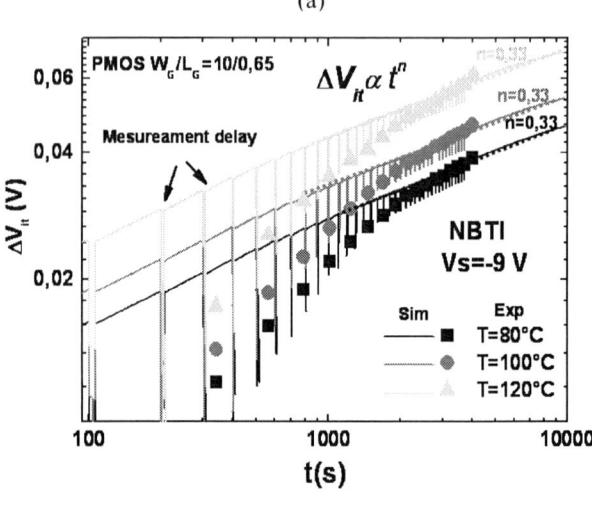

(a)

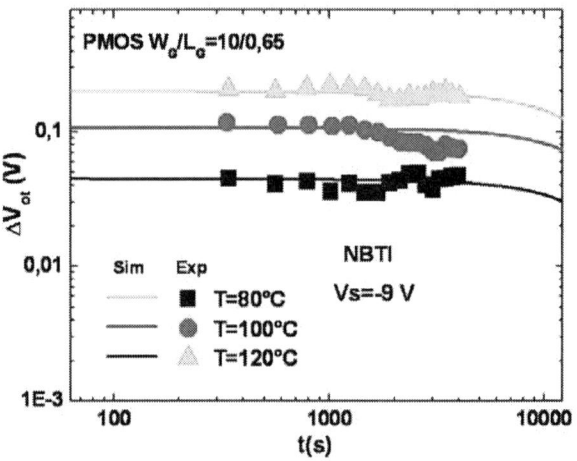

(b)

Fig.4. Comparison of experimental (Exp) extraction and numerical solution (Sim) of equations system (5)-(9) for (a) ΔV_{it} , (b) ΔV_{ot}. Note that ΔV_{it} is calculated using $\Delta V_{it} = C_{OX} \Delta N_{it}/q$ and the parameters: $N_0 = 3 \times 10^{12}$ (cm^{-2}) $k_r = 5 \times 10^{-9}$ (cm^3/s), $k_f = 10^{-2} * exp$ (-E_A/kT) (1/s) were E_A was extracted from experimental data ($E_A = 0.2eV$), $D_H = 1 \times 10^{-17}$ (cm^2/s), $D_{H2} = 1 \times 10^{-18}$ (cm^2/s), $k_H = 5e$-3 (cm^3/s), $k_{H2} = 100$ (1/s), $k_E = 10^{-20}$ (cm^3/s), $N_{E'} = 10^{13}$ (cm^{-2}), $\sigma_{HT} = 5.10^{-16}$ (cm^2), $v_{th} = 10^7$ (cm/s) and $p_h = 10^{17}$ (cm^{-3}).

hydrogen atom generated by H_2 cracking at E' center. The equation (8) corresponds to the generation of hydrogen atom by cracking H_2, were $N_{E'}$ is E' center concentration in the oxide. Whereas Eq.(9) describes the dynamic of trapping hole contribution to the threshold voltage shift (ΔV_{ot}). p_h is the inversion layer hole density, T_n is the tunnelling probability (obtained using the Wentzel-Kramers-Brillouin approximation) of hole from Si/SiO$_2$ interface to oxide trap, σ_{HT} is the capture cross section of oxide trap, ΔN_{ot} is the concentration of trapped hole. We should note that the number of notarized positive oxide charge (hole trapping) is equal to N_{H1} concentration.

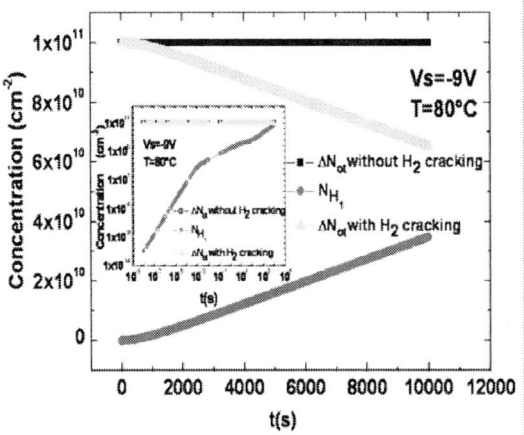

Fig.5 Simulation example of the hole trapping concentration (ΔN_{ot}) in the oxide without and with H_2 cracking for Vs=-9 V and stress temperature of 80°C. In side of this figure we presented the same figure in log-log scale

V. MODEL VALIDATION

The numerical solution of equation system (5) to (9) for stress field of 5.6 MV/cm and increasing temperature (T=80, 100, 140°C) are plotted using solid lines, in fig.4, all calculated ΔV_{it} [fig.4 (a)] and ΔV_{ot} [fig.4 (b)] are in good agreement with the experimental data. Note that the measurement delay is included in the calculation of ΔV_{it} (see fig.4 a) and ΔV_{ot}, witch is a bout 6s. Fig.4 (a) shows that the time power exponent (n) for long stress time (after 1000s) is around 0.33 (the highness of n for stress time below 1000s is due to the experimental measurement delay), this values is higher than that predicting by the R-D model developed by Islam *et a l* [6] at long stress time, with is about 0.16 due to the diffusion of H_2 in the oxide, In this model the time power exponent of 0.33 is shown in earlier times and explained by conversion of H its into H_2 However in our work, the value of 0.33 is observed at long stress time t >1000 (s). Fig .5 gives a calculated of ΔV_{ot} with/without cracking of H_2 and the hydrogen atom generated (N_{H1}) resulting by H_2 cracking at E' center which is equal to the number of neutralized positive oxide charge (hole trapping ΔN_{ot}). N_{H1} increasing with time, consequently, ΔN_{ot} decreases.

VI. CONCLUSION

We have combined the cracking of hydrogen molecule H_2 at E' center model with the reaction–diffusion (RD) framework to model the annealing of positive oxide trap (hole trapping) at long stress time under NBTI. This model can explain the reduction of NBTI-induced oxide trap at long stress time. The simulation results based on this model are in perfect agreement with the experimental data.

REFERENCES

[1] S. Fujieda, Y. Miura, M. Saitoh, E. Hasegawa, S. Koyama, and K. Ando,"Interface defects responsible for negative-bias temperature instability in plasma nitrided SiON/Si(100) systems," *Appl. Phys. Lett.*, vol. 82, no. 21, pp. 3677–3679, May 2003.

[2] S. Fujieda, Y. Miura, M. Saitoh, Y. Teraoka, and A. Yoshigoe, "Characterization of interface defects related to negative-bias temperature instability in ultrathin plasma-nitrided SiON/Si systems," *Microelectron. Reliab.*, vol. 45, no. 1, pp. 57–64, Jan. 2005.

[3] J. P. Campbell, P. M. Lenahan, A. T. Krishnan, and S. Krishnan, "Direct observation of the structure of defect centers involved in the negative bias temperature instability," *Appl. Phys. Lett.*, vol. 87, no. 20, art. no. 204106,Nov. 2005.

[4] K. O. Jeppson and C. M. Svensson, "Negative bias stress of MOS devices at high electric fields and degradation of MNOS devices," *J. Appl. Phys.*,vol. 48, no. 5, pp. 2004–2014, May 1977.

[5] S. Tsujikawa and J. Yugami, "Evidence for bulk trap generation during NBTI phenomenon in pMOSFETs with ultrathin SiON gate dielectrics," *IEEE Trans. Electron Devices*, vol. 53, no. 1, pp. 51–55, Jan. 2006.

[6] A. E. Islam, H. Kufluoglu, D. Varghese, S. Mahapatra, and M. A. Alam, "Recent issues in negative-bias temperature instability: Initial degradation, field dependence of interface trap generation, hole trapping effects,and relaxation," *IEEE Trans. Electron Devices*, vol. 54, no. 9, pp. 2143–2154, Sep. 2007.

[7] J. P. Campbell, P. M. Lenahan, A. T. Krishnan, and S. Krishnan, "Identification of atomic-scale defect structure involved in the negative bias temperature instability in plasma-nitrided devices," *Appl. Phys. Lett.*, vol. 91,no. 13, art. no. 133507, Sep. 2007.

[8] T. J. J. Ho, D. S. Ang, A. A. Boo, Z. Q. Teo, and K. C. Leong , "Are Interface State Generation and Positive Oxide Charge Trapping Under Negative-Bias Temperature Stressing Correlated or Coupled?," *IEEE Trans. Electron Devices*, vol. 59, no. 4, pp. 1013–1022, April. 2012

[9] V. Huard, M. Denais, C. Parthasarathy "NBTI degradation: From physical mechanisms to modelling," *Microelectron. Reliab.*, vol. 46, no. 1, pp. 71–81, Jan.2005

[10] G. Groeseneken, H. E. Maes, N. Bertran, and R. F. Keersmaecker, "A reliable approach to charge-pumping measurements in MOS transistors," *IEEE Trans. Electron Devices*, vol. ED-31, no. 1, pp. 42–53, Jan. 1984

[11] K.L. Brower "passivation of paramagnetic Si/SiO2 interface states with molecular hydrogen", *Appl.Phys* vol. 87, no.7, pp.4487-4489, 1992

[12] R.E.Stahlbush, A.H.Edwards, D.L.Griscom, and B.J.Mrstik, "Postirradiation cracking of H2 and formation of interface state in irradiated metaloxideseliconductor field effect transistors," *Appl.Phys* vol. 73, no.2, pp.658-667, Jan.1993.

Deep Experimental Investigation of NBTI Impact on CMOS Inverter Reliability

Amel CHENOUF*, Boualem DJEZZAR, Abdelmadjid BENADELMOUMENE & Hakim TAHI
Microelectronics & Nanotechnology Division
Centre de Développement des Technologies Avancées, CDTA
Email achenouf@cdta.dz

Abstract—Negative Bias Temperature Instability has become a major reliability concern in CMOS circuits. This is due to the fact that PMOS transistors of these circuits are submitted to many parameters shifts resulting in circuit performance degradations. In order to mitigate NBTI, IC designers need to accurately evaluate the NBTI induced performance degradations of their circuits to conduct aging simulation of their circuit lifetime. For this purpose, we present in this work a deep experimental investigation of NBTI impact on the CMOS inverter. The experimental setup of the DC NBTI has been conducted by Measure Stress Measure Method where a set of Negative stress Voltages at different temperatures were continuously applied for duration of 3600s. The results show that NBTI does shift the inverter's Voltage Transfer Curve, decreases its logic threshold, unbalance its logic states, and degrades its noise immunity. The obtained results demonstrate that these degradations worsen with both high negative stress voltage and elevated temperature and consequently impact the CMOS inverter robustness and thus its reliability.

Index Terms— NBTI stress, CMOS inverter, Noise Margins, Robustness. IC reliability.

I. INTRODUCTION

The Negative Bias Temperature Instability, as a degradation phenomenon occurring mainly in MOS Field Effect Transistors MOSFETs [1] [2], is one of the main reliability problems of CMOS Integrated Circuits that limit considerably their life time [3] [4]. It is well known that this phenomenon is ascribed to Si/SiO_2 interface states and positive charges in the oxide resulting from the breaking down of Si-H bonds at the SiO_2/Si interface of PMOS transistors under negative Bias at elevated temperatures [2]. NBTI is characterized by a positive shift in the absolute value of the PMOS threshold voltage |Vthp| leading to the decrease of drain current I_{DS} and consequently to performances degradation of CMOS Integrated Circuits [5].

At device level, many research works have been performed to study the fundamentals behind NBTI effect on PMOS transistor parameters under both DC and AC stresses and to propose models for both degradation and recovery and lifetime prediction [6] [7] [8]. Nevertheless, the focus on the low-level details does not provide insight about the circuit degradation due to NBTI on a larger scale. This is due to the fact that PMOS transistors within a circuit are interacting with other PMOS and NMOS transistors by sharing with them power

supply, and input and output signals and interconnects. So, to properly estimate NBTI degradation and carry out ageing simulation, a model that faithfully reflects the impact of NBTI on circuit performances is required. To obtain such model, an experimental NBTI characterization at circuit level is needed. In [9] an experimental characterization of NBTI has been done on CMOS inverters using stress/measure/tress (SMS) protocol conducted with one negative gate voltage at one temperature 125°C. In contrast with this work we have deeply investigate NBTI influence on CMOS inverter DC performances by applying a series of negative voltages and heating samples at different temperatures to accurately track their influence and to detect where they are more influencing the degradation.

The rest of the paper is organized as follows. In section II, the experimental DC NBTI setup details, samples, and SMS protocol are described. The obtained results are analyzed and discussed in section III. Section IV concludes the paper.

II. EXPERIMENTAL SETUP

The samples under test are CMOS inverters with aspect ratios of 1µm/0.8µm and 2µm/0.8µm for NMOS and PMOS transistors respectively. They were fabricated at ISiT (Institute for Silicon Technology) of Fraunhofer, Germany using N-well process. The gate oxide thickness is 20 nm.

As the time delay between the application of stress and measurement of degradation is quite important as it can produce a wide variation in the results, the SMS sequences have been performed using fully automated benches. The bench includes a sensitive Agilent HP 4156C for voltage measurements and triging. A hotplate is used, inside Karl Suss PA300 micro-manipulator probe station, to vary the temperature of the chip during the experiments. The test circuit chip (non-packaged) within and, the probe station were isolated from vibration and enclosed in a grounded faraday cage to avoid both RF and light effects.

The test protocol we adopted is described as follows: during stress period of 3600s, a negative voltage was applied to the inverter input that is SMU2 whereas a 0 Voltage was applied to the rest of circuit's pads (SMU1, SMU3 & SMU4) as shown in fig.1. During the measure step, a conventional DC configuration was applied, where a DC voltage source of 5V was applied by SMU2, a constant voltage of 5V by SMU1 VDD=5V and the output of the inverter was measured by

SMU4. The switch between stress and measure configurations was automated via a LabVIEW program.

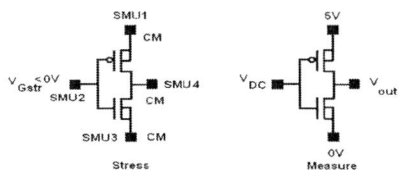

Fig.1 Measure-Stress-Measure protocol used for DC NBTI characterization

First, a measure of a virgin inverter's VTC has been done at ambient temperature, then the inverter has undergone a series of negative bias voltages of -8V, -10V, -12V and -13V for 1 hour where after each stress, the data of the VTC have been saved. This test protocol has been conducted for elevated temperatures ranging from 80°C to 140°C with a 20°C step using the hotplate in order to combine large field and elevated temperature for stress conditions.

III. RESULTS AND DISCUSSION

Fig. 2 shows the NBTI influence on the inverter voltage transfer curve VTC. It outcomes that NBTI causes a VTC shift to the left side as expected in the theory, that this VTC shift increases more and more with the increase of the applied field. This VTC shift induces a decrease of the logic threshold Vinv defined as Vout=Vin and found where the VTC crosses the X=Y line.

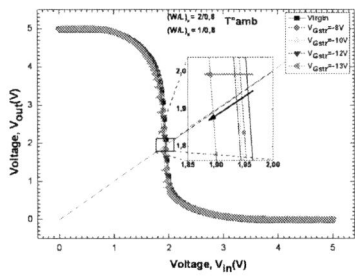

Fig. 2 Field influence on Inverter VTC under DC NBTI.

Fig. 3 shows the NBTI induced inverter VTC shift for a given stress voltage at different temperatures. It outcomes that the inverter VTC has the same trend of shift, and that the shift increases with elevated temperatures resulting in a decrease of the inverter logic threshold.

In order to demonstrate the influence of DC NBTI on the CMOS inverter, we have studied its impact on its main DC

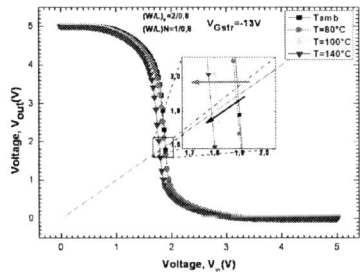

Fig. 3 Temperature influence on Inverter VTC under DC NBTI.

features which concern the logic threshold, the tolerable input and output logic levels, and its noise margins. These features qualify the robustness of the inverter and consequently its reliability. In the following subsections, we present these extracted features from the measured VTC under DC NBTI.

A. Inverter logic threshold

Fig.4.1 and fig 4.2 show the CMOS inverter logic threshold trend under DC NBTI as a function of the stress temperature

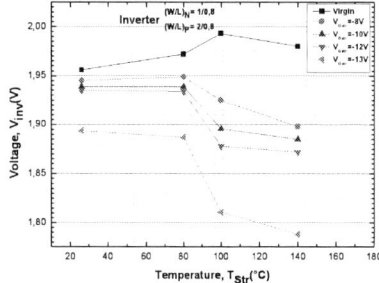

Fig. 4.1 Inverter logic threshold decrease with temperature under DC NBTI

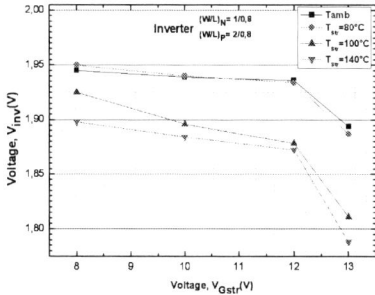

Fig. 4.2 Inverter logic threshold decrease with stress voltage under DC NBTI.

and stress voltage, respectively. Indeed, figure 4.1 shows that the logic threshold deceases with increasing temperature and the shift becomes more significant when the temperature is elevated above 80°C for a given stress voltage. We note here that logic threshold of fresh (unstressed) inverters noted by virgin in the figure has been improved and approaches the one of symmetrical VTC due to the annealing of the samples before stress.

The same shift trend of the inverter logic threshold is observed for increasing stress negative bias voltages as shown in fig 4.2 and it becomes more important when the applied field increases in its absolute value above 6.5 MV/cm corresponding to -13V for our samples of 20nm T_{ox}. Indeed, the logic threshold decreases more for negative stress voltage upper than 6MV/cm and worsen at elevated temperatures (more than 100°C).

978-1-4673-5289-5/12 $31.00 © 2012 IEEE   335

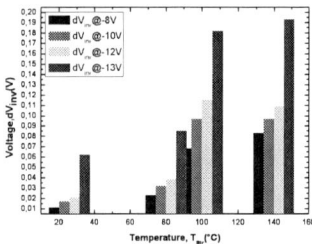

Fig. 5.1 Absolute shift of the inverter logic threshold under DC NBTI stress

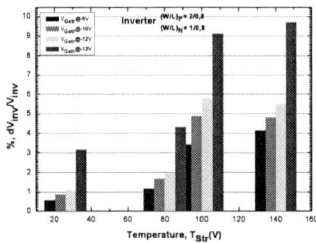

Fig. 5.2 Relative shift of the inverter logic threshold under DC NBTI stress.

The DC NBTI induced shift of the inverter's logic threshold for different temperatures is given by the flowchart in fig 5.1.

It can be seen that the shift varies from 10mV to 60mV at ambient temperature and from 80mV to 195mV at 140°C. Accordingly, the relative shift caused by high negative stress voltage is more significant and varies from 3.1% at ambient temperature to more than 9.6% at 140°C (as depicted in flowchart of fig5.2).

This shift of the logic threshold implies unequal logic states of the inverter by enlarging the voltage range interpreted as logic "1" and shrinking the one for logic "0" as will be discussed in the following section.

B. Logic voltage levels

On behalf of the logic threshold, we have studied the influence of NBTI on the logic voltage levels that define the critical voltages of the CMOS inverter VTC to know V_{IL}, V_{IH}, V_{OL} and V_{OH} which are defined where the slope of the VTC has a value of -1 as shown in fig.6. Where, the input low voltage V_{IL} represents the largest value of Vin that can be interpreted as logic "0" whereas the input high voltage V_{IH} represents the smallest value of Vin that can be interpreted as

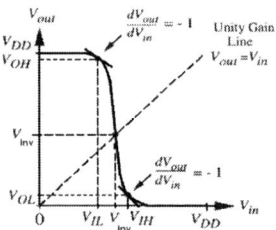

Fi.6. Critical CMOS inverter VTC voltages.

logic "1". The output low voltage V_{OL} defines the largest value of Vout's logic "0" while the output high voltage V_{OH} represents the smallest value of Vout's logic "1".

The values we extracted for these critical voltages from both fresh and stressed inverter for a series of temperatures and voltages are depicted in figure 7.

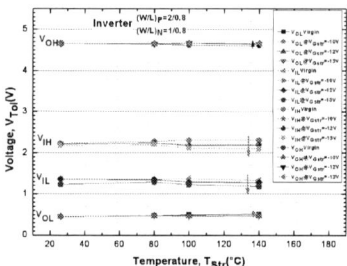

Fig.7 Measured logic voltage levels of the fresh and stressed CMOS inverters.

According to these results, we can say that NBTI has slightly influenced the CMOS inverter output levels V_{OL} and V_{OH} by decreasing the former and increasing the later. Consequently, the logic swing of the inverter defined as the difference between these two voltages is somewhat improved. In the other side, both input low and high voltages have been more affected and they have been decreased under NBTI stress. As a result, the input high voltage has been enlarged whereas its low voltage has been shrunk. Hence, the noise margins of the CMOS inverter will be obviously affected as will be demonstrated.

C. Noise immunity

The word *noise* in the context of digital circuits means "unwanted variations of voltages and currents at the logic nodes" [10]. This noise can be due to a capacitive, an inductive cross talk, or can be internally-generated power supply noise. Whatever its source, for a gate to be robust and insensitive to noise, it is essential that the "0" and "1" intervals be as large as possible. A measure of the inverter sensitivity to noise is given by the noise margins NML (noise margin low) and NMH (noise margin high), which quantize the size of the legal "0" and "1", respectively, and set a fixed maximum threshold on the noise value. So, we define the voltage noise margin for logic 1 (high) as:

$$NMH = V_{OH} - V_{IH} \qquad (1)$$

Similarly, the voltage noise margin for logic 0 (low) as:

$$NML = V_{IL} - V_{OL} \qquad (2)$$

Using these equations, we have computed the noise margins for fresh and stressed inverters and the results are depicted in fig.8. Fig 8.1 shows the influence of NBTI on noise margins with respect to temperature whereas fig 8.2 shows its influence as a function of the applied stress voltage. It is clear that both noise margins have been shifted under NBTI stress in opposite sides. While the noise margin high has been enhanced, the noise margin low has been decreased. As we are interested in NBTI induced noise margin degradation, we can see from fig 9.1 that the shift of NML caused by the application of a field of -6.5MV/cm (-13V) at 140°C is about 100mV representing a

978-1-4673-5289-5/12 $31.00 © 2012 IEEE

degradation of more than 10% of the fresh noise margin low. Thus, the noise margin at level "0" is seriously affected and also the inverter robustness.

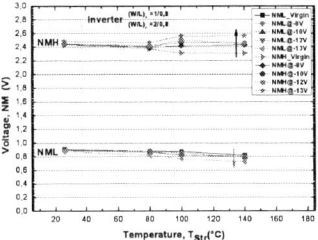

Fig.8.1 Measured inverter noise margins under NBTI stress with respect to temperature.

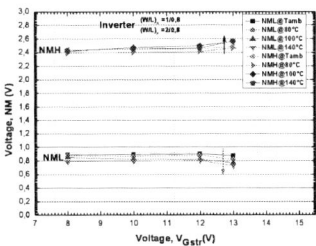

Fig.8.2 Measured inverter noise margins under NBTI stress with respect to stress voltage.

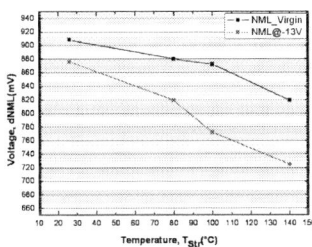

Fig.9.1 Absolute NML degradation under DC NBTI stress

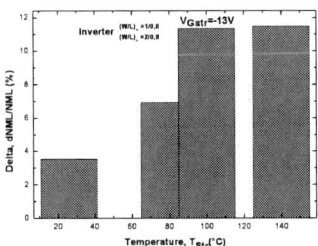

Fig.9.2 Relative NML degradation under DC NBTI stress.

IV. Conclusion

We have presented in this paper a deep investigation of NBTI impact on CMOS inverter DC performances. We have demonstrated that under NBTI, the inverter voltage transfer curve does shift to the left side as predicted by theory and that it experiences a more significant shift at higher negative voltage and worsen at elevated temperature. The results we

obtained have demonstrated that all critical voltages have been affected. Indeed, the logic threshold has shifted and undergone a degradation of more than 10% of its nominal value after one hour of stress of -6.5MV/cm at 140°C. The input low and high voltages have been oppositely affected as the logic "1" has been enlarged whereas logic "0" has been shrunk leading to more unbalanced logic states. At the same time, the output voltages have been slightly affected by NBTI. Regarding inverter noise immunity, both noise margins have been influenced where the margin noise high NMH has been increased while the noise margin low has been shrunk. This later has undergone a degradation of more than 10% of its nominal value after one hour of stress of -6.5MV/cm and just at 100°C. Consequently, the CMOS inverter has become more vulnerable to noise, and less robust. Ultimately, this work has shown that CMOS inverter reliability has been seriously impacted by DC NBTI.

As PMOS transistors within digital circuits are not continuously experiencing a DC NBTI stress, circuit lifetime prediction upon DC NBTI stress is somewhat pessimistic. Accordingly, in future work, we will continue with the experimental investigation of AC NBTI impact on CMOS inverter temporal performances namely switching times and propagation delay to more evaluate both DC and AC NBTI at circuit level in order to develop an accurate model to perform NBTI ageing simulation needed by designers to evaluate their circuits lifetimes.

REFERENCES

[1] Dieter k. schroder, "Negative bias temperature instability: What do we understand?," *Microelectronics Reliability, Elsevier*, vol. 47, 2007, pp. 841–852.

[2] V. Huard, M. Denais, C. Parthasarathy, "NBTI Degradation: From Physical Mechanisms to Modelling," *Microelectronics Reliability, Elsevier*, vol. 46, no. 1, January 2006, pp. 1-23.

[3] Sang Phill Park, Kaushik Roy & Kunhyuk Kang, "Reliability Implications of Bias-Temperature Instability in Digital ICs," *IEEE Design & Test of Computers*, vol. 26, no. 6, November/December 2009, pp.8-17.

[4] Bipul. C, Kunhyuk Kang, Haldun Kufluoglu, Muhammad A. Alam, and Kaushik Roy, "Impact of NBTI on the Temporal Performance Degradation of Digital Circuits," IEEE Electron Device Letters, vol. 26, no. 8, august 2005.pp. 560-562

[5] Xiangning Yang, Eric Weglarz, and Kewal Saluja, "On NBTI Degradation Process in Digital Logic Circuits," *20th International Conference on VLSI Design (VLSID'07)*, Bangalore, India, 6–10 January 2007, pp.723–730.

[6] Ming-Fu Li, Daming Huang, Chen Shen, "Understand NBTI Mechanism by Developing Novel Measurement Techniques," IEEE Transactions on Device and Materials Reliability, vol. 8, no. 1, march 2008, pp. 62-71.

[7] Joseph B. Bernstein, Moshe Gurfinkel, Xiaojun Li, Jörg Walters, Yoram Shapira, ,Michael Talmor, "Electronic circuit reliability modeling," *Microelectronics Reliability Elsevier*, vol. 46, no. 12, December 2006. pp. 1957–1979.

[8] Rihito Kuroda et all, "Circuit level prediction of device performance degradation due to negative bias temperature stress," *Microelectronics Reliability Elsevier*, vol 47, 2007, pp. 930–936.

[9] R. Fernandez, B. Kaczer, J Gago, R. Rodriguez, M Nafria, "Experimental characterization of NBTI effect on pMOSFET and CMOS inverter," Proceeding of the 2009 Spanish Conference on Electron Devices, Feb 11-13, 2009 Santiago de Compostela, Spain, pp. 231-233.

[10] Jan M. Rabaey, Anantha Chandrakasan , Borivoje Nikolic, "Digital Integrated Circuits: A Design Perspective," 2nd Edition, Prentice Hall, January 2003.

Electromechanical Response Simulation of Film Bulk Acoustic Wave Resonator

Rafik SERHANE
[a] Division Microélectronique et Nanotechnologie, Centre de Développement des Technologies Avancées (CDTA).
20 Août 1956, Baba Hassen, Algiers DZ-16303, BP: 17, Algeria, phone: +213 (0) 21 35 10 40, Fax: +213 (0) 21 35 10 21.
rserhane@cdta.dz

Tarek BOUTKEDJIRT
Equipe de Recherche Physique des Ultrasons, Faculté de physique, Université des Sciences et de la Technologie Houari Boumediene (USTHB), BP 32, DZ-16111, El-Alia, Bab-Ezzouar. Alger. Algeria

Abdelkader HASSEIN BEY[a,b]
[b] Micro & Nano Physics Group, Faculty of Sciences, University Saad Dahlab of Blida (USDB), BP. 270, DZ-09000 Blida. Algeria

Abstract—The FBAR (Film Bulk Acoustic Resonator) is an ultrasonic transducer consisting of a piezoelectric film which is taken as sandwich between two planar metallic electrodes. The lateral dimensions of this MEMS (Micro Electro Machinated System) are considered very large compared to its thickness [1-3].

In a first approach, an analytical one-dimensional propagation model (Mason's Model [4, 5]) is adopted. This model allowed us to determine the evolution of the electrical impedance of the transducer as a function of frequency. In a second step, we considered a FBAR transducer with 130x130x1.52 µm³ dimensions. We present the simulation of the manufactured process technology of the real structure. We perform, first, the harmonic study of the structure without embedding. Then, we study the embedding effect.

Manufacturing process steps simulation was performed under SILVACO tools (Athena, Atlas and DevEdit) [6]. Electromechanical characterization was conducted using a Finite Element Method FEM simulation [7, 8]. We perform; first, a comparison of electrical characterization results obtained using these tools with those of the one-dimensional analytical study. Then, we study the effect of embedment of the structure on the electromechanical behavior of the system.

Keywords—MEMS, FBAR, Ultrasonic transducer, piezoelectricity, FEM, SILVACO.

I. INTRODUCTION

Technology process simulation is nowadays an important tool in the research and development of industry-related manufacturing MEMS devices. Based on physical models, TCAD (Technology Computer-Aided Design) simulation can faithfully reproduce all steps of manufacturing a MEMS device and its electro-mechanical behavior. Technology process simulation takes into account the thermal oxidation, etching, deposition of thin films, etc... For 3D a complex piezoelectric structure, obtaining an analytical solution being impossible, the electrical and mechanical behaviors under different loads have been predicted by using the numerical solution of differential equations describing the electro-mechanical phenomena in these structures.

Silvaco is a set of several simulators, the two main modules of which are: Athena, for the simulation of the technological process and Atlas for electrical device simulation. For the electro-mechanical behavior, we will use a FEM simulation to deduce the electrical response of the device and to visualize the mechanical vibration modes.

II. THEORETICAL MODEL OF MASON

The piezoelectric effect is manifested directly by the appearance of electrical charges under mechanical deformation; the opposite effect results in the deformation of the solid under an electric field solicitation. These effects are in fact coupled and can be described by the following two state equations:

$$
\begin{aligned}
T_{ij} &= c_{ijkl}^{E} S_{kl} - e_{kij} E_k & T_{ij} &= c_{ijkl}^{E} \frac{\partial u_l}{\partial x_k} + e_{kij} \frac{\partial U}{\partial x_k} \\
& \quad\text{or} \\
D_j &= e_{jkl} S_{kl} + \varepsilon_{jk}^{s} E_k & D_j &= e_{jkl} \frac{\partial u_l}{\partial x_k} - \varepsilon_{jk}^{s} \frac{\partial U}{\partial x_k}
\end{aligned} \quad , (1)
$$

where D_j is the electric displacement component, E_k the electric field component, x_k the space coordinate, ε_{jk}^{S} the dielectric permittivity at constant strain, u_i the particles displacement, T_{ij} the components of the stress tensor, S_{ij} the elements of strain tensor, c_{ijkl}^{E} the elastic tensor, e_{kij} the piezoelectric tensor and U the electric potential. Using the fundamental equation of dynamics in solids, we obtain the wave equation: $\rho \partial^2 u_i / \partial t^2 = \partial T_{ij} / \partial x_j$, with ρ the material density and t the time variable. Thus :

$$
\rho \frac{\partial^2 u_i}{\partial t^2} = c_{ijkl}^{E} \frac{\partial^2 u_l}{\partial x_j \partial x_k} + e_{kij} \frac{\partial^2 U}{\partial x_j \partial x_k} . \quad (2)
$$

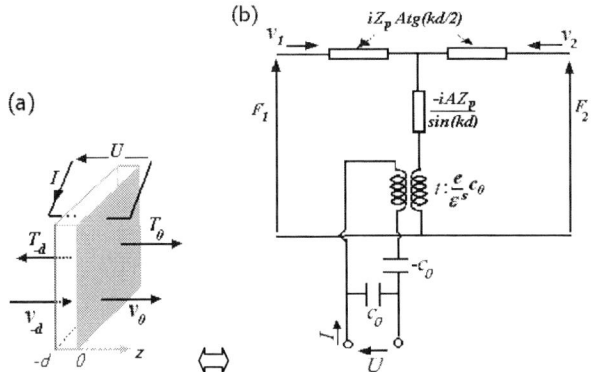

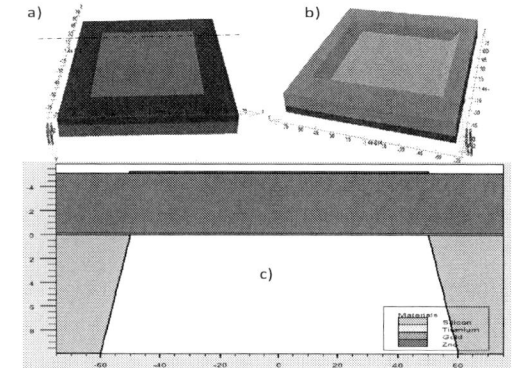

Fig. 1. (a): Piezoelectric layer, (b) Equivalent circuit of the resonator [9].

Fig. 2. Structure obtained by Silvaco simulation, (a)-Top view, (b)-Bottom view, (c)- Cross section

For a harmonic excitation, the constraints and the potential difference solutions are given by:

$$
\begin{cases}
T_0 = \left\{ Z_p \left(\dfrac{v_{-d}}{j\sin(kd)} - \dfrac{v_0}{j\,\mathrm{tg}(kd)} \right) - j\dfrac{e}{\varepsilon^s \omega}\dfrac{I_0}{A} \right\} e^{j\omega t} \\[2ex]
T_{-d} = \left\{ Z_p \left(\dfrac{v_{-d}}{j\,\mathrm{tg}(kd)} - \dfrac{v_0}{j\sin(kd)} \right) - j\dfrac{e}{\varepsilon^s \omega}\dfrac{I_0}{A} \right\} e^{j\omega t} \\[2ex]
U = \left\{ \dfrac{e}{\varepsilon^s j\omega}\left[v_{-d} - v_0 \right] + \dfrac{I_0}{j\omega A}\dfrac{d}{\varepsilon^s} \right\} e^{j\omega t}
\end{cases} \tag{3}
$$

v_0 is the vibration velocity and d the piezoectric layer thickness. k is the wave vector and ω the angular frequency. We deduce the electromechanical impedance matrix as:

$$
\begin{pmatrix} T_{-d} \\ T_0 \\ U \end{pmatrix} = je^{j\omega t}
\begin{pmatrix}
\dfrac{Z_p}{\mathrm{tg}(kd)} & \dfrac{-Z_p}{\sin(kd)} & \dfrac{e}{A\varepsilon^s \omega} \\[2ex]
\dfrac{Z_p}{\sin(kd)} & \dfrac{-Z_p}{\mathrm{tg}(kd)} & \dfrac{e}{A\varepsilon^s \omega} \\[2ex]
\dfrac{-e}{\varepsilon^s \omega} & \dfrac{e}{\varepsilon^s \omega} & \dfrac{-1}{\omega c_0}
\end{pmatrix}
\begin{pmatrix} v_{-d} \\ v_0 \\ I_0 \end{pmatrix} \tag{4}
$$

The equivalent circuit of the piezoelectric resonator of section A and acoustic impedance Z_p is represented in Fig. 1 (b). The electrical impedance is given from the previous system by $Z_e = U/I$, I being the electric current intensity. So:

$$
Z_e = \frac{1}{jc_0\omega}\left\{ 1 + \frac{K_{eff}^2}{kd}\frac{2(1-\cos(kd)) - j(z_1 + z_2)\sin(kd)}{j.(z_1 + z_2)\cos(kd) - (1 + z_1 z_2)\sin(kd)} \right\}, \tag{5}
$$

where K_{eff}^2 is the square of the electromechanical coupling coefficient. $z_1 = Z_1/Z_p$ and $z_2 = Z_2/Z_p$, the normalized mechanical impedances. In the case where the faces of the transducer are perfectly free, $z_1 = z_2 = 0$, the previous relation becomes:

$$
Z_e = \frac{1}{jc_0\omega}\left\{ 1 - K_{eff}^2 \frac{\tan(kd/2)}{kd/2} \right\} \tag{6}
$$

The electrical admittance $Y = 1/Z_e$ of the free resonator is purely imaginary. The anti-resonance ($Z = \infty$) appears at the frequencies; $f_a^n = (2n+1)f_p$, where $f_p = V_p/2d$ is the eigenfrequency, V_p the propagation velocity of the longitudinal wave in the piezoelectric layer. Each anti-resonance is preceded by a resonance at frequencies f_r^n, which are solutions of $K_{eff}^2 \tan(\pi f_r^n/2f_a) = \pi f_r^n/2f$. We also define the quality factor by the impedance phase derivation at resonance and anti-resonance respectively. That is [10]:

$$
Q_r = \frac{f_r}{2}\left.\frac{\partial \phi(Z)}{\partial f}\right|_{f=f_r} \quad \text{and} \quad Q_a = \frac{f_a}{2}\left.\frac{\partial \phi(Z)}{\partial f}\right|_{f=f_a} \tag{7}
$$

III. TECHNOLOGICAL PROCESS OF A FBAR

The manufacturing process of the FBAR consists of the following main steps [11, 12]:

1. Wafer cleaning

2. Thermal oxidation of the two faces

3. Lower electrode (20 nm Ti + 200 nm Au) deposition

4. Piezoelectric layer (1.52 µm ZnO) deposition

5. Upper electrode (20 nm Ti + 200 nm Au) deposition

6. Partial etching of the upper electrode

7. Partial etching of the backside oxide

8. Wet etching of the silicon with KOH solution to release the membrane.

The simulation of these steps is performed by Silvaco tools (Athena, Atlas and DevEdit3D). The final obtained structure is shown in Fig. 2.

When applying an electric potential $U_0 = 1V$ on the upper electrode of the structure, the static study by FEM simulation shows the appearance of a static electric charge $Q_0 = 0.976$ pC with capacitance $C_0 = Q_0/U_0 = 0.976$ pF. This value is in agreement with Mason's model [1], where $C_{0\ th} = 0.889$ pF (with $\varepsilon_{33} = 9.03$).

IV. SIMULATION OF THE ELECTROMECHANICAL BEHAVIOR OF THE RESONATOR BY FINITE ELEMENT METHOD

The geometry of the device is constituted of square blocks, so the suitable mesh type is the hexahedral one. In the thickness direction, the mesh is performed by superposing 25 piezoelectric layers. The structure contains 8 elements in the other directions which give total mesh of 1600 elements. In order to solve the constitutive equations governing the piezoelectric effect, we use a coupled-field element with 20-nodes in the hexahedral brick element. The degrees of freedom are the mechanical displacement u_x, u_y u_z and the electrical voltage U. The reaction solutions are the mechanical force and electric charge Q_0.

Fig. 3 shows the variation of the electrical admittance of the resonator versus frequency obtained by FEM method for the three first thickness vibration modes. The first column of the figure represents the amplitude of the electrical impedance, the second one its phase and the third one the Smith chart of the S_{11} Scattering parameter. The anti-resonant frequencies appear at odd multiples of 2.103 GHz. This frequency is predicted by the analytical model, which gives f_a=2.107 GHz for a thickness of 1.52 µm and a longitudinal wave velocity equal to 6405.3 m/s in ZnO (with c_{33}=210.6 GPa, e_{33}=1.321 C/m² and ρ=5665 kg/m³). Fig. 3 shows a comparison between the results obtained by Mason's model and the results of FEM simulation of a structure fixed at its ends. The simulated resonance and anti-resonance frequencies and the electromechanical coupling constant values are very close to the theoretical values (Table.1). For the real structure

(embedded), the electromechanical coupling coefficient in the three first thickness resonant modes are respectively K_{eff}=29.92%, 9.50% and 5.72%. By observing the Smith chart of these resonances in Fig. 3.(c), (f) and (i), we can evaluate qualitatively the quality factors for each mode [13, 14]. This is given by the radius of the main loop. This study shows that the best performance and best quality factor of the transducer are obtained for the fundamental resonance.

For the fundamental thickness mode, the representation in Fig. 3.b shows that the resonator is purely capacitive before resonance. It switches to pure inductive between the resonance and the anti-resonance. Then, it becomes purely capacitive after the anti-resonance frequency. The Smith chart in Fig. 3.c shows that at resonance, the device behaves like a short circuit, whereas at anti-resonance, it becomes an open circuit.

However, additional peaks appear in the curves of the electrical impedance near the resonance and anti-resonance of Fig. 3.(d) and (g), specifically, at the second and the third resonance at f=6.32 and 10.53 GHz. These peaks do not appear in the case of free ends structure. By observing the distribution of the amplitude of the mechanical displacement in the z direction (u_z) at the corresponding frequencies, we can deduce that they are due to local resonances on the surface of the transducer (Fig. 3.(d)). These spurious resonances appear on the Smith chart (Fig. 3.f and i) as miniature loops. In the capacitive region, these loops are parasitic resonances and are of mechanical origin and are caused by diffraction on the frame of the structure.

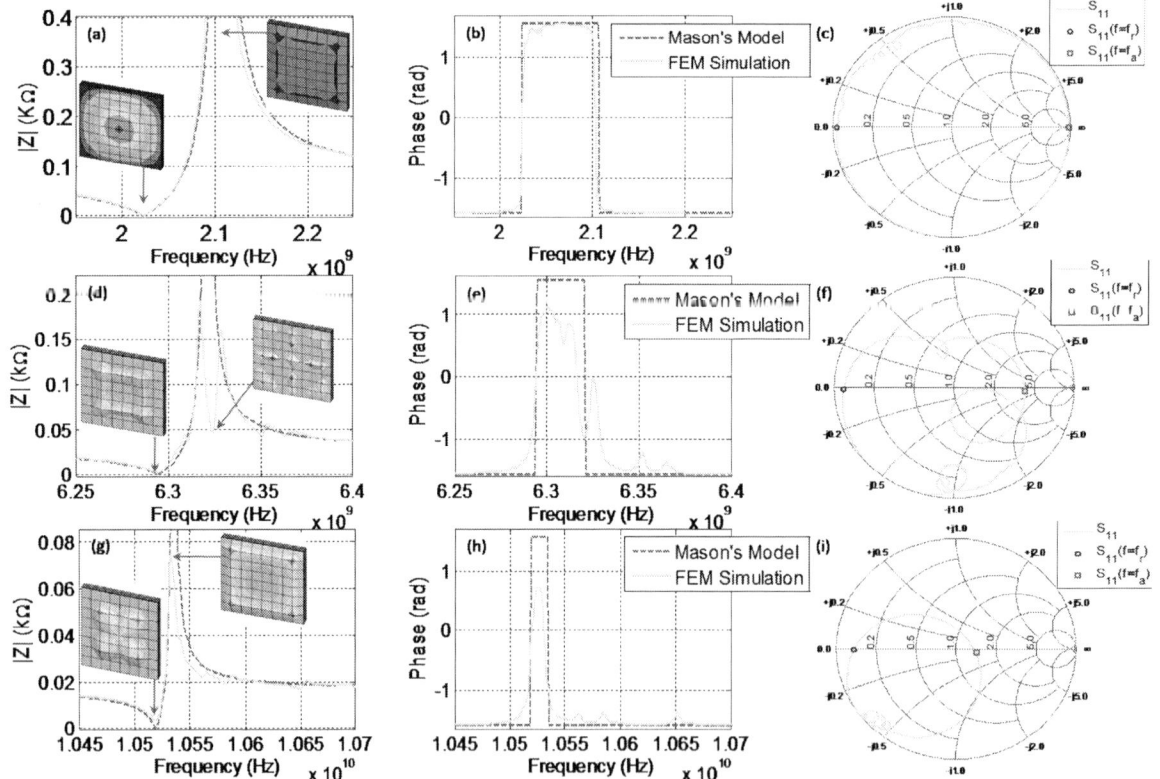

Fig. 3. Electrical input impedance of the device. (a) , (b) and (c) First thickness resonance mode, (d) , (e) and (f) Second thickness resonance mode, (g) , (h) and (i) Third thickness resonance mode. (a) , (d) and (g) Amplitude of the electrical input impedance, (b) , (e) and (h) Phase of the electrical input impedance, (c) , (f) and (i) Smith char of S_{11} scattering parameter.

TABLE I.   COMPARISON OF OBTAINED RESULTS

		Mason's model	FEM simulation (without embedding)	FEM simulation (with embedding)
C_0 (pF)		0.8896	1.1295	0.9763
f_r(GHz)	R1	2.0235	2.0240	2.0241
	R2	6.2942	6.2944	6.2941
	R3	10.519	10.519	10.52
f_a(GHz)	R1	2.107	2.1082	2.1033
	R2	6.3211	6.3216	6.3173
	R3	10.535	10.537	10.532
K_{eff}	R1	30.64 %	30.78	29.92 %
	R2	10.23 %	10.28%	9.50 %
	R3	6.12 %	6.49 %	5.72 %
Q_r	R1	∞	108 700	105 900
	R2	∞	72 240	130 570
	R3	∞	78 240	94 590
Q_a	R1	∞	124 690	116 510
	R2	∞	83 750	128 060
	R3	∞	94 860	114 870

Comparison between the results obtained by Mason's analytical model and those obtained by FEM method for free ends structure and embedded structure

V. DISCUSSION

Depending on their origins, the parasitic resonances can be classified into two main categories [15]: Resonances due to the cavity mode and those due to the lateral standing wave mode.

The cavity modes are caused by the imposed boundary conditions. The mechanical displacement is maintained at zero at the lateral ends of the structure. Mechanical waves are reflected on this interface and towards the structure. They contribute to the change in the electrical response of the device and its electrical input impedance. To avoid these parasitic modes, we use generally perfectly matched layers on the laterals ends of structure [15].

The origin of the lateral parasitic modes is given by the excitation of Lamb modes, through the coupling with the fundamental thickness mode [16]. The Lamb modes that can be piezo-electrically generated are the symmetric modes. The natural frequencies of Lamb symmetric modes are directly related to the lateral dimensions of the top electrode ($f_N=V_t$ $N/2L$).

VI. CONCLUSION

Manufacturing process steps simulation of 130x130x1.52μm FBAR, was performed under SILVACO tools. Electromechanical characterization was achieved using Finite Element Method simulation. We performed; first, a comparison of electrical characterization results obtained using these tools with those of the one-dimensional analytical study. Then, we studied the effect of embedment of the structure on the electromechanical behavior of the system.

The comparison in Fig. 3 and Table 1 shows a good correlation between results obtained by Mason's analytical model and those obtained by finite element method simulation of a structure embedded at its ends. The spurious resonances which appeared in the electrical impedance curves near the resonance and anti-resonance are due to the structure embedding. These resonances are classified into two

categories; resonances due to the cavity mode resonances and those due to the lateral standing wave mode.

FEM model simulation is a powerful tool to predict spurious modes that can appear in FBAR structure vibration. It allows solution development of complex structures. As future work, it is planed an experimental validation of the simulations with, eventually, parameters adjustment.

ACKNOWLEDGMENT

The authors wish to express their gratitude to all the MEMS & Sensors group, particularly; Hammouche KHALES, Noureddine DERGUINI, Abdelhakim ACHELI and Abderrezak SMATTI, Lamia MENASERI; Mohammed MAHDI; Nassima CHAMMA and Walid AOUIMEUR. They also thank CDS elements team: Fayçal HADJ LARBI and Abdelmadjid BENABDELMOUMENE, of Algerian Center for Development of Advanced Technologies.

REFERENCES

[1] R. Ruby, P. Merchant, "Micromachined Thin Film Bulk Acoustic Resonators," Proc. IEEE 48 Symp. on Freq. Control, 1994, pp. 135-8.

[2] P. Osbond, C.M. Beck, CJ. Brierley, M.R. Cox, S.P. Marsh, N.M. Shorrocks, "The Influence of ZnO and Electrode Thickness on the Performance of Thin Film Bulk Acoustic Wave Resonators," IEEE Ultrasonic Symp., pp. 911-914, 1999.

[3] K.M. Lakin, "Thin Film Resonators and Filters", IEEE Ultrasonic Symp., 1999, pp. 895-906.

[4] D. Royer and E. Dieulesaint, Elastic Waves in Solid II. Masson ed, vol.2, 1999.

[5] J. F. Rosenbaum, Bulk Acoustic Wave ; Theory and Devices, RTECH House, Inc. Boston London, 1988.

[6] SILVACO, International, ATLAS User's Manual Device Simulation Software. and ATHENA User's Manual 2D Process Simulation Software, Santa Clara, 2007.

[7] K.M. Lakin," Numerical analysis of two dimensional thin film resonators", IEEE Inter Freq. Contr. Symp., 1993, pp. 502-508.

[8] K. M. Lakin and K. G. Lakin, "Numerical Analysis of thin Film BAW Resonators," IEEE Ultasonic Symp., pp. 74-79, 2003.

[9] H..Lukdawala and E. S. Kim," Simple Post-processing Technique to Tune Resonant Frequency of Film Bulk Acoustic Resonators and Stacked Crystal Filters", IEEE Inter. Freq. Contr. Symp., 1998, pp.831-835.

[10] G. Caruyer, Modélisation, conception et caractérisation de résonateurs et filtres à ondes acoustique de volume pour le filtrage RF en téléphonie mobile, Phd Thesis, Lille, 2005.

[11] H. Campanella, Thin-film bulk acoustic wave resonators (FBAR): Fabrication heterogeneous integration with CMOS technologies and sensor applications, Phd Thesis, Montpelier, Décembre 2007.

[12] P. Martins, Caractérisation mécanique des matériaux pour les micro/nanosystèmes, Procédés applicables aux épaisseurs submicroniques, Phd Thesis, Lyon, 2009.

[13] G.R. Kline, K. M. Lakin, and K.T. McCarron," Overmoded High Q Resonators for microwave oscillators", IEEE Int. Freq. Contr. Symp., 1993, pp. 718-721.

[14] W. Wai hu, Y. Song and E. S. Kim," Lateral-field-excitation acoustic resonators for monolithic oscillators and filters", IEEE Inter. Freq. Contr. Symp., 1996, pp. 558-562.

[15] S. Giraud, S. Bila, M. Aubourg, D. Cros, "Bulk acoustic wave resonators 3D simulation," IEEE Inter. Freq. Contr. Symp., 2007, pp. 1147-11.

[16] J. V. Tirado, Bulk Acoustic Wave Resonators and their application to Microwave Devices, Phd Thesis, Barcelona, 2010.

Safe Operating Area of a 0.15µm GaAs PHEMT in Overdrive Operating Conditions

N. Ismail and A. Kalboussi

Laboratory of Microelectronic and Instrumentation, University of Monastir
Higher Institute of Sciences and Energy Technology of Gafsa, University of Gafsa
Tunisia
Naoufel.Ismail@iemn.univ-lille1.fr

Abstract—We present an evaluation of the safe operating area of a 0.15 µm GaAs PHEMT operating under overdrive conditions. The approach, used to delimit the device safe operating area, consists on performing on-state and off-state accelerated DC step stresses for bias conditions included in the device breakdown region. The accelerated DC stresses have induced weak degradations of the device electrical parameters attributed to trap effects.

Keywords-breakdown voltage; FETs; gallium arsenide; reliability

I. INTRODUCTION

For space applications, solid state amplifiers, based on compound semiconductor devices, can operate under critical conditions with high gain compression. These particular specifications correspond to overdrive operating conditions. When operating in these critical conditions, the device might presents degradations due to hot electrons [1], [2], [3]. This makes essential to know the safe operating area for these devices. Nowadays, the methodology to guarantee component lifetime with safety margins is based on standard specifications for space and military applications. They require to apply derating coefficient to the maximum parameters (voltage, current, temperature, …) allowed by the component manufacturer [4]. Hence, the standard specifications are not satisfactory due to cautious definition of the component max ratings. This work presents a new approach, based on results of on-state and off-state DC step stresses, to evaluate the safe operating area of the transistor, candidate to operate in overdrive conditions.

The technology under test is a GaAs PHEMT with 0.15µm long and 4x50µm wide gate. Accelerated stress tests were performed using a BILT System BE100. Current-voltage (I-V) measurements, monitored by a HP4142B semiconductor parameter analyser, were performed periodically to record the device parameters drifts.

II. THE ON-STATE SAFE OPERATING AREA

A. The Device On-State Operating Area Before Stress

Fig. 1 shows the on-state breakdown loci of the PHEMT under test, measured for several gate current conditions and using the Gate-current Extraction Technique [5]. Before performing any stress, it is necessary to determine the operating area within which the device operates without burn-out. The on-state breakdown loci are measured from low to high breakdown conditions. The operating area limit corresponds to the grey dashed line joining the exponential region extreme points of each on-state breakdown locus (Fig. 1). Operating the transistor out of this region leads to its destruction since V_{DG} values become larger than the gate-drain breakdown voltage BV_{DG}.

B. The Device On-State Safe Operating Area After Stress

In this study, the safe operating area is defined for a failure criterion of about 10% drop in the gate source current I_{GS} or the drain source current I_{DS}, corresponding to the stress bias point, for 48 hours stress. The methodology consists in applying accelerated voltage step stresses with bias conditions (the drain source voltage V_{DS}, the gate source voltage V_{GS}) defined by the following sequence: an identical value of Ids is adjusted at the beginning of each step and $|I_{GS}|$ is increased for each step until the evolution of I_{DS} or I_{GS}, corresponding to the stress bias point, exceeds 10% over one step. The duration of each step is 48 hours. Several sequences are performed for different values of I_{DS} located in the device operating area. For instance, in Fig. 1, S1 and S2 are, respectively, the first and the second step of the stress sequence performed for I_{DS}=10 mA/mm. After performing the stress sequences, the couples of points (I_{GS}; I_{DS}), which correspond to the stress conditions, allow to define the limit of the safe operating area of the device. It is important to point out that this area represents a safe operating area for non linear applications since the accelerated DC step stresses are performed in regions, which can be reached by the V_{DS} and V_{GS} sweeps in real overdrive operating conditions.

We have applied this methodology for three stress sequences with I_{DS} values of 3.3, 10 and 16.6 mA/mm respectively. The evolution of the monitored parameters during the stress is a decrease of I_{DS} and I_{GS}. The gate and drain currents, corresponding to the stress bias point, present the same evolution during each step of the on-state stress sequence.

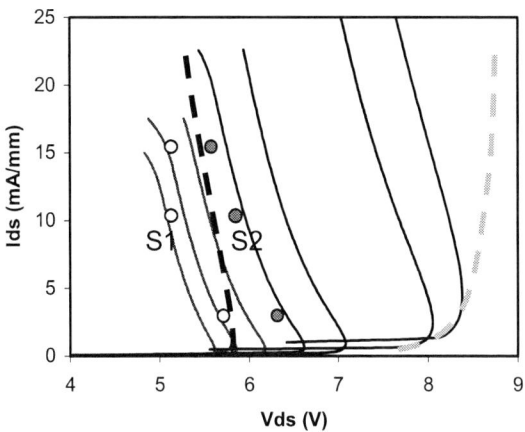

Figure. 1. On-state breakdown loci of the PHEMT under test for different breakdown conditions: I_{GS}=-0.01, -0.015, -0.025, -0.05, -0.1, -0.5 and -1 mA/mm from left to right respectively. The grey dashed line defines the limit of the device operating area before stress and the black dashed line is the limit of the device safe operating area after stress.

The evolution of I_{GS} is plotted in Fig. 2 for the stress sequence performed with I_{DS}=10 mA/mm. The drift percentage of the gate currents for each step is indicated in this figure.

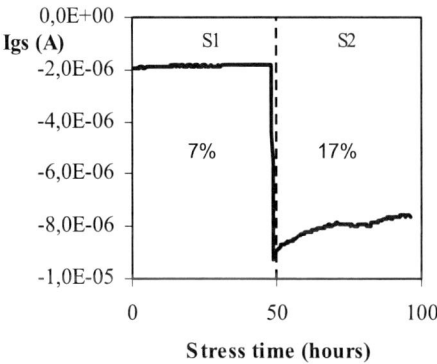

Figure. 2. Evolution of the gate-source current with time during the two steps S1 and S2 of the on-state stress sequence performed with I_{DS}=10 mA/mm.

In Fig. 1 the white dots correspond to stress conditions inducing a drift of I_{GS} or I_{DS} lower than 10%, whereas grey dots correspond to stress conditions inducing degradation higher than 10%. The contour delimiting the safe operating area is located between white and grey dots. It is important to point out that stress conditions were defined in correlation with satellite mission specifications. So, the choice of a more or less severe stress conditions leads to a shift to the left or the right of the contour delimiting the transistor safe operating area.

III. THE OFF-STATE SAFE OPERATING AREA

Since the V_{GS} sweeps can reach high negative values during the overdrive operation, we had to define the device safe operating area in the off-state operating regime.

A. The Device Off-State Operating Area Before Stress

Fig. 3 shows the off-state breakdown loci of the PHEMT under test, measured for several drain current conditions and using the Drain-Current Injection Technique [6]. The off-state breakdown loci measured from low to high breakdown conditions show that, the higher breakdown condition, the closer off-state breakdown loci. In Fig. 3, the black dots correspond to the bias conditions witch have induced the device burnout. The off-state operating area limit corresponds to the grey dashed line (Fig. 3). Operating the transistor out of this limit leads to its burnout.

B. The Device Off-State Safe Operating Area After Stress

The methodology consists in applying voltage step stresses close to the off-state breakdown regime. Step stresses are performed with bias conditions defined in a single sequence as follows: V_{GS} is held constant and V_{DS} is increased step by step for 48 hours each. Two sequences with V_{GS} value of -3.5V and -4V corresponding to the off-state breakdown regime have been performed and the stress bias conditions are plotted in Fig. 3. An identical methodology has been used for the two stress sequences. For instance, in Fig. 3, S1, S2 and S3 are, respectively, the first, the second and the third steps of the stress sequence performed for V_{GS}=-3.5V.

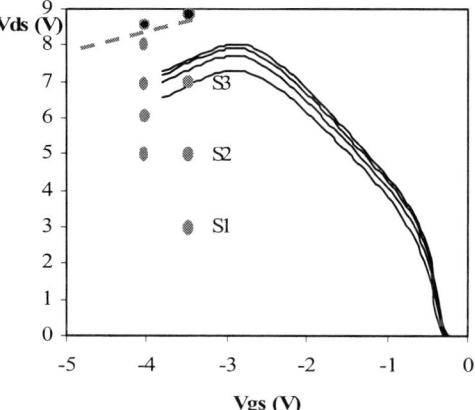

Figure. 3. Off-state breakdown loci of the PHEMT under test for different breakdown conditions: I_{DS}=0.25, 0.5, 0.75, and 1 mA/mm from bottom to top respectively. Two sequences of step stresses were performed with V_{GS} of -4V and -3.5V from left to right respectively. The grey dots correspond to stress conditions inducing a drift of I_{GS} or I_{DS} higher than 10%, whereas black dots correspond to the bias conditions witch have induced the device burnout.

The gate and drain currents, corresponding to the stress bias point, present the same evolution during each step of the off-state stress sequence. The evolution of I_{GS} is plotted in Fig. 4 for the stress sequence performed with V_{GS} of -3.5V. The drift percentage of the gate currents for each step is indicated in this figure. A decrease and stabilization of the absolute value of I_{GS} has been recorded during each step of the stress sequences. As shown in Fig. 4, the degradation percentage of I_{GS} increases with V_{DG}.

978-1-4673-5289-5/12 $31.00 © 2012 IEEE   343

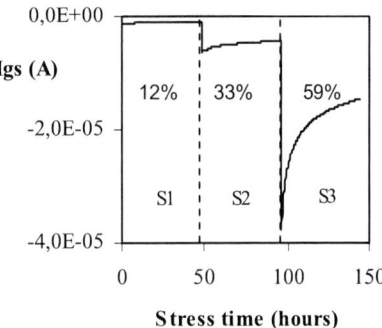

Figure. 4. Evolution of the gate-source current with time during the three steps S1, S2 and S3 of the off-state stress sequence performed with V_{GS} of -3.5V and V_{DS} of 3V, 5V and 7V respectively.

Let us notice that white dots not exist in Fig. 3 since all off-state stresses have induced degradations higher than 10%. In this case, we can not delimit the off-state safe operating area. But, we had to perform an RF stress in the area including the grey dots in order to confirm that this zone is a safe area for the off-state operating conditions of the device.

IV. DEGRADATION MODES FOR THE PHEMT AFTER THE ACCELERATED STRESSES

Fig. 5, Fig. 6, Fig. 7 and Fig. 8 present the evolution of the PHEMT characteristics after the different off-state stress steps. Let us notice that we have observed the same degradation modes after the on-state stresses.

The PHEMT presents a decrease of the saturation drain source current Idss and the threshold voltage Vt (Fig. 5), a decrease of the absolute value of the reverse gate leakage current (Fig. 6), a decrease of the gate current component related to the impact ionization mechanism (Fig. 6) and an increase of the on-state and off-state breakdown voltages (Fig. 7 and Fig. 8). Indeed, the on-state and off-state stress bias conditions correspond to high values of the drain gate voltage V_{DG}. It is generally believed that, when a transistor is submitted to high electric field, many electrons possess sufficient energy to escape from the channel and potentially become trapped in states located under the gate. These trapped electrons can increase the value of the voltage witch controls the two-dimensional electron gaz of the PHEMT [7]. As a consequence, Idss and Vt decrease.

Other electrons can be trapped in the passivation or at the interface between semiconductor and passivation over the drain access region [8]. They can even create traps themselves. The accumulation of a negative charge between the gate and the drain can relaxes the gate-drain electric field and, as a consequence, increases the on-state breakdown voltage namely "breakdown walkout" [9]. The breakdown walkout is correlated with the decrease of both the drain and gate current measured at the stress bias point. In general, a breakdown walkout is accompanied by a decrease of the impact ionization rate (Fig. 6). The increase of the off-state breakdown voltage is attributed to the decrease of the absolute value of the reverse gate leakage current.

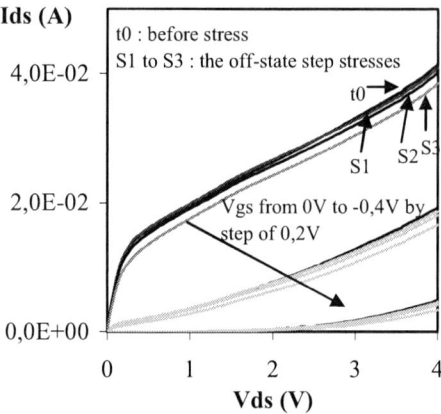

Figure. 5. Variation of the PHEMT I_{DS}-V_{DS} characteristics after the three steps of the off-state stress sequence performed with V_{GS} = -3.5V.

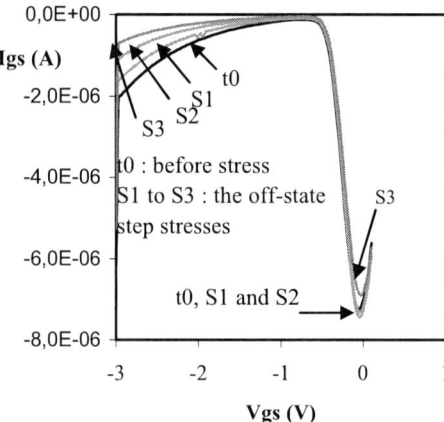

Figure. 6. Variation of the PHEMT reverse I_{GS} - V_{GS} characteristics, for V_{DS}=4V, after the three steps of the off-state stress sequence performed with V_{GS} = -3.5V.

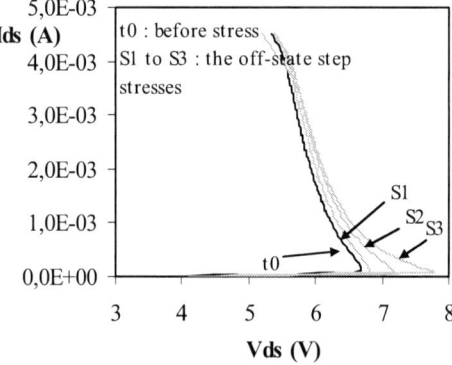

Figure. 7. Variation of the PHEMT on-state breakdown locus, for I_{GS}=-0.05 mA/mm, after the three steps of the off-state stress sequence performed with V_{GS} = -3.5V.

978-1-4673-5289-5/12 $31.00 © 2012 IEEE

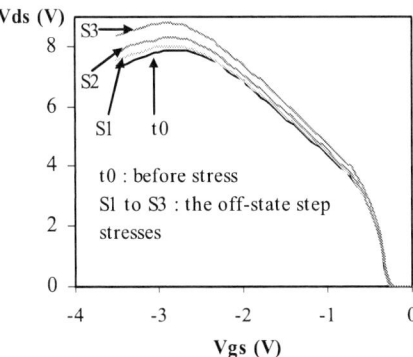

Figure. 8. Variation of the PHEMT off-state breakdown locus, for I_{DS}=1 mA/mm, after the three steps of the off-state stress sequence performed with V_{GS} = -3.5V.

In order to investigate the correlation between the weak evolution of the static characteristics of the transistor after stress and the influence of trapping states, DCTS (Drain Current Transient Spectroscopy measurements were performed before and after on-state accelerated stresses [10]. This experimental method consists in applying a voltage pulse to the gate of the transistor and measuring the drain current response. The device under test is biased in the saturation regime.

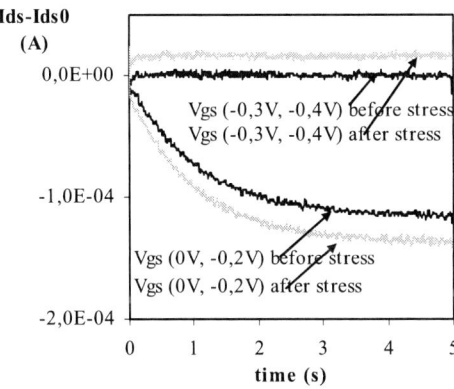

Figure. 9. Drain current transients measured with a drain bias voltage of 2V before stress and after the step S2 of the on-state stress sequence, performed with a drain-source current of 10 mA/mm, in response to gate voltage pulses of (0V, -0.2V) (pulsed from 0V to -0.2V) and (-0.3V, -0.4V) (pulsed from -0.3V to -0.4V). I_{DS0} is the initial value of I_{DS}.

As shown in Fig. 9, the gate voltage pulse (0, -0.2V) reveals an electron capture process due to surface traps. The deep levels located in the active layer and activated by the gate voltage pulse (-0.3V; -0.4V), are responsible for an electron emission process. We notice that, after stress, the DCTS characteristics shape has not changed. This suggests that no new kind of traps were created during the stresses. The slight change in the amplitude of the drain current transients, after stress, reveals a weak evolution of the density of the existing traps witch explains the weak changes of the PHEMT electrical characteristics.

V. CONCLUSION

We have presented, in this paper, an evaluation of the 0.15 μm GaAS PHEMT safe operating area for non-linear operation in overdrive conditions. The methodology consists in applying accelerated DC step stresses in the device breakdown regions. The on-state and off-state accelerated stresses have shown similar degradation modes for the PHEMT electrical characteristics: a weak decrease of the device saturation drain source current, a decrease of the absolute value of the reverse gate leakage current and an increase of the on-state and off-state breakdown voltages. DCTS measurements have demonstrated that these degradation modes are related to trap effects.

ACKNOWLEDGMENT

We want to thank N. Labat and N. Malbert for giving us the chance to perform the device electrical characterization at the IMS laboratory of the University of Bordeaux (France).

REFERENCES

[1] R. E. Leoni and J. C. M. Hwang, "Mechanisms for output power expansion and degradation of PHEMT's during high-efficiency operation," IEEE Trans. Electron Devices, vol. 46, no. 8, pp. 1608-1613, August 1999.

[2] J. C. M. Hwang, "Gradual degradation under RF Overdrive of MESFETs and PHEMTs," GaAs IC Symposium, pp. 81-84, San Diego CA, October 1995.

[3] Y. Tkachenko, A. Klimashov, C. Wei, Y. Zhao, and D. Bartle, "Comparative study of hot-electron reliability of PHEMT vs. MESFET for high efficiency power amplifiers," IEEE MTT-S Digest, pp. 799-802, Anaheim CA, June 1999.

[4] J. L. Muraro, F. Coppel, G. Gregoris, P. G. Tizien, J. L. Roux, J. Graffeuil, and R. Plana, "GaAs power MMIC: A design methodology for reliability," Microelectron. Reliab., vol. 37, no. 10-11, pp. 1651-1654, October 1997.

[5] M. H. Somerville, R. Blanchard, J. A. del Alamo, G. Duh, and P. C. Chao, "A new gate current extraction technique for measurement of on-state breakdown voltage in HEMT's," IEEE Trans. Electron Devices, vol. 19, no. 11, pp. 405-407, November 1998.

[6] S. R. Bahl and A. del Alamo, "A new drain-current injection technique for the measurement of off-state breakdown voltage in FET's," IEEE Trans. Electron Devices, vol. 40, no. 8, pp. 1558-1560, August 1993.

[7] C. Canali, P. Cova, E. De Bortoli, F. Fantini, G. Meneghesso, R. Menozzi, and E. Zanoni, "Enhancement and degradation of drain current in pseudomorphic AlGaAs/InGaAs HEMT's induced by hot-electrons," in Proc. Reliability Physics Symposium, pp. 205-211, Las Vegas, April 1995.

[8] M. Borgarino, R. Menozzi, D. Dieci, L. Cattani and F. Fantini, "Reliability physics of compound semiconductor transistors for microwave applications," Microelectron. Reliab., vol. 41, no. 1, pp. 21-30, January 2001.

[9] P. Cova, R. Menozzi, F. Fantini, M. Pavesi and G. Meneghesso, "A study of hot-electron degradation effects in pseudomorphic HEMTs," Microelectron. Reliab., vol. 37, no. 7, pp. 1131-1135, July 1997.

[10] N. Labat, N. Saysset, A. Touboul, Y. Danto, P. Cova and F. Fantini, "Analysis of hot electron degradation in pseudomorphic HEMTs by DCTS and LF noise characterization," Microelectron. Reliab., vol. 37, no. 10-11, pp. 1675-1678, October 1997.

978-1-4673-5289-5/12 $31.00 © 2012 IEEE   345

Standard CMOS Implementation of Schottky Barrier Diodes for Biomedical RFID

Sebastiao Cabral, Leonardo Zoccal, Paulo Crepaldi and Tales Pimenta

Universidade Federal de Itjauba - UNIFEI

Itajuba - Brazil

Abstract — **This paper presents and discusses the implementation of a Schottky Barrier Diode (SBD) in standard CMOS technology as a way to optimize the overall performance of a passive Radio-Frequency Identification (RFID) based biomedical implants. It is essential to limit the transmitted power in passive tags, mainly for biomedical applications in order to avoid damaging the human tissues due to local overheating. The implementation of the SBD was obtained by changing the mask flow without any modification to the CMOS fabrication process. The procedure maintains the transistors functionality and adds a new device to a standard CMOS technology. The fabricated SBD structures present a low turn on voltage of approximately 300 mV and low capacitance that are important parameters for passive RFID.**

Keywords - RFID; passive tag; SBD; Schottky diode

I. INTRODUCTION

This cost, size, lifetime and safety are important requirements on the design of an RFID based biomedical system, especially if the receiver is an implanted device. In order to attain size reduction and extended lifetime, the receiver can be implemented without batteries, thus characterizing a passive tag.

In a passive tag, the energy is transferred from the external unit to the implanted device by inductive coupling. Besides energy, the RF link is also used to make the communication path between the base unit and the transponder, so that information can be exchanged and energy can be delivered to the implant for its activation. Figure 1 shows a typical RFID topology for biomedical applications [1].

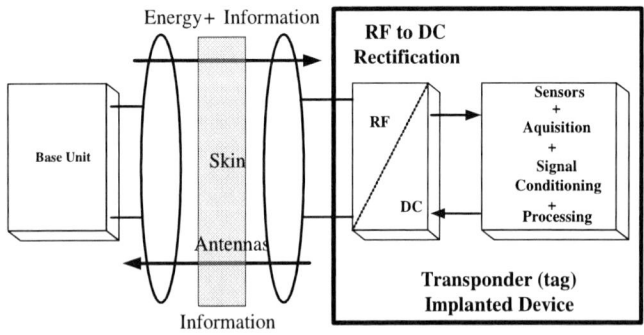

Figure 1.   Typical RFID System.

The energy is transferred by a pair of coupled coils. Nevertheless, patient safety requires keeping the induced electromagnetic fields at lower levels, in order to avoid tissue damage by raising the local temperature. The Specific Absorption Rate (SAR) represents a direct measurement of the electric field (indirect measurement of the magnetic field) and induced current density over the human tissue at the implant location. The temperature variation over the time indicates the local heating factor. Both relations are given by equations (1) and (2) [2]:

This Cost, size, lifetime and safety are important requirements on the design of an RFID based biomedical system, especially if the receiver is an implanted device. In order to attain size reduction and extended lifetime, the receiver can be implemented without batteries, thus characterizing a passive tag.

$$SAR = \frac{\sigma|E|^2}{\rho} \quad \left[W/_{Kg} \right] \qquad (1)$$

$$\frac{dT}{dt} = \frac{SAR}{c} \quad \left[{^0C}/_s \right] \qquad (2)$$

where σ. ρ and c represent the conductivity, the human tissue mass density and specific heat capacity, respectively, at the implant location. E is the incident electric field intensity (RMS). Based on equations (1) and (2), a safe value for the power transferrable by the RF link is 10mW/cm^2 [2].

Transponder front-end circuits include a rectifier used to implement the AC-DC conversion in order to provide the tag unregulated power supply. In CMOS technology, NMOS and PMOS transistors are used in different topologies to implement the rectifier circuit. These devices, however, have the disadvantage of presenting a threshold voltage (V_{th}) to turn on that can need a raising in the induced voltage at the receiver coil. Although the CMOS technology has been minimizing the transistors geometries, the V_{th} voltage does not follow the same scale of reduction. The use of a SBD is an alternative way to design the rectifier circuit in order to improve its efficiency. A more efficient rectifier will reduce the voltage drop between the tag input and the rest of the system thus reducing the power demand of the transmitter. For low current levels, as it is the case of implantable devices, the SBD voltage drop can be lower than a single V_{th} voltage.

978-1-4673-5289-5/12 $31.00 © 2012 IEEE

The SBD is not readily available in standard CMOS technology but it is possible to implement it after a few adjusts in the masking process. In this work, a mask sequence is presented to implement SBDs in CMOS process along with a simple device model. It is also presented the measurement data from a few die samples.

II. SCHOTTKY BARRIER DIODE

The metal-semiconductor contact offers some features that can find use in high frequency applications and systems that must operate at low voltage levels. These features are, basically, the low level of minority charge accumulation during commutation, which leads to high switching speeds and the low voltage drop between its terminals [3].

Lower turn on voltage, faster recovery time and lower junction capacitance are advantages offered by SBD structure when compared to other types of PN diodes. Those are the main reasons the SBDs are so popular in RF applications. As a consequence of high switching speed, ability to operate at high frequencies and low turn on voltage, SBDs are applied to RF mixer and detector diodes [4].

Considering the series resistance, the IxV function of Schottky diode can be expressed as [5]:

$$I = Is.\exp(\frac{V}{\eta Vt}) \qquad (3)$$

where, V is the bias voltage, Is is the saturation current, Vt is the thermal voltage, equals to KT/q (25,9 mV at T=300K), and η is the SBD ideality factor which can be calculated as:

$$\eta = \frac{d(\log(i))/d(v)}{\ln(10).Vt} \qquad (4)$$

The series resistance can be calculated as:

$$Rs = (\frac{\Delta V}{\Delta I}) \qquad (5)$$

The Schottky barrier height B can be calculated as:

$$\phi B = -Vt.\ln(\frac{Is}{AA^{**}.T^2}) \qquad (2)$$

where, A* is the effective Richardson constant (110 cm^{-2}K^{-2}A).

III. SBD MASK FLOW

Ideally, the SBD would be implemented by a metal layer deposition over a low doping N or P type semiconductor well as shown in Fig. 2.

In order to reduce the series resistance to improve the efficiency, the SBD, actually, is an arrangement of fingers as can be seen in Fig. 3 and Fig. 4. The whole SBD structure is surrounded by a guard ring, which is used basically to avoid latch-up and to separate the SBD from the other tag circuits that present different analog and/or digital functions.

In this design we have used 0.5µm CMOS TSMC process through MOSIS educational program.

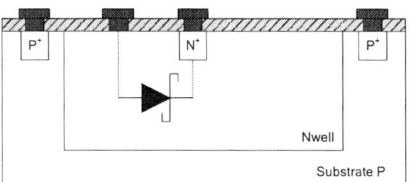

Figure 2. SBD structure cross section view.

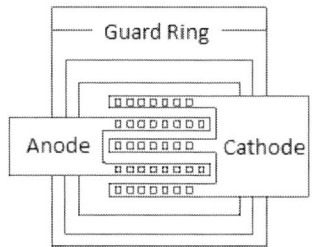

Figure 3. Five fingers SBD structure.

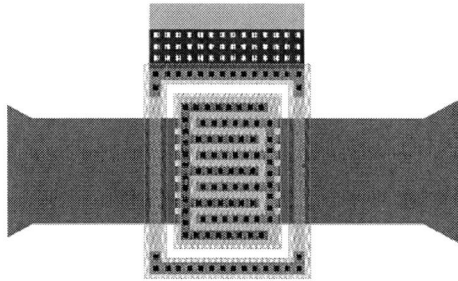

Figure 4. Nine fingers SBD layout.

In order to implement the metal-semiconductor junction it is necessary the following mask sequence. First, an NWELL (layer #42, TSMC) and an ACTIVE (#43) are used to delimit the area that will contain the multi-finger SBD and guard ring. Then an NPLUS layer (#45) is used to indicate N$^+$ regions that will be the SBD ohmic contacts (cathode). The next step is the CONTATCT layer (#48) that will be filled with METAL 1.

It is necessary to define the Schottky and guard ring contacts, and the contacts must reach the N well and the active regions directly. Others contacts must coincide with the previous N$^+$ diffusions to effectively make the ohmic contacts. With the METAL1 layer (#49), the SBD is complete. Additionally, a VIA layer (#50) is used to provide interconnections between the metal levels. Finally, the METAL 2 layer (# 51) is applied to metal 2 to provide access to the ring guard. Fig. 5 shows a 3D view of layers of the basic SBD structure without PADS and interconnections (MASK #42 to #49).

It is important to observe that it is not necessary to modify the CMOS process in terms of doping levels, metal type or other process parameters.

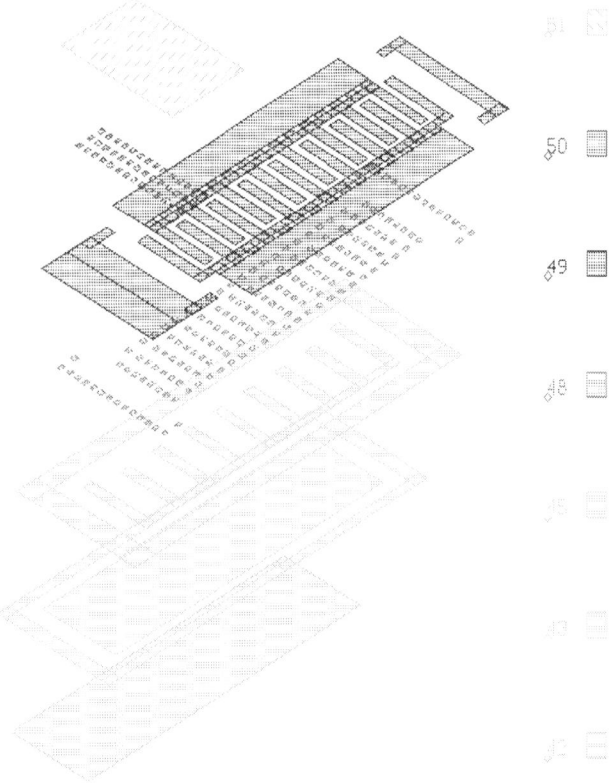

Figure 5.   3D layers view for the complete SBD structure.

Fig. 6 shows the actual silicon implementation of 5 fingers array (SBD5), 9 fingers (SBD9) and 17 fingers (SBD17).

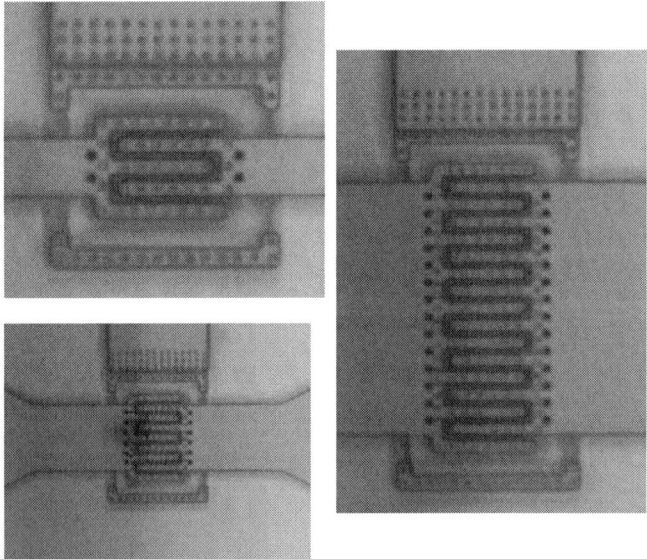

Figure 6.   Actual SBD implementation in 0.5μm technology.

Fig. 7 shows the zoom out of the 17 fingers array (SBD17) along with the dummy structures.

Figure 7.   Zoom out of the seventeen fingers SBD implementation.

IV.   THE IxV PERFORMANCE

The measured I-V curve is shown in Fig. 8 for a set of 40 samples of the nine fingers (SBD9) implementation. Fig. 9 shows the Log(I)xV measurements for the five (SBD5), nine (SBD9) and seventeen (SBD17) finger implementations. As expected, the larger number of fingers corresponds to a larger current capacity.

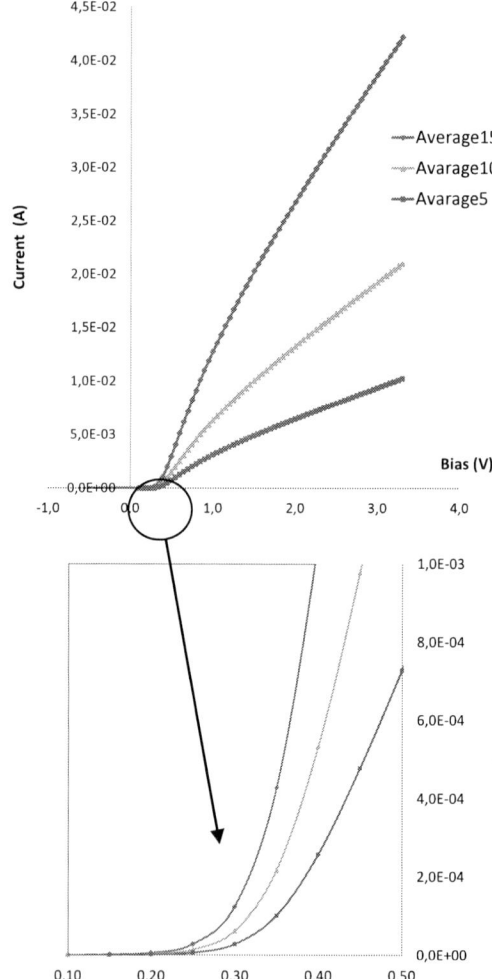

Figure 8.   IxV measurements for the three SBD topologies.

978-1-4673-5289-5/12 $31.00 © 2012 IEEE

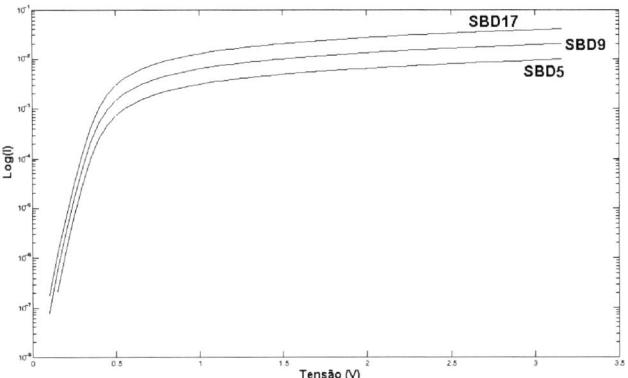

Figure 9. Log(I)xV measurements for the three SBD topologies.

By using the equations (item II) and the measured data from a set of 40 samples, the SBD parameters can be compiled as in Table 1.

TABLE I.   SBD PARAMETERS

	Area (um²)	Rs (Ohms)	Is (uA)	ΦB (eV)	η
SBD5	431	208	0.04	0.152	0.063
SBD9	753.25	117	0.7	0.101	0.13
SBD17	1.408	75	6	0.053	0.63

V.   SBD SMALL-SIGNAL MODEL

The most important difference between Schottky and PN structures is the lack junction capacitance that eliminates the electron recovery time. The equivalent small signal circuit model of the SBD, as shown in the Fig. 10.

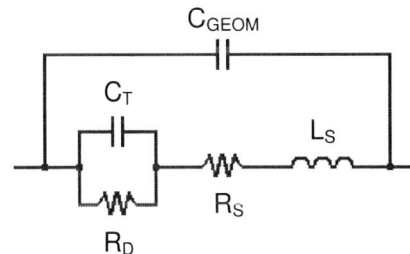

Figure 10. SBD small-signal equivalent circuit.

In SBD small-signal equivalent circuit there is the capacitance (CGEOM) that arises from the device geometry. It can range between 0.1 and 1pF [6,7].

The contact resistance R_D and capacitance C_T both come from the depletion region, and C_T can be calculated as:

$$C_T = A \left[\frac{qN_D\varepsilon_s}{2(V_{bi} - V)} \right]^{1/2} \quad (4)$$

where q is the electric charge and the V is the voltage applied across the SBD.

Finishing the model, the series resistance R_S represents the depletion region, the contacts resistance and the neutral region of the semiconductor. The parasitic impedance L_S influences the device operation at high frequencies [8,9]. Table 2 summarizes the SBD small signal parameters measured data.

TABLE II.   SBD SMALL-SIGNAL PARAMETERS.

	C_T (F)	R_D (Ohms)	R_C (Ohms)
SBD5	$4*10^{-10}$	208	1.53
SBD9	$8*10^{-10}$	117	0.81
SBD17	$1.2*10^{-9}$	75	0.41

The presented data arises from the average of 40 samples. Those parameters will be used in future work to optimize the final geometry of the SBD.

VI.   CONCLUSIONS

In this work, a diffusion of Schottky Barrier Diode in CMOS standard process is presented and discussed. This component is very attractive for RFID applications such as biomedical tags, where efficiency and power consumption are important boundary conditions to avoid patient tissue damage by overheating.

ACKNOWLEDGMENT

The authors acknowledge CAPES, CNPq and FAPEMIG for their financial support.

REFERENCES

[1] Brandl, M. et all, "Low-Cost Wireless Transponder System for Industrial and Biomedical Applications" Information, Communications and Signal Processing, 2005 Fifth International Conference on, 06-09 Dec. 2005 Page(s): 1444-1447.

[2] Pradier, A. et all, "Rigorous Evaluation of Specific Absorption Rate (SAR) Induced in a Multilayer Biological Structure" Wireless Technology, 2005. The European Conference on, 3-4 Oct. 2005 Page(s): 197-200.

[3] Janam Ku and Seonghearn Lee, "Novel SPICE Macro Modeling for an Integrated Si Schottky Barrier Diode", EGAAS 2005.

[4] Rainee N. Simons and Philip G. Neudeck, "Intermodulation-Distortion Performance of Silicon–Carbide Schottky-Barrier RF Mixer Diodes", Microwave Theory and Techniques: Jorunals, IEEE Transactions Volume 51, ISSue 2, February 2003 Page(s): 669 – 672.

[5] Rhoderick E H. Metal-semiconductor contacts. Second Edition. Oxford University Press, 1988.

[6] A. Pintar, J. Razinger, "Metalization of power schottky diodes", Vacuum, vol. 40, nº. 1/2, pp. 205-207, 1990.

[7] S. V. Averin, "Fast-Response Photo-detectors with a Large Active Area, Based on Schottky-Barrier Semiconductor Structure", Kvantovaya Electronika, vol. 23(3), 284, (1996).

[8] Streetman, "Solid State Eletronic Devices", Vol.2, pp. 185-190 (1980).

[9] Pascal Philippe, Walid El-Kamali and Vlad Pauker, "Physical Equivalent Circuit Model for Planar Schottky Varactor Diode", Microwave Theory and Techniques: Jorunals, IEEE Transactions Volume 36, Issue 2, August 2002 Page(s): 250 - 255.

978-1-4673-5289-5/12 $31.00 © 2012 IEEE

Comparative Study of Current Mode and Voltage Mode Sense Amplifier used for 28nm SRAM

Baker Mohammad[1,2], Percy Dadabhoy[2,], Ken Lin[2], Paul Bassett[2]

[1]Khalifa University of Science, Technology and Research, Abu Dhabi UAE

[2]Qualcomm Incorporated, Austin Tx, USA

Abstract—Increased process variation and reduced operating voltage present two of the main challenges in using sense amplifiers for small geometry bulk CMOS process technology. This fact coupled with the need to increase on-chip memory to reduce traffic on the bus and increase performance creates the need for a robust and reliable sense amplifier - a differentiating factor in memory area, power, and speed. We present a detailed study of the Voltage Latched Sense Amplifier (VLSA) and the Current Latched Sense Amplifier (CLSA) design in 28nm industry standard process technology [1][2]. We present results on how the two sense amplifier behave for the two design topology for low power (LP) process technology optimized for mobile low leakage application and the second one is high performance High Performance (HP) applications. Detailed Spice simulation with statistical models and Monte Carlo simulations is utilized to compare the two designs for active power, leakage power, speed, and area. Our study shows that VLSA performs better than CLSA - being 67% faster, 35% smaller area, and similar active power for the LP. The VLSA also performed better than the CLSA is the HP technology as well. The sensitivity to temperature for the LP technology node was more at high voltage but for the HP process the lower voltage perform worse.

Index Terms—Cache memory, sense amplifier, small signal array, SRAM Array

I. INTRODUCTION

EMDEDDED memory forms an integral part of the design of today's processors and SOC. The main purpose of embedded memory in general and cache in particular is to improve the instruction and data access time. Larger caches are favored since they can potentially reduce the number of time-expensive external memory accesses. The propagation delay of each cache datapath element directly contributes to the total access time. A reduction in the propagation delay of the datapath can aid in building larger caches, as large caches may have several levels of multiplexing.

One of the elements of the datapath in an SRAM design is

Manuscript received February 14, 2012. This work was supported by Qualcomm Inc.

Baker Mohammad is Assistant Professor in ECE at Khalifa University of Science, Technology, and Research, Abu Dhabi, UAE (e-mail: baker.mohammad@kustar.ac.ae)

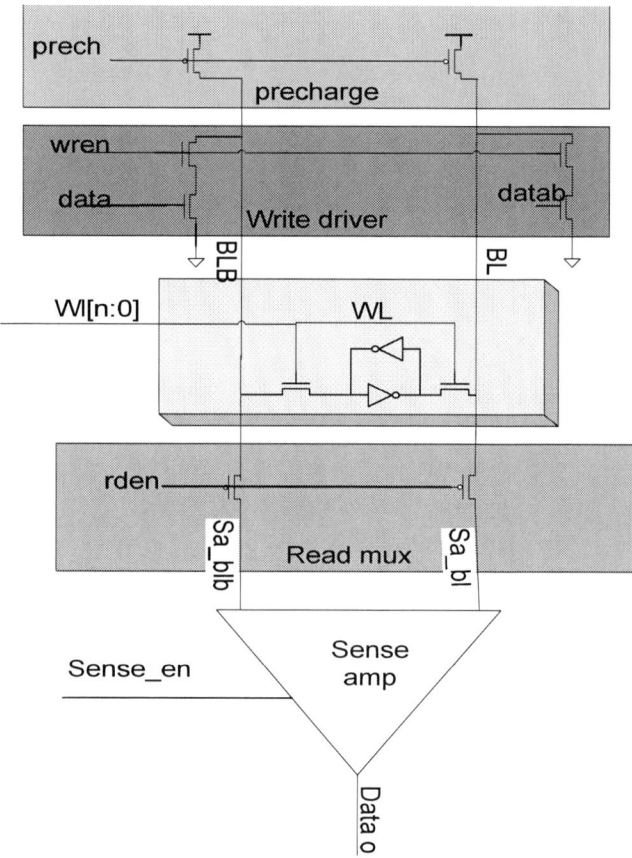

Figure 1. Basic schematic of one column for small signal memory array using 6T cell

the sense amplifier. The sense amplifier activated during a read operation is used to sense the votlage differential on the bitlines at its input, and generate a full-rail voltage swing at its output (figure 1). The reduced swing on the memory bitlines improves performance and also reduces power consumption [3].

In this paper we explore the design robustness of sense amplifiers for low voltage operation by comparing two sense amplifiers designs in two variation of 28nm process technology one is optimized for mobile application (LP) and the other is targeted for high performance circuits (HPM) [1][2]. We compare the two designs based on various factors such as power, delay, sensitivity and area. The two sense

978-1-4673-5289-5/12 $31.00 © 2012 IEEE

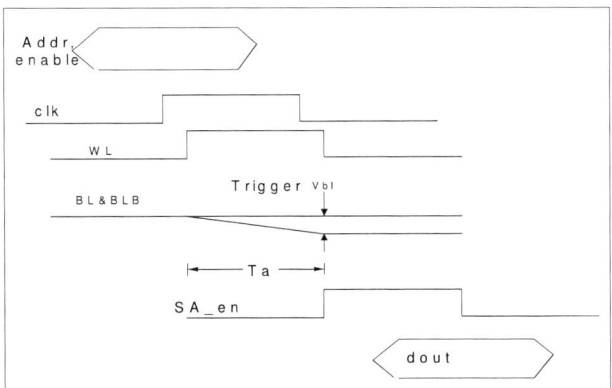

Figure 2 Timing relationship between main control signals of a 6T memory array

amplifier topologies compared are the current latched sense amplifier (CLSA), and the voltage latched sense amplifier (VLSA). The methods used for the measurement of the different comparison metrics is explained, along with a presentation of the results.

The paper is organized as follows: Section II presents an overview of the design principles of a small signal array; Section III briefly explains the topology and operation of the sense amplifiers; Section IV explains the measurement of the different comparison metrics, and Section V concludes the paper.

II. BASIC DESIGN PRINCIPLE OF SMALL SIGNAL ARRAY

Six transistor cell (6T) based memory is widely used for embedded memory due to its small area [3] and relatively fast access time. One important design parameter for memory especially when using 6T, is area utilization, which compares the actual memory cell area to the total area of the memory

which includes the memory cell area, all input, output, wordline drivers, and any other control or multiplexing circuitry. Consider a 32KB memory design where the single 6T memory cell area is $0.2um^2$, then the total cell area is

$$Total\ Cell\ Area = 0.2 * 32 * 1024 * 8 = 52428.8\ um^2$$

If the complete design of the 32KB area is 150000 um² then the utilization is 33%. Typical values for level 1 (L1) caches are 25-40% and for L2 is 60-70%. The reason for this difference between L1 and L2 is that L1 is built for performance with smaller capacity, while for L2 the emphasis is on density and less on speed. Using a sense amplifier provides the ability to have many cells in the same column (figure 1) and hence sharing of the pre-charge, write driver, and sense amplifier circuits which reduces the overhead. The bitline then does not need a full voltage swing when using a sense amplifier to transfer the data value from the cell to the output but rather a small differential (0.2*Vdd), which provides low power during read access.

The main timing signals and its relative trigger time are shown in Figure 2. The access is normally triggered from a clock edge which de-asserts the pre-charge (off) and asserts WL signal (turn it on) to access the cell. The self time Sa_en signal gets asserted through a tracking circuitry after enough voltage difference gets created between BL and BLB. The Sa_en signal enables the sense amplifier to sense the difference in voltage which is based on the data stored in the 6T. The sense amplifier also stores the data value in the latch to make it available for downstream logic. The time from WL to Sa_en (Ta) is the array access time and is programmable through a tracking circuit. This programming enables a tradeoff between timing, power and yield.

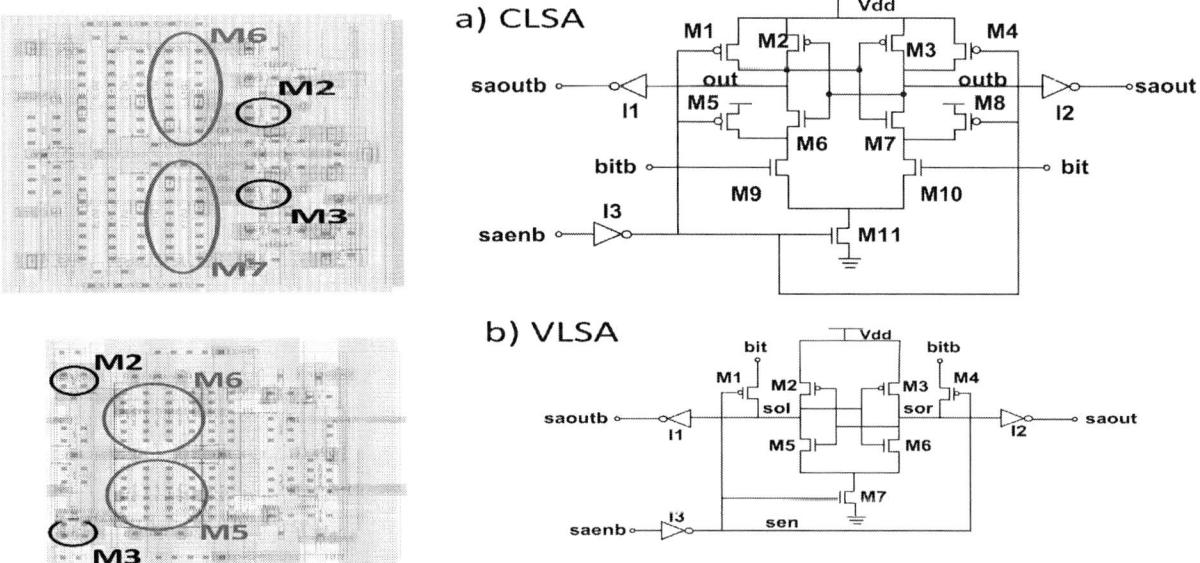

Figure 3: a) CLSA sense amplifier schematic and layout showing matching transistors b) VLSA sense amplifier schematic and layout

III.   CLSA & VLSA SENSE AMPLIFIERS

The CLSA & VLSA sense amplifier topologies are shown in figure 3a and b respectively. Both sense amplifiers have the 'bit' and 'bitb' inputs connected to the column bitlines of the SRAM. The sense input is actuated once the requisite voltage differential has developed on the bitlines. Each sense amplifier has the cross-coupled inverters that convert the voltage differential at their inputs on the bitline to a full swing at the outputs. The output inverters in each sense amplifier are used for driving the downstream logic, and also serve to isolate the internal nodes of the sense amplifier from the external load.

Figure 3a shows the CLSA sense amplifier design. Since this is a current latched design, the bitlines drive the gates of transistors M9 and M10. Transistors M1, M4, M5 and M8 are the precharge transistors. Transistors M2,M6 and M3,M7 form the inverter pair that resolves the bitline differential voltage. Traditionally this topology has been used because the memory bitlines are driving high impedance (gate) and full discharge of array bitline due to timing mismatch is not a concern.

Figure 3b shows the VLSA sense amplifier design (schematic and layout in the 28nm design rules) . M2-M5 and M3-M6 form the inverters that resolve the differential voltage on the bitlines to a full-swing at the output. The internal nodes of this design are precharged through the bitlines. The obvious advantage of this topology over the CLSA is the lower number of transistors needed which means faster access and smaller footprint. The challenge in using this topology has been the race condition for isolation signal that decouples the sense amplifier bitline (sol, sor) from the array bitline (bit, bitb). If the sense amplifier is enabled while M1 and M4 are on, the memory bitline (bit or bitb) could be discharged to logic 0. In traditional designs, a different signal other than sen (isolate) is used to control M1 and M4 which makes it hard to match sense and isolate operation, but for our design we used the same signal that enable the sense amplifier to isolate the array bitlines.

Operation

1) CLSA:

The operation of this sense amplifier design is based on the current differential produced by M9 and M10 in the two pull down branches of the sense amplifier. At the commencement of the read operation, either one of the 'bit' or 'bitb' inputs is lowered depending on the data stored in the cell. When the sense amplifier is triggered by the lowering of the input 'saenb', M11 turns on and the precharge transistors are simultaneously turned off. Since the gate voltage of M9 and M10 differs by the generated bitline differential, their channel currents are unequal. The current thus flowing in the two branches of the sense amplifier is unequal and the voltage at either 'out' or 'outb' falls faster than the other node. This difference in voltage is resolved by the cross-coupled inverters formed of M2,M6 and M3,M7.

2) VLSA:

This design operates directly based on the voltage differential developed on its internal nodes by the input bitlines. When the wordline is turned on and prior to the

TABLE I: LP TECHNOLOGY SENSE AMPLIFIER METRIC COMPARISON

Metrics	VLSA/CLSA
Area	0.65x
Bitline input capacitance	3.2x
Sense enable capacitance	1.5x

TABLE II: DELAY AND LEAKAGE COMPARESON FOR CLSA AND VLSA FOR BOTH LP AND HPM

delay	CLSA	VLSA
LP	3.03	1
HPM	1.56	0.63

Leakage	CLSA	VLSA
LP	0.28	1
HPM	3.36	1.79

triggering of the sense amplifier, transistor M7 is off and pass transistors M1 and M4 are on. As the differential develops on the bitlines, it does so too on the internal nodes of the sense amplifier 'sol' and 'sor'. When the sense signal 'saenb' is asserted, the cross-coupled inverters formed of M2-M5 and M3-M6 amplifies this differential voltage to its full-swing output.

IV.   COMPARISON METRICS

Simulations are performed on the two sense amplifier designs using full post layout netlist and same design flow used to qualify production level design.   Both design have same stimuli and drive same load for comparison. The metrics chosen for comparison help to describe all aspects of operation of the sense amplifier that are useful from a design perspective. Table I shows the comparative results of these simulations

A. Speed of Operation

Spice simulation with fully extracted post layout netlist was performed to compare the delay of both sense amplifiers. The same waveform input for both bitlines and sense enable was used. The speed of operation is determined by measuring the 50% delay that it takes for the internal nodes of the sense amplifier, once the sense amplifier is triggered.   In LP technology, the VLSA is 67% faster than the CLSA.

B. Area

Figure 3 (a) and (b) show the layouts for the CLSA and VLSA designs respectively. The transistor area for the VLS design is 35% less as compared to the CLSA design. The transistors that form the inverter pair in each sense-amp are also shown in the figure. The NMOS transistors are highlighted using red, and the PMOS transistors are shown circled in black

C. Leakage

Contemporary aggressive low power designs continuously

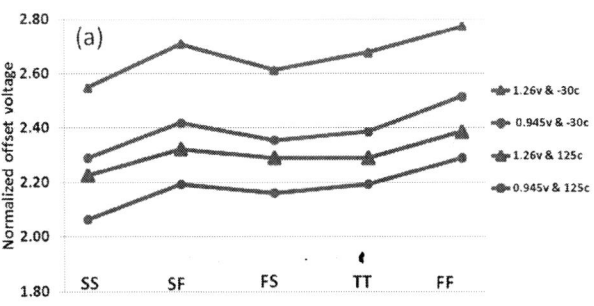

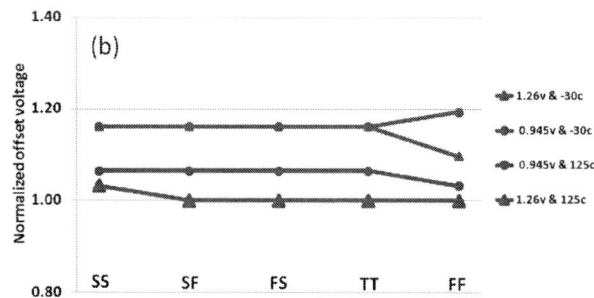

Figure 4: Normalized required offset voltage for LP process a) CLSA and b) VLSA sense amplifiers across process corners, voltages and temperatures (VLSA requires half the offset compared to CLSA at the same PVT corner)

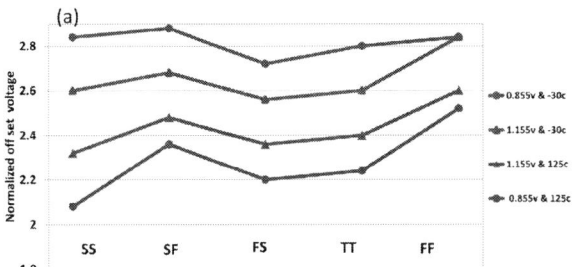

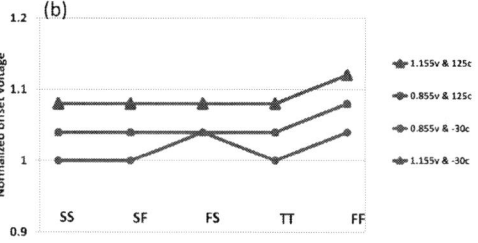

Figure 5: Normalized required offset voltage for HPM process a) CLSA and b) VLSA sense amplifiers across process corners, voltages and temperatures (VLSA requires half the offset compared to CLSA at the same PVT corner)

demand increasing performance within constant or smaller power budgets. With advancing process nodes usually characterized by higher leakage, keeping leakage power to an acceptable portion of the total power budget is a perpetual challenge.

High accuracy simulations are performed using Spice to obtain the leakage of the sense amplifiers. The simulations are performed for a long duration in order to ensure settling time and high accuracy results. The leakage profile of the sense amp is also analyzed to identify transistor channel length and threshold voltage types that can be modified to improve the leakage power consumption of the design. These changes must be made with minimal impact on the operation speed of the sense amplifier. Table 2 lists the leakage ratio for the CLSA and VLSA and shows that CLSA is less leakage due to stacking effect compare to the VLSA. As expected HPM exhibit higher leakage than the LP process.

D. Dynamic Energy

The energy consumed by the sense amplifier in operation is determined by measuring the work done by the supply voltage source. In Spice, this can be accomplished by measuring the charge that is supplied by a voltage source during the operation interval of the sense amplifier. The operation interval commences once the sense amplifier is triggered, and until the internal node of the sense-amp changes state to store the new data present at its input. Spice was used to measure the charge supplied by the voltage source during this operation interval, and the corresponding energy is then computed mathematically.

E. Input Capacitance

As mentioned earlier, the sense amplifier is one element of the SRAM datapath and hence its compatibility with upstream blocks is necessary. The input capacitance of the sense amplifier is measured using a Spice AC analysis as well as through a transient simulation

1) AC analysis: In this method a small amplitude AC source is applied to the input of the sense amplifier and Spice is used to measure the imaginary part of the current flowing through the source. This is then used to mathematically compute the impedance of the sense amplifier inputs.

2) Transient simulation: A small amplitude voltage differential is applied to the inputs of the sense amplifier, and the charge transferred from the source is measured. The measured magnitude of the charge transferred is used to mathematically determine the input capacitance of the sense amplifier.

F. Sensitivity The function of the sense amplifier is to resolve the voltage differential applied at its inputs to a full swing at the output. This important characteristic of the sense amplifier determines the minimum input voltage differential between sa_BL and sa_BLB (as shown in figure 1.) required for ensuring reliable and proper operation of the sense amplifier This minimum voltage will be referred to as offset voltage (Vdiff). During a read operation, one cell of each column in the array is turned on and it discharges a bitline dependent on the data stored in the cell. The bitline voltage differential thus developed is applied to the input of the sense amplifier. A sense amplifier with low sensitivity will require a higher voltage differential. For large cache

978-1-4673-5289-5/12 $31.00 © 2012 IEEE   353

designs that have long bitlines with many bitcells, this results in higher power dissipation and longer access time. The sensitivity of the sense amplifier thus plays an important role in determining the dynamic power dissipation of the SRAM design While it is important to use a lower bitline differential to improve power characteristics, it should be sufficiently high to ensure reliable operation. Monte Carlo simulations with full transistor spice mismatch models are performed to determine the sensitivity in the presence of process variation. A 1000 sample Monte Carlo simulation is performed for different process corners. The industry standard five corners (SS, SF, FS, TT, FF) is used. The two letters indicate the speed of NMOS and PMOS respectively for example SS referes to slow nMOS and slow PMOS while SF indicate slow NMOS and fast PMOS. The number of functional failures for a range of different bitline differential voltages (Vdiff) is observed to determine the sensitivity. We determined the required minimum voltage offset (Vdiff) to be the one where we achieve zero error for the entire 1000 samples. We simulated both design topologies on the two technologies LP and HPM. We show that VLSA design is more robust and less sensitive to process and temperature variation. Figures 4 & 5 show the variation in the minimum input offset voltage for sense-amp operation for the CLSA and VLSA designs. Note that the Y-axis has the normalized offset voltage for the CLSA & VLSA designs. We used Monte Carlo simulations with full

mismatch models at each process corner to determine the input offset voltage performed for four different combinations of operating voltages and temperatures as indicated in the figure, each line in the plot is for one particular combination of voltage and temperature. Simulations are done across different process corners (SS, SF, FS, TT, &FF)[4]–[5] as also indicated in the figures. The CLSA design shows large variation in the input offset voltage across PVT corners as compared to the VLSA design. Lower temperature exerts the worst performance with the highest voltage offset required for both cases with CLSA topology require almost twice the voltage offset compared to VLSA at the same PVT corner.

G. Technology advantage

Advanced technology nodes enable higher transistor density on the chip which allows the implementation of more functions on the chip. Smaller transistors also operate faster which enables higher operation frequencies. Simulations show that the HPM technology CLSA sense-amp is 50% faster than the design in LP technology. Similarly, the VLSA sense-ampis 37% faster. Monte Carlo simulations show that the advanced process technology node also improves the minimum required offset voltage of the sense amplifier. Lower offset voltage requirements again translate into higher operation speed benefits. Figure 6 shows the ratio by which the offset voltage of each sense amplifier topology in the LP technology is

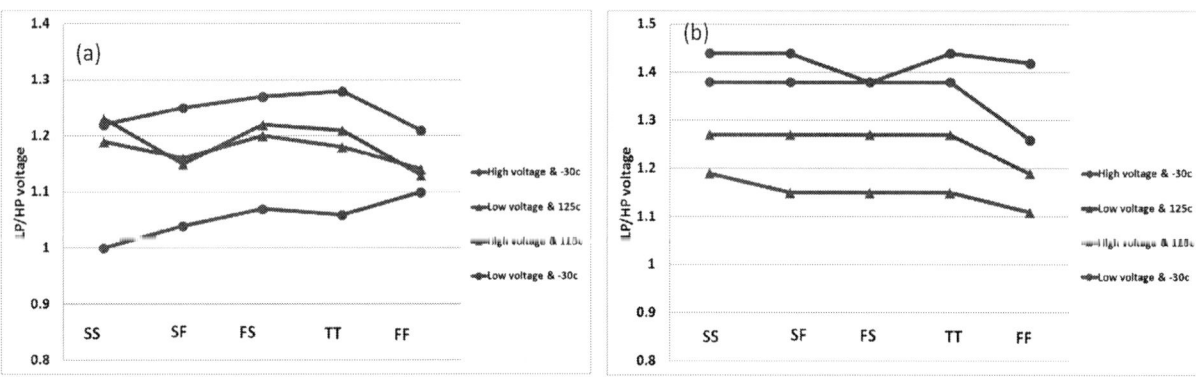

Figure 6. Increase in offset voltage for LP tech sense amp compared to HPM (a) CLSA (b) VLSA

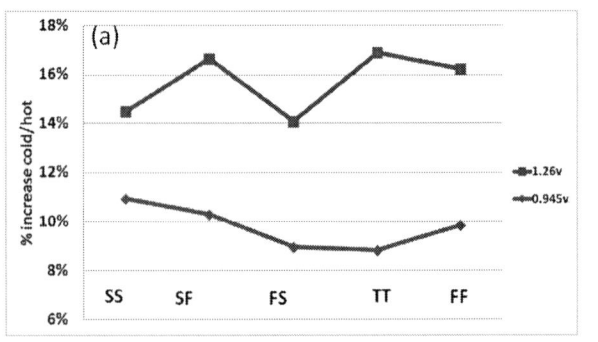
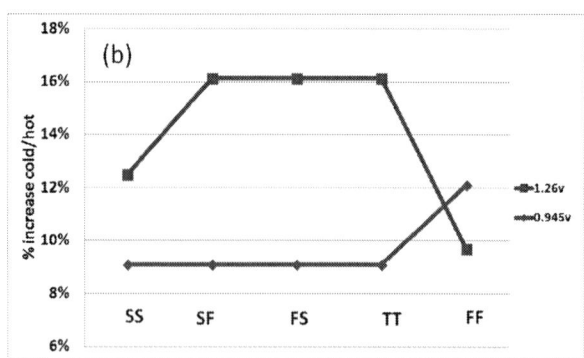

Figure 7. Increase in offset voltage for cold vs. hot temperature (a) LP CLSA (b) LP VLSA

978-1-4673-5289-5/12 $31.00 © 2012 IEEE   354

greater than that in the HPM technology. Each plot in the figure is for a particular combination of supply voltage and temperature, across the five different process corners. As can be seen, the required offset voltage for the CLSA topology is upto 30% lower in the HPM technology as compared to that in the LP technology. Similarly, the required offset voltage is about 45% greater in the LP technology for the VLSA design as compared to the same design in HPM technology.

H. Temperature Sensitivity

Increased chip transistor density in advanced process technology nodes also leads to higher chip operating temperatures. It is thus pertinent to know the impact that temperature has on the required minimum offset voltage. Figures 7 shows the extent to which the offset voltage increases at cold as compared to hot temperatures for the LP technology CLSA and VLSA designs. As can be seen, at higher voltages of 1.26v, the required offset voltage is higher by 15%-20% for the CLSA design. In the HPM technology, this trend is reversed with higher increases for the offset voltage seen for lower values of operating voltage (see Fig. 8) – at low voltages, the CLSA design offset voltage is higher by more than 35% at cold temperatures as compared to hot. There is not much variation in required offset voltage with temperature for the VLSA design in HPM technology.

V. SUMMARY & CONCLUSION

This publication describes the simulation methodology used to compare the two VLSA and CLSA sense amplifier designs. The simulations results clearly show the advantage of the VLSA design over the CLSA design. The faster speed of operation and lower input differential required by the VLSA design makes it an ideal choice for high speed, low power datapath design. Traditional design complexity arise from using VLSA has also been addressed by using one signal to enable the sense amplifier and to isolate the array bitline from sense amp bitline (as shown in figure 3b).

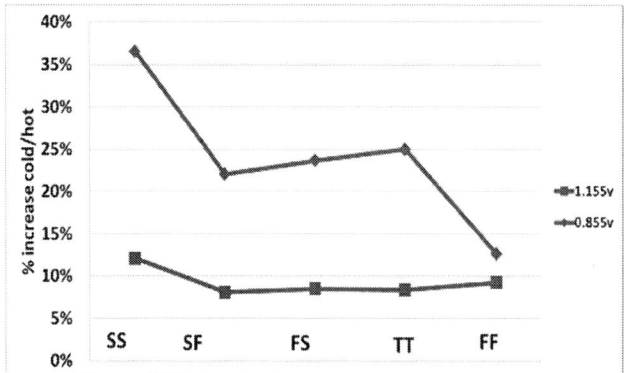

Figure 6: Percentage increase of required offset voltage for -30C compare to same 125C PV 125C (hot) Temperature sensitivity

REFERENCES

[1] J. Yuan et al., "Performance Elements for 28nm Gate Length Bulk Devices with Gate First High-k Metal Gate", Solid-State and Integrated Circuit Technology, 2010, 10th IEEE International Conference on, pp. 66–69.

[2] Wu et al., "A Highly Manufacturable 28nm CMOS Low Power Platform Technology withFully Functional 64Mb SRAM Using Dual/Tripe Gate Oxide Process", VLSI Technology, 2009 Symposium on, pp. 210–211.

[3] N. Weste and D. Harris. CMOS VLSI Design: A Circuits and Systems Perspective. Addison-Wesley, 2005

[4] International Technology Roadmap for Semiconductor (ITRS) itrs.net

[5] Baker Mohammad, Martin Saint-Laurent, Paul Bassett, and Jacob Abraham. Cache Design for Low Power and High Yield, IEEE International Symposium on Quality Electronic Design (ISQED) ,March 2008, pp 103-107, San Jose, CA, USA

[6] Baker Mohammad, Jacob Abraham; A reduced Voltage Swing Circuit Using A single Supply to Enable Lower Voltage Operation for SRAM-based Memory; Microelectronics journal, Elsevier, December 2011

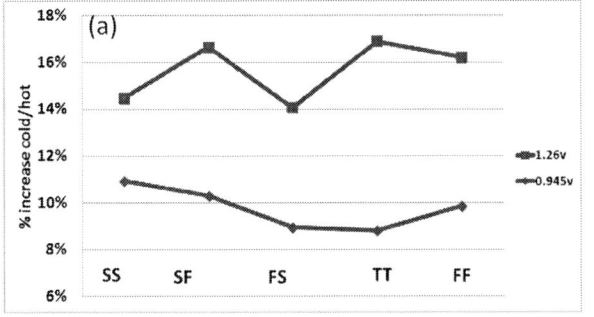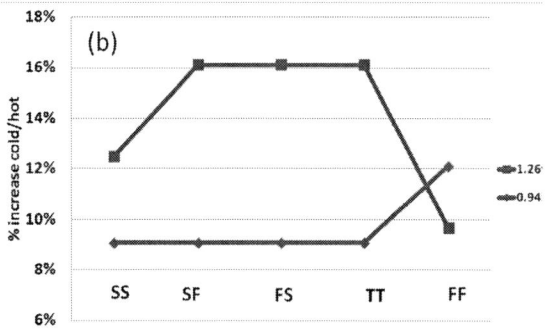

Figure 8: Increase in offset voltage for cold vs. hot temperature for LP technology (a) CLSA design (b) VLSI

A Compact Analytic Expression of the Oscillation Amplitude in MOS LC-Oscillators

Bassem Fahs[1], Adnan Harb[1], *Senior Member IEEE*, Hassan M. Bazzi[1], *Member IEEE* and Mohammad M. Mansour[2], *Senior Member IEEE*

[1] Lebanese International University (LIU), Electrical and Electronics Engineering Department, 146404 Mazraa, Beirut, Lebanon.
[2] American University of Beirut (AUB), ECE Department, Beirut, Lebanon.
bassem.fahs@liu.edu.lb

Abstract—This paper presents a compact analytic expression for the oscillation amplitude in source-tailed MOS LC-VCOs versus the oscillator bias current and LC-tank resistive losses. The developed equation is based on the use of analog saturation functions to provide a continuous analysis among the oscillator's current-limited and voltage-limited regimes, which takes also into account the oscillation start-up condition. The predictions are in good agreement with Spectre simulation results from a 0.25-μm technology LC-VCO circuit. The derived equation aims to simplify feed-forward phase noise optimization techniques by providing a simple expression relating the oscillation amplitude to the other oscillator parameters.

Index Terms—Amplitude control, MOS, LC-VCOs, phase noise.

I. Introduction

Automatic Amplitude Control (AAC) loops are often employed in modern LC-VCOs to optimize phase noise versus amplitude variations [1], [2]. Traditional AAC loops (Fig. 1a) are based on adjusting the bias current (I_0) by measuring the oscillation amplitude (A) and comparing it to a reference voltage (V_{ref}). This technique presents the drawbacks of loading the oscillator, adding more parasitic capacitance and generating additional noise. Equation-based phase-noise optimization (Fig. 1b) consists of predicting the bias current (I_0) according to the optimum oscillation amplitude regarding phase noise. This technique allows to overcome the main drawbacks of the traditional AAC technique, but, it requires at the same time an accurate knowledge of the oscillation amplitude (A) evolution versus the bias current (I_0) and some other parameters, like $\beta = \mu_0.C_{ox}.W/L$, and the equivalent parallel losses of the tank Rp. In [3], an analytic expression of the amplitude (A) has been proposed. The derived solution is shown to be in good agreement with the simulation results. However, the amplitude expression remains too complex either for hand-calculations or for direct hardware implementation.

This work proposes a simplified and compact analytic expression of the oscillation amplitude (A) versus the bias current (I_0), β, and the parallel losses (Rp). This equation is more adapted for hand-calculations and simplifies any attempt for circuit implementation.

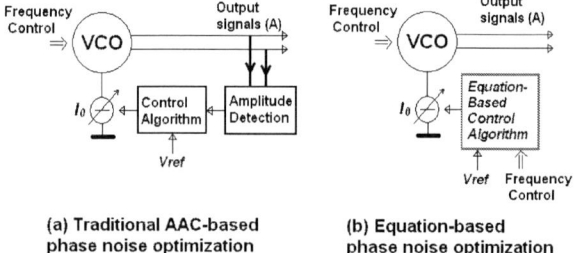

Fig. 1. (a) Traditional AAC-based phase noise optimization, and (b) proposed equation-based phase noise optimization.

II. Previuos Work In The Litterature

Previous attempts in the literature to determine the amplitude (A) versus I_0 and Rp ended usually whether with simple equations based on basic assumptions [1], [2], or a complex set of equations requiring a mathematical computer-aided resolution [4], [5]. Basic assumptions tend to separate the oscillator operation into two regimes [1], [2]: a current-limited regime and a voltage-limited regime where A is limited by the supply voltage. Although these limitations are physical, the real carrier amplitude versus I_0 behaves more like a logarithm or square-root than a linear trend which may be only asymptotically valid far away from the intersection location between the two regions, where accurate oscillator bias is generally sought out to achieve optimum phase-noise performance (Fig. 2). The second observation is that depending on Rp, an offset occurs on the bias current I_0 to get oscillation start-up. This is clearly not predictable by the linear amplitude relationship (Fig. 2).

In [3], an analytic expression of the oscillation amplitude (A) is derived based on the use of continuous saturation functions and a mathematical function approximation. The final expression describes fairly the amplitude evolution versus I_0 and the oscillator parameters (β, Rp). However, this equation is quite complex and can hardly be used by designers to predict with hand-calculations the amplitude behavior. A compact analytic expression would provide more flexibility from modeling and implementation point of views.

978-1-4673-5289-5/12 $31.00 © 2012 IEEE

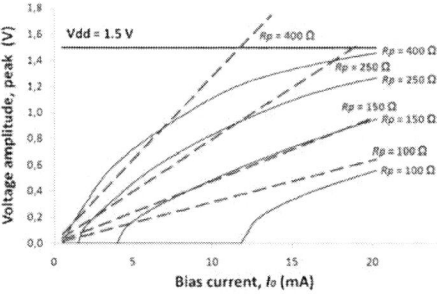

Fig. 2. Simulated (solid lines) output oscillation voltage amplitude (single-ended) with the NMOS-core oscillator given in Fig. 3 ($Vdd = 1.5$ V, $\beta = 39$ mA/V^2), and calculated (dashed lines) amplitude with $A = (1/2).2/\pi.Rp.I_0$, versus bias current (I_0) and different $Rp = 100$, 150, 250, and 400 Ω values.

III. OSCILLATION AMPLITUDE EQUATION

The simulated schematic used to support our analysis is shown in Fig. 3. L and C form the resonating tank, and the cross-coupled NMOS (M_1, M_2) is the active-core providing the negative resistance to compensate for the losses Rp. Note that the following analysis is also applicable for a PMOS cross-coupled pair architecture with appropriate sign updates. Through this analysis, we will assume that the tail current source is ideal and its output resistance is infinite. We also suppose that the back-gate effect, causing the threshold voltage V_t to be dependent on the source voltage, can be neglected ($V_t \approx V_{t0}$). Furthermore, we assume that the current full switching from a transistor to another in the active pair (M_1, M_2) occurs before that any of the transistors goes into triode region [4]. Thus, the only modes to be considered in our analysis will be the saturation and cut-off regimes. Finally, we assume that the output voltage signal (Vid) is purely sinusoidal. As represented in Fig. 3, the DC current component is driven by the inductor L, the fundamental component by Rp, and the higher harmonics are mostly driven by the capacitor C.

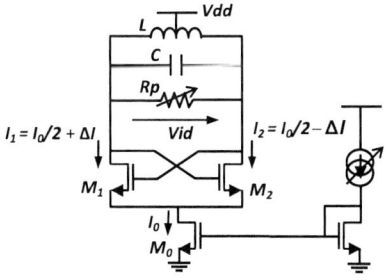

Fig. 3. Simulated NMOS oscillator schematic.

A. System Equations

According to the long-channel MOS model, the drain current in the saturation region can be expressed as $I_D = \beta/2.(V_{gs} - V_t)^2$, with $\beta = \mu_0.C_{ox}.W/L$, where μ_0 is carrier mobility, C_{ox} is the gate oxide capacitance per unit area, W is the gate width, L is the gate length, and V_{gs} is the gate-source voltage. Referring to Fig. 3, the currents in M_1, M_2 can respectively be written as

$$I_1(t) = I_0/2 + \Delta I(t) \tag{1}$$

$$I_2(t) = I_0/2 - \Delta I(t) \tag{2}$$

$$I_1(t) + I_2(t) = I_0 \tag{3}$$

$\Delta I(t)$ is the switched current in the differential pair (M_1, M_2) which can be expressed as [3]

$$\Delta I = \beta/2.V_{id}.V_c \tag{4}$$

where $V_{id} = 2.A.\cos(\omega_0 t)$ is the differential output oscillation signal, ω_0 is the angular frequency, and $V_c = (V_{gs1} + V_{gs2} - 2.V_t)/2$ is a common-mode voltage corrected by V_t, such that [3]

$$V_c = V_0\sqrt{1 - \frac{1}{4}.\frac{V_{id}^2}{V_0^2}} \tag{5}$$

$V_0 = V_{gs0} - V_t$, and V_{gs0} is the differential pair gate-source voltage at the equilibrium point.

The steady-state oscillation amplitude across the tank (A) can be obtained from

$$2.A = Rp.\Delta I_{\omega_0} \tag{6}$$

where ΔI_{ω_0} is the fundamental current component flowing into Rp at ω_0, which can be calculated as the first Fourier coefficient:

$$\Delta I_{\omega_0} = \frac{2}{T}\int_0^T \Delta I(t).\cos(\omega_0 t).dt \tag{7}$$

By applying the continuous saturation function method introduced in [3], we can substitute V_{id} and V_c by their saturated functions V_{is} and V_{cs}, such that

$$V_{is} = \sqrt{2}.V_0.\sin(\arctan(\frac{V_{id}}{\sqrt{2}.V_0})) = \frac{V_{id}}{\sqrt{1 + \frac{V_{id}^2}{2.V_0^2}}} \tag{8}$$

$$V_{cs} = V_0\sqrt{1 - \frac{1}{4}.\frac{V_{is}^2}{V_0^2}} = V_0.\sqrt{\frac{1 + V_{id}^2/4.V_0^2}{1 + V_{id}^2/2.V_0^2}} \tag{9}$$

ΔI becomes therefore

$$\Delta I = \beta/2.V_{is}.V_{cs}. \tag{10}$$

Rewriting (7) with (10), we get

$$\Delta I_{\omega_0} = \frac{\beta}{2\pi}\int_0^{2\pi} V_{is}(\theta).V_{cs}(\theta).\cos(\theta).d\theta$$

$$= \frac{2\beta}{\pi}V_0^2\alpha\int_0^{\pi/2}\frac{\cos^2(\theta).\sqrt{1 + \frac{\alpha^2.\cos^2(\theta)}{4}}}{1 + \frac{\alpha^2.\cos^2(\theta)}{2}}.d\theta$$

$$= \frac{2\beta}{\pi}V_0^2\alpha\int_0^{\pi/2} F(\alpha, \theta).d\theta \tag{11}$$

with $\theta = \omega_0 t$, and $\alpha = 2.A/V_0$. The analytic resolution of the integral in (11) provides a complex and unexploitable solution. In [3], an approximation function $P(\alpha,\theta)$ of $F(\alpha,\theta)$ was used to ease the integration of (11). $P(\alpha,\theta)$ was proposed as a polynomial function of $\theta \in [0, \pi/2]$ with α-dependent coefficients. In this work, we propose a different form for $P(\alpha,\theta)$ based on the following expression:

$$P(\alpha,\theta) = a(\alpha).sin^2(p\theta) + b(\alpha).sin^2(q\theta) + c(\alpha) \quad (12)$$

with, $\qquad p = -0.478, \ q = -0.15$

$$c(\alpha) = P(\alpha,\theta=0) = F(\alpha,\theta=0) = \frac{\sqrt{1+\alpha^2/4}}{1+\alpha^2/2}$$

$$b(\alpha) = \frac{-2.377\sqrt{1+\alpha^2/4} + 0.877)}{1+\alpha^2/2}$$

and, $\qquad a(\alpha) = \frac{1.275 - 4.98\sqrt{1+\alpha^2/4}}{1+\alpha^2/2}$

Figure 5 shows the matching between $P(\alpha,\theta)$ and $F(\alpha,\theta)$ curves for $\theta \in [0, \pi/2]$ and different α values. Note that the needed accuracy is only relative to the global integral value over the considered interval.

Hence, developing (11) with (12) leads to the following second-degree equation:

$$1 + \frac{\alpha^2}{2} = X(-1.205 + 1.703\sqrt{1+\frac{\alpha^2}{4}}) \quad (13)$$

where $X = Rp\beta V_0$.

The resolution of (13) with respect to A (with $\alpha = 2.A/V_0$) gives therefore:

$$A = \frac{V_0}{\sqrt{2}}\sqrt{(Rp\beta V_0 + 0.343)(Rp\beta V_0 - 2)} \quad (14)$$

where $V_0 = (I_0/\beta)^{1/2}$.

The amplitude A is finally obtained with a very simple expression versus I_0, Rp, and β.

Let's note that the content of the second square-root in (14) is positive only if $(Rp.\beta V_0 \geq 2)$. This can be re-written as $Rp.gm_0 \geq 2$ using $gm_0 = \beta V_0$, which is the well-known small-signal oscillator start-up condition, gm_0 being the small-signal transconductance of each transistor (M_1, M_2).

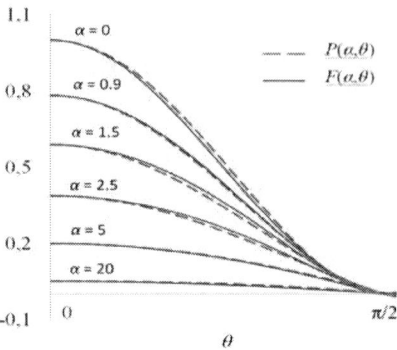

Fig. 5. Comparison between $P(\alpha,\theta)$ (solid lines) and $F(\alpha,\theta)$ (dashed lines) curves versus $\theta \in [0, \pi/2]$ for different α values.

As in [3], the large-amplitude saturation effect for levels approaching the supply voltage is modeled using the *arctan*() saturation function. This function is applied empirically on (14) using weighting coefficients. The final amplitude expression becomes thus

$$A = \frac{2}{\pi}.K(Vdd).Vdd.$$

$$arctan\{\frac{2.1}{Vdd}.\frac{V_0}{\sqrt{2}}\sqrt{(Rp\beta V_0 + 0.343)(Rp\beta V_0 - 2)}\} \quad (15)$$

where $K(Vdd)$ is a fitting coefficient with the simulation results.

This coefficient can be justified by the several neglected effects (body effect, $V_t = f(V_{source})$ ...) which can add secondary-order dependencies versus Vdd. The simulated variation range of $K(Vdd)$ is summarized in Table I.

TABLE I.   K(VDD) VERSUS VDD

Vdd (V)	1.2	1.5	2.0	2.5
K(Vdd)	1.17	1.15	1.09	1.00

B. Comparison With Simulation Results

Figures 6a, 6b, 6c and 6d provide a comparison between the predicted oscillation amplitude (A) plots from (15) and the simulated amplitude curves (in 0.25 μm CMOS) for different Rp, Vdd and NMOS-core sizes versus the oscillator dc bias current (I_0). These figures show a good fit between the plots obtained with (15) and the simulated curves over a relatively wide bias current range. The accuracy of the equation (15) starts to degrade at high Rp (> 400 Ω) and high Vdd (> 2.5V).

978-1-4673-5289-5/12 $31.00 © 2012 IEEE

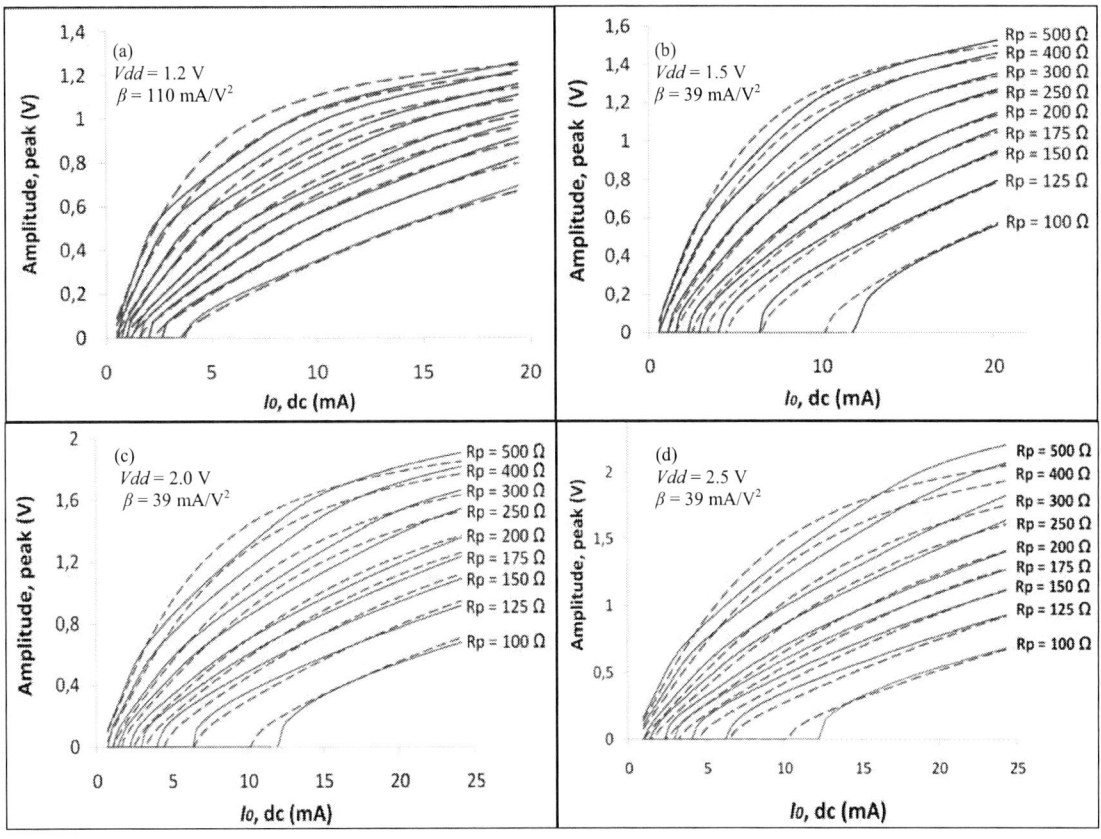

Fig. 6. Comparison between simulated amplitude curves (solid lines) and predicted curves with (15) (dashed lines) versus dc bias current (I_0) and Rp, for different Vdd values (1.2 V, 1.5 V, 2.0 V and 2.5V), and two NMOS-core sizes (β = 110 and 39 mA/V^2).

IV. CONCLUSION

In this paper, a compact and simplified analytic expression for the oscillation amplitude in a source-tailed NMOS LC-VCO has been derived. The proposed equation is based on the use of saturation functions permitting a continuous analysis over different operation regimes. Theoretical predictions have been validated with Spectre simulations. This work enables feed-forward automatic amplitude control approach to optimize phase-noise variations in wide-band LC-VCOs.

REFERENCES

[1] A. Fard, and D. Aberg, "A reconfigurable CMOS VCO with an automatic amplitude controller for multi band RF front-ends," *IEEE Proceeding of the European Conference on Circuit Theory and Design (EECTD)*, pp. 95–98, Aug. 2005.

[2] A. Berny *et al.*, "A 1.8-GHz LC VCO with 1.3-GHz tuning-range and digital amplitude calibration," *IEEE Journal of Solid-State Circuits (JSSC)*, vol. 40, no. 4, pp. 909–917, Apr. 2005.

[3] B.Fahs, P. Gamand, and C. Berland, "A Continuous Analysis of the Oscillation Amplitude in MOS LC-VCOs," *IEEE Int. Conf. Micro. (ICM)*, pp. 196−199, Dec. 2010.

[4] M. M. Mansour *et al.*, "Analysis techniques for obtainning the steady-state solution of MOS LC oscillators," *IEEE ISCAS*, pp. 512–515, May 2004.

[5] Q. Huang, "Phase noise to carrier ratio in LC oscillators," *IEEE. Trans. Circ. Syst. I (TCAS-I)*. vol. 47, pp. 965–980, July 2000.

Morphology and structural properties of nano-PZT thin films deposited on unheated substrate by RF sputtering system

Mohammed Mahdi

Devisions microélectronique et nano système(MEMS)
Centre de développent de la technologie avancée(CDTA)
Cité 20 aout 1956 Baba Hassen,
Alger, Algérie.
Tel : +213 (0) 21 35 10 18 / 40 / 75
Fax:+213 (0) 21 35 10 39
Mmahdi@cdta.dz

Mohammed Kadri

Laboratory MESM
University of sciences and technology of Oran
BP. 1505 Oran Elmnouer 31000
Algeria
Kadrimed@univ-usto.dz

Abstract—**morphology and structure of lead zirconate ceramics thin films (PZT) target were investigated. They were deposited on unheated Aluminum (Al) and silicon (Si) substrates, and then heat-treated in a tabular furnace at 200 and 400°C during 30 min in air atmosphere. Its crystallographic characteristics were determined by glancing incidences X-ray diffraction (GIXRD) analysis with Fe Kα radiation (λ=1.936A) at glancing angle of 1.5°.the micrographs of the surfaces and cross-sections of thin films were observed by using scanning electron microscopy SEM (Zeiss ultra plus) in Ultra-High resolution imaging.**

x-ray diffraction analysis showed that as-deposited PZT films, independently of the substrate nature, presents an amorphous structure and nano-crystallizes in a pure perovskite phase $PbZr_{0.44}Ti_{0.56}O_3$ when it was only deposited on Al substrate and then thermal treated at relatively low temperature 400°C.

Keyword: PZTthinfilms; nanostructure; morphology; perovskite

I. INTRODUCTION

This application of ferroelectric and piezoelectric materials to electronic devices, such as ferroelectric nonvolatile random access memories (NVRAM), integrated capacitors for dynamic random access memories (DRAMS), infrared sensors, ultrasonic transducers, piezoelectric micro-actuators and micro-electromechanical systems(MEMS) and control positioning[1.2], have been attracting great interest in recent years. Among many ferroelectric materials, lead zirconate titanate (PZT) thin films were most widely studied and have received much attention for their applications in various fields. They exhibit great piezoelectric response, large dielectric constant, low coercive field spontaneous polarization, low crystallization temperature, large electromechanical coupling coefficients, temperature stability, high driving voltage, large capacity for data storage, and high resistance to depolarization from mechanical stress [2, 3, 4, 5, 6].

There are still some serious difficulties, such as, low deposition rate, high temperature deposition, rough surface morphology, non conformal step converge, with several deposition techniques: sol gel, chemical vapor deposition (CVD) and metal organic chemical vapor deposition (MOCVD). Moreover, these processes do not have good reproducibility which is very important in the fabrication of devices [3].

As the interest for ferroelectric devices is increasing, a full knowledge of the morphological and structural properties of PZT thin films is necessary. For example, for DRAM applications surface morphology plays an important role electron transport on PZT thin film [7].

In this study, we deposited PZT thin films by RF sputtering using a ceramic PZT target previously prepared by ourselves. We investigated the surface morphology, growth rate, and structure of PZT films as function of substrate (Si, Al) nature and annealing temperature at low temperature.

II. EXPERIMENTAL SET UP

The PZT thin films were elaborated by means of conventional radio frequency (RF) sputtering using 8 cm ceramic PZT target $Pb(Zr_{0.52}Ti_{0.48})O_3$ which was already synthesized fig.1 XRD patterns of the ceramic PZT target used in our experiments are shown in Fig.1.it can be seen that the PZT crystal contains the pyrochlore phase when the target is calcined for six hours at 600°C. However, when this later was sintered at a temperature of 800°C, only perovskite structure preferentially oriented (101) was obtained. Films of 100nm thickness were deposited on both unheated aluminum (Al) and silicon Si (100) substrates.

The sputtering chamber was evacuated in the 10^{-5}torr range by an oil pump before the admission of sputtering gas Ar at a pressure of 10^{-3} torr. The RF power was set at 100W.

978-1-4673-5289-5/12 $31.00 © 2012 IEEE

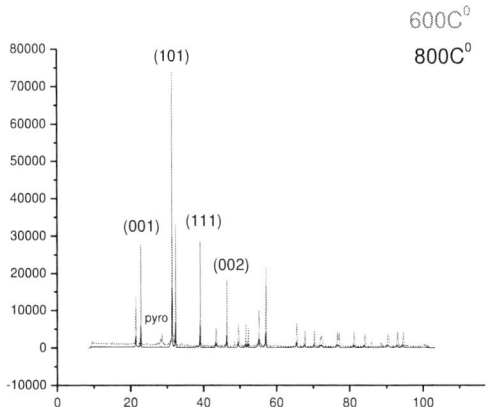

Fig.1.X-ray diffraction patterns of PZT target calcined at 600°C (red line) and sintered at 800°C (black line)

Following deposition of the PZT films and in order to induce the perovskite phase within the films, annealing treatment in a traditional furnace was carried out for 30min under air atmosphere at temperatures of 200 and 400°C. the crystalline structure of the PZT thin films was analyzed by glancing incidence X-ray diffraction (GIXRD) analysis with Fe Kα radiation (λ=1.936A) at glancing angle of 0.2, 0.5,and 1.5°. The surface morphology of the films was observed using Scanning Electron Microscope (Zeiss ultra plus), whilst PZT films thickness was measured by tilting sample.

III. RESULTS AND DISCUSSION

X-ray diffraction characteristics of PZT thin films deposited onto unheated Si (100) substrates and annealed at various temperatures are displayed in Fig.2. In all cases, spectra exhibit a broad dominant peak around 2θ=37° due to the amorphous structure [8] of the PZT films. At low annealing temperatures, the crystallinity of the PZT films (b) and (c) cannot be seen for improvement because no newly peak appears.

It is found from fig.3 that the crystallization process of the PZT films depends strongly both on the annealing temperature [9] and also on the nature of substrate.

The PZT thin films crystallization occurs at low temperature when it is deposited on Al substrate, due to its high thermal conductivity. Contrary to DRX spectra of films deposited on Si, the spectra of those deposited on Al change significantly. Indeed, spectrum of as-deposited film (a) presents a broad peak around 2θ=37°, characteristic of amorphous PZT phase as found above.

Then, after annealing at 400°C, the PZT nanocrystal phase was grown from amorphous phase and some new sharp peaks appears as shown on spectrum fig.3. X-ray diffraction intensity increased as annealing temperature rose up. The diffraction peaks of this film were all perovskite structure and strong (111) preferred orientation due to the heteronucleation [11] with no pyrochlore phase known to decrease the piezoelectric properties [10]. These peaks come from the new composition

of Pb(Zr$_{0.44}$Ti$_{0.56}$)O$_3$ thin film and aluminum substrate as indicated in fig.3. The first one peak around 2θ=12° remains unknown there are also two additional narrow peaks around 2θ=48° and 56° ascribed to Al substrate.

By applying Sherrer's equation on the peak at 2θ=39° (λ=1.936A), the crystallite diameter of the PZT thin film, deposited on silicon substrate and annealed at 400°C,was found equal to 50nm. The PZT thin film thickness was measured by SEM and equal to 90nm as shown in fig.4.

Surface morphologies of PZT thin films are shown in fig.5. It can be seen that the PZT thin film deposited on Al substrate and then annealed at 400°C, photo (C), was composed of nano-crystallite grains with an average grain size of about 60nm and the surface show free crack due the relatively thin films, less than 100nm .

However, when the film was annealed at 200°C the surface of the film was formed by irregular shapes and didn't show any grains formation photo (b) especially when the film was deposited on silicon substrate as given in photo (A).

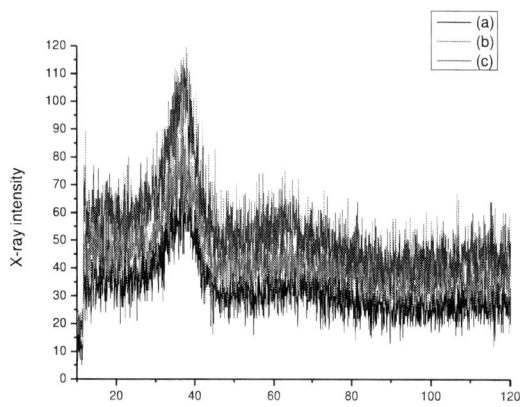

Fig.2. x-ray diffraction spectra of PZT on unheated Si (100): As deposited (a), annealed at 200°C/30min (b) and 400°C/30min(c)

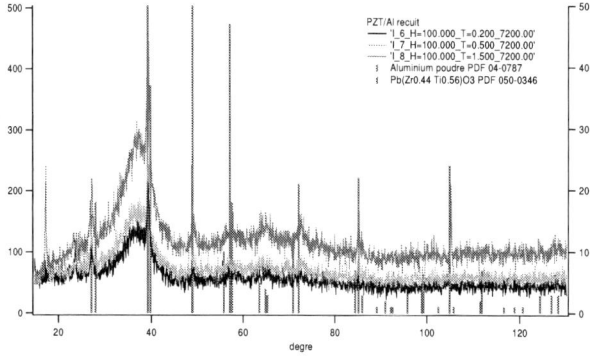

Fig.3. x-ray diffraction spectra of PZT on unheated Aluminum (Al) substrate after annealed at400°C/30min

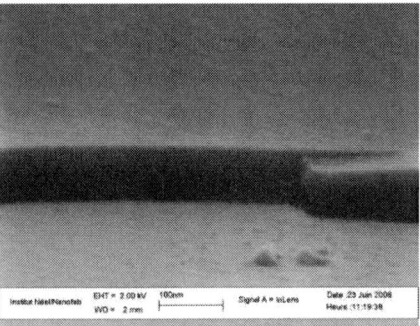

Fig.4.SEM micrograph of tilted PZT/Si (100) e=100nm

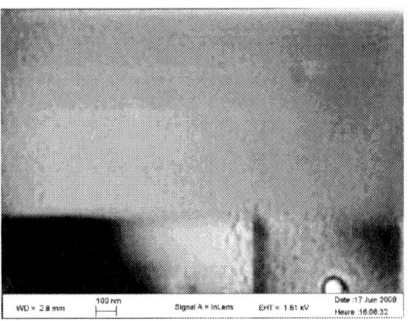

(a)

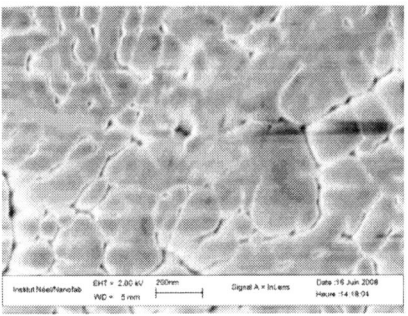

(b)

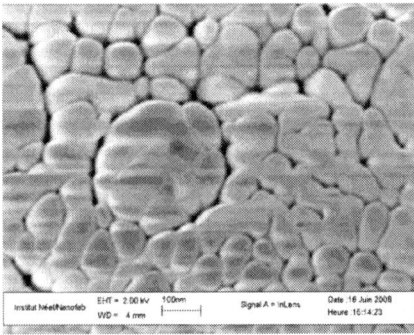

(c)

Fig.5.SEM photos of PZT thin films: (A) PZT/Si(100)

annealed at 200°C/30min, (B) PZT/Al annealed at 200°C/30min,

and (C) PZT/Al annealed at 400°C/30min

IV. Conclusion

we have deposited PZT thin films on an unheated both silicon and aluminum substrates by a RF sputtering (at low power:100W) using a ceramic PZT target and then thermal treated at very relatively low temperature, 200 and 400°C. We studied the structure and surface morphology of these films.

The results obtained from this work indicate that (1) – the analysis of the samples by x-ray glancing incidence, indicates that all as-deposited PZT films present an amorphous structure (2) – after a relatively low thermal treatment at 400°C, only films deposited on aluminum substrate exhibit a nanostructure (50nm in PZT crystallite size) and crystallized in a pure perovskite with a (111) preferred orientation and without any traces of pyrochlore phase; (3) – as shown by SEM photo, nano-crystallites shape is defined in this earlier sample.

The above mentioned results confirm that the nano-PZT film growth process depends strongly on the substrate nature and also on annealing temperatures.

ACKNOWLEDGMENT

The authors wish to thank Luc Ortéga and Khaled Ayadi of the C.N.R.S of Grenoble (France) for the access of their, respectively, glancing incidence X-ray diffraction and Scanning Electron Microscope apparatus and LMESM group Bekka Ahmed, Bentaher Mohamed of University of science and technology of Oran (Algeria).

The authors also thank, MEMS & Sensors group, and CDS group of Algerian Center for Development of Advanced Technologies.

REFERENCES

[1] C.Ruang Chalerm Wong et al, Thin solid films, Vol.517,n°24,(2009) pp.6599-6604.

[2] J.Z.Tsai et al, Sensors and actuators B: Chemical, Vol.139,2,(2009) pp.259-264.

[3] Jeong Young Lee, Byung Soo Lee, Materials Science and Engineering B79(2001) pp. 86-89.

[4] T. Masuda, Y. Miyaguchi, M.Tanimura, Y.Nishioka, K.Suu, and N.Tani, AppliedSurface Science 169-170 (2001) pp. 539-543.

[5] Ming-Ming Zhang, Ze Jia, Tian-Ling Ren, Solid-state Electronics, 53 (2009) pp. 473-477.

[6] Zongjin Li, Hongyu Gong, Yujun Zhang, Current Applied Physics 9 (2009) pp.588-591

[7] Yeau-Ren Jeng, Ping-Chi Tsai, Te-Hua Fang, Microelectrinic Engineering, 65 (2003) pp.406-415.

[8] Kazuyushi Tsuchia, Toshiaki Kitagawa and Eiji Nakmachi, Precision Enggineering 27 (2003) pp.258-264.

[9] Xin-Shan Li, Tsunehisa Tanaka and Yoshihiko Suzuki, Thin Solid films 375 (2000) pp.91-94.

[10] C.Zink, D. Pinceau, E.Delovoye and Delovoye and D.Barbier, Sensors and Actuators A115 (2004) pp.484-489.

[11] W.Huang, S.W.Jiang et al, Thin solide films 500(2006) pp.138-143

Characterization by SEM and FTIR of B-LPCVD polysilicon films after thermal oxidation

Moufida BOUZERDOUM
Department of Electronics, University of Jijel,
Ouled Aïssa Zone, POB 98, Jijel, Algeria
m83.bouzerdoum@yahoo.com

Boubekeur BIROUK
Department of Electronics, University of Jijel,
Ouled Aïssa Zone, POB 98, Jijel, Algeria
bbirouk@univ-jijel.dz

Abstract— **The aim of this work is the study and analysis of the properties of polycrystalline silicon thin films that have been deposited on oxidized monocrystalline silicon for a use as MOS structures gates. The polysilicon films were elaborated by chemical vapour deposition at low pressure (LPCVD) from silane (SiH$_4$) and boron trichloride (BCl$_3$) gas mixture, at temperature 605 °C. They have been subjected to thermal annealing in dry oxidizing atmosphere at temperatures ranged between 850 and 1100°C. The analysis of the specific behavior of such a structure towards oxidation, was based on two methods of investigation namely scanning electron microscopy (SEM) and Fourier transform infrared spectroscopy (FTIR). The SEM images show an increase in grain size and crystallinity when the doping level and temperature of oxidation increase. FTIR results analysis highlights the various chemical bonds existing in films such as Si-H and Si-O-Si and reveals the breaking of some bonds at high oxidation temperature, which gives rise to more trapping states.**

Keywords- Polysilicon, LPCVD deposit, boron, thermal oxidation, grains, grains boundary, SEM, FTIR.

I. INTRODUCTION

The polycrystalline silicon is a semiconductor material possessing many applications. It allows the increasing of integration density and the integrated circuits performance. It is specially used as resistance in circuits such as static memory (Si-poly weakly doped), as gate electrode and interconnections in MOS structures (Si-poly highly doped), and also in the fabrication of flat monitors and solar cells (Si-poly with large grains). Most of these applications necessitate good microstructural properties of polycrystalline silicon films.

The aim of this work is to study the microstructural and physico-chemical properties of Si-LPCVD films used as gate of MOS structures, by means of SEM and FTIR techniques.

II. EXPERIMENTAL PROCEDURE

A. Sample preparation

The polysilicon films are elaborated by chemical vapour deposition at low pressure (LPCVD) from silane (SiH$_4$) and boron trichloride (BCl$_3$) gas mixture, at temperature 605°C. They have been subjected to thermal annealing in dry oxidizing atmosphere at temperatures ranged between 850 and 1100°C.

Samples destined to SEM observations must undergo a preparation including a cleaning procedure thanks to a

solution "SECOO" [1]. The latter is composed of two volumes of HF (40%) for a volume of K$_2$Cr$_2$O$_7$ (44g/litre), and aims to remove the amorphous silicon surrounding the crystallites so that to highlight them. The samples were soaked in this solution for no longer than few seconds. After drying, the sample is immediately deposited into the microscope chamber to be observed.

B. Structural measurements

After removal of the formed silicon oxide layer, photographic views of the Si-LPCVD films structure were taken on the surface using a scanning electron microscope Philips XL 30 type, allowing a magnification of 400 000 times under acceleration voltage of 30 KV.

Investigation of some physico-chemical properties brought into play in this material type, allows the identification of the mechanisms influencing films conduction and microstructure. To this end, we performed analysis by infrared spectroscopy with Fourier transform (FTIR). Our spectra were carried out between 400cm^{-1} and 4000cm^{-1} with an infrared spectrometer Affinity-1 type.

III. RESULTS

A. Characterization by SEM

1. Evolution of grain size as oxidation temperature function

Figure 1 shows the effect of thermal oxidation process on the evolution of grains size. It includes, in fact, different SEM photographs on which a comparison can be made between three films highly in situ boron doped in the order of 2×10^{20} cm^{-3}, deposited on oxidized substrate, and annealed at oxidation temperature of 850, 1000 or 1100° C.

According to Figure 1, we note that the microstructure of polycrystalline silicon layers is presented as a set of tiny crystals or grains, separated from each other by highly disordered zones, called grain boundaries. Crystalline regions are represented by enlightened regions on the SEM photos. Indeed, the crystallites created at the deposition beginning tend to grow and develop in grains and then merging themselves by coalescence. The medium grain size is estimated at 450 nm for the oxidation temperature of 850 °C, 900 nm for 1000°C and 1500 nm for 1100°C.

The thermal annealing is one of the treatments modifying the grain size and crystallinity of our thermo-oxidized Si-LPCVD deposits. This appears clearly from our analyzed SEM photographs taken on films after removing the silicon dioxide layer.

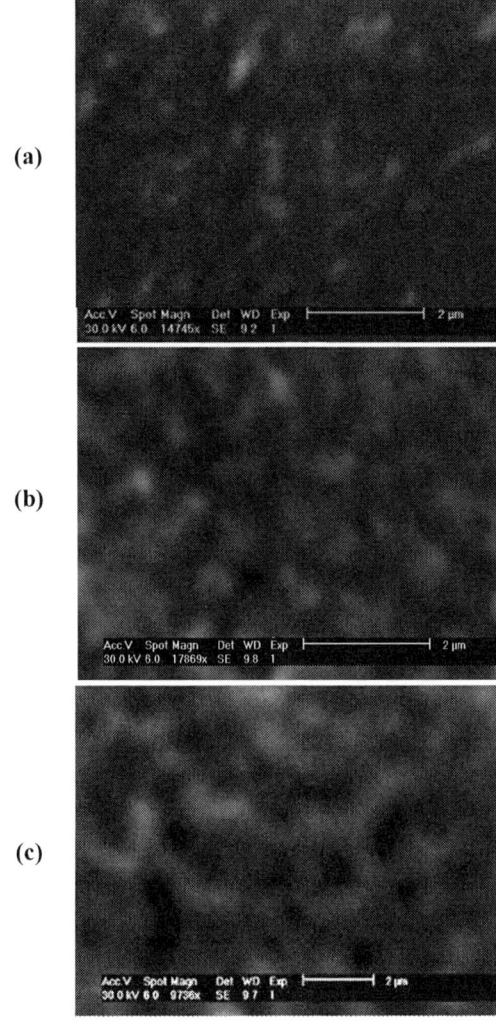

Figure 1. SEM observation of Si-LPCVD films oxidized at temperatures: (a) 850 °C, (b) 1000 °C and (c) 1100 °C

It is found that the grain size increases with oxidation temperature. The temperature rise, in any diffusion mechanism, leads to grain growth and reduced internal stresses [2]. Indeed, the elementary process of grain growth in our Si-LPCVD films is attributed to the diffusion of silicon atoms through the grain boundaries. Thus, under the effect of a heat source sufficiently powerful, silicon atoms can migrate from one crystallite to another through the grain boundaries, some grains will therefore find themselves rich at the expense of other grains, this gives rise to the merger of two or more grains, and consequently the disappearance of some grain boundaries. Therefore the total volume of grain boundaries becomes smaller. The increase in average grain size is in the opposite direction from that of the number of these grains because each grows in depends on other [3].

2. Evolution of grain size with doping level

Figure 2 illustrates the effect of doping level on the evolution of the crystallites size. It shows SEM photographs from which we can compare three in situ doped films with boron concentrations of $4x10^{19}$, $1x10^{20}$ and $2x10^{20}$ cm^{-3}, and annealed in oxygen at 950 °C for 15 min.

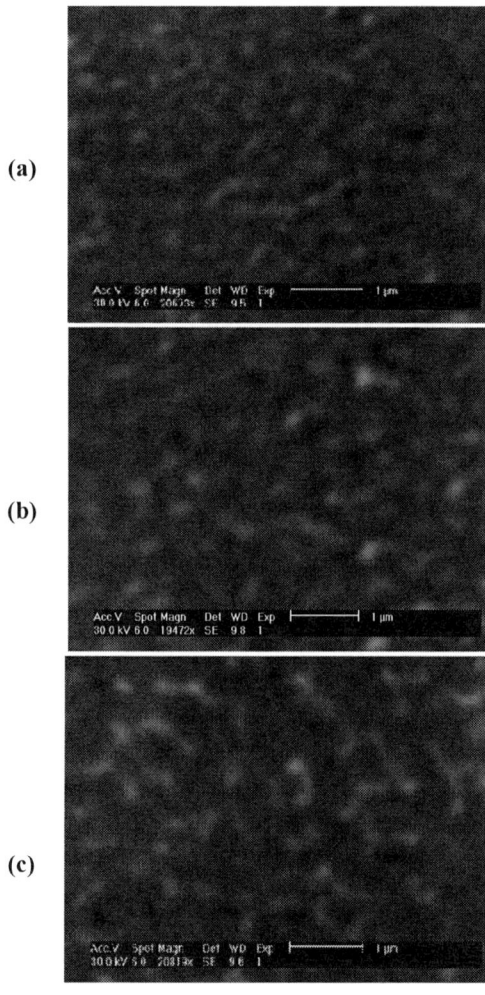

Figure 2. SEM observation of Si-LPCVD films doped with boron at level: (a) $4x10^{19}cm^{-3}$ (b) $1x10^{20}cm^{-3}$ and (c) $2 x10^{20}cm^{-3}$

Grains undergo a clear increase. It results from measured average size, values of 400 nm for the doping level of $4x10^{19}$ cm^{-3}, 550 nm for the doping level of $1x10^{20}$ cm^{-3}, and 650 nm for the doping level of $2x10^{20}$ cm^{-3}. Thus, the effects of successive high temperature annealing and high doping, first result in an improvement in crystallinity and an increase in grain size [4].

Given these various observations, we can first say that the annealing temperature in oxygen ambient, independent of doping, acts significantly on grain size. As for the effect of the

978-1-4673-5289-5/12 $31.00 © 2012 IEEE

dopant, increasing the size of the crystallites is activated by the presence of boron [5]. Both of them enhance the film conductivity, knowing that the amorphous region (grain boundaries) becomes weaker.

From these results, we can notice that the oxidation temperature and doping level, significantly affect the crystallinity and the grain size. This allows us to say that the characteristics of these films are closely related to experimental conditions under which the deposit and the oxidation were made.

B. Characterization by FTIR

1. Evolution of FTIR spectra with oxidation temperature

FTIR absorption spectra analysis of polysilicon films doped with boron and deposited on oxidized monocrystalline silicon substrate, will allow us to identify the chemical bonds existing between atoms of the structure.

Figure 3, shows the absorption spectra raised up in function of wave number, for a sample doped with boron concentration $4 \times 10^{19} cm^{-3}$, oxidized at different temperatures. The peaks appearing at 605, 733, 886 and 1105 cm^{-1} correspond to the Si-Si [6], Si-O [6, 7], Si-H [8] and the Si-O-Si [9] bonds, respectively.

The peak appeared in the band [2330-2360cm^{-1}] belong to SiH$_x$ (x = 1, 2, 3) group [6]. This absorption band seems to consist of several sub-bands that correspond to different chemical environments of the Si-H bonds.

The appearance of the Si-Si bond at wave number 605 cm^{-1} is due to the monocrystalline material. The appearance of Si-H bond at 870 cm^{-1} for low wave numbers and SiH$_x$ at 2350 cm^{-1} for high wave numbers [6], is due to the process of preparing of Si-LPCVD film: the use of silane SiH$_4$ during the deposition phase of our films. On the other hand, the appearance of Si-O and Si-O-Si bonds at wave numbers 740 and 1100 cm^{-1} respectively, is of course due to the process of annealing of polysilicon dry oxygen (thermal oxidation) and also for a great part, to the oxide layer that sits between the two monocrystalline and polycrystalline layers of our structures.

We have attempted to estimate the area under the peaks corresponding to Si-O (740cm^{-1}) and Si-O-Si (1100cm^{-1}) bonds depending on the annealing temperature in dry oxygen. We note an increase in the intensity which means an increase in oxygen concentration with oxidation rate in the annealed films.

The high-temperature annealing allows the breaking of Si-Si and Si-H bonds. Indeed, after annealing at 1100 °C (see figure 3), a decrease in the intensity of the Si-H bond, reveals a probable dissociation of hydrogen atom from silicon, at high temperatures. So it would be tempting to conclude that high temperatures lead to releasing of hydrogen, and increasing of some free bonds which act as trapping states for oxygen atoms.

The variation of the annealing temperature will change the location of certain positions of the spectra. For example, we find that the Si-O-Si appears at wave number 1098 cm^{-1} for the oxidation temperature of 850 °C and at 1105 cm^{-1} for the oxidation temperature of 1100 ° C. So there is shift of the peak of Si-O-Si to larger values of wave number when the oxidation temperature increases.

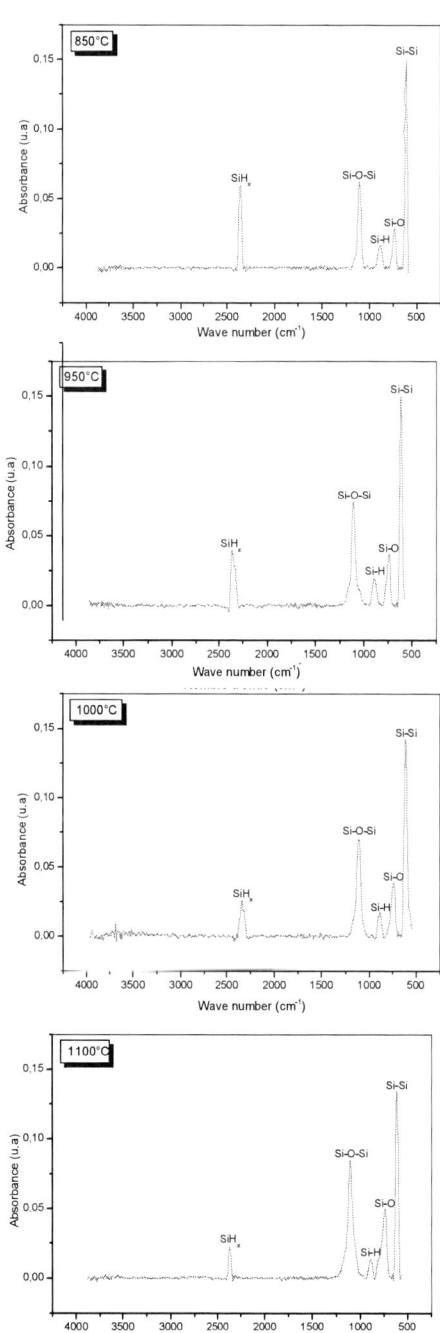

Figure 3. FTIR absorption spectra of Si-LPCVD films oxidized at different temperatures

2. Evolution of FTIR spectra with doping level

The electrical properties of these films are highly dependent from effects relative to grains boundary, involving specific silicon-boron interactions, especially at high doping levels (in the order of 10^{20} cm^{-3}) [5].

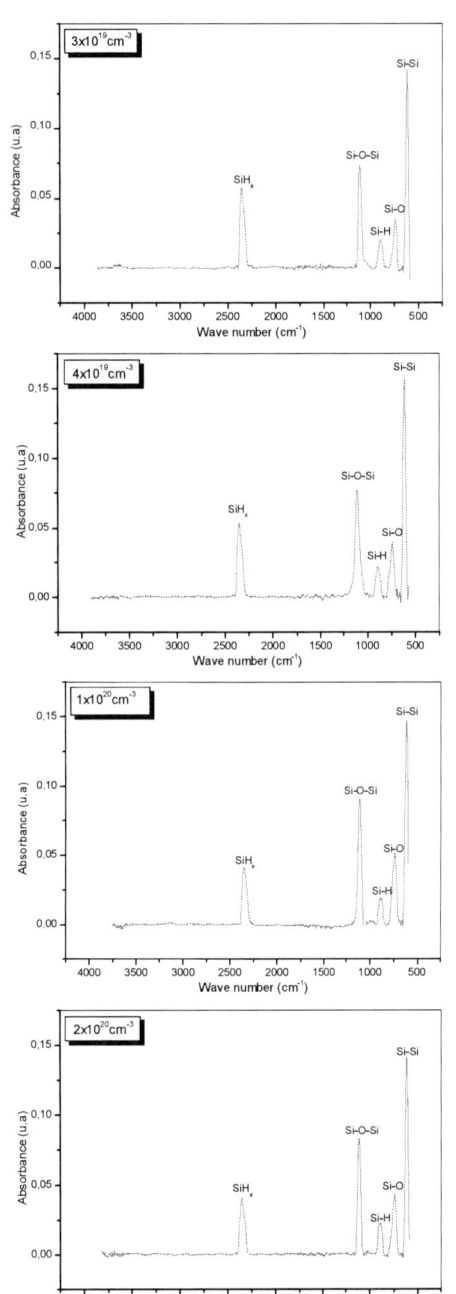

Figure 4. FTIR absorption spectra of Si-LPCVD films doped with boron at different doping levels.

Figure 4 shows the absorption spectra of LPCVD silicon deposits, oxidized at temperature 850°C and highly boron doped with concentration levels of: 3×10^{19}, 4×10^{19}, 1×10^{20} and 2×10^{20} cm^{-3}.

As observed before, we note on the absorption spectra showing the chemical composition of our films at different doping levels. These spectra are noisy due to atmosphere molecules (air of the vacuum) and we met great difficulties in extracting information about the Si-HB (1899cm^{-1}) bond [10] which should appear on the spectrum but it is masked by the noisy zone. From this figure, although the spectrum of boron does not appear, it is clear that the doping level has an effect on the absorption spectra either in intensity or location of the spectra.

Knowing that the studied films are highly doped with boron at solid solubility limit, it is estimated that the additional dopant precipitates and forms clusters or silicon compounds, such as: SiB_x (SiB_4, SiB_6...) electrically inactive and immobile. The formation of these clusters by segregation of the dopant in interstitial defects of grains and grain boundaries, have probably reduces the diffusion coefficient of the dopant [5].

IV. CONCLUSION

In this study, the experimental results allowed us to determine the microstructure and the chemical composition of thin Si-LPCVD films, by using SEM and FTIR techniques. The SEM images show an increase in grain size and crystallinity enhancement, when the doping level and oxidation temperature increase. FTIR analysis shows the various bonds existing in films such as Si-H and Si-O-Si. It reveals the breaking of Si-H bonds at high oxidation temperature, meanwhile the Si-O ones increase, thanks to the formation of trapping states.

REFERENCES

[1] M.BONNEL, N.DUHAMEL,M.GUENDOUZ,L.HAJI, B.LOISEL and P.RUAULT, "Poly-Si thin film transistors fabricated with rapid thermal annealed silicon films", Journal of Appl. Phys, 30 (11B), p.L 1924, November 1991.

[2] M.LEMITI, S.AUDISIOP, C.MAI, B.BALLAUD, "Evolution of the grain size of the polycrystalline silicon during heat treatment or oxidation",Rev. Phys.App (1989), 24, 133-41.

[3] B.E.DEAL and M.SKLAR, "Thermal oxidation of heavily doped silicon", J. Electrochem; Soc,112, 1965, p.430.

[4] WON-JU CHO, EUNG-SOO KIM, JUN-JIN KANG, KWAN-GOO RHA and HONG-SEOK KIM, "Annealing effects of polycristalline silicon gate on electrical properties of thin gate oxide", Solid-State Electronics, vol.42.No.4,pp.557-566.1998.

[5] BOUKEZZATA M, BIROUK B, MANSOR F AND BIELLE-DASPET D. "structural and Electrical changes in polycrystalline silicon thin films that are heavily in-situ boron-doped and thermally oxidized with dry oxygen".Chem. Vap.Deposition, 3, N°5, p.271-279, (1997).

[6] M. A VASQUEZ-A, "FTIR and photoluminescence studies of porus silicon layers oxidized in controlled water vapour conditions", Mexico, November 2007.

[7] ROSS BOYLE, "FTIR Measurement of Interstitial Oxygen and Substitutional Carbon in Silicon Wafers", Thermo Fisher Scientific, Madison, WI, USA, 2008.

[8] J.LUNA-LOPEZ,J. CARRILLO-LOPEZ, "FTIR and photoluminescence of annealed silicon rich oxide films", Mexico, Mars 2009.

[9] S. P. KIMA and S. K. CHOI, "Effect of water absorption on the residual stress in fluorinated silicon-oxide thin films fabricated by electron-cyclotron-resonance plasma-enhanced chemical-vapor deposition", Applied Physics Letters volume 79, number 2, Korea, 2001.

[10] H. FUJIWARA, M. KONDO, AND A. MATSUDA, Appl. Phys. Lett. 82, 1227 (2003).

A low power thermal protection topology

Alex Pivoto, Paulo Crepaldi and Tales Pimenta
Universidade Federal de Itajuba - UNIFEI
Itajuba, Brazil

Abstract — **Many circuits are subject to excessive heating such as digital and power switching circuits. Overheating may cause permanent damage to the circuits and devices. In this article we present a temperature monitor the temperature variation and once a certain limit is reached, a protection mechanism could shut off the circuit operation or at least signalize for a specific operation. The circuit offers linear operation from -25 ºC to +120ºC, and it can be adjusted to detect any temperature in that range. The proposed circuit was developed for low voltage and low power operation, and the prototyped was designed in TSMC 0.35um technology.**

I. INTRODUCTION

Heating is the largest bottleneck of power and digital switching circuits. In fact heating is the cause of concern in most circuits.

Currently, power management circuits are used in most integrated circuits applied to power switching. Thermal protection or thermal shutdown circuit is one of them, and in this paper we propose an alternative topology.

The proposed topology can be used in any circuit that requires temperature management, in order to identify and flag excessive temperatures, thus avoiding irreversible damage to the component.

The proposed topology of thermal protection can be used in any integrated circuit, including industrial and remote applications, where low power consumption is necessary.

II. CONVENTIONAL CIRCUITS

The present thermal protection circuits currently use comparators. Fig. 1 shows a widely used thermal protection circuit [1].

From the analysis of the circuit in Fig. 1 it is possible to verify that Vbe of transistor Q1 is responsible for sensing the temperature variations. It is known that a temperature variation will cause a Vbe variation on bipolar transistors. Therefore if it is known the Vbe variation for the desired temperature variation, resistors R1 and R2 can be selected so that the comparator toggle once the desired temperature has been reached. Transistor M1 is responsible for the generation of a hysteresis required by the comparator.

Depending on the Vbe variation and the comparator response time, that topology can be used on the detection of high temperatures and can present good stability [1].

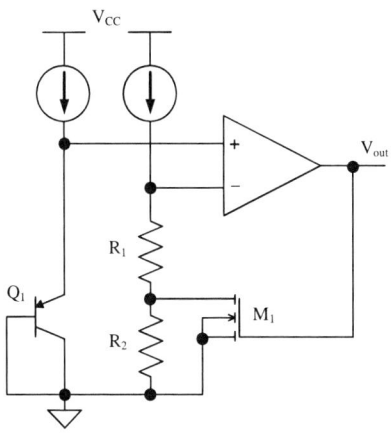

Figure 1. Conventional thermal protection circuit.

Nevertheless, that topology suffers two disadvantages. The current supplied by the Q1 branch is not stable for different power supply voltages. Additionally, the bipolar transistors also dissipate power.

III. PROPOSED CIRCUIT

The proposed circuit was developed using CMOS transistor technology that allows low power consumption, small layout area and low voltage operation.

The proposed circuit will be dived into smaller block in order to ease its understanding.

A. Biasing and PTAT

Fig. 2 shows the PTAT circuit. It is implemented using both single and composed, NMOS and PMOS transistors. The circuit is responsible to provide a voltage on C4 that is linearly dependent on the temperature variation.

The composite MOS transistor used in the PTAT circuit offers an expressive increase in the output impedance (or correspondingly decrease in the conductance) as compared to a saturated simple transistor. At the same time, the composite transistor requires the same biasing as the simple transistor, and thus does not require extra quiescent current.

Consequently, the composite transistor offers a behavior close to the ideal transistor, modeled as an ideal current source, as compared to the simple transistor [2].

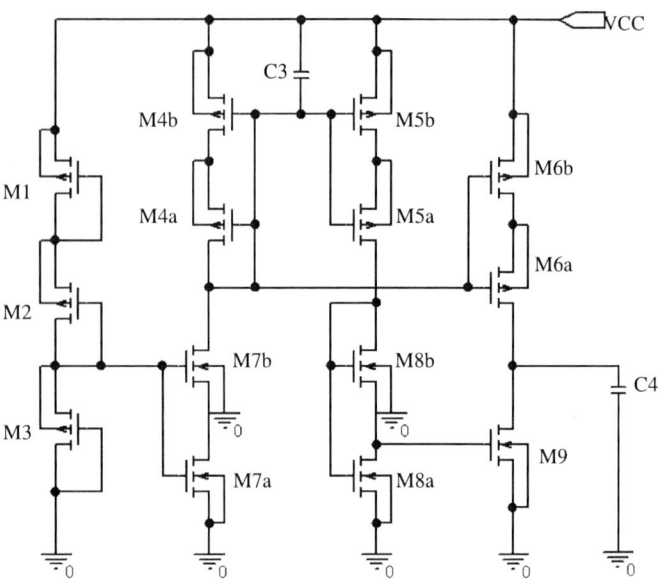

Figure 2.   Biasing and PTAT circuit.

Under a 1V power supply voltage the circuit requires current in the range of few nA. The circuit offers a highly linear behavior to temperature variations. Transistors M1, M2 and M3are responsible to provide a reference voltage of 333mV at the gate of composite transistor M3, thus forcing it to weak inversion operation. The composite transistors M4, M5 and M6 are current mirrors responsible to provide enough current to the PTAT circuit, as its biasing current. That current is adjusted by the composite transistor M7 ratio. The values of (W/L) for M7a and M7b were calculated and adjusted for a biasing current of approximately 72nA.

In order to maintain M7a under weak inversion saturation, the voltage V_{DS}M7a must be higher than 3kT/q [2]. Consequently, the dimensions of transistor M7b required to maintain M7a under weak inversion saturation are given by equations (1) and (2).

$$\left(\frac{W}{L}\right)_b \geq \left(e^3 - 1\right)\left(\frac{W}{L}\right)_a \tag{1}$$

$$\left(\frac{W}{L}\right)_b \geq 19\left(\frac{W}{L}\right)_a \tag{2}$$

Nevertheless, it is required a large area (W-L) for transistor M7b in order to maintain M7a under weak inversion saturation. The same rule should also be valid for the other mirror transistors.

Transistors composite M8a and M8b work as a low power thermistor, so that V_{GS}M9 varies linearly with the temperature. The circuit also requires a simple current source, obtained by the previous current mirror. Nevertheless the current source does not require special attention since V_{GS}M9 does not depend on that current [9].

Transistors implemented in the same subtract are subject to expression (3), and transistors implemented in different substrates are subject to expression (4).

$$V_{DSa} = \frac{kT}{q} \ln\left(1 + \left(\frac{\left(\frac{W}{L}\right)_a}{\left(\frac{W}{L}\right)_b}\right)\right) \tag{3}$$

$$V_{DSa} \approx \frac{kT}{q} \ln\left(1 + \left(\frac{\left(\frac{W}{L}\right)_a}{\left(\frac{W}{L}\right)_b}\right)^n\right) \tag{4}$$

The single transistor M9 is responsible to elevate the value of PTAT, capacitor C3 is used for the start up and capacitor C4 is used to eliminate or at least reduce the influences of ripple or any power supply voltage variation.

B.   Voltage Reference

In the proposed circuit, the voltage reference is implemented by stacking PMOS transistors, as shown in Fig. 3. The stacking is necessary in order to provide a reference value for comparison with the PTAT, which refers to the temperature variation. The transistor stacking was designed in order to obtain a reference capable of detecting whenever the temperature reaches 100°C.

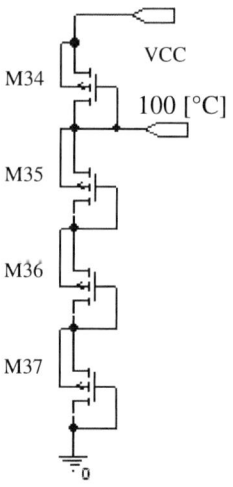

Figure 3.   Voltage reference.

We are currently working on a programmable reference so that few temperatures can be selected.

C.   Buffer

Fig. 4 shows the buffer implementation used to isolate the comparator reference source. It was used an OTA Miller, one of the most popular operational amplifier architectures.

The operational amplifier in a voltage follower connection or buffer isolates the input signal from the load by a unit gain stage. It provides no phase inversion or signal shifting, high input impedance and low output impedance.

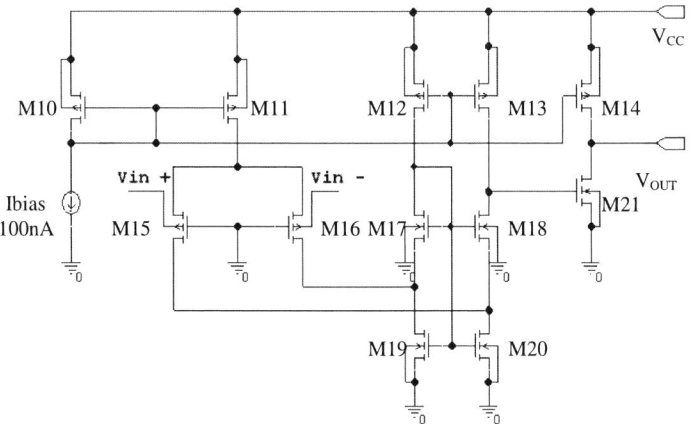

Figure 4. Bulk driven buffer.

The operational amplifier in a voltage follower connection or buffer isolates the input signal from the load by a unit gain stage. It provides no phase inversion or signal shifting, high input impedance and low output impedance.

The bulk driven unit gain OTA Miller allows a rail-to-rail swing with shutting off the transistors [1].

D. Voltage Comparator

Current and voltage comparators are used whenever a signal needs to be compared to another one or to a specific value.

Fig. 5 shows the voltage comparator that will be used to compare the PTAT signal that varies according to the temperature variation and the reference corresponding to the desired temperature.

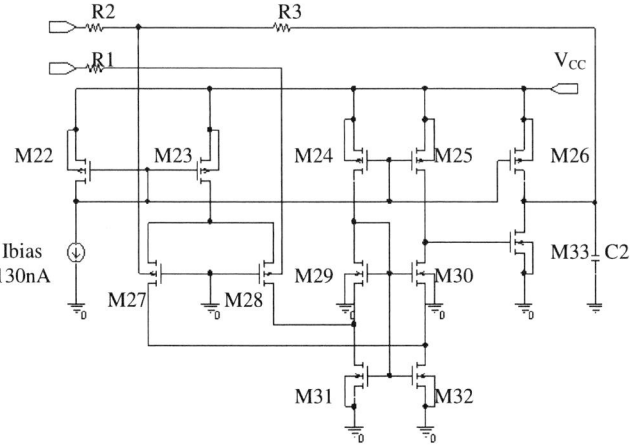

Figure 5. Voltage comparator.

It was used a bulk driven OTA Miller capable of 5uA output current. The comparator presents a 4°C hysteresis that was obtained by expressions (5) and (6). The hysteresis corresponds to 11mV.

$$H = \frac{1}{n}\left[+V_{SAT} - \left(-V_{SAT}\right)\right] \qquad (5)$$

$$H \approx \frac{2}{n}\left|V_{SAT}\right| \qquad (6)$$

IV. SIMULATION RESULTS

Figure 6 shows the simulation results of PTAT against the temperature. It can be observed a linear behavior, ranging from -20 °C to 120 °C.

$V_{DS}M_{8a}$

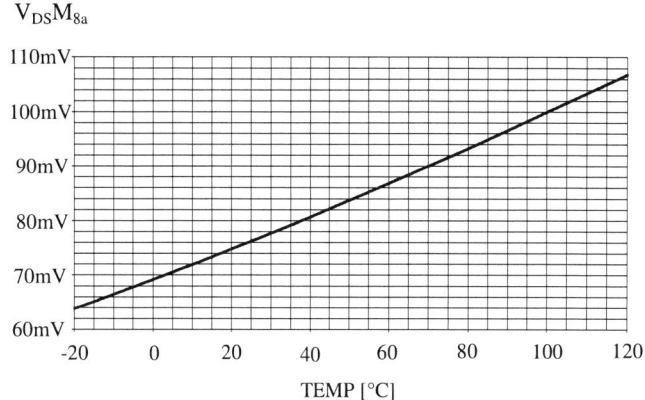

Figure 6. PTAT simulation response.

As can be observed from Fig. 06, the PTAT voltage variation is limited to a few tens of millivolts. Therefore it is necessary to raise the PTAT signal before the comparator. It was used just a PMOS transistor to implement it. Fig. 7 shows the PTAT signal at C4, which is after the voltage raise.

$V_{DS}M_9$

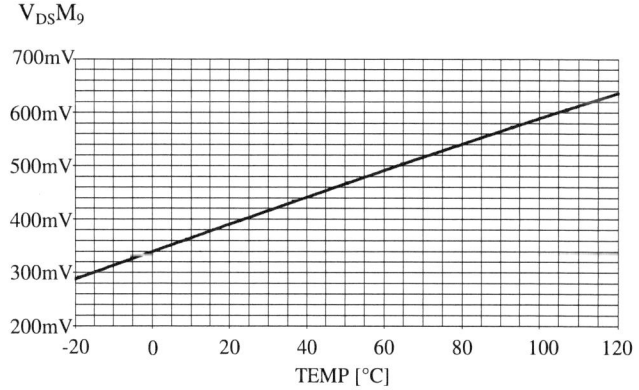

Figure 7. PTAT at C4.

Fig. 8 presents the circuit response to biasing at start up at the gate of M3 & M4 (a), at the gate of M8 (b) and at the source of M8 (c).As it can be observed, the circuit takes approximately 150us to stabilize. The delay is promoted by the capacitor C3.

Fig. 9 illustrates the global operation of the thermal sensor. In this work, the voltage reference was set for the operation at 100°C (Fig 9.a). If the temperature is under 100°C, the comparator output is at logic level 1, and if the temperature raises over 100°C, the comparator toggles to 0. The signal from the PTAT (Fig. 9.b), already raised, continues to vary along with the temperature, even after the comparison point.

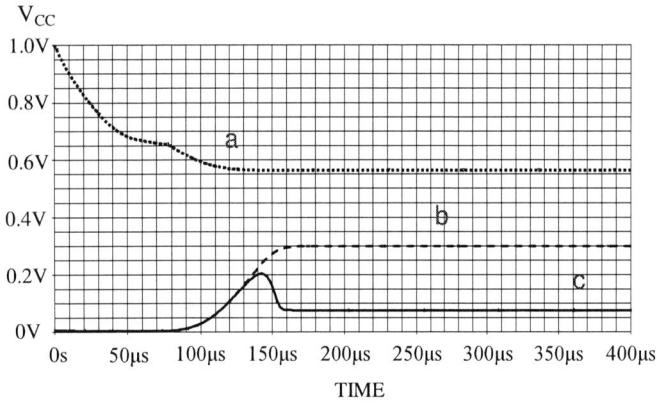

Figure 8. Circuit biasing at start up: (a) Gate of M4 & M5, (b). Gate of M8a and M8b, (c) Drain of M8a.

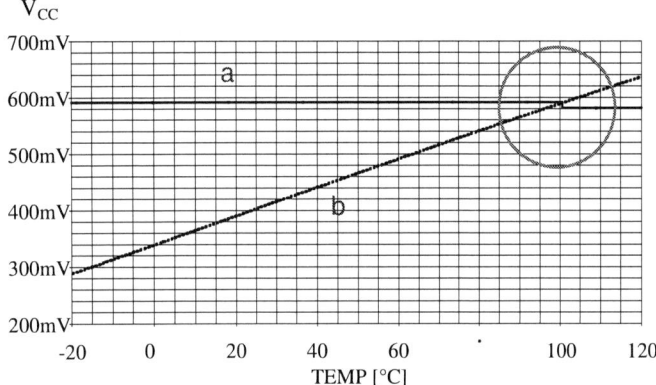

Figure 9. Global circuit operation: (a) Voltage from the reference circuit, (b) Raised PTAT signal.

We are currently working on a circuit in which the operation point can be adjusted to the desired temperature, on a limited range. Fig. 10 shows both the PTAT signal and the comparator output along with the reference voltage used to set up the temperature operation at 100°C. Fig. 10.a and Fig. 10.b are basically the same as Fig 9.a and Fig. 9.b, respectively. Fig. 10.c corresponds to the output of the comparator. In this case, it was used a 280K and 15pF to represent a load corresponding to the next stage considered in the simulation.

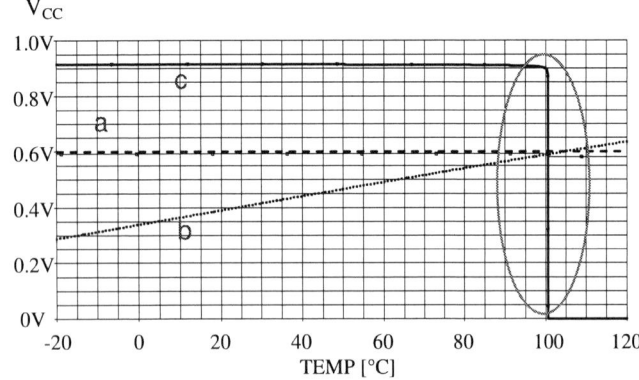

Figure 10. PTAT circuit operation: (a) Voltage from the reference circuit, (b) Raised PTAT signal. (c) Comparator output.

Fig. 11 presents the layout of the entire PTAT circuit. It was implemented in standard TSMC 0.35um technology.

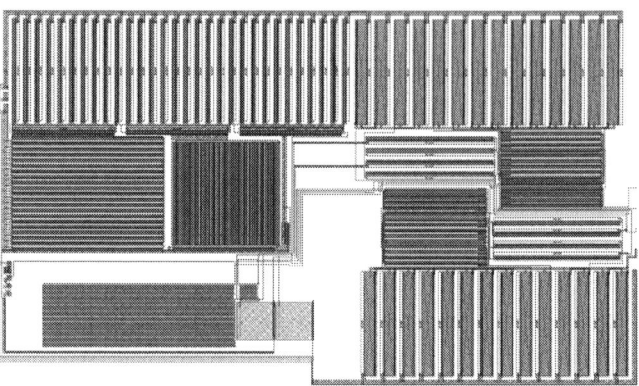

Figure 11. Layout of the global circuit.

ACKNOWLEDGMENT

The authors acknowledge CAPES, CNPq and FAPEMIG for their financial support.

REFERENCES

[1] Zhang Bin and Feng Quan-yuan, "A Novel thermal-shutdown Protection Circuit", 3rd International Conference on Anti-counterfeiting, Security, and Identification in Communication, 2009.

[2] L. H. Ferreira, T. Pimenta and R. Moreno, "An Ultra-Low-Voltage Ultra-Low-Power Weak Inversion Composite MOS Transistor: Concept and Applications", IEICE Transactions on Electronics, v. E91-C, p. 662-665, 2008.

[3] Gerard C. M. Meijer and A. W. Herwaarden, Thermal Sensors, Taylor & Francis; 1st edition, 1994.

[4] L.H.C. Ferreira and T.C. Pimenta, "A CMOS voltage reference for ultra low-voltage applications," 12th IEEE International Conference on Electronics, Circuits and Systems, Dec. 2005.

[5] David J. Comer, d Donald T. Comer, "Operation of Analog MOS Circuits in the Weak or Moderate Inversion Region", IEEE Transactions on Education, Vol. 47, No. 4, November 2004.

[6] L. H. C. Ferreira, T. C. Pimenta, and R. L. Moreno, "An ultra-low-voltage ultra-low-power CMOS miller OTA with rail-to-rail input/output swing", IEEE Trans. Circuits Syst. II, 843–847, Oct. 2007.

[7] Ken Ueno, Tetsuya Hirose, Tetsuya Asai and Yoshihito Amemiya, "Ultra low-Power Smart Temperature Sensor with Subthreshold CMOS Circuits", International Symposium on Intelligent Signal Processing and Communication Systems – ISPACS, 2006.

[8] Y. Cheng and C. Hu, *MOSFET Modeling & BSIM3 User's Guide*, Kluwer, 1999.

[9] Micheal A.P. Pertijs and Johan Huijsing, Precision Temperature Sensors in CMOS Technology, Springer, 1st ed, Nov, 2010.

[10] P. Crepaldi, T. Pimenta and R. Moreno, " A CMOS low-voltage low-power temperature sensor. Microelectronics", Microelectronics Journal), v. 41, p. 594-600, 2010.

[11] P. Crepaldi, R. Moreno, T. Pimenta, "Low-voltage, low-power, high linearity front-end thermal sensing element",. El Letters, v. 46, p. 1271.

[12] P. Crepaldi, L. H. Ferreira, R. Moreno, Leonardo B. Zoccal; T. Pimenta, "A Vt Independent Voltage Reference Based on Composite Transistor Operating in Weak Inversion", 10th IEEE Faible Tension Faible Consommation – FTFC, 2011, Marrakech - Marrocos. p. 12-14.

Boron Redistribution in Strongly Doped Silicon Thin Bi-layers Gates

Salah ABADLI
Department of Electrical Engineering,
University 20 Août 1955, Skikda, 21000
Skikda, Algeria
abadli.salah@gmail.com

Farida MANSOUR
LEMEAMED Laboratory, Department of Electronics,
University of Mentouri, Constantine, 25000
Constantine, Algeria
farida.mansour@yahoo.fr

Abstract—**We have investigated and modeled the complex phenomenon of boron (B) redistribution process in strongly doped silicon (Si) bi-layers structure. A two stream transfer model well adapted to the particular structure of bi-layers and to the effects of strong-concentrations has been developed. This model takes into account the instantaneous kinetics of B transfer, trapping, clustering and segregation during the thermal B activation annealing. The used Si bi-layers have been obtained by low pressure chemical vapor deposition (LPCVD) method, using in-situ nitrogen-doped-silicon (NiDoS) layer and strongly B doped poly-Si (P$^+$) layer. To avoid long redistributions, thermal annealing was carried out at relatively low-temperatures (700 and 850°C) for various times ranging between 15 minutes and 2 hours. The good adjustment of the simulated profiles with the experimental secondary ion mass spectroscopy (SIMS) profiles allowed a fundamental understanding about the instantaneous physical phenomena giving and disturbing the complex B redistribution profiles-shoulders kinetics.**

Keywords-boron; nitrogen; diffusion; polysilicon; activation annealing; bi-layers

I. INTRODUCTION

The strongly doped polycrystalline-silicon (P$^+$ poly-Si) is a key element of today's advanced very-large-scale-integration (VLSI) technology. Boron (B) is the most widely used P-type dopant in recent integrated devices due its high solid solubility in Si. It is usually introduced into Si layers, followed by rapid thermal annealing which electrically activates the B atoms. To continue scaling down P$^+$ poly-Si gates of metal-oxide-semiconductor (MOS) integrated circuits, it is necessary to create very shallow-junctions with strong-concentrations of electrically active boron (B). Two related processes limit the realization of this goal: (i) the enhanced redistribution of the B during the thermal dopant activation annealing, which causes B penetration through thin oxides from the P$^+$ poly-Si gate into the underlying layers, and (ii) the formation of electrically inactive B clusters and B precipitates, which decreases the B activation rate. The use of low-energy doping methods, co-doping techniques, low thermal annealing temperatures, short annealing times, amorphous-Si layers, and thin Nitrogen-Doped-Silicon (NiDoS) layers have been practiced to avoid the doping depletion of P$^+$ poly-Si gate at the oxide interface. This last effect is one of the major performance limitations in the standard advanced MOS circuits. Currently, significant research efforts are focused on improving the B activation; problem common to all the practiced methods and techniques.

In this work, based on the approach of using thin bi-layers gate, we investigate NiDoS layers for reducing B enhanced redistribution or gate depletion and for improving B activation. Fast B transfer is reduced by increasing nitrogen atoms. The goal is to develop a fundamental understanding about the instantaneous kinetics of complex B transfer process into strongly in-situ doped P$^+$ poly-Si/NiDoS bi-layers.

II. EXPERIMENTAL DETAILS

The used simples, obtained in laboratory LAAS-CNRS of Toulouse (France), consist of two-layers (bi-layers) deposited at 480°C by thermal low-pressure chemical vapor deposition (LPCVD) method. The first layer, deposited on oxidized single-crystal Si substrates with thermal oxide SiO$_2$ of 25 nm-thickness, consist of 200 nm-thickness of nitrogen-doped-silicon (NiDoS); deposited with 1% of nitrogen content in disilane (Si$_2$H$_6$) and ammonia (NH$_3$) gases mixture. The second layer, deposited on the NiDoS layer, consists of about 110 nm-thickness of in-situ doped P$^+$ poly-Si; deposited from Si$_2$H$_6$/BCl$_3$ gases mixture. Fig. 1 illustrates a schematic representation of the investigated samples structure.

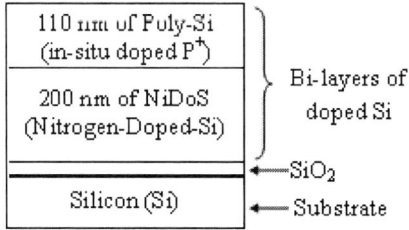

Figure 1. Schematic representation of the investigated P$^+$ doped poly-Si/NiDoS bi-layers structure.

In order to avoid long-rage redistributions, thermal B activation anneals were carried out at relatively low-temperatures (700 and 850°C) for various periods ranging between 15 min and 2 hours. The experimental B redistribution profiles in the bi-layers have been obtained using secondary ion mass spectroscopy SIMS technique, with a CAMECA IMS4F6 micro-analyzer. Fig. 2 shows superposition of SIMS profiles obtained before and after annealing at 700 and 850°C, for various annealing times. All the SIMS profiles show a concentration peak appearance, in the NiDoS region, near the bi-layers interface, which indicates important B transfer from the poly-Si layer and trapping at NiDoS. The B concentration in the trapping peak increases with increasing annealing time and temperature, in presence of opposite concentration gradient and nitrogen atoms. The profiles peak evolution is similar to a Gaussian in the NiDoS region. The curves are not very abrupt at the interface and they differ clearly to those observed in poly-Si/mono-Si interfaces [1,2].

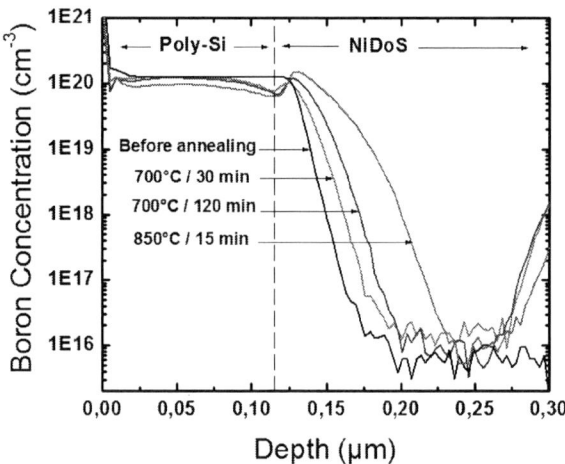

Figure 2. Comparison between SIMS diffusion profiles before and after annealing for various temperatures and times.

III. TWO-STREAM TRANSFER MODEL

The model parameters published in our previous work [3] and some other set of parameters have been used. To explain B redistribution in poly-Si/NiDoS bi-layers, the total B population is divided between the poly-Si region and NiDoS region, coupled together by effective B transfer or exchange between the two regions; named a two-stream transfer process. The B redistribution within poly-Si region differs to that within NiDoS region. Nitrogen (N) concentration in NiDoS region is almost uniform and so it is not necessary to use co-diffusion equations. The effect of N population on B redistribution has been only taken as a calibrating parameter. The B transfer is strongly affected by the kinetics of trapping, peak increasing, segregation, and solubility limit motion; since the solubility limit can be enhanced [4]. So, five transport mechanisms for B-atoms can be considered: (a) effective transfer in poly-Si region, (b) effective transfer in NiDoS region, (c) effective segregation near the interface, (d) clustering effects in the two regions, and (e) peak concentration appearance and evolutions. Moreover, effects of strong B concentrations such as that of the internal electric field, the charged vacancies, and the solubility solid limit excess are also considered [3]. Fig. 3 illustrates the particular transfer kinetics in Si bi-layers structure during B activation annealing. B atoms diffuse with two effective diffusion coefficients in P+ doped-Si/NiDoS structure.

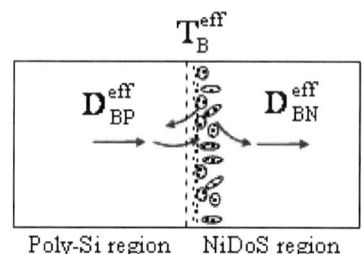

Figure 3. Schematic of effective B redistribution and transfer process in P+ doped poly-Si / NiDoS bi-layers structure.

With a two-stream transfer process, the established model is given by the coupled continuity equations (1) and (2) for the two B populations in the Poly-Si region C_{BP} and the NiDoS region C_{BN}.

$$\frac{\partial C_{BP}}{\partial t} = \nabla \left(D_{BP}^{eff} \, \nabla C_{BP} \right) - T_{B}^{eff} \qquad (1)$$

$$\frac{\partial C_{BN}}{\partial t} = \nabla \left(D_{BN}^{eff} \, \nabla C_{BN} \right) + T_{B}^{eff} \qquad (2)$$

$$T_{B}^{eff} = k_{t}^{eff} \left(C_{BP}^{int} - \frac{C_{BN}^{int}}{k_{seg}^{eff}} \right) \qquad (3)$$

T_{B}^{eff} in equation (3) describes the effective transfer or exchange of B-atoms at the interface, between Poly-Si and NiDoS regions. It is obtained by use of effective transfer rate k_{t}^{eff} and effective B segregation coefficient k_{seg}^{eff} [3]. D_{BP}^{eff} and D_{PN}^{eff} are respectively, the effective B diffusion coefficients in the Poly-Si region and NiDoS region [3]. As a result, the specified effective B redistribution coefficients in strongly doped Si bi-layers are modulated during annealing by the instantaneous concentrations and physical parameter.

IV. RESULTS AND DISCUSSION

The SIMS profiles after annealing have been simulated by using the initial profile before annealing as initial condition in redistribution model, and then calibrating independently values of sixe set of physical parameters; each of which has a clear physical meaning. Indeed, each of these parameters has a leading effect on a particular part of the redistribution profile where the associated phenomenon is dominating. The initial B concentration profile in Si bi-layers before annealing has been simulated without any difficulties by reproducing the SIMS profile, using theoretical expressions and adjustment method. This first simulated profile has been used as initial condition during simulation of redistribution profiles after annealing. We have then deduced two individual B population profiles, according to the depth: B population in the Poly-Si region C_{BP} and B population in the NiDoS region C_{BN}. The simulation which reproduces very well the investigated SIMS profiles, shown in Fig. 4, illustrates the significant roles of the instantaneous kinetics of B transfer, clustering, clusters evolution, and that of B segregation for the precise reproduction of the redistribution profiles. The simulations

978-1-4673-5289-5/12 $31.00 © 2012 IEEE

accurately predict B redistribution in Poly-Si region, transfer process within Poly-Si/NiDoS interface, concentration peak evolutions, and redistribution within NiDoS region, with increasing annealing time and temperature.

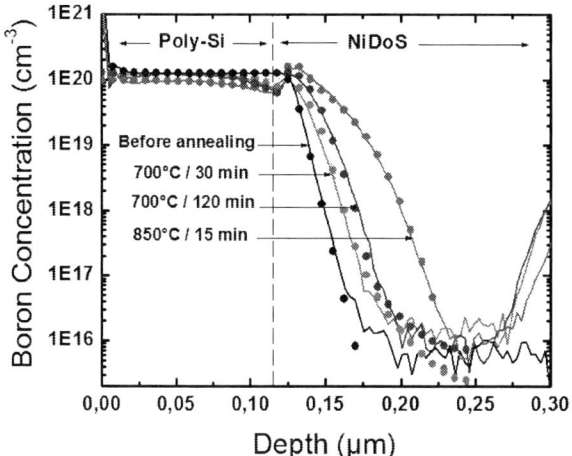

Figure 4. Simulated profiles (symbols) and SIMS profiles (lines) of B concentrations after annealing at: 700°C/2h, 700°C/30min, and 850°C/15min.

The effect of B-atoms interactions with charged vacancies, that enhances the B diffusivity, has been evaluated by varying the ratio of the intrinsic diffusivity induced by the positively charged vacancies on the global diffusivity induced by neutral vacancies D_i^+/D_i^0 [3]. The optimal value of this ratio obtained after best fitting is between 0.13 and 0.14, which is in good agreement with that reported by Mansour et al. [5]. The medium number of atoms to be trapped in formed small Si–B and B–B clusters in poly-Si region takes the value of 2. This value is similar to that obtained by Uematsu [6] for strongly doped single-crystal Si layers. However, the medium number of atoms to be trapped in formed small B–Si, B–B, N–Si, B–N, and B–N–B clusters in NiDoS region takes the value of 3; which lead to the best fitting for all the investigated profiles. Below the B solubility limit, strong-concentration B will lead to the formation of small Si–B and B–B clusters that enhances the B diffusivity. This enhancement diminishes with the clusters evolution during annealing [3,6-7]. The B solubility limit has been found to be enhanced in our investigated bilayers. The used expression given by Solmi et al. [8] was multiplied by a factor of about 4 to simulate the profiles inclinations or trend, which is in good agreement with the results in literature [4]. This means the excess of the produced complexes. Fig. 5 shows the instantaneous evolution of the B concentration in Poly-Si region, for different depths before bi-layers interface, for annealing at 700°C. We notice that the instantaneous B concentration is more and more influenced near the bi-layers interface. Its variation depends also on the annealing temperature. Near the bi-layers interface, the behavior of the instantaneous effective B diffusivity in Poly-Si side is not very different to that in NiDoS side. It is transient because it firstly increases and then decreases with increasing annealing time. The B transient enhanced diffusion in NiDoS region is obviously related to that in poly-Si region. This can be explained by the effective B transfer process starting from the Poly-Si region and the involvement of the same physical phenomena in the two regions (such as evolution of trapping, emission and crystallization). The same effects have been observed for annealing at 850°C but they are only activated. The instantaneous contributions of the physical phenomena in the two regions are dependent on annealing temperature.

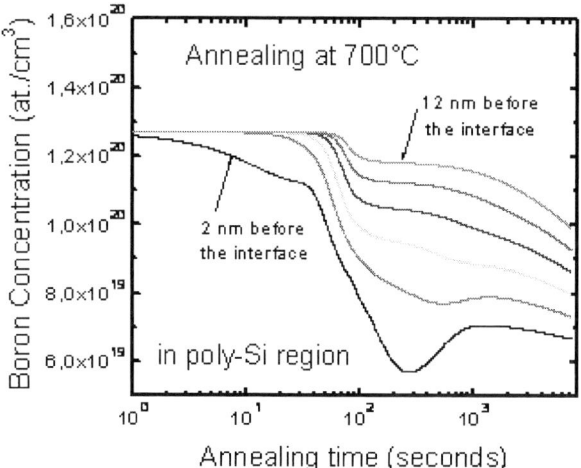

Figure 5. The instantaneous evolutions of the B concentrations near the bi-layers interface.

V. CONCLUSION

We have developed a redistribution model well adapted to the strong impurities-concentrations and to the particular structures of doped poly-Si/NiDoS bi-layers. This model takes into account technique of two-stream transfer via the interface, due to the dissimilar bi-layers. It has been revealed that B transient enhanced redistribution in NiDoS layer is almost related to that in poly-silicon layer. The formation of small B–Si and B–Si–N clusters enhances the diffusivity whereas the evolution of these clusters reduces this enhancement. We have concluded that B peak-concentration evolution dose not depend only on transfer kinetics from poly-Si region, but also on the trapping/segregation to the NiDoS region; under the effects of N concentrations and clustering kinetics.

REFERENCES

[1] S. Batra, M. Manning, C. Dennison, A. Sultan, S. Bhattacharya, K. Park, S. Banerjee, M. Lobo, G. Lux, C. Kirschbaum, J. Noberg, T. Smith, B. Mulvaney, J. Appl. Phys. 73, (1993) 3800.

[2] A. Merabet, J. Marcon, Nucl. Instrum. Methods B 253, (2006) 122.

[3] S. Abadli, F. Mansour, E. Bedel Pereira, Cryst. Res. Technol. 47, DOI 1002/crat.201200199, (2012) 1047.

[4] L. Shao, J. Zhang, J. Chen, D. Tang, P. Thompson, S. Patel, X. Wang, H. Chen, J. Liu, and W. Chu, Appl. Phys. Lett. 84, (2004) 3325.

[5] F. Mansour, R. Mahamdi, and L. Jalabert, P.T. Boyer, Thin Solid Films 434, (2003) 152.

[6] Masashi Uematsu, Jpn. J. Appl. Phys. 38, (1999) 3433.

[7] S. Abadli, F. Mansour, Thin Solid Films 517, (2009) 1961.

[8] S. Solmi, F. Baruffaldi, R. Canteri, J. Appl. Phys. 69, (1991) 2135.

FPGA-Based implementation of IncCond algorithm for photovoltaic applications

N. Chettibi, A. Mellit

Faculty of Science and Technology, Renewable Energy Laboratory, Jijel University
Jijel, Algeria

M. Drif

Electronics department, M'sila University,
M'sila, Algeria

Abstract— **Recent advances in control algorithms embedded into a Field Programmable Gate Array (FPGA) allowed the application of such algorithms in real engineering problems (robotic, image and signal processing, control, power electronics, etc.), however, the application of such technologies in the solar energy field is very limited. The embedded advanced algorithm into FPGA can play a very important role in renewable energy systems for control, monitoring, supervision, etc. FPGA technology was employed due to its development, flexibility and low cost. In this paper, the incremental conductance (IncCond) algorithm is implemented on FPGA for tracking the Maximum Power Point (MPP) of a photovoltaic (PV) array. The IncCond algorithm has been designed using a description language (VHSIC standing for very-high-speed integrated circuits) then implemented on Xilinx Virtex-II-Pro(xc2v1000-4fg456. Modelsim-based simulation results confirm the good tracking efficiency (96%) and rapid response time (2ms). It has been confirmed that the implemented algorithm performs better than the well known Perturb and Observe (P&O) algorithm.**

Index Terms— **Photovoltaic, IncCond algorithm, MPPT, VHDL, FPGA**

I. INTRODUCTION

In Algeria, the potential for renewable energy is strongly dominated by solar energy. Algeria considers this source of energy as an opportunity and a lever for economic and social development, particularly through the establishment of wealth and job-creating industries.

The photovoltaic (PV) generation system is one of the most promising renewable energy sources, as it has the advantages of being safe, inexhaustible, pollution free and requiring little maintenance. Recently, the photovoltaic energy source has received much attention as alternative means of generating electricity. The photovoltaic systems are used today in many applications, which can be classified into two main categories: the stand-alone (grid-off) PV system and the grid-connected PV system. In remote rural areas where the grid connection is

impossible, the stand-alone PV systems are used with a battery bank for the energy storage. On the other hand, to answer the need for alternative energy, the grid connected systems are used. In these systems, the PV panels are connected to the utility grid without the employ of battery bank such that the PV electricity produced is injected into the utility grid.

The amount of power generated by photovoltaic modules depends mostly on the atmospheric conditions (irradiation and air temperature). In order to extract maximum power from a photovoltaic (PV) module for a given climatic conditions, a maximum-power-point tracker (MPPT) is used. This control unit is usually associated with a DC/DC converter permitting the power transfer from PV generator to DC load.

Different MPPT algorithms have been proposed in literature [1,2], Perturb and observe (P&O), incremental conductance (IncCond), parasitic capacitance, constant voltage, artificial intelligence techniques, etc [3]. Conventional algorithms, as P&O, guarantee acceptable performances, and are easy to implement. On the other hand, artificial intelligent methods perform better but are generally more complicated to implement; they also require relatively high performance processor. Since the functions of various components can be integrated onto the same chip, FPGAs offer lower implementation cost than microcontrollers and DSPs. FPGAs can provide equivalent or higher performance with the customization potential of an ASIC. As FPGAs can also be reprogrammed at any time, repairs can be performed in-situ while the system is running providing a high degree of robustness. Beside robustness, this re-programmability can also provide a high level of flexibility [4]. Therefore, to meet the required performance, FPGAs are desirable since their performance can easily surpass the performance of microcontrollers and DSPs.

Due to its performance, simplicity and easy implementation, the well-known IncCond algorithm is used for the MPP tracking. Thus, the main objective of this paper is to implement the IncCond algorithm on FPGA board for possible PV system application. The used grid-off PV system is installed on the rooftop of our Renewable Energy Laboratory, at Jijel University.

This paper is organized as follows: The IncCond algorithm description is provided in the next section. MPPT algorithm

978-1-4673-5289-5/12 $31.00 © 2012 IEEE

implementation as well as simulation results are given in section IV. Conclusion and perspectives are presented in the final section.

II. THE INCCOND ALGORITHM

To overcome the P&O algorithm problem in the case of sudden variation of irradiance, the IncCond algorithm was developed by Hussein et al. [5]. This method can determine the right direction in which voltage change should be done even under rapidly changing conditions.

The IncCond [1,5] algorithm is based on the fact that the slope of the PV array power curve (Fig. 1) is zero at the MPP, positive on the left of the MPP, and negative on the right, as given by

$$
\begin{cases}
dP/dV = 0, & \text{at MPP} \\
dP/dV > 0, & \text{left of MPP} \\
dP/dV < 0, & \text{right of MPP.}
\end{cases} \tag{1}
$$

Since

$$
\frac{dP}{dV} = \frac{d(IV)}{dV} = I + V\frac{dI}{dV} \cong I + V\frac{\Delta I}{\Delta V} \tag{2}
$$

Eq. 1 can be rewrite as:

$$
\begin{cases}
\Delta I/\Delta V = -I/V, & \text{at MPP} \\
\Delta I/\Delta V > -I/V, & \text{left of MPP} \\
\Delta I/\Delta V < -I/V, & \text{right of MPP.}
\end{cases} \tag{3}
$$

The MPP can thus be tracked by comparing the instantaneous conductance $(G=I/V)$ to the incremental conductance $(\Delta G=\Delta I/\Delta V)$ as shown in the flowchart in Fig. 2. V_{ref} is the reference voltage at which the PV array is forced to operate. At the MPP, V_{ref} equals to V_{MPP}.

Once the MPP is reached, the operation of the PV array is maintained at this point unless a change in ΔI is noted, indicating a change in atmospheric conditions and the MPP. The algorithm decrements or increments V_{ref} to track the new MPP.

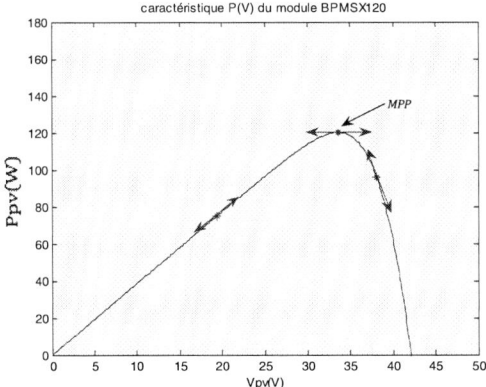

Fig.1 Power-Voltage curve under constant irradiance and temperature.

The tracking efficiency (η_{MPPT}) is an important parameter of an MPPT algorithm. This value is calculated as:

$$
\eta_{PV} = \frac{\int_0^t P_{MPPT}(t)dt}{\int_0^t P_{max}(t)dt} \tag{4}
$$

Where P_{MPPT} represents the output power of PV system with MPPT, and P_{max} is the output power at true MPP [6].

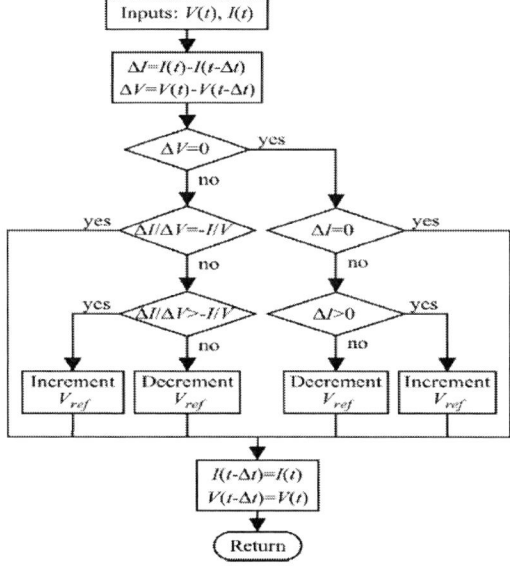

Fig.2 the flowchart of IncCond algorithm [1]

III. BOOST CONVERTER

In this work, the grid-off PV system adopts step up DC/DC converter topology in order to adapt the PV panel output terminals to a large voltage range. Therefore, the boost converter performs the tracking of the optimum operating point. For a PWM boost converter, the relationship of output voltage to input voltage can be determined as follows:

$$
\frac{V_{out}}{V_{in}} = \frac{1}{1-D} \Rightarrow V_{in} = (1-D)V_{out} \tag{5}
$$

The IncCond MPPT controller adjusts the PV panel voltage to follow the reference voltage by regulating the duty cycle (D) of the DC/DC converter control signal.

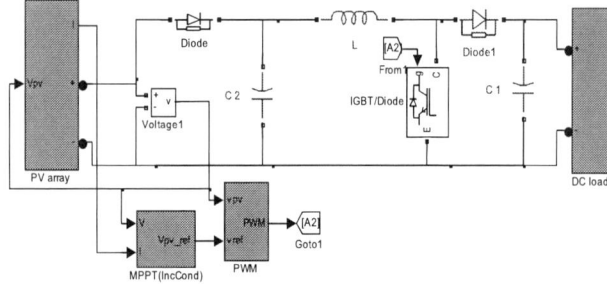

Fig.3 The stand-alone PV system with IncCond control method

IV. ALGORITHM IMPLEMENTATION AND RESULTS

Fig. 4 shows the considered PV system which consists of four main blocks:

1. A circuit which adapts the current and the voltage of the PV array;
2. A DC/DC boost converter;
3. An ADC converter which converts the analogue current and voltage to digital data;
4. A control tool (IncCond algorithm) implemented on the FPGA device.

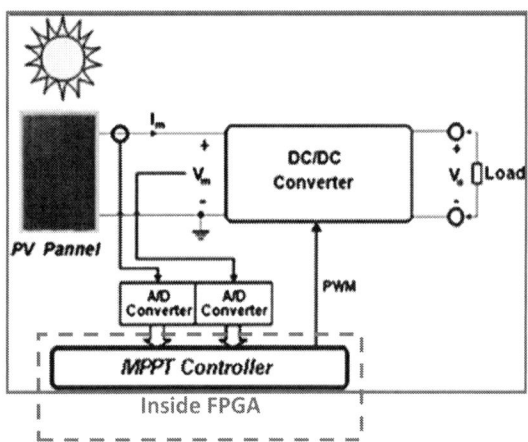

Fig.4 The considered PV system with IncCond algorithm integrated on FPGA

The simulation was carried out in a manner such that to each value of the duty cycle ratio (D) the signal generated at the output of the controller, corresponds a single point with (I, V) coordinates on the memorized I–V characteristic of the considered PV panel (BP SX120).

During each acquisition period, the controller generates the corresponding new duty cycle ratio (D) signal which sets the new position of the tracked MPP and vice versa.

The IncCond algorithm has been simulated in the ModelSim environment. Fig. 5(a) shows the convergence of the IncCond algorithm at STC (G=1000W/m², and T=25°C). As can be seen the simulated algorithm converge to the MPP after 2ms. The efficiency of the system 95.93% for

perturbation step ΔD= 0.5. With reference to our paper published in [3] the implemented IncCond performs better than Perturb & Observe (P&O) algorithm.

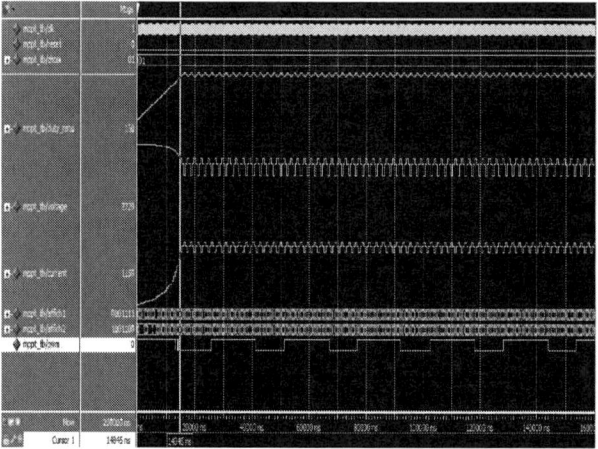

Fig.5.a Convergence of IncCond algorithm using ModelSim

Fig. 5.b depicts the RTL view of the synthesize IncCond algorithm by using the ISE software.

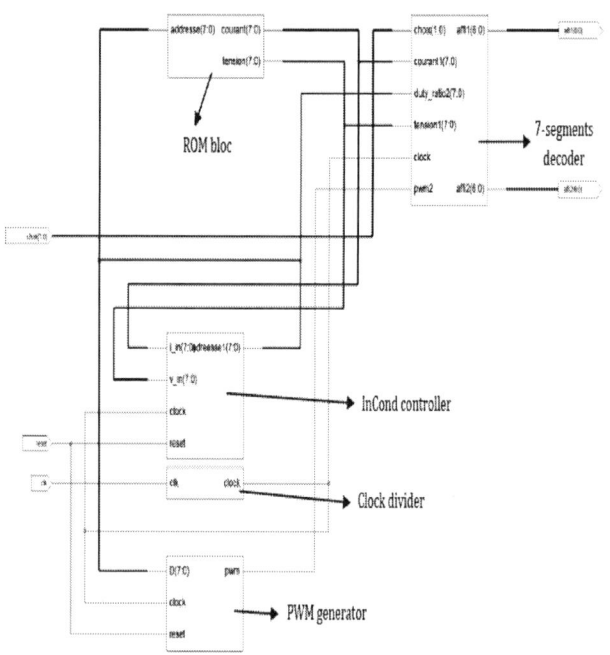

Fig. 5.b the RTL view of the synthesize algorithm

The different sub-units described above were implemented separately on a Virtex II (XC2v1000-4fg456) FPGA chip from Xilinx, shown in Fig.5.c. Table II reports the FPGA (xc2v1000-4fg456) resources summary for the designed MPPT-controller. With reference to fig.5.c and Table II, this chip was largely sufficient to implement all the constituents of the MPPT controller addressed in this work.

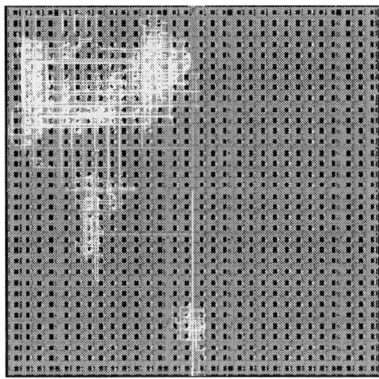

Fig.5.c Floor planning of the FPGA implementing the IncCond algorithm

TABLE II DEVICE UTILIZATION SUMMARY (XC2V1000-4FG456)

Selected devices	xc2v1000-4fg456
Number of slices	140 out of 1536
Number of slice flip flops	165 out of 3072
Number of 4 input LUTs	182 out of 3072
Number of bonded IOBs	60 out of 200
RAMs	2 out of 24
Number of MULT18X18s	1 out of 24
Number of GCLKs 1	1 out of 16
Maximum frequency	90

V.   CONCLUSION

In this paper, the IncCond algorithm has been implemented on a FPGA chip for real time simulation of the MPPT for photovoltaic applications. The efficiency of the implemented IncCond controller is 96% and the time response is 2ms. It has been demonstrated that the implemented algorithm can quickly reach the MPP. It has been also shown that the used FPGA Virtex II is sufficient for this application; therefore, the implementation algorithm into FPGA for the tracking of the MPP is very promising for photovoltaic applications.

Our future action consists to make an experimental realization of the implemented algorithm for real-time photovoltaic applications by taking in account the expenditure required.

ACKNOWLEDGMENT

The second author expresses a special acknowledgment to the International Centre for Theoretical Physics (ICTP), Italy for his helpful in this work. This work was supported by the TWAS under grant Ref. 09-108 RG/REN/AF/AC_C: UNESCO FR: 3240231224.

VI.   REFERENCES

[1] T. Esram and P.L Chapman, "Comparison of PV array maximum power point tracking techniques" IEEE Transactions on energy conversion, vol. 22, no. 2, June 2007, pp.439-449

[2] N. Khaehintung T. Wiangtong P. Sirisuk "FPGA implementation of MPPT using variable step-size P&O algorithm for PV applications". In:, IEEE-ISCIT'06; 2006. pp. 212-215.

[3] A. Mellit, H. Rezzouk A. Messai and B. Medjahed, "FPGA-based real time implementation of MPPT controller for photovoltaic systems", Renewable energy, vol.36, no.5, 2011, pp.1652-1661.

[4] A. Messai A. Mellit, A. Guessoum, and S.A. Kalogirou, "Maximum power point tracking using a GA optimized fuzzy logic controller and its FPGA implementation", Solar energy, vol.85, pp.265–277, 2011 .

[5] K. Hussein, I. Muta, T. Hoshino, M. Osakada, "Maximum photovoltaic power tracking: an algorithm for rapidly changing atmospheric conditions" IEE Generation, Transmission and Distribution, vol.142, 1995, pp. 59-64.

[6] V. Salas, M. Alonso-Abella, F. Chenlo, and E. Olıas, "Analysis of the maximum power point tracking in the photovoltaic grid inverters of 5kW," Renewable Energy, vol. 34, no. 11, 2009, pp. 2366–2372.

Comparative study on Water Max A 64 DC pump performances based Photovoltaic Pumping System design to select the optimum heads in arid area

Azzedine Boutelhig*
PV water pumping laboratory
Applied Research Unit for Renewable Energies - URAER
Ghardaia, Algeria
boutelhig@yahoo.com

Yahia Bakelli
PV water pumping laboratory
Applied Research Unit for Renewable Energies - URAER
Ghardaia, Algeria
bakelliy@gmail.com

Abstract—Stand Alone Solar Pumping System is being attracting the citizen of remote desert locations, especially, those far away from electric grid. Among them there are many owners of small scale agricultural fields which need a permanent water supply for irrigation and livestock watering as well as for domestic usage. In Direct Coupling Photovoltaic Pumping System (DC/PVPS), the PV array constitutes the expansiveness part of the whole installation; however, an accurate sizing of such system is inevitably to satisfy the customer's request on daily required quantity of water with a minimum low erection cost. Throughout this work, a proposed comparative study based on Water Max A 64 DC pump (MPP=300W) performances, PV powered by four Isofoton (110W/24V) modules, has been carried out. The main goal is to determine with a selected PV generator design, the optimum total mano-metric height (TMH), suitable to drawdown the demanded quantity of water, on different boreholes of Ghardaia region. Under the real outdoor conditions of the site, the pump has been put into test, at different heads by mean of the PV pumping test bench. Following the treatment of the experimental results, the pump efficiency and the daily cumulative water corresponding to each head were calculated. A data sheet involves the appropriate region and the corresponding optimum head, suitable to extract the demanded daily volume of water has been concluded.

Index Terms—: PV pumping, expansiveness, sizing, TMH, efficiency.

I. INTRODUCTION

Around 80% of the Algerian total area is classified as arid and semi-arid regions. A substantial potential of solar energy source is covering the large south of the country with an average yearly solar radiation, received on horizontal surface ranges between 5kWh/m^2/day and 7 kWh/m^2/day. The area is characterized by a high flow of ground water resources, where it can be found at low drilling heads. The borehole and well average heads vary from 5m until 20m, in Adrar and Meniaa basin and from 20m until 50m in Metlili and Sebseb basin. The availability of these two factors (solar and water sources) in M'Zab Valleys of Ghardaia region, makes the Photovoltaic

water pumping system (PVPS) an appropriate solution to supply water, especially for remote areas.

Many works to implement optimum PVPS installations have been investigated, in this region [1]. The promotion of PV water pumping systems in irrigation is today a necessity to ovoid high cost of the grid energy invoice [2]. Based on real data, obtained through different experimental tests, which have been carried out in our PV pumping facility, different configurations of PVPS have been studied [3]. The recent works have been concentrated on the development of software as a yet simple and cost effective, especially a considerable interest in the development of the Fuzzy Logic Controllers (FLC) hardware implementations [4];[5]. However, the local soil specifications should be taken into consideration, as the decrease of the rainfall which affects on the groundwater table and the water depth variation, from a valley to another. In this view, a comparative study has been carried out to determine optimum well heads, suitable to drawdown the required volume of water by mean of the selected PV water pumping system design. The concluded results have been involved in data sheet, which serves as tool to be consulted, whenever, the PV pumping system design is selected to be installed in the appropriate region.

II. CHARACTERISTICS OF GHARDAÏA REGION

- Location: 600 km south of the Mediterranean Sea
- Latitude: 32° 36' N
- Longitude: 3° 81'E
- Altitude: 450 m above the sea level
- Rate of sunny days per year: 77%
- Annual daily average of global solar irradiance ranges between 5 kWh/m^2 and 7 kWh/m^2 horizontal surfaces.

The geographic situation of the Ghardaia province area is shown in the map of the figure1.

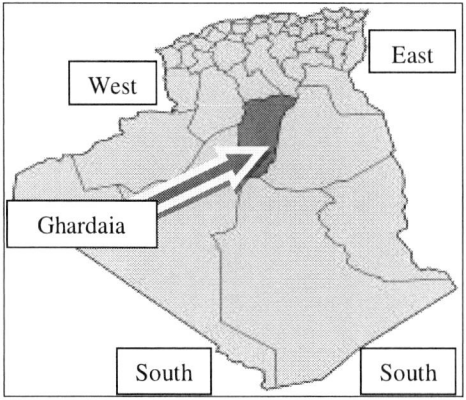

Figure 1: Geographic situation of Ghardaia

III. MATERIALS

A. Stationary PV pumping system

The stationary PV water pumping lab installed at URAER/Ghardaïa site consists of a complete test bench assembled by the following parts:

Inside the lab:

- Stainless steel tank (artificial well), type acerinox 1.4301 2B / 034DC7, completed by hydraulic system which involve two flow meters, two pressure sensors and control valve to adjust the pumped water pressure.
- MPPT (300W) for low power
- DC/AC inverter for three phases pumps.
- Electrical panel display which displays the following parameters: Q (m^3/h), I(A), V(V),E(W/m^2), TMH(m).
- Connexion box to select the different configuration (DC pump, three phase pump or DC pump via the MPPT).
- Data acquisition connected to PC

Outside the lab:

- PV generator composed of 25 Isofoton (110 W/24V) PV modules, implemented about 40 m away from the lab, direction full south, with optimal inclination angle of 32°.
- Earth installation.

The global PV pumping test bench complete with the accessories is shown in the figure.2;

Figure 2: The PV pumping test facility at URAER

B. Description of the selected PV pumping configuration

The Water Max A64 DC pump with nominal characteristics (I_m = 4.6 A, V_m = 64V, P_m = 300 W), is PV powered by a selected PV array configuration, comprised of two paralleled raw with two serial Isofoton (110 W) PV modules each (2X2), has been put into test. The figure.3 shows the selected PV array:

Figure 3: The (2 X 2) Isofoton (110/24) selected PV array

The designed PV array consists of four mono-crystalline silicon Isofoton (110/24) module (2X2) with nominal characteristics (I_{sc} = 3.5 A, V_{oc} = 40 V, P_m =110 W), each. The global PV array nominal characteristics are: (I_{sc}= 7 A, V_{oc} = 80 V, P_m = 440 W) and the total area is 3.328 m². The mentioned PV array has been tested, under the outdoor conditions of the site in the purpose to determine its maximum power point (MPP). The MPP is the product of the maximum possible voltage that the PV total cell area can produce in sunlight and the maximum possible current that the PV total cell area can produce at a given level of irradiance. Then it is given by the equation 1.

$$P_m = V_m * I_m \qquad (1)$$

The performances of the PV generator is characterized by the quality of the (I,V) sharpness curve, which consists of the percentage rate of the power that the PV array design can provide from its total power capacity rate to feed such load. It is calculated by the formula.2.

$$FF = \frac{V_m * I_m}{V_{oc} * I_{sc}} \qquad (2)$$

Where; V_m, I_m ; are the maximum power and current. V_{oc} , I_{sc} ; are the open circuit voltage and the short circuit current.

The efficiency of the PV array is given by the formula.3:

$$E_{PV} = \frac{V_m * I_m}{E * A_{pv}} \qquad (3)$$

Where; A_{pv} is the total PV cell area (A_{pv} =2.88 m²). E is the daily average irradiance in (W/m²).

IV. METHOD

Under the real weather conditions, the pump is tested for different heads, during 8 hours of straight pumping per day. The results were recorded directly in the PC via the Agilent 34970A acquisition data.
Through the obtained data, the daily hydraulic energy and the daily electric energy demanded by the pump were calculated by mean of the (equation.4) and the (equation.5), respectively. The efficiency of the pump was concluded by mean of the (equation.6), for each dynamic head.

$$E_h = Ch * H * \int_{8h} Qdt \qquad (4)$$

Where, Ch (Ch = 2.725) is the hydraulic constant, H (m) is the head and Q (m³/day) is the daily cumulative water.

$$E_e = \int_{8h} Pdt \qquad (5)$$

P (W) is the instantaneous power consumed by the pump. The efficiency is then calculated by the comparison of the hydraulic energy provided by the pump and the required energy coming from the PV array.

$$E_p = \frac{E_h}{E_e} \qquad (6)$$

V. RESULTS AND DISCUSSION

Following recurrent characterization tests, carried out on the selected different PV array configurations, the obtained (I, V)

characteristic results were treated and then the fill factor and the efficiency of the PV array were calculated. The following table.1 concluded the PV array characteristics in Standard Test Conditions (STC) and the corresponding obtained Fill Factor and efficiency, under the local outdoor conditions:

TABLE I. TABLE TYPE STYLES

PVG Design	Nominal values (STC)			Area	FF	Efficiency
P X S	I_{sc}(A)	V_{oc}(V)	P(W)	A_{pv}(m²)	(%)	E_{pv}(%)
1 module	3.5	40	110	0.832	72	12
(2x1)	7.0	40	220	1.664	65	12
(1X2)	3.58	80	220	1.664	63	11.4
(2X2)	7.0	80	440	3.328	63	12.5

The selected PV array configuration (2X2) efficiency rating is about (12.5%). The plots showed by the figure 4, averred a good rate of Maximum power point (MPP) produced at noon of a cool clear sky day. The figure.5 shows the (MPP) line produced during one day at different solar radiation levels;

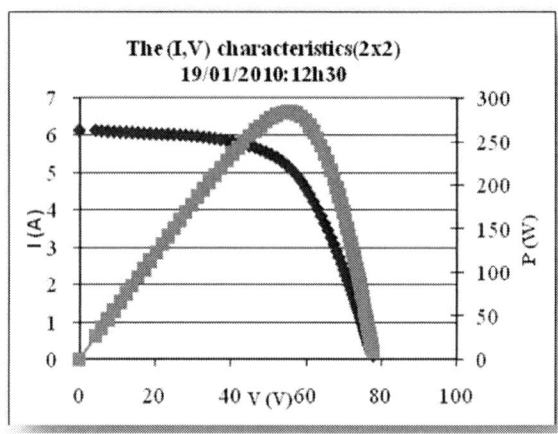

Figure.4: (I,V) characteristics and MPP rating of (2X2) PV array

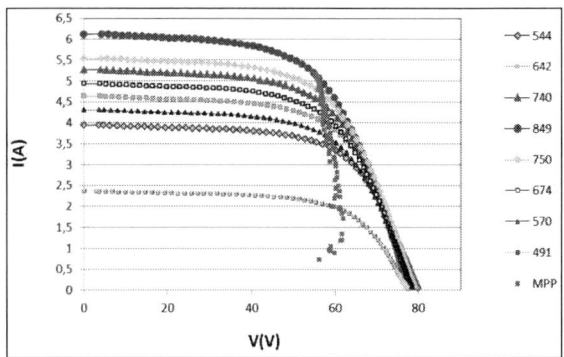

Figure.5: The MPP line in red produced by the (2X2) PV array

The following curves ranges from figure 6 until figure.9 illustrate the pump efficiency for each head.

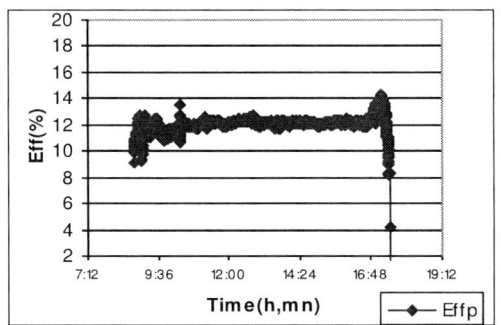

Figure 6 Water Max A64 pump efficiency for head=10m

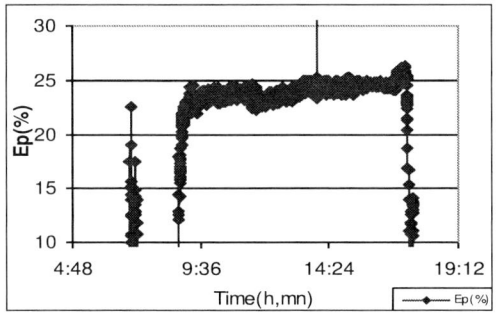

Figure.7 Water Max A64 pump efficiency for head=20m

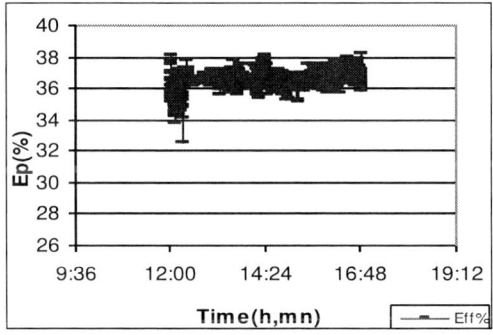

Figure.8 Water Max A64 pump efficiency for head=30m

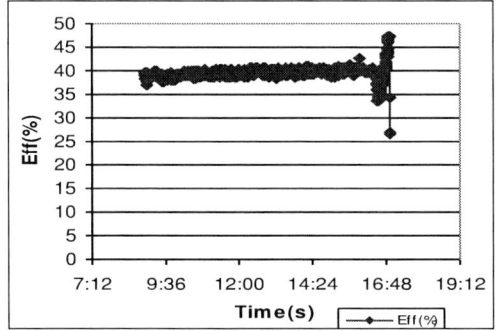

Figure.9 Water Max A64 pump efficiency for head=40m

The table.2 shows the appropriate sites suitable to the mentioned PVPS configuration according to the daily required average volume of water:

TABLE II. DATA SHEET OF THE APPROPRIATE ZONE

Head (m)	Daily cumulative water	Efficiency of the pump (%)	Suitable region
10	7.55	12.04	Menia
15	7.04	22.35	Sebseb
20	6.75	24.018	Sebseb
25	6.68	32.25	Mansoura
30	6.50	36.50	Metlili
40	6.28	39.52	Metlili

VI. CONCLUSION

The analysis study showed that the selected direct-coupling PV pumping configuration fit well with the well optimum heads of the region, range from 20 m until 40 m, taking into consideration the average daily volume of water needed by each citizen, which ranges from 6 m^3 until 8 m^3. The selection of the optimum head for this current PVPS design is based on the daily water needed and on the efficiency of the pump. In order to protect the pump from an early damage, it was advised to use the pump at its minimum optimal efficiency. The average efficiency of the pump is about 45%, and then it will be very suitable to select the heads corresponding to the efficiency between 30% and 40%.

REFERENCES

[1] Azzedine BOUTELHIG [1], Yahia BAKELLI [1]and Amar HADJ, Promotion of Photovoltaic Water Pumping System (PVPS) for irrigation in Desert Regions, Journal of Agriculture machinery Science, Turkey; Volume4; 2008.

[2] Azzedine BOUTELHIG [1], Yahia BAKELLI [1]and Amar HADJ, Study and Implementation of a Stand Alone Photovoltaic water pumping system (PVPS) in Desert Region, Revue des energies renouvelables, CDER, 2008.

[3] Azzedine BOUTELHIG[1], Yahia BAKELLI[1]. HADJMAHAMMED, and Amar HADJARAB, "Performances study of PV powered DC pump Shurflo model for an optimum energy use for different heads under the outdoor conditions of a desert area, Elsevier Journal of Energy volume 39, Mars 2012..

[4] A. Messai [a,d], A. Mellit [b,d,1], A. Massi Pavan [c], A. Guessoum [d], H. Mekki [a,d]; FPGA-based implementation of a fuzzy controller (MPPT) for photovoltaic module; Energy Conversion and Management 52 (2011) 2695‐2704 .

[5] Mohamed M. Algazar [a], Hamdy AL-monier [b], Hamdy Abd EL-halim [a], Mohamed Ezzat El Kotb Salem [c]; Maximum power point tracking using fuzzy logic control; Electrical Power and Energy Systems 39 (2012) 21‐28.

Optimal Model Selection For PV Module Modeling

I.Hadj Mahammed, Y. Bakelli, S. Hamid Oudjana
Applied Research Unit on Renewable Energies
URAER
Ghardaïa, Algeria
E-mail: hmidriss65@yahoo.fr

A. Hadj Arab
Center of Renewable Energies Development
CDER
Algiers, Algeria
E-mail: hadjarab@hotmail.com

S. Berrah
Electronics Lab/ Faculty of Electronics,
A. Mira University
Bejaia, Algeria
E-mail: sm_berrah@yahoo.fr

Abstract— In the current work, five parameters model (FPM) has been used to improve the accuracy of the I-V PV module characteristics curves. This model has been applied to characterize five PV modules of different technologies. These are Mono-crystalline silicon, polycrystalline silicon, triple –junction Amorphous and Thin-film (CIS). Characterization was carried out at outdoor conditions of an arid area located in Ghardaia (Algeria). Through analysis of the recorded results using RMSE (root mean square error) method, it has been found that the square error is kept to minimum. Taking into account errors of different origins and the measurement method it can be concluded that the mentioned model fits better with accuracy to characterize PV modules. The proposed method is also discussed.

Keywords—Photovoltaic module, characterization, outdoor measurement, modeling simulation.

I. INTRODUCTION

The research of alternative sources of energy is becoming increasingly important as the prices of electricity from national grid are getting higher and higher. Moreover, the establishment of an electrical power network to supply remote residential area is practically complicated and difficult task and not cost effective solution. In this context, electricity generated through photovoltaic technology constitutes an appropriate alternative regarding the abundance of solar radiations.

The performances of PV module are directly influenced by solar irradiance and module temperature and by several other parameters such as tilt and azimuth angles. As a result of this understanding, the PV module behavior under different conditions stills of great concern of photovoltaic research field.

An extensive works on PV module characterization had been carried out based upon the four most important electrical characteristics of PV module which are short circuit current (I_{SC}), open circuit voltage (V_{OC}), maximum current (I_m) and voltage maximum point (V_m). The present work of characterization is based on five parameters model since it gives a better accuracy. Characterization tests have carried out on four PV modules of different technologies namely, Mono-crystalline silicon, Poly-crystalline silicon, triple-junction Amorphous and Thin film (CIS).

The modules were subjected to outdoor conditions of a semi-arid area located in Ghardaia (Algeria). The recorded Data allowed obtaining the correlation between model parameters in function of irradiance and temperature.

II. MATERIALS AND METHOD

A. Characteristics of the site:

Characterization tests were performed at outdoor conditions of the testing site (Applied Research Unit for Renewable Energies). The site is located in Ghardaia about 600 km south of Algiers. Having the following coordinates 32°36' N, Longitude: 3°81' E and elevation of 450 m above sea level, this area is classified as an arid region. Based upon meteorological data the solar resources and temperatures are as follows: rate of sunny days varies between 77% and 80% per year; the average daylight hours is approximately 5 hours in winter and up to 10 hours over other seasons (summer, spring, autumn). The daily average irradiation on a horizontal surface varies between 6 kW/m^2/day and 7 kW/m^2/day. The minimum and maximum summer average temperatures are 26°C and 42°C respectively. Agriculture activities are resumed mainly in growing palm trees, livestock and vegetables.

B. Theory of five parameters model

The behavior of a photovoltaic generator may be closely represented by the five parameter model based on one diode equivalent circuit of a solar cell; it consists of a diode, a current source, a series resistance, and a parallel resistance. [1-6]The current source generates photocurrent (I_{ph}) which is a function of incident solar irradiation and cell temperature. The diode represents p–n junction of the solar cell. At real solar cells, the voltage loss on the way to the external contacts is observed. This voltage loss is expressed by a series resistance (R_s). Furthermore, leakage currents are described by a shunt

978-1-4673-5289-5/12 $31.00 © 2012 IEEE

resistor (R_{sh}). Using Kirchhoff's first law, the equation for the extended I–V curve is derived as shown by equation 1:

$$I = I_L - I_0 \left[\exp\left(\frac{V + IR_s}{m \, V_t} \right) - 1 \right] - \frac{V + IR_s}{R_{sh}} \qquad (1)$$

Where I is the output current of PV module, V is the terminal voltage of the cell or module, q is the electric charge ($1.6 \ 10^{-19}$ C), k is the Boltzmann constant ($1.38 \ 10^{-23}$ J/K), and T is the cell temperature (K). m is the ideality factor, I_L the photogenerated current under insolation, R_{sh}: shunt resistor; R_s: series resistor and I0 is the diode saturation current. The five equivalent circuit parameters are given in [2, 7-8].

After the parameters calculation their values were substituted at equation 1 and the I-V characteristic curve (current versus voltage) for the photovoltaic module was simulated. The values of V_{OC}, I_{SC}, V_m, I_m, are given by the manufactures and were used in the simulation.

One file with all data measured was created and one can simulate different photovoltaic modules to any given solar radiation and cell temperature.

C. Equipement

The instrument for recording IV-characteristics is a PV Photovoltaic Field Array Tracer (PVPM) 250V 40A model. It is a portable array tester and it uses a discharged capacitor that functions as 'load resistor'. Current and voltage are measured during the process of charging the capacitor. (Figure 1)

A calibrated reference solar cell was used for measuring irradiance and cell temperature. It consists of m-Si cell. A Pt1000 temperature sensor is glued to the backside of the centre cell. The Pt1000 sensor gives directly the cell temperature.

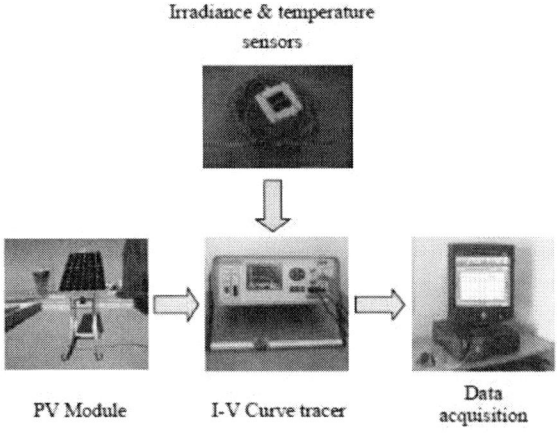

Figure 1. Synoptic of the test bench

Under the local outdoor test condition, the five parameter model has been applied on the obtained results of the four different mentioned PV modules. The standard characteristics of the subjected modules are shown in Table1.

TABLE I.   STANDARD CONDITIONS TESTED MODULES CHARACTERISTICS

Module	ASE100 GTFT95	BP3160	US64	ST40
Technologies	mc-Si	pc-Si	TJ a-Si	TF (CIS)
I_{SC} (A)	3.2	4.8	5.1	2.68
V_{OC} (V)	42.3	44.2	21.3	23.3
Im (A)	2.8	4.55	4.1	2.41
Vm (V)	34.1	35.1	15.6	16.6
P (W)	95	160	64	40

III. RESULTS AND DISCUSSION

The analysis of the recorded results by mean of RMSE (Root Mean Square Error)

With all the necessary data and parameters for plotting the curves (I_L, I R_S, V_m, V_{OC}, I_{SC} and I_m), it was possible to simulate, the desired photovoltaic module. Figure 2 show an example of the simulation output that are the graphs of voltage versus current and measured curve for the photovoltaic ST40 module, for different radiation levels and different temperatures

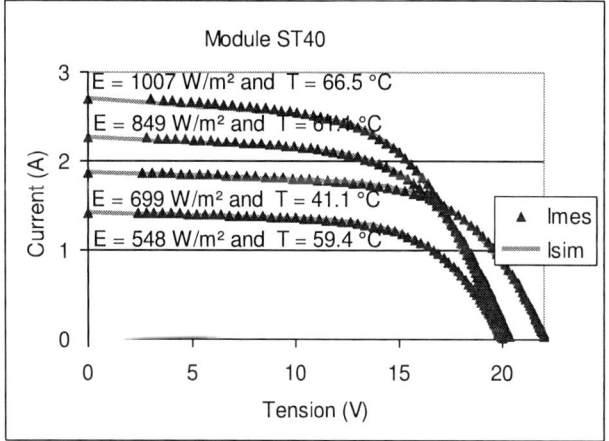

Figure 2.   IV characteristics measured and simulated under different test conditions for ST40.

The statistic indicators used to test the quality of the fit to the experimental data are the Root Mean Square Error (RMSE) equation 2

$$RMSE = \left[\frac{\sum \left(I_{cal} - I_{exp} \right)^2}{N} \right]^{0,5} \qquad (2)$$

Where Ical and I_{exp} are respectively, the calculated and measured currents and N: number of measured points.

Comparing the simulated values with the measured values, it is noted that the results are practically identical.

The maximum RMSE between the simulated and observed values, supplied by the measured, was presented at below. This indicates that the model used for simulation of the four type photovoltaic modules is valid.

TABLE II.   MAXIMUM RMSE BETWEEN THE SIMULATED AND MEASURED VALUES

Module	RMSEm	Nb PV test
ASE95	0.053767	210
BP3160	0.079658	285
US64	0.099373	259
ST40	0.046094	253

IV. CORRELATION OF THE FIVE PARAMETERS

To characterize a PV module as a power source in performance analysis, it is very important to take into consideration the dependence of all equivalent circuit parameters of PV module on irradiation and cell temperature, to be able to obtain the changing of the parameters over the whole range of operating condition. For the determination of the veracity of the simulated I-V characteristic curves, the curves supplied by measured data. Simple correlations (equations (3-5)) [9-12], were used to compare, and quantify errors between the simulated curves with the ones provided by measured ones by the correlation coefficient which is given by equation 6:

$$y = a * \frac{E}{1000}\left(1 + b*(T-25)\right) \quad (3)$$

$$y = a * \frac{E}{1000} + b*(T-25) + c \quad (4)$$

$$y = a * \log\left(\frac{E}{1000}\right) + b*(T-25) + c \quad (5)$$

$$CC = \frac{\sum\left(I_{exp} - \overline{I_{exp}}\right)\left(I_{cal} - \overline{I_{cal}}\right)}{\sqrt{\sum\left(I_{exp} - \overline{I_{exp}}\right)^2 \sum\left(I_{cal} - \overline{I_{cal}}\right)^2}} \quad (6)$$

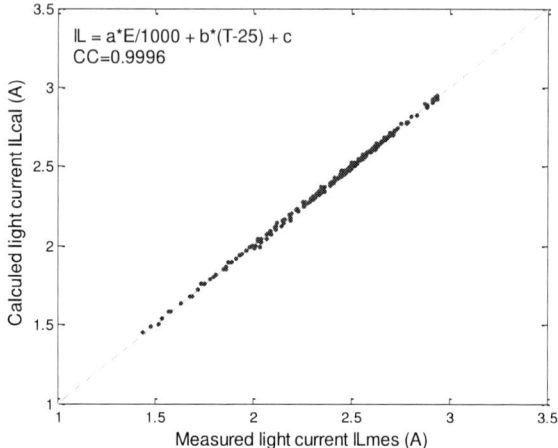

Figure 3.   Correlation of light current as function of irradiance and temperature for ST40.

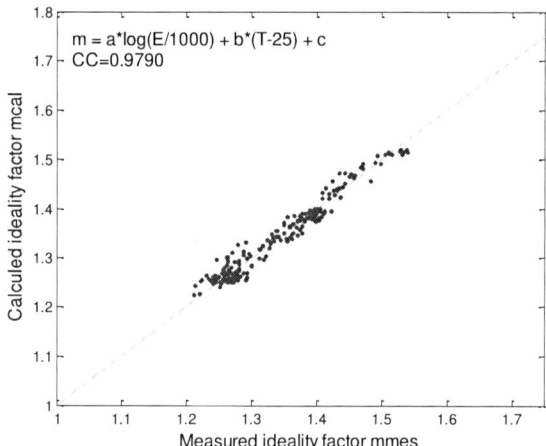

Figure 4.   Correlation of ideality factor as function of irradiance and temperature for ST40.

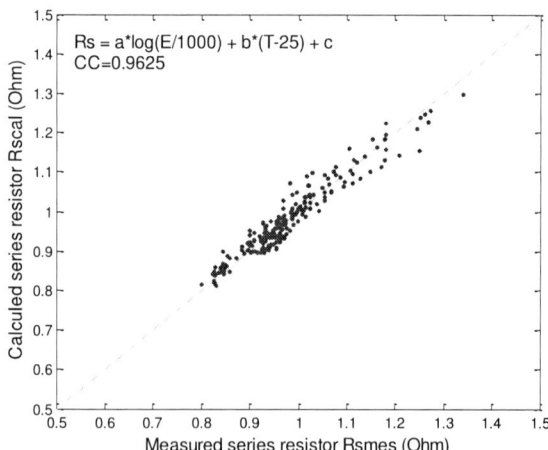

Figure 5.   Correlation of series resistor as function of irradiance and temperature for ST40.

978-1-4673-5289-5/12 $31.00 © 2012 IEEE

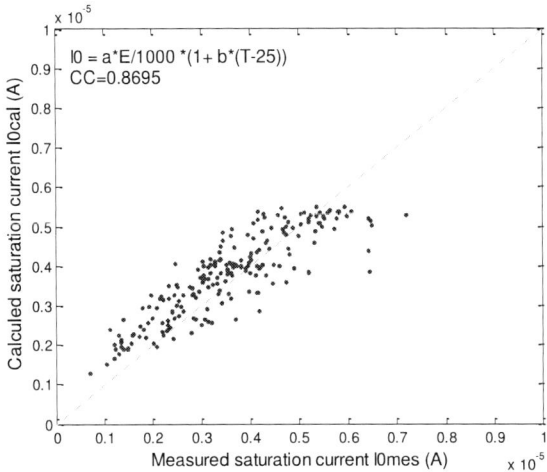

Figure 6. Correlation of saturation current as function of irradiance and temperature for ST40.

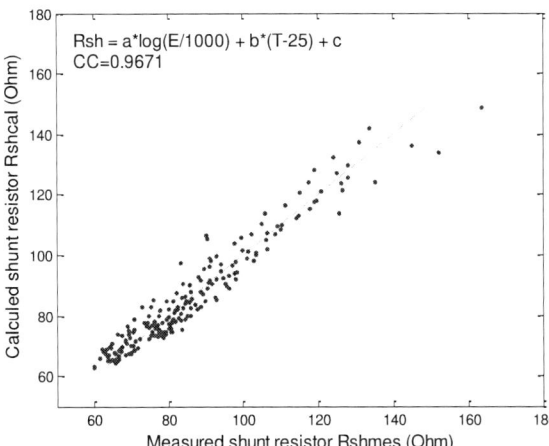

Figure 7. Correlation of shunt resistor as function of irradiance and temperature for ST40.

TABLE III. CORRELATION COEFFICIENTS OF THE FOUR PV MODULE CHARACTERISTICS:

Param.	Equ.	Module			
		ASE95	BP3160	US64	ST40
P	(3)	0.9983	0.9984	0.9934	0.9964
I_{SC}	(4)	0.9982	0.9984	0.9980	0.9995
V_{OC}	(5)	0.9172	0.9711	0.9675	0.9830
FF	(4)	0.9191	0.6733	0.8549	0.9756
I_m	(4)	0.9978	0.9985	0.9966	0.9975
V_m	(5)	0.8879	0.9259	0.9598	0.9803
I_L	(4)	0.9981	0.9985	0.9979	0.9996
m	(5)	0.7025	0.0519	0.6911	0.9790
R_s	(5)	0.9694	0.7153	0.7441	0.9625
I_0	(3)	0.7594	0.0981	0.4281	0.8695
R_{sh}	(5)	0.7937	0.4945	0.5941	0.9671

From Table III, it is obviously showed the better correlation of the IV characteristics of the ST40 module. A bad correlation for I_0 is remarked for non-existed linear correlation.

V. CONCLUSION

In this paper, we have investigated the IV characteristics performance of four different PV module technologies by using the five parameters model

Simple correlations have been used in order to illustrate the relation between the five parameters in function of the insolation and the PV module temperature. The obtained results averred that the ST40 module presents a better correlation than the others, thus, it can be used for further IV characteristics translation under different operating conditions.

REFERENCES

[1] J. P. Charles, M. Abdelkrim, Y. H. Muoy and P. Mialhe "A Practical Method Of Analysis Of The Current Voltage Characteristics Of Solar Cells," Solar Cells, 4, pp. 169 – 178, 1981

[2] D. S. H. Chan, J. R. Phillips and J. C. H. Phang "A Comparative Study Of Extraction Methods For Solar Cell Model Parameters," Solid-State Electromcs Vol. 29, No. 3, pp. 329-337, 1986.

[3] Z. Ouennoughi and M. Chegaar "A simpler method for extracting solar cell parameters using the conductance method," Solid-State Electronics 43, pp. 1985-1988, 1999

[4] K. Bouzidi, M. Chegaar and A. Bouhemadou, "Solar cells parameters evaluation considering the series and shunt resistance," Solar Energy Materials & Solar Cells 91, pp. 1647–1651, 2007

[5] M. Bashahu a, P. Nkundabakura "Review and tests of methods for the determination of the solar cell junction ideality factors" Solar Energy 81, pp. 856‑863, 2007

[6] I. Hadj Mahammed, A. Hadj Arab, F. Youcef Ettoumi, Y. Bakelli, S. Semaoui,. "A comparative study of photovoltaic generator models" 3rd International Convention on the Renewable Energies and the environment, November 6-8, 2006, Mahdia, Tunisia,

[7] A. Hadj Arab, F. Chenlo, K. Mukadam, and J. L. Balenzategui, "Performance of PV water pumping systems ", Renewable Energy, vol. 18, no. 2, p. 191–204, Oct 1999

[8] Hadj Mahammed, A. Hadj Arab, F. Youcef Ettoumi, Y. Bakelli, S. Semaoui " comparative Study of module photovoltaic models at the end of the study in situ for different photovoltaic module technologies " international Symposium of the Renewable energies (CER), May 04-05 2007 Oujda, Morocco,

[9] E. Skoplaki and J. A. Palyvos, " On the temperature dependence of photovoltaic module electrical performance: A review of efficiency/power correlations ", Solar Energy, vol. 83, no. 5, p. 614–624, May 2009.

[10] G. N. Tiwari and S. Dubey, Fundamentals of Photovoltaic Modules and Their Applications, RSC Publishing. 2010

[11] E. Q. B. Macabebe, C. J. Sheppard, et E. E. van Dyk, « Parameter extraction from I–V characteristics of PV devices », Solar Energy, vol. 85, n°. 1, p. 12–18, janv. 2011.

[12] Y. Tsuno, Y. Hishikawa, et K. Kurokawa, « Modeling of the I–V curves of the PV modules using linear interpolation/extrapolation », Solar Energy Materials and Solar Cells, vol. 93, n°. 6–7, p. 1070–1073, juin 2009.

Photovoltaics in Italy:
toward grid parity in the residential electricity market

Alessandro Massi Pavan and Vanni Lughi

Department of Engineering and Architecture, University of Trieste Via A. Valerio 6/A Trieste (Italy)

Abstract – The Italian photovoltaic market, since 2011 the world's largest, represents a success story having attained grid parity for the residential market of electricity – thus setting the basis for surviving without subsidies. The Levelized Cost Of Energy (LCOE) is calculated for three representative locations in Northern, Central, and Southern Italy, and compared with the residential end-user electricity price. It is shown that grid parity is attained under most conditions for the Italian residential market, thus laying the basis for its survival without subsidies.

Index Terms — Photovoltaic, grid parity, subsidies, incentives, LCOE, Italy.

I. INTRODUCTION

Historically, demand for photovoltaic (PV) systems and products in grid-connected applications has been driven by incentives [1]. In the history of PV, three major inventions led to sustainable markets for PV systems [2]. The first phase started after introducing PV power supply in space; as a consequence of the oil-price crises in in 1970s, PV applications started the second phase with off-grid applications; the third (current) phase has been enabled by the political invention of rooftop plans and feed-in tariff laws [3]. During this third phase a fundamental milestone in the history of PV market can be observed: the grid parity in many countries across the world, defined as the point where the price of electricity for the end consumer equals generation cost from PV. These considerations perfectly apply to the Italian market, which in 2011 became the world largest. The reduction of PV cost in Italy – which dropped by the 70% over the last decade – in conjunction with the good level of irradiation, led to the most important result achieved by this technology in its entire life: the realization of grid parity in the residential electricity market.

In this paper, starting from the cost of a domestic PV plant in Italy and its productivity, the photovoltaic Levelized Cost of Energy (LCOE) is calculated and compared with the utility price of electricity for residential customers. The comparison shows how incentives in the photovoltaic industry have been able to guide a niche market to its maturity, where subsidies will not be needed the near future. This paper is organized as follows: in the next section an overview of the Italian market is given, while Section III deals with the cost of electricity and Section IV with the yield of a PV plant. Finally, Section V discusses grid parity in Italy and Section VI presents conclusions and perspectives.

II. THE ITALIAN PV MARKET

Italy has become the largest PV market worldwide. Figures 1 and 2 show the installed PV power in Italy and Germany: in 2011, the PV power installed in Italy ($9250MW_p$) exceeded that installed in Germany ($7500MW_p$), which has historically been the market leader and in general the country that has most invested and believed in PV [4-5]. These results led to a cumulative installed power, at the end of 2011, of about 13 and $25GW_p$ in Italy and Germany, respectively.

This result supports the original core assumption of the PV incentive policies, i.e. that once the demand has been stimulated by incentives, the market starts growing at a rapid pace: in Italy the PV industry has seen a growth of more than 350% in one single year.

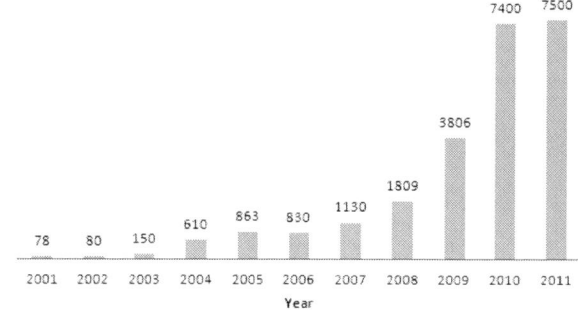

Fig. 1. Germany. Installed PV power in MW_p.

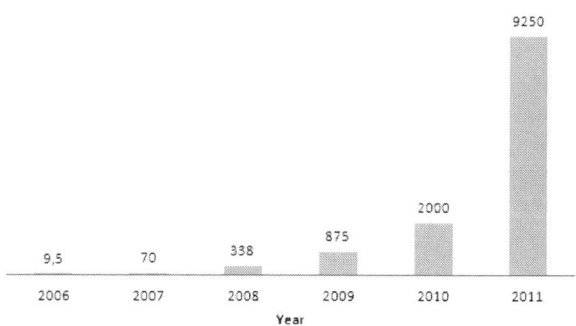

Fig. 2. Italy. Installed PV power in MW_p.

This growth is undoubtedly associated to the introduction of the feed-in tariff program in 2005. However, the incentives accorded through the feed-in tariffs programs, in the form of a reward paid for each kWh of electricity produced by PV plants, have been continuously reduced over the years and it is likely that after 2013 this type of subsidies will be discontinued. Therefore, survival of the PV market in Italy depends upon its ability to reach full maturity, by attaining grid parity in the residential segment first (where the cost of electricity is highest), followed by grid parity for commercial and industrial electricity, eventually reaching the so-called fuel parity – that corresponds to a scenario where the LCOE of PV equals that of conventional energy sources.

In this paper, we calculate current values of LCOE for the residential market, showing that for this segment, grid parity is already a reality.

III. THE COST OF ELECTRICITY

In this section, the methodology for the calculation of the cost of energy produced by a PV plant is given. A summary of the electricity price scheme for the final residential consumer is also given.

A. LCOE of a PV plant

The Simple Levelized Cost of Energy (LCOE) for renewable energies can be calculated as [6]:

$$LCOE = \frac{OCS \times CRF + FO\&MC}{8760 \times CF} + VO\&MC \qquad (1)$$

Where:

- OCS [€] is the overnight capital cost, i.e. the upfront cost of the plant;
- CRF [ ] is the capital recovery factor;
- FO&MC [€/year] are the fixed operation and maintenance costs;
- CF [ ] is the capacity factor, a function of the amount of time that the plant is operational;
- VO&MC [€/year] are the variable operation and maintenance costs.

The capital recovery factor is expressed as:

$$CRF = \frac{i \times (i+1)^N}{(i+1)^N - 1} \qquad (2)$$

Where:

- i [%] is the interest rate;
- N [ ] is the number of annuities received.

In the case of a PV plant, the fuel cost is zero; the variable operation and maintenance costs are usually set to zero, assuming to include the substitution of the inverter, once in the life of the system, in the fixed O&M costs. Moreover, the denominator of Equation 1 for a solar plant can be written as:

$$\frac{E_0}{N} \times \sum_{k=1}^{N} \left(1 - \frac{d_r \times (k-1)}{100}\right) \qquad (3)$$

Where:

- E_0 [kWh/year] is the yield of the plant over the first year of operation;

- d_r [%/year] is the degradation rate of the PV modules.

Thus, the final expression for the LCOE of a PV plant may be written as:

$$LCOE = \frac{OCS \times CRF + FO\&MC}{\frac{E_0}{N} \times \sum_{k=1}^{N} \left(1 - \frac{d_r \times (k-1)}{100}\right)} \qquad (4)$$

B. The price of electricity in Italy

The end-customer price of electricity in the Italian market depends on several factors as: the amount of kWh consumed in a year, the time of the day the electricity is used, the type of contract, etc. However, the Regulatory Authority for Electricity and Gas (AEEG), the independent body which regulates, controls, and monitors the electricity and gas markets in Italy, provides the mean price of electricity as a function of the amount of energy consumed in a year. Table I reports the mean price for electricity for the residential market in 2010 [7]. We assumed that current prices have remained constant.

TABLE I. MEAN ELECTRICITY PRICE FOR RESIDENTIAL CUSTOMERS IN ITALY

Energy [kWh/year]	Price [€]
< 1.000	0,28
1.000 – 2.500	0,16
2.500 – 5.000	0,20
5.000 – 15.000	0,25
> 15.000	0,28

IV. YIELDS

The LCOE of an Italian residential PV plant will be calculated considering three different locations that represent three common case studies for Italy (because of their different latitude and different level of available solar irradiance):

- Trieste in Northern Italy;
- Rome in Central Italy;
- Palermo in Southern Italy.

With reference to Table II and III, for these three cities two different yields have been used:

1. The *mean yield* [kWh/kWp/year] provided by GSE, the state-owned company who grants the incentives for the energy produced by the PV plants. The mean yield for a given location is the mean value of electricity injected into the electrical grid by the PV plants operating in that area;

2. The *optimal yield* [kWh/kWp/year] calculated as the product between the Balance Of the System (BOS) and the optimal solar irradiation for the specific location.

TABLE II. MEAN YIELDS FOR THREE REPRESENTATIVE CITIES IN ITALY

City	Yield [kWh/kW$_p$/year]
Trieste (North)	1.139
Rome (Center)	1.278
Palermo (South)	1.371

TABLE III. BOS, SOLAR IRRADIATIONS AND OPTIMAL YIELDS FOR THREE REPRESENTATIVE CITIES IN ITALY

City	η_{BOS} [%]	Irradiation [kWh/m²]	Yield [kWh/kW$_p$/year]
Trieste (North)	84	1.460	1.235
Rome (Center)	82	1.680	1.378
Palermo (South)	80	1.870	1.496

The power produced by a PV plant is [8]:

$$P = A \times \eta_{STC} \times \eta_{BOS} \times G \qquad (5)$$

Where:

- A [m²] is the PV array area;
- η_{STC} [%] is the PV module efficiency at STC (Standard Test Conditions);
- η_{BOS} [%] is the Balance Of the System;
- G [kW/m²] is the solar irradiance.

As for a 1kW$_p$ PV array it can be written:

$$A = \frac{1}{\eta_{STC}} \qquad (6)$$

The power produced by a 1kWp PV plant is:

$$P = \eta_{BOS} \times G \qquad (7)$$

Finally, substituting in Equation 7 the solar irradiance G with the solar irradiation I [kWh/m²], the yield of a 1kWp PV plant is:

$$E_{1kWp} = \eta_{BOS} \times I \qquad (8)$$

The values for the solar irradiation reported in Table III have been calculated with the Photovoltaic Geographical Information System (PVGIS) developed for the European Commission, which provides a map-based inventory of the solar energy resource [9]. Irradiances have been calculated assuming optimal tilt (35° for Trieste, 33° for Roma and 32° for Palermo) and azimuth (corresponding to the South direction for each of the three cities) angles. Finally, the BOSs have been calculated with the software PVSYST [10]. The different BOSs are due to different yearly average ambient temperatures, and therefore different temperature losses, in the considered locations.

V. GRID PARITY

In order to calculate the LCOE for a residential Italian PV plant, the following assumptions have been applied:

- The overnight capital cost is in the range of 2.200-2.400€/kWp plus VAT (10%), that represents a fair and widely applied price for a residential 3kWp turn-key plant in Italy;
- The number of annuities is 30 that corresponds to widely accepted lifespan for a PV plant;

- The interest rate is 5.2% that is the typical interest rate for a similar investment in terms of duration, size and security in Italy (e.g. the state bonds);
- The fixed operation and maintenance costs are calculated as the 1.5% of the overnight capital cost. This includes the substitution of the inverter once in the entire life of the plant, the ordinary maintenance of the electrical plant and damage insurance;
- The degradation factor is 0.7%/year, corresponding to the standard warranty on the power output of a PV module. As an example, the PV module Q.Pro G2 is guaranteed to produce at least 83% of the initial power after 25 years of operation [11.

The calculated LCOEs for the three locations and for two different OCSs are reported in Figures 3 and 4 in the case of optimal and mean yield, respectively, and compared with the average utility electricity price for the main consumption ranges.

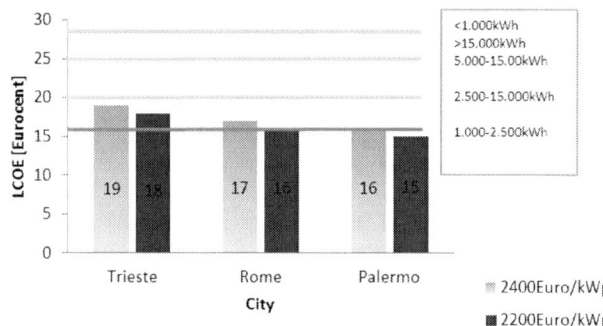

Fig. 3. LCOEs for PV plants with an optimal yield

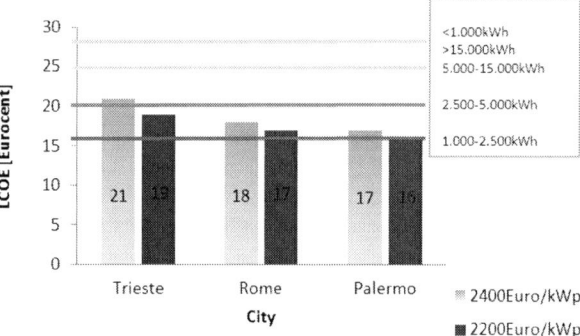

Fig. 4. LCOEs for PV plants with an average yield

The results show that:

- In Palermo (Southern Italy), grid parity is attained under any condition, except when the up-front cost of the PV system (OCS) approaches the upper extreme of the range (2.400 €/kWp)
- Grid parity is attained for any latitude and any electricity consumption rate above 2.500 kWh/year as well as for very low (< 1000 kWh/year) consumption

rates. The only exception is for the location of Trieste (Northern Italy) in the case of a high up-front cost of the PV system and low consumption rates (1.000-2.500 kWh/year); in this case the grid parity condition is nearly approached but not rigorously fulfilled.

It is worth noting that, since the average household consumption in Italy is 2.700kWh/year [7], the grid parity condition is fulfilled for the average Italian family.

Another key parameter is the payback time for the investment on the PV plant. The cash flow analysis has been performed for plants with an OCS of 2.200€/kWp, with an inflation rate set to 2%, for the three locations and the six yields. Table IV reports the payback time (PBT) for the six cases considered in this report. As an example, Figure 3 reports the cash flow for a 3kWp PV plant built in Rome where an initial yield of 1.378kWh/year and a price of electricity for the average Italian family (€0,20) are considered.

TABLE IV. PBTs FOR A TYPICAL DOMESTIC ITALIAN PV PLANT WITHOUT INCENTIVES

City	PBT mean yield	PBT optimal yield
Trieste	13,0	12,0
Rome	11,5	10,5
Palermo	10,5	9,5

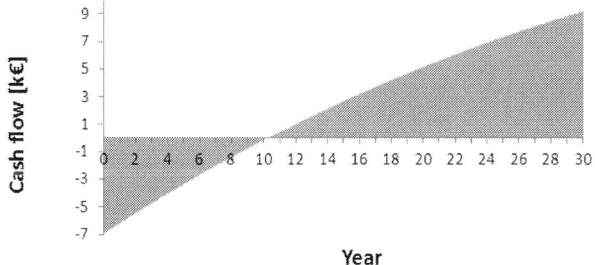

Fig. 5. Cash flow analysis for a 3kWp PV plant installed in Rome without incentives

VI. CONCLUSIONS AND PERSPECTIVES

The key observation is that, essentially, grid parity has been attained in Italy for most residential PV applications. The exceptions that have been pointed out only apply to consumption rates below average and for unfavorable locations, and – considering the market trends - in most cases are nevertheless rapidly approaching the grid parity condition.

No incentives or subsidies have been considered, demonstrating the maturity of the residential market in Italy. However, this remarkable result has only been possible thanks to an incentive scheme that has proven very effective.

The next step for the Italian PV market is the attainment of grid parity for commercial and industrial applications – already happening - and ultimately the attainment of fuel parity. Of the factors that determine the LOCE, some are set by external conditions – such as the interest rate – but most can be influenced by technological advances as well as by commercial and industrial policies. Research and development, for the current situation especially in the field of BOS, still remain key factors in making of PV one of the key player in the future energy scenarios.

As a final consideration we point out that, since irradiance is among the key factors in determining the LOCE, one important global strategy for the future is to push the PV market – profiting from Italian experience – in other favorable locations. Of particular interest are countries where high levels of average irradiance are combined with an emerging economy and therefore increasing energy demand combined. A number of countries of the MENA (Middle East – North Africa) region fit this profile: Algeria can be considered an important example, as PV yields are on average about 20% higher than in Central Italy, and a threefold increase of the energy demand is expected over the next 20 years.

VII. ACKNOWLEDGMENTS

One of the authors (A.M.P.) is grateful for the support provided by the Regione Friuli-Venezia Giulia under the GreenBoat POR project.

VIII. REFERENCES

[1] P. Mints, "The history and future of incentives and the photovoltaic industry and how demand is driven", Progress in photovoltaics: research and applications, vol. 20, pp. 711-716, 2012.

[2] Breyer C, Gerlach A, "Global overview on grid-parity", Progress in photovoltaics: research and applications, in press.

[3] Fell HJ, "Will this work? Is it realistic? Thoughts and acts of a political practitioner with a solar vision", Power for the world: the emergency of electricity from the sun, Pan Stanford Publishing, Singapore, 2010.

[4] www.gse.it

[5] International Energy Agency. PVPS annual report, 2011

[6] http://www.nrel.gov/analysis/tech_lcoe_documentation.html

[7] http://www.autorita.energia.it/it/inglese/index.htm

[8] A. Mellit, A. Massi Pavan, "A 24-hours forecast of solar irradiance using artificial neural network: application for performance prediction of a grid-connected PV plant at Trieste, Italy," Solar Energy, vol. 84, pp. 807–821, 2010.

[9] http://www.re.jrc.ec.europa.eu/pvgis/app4/pvest.php#

[10] http://www.pvsyst.com/

[11] http://www.q-cells.com

Modeling and Simulation of a PV Pumping System Under Real Climatic Conditions

BENDIB Douadi, MAHRANE Achour, AYAD Mohmaed, CHIKH Madjid

Unité de Développement des Equipements Solaires .

UDES, RN N°11, BP 386, Bou -Is/mail, 42415 Wilaya de Tipaza, Algéri e

bendib.douadi@yahoo.fr

Abstract—This article presents the modeling and simulation of a PV pumping system operating over the sun which has been developed under PSIM environment. The main objective is to optimize the PV pumping system through the introduction of two stages. The first one concerns the insertion of a DC-DC converter working with an MPPT in order to get the maximum power from the PV array. The second concerns the optimization of the DC-AC inverter which is controlled through a PWM control in order to get good voltage and current outputs spectrum with reduced losses and harmonics rate. The improvements brought by the MPPT and the PWM on the PV pumping system are shown through the example treated under real climatic conditions.

Keywords: Pv pumping, solar radiation, temperature, invert er, dc-dc converter, mppt , flow rate,PSIM.

I. INTRODUCTION

Photovoltaic pumping systems are considered as one of the most important applications of photovoltaic standalone systems, especially in areas with considerable amount of solar radiation and with no access to the utility grid. In rural areas with no access to grid power, national water authorities and private farmers have to rely on hand pumps and diesel -driven pumps. Hand-operated pumps are the least-cost option for low consumption rates and low pumping heads. Wind pumps are effective only if the wind resource is adequate. If hand pumps cannot satisfy the demand, diesel -driven pumps are commonly used for drinking and irrigation water supply. These pumps stand in competition with the PV water pumps, which present themselves as a reliable, cost effective and environmentally - sound alternative means of water delivery.

This work presents the modeling and simulation of a PV pumping system operating over the sun. The main objective is to optimize the PV pumping system through the introduction of two stages. The first one concerns the insertion of a DC -DC converter working with P&O MPPT in order to get the maximum power from the PV array . The second concerns t he optimization of the DC-AC inverter. The latter is controlled through a PWM control in order to get a high quality voltage and current spectrum with reduced losses and harmonics rate. The improvements brought by the MPPT and the PWM on the PV pumping sys tem are shown through the example treated under real climatic conditions.

II. PV PUMPING SYSTEM DESCRIPYION

A typical photovoltaic pumping system is composed as shown on fig.1 by a PV solar generator which provides DC electricity converted through the po wer inverter in AC and then used for driving a motor pump, which in turn pumps the water. The water stored into tanks allows to bridge night -time periods and cloudy days. The water can flow from the tanks to public water taps and watering points for livest ock or to the irrigation system.

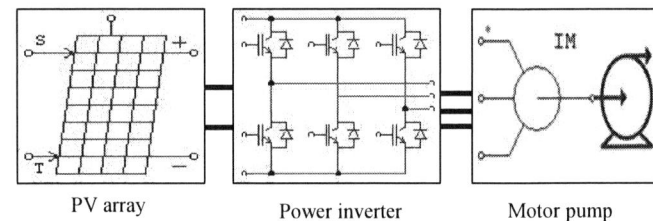

PV array   Power inverter   Motor pump

Fig. 1. Synopsis of a PV pumping system

One major advantage of solar pumps is that they do not need a lot of maintenance. The only moving part of the system, the submersible motor pump, has to be check ed every 3 to 5 years. The PV pumping system under study is installed in our research unit 'Solar equipment Development Unit' (UDES). As shown on figure 1, it is composed by a PV array of 1.1KWp connected to a submersible Grandfos moto -pump through Omron S ysdrive inverter.

III. PV PUMPING SYSTEM MODELING

This section presents the models used for each stage of the studied pumping system in order to simulate its behavior and operation and try to optimize its performance.

A. Solar PV generator modeling

The PV generat or is a combination in series and in parallel of PV panels. Its model is based on the model used for a cell.   The model chosen is the s ingle-diode model, which is represented as shown on fig.2 by a diode, a current source, a series resistance and a shunt re sistance.

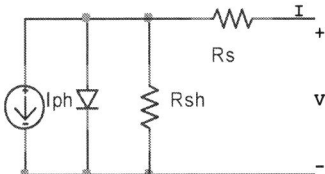

Fig. 2. Electric model of a PV cell

The current source generates a photocurrent which is a function of solar irradiance and temperature of the cell. The diode represents the PN junction solar cell. The I-V characteristic of a solar panel has the following expression [1][5][6]:

$$I = I_{ph} - I_0 \left[\exp\left(\frac{q(V + IR_s)}{mNs\ kT_C} \right) - 1 \right] - \frac{V + IR_s}{R_{sh}} \quad (1)$$

Where I is the current output of the module, Iph is the photo current, Io the diode saturation current, NS is the number of the solar cells in series in the module, V is the voltage output of the module, q is the electric charge (1.6x10-19C), k is the Boltzmann constant (1.38x10-23J/K), TC the cell temperature in degrees Kelvin and m is the ideality factor comprised between 1 and 2 [5].

B. P&O MPPT controller

In the case of a direct connection between the pv array and the load, the system operates at the intersection of the current – voltage curve of the pv array and the load-line. This operation point may be far from the maximum power point of the pv array. As a result a significant part of the pv power is wasted.
As the investment in the PV generator is important, we put our best to get the optimum energy from the PV array. As the maximum power produced by the PV generator is fluctuating with the solar irradiation, tracking the maximum power point of a PV array is usually an essential part of a PV system.

Many MPPT tracking methods have been developed and implemented. The methods vary in complexity, sensors required, convergence speed, cost, range of effectiveness and implementation hardware. It exists many types of MPPT algorithms; the most common are P&O, hill climbing, incremental conductance, fuzzy logic and neural network controllers [2]. In our study we have used the simplest one which is the Perturb and Observe (P&O) controller. The P&O algorithms operate by periodically perturbing (i.e. incrementing or decrementing) the array terminal voltage or current and compare the PV output power with that of the previous perturbation cycle.
If the PV array operating voltage changes and power increases (dP/dVPV>0), the control system moves the PV array operating point in that direction, otherwise the operating point is moved in the opposite direction. In the next perturbation cycle the algorithm continues in the same way (cf.fig. 3).

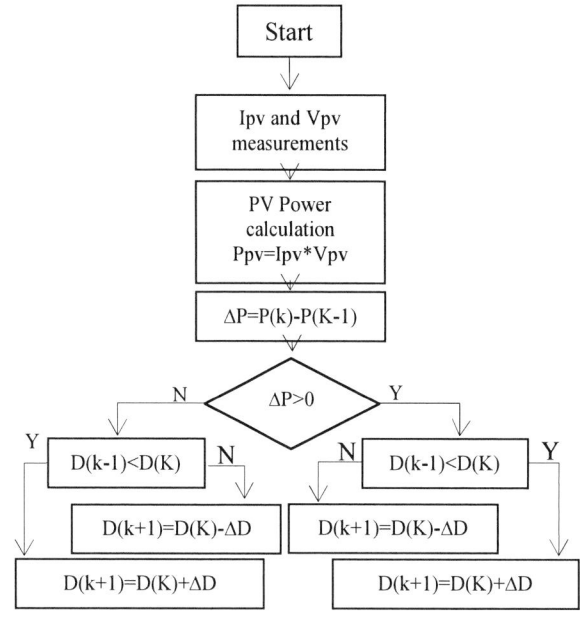

Fig. 3. MPPT P&O algorithm

C. Inverter modeling

As the PV generator produces dc electricity, in most of the cases we need to use an inverter in order to feed an AC voltage to the PV pump. In our case, as we use a three phase immerged PV pump, a standard three-phase voltage source inverter composed of six controlled switches such as IGBT has been used. To avoid harmonics for the output signals (voltages and currents), PWM techniques are used. There are various PWM techniques. The well-known among them are the Sinusoidal PWM (SPWM), the hysteresis PWM, the Space Vector Modulation (SVM) and "optimal" PWM technique which is based on the optimization of certain performance criteria as selective harmonic elimination, increasing efficiency, and minimization of torque pulsation.
The phase voltages of the inverter are given by the following expression:

$$\begin{bmatrix} v_a \\ v_b \\ v_c \end{bmatrix} = \frac{V_{in}}{3} \begin{bmatrix} 2 & -1 & -1 \\ -1 & 2 & -1 \\ -1 & -1 & 2 \end{bmatrix} \begin{bmatrix} f_a \\ f_b \\ f_c \end{bmatrix} \quad (2)$$

Where V_{in} is the input voltage, f_a, f_b and f_c are the control signals which depend on the control strategy adopted. In our case we have used the SPWM in which the generation of the desired output voltage is achieved by comparing the desired reference waveform (modulating signal) with a high-frequency triangular 'carrier' wave as depicted schematically in Fig.4. Depending on whether the signal voltage is larger or smaller than the carrier waveform, either the positive or negative dc bus voltage is applied at the output, and over the period of one triangle wave, the average voltage applied to the load is proportional to the amplitude of the signal (assumed constant) during this period.

The resulting chopped square waveform contains a replica of the desired waveform in its low frequency components, with the higher frequency components being at frequencies close to the carrier frequency. The root mean square value of the ac voltage waveform is still equal to the dc bus voltage, and hence the total harmonic distortion is not affected by the PWM process. The harmonic components are merely shifted into the higher frequency range.

When the modulating signal is a sinusoid of amplitude A_m, and the amplitude of the triangular carrier is A_c, the ratio $m=A_m/A_c$ is known as the modulation index.

However, a higher carrier frequency does result in a larger number of switching per cycle and hence in an increased power loss [3] [7].

D. Motor pump modeling

In order to simulate the operation of the motor pump of the PV pumping system, we have chosen from the Psim library an induction motor we connected a pump as a load. The Psim software provides an opportunity to model the different mechanical loads. For centrifugal pumps and screw propeller, centrifugal compressors, fans and blowers, are modeled by a mechanical load whose torque is proportional to the square of the velocity [8]. The PV pumping chain under Psim environment is shown on the figure 4.

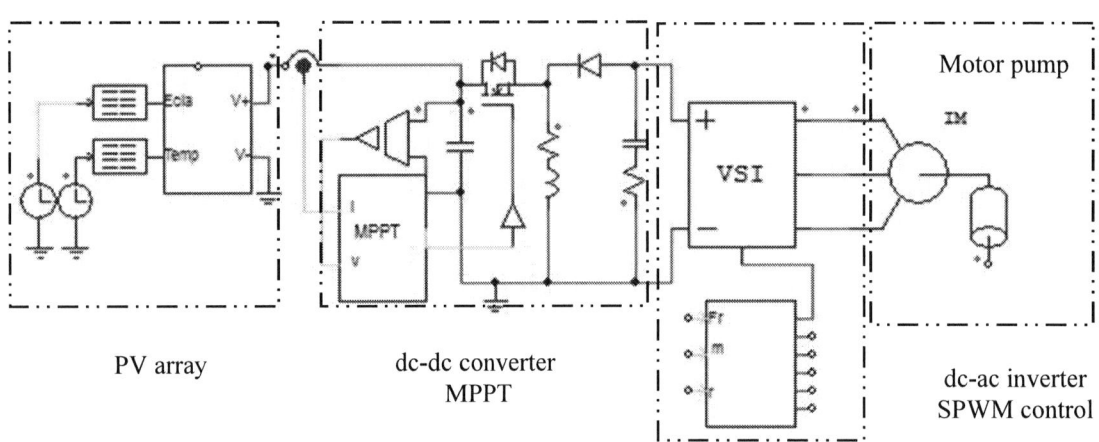

Fig. 4. PV pumping scheme using PSIM

IV. SIMULATION RESULTS AND DISCUSSIONS

After having presented the models used for each stage of the PV pumping chain , we will now simulate the operation of the PV pumping bench installed at UDES (latitude of 36.80250 and longitude of 2.919440) using the experimental data that have been measured at UDES for the 23 march 2011) (cf. fig.5).

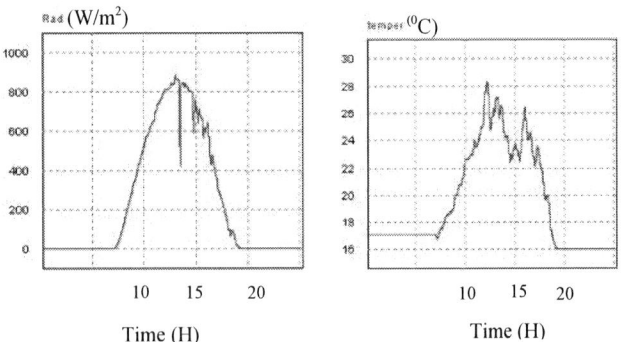

Fig. 5. Experimental solar radiation and temperature data at UDES (Bou-Ismail) for the day march 23rd 2011

Figure 6 shows the improvements introduced by the MPPT control and its effect on the power generated by the PV array.

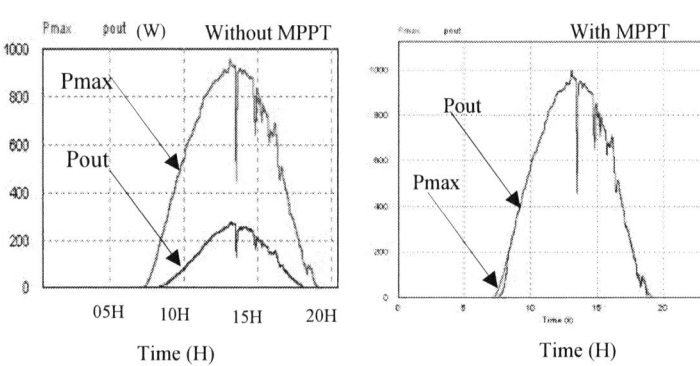

P_{max} : The maximum power received by the PV array.
P_{out} : The power delivered by the PV array

Fig. 6. Electrical power got from the PV array without and with the MPPT

The solar array with the MPPT was connected to a three-phase PWM inverter. The latter feeds the motor-pump unit. At the output of the inverter, we have obtained voltages and currents with low harmonics, which means less waste. Figure 7 shows the power got from the PV array and the electrical power at the output of the motor-pump.

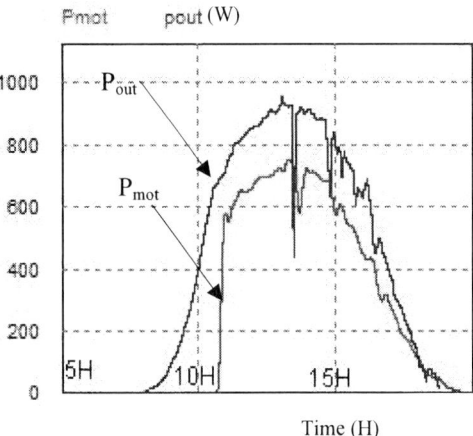

Fig. 7. Motor-pump power and PV array power

The indicator of the optimization of a PV pumping system is undoubtedly the flow of water which is, from the relationship 1, proportional to the transmitted power to the motor-pump. The motor-pump performance is given by the equations 1and 2 [4] [6].

$$Q = \frac{P_{hydr}}{\rho * g * H} \qquad (3)$$

$$P_{hydr} = \eta * P_{in,motor} \qquad (4)$$

Where $P_{in,motor} = P_{out,inverter}$,

Q (m^3.h^{-1}) is the flow rate,

P_{hydr} (W) the Hydraulic power,

ρ (Kg/m^3) the water density,

g (m/s^2) the gravity acceleration,

H *(m)*the total H*ead.*

From these equations we have determined the daily value of the water flow rate shown in fig.8.The amount of the pumped water can be obtained by integrating the flow curve on the day of 23 March 2011. The result obtained is equal to 2.5 m^3.h^{-1} without the MPPT and 4.56 m^3.h^{-1} with the MPPT, which represents a rate of 54%.

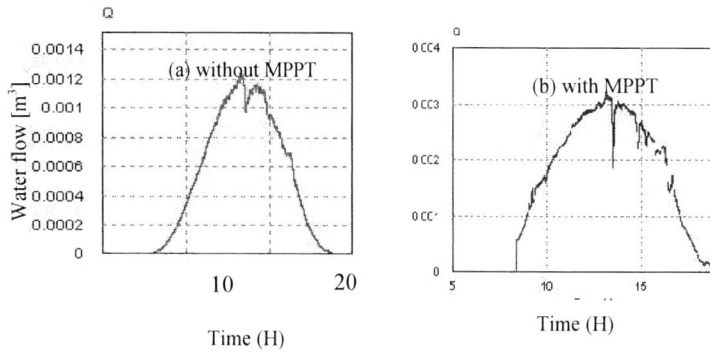

Fig. 8. Simulated water flow-rate during the 23rd march 2011

V. CONCLUSION

This article presented the simulation of a pv pumping system with real values of climatic conditions to verify the effect of the climate and geographical situation. As a first step, simple models are used for each stage of the chain in order to simulate the operation of the whole PV pumping system. Some techniques, the P&O and SPWM, have been introduced in order to optimize the performances of the system. These improvements were quantified through the flow of water which is increased 54% depending on the meteorological conditions. The next step is to validate the simulations results with the experience and improve the models in order to get more precise performance prediction.

REFERENCES

[1] Safari A., Mekhilef S." Simulation and hardware implemenation of incremental conductance MPPT with direct control method using cuk converter", IEEE Trans. on Industrial Electronics ,vol 58, Issue 4, 2011, 1154-1161.

[2] N.SenthilMurugan, N.SenthilMurugan, C.Sharmeela, , "Modeling of Photo Voltaic Arrays with Soft Switching Converter Design and Simulation for Maximum Power Point Tracking", International Journal of Computer Applications (0975 – 8887) Vol 8– No.13,pp. 43-49 October 2010

[3] N. Hamrouni, Moncef Jraidi, Adnène Chérif "Theoretical and experimental analysis of the behaviour of a photovoltaic pumping system", Solar Energy 83 (2009) 1335–1344

[4] N. Hamrouni, Moncef Jraidi, Adnène Chérif Cherif, Dhouib, A. "Measurements and simulation of a PV pumping systems parameters using MPPT and PWM control strategies" IEEE MELECON, pp. 885-888, May 2006.

[5] Mummadi Veerachary "PSIM circuit-oriented simulator model for the nonlinear photovoltaic sources", IEEE Trans. on Aerospace and Electronic systems vol. 42, no. 2, pp.735-74, april 2006.

[6] I. Odeha, Y.G. Yohanisb, B. Nortonc , "Influence of pumping head, insolation and PV array size on PV water pumping system performance", Solar Energy 80 (2006) 51–64.

[7] N.Mohan, M.Tore and P.William, "Power electronics. converters, applications and design", 2nd Ed. NewYork, JOHN WILEY & SONS, 2003 802p.

[8] http://sti.ac-dijon.fr/IMG/diapo2.pdt

New approach for modeling effect of phosphorous diffusion on minority carrier lifetime in multicrystalline silicon wafers

Bentalhoda Soleimany[*], Anahita Shojai Hashemi, Ebrahim Asl-Soleimani
Department of Electrical and Computer Engineering
University of Tehran
Tehran, Iran
[*]h.soleimany@ut.ac.ir

Abstract— A model has been developed for the joint effect of multicrystalline silicon (mc-Si) wafer characteristics and diffusion parameters on minority carrier lifetime. Numerical simulations of mc-Si lifetime based on drift-diffusion model involve many simplifying assumptions, leading to non-practical results for optimizing gettering process. To overcome this deficiency, the proposed model has a completely different approach to estimate carrier lifetime enhancement for various mc-Si wafers processed under different diffusion conditions. SVM regression, which is a machine learning-based procedure, has been employed to foretell the efficiency of gettering through predicting the average lifetime after diffusion. Experimental data from processed wafers have been used to train the model. Examining test data using mean square error (MSE) confirms that Gaussian model has the best accuracy.

Index Terms—Multicrystalline silicon, lifetime, diffusion, regression, SVM.

I. INTRODUCTION

Multicrystalline silicon (mc-Si) wafers are widely used as a preferred material by the manufactures of solar cells [1]. Defects and metal impurities present in the feedstock or introduced during crystal growth, degraded electrical performance of such materials, i.e. mainly through their minority carrier lifetime [2].The enhancement of minority carrier lifetime due to phosphorous diffusion gettering is a well known phenomenon [3, 4]. Modeling the efficiency of such gettering through lifetime improvement is a complicated procedure because of incomplete understanding of this phenomenon and its dependence on both material characteristics and diffusion parameters [5, 6]. Material parameters such as grain boundaries, dislocation densities and impurities concentration differ for each mc-Si wafer; on the other hand, amount of the effect of diffusion on lifetime improvement depends on these parameters.

Nowadays, in order to estimate mc-Si wafer's lifetime one has to implement standard drift-diffusion model. Simulations based on this model available in the literature [7, 8, 9] have different simplification assumptions which make their result

not completely applicable to estimate enhancement of carrier lifetime after diffusion for different mc-Si wafers. For example in a recent work, K. Metzger [7] proposed a two-dimensional simulation of different lifetime measurement techniques and solar cell performance on silicon model that incorporate only lateral variation in SRH lifetime through one or two vertical grain-boundaries while when considering the emitter junction the model assumes a uniform lifetime throughout the device. In another recent work, M. Schubert et al. [10] model impurity redistribution and consequent lifetime enhancement with respect to phosphorus diffusion gettering but as they mentioned they adjust their model for monocrystalline silicon. In another work, G. Stokkan et al. [9] developed a model for the combined effect of dislocations and grain boundaries on minority carrier lifetime but their model consist of mc-Si wafer without emitter junction and needs a complete measurement of dislocation density and grain boundary misorientation maps as input data for each individual wafer. As illustrated by these examples which are of the recent and best works, establishing a physical model which can combine completely the varying mc-Si characteristics of different wafers and also different diffusion parameters to examine the effect of gettering process on lifetime enhancement is very complicated and still out of hand.

In this paper, a completely different approach to traditional models is used to estimate carrier lifetime enhancement for different mc-Si wafers processed with different diffusion conditions. A machine learning procedure is employed in the proposed method to foretell the efficiency of gettering through predicting the average lifetime after diffusion. Observing several samples, the estimator learns the approximate relationship between final lifetime and characteristics of initial mc-Si wafer and diffusion process. These features include initial lifetime and resistivity of the wafer and diffusion parameters such as temperature and process time. The initial lifetime is supposed as an indicator of all stochastic wafer parameters, i.e. dislocation density, grain boundaries structure and impurities concentration. In this way, the estimator can predict the approximate lifetime after diffusion for diverse new multicrystalline silicon samples and for different diffusion conditions.

II. FUNDAMENTALS

A. Gettering process and lifetime

As mentioned in Section 1, structural defects, such as grain boundaries and dislocations, and high concentration of transition metal impurities exist in the mc-Si wafers. These impurities may exist in different states, for example, interstitially dissolved, as metal-silicide precipitates, or as larger micron-sized particles [11]. Energy distribution of trapping states induced by these defects at grain boundaries has a Gaussian distribution [13, 14]. As known, recombination of minority carriers increase in regions containing high dislocation density and impurities and also around grain boundaries and ultimately limit the electrical performance of mc-Si solar cells [5].

Impurity gettering through P diffusion process extensively employed to increase minority carrier lifetime. According to a review study by Myers et al. [15], impurities gettering can be classified into five categories based on the mechanisms involved: (i) precipitation of metal-silicides at energetically favorable nucleation sites, such as at the surface, at grain boundaries or dislocations within the Si material; (ii) atomic trapping at lattice defects without formation of a second phase; (iii) segregation into a second phase due to the impurity solubility difference in different phases; (iv) interaction with electronic dopants due to electrostatic attraction between ionized dopant atoms and oppositely charged metals; (v) P diffusion gettering which combines all or some of the above mechanisms. However, physics of the effect of P diffusion process in mc-Si wafers is very complicated. In spite of the vast studies on P diffusion process [4, 6, 10, 12], there is no analytical relationship between diffusion parameters such as temperature and drive-in or gettering time with the amount of enhancement of minority carrier lifetime. The relationship between intrinsic diffusivities of the impurity atoms or impurity-induced defects and diffusion temperature can be expressed as:

$$D = D_0 exp\left(-\frac{E_a}{k_B T}\right) \quad (1)$$

where D is the intrinsic diffusivity of the impurities or defects, D_0 corresponds to diffusion constant, E_a is the activation energy required to activate the impurities or defects to diffuse, k_B is Boltzmann constant and T is the diffusion gettering temperature, in which D_0 and E_a is changing for different impurities. The relationship between minority carrier diffusion length, minority carrier lifetime and diffusivity can be described as:

$$L = \sqrt{D\tau} \quad (2)$$

Applying this equation to multicrystalline silicon wafers is very complicated since carrier lifetime and diffusion length vary in different grains with different dislocation density and impurity concentration and also around grain boundaries. In addition, there is no analytical relationship between diffusion time and minority carrier lifetime for mc-Si wafers [5].

Therefore, even recent numerical simulation of carrier lifetime or P diffusion process available in the literature [7, 8, 9 and 10] based on physical models cannot combine the varying mc-Si characteristics of different wafers and diffusion parameters.

B. Regression with support vector machines

Suppose a set of data in the form of $\{(x_i, y_i)\}_{i=1}^n$ where x_i is a vector of l inputs and y_i is the corresponding output. In many applications, it is required to find a reasonable relationship between $x_i's$ and $y_i's$, which are called training data, so that for any new input vector, the output can be efficiently predicted. This is achieved by a procedure called regression. There exist several regression methods, each having its own pros and cons. A famous scheme, started by Vapnik et. al [16] in 1979 and made close to its current form by Boser, Guyon and Vapnik [17] in 1992, is support vector machine (SVM). This method has received increasing attention during the last decade especially due to its potential for generalization.

In case when there is nearly linear relationship between the inputs and the output, SVM tries to find a hyper-plane representing this relationship with feasible error. If the predicted output for x_i is presented by $\hat{y}_i$, the first step is to choose an upper limit, such as ε , for deviations$|\hat{y}_i - y_i|$, where y_i is the actual output. In this way, the regression will be so-called ε-sensitive. The SVM gives out a hyperplane in such a way that the deviations for the training data do not exceed ε as far as possible. The performance of the trained SVM in prediction of outputs is evaluated by feeding the SVM with a new set of input vectors $x_i's$, the output of which are already known, and comparing the predicted outputs with the actual ones. This new set of data used for the purpose of evaluation is called testing data. In order to have a well-trained SVM, the number of training data has to be at least a few times of the number of the input vector elements. In addition, there should be enough testing data in evaluation phase.

The priority of SVM over other regression algorithms becomes evident when the relationship between the inputs and the output is far from linear. In this situation, the input vectors are transformed to another space, which is obviously of higher dimension. In this space, the relationship will be closer to linear and the SVM will find the best hyper-plane. The significant point in this regard is, however, the fact that the mentioned transform is not needed to acquire the predicted output. To be more specific, only the equation for the inner product of two vectors in the new space is required, which is a function of the elements of the corresponding input vectors. This function is called a 'kernel'. In order to achieve the best predictor, different kernels, which are equivalent of different vector space transforms, are tested to find the one resulting in the least prediction error. Two renowned kernels are polynomial and Gaussian radial based function (Gaussian RBF).

III. NUMERICAL SIMULATIONS

In order to predict final lifetime, non-linear SVM has been used as the function estimator. The input vector for each sample consists of the initial average lifetime and resistivity of the wafer as well as diffusion temperature and time, while the output is the final average lifetime. To increase the efficiency of learning procedure, outlier data were omitted before feeding training data to the SVM. Moreover, in order for all inputs to be of the same significance to the estimator despite their different ranges of values, all train and test data were normalized. It should be noted that the real value of predicted final lifetime is achieved by performing the reverse procedure, i.e. denormalization.

Experimental data required for the simulation have been extracted from [18-26] and contains 75 wafers data, respectively. To train the SVM, 55 samples were used and the rest were employed in testing stage to evaluate the performance of the trained SVM estimator, which is the ultimate desired model. In training phase, the output of each sample, which is the final lifetime after diffusion, should be fed to the SVM. The measured final lifetimes of train wafers come into application in this step.

The model gets the initial average lifetime and resistivity of test wafers in addition to the diffusion temperature and time of the corresponding gettering process, and gives out the predicted final average lifetime of the wafers. The evaluation of the model has been based on mean square error (MSE). That is, the measured final lifetimes of the test wafers are subtracted from the predicted ones and the result is divided to the number of the test wafers.

IV. RESULTS AND DISCUSSIONS

Two factors affecting the performance of the model are the kernel of the SVM and the data normalization method. In order to get the most accurate predictions, i.e. the least MSE, different kernels, including linear, polynomial and Gaussian RBF, were employed and two different normalization schemes were implemented. In the first normalization method, the difference between the maximum and the minimum of each input among train data is obtained and after subtracting the minimum from the corresponding element in each train or test vector, the result is divided by this value. The same was done for the output. In the second approach, the mean and the variance of each input among train data are computed. The mean is subtracted from the corresponding element in each train or test vector and the result is divided by the variance.

Table I shows rang of input values for all wafers and figure 1 contains the measured final lifetimes after diffusion. The horizontal axis indicates the sample number. For the three mentioned kernels, the predicted and measured values of final lifetime for test wafers have been shown in figures 2 and 3. The predictions in figure 2 are based on the first normalization scheme, whereas those of figure 3 are obtained through the second normalization method.

Table II demonstrates the mean square error (MSE) of predicted final lifetimes obtained from different simulations. It can be seen that normalization algorithm does not have considerable effect on the prediction accuracy, while the type of kernel can highly affect the performance of the estimator and RBF kernel gives the most accurate results. Consequently,

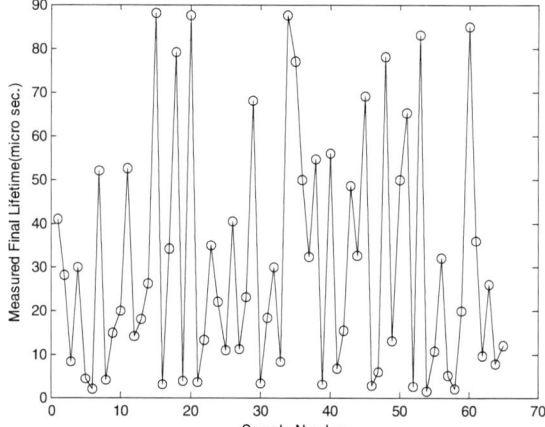

Fig. 1. Final measured lifetime after different diffusion processes for all wafers

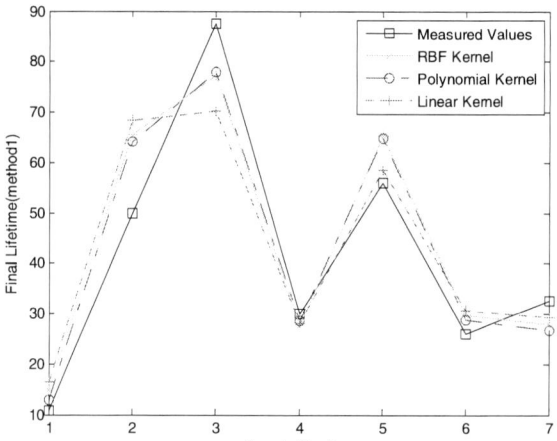

Fig. 2. Final lifetime predicted by trained model for different kernels based on normalization method 1.

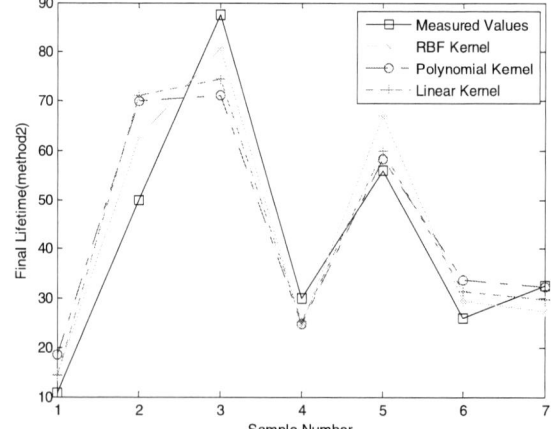

Fig. 3. Final lifetime predicted by trained model for different kernels based on normalization method 2.

TABLE I. RANGE OF INPUT VALUES

Input Parameter	Experimental Rang	Unit
Initial Lifetime	0.2 - 73	min
Resistivity	0.6 - 2	Ωcm
Diffusion Temperature	840 - 990	°C
Diffusion Time	4 – 300	min

RBF kernel along with the first normalization method leads into the least MSE.

As illustrated in table II, linear kernel results in the highest MSE. This implies that the relationship between final lifetime, diffusion parameters and wafer initial characteristics is not close to linear. The non linear RBF kernel representing Gaussian function has the least MSE which is in correspondence with the Gaussian relationship of several physical phenomena. For instance, the least MSE of RBF kernel may deduce that the Gaussian distribution of traps around grain boundaries, mentioned in section 2.1, leads to Gaussian relationship between final lifetime, diffusion parameters and wafer initial characteristics. It should be noted that the numbers of train and test sample has a great influence on prediction accuracy. In other words, the more are training samples, the better the SVM estimator learns the relationship between the inputs and the output. In addition, increasing test wafers will also expand sample space and consequently gives more reliable predictions.

TABLE II. PREDICTION PERFORMANCE OF SVM MODEL BASED ON DIFFERENT KERNELS AND NORMALIZATION METHOD

Normalization Method	Method 1			Method 2		
SVM Kernel	RBF	Polynomial	Linear	RBF	Polynomial	Linear
MSE	6.3861	7.0987	7.6056	6.8162	7.7462	8.5083

V. CONCLUSIONS

A model was developed that implicates the effect of mc-Si wafer characteristics and diffusion parameter on minority carrier lifetime enhancement due to gettering process. Since grain boundaries, dislocation density and impurities concentration vary for different mc-Si wafers, physical models based on drift-diffusion equation require several simplification assumptions that leads to non-practical results for average lifetime of unprocessed mc-Si wafers. Phosphorous diffusion process adds more complexity due to the incomplete understanding of this phenomenon.

On the contrary, this machine learning-based model employs SVM regression to model the lifetime of mc-Si wafer after diffusion and consequently, predicts the efficiency of gettering process with acceptable accuracy. The inputs of the SVM for each wafer are the initial average lifetime and resistivity of the wafer as well as diffusion temperature and time, while the output is the final average lifetime.

The performance of the model has been evaluated through MSE. The results imply that the relationship between final lifetime, diffusion parameters and wafer initial characteristics is far from linear, while Gaussian RBF best represents this relationship. This has something to do with the Gaussian distribution of several physical phenomena, such as the distribution of traps around grain boundaries.

REFERENCES

[1] A. Goetzberger, V. U. Hoffmann, "Photovoltaic solar energy generation, Springer," 2005, 4-5
[2] T. Buonassisi, A. Istratov, T.F. Ciszek, R.F. Clark, D.W. Cunningham, A.M. Gabor, R. Jonczyk, S. Narayanan, E. Sauar, E.R.Weber, Prog. Photovoltaic: Res. Appl., 2006, 14, 513.
[3] A. Bentzen, A. Holt, et. Al. "Gettering of transition metal impurities during phosphorus emitter diffusion in multicrystalline silicon solar cell processing," J. Appl. Phys., 2006, 99, 093509
[4] S. Dubois, N. Enjalbert, F. Warchol, S. Martinuzzi, "Is impurity gettering or passivation by hydrogen the improvement key of mc-Si solar cells during processing steps?," Mater. Sci. Eng. B, 2009, 159–160, 239–41
[5] A. Bentzen, A. Holt, "Overview of phosphorus diffusion and gettering in multicrystalline silicon," Mater. Sc. & Eng. B, 2008, 159–160, 228–34.
[6] H.B. Xu, R.J. Hong, B. Ai, L. Zhuang, H. Shen," Application of phosphorus diffusion gettering process on upgraded metallurgical grade Si wafers and solar cells," Applied Energy, 2010, 87, 3425–3430
[7] W. K. Metzger, "How lifetime fluctuations, grain-boundary recombination, and junctions affect lifetime measurements and their correlation to silicon solar cell performance," Solar Energy Materials & Solar Cells, 2008, 92, 1123– 1135
[8] W. K. Metzger, R. Ahrenkiel, J. Dashdorj, "Analysis of charge separation dynamics in a semiconductor junction," Phys. Rev. B, 2005, 71, 035301
[9] G. Stokkan, S. Riepe, O. Lohne, W. Warta, "Spatially resolved modeling of the combined effect of dislocations and grain boundaries on minority carrier lifetime in multicrystalline silicon," J. Appl. Phys., 2007, 101, 053515
[10] M.C. Schubert, W.Kwapil, et. Al. "Analysis of performance limiting material properties of multicrystalline silicon," Solar Energy Materials & Solar Cells, 2010, 94, 1451–1456
[11] J. Jourdan, S. Dubois, R. Cabal, Y. Veschetti, "Electrical properties of n-type multicrystalline silicon for photovoltaic application-Impact of high temperature boron diffusion," Material Sci. and Eng. B, 2009, 159, 305-308
[12] A. Bentzen, A. Holt, JS. Christensen, BG. Svensson, "High concentration in diffusion of phosphorus in Si from a spray-on source," J. Appl. Phys., 2006
[13] K. Sharma, D. P. Joshi, "A model of electrical conduction across the grain boundaries in polycrystalline silicon thin film transistors and metal oxide semiconductor field effect transistors," J. Appl. Phys., 2009, 106
[14] I. Yamamoto, H. Kuwano, Y. Saito, "Energy distribution of trapping states at grain boundaries in polycrystalline silicon, J. Appl. Phys., 1992, 71
[15] S.M. Myers, M. Seibt,W. Schröter, "Mechanisms of transition-metal gettering in silicon," J. Appl. Phys., 2000, 88, 3795–3819
[16] V. Vapnik, "Estimation of Dependences Based on Empirical Data," Nauka, Moscow, 1979, English translation: Springer Verlag, New York, 1982
[17] B. E. Boser, I. M. Guyon, and V. N. Vapnik, "A training algorithm for optimal margin classiers," In D. Haussler, editor, 5th Annual ACM Workshop on COLT, pages144-152, Pittsburgh, PA, 1992, ACM Press
[18] M. Sheoran, A. Upadhyaya, A. Rohatgi, "Bulk lifetime and efficiency enhancement due to gettering and hydrogenation of defects during cast multicrystalline silicon solar cell fabrication," Solid-State Electronics, 2008
[19] A. Bentzen, E. S. Marstein, R. Kopecek, A. Holt, "Phosphorus diffusion and gettering in multicrystalline silicon solar cell processing," 19th European Photovoltaic Solar Energy Conference, 7-11 June 2004, Paris, France
[20] J. H. arkonen, V.P. Lempinen, T. Juvonen, J. Kyaluoma, "Recovery of minority carrier lifetime in low-cost multicrystalline silicon," Solar Energy Materials & Solar Cells, 200, 73, 125–130
[21] D. H. Macdonald, "Recombination and trapping in multicrystalline silicon solar cells," Ph.D. thesis, Australian National University, 2001
[22] J. H. Tan, "Overcoming Performance Limitations of Multi-Crystalline Silicon Solar Cells," Ph.D. thesis, Australian National University, 2007
[23] O. Schultz, "High-Efficiency Multicrystalline Silicon Solar Cells," Ph.D. thesis, University of Konstanz, Germany, 2005
[24] D. Bouhafsi, M. Boumaour, "Improvement of charge carrier lifetime in heat exchange method multicrystalline silicon wafers by extended phosphorous gettering process," Revue des Energies Renouvelables Vol. 14 N°4, 2011, 665 – 674
[25] K. Nakayashiki, "Understanding of defect passivation and its effect on multicrystalline silicon solar cell performance," Ph.D. thesis, Georgia Institute of Technology, 2007
[26] K. Austad, "Characterization of electrical activity and lifetime in compensated multicrystalline silicon," Ph.D. thesis, Norwegian University of Science and Technology, 2011

Compact Multi-band Rectangular slotted Antenna for Global Navigation Satellite Systems (GNSS)

Mustapha Djebari, Amine Abdelhadi
Department of Electronic, Faculty of Technology
University Saad Dahlab of Blida
Blida, Algeria
djebari_m@yahoo.fr, abdelaminester@gmail.com

Abstract— A compact multi-band rectangular microstrip antenna with two different slots suitable for future Global Navigation Satellite Systems (GNSS) applications is proposed. The antenna operates at multiple frequency bands which are L1, L2, L5 for GPS and G1 G2, G3 for GLONASS and E1,E6, E5a&E5b for GALILEO. The proposed antenna is excited by a microstrip feed line and has three frequency bands through loading an U-shaped slot and an inverted H-shaped slot. It achieved good performance within a large 3-dB beamwidth, which will enable the antenna to receive signals from several GNSS satellites. The obtained results of the return loss, radiation patterns and peak antenna gains are presented using computer simulations.

Key words: compact, microstrip, antenna GPS, Galileo, Glonass

I. INTRODUCTION (HEADING 1)

In the coming years more than 60 satellites with multiple rangingsignals will be available from Global Navigation Satellite Systems (GNSSs) consisting of the modernized American GPS (L2C and L5), updated GLONASS (L1 and L2), European Galileo (E5a, E5b, and E6) and other regional systems [1, 2]. Therefore, the future navigation receivers should be capable of working for all the GNSS frequencies. This interoperability between the GNSS systems will help in overcoming some of the shortfalls of individual navigation systems such as service guarantees, integrity monitoring, and improved service performance [3]. Recently, the ability to integrate more than one communication standard into a single system has become an increasing demand for a modern wireless communication device. Due to the limited space, it often requires the antenna can work at several frequencies simultaneously [4]. Among the various types of antennas, circularly polarized (CP) antennas are the most desired ones, owing to their inevitable merits like reducing polarization mismatch and multipath fading. To benefit from broadband and low profiles, various shapes and designs of multi-bands circularly polarized slot antennas have been developed to overcome both the narrow impedance and axial-ratio

bandwidths by applying different techniques on patch and ground structures.

Therefore, there are various multi-band antennas that have been developed over the years, which can be utilized to achieve multi-band operations [5, 6, 7], such as in our article, a compact patch antenna consisting of a rectangular slotted patch with a size of 80mm x75mm x1.6mm and two different shaped slots fed by a 50 Ω microstrip feed line is presented in detail.

The inverted H- and U-shaped slots incise and change the path of the surface current on the rectangle patch.

Furthermore, the path of the surface current is also expanded due to the two slots. Thus, three resonant modes are excited, and the dimensions of the slots have a great effect on the matching performance of the proposed antenna. The designed antenna is achieved which operates in multi-frequency bands for future GNSS applications.

The simulation software CST studio is used in the design and simulation processes of the designed antenna. The simulated results on the radiation patterns and return loss indicate good agreements with each other are fully explained in the following sections.

II. ANTENNA DESIGN

The proposed rectangular slotted antenna is printed on the substrate with relative dielectric constant of ε_r = 4.3 and thickness of 1.6 mm. On the top of the substrate, a rectangular monopole antenna consisting of a rectangular patch with the inverted-H and U-shaped slots is printed to create three major frequency bands. This structure is fed by a single microstrip line of 50 Ω.

On the opposite side of the substrate, a conducting partial ground plane of width Wg and length Lg is placed. The inverted H-shaped slot deals with the bands of (1.24 – 1.30 GHz) and (1.575 - 1.602 GHz), while the inverted U-shaped slot deals with the bands of (1.176 GHz) and (1.30 GHz).

978-1-4673-5289-5/12 $31.00 © 2012 IEEE

The final design of the multi-band antenna with a size of 75 mm x 80 mm x 1.6 mm is illustrated in "Figure 2.1". The optimized parameters of the antenna are presented in Table 1.

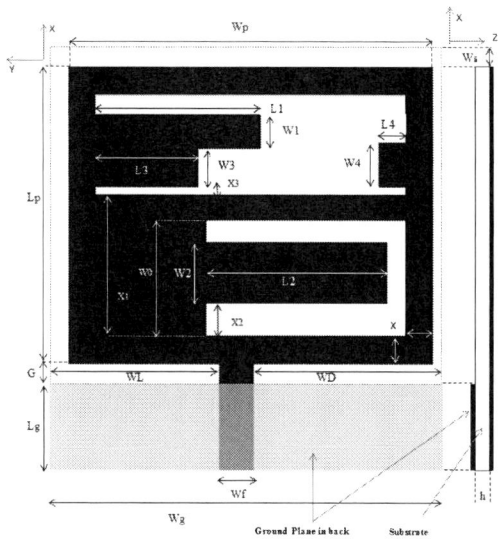

Fig. 2.1 Geometry and dimensions of proposed antenna
(75 mm x 80 mm x 1.6 mm)

TABLE I. Optimized parameters of the a Antenna (in mm)

L_p	W_p	L_g	W_g	W_L	W_D	L_f	W_f
53.59	72.69	12.43	75.88	42.94	37.94	15.87	6.35
h	G	Ws	L1	L2	L3	L4	W_0
1.6	3.72	6.8	30.74	33.48	19.5	5	21
X	X1	X2	W1	W2	W3	W4	
5	25.59	6	6	11	7	8	

III. EFFECTS OF THE PARAMETERS OF THE ANTENNA

An important feature of the proposed antenna is the capability of impedance matching at multiple resonate frequencies using the two slots which are presented above "Figure 1".

A. Effect of the length L1

The simulated 10 dB return loss for different lengths of the L1 parameter of the antenna is presented in "Figure 3.1". The optimized value of the L1 is given in Table 1. It can be observed that decreasing the length L1, decreases the band., but don't change noticeable the first and the second band.

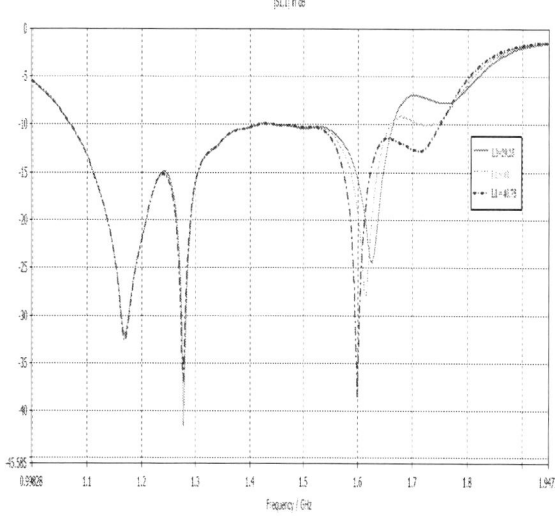

Fig 3.1 Simulated return loss for the antenna with various lengths of L1

B. Effect of the length L4

To highlight these variations, different values of L4 are studied. As presented in "Figures 3.2", by increasing and decreasing L4, there no change in the first and third band frequencies

Picking L4 = 5mm (optimized value), enhances the antenna with maximum bandwidth of 12 % for the second band,small changes in this parameter (L1) leads to considerable variations in both impedance matching and central frequency of the second band

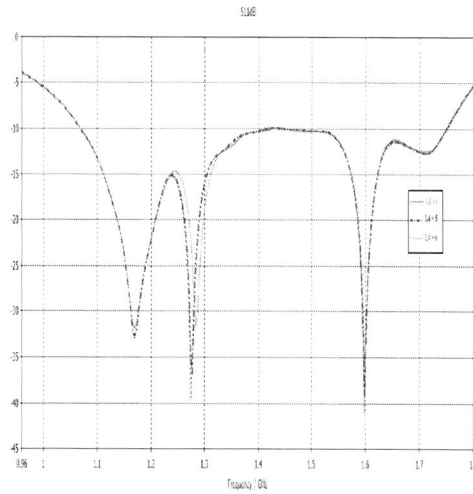

Fig. 3.2 Simulated return loss for the antenna with various lengths of L4

C. Effect of the length L2

Through extensive simulations and experiments, it was found that length of the L2 trip should be selected as L2=33.48 mm , small changes in this parameter will cause a slight shift to the left and the right in the middle resonance frequency of the antenna in the second band and decrease impedance matching of the first band as shown in "Figure 2.3".

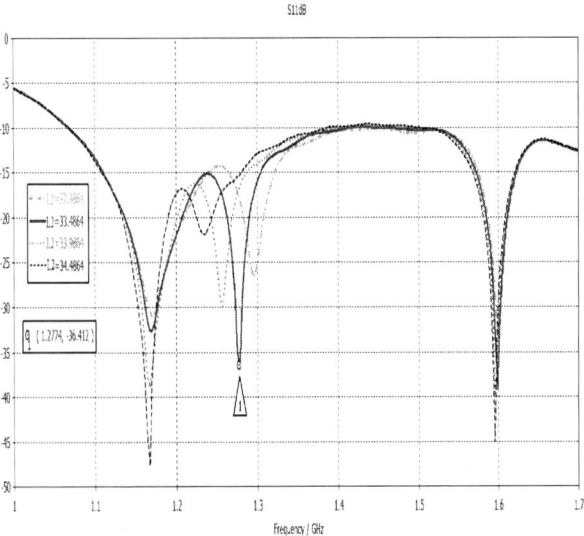

Fig. 3.3 Simulated return loss for the antenna with various lengths of L2

IV. ANTENNA POLARIZATION AND RADIATION PATTERN

To illustrate the circular polarization mechanism, which requires modes of equal magnitude that are in opposite phase, the simulated surface current distributions viewed from the microstrip side are illustrated in "Figure 4.1".

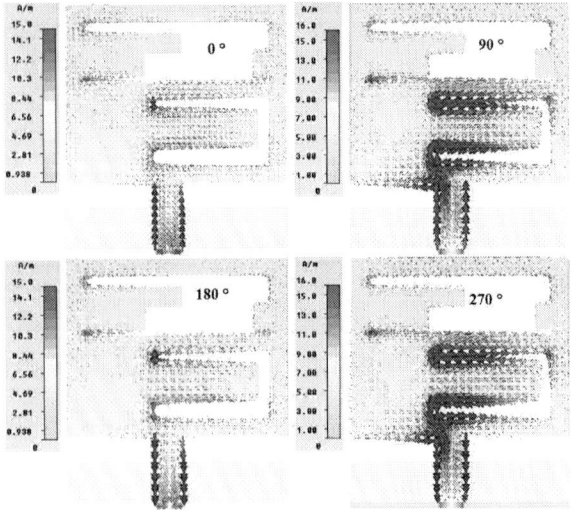

Fig. 4.1 Simulated Distribution of the surface current on the feed and ground of the antenna at 1.278 GHz in 0±, 90±, 180±, and 270± phase.

The direction of the surface currents on the antenna slots and the microstrip feed network is presented at 1.278 GHz as the phase changes from 0 through 270 degrees. It is observed that the surface current distribution in 90 and 270 degrees are equal in magnitude and opposite in phase as in 0 and 180 degrees, as shown in "figure 4.1".

The Radiation patterns and directivity of the antenna at typical frequencies of 1.176 GHz and1.602 GHz are presented in "Figures" 4.2 and 4.3, respectively.

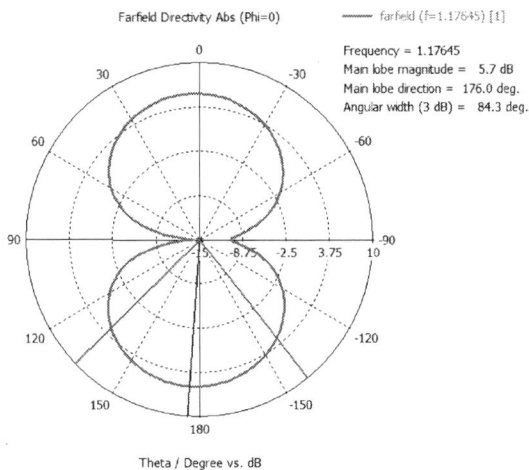

Fig. 4.2 Simulated radiation patterns at 1.176GHz (Phi=0)

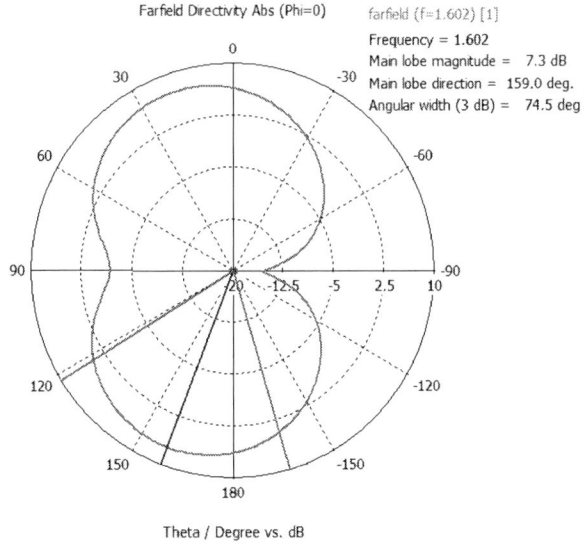

Fig. 4.3 Simulated radiation patterns at 1.602 GHz (Phi == 0)

978-1-4673-5289-5/12 $31.00 © 2012 IEEE   400

V. RETURN LOSS AND ANTENNA BANDWIDTH

The result of the simulated return loss can observed in "Figure 5.1" and we can see that the antenna resonate well at all frequencies. All the parameters of the return loss, resonating frequencies and impedances are resumed in table 2.

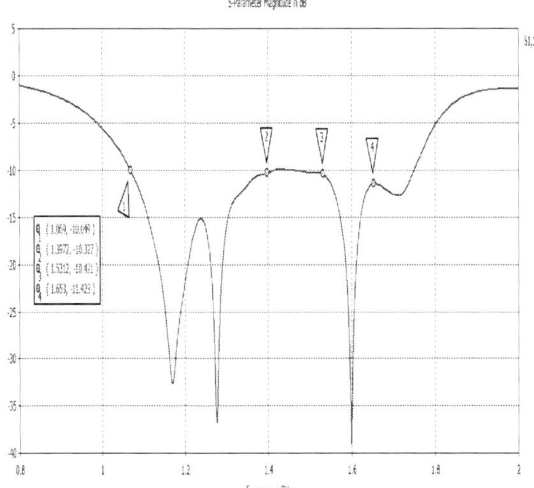

Fig. 5.1 Simulated return loss of the antenna

TABLE II. Resumed characteristics of the antenna

Resonating frequency (GHz)	impedance bandwidth (%)	Return loss (dB)
1.176	11 (at -15 dB)	-32.65
1.278	5 (at -15 dB)	-36.80
1.575	30 (at -15 dB)	-15.55
1.602	30 (at -15 dB)	-31.08

VI. CONCLUSION

In this paper, we have investigated printed rectangular slotted antennas, which is basically a printed microstrip antenna with two different slotted shapes multi-band GNSS applications.

The effects of different dimensions of the slots on the feature of the proposed antenna have also been discussed. Good antenna performances of the operating frequencies across the three operating bands have been obtained.

The obtained three operation bands of the proposed antenna are ranging from (1.06-1.23 GHz), (1.23-139 GHz) and (1.53-1.65 GHz) respectively, which are wide enough to cover the required bandwidths of all the GNSS frequencies.

The antenna gains at 1.176/1.278/1.575/1.602 GHz are 2.7dBi, 2.7 dBi,4.2 dBi and 4.3dBi respectively.

REFERENCES

[1] P. Kovar, P. Puricer, P. Kacmarik, and F. Vejrazka, "Augmentation methods for GNSS integrity and precisionenhancement in di±cult environment," Proceedings of TimeNav07, ENC-GNSS, European Navigation Conference, 107{114, ThePrinting House Inc., Stoughton, 2007.

[2] C. Rizos., et al., "New GNSS developments and their impact onproviders and users spatial information," http://www.gmat.un-sw.edu.au/snap/publications/rizosetal 2005a.pdf.

[3] A. Constantinescu and R. J. Landry, "GPS/Galileo/GLONASS hybrid satellite constellation simulator, GPS constellation validation analysis," The Institute of Navigation 61st AnnualMeeting, 733{737, Cambridge, MA, USA, 2005.

[4] W. C Liu., C. M. Wu, and N. C. Chu, "A compact CPW-fed slotted patch antenna for dual-band operation," IEEE Antennas and Wireless Propagation Letters, Vol. 9, 110{113, 2010.

[5] S. A. Rezaeieh and M. Kartal, "A new triple band circularly polarizedsquare slot antenna design with crooked tand f-shape strips for wireless applications," Progress In Electromagnetics Research, Vol. 121, 2011.

[6] A. Kumar, A D Sarma, A. K. Mondal, K. Ycdukondalu, "A wide band antenna for multi-constellationgnss and augmentation systems," Progress In Electromagnetics Research M, Vol. 11, 2010.

[7] S. M. Zhang*, F. S. Zhang, W.M. .Li, W. Z. Li, and H. Y. Wu, "A multi-band monopole antenna with twodifferent slots for wlan and wimax applications," Progress In Electromagnetics Research Letters, Vol. 28, 2012.

2018 IEEE 30th International Symposium on Power Semiconductor Devices and ICs (ISPSD 2018)

Chicago, Illinois, USA
13 – 17 May 2018

IEEE Catalog Number: CFP18ISP-POD
ISBN: 978-1-5386-2928-4